USMG NATO ASI

# LES HOUCHES
## Session XXXVII
## 1981

THÉORIES DE JAUGE
EN PHYSIQUE DES HAUTES ÉNERGIES

GAUGE THEORIES
IN HIGH ENERGY PHYSICS

## CONFÉRENCIERS

Richard C. BROWER
Sidney COLEMAN
John ELLIS
Leon M. LEDERMAN
Chris QUIGG
Chris T.C. SACHRAJDA
Julius WESS
Björn WIIK

Gordon L. Kane, David N. Schramm,
Larry R. Sulak, Michael S. Turner.

USMG NATO ASI

# LES HOUCHES

SESSION XXXVII

3 Août–11 Septembre 1981

# THÉORIES DE JAUGE

## EN PHYSIQUE DES HAUTES ÉNERGIES

# GAUGE THEORIES

## IN HIGH ENERGY PHYSICS

*édité par*

MARY K. GAILLARD *et* RAYMOND STORA



## PART I

1983
NORTH-HOLLAND PUBLISHING COMPANY
AMSTERDAM · NEW YORK · OXFORD

ISBN 0 444 86543 8 Set
0 444 86722 8 Part I
0 444 86723 6 Part II

Published by:
NORTH-HOLLAND PUBLISHING COMPANY
AMSTERDAM · NEW YORK · OXFORD

Sole distributors for the USA and Canada:
ELSEVIER SCIENCE PUBLISHING COMPANY, INC.
52 VANDERBILT AVENUE
NEW YORK, N.Y. 10017

**Library of Congress Cataloging in Publication Data**
Main entry under title:

Théories de jauge en physique des hautes énergies
Gauge theories in high energy physics.

English and French.
At head of title: USMG, NATO ASI.
Lectures delivered at Les Houches, École d'été de physique théorique, session XXXVII.
Includes bibliographies.
1. Gauge fields (Physics)--Addresses, essays, lectures. 2. Particles (Nuclear physics)--Addresses, essays, lectures. 3. Quantum chromodynamics--Addresses, essays, lectures. I. Gaillard, Mary K. II. Stora, Raymond, 1930- . III. NATO Advanced Study Institute. IV. Université seientifique et médicale de Grenoble. École d'été de physique théorique. V. Gauge theories in high energy physics.
QC793.3.F5T46 1983 539.7'6'01 83-13083
ISBN 0-444-86543-8 (Elsevier Science)

Printed in The Netherlands

# LES HOUCHES
# ÉCOLE D'ÉTÉ DE PHYSIQUE THÉORIQUE

ORGANISME D'INTÉRÊT COMMUN DE L'UNIVERSITÉ SCIENTIFIQUE ET MÉDICALE DE GRENOBLE ET DE L'INSTITUT NATIONAL POLYTECHNIQUE DE GRENOBLE
AIDÉ PAR LE COMMISSARIAT A L'ÉNERGIE ATOMIQUE



*Directeur*: Raymond Stora, Division Théorique, CERN, CH 1211 Genève 23

## SESSION XXXVII

INSTITUT D'ÉTUDES AVANCÉES DE L'OTAN
NATO ADVANCED STUDY INSTITUTE

3 Aout–11 Septembre 1981

*Directeur scientifique de la session*: Mary K. Gaillard, LAPP, Chemin de Bellevue, BP no. 909, F 74019 Annecy-le-Vieux (nouvelle adresse: Physics Department, University of California, Berkeley, CA 94720, USA)

# SESSIONS PRÉCÉDENTES

| | | |
|---|---|---|
| I | 1951 | Mécanique quantique. Théorie quantique des champs |
| II | 1952 | Quantum mechanics. Mécanique statistique. Physique nucléaire |
| III | 1953 | Quantum mechanics. Etat solide. Mécanique statistique. Elementary particles |
| IV | 1954 | Mécanique quantique. Théorie des collisions; two-nucleon interaction. Electrodynamique quantique |
| V | 1955 | Quantum mechanics. Non-equilibrium phenomena. Réactions nucléaires. Interaction of a nucleus with atomic and molecular fields |
| VI | 1956 | Quantum perturbation theory. Low temperature physics. Quantum theory of solids; dislocations and plastic properties. Magnetism; ferromagnetism |
| VII | 1957 | Théorie de la diffusion; recent developments in field theory. Interaction nucléaire; interactions fortes. Electrons de haute énergie. Experiments in high energy nuclear physics |
| VIII | 1958 | Le problème à *N* corps (Dunod, Wiley, Methuen) |
| IX | 1959 | La théorie des gaz neutres et ionisés (Hermann, Wiley)* |
| X | 1960 | Relations de dispersion et particules élémentaires (Hermann, Wiley)* |
| XI | 1961 | La physique des basses températures. Low-temperature physics (Gordon and Breach, Presses Universitaires)* |
| XII | 1962 | Géophysique extérieure. Geophysics: the earth's environment (Gordon and Breach)* |
| XIII | 1963 | Relativité, groupes et topologie. Relativity, groups and topology (Gordon and Breach)* |
| XIV | 1964 | Optique et électronique quantiques. Quantum optics and electronics (Gordon and Breach)* |
| XV | 1965 | Physiques des hautes énergies. High energy physics (Gordon and Breach)* |
| XVI | 1966 | Hautes énergies en astrophysique. High energy astrophysics (Gordon and Breach)* |
| XVII | 1967 | Problème à *N* corps. Many-body physics (Gordon and Breach)* |
| XVIII | 1968 | Physique nucléaire. Nuclear physics (Gordon and Breach)* |
| XIX | 1969 | Aspects physiques de quelques problèmes biologiques. Physical problems in biology (Gordon and Breach)* |
| XX | 1970 | Mécanique statistique et théorie quantique des champs. Statistical mechanics and quantum field theory (Gordon and Breach)* |
| XXI | 1971 | Physique des particules. Particle physics (Gordon and Breach)* |
| XXII | 1972 | Physique des plasmas. Plasma physics (Gordon and Breach)* |
| XXIII | 1972 | Les astres occlus. Black holes (Gordon and Breach)* |
| XXIV | 1973 | Dynamique des fluides. Fluid dynamics (Gordon and Breach) |
| XXV | 1973 | Fluides moléculaires. Molecular fluids (Gordon and Breach)* |
| XXVI | 1974 | Physique atomique et moléculaire et matière interstellaire. Atomic and molecular physics and the interstellar matter (North-Holland)* |
| June Inst. | 1975 | Structural analysis of collision amplitudes (North-Holland) |
| XXVII | 1975 | Aux frontières de la spectroscopie laser. Frontiers in laser spectroscopy (North-Holland)* |
| XXVIII | 1975 | Méthodes en théorie des champs. Methods in field theory (North-Holland)* |

| | | |
|---|---|---|
| XXIX | 1976 | Interactions électromagnétiques et faibles à haute énergie. Weak and electromagnetic interactions at high energy (North-Holland)* |
| XXX | 1977 | Ions lourds et mésons en physique nucléaire. Nuclear physics with mesons and heavy ions (North-Holland)* |
| XXXI | 1978 | La matière mal condensée. Ill-condensed matter (North-Holland)* |
| XXXII | 1979 | Cosmologie physique. Physical cosmology (North-Holland)* |
| XXXIII | 1979 | Membranes et communication intercellulaire. Membranes and intercellular communication (North-Holland)* |
| XXXIV | 1980 | Interaction laser–plasma. Laser–plasma interaction (North-Holland) |
| XXXV | 1980 | Physique des défauts. Physics of defects (North-Holland)* |
| XXXVI | 1981 | Comportement chaotique des systèmes déterministes. Chaotic behaviour of deterministic systems (North-Holland)* |
| XXXVII | 1981 | Théories de jauge en physique des hautes énergies. Gauge theories in high energy physics (North-Holland)* |

*En préparation*

| | | |
|---|---|---|
| XXXVIII | 1982 | Tendances Actuelles en Physique atomique. New trends in atomic physics (North-Holland)* |
| XXXIX | 1982 | Développements récents en théorie des champs et mecanique statistique. Recent developments in field theory and statistical mechanics (North-Holland) |

*Sessions ayant reçu l'appui du Comité Scientifique de l'OTAN.

## LECTURERS

*BROWER, Richard*, Physics Department, Natural Sciences 2, University of California, Santa Cruz, CA 95064, USA.

*COLEMAN, Sidney*, Theoretical Physics Group, Lyman Laboratory, Harvard University, Cambridge, MA 02138, USA.

*ELLIS, John*, Theory Division, CERN, CH-1211 Genève 23, Switzerland.

*KANE, Gordon*, Randall Laboratory of Physics, University of Michigan, Ann Arbor, MI 46109, USA.

*LEDERMAN, Leon*, Fermi National Accelerator Laboratory, PO Box 500, Batavia, IL 60510, USA.

*QUIGG, Chris*, Fermi National Accelerator Laboratory, PO Box 500, Batavia, IL 60510, USA.

*SACHRAJDA, Chris*, Department of Physics, University of Southampton, Southampton SO9 5NH, UK.

*SCHRAMM, David*, Astrophysical and Astronomical Center 100, University of Chicago, 5840 S. Ellis Avenue, Chicago, IL 60637, USA.

*SULAK, Larry*, Randall Laboratory of Physics, University of Michigan, Ann Arbor, MI 46109, USA.

*TURNER, Michael*, Astrophysical and Astronomical Center, University of Chicago, 5640 Ellis Avenue, Chicago, IL 60637, USA.

*WESS, Julius*, Institut für Theoretische Physik, Universität Karlsruhe, Kaiserstrasse 12, Postfach 6380, D-7500 Karlsruhe 1, FRG.

*WIIK, Björn*, II Institut für Experimentalphysik, Universität Hamburg, Luruper Chaussee 149, 2000 Hamburg 50, FRG.

# PARTICIPANTS

*Alvarez, Enrique*, Departamento de Fisica Teórica, CXI, Universidad Autónoma de Madrid, Canto Blanco, Madrid 34, Spain.

*Bagger, Jonathan A.*, Department of Physics, Princeton University, Princeton, NJ 08544, USA.

*Bailey, David*, High Energy Physics, Rutherford Physics Bldg., McGill University, 3600 University, Montreal, Québec, Canada.

*Bandelloni, Giuseppe*, Istituto di Scienze Fisiche, Viale Benedetto XV-5, I-16132 Genova, Italy.

*Barate, Robert*, DPhPE/SEE, CEN Saclay, BP 2, 91191 Gif-sur Yvette, France.

*Carpenter, David B.*, Physics Department, Westfield College, Kidderpore Ave, London NW3, England.

*Chiu, Ting-Wai*, Physics Department, University of California, Irvine, CA 92717, USA.

*Cohen, Eyal*, Department of Nuclear Physics, Weizmann Institute of Science, Rehovot, Israel.

*Coquereaux, Robert*, CPT-CNRS, Case 907, Luminy, 13288 Marseille cedex 2, France.

*Dawson, Sara*, Theory Group, Box 500, Fermilab, Batavia, IL 60510, USA.

*Dhar, Avinash*, Theory Group, T.I.F.R., Bombay 400005, India.

*DiLieto, Christine*, Blackett Laboratory, Imperial College, Prince Consort Road, London SW7, England. From Oct.: SLAC Theory Division, Stanford, CA, USA.

*El Hassouni, Abdellah*, Laboratoire de Physique Théorique, Département de Physique, Faculté des Sciences de Rabat, BP 1014, Rabat, Maroc.

*Fayard, Louis*, Laboratoire de l'Accélérateur Linéaire, Université Paris 11, 91405 Orsay, France.

*Gavai, Rajiv V.*, Fakultät für Physik, Universität Bielefeld, Postfach 8640, D-4800 Bielefeld 1, FRG.

*Gavela Legazpi, Maria Belén*, Laboratoire de Physique des Particules, Chemin de Bellevue, BP 909, 74019 Annecy-le-Vieux cedex, France.

*Gelmini, Graciela*, Sektion Physik der Universität München, Theoretische Physik, Theresienstrasse 37, 8000 München 2, FRG.

*Goossens, Michel*, Division EP, CERN, CH-1211 Genève 23, Switzerland.

*Govaerts, Jan*, Université Catholique de Louvain, Institut de Physique Théorique, Chemin du Cyclotron 2, 1348 Louvain-la-Neuve, Belgium.

*Haggerty, John*, Physics Department, Fermilab, PO Box 500, Batavia, IL 60510, USA

*Hasenfratz, Anna*, CRIP, PO Box 114, Budapest, Hungary.

*Hornaes, Arne*, Fysisk Institutt Allégt. 55, 5014-U, Bergen, Norway.

*Jonsson, Thordur*, Raunvisindastofnun Haskolans, Dunhaga 3, 107 Reykjavik, Iceland.

*Kaplunovsky, Vadim*, Department of Physics and Astronomy, Tel-Aviv University, Ramat-Aviv, Tel-Aviv, Israel.

*Karasinski, Piotr*, Theory group, Gibbs Research Laboratory, Yale University, New Haven, CT 06520, USA

*Kleiss, Ronald*, Lorentz Instituut, Niewsteeg 18, 2311 BS Leiden, The Netherlands.

*Klinkhamer, Frans*, Astronomisch Instituut, PO Box 9513, 2300 Leiden, The Netherlands.

*Lauwers, Paul*, Niels Bohr Institutet, Blegdamsvej 17, DK-2100 København Ø, Denmark.

*Leung, Chung Ngoc*, School of Physics and Astronomy, University of Minnesota, 116 Church Street SE, Minneapolis, MN 55455, USA.

*Machet, Bruno*, CNRS, Centre de Physique Théorique, Section II, Luminy, Case 907, 13288 Marseille cedex 2, France.

*Maciel, Arthur*, Rua viúva Lacerda 433/401, Humaitá, Rio de Janeiro, Brasil.

*McBride, Patricia*, 5th floor JWG, Department of Physics, Box 6666, 260 Whitney Ave, New Haven, CT 06511, USA

*Mukhi, Sunil*, ICTP, PO Box 586, Miramare, 34100 Trieste, Italy.

*Nadkarni, Sudhir*, Yale University, Department of Physics, 217 Prospect Street, PO Box 666, New Haven, CT 06511, USA.

*Napoly, Olivier*, Service de Physique Théorique, CEN Saclay, BP 2, 91191 Gif-sur-Yvette, France.

*Neufeld, Helmut*, Institut für Theoretische Physik der Universität Wien, Boltzmanngasse 5, A-1090 Wien, Austria.

*Oudrhiri Safiani, El Ghali*, Laboratoire de Physique Théorique, Département de Physique, Faculté des Sciences de Rabat, BP 1014, Rabat, Maroc.

*Petcov, Sergey*, Institute of Nuclear Research and Nuclear Energy, Bulgarian Academy of Sciences, Boul. Lenin 72, 1184 Sofia, Bulgaria.

*Pitman, Dale*, Physics Department, University of Toronto, Toronto, Ontario M5S 1A7, Canada.

*Polley, Lutz*, Institut f. Kernphysik, Technische Hochschule Darmstadt, Schlossgarten Str. 9, D-6100 Darmstadt, FRG.

*Roudeau, Patrick*, L.A.L., Université Paris-Sud, Bât. 200, 91405 Orsay cedex, France.

*Sever, Ramazan*, Middle East Technical University, Physics Department, Ankara, Turkey.

*Sjöstrand, Torbjörn*, Department of Theoretical Physics, University of Lund, Sölvegatan 14 A, S-223 62 Lund, Sweden.

*Sodano, Pasquale*, Istituto di Fisica, Universitá di Salerno, 84100 Salerno, Italy.

*Van Ramshorst, Jan Gerrit*, Instituut voor Theoretische Natuurkunde der R.U.U., PO Box 80006, 3508 TA Utrecht, The Netherlands.

*Verschelde, Henri*, Seminarie voor Theoretische Vaste Stof, en lage Energie Kern-Fysica R.U.G., Krijgslaan 271,S9, 9000 Gent, Belgium.

*Wheater, John*, Department of Theoretical Physics, University of Oxford, 1 Keble Road, Oxford, UK.

*Wrigley, James C.*, HEP Group, Cavendish Laboratory, Madingley Road, Cambridge, UK.

*Zaks, Alexander*, Tel-Aviv University, Ramat-Aviv, Israel.

*Zhang, Yi-Cheng*, SISSA (I.C.T.P.), Trieste, Italy.

*Zoupanos, Georges*, Theoretical Physics, National Technical University of Athens, Zografou Campus, Athens 624, Greece.

# PRÉFACE

Depuis la session d'été de 1976 à l'Ecole des Houches sur la physique des particules à haute énergie, les théories de jauge se sont imposées comme le cadre commun à l'intérieur duquel les théoriciens tentent de décrire et de prévoir les phénomènes, et avec lequel les expérimentateurs confrontent leurs données.

Les prédictions du modèle standard "SU(2)×U(1)" des interactions électromagnétiques et faibles ont maintenant trouvé un succès incontesté, la chromodynamique quantique est devenue le language communément accepté pour l'étude des hadrons et de leurs interactions fortes, et, même les théories de jauge qui tentent d'unifier les trois forces fondamentales de la physique des particules à haute énergie ont reçu un soutien phénoménologique encourageant. Il a donc semblé approprié de présenter une revue exhaustive des théories de jauge: leur motivation et les principes sous-jacents, l'évolution rapide des développements techniques des traitements aussi bien perturbatifs que non perturbatifs de ces théories, leur large spectre d'applications à la physique des particules aussi bien qu'à l'astrophysique et à la cosmologie, leur succès sur le plan expérimental et les possibilités de futurs tests de leur structure qui seront offertes par les nouveaux équipements expérimentaux.

Le volume que nous proposons inclut la rédaction de douze articles écrits à partir de huit séries de cours et de quatre séries de séminaires.

Les théories de jauge ont été introduites dans les conférences de Julius Wess, avec une focalisation particulière sur la construction des théories de jauge supersymétriques qui récemment ont été largement reconnues comme susceptibles d'offrir une approche prometteuse à la résolution de certaines des difficultés rencontrées dans les modèles réalistes. Ces conférences comprennent aussi une introduction à la théorie de la gravitation et à sa formulation supersymétrique.

Dans une série de cours simultanément proposés aux étudiants, Chris Sachrajda a décrit l'approche perturbative à la chromodynamique quantique, sujet qui a connu un considérable développement technique au cours des cinq dernières années avec un élargissement du domaine d'applicabilité permettant d'inclure certains phénomènes exclusifs en plus du régime standard des processus inclusifs profondément inélastiques.

John Ellis a passé en revue les applications des théories de jauge aux interactions électromagnétiques et faibles, commençant par la théorie électrofaible maintenant standard, développant ensuite les idées qui cherchent à unifier cette théorie avec la chromodynamique quantique, et offrant pour terminer certaines spéculations sur une ultime unification avec la gravité.

Ces conférences ont été complètées par des revues de l'interpénétration entre la physique des particules et la cosmologie par David Schramm et Michael Turner, une discussion par Gordon Kane des particules scalaires insaisissables que l'on doit introduire dans la théorie pour comprendre la brisure spontanée des théories de jauge, et une revue par Larry Sulak des expériences à venir destinées à mesurer la vie moyenne du proton.

Il y a eu aussi d'importants développements dans le traitement non perturbatif des théories de jauge. Sidney Coleman a décrit des configurations classiques appelées monopôles et les solutions correspondantes qui doivent apparaître dans les théories de jauge spontanément brisées utilisées dans les descriptions unifiées des interactions entre particules. Un autre sujet auquel s'est appliquée une activité d'importance majeure de la part des théoriciens des particules, avec l'espoir de pénétrer les mécanismes du phénomène de confinement, est l'étude des théories de jauge sur réseau qui a été le sujet des conférences de Richard Brower.

Tandis que la plupart des théoriciens travaillent aujourd'hui dans le language des quarks et des gluons, les propriétés de ces objets se manifestent par l'intermédiaire de l'étude expérimentale des hadrons et l'une des questions fondamentales est la description quantitative des hadrons eux-mêmes à partir de la théorie quantique des champs sous-jacente. L'état actuel de cet art a été décrit dans les conférences de Chris Quigg.

Finalement, le sort ultime de toute théorie repose sur sa vérification expérimentale. Leon Lederman a donné une description personnalisée des pièges, mais aussi des heures de gloire de la physique expérimentale et décrit le travail qui reste à faire pour étudier plus avant la structure des interactions entre particules aussi bien à l'aide des accélérateurs de protons sur cibles fixes que des anneaux de collision proton–anti-proton. Björn Wiik a décrit les accélérateurs actuels en insistant sur les anneaux de collision électron–positron et sur la physique qu'ils ont produit jusqu'à maintenant. Il a aussi passé en revue les résultats sur la lepto-production et conclu avec une perspective des équipements à venir pour les interactions de très haute énergie induites par des leptons.

Outre les conférences reproduites dans ce volume, des séminaires ont été présentés sur des aspects plus spécialisés de chacun des sujets, aussi bien par des étudiants que par des orateurs invités. Comme ils traitent de travaux déjà publiés, à paraître, ou dans un état préliminaire, ces séminaires ne sont pas reproduits, mais répertoriés à la place qui leur est appropriée.

Cependant, les qualités dynamiques de l'Ecole ne peuvent être reproduites par l'imprimerie. Elles doivent aussi bien au dévouement des conférenciers qui ont encouragé le dialogue et fréquemment participé aux sessions de discussion, qu'à l'enthousiasme des étudiants qui ont organisé les discussions non seulement dans un but de clarification et d'approfondissement des sujets discutés dans les conférences, mais aussi sur de nombreux autres sujets connexes qui les intéressaient spécialement. Nous désirons remercier ici tous les participants, conférenciers, orateurs de séminaires pour leur contribution à cette magnifique session de six semaines à l'Ecole de Physique Théorique des Houches, la 37ème depuis la création de l'Ecole, qui a regroupé 51 étudiants appartenant à plus de 20 nations, et 27 conférenciers et spécialistes.

*Remerciements*

La réalisation de la session XXXVII de l'Ecole des Houches et la publication de ce volume sont le résultat de nombreuses contributions:

–le soutien financier de l'Université Scientifique et Médicale de Grenoble, et les subventions de la Division des Affaires Scientifiques de l'OTAN, qui a inclus cette session dans son programme d'Instituts d'Etudes Avancées, et du Commissariat à l'Energie Atomique;

–l'orientation et le soutien effectif du Conseil de l'Ecole;

–le soin apporté par Valérie Lecuyer dans la préparation et la frappe des manuscrits;

–la coopération de tous les participants qui ont contribué de façon inestimable à la richesse du programme scientifique, par leur aide active au cours de l'élaboration des notes de cours et les compléments sous forme de séminaires;

–l'aide d'Henri et Nicole Coiffier, d'Anny Battendier, et de toute l'équipe qui a rendu la vie de tous les jours aussi confortable que possible;

—la direction enthousiaste et communicative des opérations photographiques assurée par Piotr Karasinski.

Nous tenons finalement à remercier les conférenciers pour avoir donné leur temps à la préparation et à la rédaction des cours.

Mary K. Gaillard
Raymond Stora

# PREFACE

Since the 1976 Les Houches summer school on high energy particle physics, gauge theories have asserted themselves as the standard framework within which theorists attempt to describe and predict phenomena, and against which experimenters confront their data.

The predictions of the standard SU(2)×U(1) model of electromagnetic and weak interactions have by now met with uncontested success, quantum chromodynamics has become the accepted language for the study of hadrons and their strong interactions, and even those gauge theories which attempt to unify the three basic forces of high energy particle physics have received encouraging phenomenological support. It therefore appeared appropriate to present a comprehensive survey of gauge theories: their motivation and underlying principles, the rapidly evolving technical developments in both perturbative and non-perturbative treatments of these theories, their wide range of applications to particle physics as well as to astrophysics and cosmology, their experimental successes, and the possibilities for future tests and probes of their structure, which will be offered by new experimental facilities.

The present volume includes twelve articles, written from eight sets of lectures and four series of seminars.

Gauge theories were introduced in the lectures of Julius Wess, with particular emphasis on the construction of supersymmetric gauge theories which have recently become widely recognized as providing a promising approach to the resolution of some of the difficulties confronting realistic models. These lectures also include an introduction to the theory of gravitation and to its supersymmetric formulation.

In a concurrent series of lectures, Chris Sachrajda described the perturbative approach to quantum chromodynamics, a field which has undergone considerable technical development over the past five years, widening its domain of applicability to include certain exclusive phenomena in addition to the standard regime of deep inelastic inclusive processes.

John Ellis reviewed the applications of gauge theories to non-strong interactions, starting with the by now "standard" electroweak theory, developing the ideas which seek to unify this theory with quantum chromodynamics, and finally offering some speculations on an ultimate unification with gravity. These lectures were supplemented with reviews on the interplay between particle physics and cosmology by David Schramm and Michael Turner, a discussion by Gordon Kane of the properties of the elusive scalar particles which must be introduced into the theory to understand the spontaneous breaking of gauge theories, and an overview by Larry Sulak of the forthcoming experiments designed to measure the proton lifetime.

There have also been important developments in the non-perturbative treatment of gauge theories. Sidney Coleman described those classical configurations known as monopoles and the corresponding solutions which are expected to appear in the spontaneously broken quantum gauge theories used in unified descriptions of particle interactions. Another field which has become a major activity among particle theorists, with the hope of gaining insight into the phenomenon of confinement, is the study of lattice gauge theories which was the subject of the lecture series by Richard Brower.

While most theorists today work in the language of quarks and gluons, the properties of these objects are discerned through the experimental study of hadrons, and one of the outstanding issues is the quantitative description of the hadrons themselves in terms of the underlying quantum field theory. The present state of this art was described in the lectures of Chris Quigg.

Finally, the ultimate fate of any theory rests on its experimental verification. Leon Lederman gave a personalized account of the pitfalls and rewards of experimental physics and described the work still to be done by both fixed-target proton accelerators and proton–antiproton colliding rings in probing further the structure of particle interactions. Björn Wiik described existing accelerators with emphasis on electron–positron colliding rings and the physics they have so far produced. He also reviewed leptoproduction results and concluded with a perspective of future facilities for very high energy lepton interactions.

In addition to the lectures reproduced in this volume seminars were presented by both students and outside speakers on more specialized aspects of each topic. As they cover work that is published, soon to be published or still in a preliminary stage, these seminars are not reproduced, but are listed at the end of the appropriate section. However, the dynamic quality of the School cannot be reproduced in print. This was due both to the dedication of the lecturers who encouraged a dialogue and participated frequently in discussion sessions, and to the enthusiasm of the students who organized on-going discussions, not only for clarification of and further insight into subjects covered by the lectures, but also on many other related topics of special interest to themselves. We wish to thank here all the participants, lecturers and seminar speakers for their contribution to a rewarding six weeks session at the Les Houches summer school, the 37th since its creation, which gathered 51 students from more than 20 nations, and 27 lecturers and seminar speakers.

*Acknowledgements*

The XXXVIIth Session of the Les Houches summer school and the publication of this volume of lecture notes would not have been possible without:

–the financial support from the Université Scientifique et Médicale de Grenoble, the NATO Scientific Affairs Division (which included this session in its Advanced Studies Institute Programme) and the Commissariat à l'Energie Atomique;

–the guidance of the school board;

–the careful preparation and typing of the manuscripts by Valérie Lecuyer;

–the cooperation of all the participants who contributed in an invaluable manner to the whole scientific programme, including substantial help in the preparation of many of the lecture notes and providing complements to the courses in the form of carefully chosen seminars;

–the help of Henri and Nicole Coiffier, Anny Battendier and the whole team, in making everyday life as comfortable as possible;

–Piotr Karasinski and his enthusiastic and communicative direction of the photographic operations.

Special thanks are due to the lecturers for giving so much of their time in preparing the lectures, and writing them down.

Mary K. Gaillard
Raymond Stora

# CONTENTS

# PART II

COURSE 1

# INTRODUCTION TO GAUGE THEORY

Julius WESS

*Institut für Theoretische Physik, Universität Karlsruhe*
*D-7500 Karlsruhe 1, Germany*

Lecture notes by

Jonathan BAGGER

*Department of Physics, Princeton University*
*Princeton, NJ 08544, USA*

and

Julius WESS

*M.K. Gaillard and R. Stora, eds.*
*Les Houches, Session XXXVII, 1981*
*Théories de jauge en physique des hautes énergies / Gauge theories in high energy physics*

## Contents

# 1. Symmetries

In these lectures we shall study gauge theories and their symmetries. In particular, we shall discuss Poincaré, internal and super symmetries.

Poincaré symmetry is the underlying symmetry of spacetime. It is an unbroken symmetry of nature.

Internal symmetries are compact, semisimple Lie groups which act in some internal space. They are often broken spontaneously in particle phenomenology.

Supersymmetries are symmetries in which anticommutators as well as commutators are used in the defining relations. Supersymmetries must be broken in realistic applications to particle physics.

There are several "no-go" theorems which restrict the symmetries of the $S$-matrix to those mentioned above. They tell us that the only symmetries of the $S$-matrix consistent with conventional quantum field theory are direct products of the Poincaré group with compact internal symmetry groups (if we do not allow anticommutators) and supersymmetries (if we do).

Gauging internal symmetries leads to the usual gauge theories, such as QED, QCD, the Glashow–Salam–Weinberg model and grand unified theories. These are well-behaved quantum field theories; they have been quite successful in particle phenomenology.

Gauging the Poincaré group yields a theory which is invariant under general coordinate transformations. This theory contains Einstein's theory of gravitation. Einstein's theory provides a very successful description of classical long-range gravitational effects. When quantized, however, it becomes extremely nonrenormalizable.

Gauging supersymmetry gives a supergravity theory which might be able to produce a renormalizable theory of gravitation.

We begin by studying the spontaneous breakdown of internal symmetry. In particular, we assume that the internal symmetry group has $n$ generators $T^a$ which satisfy the following commutation relations:

$$[T^a, T^b] = \mathrm{i}t^{abc}T^c. \tag{1.1}$$

The regular representation is $n$-dimensional. We normalize the generators such that

$$\operatorname{tr} T^a T^b = 2\delta^{ab} \tag{1.2}$$

in the regular representation. This is possible because the group is compact and semisimple. With this normalization, the $t^{abc}$ are totally antisymmetric.

We may always associate the following matrix with any vector field $v_m^a$ in the regular representation:

$$v_m = v_m^a T^a.$$

When we do this we say that the matrix $v_m$ is Lie algebra valued.

The simplest model which exhibits spontaneous symmetry breaking is described by the Lagrangian

$$\mathfrak{L} = A^* \Box A - V(A^* A). \tag{1.3}$$

It is invariant under phase transformations:

$$A \to e^{i\lambda} A. \tag{1.4}$$

Renormalizability restricts the potential to the following form:

$$V(A^* A) = m^2 A^* A + g(A^* A)^2. \tag{1.5}$$

The coupling constant $g$ must be positive to exclude solutions with arbitrarily negative energy. The sign of $m^2$ is free as long as $g \neq 0$. When $m^2 > 0$ the potential has a symmetric minimum at the origin. When $m^2 < 0$ the origin is no longer a minimum, so perturbation expansions about this point lead to unstable solutions.

To find a stable solution we must first find a minimum of the potential (1.5),

$$\partial V / \partial A^* = A(m^2 + 2gAA^*) = 0. \tag{1.6}$$

For $m^2 < 0$, we see that minima must satisfy the following condition:

$$AA^* = -m^2/2g > 0. \tag{1.7}$$

This shows that the minima are undetermined up to a phase. Because of this symmetry, we are free to choose $A = (-m^2/2g)^{1/2}$ as the point around which to expand our fields,

$$A = (-m^2/2g)^{1/2} + \tilde{A}. \tag{1.8}$$

We assume that $\tilde{A} \to 0$ as $|\boldsymbol{x}| \to \infty$. Of course, any other point satisfying eq. (1.7) yields an equivalent theory.

Expanding the potential around this minimum,

$$V = g\left[\frac{-m^2}{2g} + \left(\frac{-m^2}{2g}\right)^{1/2}(\tilde{A}+\tilde{A}^*)+\tilde{A}\tilde{A}^*\right]$$
$$\times\left[\frac{m^2}{2g} + \left(\frac{-m^2}{2g}\right)^{1/2}(\tilde{A}+\tilde{A}^*)+\tilde{A}\tilde{A}^*\right], \tag{1.9}$$

and decomposing the field $\tilde{A}$ into its real and imaginary parts,

$$u = \frac{1}{\sqrt{2}}(\tilde{A}+\tilde{A}^*), \qquad v = -\frac{\mathrm{i}}{\sqrt{2}}(\tilde{A}-\tilde{A}^*), \tag{1.10}$$

we find that $u$ and $v$ enter the Lagrangian in the following way:

$$\mathfrak{L} = \tfrac{1}{2}u\Box u + \tfrac{1}{2}v\Box v + m^2u^2 + \mathfrak{L}_{\text{int}}. \tag{1.11}$$

We have only written the free Lagrangian explicitly because it contains all we need to know about the mass spectrum. We see that the theory contains two scalar fields, one massless, the other of mass $(-2m^2)^{1/2}$. The massless field $v$ is called the Goldstone field. We shall see that in general there are as many massless Goldstone fields as there are spontaneously broken symmetries.

Under the internal symmetry group, the fields $u$ and $v$ transform as follows:

$$u' = u\cos\lambda - v\sin\lambda + (-m^2/g)^{1/2}(\cos\lambda - 1),$$
$$v' = u\sin\lambda + v\cos\lambda + (-m^2/g)^{1/2}\sin\lambda. \tag{1.12}$$

In the neighborhood of the origin, we find

$$u' = u, \qquad v' = v + (-m^2/g)^{1/2}\lambda \tag{1.12a}$$

for infinitesimal transformations $\lambda$. This shift of origin signals a truly nonlinear transformation law. No redefinition of the fields $u$ and $v$ can ever reduce eq. (1.12) to a linear transformation. This is a special case of a general theorem. Let $P$ be a point with neighborhood $U$, and let $G$ be a group of transformations. Then it is possible to find coordinates in $U$ which transform linearly under $G$ if and only if $P$ is left invariant under transformations in $G$. Therefore from eq. (1.12a) we learn that eq. (1.12) describes an inherently nonlinear transformation law. Furthermore, since Goldstone fields transform nonlinearly, we identify $v$ as the Goldstone field.

Instead of $u$ and $v$, we could have chosen a different parametrization for our fields:

$$A = \frac{1}{\sqrt{2}} \rho e^{i\phi}. \tag{1.13}$$

This parametrization is not invertible at $A = 0$, but it is invertible about the point $A^*A = \frac{1}{2}\rho^2 = -m^2/2g$. It may therefore be used for an expansion around the minimum of the potential. In terms of $\rho$ and $\phi$, the Lagrangian becomes

$$\mathcal{L} = -\tfrac{1}{2}\left(\partial_m \rho\, \partial^m \rho + \rho^2 \partial_m \phi\, \partial^m \phi\right) - \tfrac{1}{2}\rho^2\left(m^2 + \tfrac{1}{2}g\rho^2\right). \tag{1.14}$$

Again, the potential has a minimum at $\rho = (-m^2/g)^{1/2}$. Normalizing $\phi$ and shifting $\rho$,

$$\tilde{\phi} = \left(-m^2/g\right)^{1/2}\phi, \qquad \rho = \left(-m^2/g\right)^{1/2} + \tilde{\rho}, \tag{1.15}$$

we find

$$\mathcal{L} = -\tfrac{1}{2}\left(\partial_m \tilde{\rho}\, \partial^m \tilde{\rho} + \partial_m \tilde{\phi}\, \partial^m \tilde{\phi}\right) + m^2 \tilde{\rho}^2 + \mathcal{L}_{\text{int}}. \tag{1.16}$$

This Lagrangian has the same degrees of freedom and the same spectrum as (1.11).

This is a consequence of a general theorem which states that the $S$-matrix does not depend on one's choice of interpolating field, as long as the interpolating fields are locally connected and have nonvanishing matrix elements between the vacuum and identical one-particle states. For the classical approximation which we are discussing, this means that the same results will be obtained from any set of fields

$$A'_i(x) = F_i\left(A_j(x)\right), \tag{1.17}$$

as long as the transformation is invertible, that is to say, as long as

$$\det \partial F_i / \partial A_j \neq 0. \tag{1.18}$$

Another model which exhibits spontaneous symmetry breaking is given by

$$\mathcal{L} = \tfrac{1}{2}\boldsymbol{A} \cdot \Box \boldsymbol{A} - \tfrac{1}{2}m^2 \boldsymbol{A}^2 - g\left(\boldsymbol{A}^2\right)^2. \tag{1.19}$$

This Lagrangian is invariant under SO(3). For $m^2 < 0$, the minimum of the potential is at

$$\boldsymbol{A}^2 = -m^2/4g. \tag{1.20}$$

Because of the SO(3) invariance, we may choose the point $\boldsymbol{A} =$

$(-m^2/4g)^{1/2}\hat{\boldsymbol{e}}_3$ as our vacuum. Here $\hat{\boldsymbol{e}}_3$ denotes the unit vector in the 3-direction. Note that this choice of ground state is still invariant under rotations about the 3-axis. Expanding the potential about the minimum,

$$\boldsymbol{A} = \left(-m^2/4g\right)^{1/2}\hat{\boldsymbol{e}}_3 + \tilde{\boldsymbol{A}}, \tag{1.21}$$

we find that $\tilde{A}_3$ acquires a mass $(-2m^2)^{1/2}$, while $\tilde{A}_1$ and $\tilde{A}_2$ become massless.

As a final example, we consider a model invariant under SU(2) in which the potential breaks the symmetry completely:

$$\varphi = \begin{pmatrix} \varphi_1 \\ \varphi_2 \end{pmatrix}, \qquad \mathcal{L} = \varphi^\dagger \Box \varphi - \varphi^\dagger\varphi\left(m^2 + g\varphi^\dagger\varphi\right). \tag{1.22}$$

The potential has a minimum at

$$\varphi^\dagger\varphi = -m^2/2g. \tag{1.23}$$

Any choice of vacuum completely breaks the SU(2) symmetry. In particular, the point $(0,(-m^2/2g)^{1/2})^{\mathrm{T}}$ is left invariant only by the unit element. Expanding about this point,

$$\varphi = \begin{pmatrix} 0 \\ \left(-m^2/2g\right)^{1/2} \end{pmatrix} + \begin{pmatrix} \tilde{\varphi}_1 \\ \tilde{\varphi}_2 \end{pmatrix}, \tag{1.24}$$

we find that $\varphi_1$ and $\mathrm{Im}\,\varphi_2$ become the massless Goldstone fields, while $\mathrm{Re}\,\varphi_2$ acquires a mass $(-2m^2)^{1/2}$. The three Goldstone fields correspond to the three broken symmetries.

This is a very general phenomenon. In fact, the Goldstone theorem states that there are always as many massless particles as there are broken symmetries. To prove this, we consider a potential invariant under the action of an internal symmetry group:

$$\delta V = \frac{\partial V}{\partial \varphi_i}\delta\varphi_i = \frac{\partial V}{\partial \varphi_i} T^a_{ik}\varphi_k = 0, \tag{1.25}$$

$$\delta\varphi_i = T^a_{ik}\varphi_k. \tag{1.26}$$

We assume that the potential has a minimum at $\varphi_i = c_i$:

$$\left.\frac{\partial V}{\partial \varphi_i}\right|_{\varphi_i = c_i} = 0. \tag{1.27}$$

Expanding around this minimum,

$$\varphi_i = c_i + \tilde{\varphi}_i, \tag{1.28}$$

we obtain the mass matrix associated with the shifted fields,

$$V = \tfrac{1}{2} m_{ij} \tilde{\varphi}_i \tilde{\varphi}_j, \tag{1.29}$$

where

$$m_{ij} = \left. \frac{\partial^2 V}{\partial \varphi^i \partial \varphi^j} \right|_{\varphi^l = c^l}. \tag{1.30}$$

Differentiating eq. (1.25), setting $\varphi_i = c_i$, and invoking eq. (1.27), we find

$$\left. \frac{\partial^2 V}{\partial \varphi^i \partial \varphi^j} \right|_{\varphi^l = c^l} T^a_{jk} c_k = 0. \tag{1.31}$$

This leaves two possibilities. Either $T^a_{jk} c_k$ is zero, in which case the vacuum is left invariant by $T^a$, or $T^a_{jk} c_k$ is nonzero, in which case it is an eigenvector of the mass matrix with eigenvalue zero. Thus we see that there exists a massless eigenvector for each broken symmetry. With a little work, the reader may verify that the zero mass eigenvectors are linearly independent, completing the proof of the theorem for the classical theory.

## 2. Gauging internal symmetries

When we gauge an internal symmetry we render a theory invariant under symmetry transformations with $x$-dependent parameters. This requires the introduction of vector fields which transform inhomogeneously under the internal symmetry group.

One well-known example is provided by scalar electrodynamics. In this theory, the scalar field transforms homogeneously under the internal symmetry group,

$$A \to e^{i\lambda(x)} A, \tag{2.1}$$

but derivatives of this field transform inhomogeneously,

$$\partial_m A \to e^{i\lambda} \partial_m A + i \partial_m \lambda(x) e^{i\lambda} A. \tag{2.2}$$

A vector field is introduced to compensate for the inhomogeneous term:

$$v_m \to v_m + \partial_m \lambda(x). \tag{2.3}$$

The vector field guarantees that the full theory can be made invariant under gauge transformations. With the help of the vector field, the covariant derivative

$$\mathcal{D}_m A = (\partial_m - \mathrm{i} v_m) A \tag{2.4}$$

transforms as does $A$.

It is not difficult to generalize this process to non-Abelian groups,

$$A \to \mathrm{e}^{\mathrm{i}\Lambda(x)} A, \tag{2.5}$$

where $\Lambda = T^a \lambda^a(x)$. We introduce the Lie algebra valued vector field

$$v_m = T^a v_m^a \tag{2.6}$$

and assign to it the following transformation law:

$$v_m' = \mathrm{e}^{\mathrm{i}\Lambda} v_m \mathrm{e}^{-\mathrm{i}\Lambda} + \mathrm{i}\mathrm{e}^{\mathrm{i}\Lambda} \partial_m \mathrm{e}^{-\mathrm{i}\Lambda}. \tag{2.7}$$

Gauge transformations leave the vector field Lie algebra valued because only commutators enter into the right hand side of eq. (2.7). This guarantees that the transformation law for $v_m^a$ is independent of the representation chosen for the vector field. The transformation law (2.7) insures that the covariant derivative

$$\mathcal{D}_m A = (\partial_m - \mathrm{i} v_m) A \tag{2.8}$$

transforms as follows:

$$\mathcal{D}_m A \to \mathrm{e}^{\mathrm{i}\Lambda} \mathcal{D}_m A \tag{2.9}$$

under the internal symmetry group.

Covariant derivatives make it very easy to gauge Lagrangians invariant under rigid transformations. One simply replaces all ordinary derivatives by covariant derivatives:

$$\mathcal{L}_0 = -\mathcal{D}_m A^* \mathcal{D}^m A - m^2 A^* A. \tag{2.10}$$

Note that it is possible to partially integrate covariant derivatives. This follows from the fact that covariant derivatives satisfy the product rule,

$$\mathcal{D}_m (A \times B) = \mathcal{D}_m A \times B + A \times \mathcal{D}_m B, \tag{2.11}$$

and from the observation that the covariant derivative of an invariant is identical to its ordinary derivative.

To formulate the dynamics of the vector field, we need a linearly transforming object with derivatives of $v_m$. For the Abelian case,

$$F_{mn} = \partial_m v_n - \partial_n v_m \tag{2.12}$$

is such an object. In fact, $F_{mn}$ is a gauge invariant tensor, and

$$\mathcal{L} = -(1/4e^2)F_{mn}F^{mn} \tag{2.13}$$

is the gauge invariant Lagrangian for the vector field. If we commute two covariant derivatives, we find

$$[\mathcal{D}_m, \mathcal{D}_n]A = -\mathrm{i}F_{mn}A. \tag{2.14}$$

We shall use this expression to define the Yang–Mills field strength $F_{mn}$ in the non-Abelian case as well:

$$\begin{aligned}[\mathcal{D}_m, \mathcal{D}_n]A &= -\mathrm{i}F_{mn}A \\ &= -\mathrm{i}(\partial_m v_n - \partial_n v_m - \mathrm{i}[v_m, v_n])A. \end{aligned} \tag{2.15}$$

The transformation law for $F_{mn}$ follows immediately from (2.14):

$$\begin{aligned}&[\mathcal{D}_m, \mathcal{D}_n]A \to \mathrm{e}^{\mathrm{i}\Lambda}[\mathcal{D}_m, \mathcal{D}_n]A \\ \Rightarrow \quad & F_{mn} \to \mathrm{e}^{\mathrm{i}\Lambda}F_{mn}\mathrm{e}^{-\mathrm{i}\Lambda}. \end{aligned} \tag{2.16}$$

With this assignment,

$$\mathcal{L} = -(1/8e^2)\operatorname{tr} F_{mn}F^{mn} \tag{2.17}$$

is the gauge invariant generalization of eq. (2.13) for non-Abelian fields.

The field strength tensor satisfies a cyclic identity which we shall later identify as one of the Bianchi identities. In the Abelian case, we find

$$\partial_m F_{nl} + \partial_n F_{lm} + \partial_l F_{mn} = 0, \tag{2.18}$$

while for the non-Abelian case,

$$\mathcal{D}_m F_{nl} + \mathcal{D}_n F_{lm} + \mathcal{D}_l F_{mn} = 0. \tag{2.19}$$

This latter relation follows from

$$[\mathcal{D}_k, [\mathcal{D}_l, \mathcal{D}_m]]A = -\mathrm{i}(\mathcal{D}_k F_{lm})A \tag{2.20}$$

and from the fact that covariant derivatives satisfy the Jacobi identity:

$$[\mathcal{D}_k, [\mathcal{D}_l, \mathcal{D}_m]] + [\mathcal{D}_l, [\mathcal{D}_m, \mathcal{D}_k]] + [\mathcal{D}_m, [\mathcal{D}_k, \mathcal{D}_l]] = 0. \tag{2.21}$$

We are now prepared to study gauge theories with spontaneously broken symmetries. We shall notice a new phenomenon through which Goldstone fields disappear and gauge fields acquire mass. This is called the Higgs–Kibble effect. It provides an important link between gauge theories and particle phenomenology.

We shall first study the simplest model in which this occurs. This model is invariant under a U(1) gauge group:

$$\mathcal{L} = -(1/4e^2)F_{mn}F^{mn} - \mathcal{D}_m A^* \mathcal{D}^m A - (m^2 + gA^*A)A^*A. \tag{2.22}$$

The covariant derivative is given by

$$-\mathcal{D}_m A^* \mathcal{D}^m A = -(\partial_m + \mathrm{i}v_m)A^*(\partial^m - \mathrm{i}v^m)A. \tag{2.23}$$

Expanding the field $A$ around the minimum of the potential,

$$A = \left(\frac{-m^2}{2g}\right)^{1/2} + \tilde{A}, \qquad \tilde{A} = \frac{1}{\sqrt{2}}(u + \mathrm{i}v), \tag{2.24}$$

we find

$$\begin{aligned} -\mathcal{D}_m A^* \mathcal{D}^m A &= -\tfrac{1}{2}\left[\partial_m(u - \mathrm{i}v) + \mathrm{i}v_m\left(u - \mathrm{i}v + (-m^2/g)^{1/2}\right)\right] \\ &\quad \times \left[\partial^m(u + \mathrm{i}v) - \mathrm{i}v^m\left(u + \mathrm{i}v + (-m^2/g)^{1/2}\right)\right] \\ &= -\tfrac{1}{2}\Big\{(\partial_m u + v_m v)^2 \\ &\quad + \left[\partial_m v - v_m u - v_m(-m^2/g)^{1/2}\right]^2\Big\}. \end{aligned} \tag{2.25}$$

This expression contains field mixings of the form $v_m \partial^m v$. These may be removed by diagonalizing the kinetic terms. It is much simpler, however, to observe that there exists a certain gauge in which the mixings do not occur. This gauge is found from the transformation laws for $u$ and $v$:

$$\begin{aligned} u' &= u\cos\lambda(x) + v\sin\lambda(x) + (-m^2/g)^{1/2}(\cos\lambda(x) - 1), \\ v' &= -u\sin\lambda(x) + v\cos\lambda(x) - (-m^2/g)^{1/2}\sin\lambda(x). \end{aligned} \tag{2.26}$$

These tell us that for any $u$ and $v$, it is always possible to find a $\lambda(x)$ such that $v'(x) = 0$. In other words, these equations tell us that $v(x) = 0$ is an admissible gauge. In this gauge, (2.25) simplifies considerably:

$$\begin{aligned} -\mathcal{D}_m A^* \mathcal{D}^m A &= -\tfrac{1}{2}\partial_m u \partial^m u - \tfrac{1}{2}(-m^2/g)v_m v^m \\ &\quad + [\text{interaction terms}]. \end{aligned} \tag{2.27}$$

The mixings are gone, the scalar field $v$ disappears, and the vector field $v_m$ acquires the mass $(-m^2/g)^{1/2}$. The spontaneous symmetry breaking also gives the scalar field $u$ the mass $(-2m^2)^{1/2}$. Note that the number of degrees of freedom does not change with the symmetry breaking. We start with one massless vector field and one complex scalar field, with a total of four degrees of freedom, and we end with one massive vector field and one real scalar field, again with four degrees of freedom.

An example of spontaneous symmetry breaking and the Higgs–Kibble mechanism for a non-Abelian gauge group is given by

$$\mathcal{L} = -(1/8e^2)\mathrm{tr}\,\boldsymbol{F}_{mn}\cdot\boldsymbol{F}^{mn} - \tfrac{1}{2}\mathcal{D}_m\boldsymbol{A}\cdot\mathcal{D}^m\boldsymbol{A} - \boldsymbol{A}^2(m^2/2+g\boldsymbol{A}^2). \tag{2.28}$$

This Lagrangian is invariant under O(3) transformations:

$$\begin{aligned}\delta\boldsymbol{A} &= \boldsymbol{\lambda}(x)\times\boldsymbol{A}(x),\\ \delta\boldsymbol{v}_m &= \boldsymbol{\lambda}(x)\times\boldsymbol{v}_m(x)+\partial_m\boldsymbol{\lambda}(x).\end{aligned} \tag{2.29}$$

The covariant derivative

$$\mathcal{D}_m\boldsymbol{A} = \partial_m\boldsymbol{A} - \boldsymbol{v}_m\times\boldsymbol{A} \tag{2.30}$$

transforms homogeneously under the gauge group. As before, we expand $\boldsymbol{A}$ around the minimum of the potential:

$$\boldsymbol{A} = (-m^2/4g)^{1/2}\hat{\boldsymbol{e}}_3 + \tilde{\boldsymbol{A}}. \tag{2.31}$$

From the transformation law for $\tilde{\boldsymbol{A}}$, we see that $A_1 = A_2 = 0$ is an admissible gauge. In this gauge, the Lagrangian takes a simple form:

$$\begin{aligned}-\tfrac{1}{2}\mathcal{D}_m\boldsymbol{A}\cdot\mathcal{D}^m\boldsymbol{A} &= -\tfrac{1}{2}\partial_m\tilde{A}_3\partial^m\tilde{A}_3 - \tfrac{1}{2}(-m^2/4g)\left[(v_1^m)^2+(v_2^m)^2\right]\\ &\quad+[\text{interaction terms}].\end{aligned} \tag{2.32}$$

Again we see that the Goldstone boson fields $\tilde{A}_1$, $\tilde{A}_2$ disappear and that the vector fields $v_1^m$ and $v_2^m$ acquire the mass $(-m^2/4g)^{1/2}$.

A final example is given by the Lagrangian

$$\mathcal{L} = -(1/8e^2)\,\mathrm{tr}\,\boldsymbol{F}_{mn}\cdot\boldsymbol{F}^{mn} - \mathcal{D}_m\varphi^\dagger\mathcal{D}^m\varphi - \varphi^\dagger\varphi(m^2+g\varphi^\dagger\varphi). \tag{2.33}$$

The transformation laws for the fields

$$\varphi' = \mathrm{e}^{(\mathrm{i}/2)\boldsymbol{\tau}\cdot\boldsymbol{\lambda}(x)}\varphi, \qquad \varphi^{\dagger\prime} = \varphi^\dagger\mathrm{e}^{-(\mathrm{i}/2)\boldsymbol{\tau}\cdot\boldsymbol{\lambda}(x)} \tag{2.34}$$

lead to the following expressions for the covariant derivatives:

$$\mathcal{D}_m\varphi = (\partial_m+\tfrac{1}{2}\mathrm{i}\boldsymbol{\tau}\cdot\boldsymbol{v}_m)\varphi, \qquad \mathcal{D}_m\varphi^\dagger = \varphi^\dagger(\overleftarrow{\partial}_m-\tfrac{1}{2}\mathrm{i}\boldsymbol{\tau}\cdot\boldsymbol{v}_m). \tag{2.35}$$

Expanding around the minimum,

$$\varphi = \left(\frac{-m^2}{2g}\right)^{1/2}\begin{pmatrix}0\\1\end{pmatrix} + \tilde{\varphi}, \tag{2.36}$$

and fixing the gauge,

$$\tilde{\varphi}_1 = 0, \qquad \tilde{\varphi}_2 = \frac{1}{\sqrt{2}} u \quad (u \text{ real}), \tag{2.37}$$

we find

$$-\mathfrak{D}_m \varphi^\dagger \mathfrak{D}^m \varphi = -\tfrac{1}{2} \partial_m u \partial^m u - \tfrac{1}{8}(m^2/g)\left(v_x^2 + v_y^2 + v_z^2\right) + [\text{interaction terms}]. \tag{2.38}$$

The three vector fields acquire the mass $(-m^2/4g)^{1/2}$ by absorbing three scalar fields. The remaining scalar field $u$ receives the mass $(-2m^2)^{1/2}$.

We shall conclude this section with a few general remarks on gauge fixing. The theories which we have examined are invariant under gauge transformations. This means that from one solution many others may be generated by gauge transformations. These solutions are equivalent and describe the same physical state. We have used this gauge freedom extensively throughout this chapter. Among all gauge-equivalent solutions we have chosen those in terms of which the Lagrangians took particularly simple forms. To specify these solutions we have required that the fields satisfy equations of the form

$$F\left(v_i^m, A_j\right) = 0. \tag{2.39}$$

These gauge fixings are admissible if and only if there exist gauge transformations from arbitrary fields $v_i^m$, $A_j$ to fields $v_i^{m\prime}$, $A_j'$ such that

$$F\left(v_i^{m\prime}, A_j'\right) = 0. \tag{2.40}$$

Consider, for example, the Lorentz gauge in the Abelian case,

$$\partial_m v^m = 0. \tag{2.41}$$

Starting from an arbitrary field $v_m$, we must transform it to the field $v_m'$,

$$v_m' = v_m + \partial_m \lambda, \tag{2.42}$$

requiring

$$0 = \partial^m v_m' = \partial^m v_m + \Box \lambda. \tag{2.43}$$

This tells us that the Lorentz gauge is admissible if and only if eq. (2.43) has a solution $\lambda$ for arbitrary $v_m$. This equation has a solution so the Lorentz gauge is admissible.

Other well-known admissible gauges include the Coulomb gauge ($\nabla \cdot \boldsymbol{v} = 0$), the Hamilton gauge ($v^0 = 0$) and the axial gauge ($v^3 = 0$).

We have also used gauges which were formulated in terms of matter fields, e.g. $v = 0$. From the transformation law (2.26),

$$v' = -u\sin\lambda + v\cos\lambda - \left(-m^2/g\right)^{1/2}\sin\lambda, \tag{2.44}$$

we see that this is admissible at least for small $u$, $v$ and $\lambda$:

$$v = \left[u + \left(-m^2/g\right)^{1/2}\right]\lambda. \tag{2.45}$$

In general, it is only easy to show that gauges are admissible for small values of the fields. For small fields, we expand $F(v_i^{m\prime}, A_j')$ around $v_i^m$ and $A_j$:

$$\begin{aligned} F\left(v_i^{m\prime}, A_j'\right) = {} & F\left(v_i^m, A_j\right) \\ & + \int \mathrm{d}y \left[\frac{\delta F\left(v_i^m(x), A_j(x)\right)}{\delta v_k^n(y)}\delta v_k^n(y)\right. \\ & \left. + \frac{\delta F\left(v_i^m(x), A_j(x)\right)}{\delta A_k(y)}\delta A_k(y)\right] + \cdots . \end{aligned} \tag{2.46}$$

We know that

$$\begin{aligned} \delta A_j(y) &= \mathrm{i}\Lambda(y)A_j(y), \\ \delta v_m(y) &= \partial_m^y\Lambda(y) - \mathrm{i}\left[v_m(y), \Lambda(y)\right]. \end{aligned} \tag{2.47}$$

Inserting these expressions into the above equation allows us to write the integral as a differential operator $M_F(x, y)$ acting on $\Lambda$. When we do this, we find that a gauge condition is admissible only if

$$\int \mathrm{d}y\, M_F(x, y)\Lambda(y) = F\left(v_i^m(x), A_j(x)\right) \tag{2.48}$$

can be solved for $\Lambda$. A necessary condition for this is

$$\det M_F(x, y) \neq 0. \tag{2.49}$$

For the sake of illustration, we return to the Lorentz condition (2.41):

$$0 = \partial_m v^{m\prime}$$

$$= \partial_m v^m + \int \mathrm{d}y \frac{\partial}{\partial x^m}\left[\frac{\delta v^m(x)}{\delta v^n(y)}\left(\frac{\partial}{\partial y_n}\Lambda(y) - \mathrm{i}[v^n(y), \Lambda(y)]\right)\right]$$

$$= \partial_m v^m + \frac{\partial}{\partial x^m}\left(\frac{\partial}{\partial x_m}\Lambda(x) - \mathrm{i}[v^m(x), \Lambda(x)]\right)$$

$$= \partial_m v^m + \Box\Lambda(x) - \mathrm{i}\partial_m[v^m, \Lambda]. \tag{2.50}$$

Since $\Box$ is invertible, this equation may be solved for $\Lambda$.

A second example is provided by the case where $v = 0$:

$$0 = v' = v + \int \mathrm{d}y \frac{\delta v(x)}{\delta v(y)}\left(-\lambda(y)\left[u(y) + (-m^2/g)^{1/2}\right]\right)$$

$$= v(x) - \lambda(x)\left[u(x) + (-m^2/g)^{1/2}\right]. \tag{2.51}$$

This is precisely eq. (2.45).

## 3. The Glashow–Salam–Weinberg model

The Glashow–Salam–Weinberg model uses spontaneous symmetry breaking and the Higgs–Kibble mechanism to describe the weak and electromagnetic interactions of quarks and leptons. We shall consider a simpler version of this model, treating only the electron and its neutrino, $\psi_e$ and $\psi_\nu$. This minimal model demonstrates many of the features of the full standard model.

The electron and neutrino fields may be decomposed into their chiral components through the chiral projection operators $\frac{1}{2}(1 \pm \gamma_5)$:

$$e_L = \tfrac{1}{2}(1+\gamma_5)\psi_e, \qquad e_R = \tfrac{1}{2}(1-\gamma_5)\psi_e,$$

$$\nu_L = \tfrac{1}{2}(1+\gamma_5)\psi_\nu, \qquad \nu_R = \tfrac{1}{2}(1+\gamma_5)\psi_\nu = 0. \tag{3.1}$$

These fields enter into the observed weak and electromagnetic currents:

$$J^m_{EM} = \bar{\psi}_e\gamma^m\psi_e = \bar{e}_R\gamma^m e_R + \bar{e}_L\gamma^m e_L,$$

$$J^{(-)m}_W = \bar{e}_L\gamma^m\nu_L, \qquad J^{(+)m}_W = \bar{\nu}_L\gamma^m e_L. \tag{3.2}$$

Grouping the left-handed fields into a doublet,

$$L = \begin{pmatrix} \nu_L \\ e_L \end{pmatrix}, \tag{3.3}$$

we may write the currents in terms of the Pauli matrices,

$$\begin{aligned} J^m_{EM} &= \bar{e}_R \gamma^m e_R + \bar{L}\gamma^m \tfrac{1}{2}(1-\tau_3)L, \\ J^{(-)m}_W &= \bar{L}\gamma^m \tau_- L, \qquad J^{(+)m}_W = \bar{L}\gamma^m \tau_+ L, \end{aligned} \tag{3.4}$$

where

$$\begin{aligned} \tau_3 &= \begin{pmatrix} 1 & 0 \\ 0 & -1 \end{pmatrix}, \\ \tau_- &= \tfrac{1}{2}(\tau_1 - i\tau_2) = \begin{pmatrix} 0 & 0 \\ 1 & 0 \end{pmatrix}, \qquad \tau_+ = \tfrac{1}{2}(\tau_1 + i\tau_2) = \begin{pmatrix} 0 & 1 \\ 0 & 0 \end{pmatrix}. \end{aligned} \tag{3.5}$$

From this we see that the observed weak and electromagnetic currents are linear combinations of SU(2)×U(1) currents:

$$\boldsymbol{J}^m = \bar{L}\gamma^m \boldsymbol{\tau} L, \qquad J^m = \tfrac{1}{2}\bar{L}\gamma^m L + \bar{e}_R \gamma^m e_R. \tag{3.6}$$

These currents generate the group SU(2)×U(1). They could have been obtained as the Noether currents associated with the following field transformations:

$$\delta L = -\tfrac{1}{2}i\boldsymbol{\tau}\cdot\boldsymbol{\lambda}L - \tfrac{1}{2}i\eta L, \qquad \delta e_R = -i\eta e_R. \tag{3.7}$$

Since we plan to break this symmetry spontaneously we must also introduce a Higgs multiplet. The simplest choice is a Higgs doublet transforming as follows under the symmetry group:

$$\delta\varphi = -\tfrac{1}{2}i\boldsymbol{\tau}\cdot\boldsymbol{\lambda}\varphi + \tfrac{1}{2}i\eta\varphi. \tag{3.8}$$

Having specified the matter content of our theory, we are now prepared to construct the corresponding gauge invariant Lagrangian:

$$\begin{aligned} \mathcal{L} = &-\tfrac{1}{8}\mathrm{tr}\,\boldsymbol{F}_{mn}\cdot\boldsymbol{F}^{mn} - \tfrac{1}{4}F_{mn}F^{mn} + i\bar{L}\gamma^m\mathcal{D}_m L + i\bar{e}_R\gamma^m\mathcal{D}_m e_R \\ &- \mathcal{D}_m\varphi^\dagger\mathcal{D}^m\varphi - (m^2 + \lambda^2\varphi^\dagger\varphi)\varphi^\dagger\varphi. \end{aligned} \tag{3.9}$$

This Lagrangian is invariant under local SU(2)×U(1) transformations. It describes three fermion fields, two complex Higgs scalars and four gauge vector bosons.

From our previous discussion we know that the minimum of the potential is at $\varphi = (0,(-m^2/2\lambda^2)^{1/2})^T$. As usual, this requires us to shift the Higgs field:

$$\varphi = \begin{pmatrix} 0 \\ (-m^2/2\lambda^2)^{1/2} \end{pmatrix} + \tilde{\varphi}. \tag{3.10}$$

The minimum of the potential at $\varphi = (0, (-m^2/2\lambda^2)^{1/2})^{\mathrm{T}}$ is left invariant by a gauge transformation with $\lambda_3 = -\eta$, $\lambda_1 = \lambda_2 = 0$. This symmetry is unbroken, so we identify it with the unbroken gauge symmetry of electrodynamics. Under this transformation, we find

$$\delta L = -\tfrac{1}{2}\mathrm{i}\eta(1-\tau_3)L, \qquad \delta e_{\mathrm{R}} = -\mathrm{i}\eta e_{\mathrm{R}}, \qquad \delta\varphi = \tfrac{1}{2}\mathrm{i}\eta(1+\tau_3)\varphi. \tag{3.11}$$

This defines the charges of the respective fields.

We shall now study the effect of spontaneous symmetry breaking on the vector fields. We begin by examining the covariant derivative of the Higgs field:

$$\mathcal{D}_m\varphi = \partial_m\varphi - \tfrac{1}{2}\mathrm{i}g\boldsymbol{\tau}\cdot\boldsymbol{A}_m\varphi + \tfrac{1}{2}\mathrm{i}g'B_m\varphi. \tag{3.12}$$

As before, we are free to choose

$$\tilde{\varphi}_1 = 0, \qquad \tilde{\varphi}_2 = u/\sqrt{2} \tag{3.13}$$

as our gauge condition. In this gauge, (3.12) becomes

$$\mathcal{D}_m\varphi = \frac{1}{\sqrt{2}}\begin{pmatrix} 0 \\ \partial_m u \end{pmatrix} - \tfrac{1}{2}\mathrm{i}g\left(\frac{-m^2}{2\lambda^2}\right)^{1/2}\begin{pmatrix} A_m^1 + \mathrm{i}A_m^2 \\ -A_m^3 \end{pmatrix}$$
$$+ \tfrac{1}{2}\mathrm{i}g'\left(\frac{-m^2}{2\lambda^2}\right)^{1/2}\begin{pmatrix} 0 \\ B_m \end{pmatrix} + [\text{higher order terms}]. \tag{3.14}$$

The mass spectrum of the vector bosons is obtained from the kinetic part of the Higgs Lagrangian:

$$-\mathcal{D}_m\varphi^\dagger\mathcal{D}^m\varphi = -\tfrac{1}{2}\partial_m u\partial^m u + \tfrac{1}{8}g^2\frac{m^2}{\lambda^2}\left[(A_m^1)^2 + (A_m^2)^2\right]$$
$$+ \tfrac{1}{8}\frac{m^2}{\lambda^2}\left[gA_m^3 + g'B_m\right]^2 + \cdots. \tag{3.15}$$

This expression is diagonalized by the following linear combinations:

$$Z_m = (g^2 + g'^2)^{-1/2}(gA_m^3 + g'B_m),$$
$$A_m^{\mathrm{EM}} = (g^2 + g'^2)^{-1/2}(-g'A_m^3 + gB_m). \tag{3.16}$$

These combinations rotate $A^3$ and $B$ by an angle $\theta_W$, the Weinberg angle:

$$g/(g^2+g'^2)^{1/2} = \cos\theta_W, \qquad g'/(g^2+g'^2)^{1/2} = \sin\theta_W. \tag{3.17}$$

Defining

$$W_m^{\pm} = \frac{1}{\sqrt{2}}(A_m^1 \pm \mathrm{i}A_m^2), \tag{3.18}$$

we write (3.15) in the following form:

$$-\mathcal{D}_m\varphi^{\dagger}\mathcal{D}^m\varphi = -\tfrac{1}{2}\partial_m u\partial^m u + \tfrac{1}{4}g^2\frac{m^2}{\lambda^2}W_m^+W_m^- + \tfrac{1}{8}\frac{m^2}{\lambda^2}(g^2+g'^2)Z_m^2 + \cdots. \tag{3.19}$$

From this we see that the photon $A_m^{\mathrm{EM}}$ remains massless, but that the fields $W_m^+$, $W_m^-$ and $Z_m$ all acquire mass:

$$m_W^2 = -\tfrac{1}{4}g^2m^2/\lambda^2, \qquad m_Z^2 = -\tfrac{1}{4}(g^2+g'^2)m^2/\lambda^2. \tag{3.20}$$

Note that $m_Z^2 > m_W^2$ for $\theta_W \neq 0$.

The weak and electromagnetic currents are obtained by gauging the fermionic part of the Lagrangian:

$$\mathrm{i}\bar{L}\gamma^m\mathcal{D}_mL + \mathrm{i}\bar{e}_R\gamma^m\mathcal{D}_me_R = \mathrm{i}\bar{L}\gamma^m[\partial_m - \tfrac{1}{2}\mathrm{i}g\boldsymbol{\tau}\cdot\boldsymbol{A}_m - \tfrac{1}{2}\mathrm{i}g'B_m]L + \mathrm{i}\bar{e}_R\gamma^m[\partial_m - \mathrm{i}g'B_m]e_R. \tag{3.21}$$

This gives rise to interactions between the leptons and the physical vector bosons:

$$\begin{aligned}&\frac{1}{\sqrt{2}}g\bar{L}\gamma^m\tau_+LW_m^+ + \frac{1}{\sqrt{2}}g\bar{L}\gamma^m\tau_-LW_m^-\\ &+(g^2+g'^2)^{-1/2}\left[\tfrac{1}{2}g^2\bar{L}\gamma^m\tau_3L + g'^2(\tfrac{1}{2}\bar{L}\gamma^mL + \bar{e}_R\gamma^me_R)\right]Z_m\\ &+\left[gg'/(g^2+g'^2)^{1/2}\right]\left[\bar{e}_R\gamma^me_R + \bar{L}\gamma^m\tfrac{1}{2}(1-\tau_3)L\right]A_m^{\mathrm{EM}}.\end{aligned} \tag{3.22}$$

These interactions allow us to identify the electric charge

$$e = gg'/(g^2+g'^2)^{1/2} = g\sin\theta_W = g'\cos\theta_W \tag{3.23}$$

as well as the Fermi coupling constant

$$\frac{1}{\sqrt{2}}G_F = \frac{1}{8}\frac{g^2}{m_W^2} = -\frac{1}{2}\frac{\lambda^2}{m^2}. \tag{3.24}$$

From these relations follow

$$g > e, \qquad g' > e, \tag{3.25}$$

giving

$$m_{\mathrm{W}}^2 > \tfrac{1}{8}\sqrt{2}\, e^2/G_{\mathrm{F}} \approx (75\ \mathrm{GeV})^2. \tag{3.26}$$

The Lagrangian as given in eq. (3.9) contains no fermion mass terms. These too may be generated by the spontaneous symmetry breaking. In particular, a fermion–Higgs coupling of the form

$$f\left(\bar{e}_{\mathrm{R}}\varphi^{\dagger}L + \bar{L}\varphi e_{\mathrm{R}}\right) \tag{3.27}$$

gives rise to an electron mass

$$m_{\mathrm{e}} = f\left(-m^2/2\lambda^2\right)^{1/2} = f/\left(2\sqrt{2}\, G_{\mathrm{F}}\right)^{1/2} \tag{3.28}$$

after spontaneous symmetry breaking. Neutrinos, however, remain massless. Note that the parameter $f$ is completely arbitrary; it allows us to adjust the electron mass at will.

## 4. Weyl, Majorana, and Dirac spinors

Our discussion of supersymmetry in the coming sections will be phrased in terms of two-component spinors. Before proceeding, therefore, we first introduce Weyl spinors in the Van der Waerden notation.

We begin by examining the connection between the Lorentz group and SL(2,$\mathbb{C}$), the group of two-by-two complex matrices of determinant one. If $M$ is an element of SL(2,$\mathbb{C}$), it is easy to see that $M^*$ (complex conjugate), $(M^{\mathrm{T}})^{-1}$ (transpose inverse) and $(M^{\dagger})^{-1}$ (hermitian conjugate inverse) all represent the same group. Spinors which transform under such representations are characterized by upper or lower, dotted or undotted indices:

$$\psi'_{\alpha} = M_{\alpha}{}^{\beta}\psi_{\beta}, \qquad \bar{\psi}'_{\dot{\alpha}} = (M^*)_{\dot{\alpha}}{}^{\dot{\beta}}\bar{\psi}_{\dot{\beta}},$$

$$\psi'^{\alpha} = (M^{-1})_{\beta}{}^{\alpha}\psi^{\beta}, \qquad \bar{\psi}'^{\dot{\alpha}} = (M^{*\,-1})_{\dot{\beta}}{}^{\dot{\alpha}}\bar{\psi}^{\dot{\beta}}. \tag{4.1}$$

The restriction to $M$ unimodular, $\det M = 1$, is easily expressed in terms of the antisymmetric $\varepsilon$-symbol,

$$\varepsilon^{\gamma\delta} = \varepsilon^{\alpha\beta} M_{\alpha}{}^{\gamma} M_{\beta}{}^{\delta}, \tag{4.2}$$

where $\varepsilon^{12} = 1$, $\varepsilon^{21} = -1$, $\varepsilon^{11} = \varepsilon^{22} = 0$. From this we see that the $\varepsilon$-symbol

$\varepsilon^{\alpha\beta}$ and its inverse $\varepsilon_{\alpha\beta}$ ($\varepsilon_{\alpha\beta}\varepsilon^{\beta\gamma} = \delta_{\alpha}{}^{\gamma}$) transform as tensors. They may therefore be used to raise and lower spinor indices:

$$\psi^{\alpha} = \varepsilon^{\alpha\beta}\psi_{\beta}, \qquad \psi_{\alpha} = \varepsilon_{\alpha\beta}\psi^{\beta}. \tag{4.3}$$

A similar analysis holds for dotted indices. Hence $M$ and $(M^{\mathrm{T}})^{-1}$ [and $M^*$ and $(M^{\dagger})^{-1}$] form equivalent representations of SL(2,$\mathbb{C}$).

The connection between SL(2,$\mathbb{C}$) and the Lorentz group is established through the $\sigma$-matrices,

$$\begin{aligned} \sigma^0 &= \begin{pmatrix} -1 & 0 \\ 0 & -1 \end{pmatrix}, \qquad \sigma^1 = \begin{pmatrix} 0 & 1 \\ 1 & 0 \end{pmatrix}, \\ \sigma^2 &= \begin{pmatrix} 0 & -\mathrm{i} \\ \mathrm{i} & 0 \end{pmatrix}, \qquad \sigma^3 = \begin{pmatrix} 1 & 0 \\ 0 & -1 \end{pmatrix}, \end{aligned} \tag{4.4}$$

in complete analogy with SU(2) and the rotation group. These matrices form a basis for two-by-two hermitian matrices $P$:

$$P = \sigma^m P_m = \begin{pmatrix} -P_0 + P_3 & P_1 - \mathrm{i}P_2 \\ P_1 + \mathrm{i}P_2 & -P_0 - P_3 \end{pmatrix}. \tag{4.5}$$

Additional hermitian matrices may always be generated by the following transformation:

$$P' = MPM^{\dagger}. \tag{4.6}$$

Both $P$ and $P'$ have expansions in $\sigma$:

$$M\sigma^m P_m M^{\dagger} = \sigma^m P'_m. \tag{4.7}$$

Taking determinants, we find $P_0^2 - \boldsymbol{P}^2 = P_0'^2 - \boldsymbol{P}'^2$ since $M$ has determinant one. This shows that $P_m$ and $P'_m$ are connected by a Lorentz transformation for $M \in$ SL(2,$\mathbb{C}$).

Equation (4.7) also shows that $\sigma$ has the following index structure:

$$\sigma^m_{\alpha\dot{\alpha}}. \tag{4.8}$$

These indices may be raised by the $\varepsilon$-tensor:

$$\bar{\sigma}^{m\dot{\alpha}\alpha} = \varepsilon^{\alpha\beta}\varepsilon^{\dot{\alpha}\dot{\beta}}\sigma^m_{\beta\dot{\beta}}. \tag{4.9}$$

Evaluating this expression, we find

$$\bar{\sigma}^0 = \sigma^0, \qquad \bar{\sigma}^i = -\sigma^i \quad \text{for } i = 1,2,3. \tag{4.10}$$

The $\sigma$'s and $\bar{\sigma}$'s may be multiplied covariantly, and their products satisfy the following relations:

$$\sigma^m\bar{\sigma}^n + \sigma^n\bar{\sigma}^m = -2\eta^{mn}, \qquad \bar{\sigma}^m\sigma^n + \bar{\sigma}^n\sigma^m = -2\eta^{mn},$$

$$\operatorname{tr}\sigma^m\bar{\sigma}^n = -2\eta^{mn}, \qquad \sigma^m_{\alpha\dot{\alpha}}\bar{\sigma}^{m\dot{\beta}\beta} = -2\delta_\alpha{}^\beta\delta_{\dot{\alpha}}{}^{\dot{\beta}}. \tag{4.11}$$

These relations lead immediately to a representation of the Dirac $\gamma$-matrices:

$$\gamma^m = \begin{pmatrix} 0 & \sigma^m \\ \bar{\sigma}^m & 0 \end{pmatrix}, \qquad \gamma_5 = i\gamma^0\gamma^1\gamma^2\gamma^3 = \begin{pmatrix} 1 & 0 \\ 0 & -1 \end{pmatrix},$$
$$\{\gamma^m, \gamma^n\} = -2\eta^{mn}. \tag{4.12}$$

We shall call this the Weyl basis. In this basis, Dirac spinors are represented in terms of two Weyl spinors as follows,

$$\psi = \begin{pmatrix} \varphi_\alpha \\ \bar{\chi}^{\dot{\alpha}} \end{pmatrix}, \tag{4.13}$$

where

$$\tfrac{1}{2}(1+\gamma_5)\psi = \begin{pmatrix} \varphi_\alpha \\ 0 \end{pmatrix}, \qquad \tfrac{1}{2}(1-\gamma_5)\psi = \begin{pmatrix} 0 \\ \bar{\chi}^{\dot{\alpha}} \end{pmatrix}. \tag{4.14}$$

As usual, we denote the spinor transforming inversely to $\psi$ by $\bar{\psi}$. In the Weyl basis, we find

$$\bar{\psi} = (\bar{\varphi}_{\dot{\alpha}} \ \ \chi^\alpha)\begin{pmatrix} 0 & -1 \\ -1 & 0 \end{pmatrix} = (-\chi^\alpha, -\bar{\varphi}_{\dot{\alpha}}). \tag{4.15}$$

This renders the combination

$$\bar{\psi}\psi = -\chi^\alpha\varphi_\alpha - \bar{\varphi}_{\dot{\alpha}}\bar{\chi}^{\dot{\alpha}} \tag{4.16}$$

invariant under Lorentz transformations. To simplify our notation, we shall always employ the following summation convention

$$\varphi\chi = \varphi^\alpha\chi_\alpha, \qquad \bar{\chi}\bar{\varphi} = \bar{\chi}_{\dot{\alpha}}\bar{\varphi}^{\dot{\alpha}}. \tag{4.17}$$

With this convention, we have

$$\chi\varphi = \varphi\chi, \qquad \bar{\chi}\bar{\varphi} = \bar{\varphi}\bar{\chi} \tag{4.18}$$

and

$$(\chi\varphi)^* = \bar{\chi}\bar{\varphi}. \tag{4.19}$$

The Dirac current may also be written in the two-component notation:

$$\bar{\psi}\gamma^m\psi = -\chi\sigma^m\bar{\chi} - \bar{\varphi}\bar{\sigma}^m\varphi. \tag{4.20}$$

Combining eqs. (4.16) and (4.20), we find the free Dirac Lagrangian:

$$\begin{aligned}\mathcal{L} &= \mathrm{i}\bar{\psi}\gamma^m\partial_m\psi - m\bar{\psi}\psi \\ &= -\mathrm{i}(\bar{\varphi}\bar{\sigma}^m\partial_m\varphi + \chi\sigma^m\partial_m\bar{\chi}) + m(\varphi\chi + \bar{\chi}\bar{\varphi}).\end{aligned} \tag{4.21}$$

This Lagrangian is invariant under the phase transformation

$$\psi \to \mathrm{e}^{\mathrm{i}\lambda}\psi, \tag{4.22}$$

or

$$\varphi \to \mathrm{e}^{\mathrm{i}\lambda}\varphi, \qquad \chi \to \mathrm{e}^{-\mathrm{i}\lambda}\chi. \tag{4.23}$$

It is precisely this symmetry which leads to conservation of the Dirac current.

Charge conjugation is obtained by a matrix $C$ which transforms the $\gamma$-matrices in the following way:

$$C\gamma^m C^{-1} = -(\gamma^m)^{\mathrm{T}}. \tag{4.24}$$

The matrix $C$ relates Dirac spinors to their charge conjugates:

$$\psi^{\mathrm{C}} = -C\bar{\psi}^{\mathrm{T}}. \tag{4.25}$$

In the Weyl basis, $C$ may be represented in terms of the $\varepsilon$-tensors:

$$C = \begin{pmatrix} \varepsilon_{\alpha\beta} & 0 \\ 0 & \varepsilon^{\dot{\alpha}\dot{\beta}} \end{pmatrix}. \tag{4.26}$$

In this basis, charge conjugation exchanges the two Weyl spinors:

$$\psi = \begin{pmatrix} \varphi_\alpha \\ \bar{\chi}^{\dot{\alpha}} \end{pmatrix}, \qquad \psi^{\mathrm{C}} = \begin{pmatrix} \chi_\alpha \\ \bar{\varphi}^{\dot{\alpha}} \end{pmatrix}. \tag{4.27}$$

From this we see that

$$\bar{\psi}^{\mathrm{C}}\psi^{\mathrm{C}} = -\varphi\chi - \bar{\chi}\bar{\varphi} = \bar{\psi}\psi, \tag{4.28}$$

while

$$\bar{\psi}^{\mathrm{C}}\gamma^m\psi^{\mathrm{C}} = -\varphi\sigma^m\bar{\varphi} - \bar{\chi}\bar{\sigma}^m\chi = \bar{\varphi}\bar{\sigma}^m\varphi + \chi\sigma^m\bar{\chi} = -\bar{\psi}\gamma^m\psi. \tag{4.29}$$

This justifies the name charge conjugation.

Majorana spinors are defined through the relation

$$\psi^{\mathrm{C}} = \psi. \tag{4.30}$$

This implies that $\varphi^\alpha = \chi^\alpha$ for the associated Weyl spinors, or, equivalently, that a Majorana spinor is represented by only one Weyl spinor:

$$\psi = \begin{pmatrix} \chi_\alpha \\ \bar{\chi}^{\dot\alpha} \end{pmatrix}. \tag{4.31}$$

In fact, the four components of a Majorana spinor may be viewed as the real and imaginary parts of a two-component Weyl spinor. This is easily seen in the Majorana representation:

$$(\gamma_{\mathrm{M}}^m)^* = -\gamma_{\mathrm{M}}^m,$$

$$\gamma_{\mathrm{M}}^0 = \begin{pmatrix} 0 & -\sigma^2 \\ -\sigma^2 & 0 \end{pmatrix}, \qquad \gamma_{\mathrm{M}}^1 = \begin{pmatrix} 0 & \mathrm{i}\sigma^3 \\ \mathrm{i}\sigma^3 & 0 \end{pmatrix},$$

$$\gamma_{\mathrm{M}}^2 = \begin{pmatrix} \mathrm{i} & 0 \\ 0 & -\mathrm{i} \end{pmatrix}, \qquad \gamma_{\mathrm{M}}^3 = \begin{pmatrix} 0 & -\mathrm{i}\sigma^1 \\ -\mathrm{i}\sigma^1 & 0 \end{pmatrix}. \tag{4.32}$$

This representation is obtained from the Weyl representation by a similarity transformation:

$$\gamma_{\mathrm{M}}^m = \mathrm{Y}\gamma_{\mathrm{W}}^m\mathrm{Y}^{-1}, \qquad \mathrm{Y} = \frac{1}{\sqrt{2}}\begin{pmatrix} 1 & \varepsilon_{\dot\alpha\dot\beta} \\ -\mathrm{i} & \mathrm{i}\varepsilon_{\dot\alpha\dot\beta} \end{pmatrix}. \tag{4.33}$$

Majorana fields in the Weyl basis

$$\psi = \begin{pmatrix} \chi_\alpha \\ \bar{\chi}^{\dot\alpha} \end{pmatrix} \tag{4.34}$$

become

$$\psi_{\mathrm{M}} = \mathrm{Y}\psi = \frac{1}{\sqrt{2}}\begin{pmatrix} \chi_\alpha + \bar{\chi}_{\dot\alpha} \\ -\mathrm{i}(\chi_\alpha - \bar{\chi}_{\dot\alpha}) \end{pmatrix} = \sqrt{2}\begin{pmatrix} \mathrm{Re}\,\chi_\alpha \\ \mathrm{Im}\,\chi_\alpha \end{pmatrix} \tag{4.35}$$

in the Majorana basis.

## 5. Supersymmetry

The supersymmetry algebra is defined through the following structure relations:

$$\{Q_\alpha^L, \bar{Q}_{K\dot\alpha}\} = 2\sigma_{\alpha\dot\alpha}^m P_m \delta^L{}_K,$$

$$\{Q_\alpha^L, Q_\beta^K\} = \{\bar{Q}_{L\dot\alpha}, \bar{Q}_{K\dot\beta}\} = 0,$$

$$[Q_\alpha^L, P_m] = [\bar{Q}_{K\dot\alpha}, P_m] = 0. \tag{5.1}$$

The spinorial charges $Q$ and $\overline{Q}$ are labelled by an index $K$ or $L$. This index runs from 1 to $N$ and denotes the number of spinorial charges. When $N > 1$, the algebra is called extended supersymmetry. Note that spinorial charges commute with the four-momentum $P_m$.

The algebra (5.1) leads to several immediate consequences. In particular, it tells us that supersymmetric theories must contain equal numbers of fermionic and bosonic degrees of freedom. To see this, we introduce a fermionic number operator $N_{\mathrm{F}}$, such that

$$N_{\mathrm{F}}|\Omega\rangle = n_{\mathrm{F}}|\Omega\rangle, \tag{5.2}$$

when the state $|\Omega\rangle$ contains $n_{\mathrm{F}}$ fermions. Since $Q$ and $\overline{Q}$ create and annihilate fermions, we have immediately that

$$(-1)^{N_{\mathrm{F}}}Q = -Q(-1)^{N_{\mathrm{F}}}. \tag{5.3}$$

Taking the trace,

$$\begin{aligned} \operatorname{tr}(-1)^{N_{\mathrm{F}}}\{Q,\overline{Q}\} &= \operatorname{tr} Q(-1)^{N_{\mathrm{F}}}\overline{Q} + \overline{Q}(-1)^{N_{\mathrm{F}}}Q \\ &= -\operatorname{tr}(-1)^{N_{\mathrm{F}}}\{Q,\overline{Q}\} = 0, \end{aligned} \tag{5.4}$$

and using the supersymmetry algebra,

$$0 = \operatorname{tr}(-1)^{N_{\mathrm{F}}}\{Q,\overline{Q}\} = 2\sigma^m P_m \operatorname{tr}(-1)^{N_{\mathrm{F}}}, \tag{5.5}$$

we find

$$\operatorname{tr}(-1)^{N_{\mathrm{F}}} = 0. \tag{5.6}$$

This shows that the fermionic and bosonic modes cancel in trace over states. As we shall see, this complicates the discussion of supersymmetry phenomenology.

To formulate a supersymmetric field theory, we must represent the supersymmetry generators $P_m$, $Q$ and $\overline{Q}$ as conserved charges derived from local Noether currents. The supersymmetry algebra may then be viewed as a direct consequence of the field commutators. The Noether currents are associated with supersymmetry transformations which act linearly on field multiplets and leave the action invariant.

To find such field multiplets we first represent the supersymmetry algebra in terms of differential operators. For $N = 1$, these operators act in a space called superspace:

$$z^M = \left(x^m, \theta^\mu, \bar{\theta}_{\dot{\mu}}\right). \tag{5.7}$$

The variables $x^m$ are spacetime coordinates while $\theta^\mu$ and $\bar{\theta}_{\dot{\mu}}$ are anticommuting spinors:

$$\{\theta^\alpha, \theta^\beta\} = \{\bar{\theta}_{\dot{\alpha}}, \bar{\theta}_{\dot{\beta}}\} = \{\theta^\alpha, \bar{\theta}_{\dot{\alpha}}\} = 0,$$

$$[x^m, \theta^\alpha] = [x^m, \bar{\theta}_{\dot{\alpha}}] = 0. \tag{5.8}$$

One may easily verify that the differential operators

$$Q_\alpha = \frac{\partial}{\partial \theta^\alpha} - \mathrm{i}\sigma^m_{\alpha\dot{\alpha}}\bar{\theta}^{\dot{\alpha}}\partial_m, \qquad \bar{Q}_{\dot{\alpha}} = -\frac{\partial}{\partial \bar{\theta}^{\dot{\alpha}}} + \mathrm{i}\theta^\alpha\sigma^m_{\alpha\dot{\alpha}}\partial_m, \tag{5.9}$$

satisfy the following algebra:

$$\{Q_\alpha, \bar{Q}_{\dot{\alpha}}\} = 2\mathrm{i}\sigma^m_{\alpha\dot{\alpha}}\partial_m, \qquad \{Q_\alpha, Q_\beta\} = \{\bar{Q}_{\dot{\alpha}}, \bar{Q}_{\dot{\beta}}\} = 0. \tag{5.10}$$

In addition, there exist differential operators

$$D_\alpha = \frac{\partial}{\partial \theta^\alpha} + \mathrm{i}\sigma^m_{\alpha\dot{\alpha}}\bar{\theta}^{\dot{\alpha}}\partial_m, \qquad \bar{D}_{\dot{\alpha}} = -\frac{\partial}{\partial \bar{\theta}^{\dot{\alpha}}} - \mathrm{i}\theta^\alpha\sigma^m_{\alpha\dot{\alpha}}\partial_m, \tag{5.11}$$

which anticommute with $Q$ and $\bar{Q}$

$$\{Q_\alpha, D_\beta\} = \{\bar{Q}_{\dot{\alpha}}, \bar{D}_{\dot{\beta}}\} = \{Q_\alpha, \bar{D}_{\dot{\alpha}}\} = \{\bar{Q}_{\dot{\alpha}}, D_\alpha\} = 0, \tag{5.12}$$

and which satisfy

$$\{D_\alpha, \bar{D}_{\dot{\alpha}}\} = -2\mathrm{i}\sigma^m_{\alpha\dot{\alpha}}\partial_m, \qquad \{D_\alpha, D_\beta\} = \{\bar{D}_{\dot{\alpha}}, \bar{D}_{\dot{\beta}}\} = 0. \tag{5.13}$$

These $D$-operators are called covariant derivatives.

We are now prepared to define superfields. Superfields are functions of superspace which must be understood in terms of their power series expansions in $\theta$ and $\bar{\theta}$:

$$\begin{aligned} F(x, \theta, \bar{\theta}) = {} & f(x) + \theta\varphi(x) + \bar{\theta}\bar{\chi}(x) + \theta\theta m(x) + \bar{\theta}\bar{\theta}n(x) \\ & + \theta\sigma^m\bar{\theta}v_m(x) + \theta\theta\bar{\theta}\bar{\lambda}(x) + \bar{\theta}\bar{\theta}\theta\psi(x) \\ & + \theta\theta\bar{\theta}\bar{\theta}d(x). \end{aligned} \tag{5.14}$$

This is the most general expansion because all higher powers of $\theta$ and $\bar{\theta}$ vanish and because

$$\theta^\alpha\theta^\beta = -\tfrac{1}{2}\varepsilon^{\alpha\beta}\theta\theta, \qquad \bar{\theta}_{\dot{\alpha}}\bar{\theta}_{\dot{\beta}} = -\tfrac{1}{2}\varepsilon_{\dot{\alpha}\dot{\beta}}\bar{\theta}\bar{\theta}. \tag{5.15}$$

Supersymmetry transformations are generated by the differential operators $Q$ and $\bar{Q}$,

$$\begin{aligned}\delta_\xi F &= (\xi Q + \bar{\xi}\bar{Q})F \\ &= \delta_\xi f(x) + \theta\delta_\xi\varphi(x) + \bar{\theta}\delta_\xi\bar{\chi}(x) \\ &\quad + \theta\theta\delta_\xi m(x) + \bar{\theta}\bar{\theta}\delta_\xi n(x) + \theta\sigma^m\bar{\theta}\delta_\xi v_m(x) \\ &\quad + \theta\theta\bar{\theta}\delta_\xi\bar{\lambda}(x) + \bar{\theta}\bar{\theta}\theta\delta_\xi\psi(x) + \theta\theta\bar{\theta}\bar{\theta}\delta_\xi d(x).\end{aligned} \tag{5.16}$$

Here $\xi$ and $\bar{\xi}$ are anticommuting parameters. The transformation laws for the component fields $(f, \varphi, \bar{\chi}, \ldots)$ may be found from (5.16) by comparing powers of $\theta$ and $\bar{\theta}$. Note that the highest component of a superfield always transforms into a spacetime derivative.

Because of eq. (5.10) the commutator of two supersymmetry transformations yields

$$(\delta_\eta\delta_\xi - \delta_\xi\delta_\eta)f(x) = -2\mathrm{i}(\eta\sigma^m\bar{\xi} - \xi\sigma^m\bar{\eta})\,\partial_m f(x) \tag{5.17}$$

on any component field. A set of fields obeying this property is called a component multiplet. Every superfield yields a component multiplet and it may be shown that every component multiplet comes from a superfield.

Superfields do not in general yield irreducible component multiplets. Such multiplets are obtained by imposing appropriate restrictions on the superfields. These restrictions must respect supersymmetry, but otherwise there are no general formulae to indicate the proper constraints. Two sets of constraints have proven particularly useful. The first defines a chiral multiplet

$$\bar{D}_{\dot{\alpha}}\Phi = 0, \tag{5.18}$$

while the second specifies a vector multiplet

$$V^\dagger = V. \tag{5.19}$$

The reasons for these names will become apparent in the coming lectures.

We shall first investigate the chiral superfield. It is easy to see that

$$\bar{D}_{\dot{\alpha}}\theta^\alpha = 0, \qquad \bar{D}_{\dot{\alpha}}y^m = \bar{D}_{\dot{\alpha}}(x^m + \mathrm{i}\theta\sigma^m\bar{\theta}) = 0. \tag{5.20}$$

Any function of these variables satisfies eq. (5.18):

$$\begin{aligned}\Phi(y,\theta) &= A(y) + \sqrt{2}\,\theta\psi(y) + \theta\theta F(y) \\ &= A(x) + \mathrm{i}\theta\sigma^m\bar{\theta}\partial_m A(x) + \tfrac{1}{4}\theta\theta\bar{\theta}\bar{\theta}\Box A(x) + \sqrt{2}\,\theta\psi(x) \\ &\quad - (\mathrm{i}/\sqrt{2})\,\theta\theta\partial_m\psi(x)\sigma^m\bar{\theta} + \theta\theta F(x).\end{aligned} \tag{5.21}$$

This is the most general chiral superfield. The transformation law for the chiral multiplet may be obtained from eq. (5.16):

$$\delta_\xi A = \sqrt{2}\,\xi\psi, \qquad \delta_\xi \psi = \sqrt{2}\,\mathrm{i}\sigma^m\bar{\xi}\partial_m A + \sqrt{2}\,\xi F$$

$$\delta_\xi F = \sqrt{2}\,\mathrm{i}\bar{\xi}\bar{\sigma}^m\partial_m\psi. \tag{5.22}$$

As expected, the highest component transforms into a spacetime derivative.

An important advantage of superfields stems from the fact that sums and products of superfields are again superfields. This simplifies the construction of invariant actions. Furthermore, sums and products of chiral superfields remain chiral because $\bar{D}_{\dot{\alpha}}$ is a linear differential operator:

$$\bar{D}_{\dot{\alpha}}(\Phi_i\Phi_j) = 0, \qquad \bar{D}_{\dot{\alpha}}(\Phi_i\Phi_j\Phi_k) = 0,$$

$$\begin{aligned}
\Phi_i\Phi_j &= A_i(y)A_j(y) + \sqrt{2}\,\theta\big[A_i(y)\psi_j(y) + A_j(y)\psi_i(y)\big] \\
&\quad + \theta\theta\big[A_i(y)F_j(y) + A_j(y)F_i(y) - \psi_i(y)\psi_j(y)\big], \\
\Phi_i\Phi_j\Phi_k &= A_i(y)A_j(y)A_k(y) + \sqrt{2}\,\theta\big[\psi_iA_jA_k + \psi_jA_kA_i + \psi_kA_iA_j\big] \\
&\quad + \theta\theta\big[F_iA_jA_k + F_jA_kA_i + F_kA_iA_j \\
&\qquad - \psi_i\psi_jA_k - \psi_j\psi_kA_i - \psi_k\psi_iA_j\big].
\end{aligned} \tag{5.23}$$

The $\theta\theta$-components of $\Phi_i\Phi_j$ and $\Phi_i\Phi_j\Phi_k$ transform as the $F$-components of chiral multiplets. Since $F$-components transform into spacetime derivatives, the $\theta\theta$-components of $\Phi_i\Phi_j$ and $\Phi_i\Phi_j\Phi_k$ may be used to construct invariant actions.

To obtain the kinetic part of the Lagrangian, it is necessary to compute $\Phi_i^\dagger\Phi_i$. This is not a chiral multiplet because $\bar{D}_{\dot{\alpha}}\Phi = 0$, whereas $D_\alpha\Phi^\dagger = 0$. Only the $\theta\theta\bar{\theta}\bar{\theta}$-component of this product transforms into a spacetime derivative:

$$\begin{aligned}
\Phi_i^\dagger\Phi_i|_{\theta\theta\bar{\theta}\bar{\theta}} &= F_i^*F_i + \tfrac{1}{4}A_i^*\Box A_i + \tfrac{1}{4}A_i\Box A_i^* \\
&\quad - \tfrac{1}{2}\partial_m A_i^*\partial^m A_i + \tfrac{1}{2}\mathrm{i}\,\partial_m\bar{\psi}_i\bar{\sigma}^m\psi_i - \tfrac{1}{2}\mathrm{i}\,\bar{\psi}_i\bar{\sigma}^m\partial_m\psi_i.
\end{aligned} \tag{5.24}$$

With this result, we are able to write the most general supersymmetric renormalizable Lagrangian involving only spin-0 and spin-1/2 component fields:

$$\mathcal{L} = \Phi_i^\dagger \Phi_i|_{\theta\theta\bar{\theta}\bar{\theta}} + \left(\lambda_i \Phi_i + \tfrac{1}{2} m_{ij} \Phi_i \Phi_j + \tfrac{1}{3} g_{ijk} \Phi_i \Phi_j \Phi_k\right)|_{\theta\theta}$$
$$+ \left(\lambda_i^* \Phi_i^\dagger + \tfrac{1}{2} m_{ij}^* \Phi_i^\dagger \Phi_j^\dagger + \tfrac{1}{3} g_{ijk}^* \Phi_i^\dagger \Phi_j^\dagger \Phi_k^\dagger\right)|_{\bar{\theta}\bar{\theta}}. \quad (5.25)$$

The mass matrix $m_{ij}$ and the coupling constant $g_{ijk}$ are completely symmetric in their indices.

In components, the Lagrangian becomes

$$\mathcal{L} = \mathrm{i}\partial_m \bar{\psi}_i \bar{\sigma}^m \psi_i + A_i^* \Box A_i + F_i^* F_i$$
$$+ \left[\lambda_i F_i + m_{ij}\left(A_i F_j - \tfrac{1}{2}\psi_i \psi_j\right)\right.$$
$$\left. + g_{ijk}\left(A_i A_j F_k - \psi_i \psi_j A_k\right) + \text{h.c.}\right]. \quad (5.26)$$

The fields $F_i$ enter this expression without derivatives. They are auxiliary fields and may be eliminated through their Euler equations:

$$\partial\mathcal{L}/\partial F_i^* = F_i + \lambda_i^* + m_{ik}^* A_k^* + g_{ijk} A_j^* A_k^* = 0,$$
$$\partial\mathcal{L}/\partial F_i = F_i^* + \lambda_i + m_{ik} A_k + g_{ijk} A_j A_k = 0. \quad (5.27)$$

Substituting for the auxiliary fields, we find

$$\mathcal{L} = \mathrm{i}\partial_m \bar{\psi}_i \bar{\sigma}^m \psi_i + A_i^* \Box A_i$$
$$+ \left[-\tfrac{1}{2} m_{ij} \psi_i \psi_j - g_{ijk} A_i \psi_j \psi_k + \text{h.c.}\right] - F_i^* F_i. \quad (5.28)$$

In this expression $F_i$ and $F_i^*$ are solutions to eq. (5.27). The potential $V = F_i^* F_i$ is always greater than or equal to zero. Points where $F_i = 0$ are absolute minima of the potential.

In realistic applications to physics, the superfields $\Phi_i$ span a representation of a compact Lie group:

$$\Phi_i' = \left(\mathrm{e}^{\mathrm{i}T^r \lambda^r}\right)_{ik} \Phi_k \quad (5.29)$$

with $\lambda^r \in \mathbb{R}$. If $\lambda_i$, $m_{ij}$ and $g_{ijk}$ are invariant tensors, the Lagrangian (5.25) is invariant under the compact symmetry group. Note, however, that the part of the Lagrangian without the kinetic term $\Phi_i^\dagger \Phi_i$ is invariant under transformations of the type (5.29) with *complex* parameters $\lambda^r$. Points where $F_i = 0$ are also invariant under this complex extended symmetry. This gives supersymmetric ground states a larger degeneracy than might have been expected from the symmetry group itself.

## 6. Supersymmetric gauge theories

Gauge theories require vector fields to insure invariance under local symmetry groups. Similarly, supersymmetric gauge theories require vector superfields, defined by the condition

$$V^\dagger = V. \tag{6.1}$$

Such superfields contain ordinary vector fields as well as their supersymmetric partners:

$$\begin{aligned} V = {} & C(x) + \mathrm{i}\theta\chi(x) - \mathrm{i}\bar{\theta}\bar{\chi}(x) + \tfrac{1}{2}\mathrm{i}\theta\theta[M(x)+\mathrm{i}N(x)] \\ & - \tfrac{1}{2}\mathrm{i}\bar{\theta}\bar{\theta}[M(x)-\mathrm{i}N(x)] - \theta\sigma^m\bar{\theta}v_m(x) \\ & + i\theta\theta\bar{\theta}[\bar{\lambda}(x) + \tfrac{1}{2}\mathrm{i}\bar{\sigma}^m\partial_m\chi(x)] - \mathrm{i}\bar{\theta}\bar{\theta}\theta[\lambda(x) + \tfrac{1}{2}\mathrm{i}\sigma^m\partial_m\bar{\chi}(x)] \\ & + \tfrac{1}{2}\theta\theta\bar{\theta}\bar{\theta}[D(x) + \tfrac{1}{2}\Box C(x)]. \end{aligned} \tag{6.2}$$

The fields $C$, $M$, $N$ and $D$ are all real. We have chosen a very special decomposition in terms of component fields. This decomposition is suggested by the hermitian field $\Phi + \Phi^\dagger$, where $\Phi$ and $\Phi^\dagger$ are chiral superfields:

$$\begin{aligned} \Phi + \Phi^\dagger = {} & A + A^* + \sqrt{2}(\theta\psi + \bar{\theta}\bar{\psi}) + \theta\theta F + \bar{\theta}\bar{\theta}F^* \\ & + \mathrm{i}\theta\sigma^m\bar{\theta}\partial_m(A - A^*) + (\mathrm{i}/\sqrt{2})\theta\theta\bar{\theta}\bar{\sigma}^m\partial_m\psi \\ & + (\mathrm{i}/\sqrt{2})\bar{\theta}\bar{\theta}\theta\sigma^m\partial_m\bar{\psi} + \tfrac{1}{4}\theta\theta\bar{\theta}\bar{\theta}\Box(A + A^*). \end{aligned} \tag{6.3}$$

From this it is easy to see that

$$V \to V + \Phi + \Phi^\dagger \tag{6.4}$$

is a supersymmetric generalization of a gauge transformation. In component fields, this becomes

$$\begin{aligned} C &\to C + A + A^*, & \chi &\to \chi - \mathrm{i}\sqrt{2}\psi, \\ M + \mathrm{i}N &\to M + \mathrm{i}N - 2\mathrm{i}F, & v_m &\to v_m - \mathrm{i}\partial_m(A - A^*), \\ \lambda &\to \lambda, & D &\to D. \end{aligned} \tag{6.5}$$

Our choice of component fields renders $\lambda$ and $D$ gauge invariant. There is a special gauge, the WZ gauge, in which $C$, $\chi$, $M$ and $N$ are all zero. This gauge breaks supersymmetry but still allows the usual gauge transformations $v_m \to v_m + \partial_m a$.

The vector field $V$ may be viewed as a supersymmetric vector potential. To formulate a theory which is both supersymmetric and gauge invariant, we must incorporate this vector potential into a supersymmetric gauge invariant field strength:

$$W_\alpha = -\tfrac{1}{4}\bar{D}\bar{D}D_\alpha V, \qquad \bar{W}_{\dot\alpha} = -\tfrac{1}{4}DD\bar{D}_{\dot\alpha}V. \tag{6.6}$$

These expressions are gauge invariant because

$$\bar{D}\bar{D}D_\alpha(\Phi+\Phi^\dagger) = \bar{D}_{\dot\alpha}\{\bar{D}^{\dot\alpha}, D_\alpha\}\Phi = 2\mathrm{i}\sigma^m_{\alpha\dot\alpha}\bar{D}^{\dot\alpha}\partial_m\Phi = 0. \tag{6.7}$$

Furthermore, they satisfy the constraint equations

$$\bar{D}_{\dot\alpha}W_\beta = 0, \qquad D_\alpha\bar{W}_{\dot\beta} = 0, \qquad D^\alpha W_\alpha = \bar{D}_{\dot\alpha}\bar{W}^{\dot\alpha}. \tag{6.8}$$

The superfields $W_\alpha$ and $\bar{W}^{\dot\alpha}$ may be expanded in components using the decomposition of $V$:

$$W_\alpha = -\mathrm{i}\lambda_\alpha(y) + \left[\delta_\alpha{}^\beta D(y) - \tfrac{1}{2}\mathrm{i}(\sigma^m\bar\sigma^n)_\alpha{}^\beta v_{mn}(y)\right]\theta_\beta + \theta\theta\sigma^m_{\alpha\dot\alpha}\partial_m\bar\lambda^{\dot\alpha}(y),$$

$$y^m = x^m + \mathrm{i}\theta\sigma^m\bar\theta, \qquad v_{mn} = \partial_m v_n - \partial_n v_m. \tag{6.9}$$

It is possible to show that this expansion is the most general solution to the constraints (6.8). In particular, the constraints imply that $D = D^*$ and that $\partial_m v_{lk} + \partial_l v_{km} + \partial_k v_{ml} = 0$.

It is easy to form a supersymmetric gauge invariant Lagrangian from the field strengths (6.6):

$$\mathcal{L} = \tfrac{1}{4}\left(W^\alpha W_\alpha|_{\theta\theta} + \bar{W}_{\dot\alpha}\bar{W}^{\dot\alpha}|_{\bar\theta\bar\theta}\right). \tag{6.10}$$

In components, this becomes

$$\mathcal{L} = \tfrac{1}{2}D^2 - \tfrac{1}{4}v^{mn}v_{mn} - \mathrm{i}\lambda\sigma^m\partial_m\bar\lambda, \tag{6.11}$$

where we have freely integrated by parts.

We are now able to gauge any Abelian symmetry group,

$$\Phi_l \to \mathrm{e}^{-\mathrm{i}t_l\lambda}\Phi_l, \tag{6.12}$$

by promoting $\lambda$ to a full scalar multiplet. This insures that $\Phi$ remains chiral after a gauge transformation:

$$\Phi_l \to \mathrm{e}^{-\mathrm{i}t_l\Lambda}\Phi_l, \qquad \bar{D}_{\dot\alpha}\Lambda = 0. \tag{6.13}$$

If $V$ transforms as

$$V \to V + \mathrm{i}(\Lambda - \Lambda^\dagger), \tag{6.14}$$

the Lagrangian

$$\mathcal{L} = \tfrac{1}{4}\big(W^\alpha W_\alpha|_{\theta\theta} + \overline{W}_{\dot\alpha}\overline{W}^{\dot\alpha}|_{\bar\theta\bar\theta}\big) + \Phi_i^\dagger e^{t_i V}\Phi_i$$
$$+ \big[\big(\tfrac{1}{2}m_{ij}\Phi_i\Phi_j + \tfrac{1}{3}g_{ijk}\Phi_i\Phi_j\Phi_k\big)|_{\theta\theta} + \text{h.c.}\big] \tag{6.15}$$

is gauge invariant, provided, of course, that the mass and coupling terms are themselves invariant under rigid transformations.

The supersymmetric generalization of electrodynamics requires two chiral superfields. These superfields carry opposite $U(1)$ charges:

$$\Phi'_\pm = e^{\mp ie\Lambda}\Phi_\pm. \tag{6.16}$$

The invariant Lagrangian

$$\mathcal{L}_{\text{QED}} = \tfrac{1}{4}\big(WW|_{\theta\theta} + \overline{W}\,\overline{W}|_{\bar\theta\bar\theta}\big) + \Phi_+^\dagger e^{eV}\Phi_+|_{\theta\theta\bar\theta\bar\theta}$$
$$+ \Phi_-^\dagger e^{-eV}\Phi_-|_{\theta\theta\bar\theta\bar\theta} + m\big(\Phi_+\Phi_-|_{\theta\theta} + \Phi_+^\dagger\Phi_-^\dagger|_{\bar\theta\bar\theta}\big), \tag{6.17}$$

has the following decomposition in terms of component fields:

$$\mathcal{L}_{\text{QED}} = \tfrac{1}{2}D^2 - \tfrac{1}{4}v_{mn}v^{mn} - i\lambda\sigma^n\partial_n\bar\lambda$$
$$+ F_+F_+^* + F_-F_-^* + A_+^*\Box A_+ + A_-^*\Box A_-$$
$$+ i\big(\partial_n\bar\psi_+\bar\sigma^n\psi_+ + \partial_n\bar\psi_-\bar\sigma^n\psi_-\big)$$
$$+ ev^n\big(\tfrac{1}{2}\bar\psi_+\bar\sigma^n\psi_+ - \tfrac{1}{2}\bar\psi_-\bar\sigma^n\psi_-$$
$$+ \tfrac{1}{2}iA_+^*\partial_nA_+ - \tfrac{1}{2}i\partial_nA_+^*A_+ - \tfrac{1}{2}iA_-^*\partial_nA_- + \tfrac{1}{2}i\partial_nA_-^*A_-\big)$$
$$- \tfrac{1}{2}ie\sqrt{2}\big(A_+\bar\psi_+\bar\lambda - A_+^*\psi_+\lambda - A_-\bar\psi_-\bar\lambda + A_-^*\psi_-\lambda\big)$$
$$+ \tfrac{1}{2}eD\big(A_+^*A_+ - A_-^*A_-\big) - \tfrac{1}{4}e^2v_nv^n\big(A_+^*A_+ + A_-^*A_-\big)$$
$$+ m\big(A_+F_- + A_-F_+ - \psi_+\psi_- - \bar\psi_+\bar\psi_- + A_+^*F_-^*$$
$$+ A_-^*F_+^*\big). \tag{6.18}$$

From eq. (6.18) we see that the two Weyl spinors $\psi_+$ and $\psi_-$ combine to form one massive Dirac spinor, the electron.

It is not difficult to extend this formalism to non-Abelian groups. The transformation laws for the chiral and vector superfields become

$$\Phi' = e^{-i\Lambda}\Phi, \qquad \Phi'^\dagger = \Phi^\dagger e^{i\Lambda^\dagger}, \qquad e^{V'} = e^{-i\Lambda^\dagger}e^{V}e^{i\Lambda}. \tag{6.19}$$

Here $\Phi$ is a column vector and $\Lambda$ is a Lie algebra valued chiral superfield:

$$\Lambda = T^a\Lambda^a. \tag{6.20}$$

With these transformation laws, the combination

$$\Phi^{\dagger} \mathrm{e}^{V} \Phi \tag{6.21}$$

is gauge invariant.

It should be noted that the transformation law for the vector superfield is independent of the representation chosen for the generators $T^a$. This is because only commutators enter into eq. (6.19). Furthermore, the transformation law starts with a term independent of $V$,

$$V' = V + \mathrm{i}(\Lambda - \Lambda^{\dagger}) + \cdots, \tag{6.22}$$

so it is always possible to go to the WZ gauge. The generalization of the Yang–Mills field strength

$$W_{\alpha} = -\tfrac{1}{4} \bar{D}\bar{D} \mathrm{e}^{-V} D_{\alpha} \mathrm{e}^{V} \tag{6.23}$$

transforms covariantly under gauge transformations:

$$W'_{\alpha} = \mathrm{e}^{-\mathrm{i}\Lambda} W_{\alpha} \mathrm{e}^{\mathrm{i}\Lambda}. \tag{6.24}$$

This may be seen as follows:

$$\begin{aligned} W'_{\alpha} &= -\tfrac{1}{4} \bar{D}\bar{D} \mathrm{e}^{-\mathrm{i}\Lambda} \mathrm{e}^{-V} \mathrm{e}^{\mathrm{i}\Lambda^{\dagger}} D_{\alpha} (\mathrm{e}^{-\mathrm{i}\Lambda^{\dagger}} \mathrm{e}^{V} \mathrm{e}^{\mathrm{i}\Lambda}) \\ &= -\tfrac{1}{4} \mathrm{e}^{-\mathrm{i}\Lambda} \bar{D}\bar{D} \mathrm{e}^{-V} \left[ (D_{\alpha} \mathrm{e}^{V}) \mathrm{e}^{\mathrm{i}\Lambda} + \mathrm{e}^{V} D_{\alpha} \mathrm{e}^{\mathrm{i}\Lambda} \right] \\ &= \mathrm{e}^{-\mathrm{i}\Lambda} W_{\alpha} \mathrm{e}^{\mathrm{i}\Lambda}. \end{aligned} \tag{6.25}$$

With these assignments, the Lagrangian

$$\begin{aligned} \mathcal{L} &= \tfrac{1}{8} \operatorname{tr} \left( W^{\alpha} W_{\alpha}|_{\theta\theta} + \bar{W}_{\dot{\alpha}} \bar{W}^{\dot{\alpha}}|_{\bar{\theta}\bar{\theta}} \right) + \Phi^{\dagger} \mathrm{e}^{V} \Phi|_{\theta\theta\bar{\theta}\bar{\theta}} \\ &\quad + \left[ \left( \tfrac{1}{2} m_{ij} \Phi_i \Phi_j + \tfrac{1}{3} g_{ijk} \Phi_i \Phi_j \Phi_k \right)|_{\theta\theta} + \text{h.c.} \right] \end{aligned} \tag{6.26}$$

is gauge invariant. The mass and the coupling terms are totally symmetric in their indices. They must also be invariant under rigid symmetry transformations.

As an example of a supersymmetric Yang–Mills theory, we shall consider a possible supersymmetric generalization of the Glashow–Salam–Weinberg model. The left-handed lepton doublet, the right-handed electron singlet and the Higgs doublet all become chiral superfields, transforming exactly as in the standard model:

$$\begin{aligned} \delta\Phi_{\mathrm{L}} &= -\tfrac{1}{2} \mathrm{i} \boldsymbol{\tau} \cdot \boldsymbol{\Lambda} \Phi_{\mathrm{L}} - \tfrac{1}{2} \mathrm{i} \Lambda \Phi_{\mathrm{L}}, \qquad \delta\Phi_{\mathrm{R}} = \mathrm{i} \Lambda \Phi_{\mathrm{R}}, \\ \delta\Phi_{\mathrm{H}} &= -\tfrac{1}{2} \mathrm{i} \boldsymbol{\tau} \cdot \boldsymbol{\Lambda} \Phi_{\mathrm{H}} - \tfrac{1}{2} \mathrm{i} \Lambda \Phi_{\mathrm{H}}. \end{aligned} \tag{6.27}$$

Their only invariant interaction is given by

$$g^{ab}\Phi_{\mathrm{H}a}\Phi_{\mathrm{L}b}\Phi_{\mathrm{R}} + \text{h.c.}, \tag{6.28}$$

where the antisymmetric two-by-two tensor $g^{ab}$ combines two doublets into a singlet. This interaction does not break the SU(2)×U(1) symmetry. To do this we must add additional chiral superfields. One possible choice is to add a doublet $\Phi_{\mathrm{T}}$ and a singlet $\Phi_{\mathrm{S}}$:

$$\delta\Phi_{\mathrm{T}} = -\tfrac{1}{2}\mathrm{i}\boldsymbol{\tau}\cdot\boldsymbol{\Lambda}\Phi_{\mathrm{T}} + \tfrac{1}{2}\mathrm{i}\Lambda\Phi_{\mathrm{T}}, \qquad \delta\Phi_{\mathrm{S}} = 0. \tag{6.29}$$

Imposing the discrete symmetries

$$\begin{aligned} &\Phi_{\mathrm{L}} \to -\Phi_{\mathrm{L}}, \qquad \Phi_{\mathrm{R}} \to -\Phi_{\mathrm{R}}, \qquad \Phi_{\mathrm{H}} \to \Phi_{\mathrm{H}}, \\ &\Phi_{\mathrm{T}} \to \Phi_{\mathrm{T}}, \qquad \Phi_{\mathrm{S}} \to \Phi_{\mathrm{S}}, \end{aligned} \tag{6.30}$$

we find the following gauge invariant chiral interactions:

$$\begin{aligned} &Yg^{ab}\Phi_{\mathrm{H}a}\Phi_{\mathrm{L}b}\Phi_{\mathrm{R}} + Zg^{ab}\Phi_{\mathrm{H}a}\Phi_{\mathrm{T}b}\Phi_{\mathrm{S}} \\ &\quad + mg^{ab}\Phi_{\mathrm{H}a}\Phi_{\mathrm{T}b} + W\Phi_{\mathrm{S}} \\ &\quad + \tfrac{1}{2}\mu\Phi_{\mathrm{S}}^2 + \tfrac{1}{3}f\Phi_{\mathrm{S}}^3 + \text{h.c..} \end{aligned} \tag{6.31}$$

In the next section we shall see that these interactions spontaneously break the internal symmetry.

To complete the theory, we also introduce four vector multiplets $\boldsymbol{V}$, $V$, along with their associated transformation laws:

$$\begin{aligned} &\mathrm{e}^{\frac{1}{2}\boldsymbol{\tau}\cdot\boldsymbol{V}'} = \mathrm{e}^{-\frac{1}{2}\mathrm{i}\boldsymbol{\tau}\cdot\boldsymbol{\Lambda}^{\dagger}}\mathrm{e}^{\frac{1}{2}\boldsymbol{\tau}\cdot\boldsymbol{V}}\mathrm{e}^{\frac{1}{2}\mathrm{i}\boldsymbol{\tau}\cdot\boldsymbol{\Lambda}} \\ &V' = V + \mathrm{i}(\Lambda - \Lambda^{\dagger}). \end{aligned} \tag{6.32}$$

These fields render the full Lagrangian gauge invariant:

$$\begin{aligned} \mathcal{L}_{\mathrm{GWS}} = {}& \tfrac{1}{4}\big[\big(\boldsymbol{W}^{\alpha}\cdot\boldsymbol{W}_{\alpha} + W^{\alpha}W_{\alpha}\big)|_{\theta\theta} + \big(\overline{\boldsymbol{W}}_{\dot\alpha}\cdot\overline{\boldsymbol{W}}^{\dot\alpha} + \overline{W}_{\dot\alpha}\overline{W}^{\dot\alpha}\big)|_{\bar\theta\bar\theta}\big] \\ & + \big[\Phi_{\mathrm{L}}^{\dagger}\exp\big(\tfrac{1}{2}\boldsymbol{\tau}\cdot\boldsymbol{V} + \tfrac{1}{2}V\big)\Phi_{\mathrm{L}} + \Phi_{\mathrm{R}}^{\dagger}\mathrm{e}^{-V}\Phi_{\mathrm{R}} + \Phi_{\mathrm{S}}^{\dagger}\Phi_{\mathrm{S}} \\ & \quad + \Phi_{\mathrm{H}}^{\dagger}\exp\big(\tfrac{1}{2}\boldsymbol{\tau}\cdot\boldsymbol{V} + \tfrac{1}{2}V\big)\Phi_{\mathrm{H}} \\ & \quad + \Phi_{\mathrm{T}}^{\dagger}\exp\big(\tfrac{1}{2}\boldsymbol{\tau}\cdot\boldsymbol{V} - \tfrac{1}{2}V\big)\Phi_{\mathrm{T}}\big]|_{\theta\theta\bar\theta\bar\theta} \\ & + \big[\big(W\Phi_{\mathrm{S}} + \tfrac{1}{2}\mu\Phi_{\mathrm{S}}^2 + \tfrac{1}{3}f\Phi_{\mathrm{S}}^3 + mg^{ab}\Phi_{\mathrm{H}a}\Phi_{\mathrm{T}b} \\ & \quad + Yg^{ab}\Phi_{\mathrm{H}a}\Phi_{\mathrm{L}b}\Phi_{\mathrm{R}} + Zg^{ab}\Phi_{\mathrm{H}a}\Phi_{\mathrm{T}b}\Phi_{\mathrm{S}}\big)|_{\theta\theta} + \text{h.c.}\big]. \end{aligned} \tag{6.33}$$

This is the most general supersymmetric gauge invariant Lagrangian consistent with the discrete symmetries (6.30). We leave to the reader the task of expressing this Lagrangian in terms of component fields.

## 7. Spontaneous symmetry breaking

In supersymmetric gauge theories both gauge symmetry and supersymmetry may be broken spontaneously. We shall discuss examples of each later in this section. Before we do this, however, we first discuss some general properties of supersymmetry breaking. These rest on the fact that $H$ takes the following form:

$$H = \tfrac{1}{4}(\bar{Q}_1 Q_1 + Q_1 \bar{Q}_1 + \bar{Q}_2 Q_2 + Q_2 \bar{Q}_2) \tag{7.1}$$

in supersymmetric theories. This is a direct consequence of the supersymmetry algebra. It tells us that $\langle \Psi | H | \Psi \rangle \geqslant 0$ for every state $|\Psi\rangle$. It also tells us that states with vanishing energy are supersymmetric ground states of the theory. These states are ground states because there are no negative eigenvalues of $H$. Furthermore, they are supersymmetric since $\langle 0 | H | 0 \rangle = 0$ implies $Q|0\rangle = 0$ from eq. (7.1). To break supersymmetry we must find ground states with positive energy density.

Fayet and Iliopoulos have shown how to break supersymmetry spontaneously in Abelian gauge theories. They note that the $\theta\theta\bar{\theta}\bar{\theta}$-component of the vector superfield is both supersymmetric and gauge invariant. They add this $D$-term to the interaction Lagrangian and find that it spontaneously breaks supersymmetry.

To see this, we return to the supersymmetric extension of electrodynamics:

$$\begin{aligned} \mathcal{L} &= \tfrac{1}{4}\left[ W^\alpha W_\alpha |_{\theta\theta} + \bar{W}_{\dot\alpha} \bar{W}^{\dot\alpha} |_{\bar\theta\bar\theta} \right] \\ &\quad + \left( \Phi_+^\dagger \mathrm{e}^{eV} \Phi_+ + \Phi_-^\dagger \mathrm{e}^{-eV} \Phi_- + 2\kappa V \right)|_{\theta\theta\bar\theta\bar\theta} \\ &\quad + m\left( \Phi_+ \Phi_- |_{\theta\theta} + \Phi_+^\dagger \Phi_-^\dagger |_{\bar\theta\bar\theta} \right). \end{aligned} \tag{7.2}$$

In this model, the potential is given by

$$V = \tfrac{1}{2} D^2 + F_+ F_+^* + F_- F_-^*, \tag{7.3}$$

where $D$, $F_+$ and $F_-$ are solutions to the Euler equations:

$$\begin{aligned} &D + \kappa + \tfrac{1}{2} e (A_+^* A_+ - A_-^* A_-) = 0, \\ &F_+ + m A_-^* = 0, \qquad F_- + m A_+^* = 0. \end{aligned} \tag{7.4}$$

Substituting for the auxiliary fields, we find

$$\begin{aligned} V &= \tfrac{1}{2}\kappa^2 + \left( m^2 + \tfrac{1}{2} e\kappa \right) A_+^* A_+ \\ &\quad + \left( m^2 - \tfrac{1}{2} e\kappa \right) A_-^* A_- + \tfrac{1}{8} e^2 (A_+^* A_+ - A_-^* A_-)^2. \end{aligned} \tag{7.5}$$

When $m^2 + \frac{1}{2}e\kappa > 0$ and $m^2 - \frac{1}{2}e\kappa > 0$, this expression is positive definite, so supersymmetry is spontaneously broken. The masses of the scalar particles become $m_+^2 = m^2 + \frac{1}{2}e\kappa$ and $m_-^2 = m^2 - \frac{1}{2}e\kappa$, while the spinor retains its mass $m$. Note that $m_+^2 + m_-^2 = 2m^2$.

Since supersymmetry is spontaneously broken, we expect to find a massless Goldstone fermion associated with the symmetry breaking. From the transformation law for $\lambda$, the spinorial partner of the vector field,

$$\delta_\xi \lambda = \mathrm{i}\xi D + \sigma^{mn}\xi v_{mn}, \tag{7.6}$$

we see that $\lambda$ transforms inhomogeneously as soon as $D$ acquires a vacuum expectation value:

$$\delta_\xi \lambda = -\mathrm{i}\xi\kappa + \cdots. \tag{7.7}$$

Such a transformation law is incompatible with an invariant mass, so we identify $\lambda$ as the Goldstone field.

As a second example of symmetry breaking we consider a model in which the gauge symmetry is broken spontaneously. We start again from supersymmetric QED, but this time we add one neutral chiral superfield. We take

$$\mathcal{L}_{\mathrm{PE}} = \left(\tfrac{1}{2}\Phi^2 + \lambda\Phi + \tfrac{1}{3}f\Phi^3 + \mu\Phi_+\Phi_- + g\Phi\Phi_+\Phi_-\right)|_{\theta\theta} + \mathrm{h.c.} \tag{7.8}$$

as our interaction Lagrangian. As usual, the potential energy is expressed in terms of the auxiliary fields,

$$V = \tfrac{1}{2}D^2 + F_+F_+^* + F_-F_-^* + FF^*, \tag{7.9}$$

where the auxiliary fields are solutions to the Euler equations.

Supersymmetric ground states are those for which

$$D = F_+ = F_- = F = 0. \tag{7.10}$$

The $F$ terms lead to the following equations:

$$\begin{gathered} \lambda + m^2 A + gA_+A_- + fA^2 = 0, \\ A_-\left(\mu^2 + gA\right) = 0, \qquad A_+\left(\mu^2 + gA\right) = 0. \end{gathered} \tag{7.11}$$

These equations have two solutions:

$$A_+ = A_- = 0, \quad \lambda + m^2 A + fA^2 = 0, \tag{7.12a}$$

$$A = \frac{-\mu^2}{g}, \quad A_+A_- = -\frac{1}{g}\left(\lambda - \frac{m^2\mu^2}{g} + \frac{f\mu^4}{g^2}\right). \tag{7.12b}$$

The first preserves the U(1) symmetry, but the second breaks it spontaneously. In the second case only the product $A_+A_-$ is determined. This is a consequence of the fact that $\mathcal{L}_{PE}$ is invariant under the complex extension of U(1).

The term $\frac{1}{2}D^2$ in the potential (7.9) lifts this degeneracy. From the Euler equation

$$D = \tfrac{1}{2}e(A_+A_+^* - A_-A_-^* + 2\kappa/e), \tag{7.13}$$

we see that it is always possible to choose $A_+$ and $A_-$ such that $D = 0$. In this model there exists a supersymmetric minimum for any choice of $\kappa$. This illustrates the difficulty of simultaneously breaking both gauge symmetry and supersymmetry.

As a final example of symmetry breaking we discuss the ground states of the supersymmetric Glashow–Salam–Weinberg model introduced at the end of the previous section. Because the term $mg^{ab}\Phi_{\mathrm{H}a}\Phi_{\mathrm{T}b}$ may be obtained by a shift in the field $\Phi_{\mathrm{S}}$, we shall only consider the case $m = 0$. With this restriction, the potential takes the form

$$V = \tfrac{1}{2}D^2 + \tfrac{1}{2}\boldsymbol{D}^2 + F_{\mathrm{S}}^*F_{\mathrm{S}} + F_{\mathrm{R}}^*F_{\mathrm{R}} + F_{\mathrm{L}}^\dagger F_{\mathrm{L}} + F_{\mathrm{H}}^\dagger F_{\mathrm{H}} + F_{\mathrm{T}}^\dagger F_{\mathrm{T}}, \tag{7.14}$$

where the $F$'s and $D$'s are given by their Euler equations:

$$\begin{aligned}
&F_{\mathrm{S}}^* + W + \mu A_{\mathrm{S}} + fA_{\mathrm{S}}^2 + Zg^{ab}A_{\mathrm{H}a}A_{\mathrm{T}b} = 0,\\
&F_{\mathrm{R}}^* + Yg^{ab}A_{\mathrm{H}a}A_{\mathrm{L}b} = 0,\\
&F_{\mathrm{L}}^{\dagger r} - Yg^{ra}A_{\mathrm{H}a}A_{\mathrm{R}} = 0,\\
&F_{\mathrm{H}}^{\dagger r} + Zg^{rb}A_{\mathrm{T}b}A_{\mathrm{S}} + Yg^{rb}A_{\mathrm{L}b}A_{\mathrm{R}} = 0,\\
&F_{\mathrm{T}}^{\dagger r} - Zg^{rb}A_{\mathrm{H}b}A_{\mathrm{S}} = 0,\\
&D + \tfrac{1}{2}\left(\tfrac{1}{2}A_{\mathrm{L}}^\dagger A_{\mathrm{L}} + \tfrac{1}{2}A_{\mathrm{H}}^\dagger A_{\mathrm{H}} - \tfrac{1}{2}A_{\mathrm{T}}^\dagger A_{\mathrm{T}} - A_{\mathrm{R}}^*A_{\mathrm{R}}\right) = 0,\\
&\boldsymbol{D} + \tfrac{1}{2}\left(A_{\mathrm{L}}^\dagger\boldsymbol{\tau}A_{\mathrm{L}} + A_{\mathrm{H}}^\dagger\boldsymbol{\tau}A_{\mathrm{H}} + A_{\mathrm{T}}^\dagger\boldsymbol{\tau}A_{\mathrm{T}}\right) = 0.
\end{aligned} \tag{7.15}$$

Supersymmetric minima correspond to solutions with $F_{\mathrm{S}} = F_{\mathrm{R}} = F_{\mathrm{L}} = F_{\mathrm{H}} = F_{\mathrm{T}} = D = \boldsymbol{D} = 0$. We first set the $F$'s to zero. The resulting equations

$$\begin{aligned}
&W + \mu A_{\mathrm{S}} + fA_{\mathrm{S}}^2 + Zg^{ab}A_{\mathrm{H}a}A_{\mathrm{T}b} = 0,\\
&Yg^{ab}A_{\mathrm{H}a}A_{\mathrm{L}b} = 0, \qquad Yg^{ra}A_{\mathrm{H}a}A_{\mathrm{R}} = 0,\\
&Zg^{rb}A_{\mathrm{T}b}A_{\mathrm{S}} + Yg^{rb}A_{\mathrm{L}b}A_{\mathrm{R}} = 0, \qquad Zg^{rb}A_{\mathrm{H}b}A_{\mathrm{S}} = 0,
\end{aligned} \tag{7.16}$$

have two sets of solutions:

1) $A_S \neq 0$, which implies

$$A_H = 0, \qquad A_T = -(YA_R/ZA_S)A_L, \qquad W + \mu A_S + fA_S^2 = 0, \tag{7.17}$$

and

2) $A_S = 0$, which implies

$$g^{ab}A_{Ha}A_{Tb} = -W/Z, \qquad g^{ab}A_{Ha}A_{Lb} = 0, \qquad A_R = 0. \tag{7.18}$$

The second solution gives a nonzero vacuum expectation value to the Higgs field. By an SU(2) rotation we are free to choose

$$A_H = \begin{pmatrix} 0 \\ a \end{pmatrix}, \tag{7.19}$$

giving

$$A_T = \begin{pmatrix} W/aZ \\ t \end{pmatrix}, \qquad A_L = \begin{pmatrix} 0 \\ b \end{pmatrix}, \tag{7.20}$$

in accord with (7.18).

We now set the $D$'s to zero:

$$A_L^\dagger A_L + A_H^\dagger A_H - A_T^\dagger A_T - 2A_R^* A_R = 0,$$
$$A_L^\dagger \tau A_L + A_H^\dagger \tau A_H + A_T^\dagger \tau A_T = 0. \tag{7.21}$$

Substituting eqs. (7.19) and (7.20), we find supersymmetric minima for those values of $a$, $b$ and $t$ that satisfy:

$$bb^* + aa^* - tt^* - \frac{1}{aa^*}\frac{WW^*}{ZZ^*} = 0,$$
$$-bb^* - aa^* - tt^* + \frac{1}{aa^*}\frac{WW^*}{ZZ^*} = 0,$$
$$\frac{tW^*}{a^*Z^*} + \frac{t^*W}{aZ} = 0, \qquad \frac{tW^*}{a^*Z^*} - \frac{t^*W}{a^*Z^*} = 0. \tag{7.22}$$

These equations force $t = 0$. They also fix $aa^*$ in terms of $b$:

$$aa^*(aa^* + bb^*) = WW^*/ZZ^*. \tag{7.23}$$

Equation (7.23) gives a zero of the potential for any choice of $b$. These minima preserve supersymmetry but break SU(2)×U(1). If we set $b = 0$, we obtain a model which resembles the standard Glashow–Salam–Weinberg model. We have, however, many more particles, including the supersymmetric partners of the observed leptons. Because supersymmetry is unbroken, the leptons and their partners have identical mass. This is phenomenologically unacceptable, so supersymmetry must be broken. But how?

## 8. Differential forms

In this section we shall introduce the basic concepts of differential forms. We shall use differential forms to formulate gauge theories, supersymmetric gauge theories, general relativity and supergravity. This formalism is the most systematic way to formulate supergravity, and it illustrates quite clearly the close connections between supergravity and other gauge theories.

We start with the definition of the *exterior product*. Let $L$ be an $n$-dimensional vector space over $\mathbb{C}$ with elements $\alpha$, $\beta$,... and basis $\sigma^1,\dots,\sigma^n$. The exterior product of two elements of $L$ is defined through the following relations:

$$1)\ \alpha \wedge \beta = -\beta \wedge \alpha,$$

$$2)\ (a_1\alpha_1 + a_2\alpha_2) \wedge \beta = a_1\alpha_1 \wedge \beta + a_2\alpha_2 \wedge \beta, \tag{8.1}$$

where $a_1, a_2 \in \mathbb{C}$. The exterior product of all the vectors of $L$ forms a vector space $\Lambda^2$:

$$\lambda \in \Lambda^2, \quad \lambda = \frac{1}{2}\sum_{i,j} a_{ij}\sigma^j \wedge \sigma^i, \quad a_{ij} = -a_{ji}. \tag{8.2}$$

This space has dimension $n(n-1)/2$.

Definition (8.1) may be easily generalized to exterior products of $p$ elements of $L$. Such products span $\binom{n}{p}$-dimensional vector spaces $\Lambda^p$:

$$\lambda \in \Lambda^p, \quad \lambda = \frac{1}{p!}\sum_{k_1,\dots,k_p} b_{k_1\cdots k_p}\sigma^{k_p} \wedge \cdots \wedge \sigma^{k_1}. \tag{8.3}$$

The coefficients $b_{k_1\cdots k_p}$ are totally antisymmetric in their indices. Note that $\Lambda^p = 0$ for $p > n$.

Defining $\Lambda^0 = \mathbb{C}$, $\Lambda^1 = L$, we can now multiply elements of $\Lambda^p$ and $\Lambda^q$ for arbitrary $p$ and $q$:

$$\lambda \in \Lambda^p, \qquad \mu \in \Lambda^q, \qquad \lambda \wedge \mu \in \Lambda^{p+q}. \tag{8.4}$$

This multiplication obeys the following properties:

$$\lambda \wedge \mu = (-1)^{pq} \mu \wedge \lambda, \tag{8.5a}$$

$$(a_1\lambda_1 + a_2\lambda_2) \wedge \mu = a_1\lambda_1 \wedge \mu + a_2\lambda_2 \wedge \mu, \tag{8.5b}$$

$$\lambda \wedge (\mu \wedge \omega) = (\lambda \wedge \mu) \wedge \omega. \tag{8.5c}$$

These properties may be derived directly from the definitions (8.1) and (8.3).

*Differential forms* are elements of $\Lambda^p$ defined on an $n$-dimensional manifold $\mathfrak{M}$. The antisymmetric coefficients are smooth functions on $\mathfrak{M}$. In the neighborhood of any point they are functions of the $n$-dimensional euclidian space $E^n$. At each point, $\Lambda^1$ is spanned by the differentials $dx^1,\ldots,dx^n$. The spaces $\Lambda^p$ are built directly from these differentials:

$$\text{0-form:} \quad a(x^1,\ldots,x^n),$$

$$\text{1-form:} \quad \sum_i dx^i a_i(x^1,\ldots,x^n),$$

$$p\text{-form:} \quad \frac{1}{p!} \sum_{i_1,\ldots,i_p} dx^{i_p} \wedge \cdots \wedge dx^{i_1} a_{i_1 \cdots i_p}(x^1,\ldots,x^n). \tag{8.6}$$

As before, $\Lambda^p = 0$ for $p > n$.

Having introduced differential forms, we are now prepared to define *exterior derivatives*. Exterior derivatives map $p$-forms into $(p+1)$-forms according to the following properties:

$$d(\omega + \mu) = d\omega + d\mu, \tag{8.7a}$$

$$d(\omega \wedge \mu) = \omega \wedge d\mu + (-1)^p d\omega \wedge \mu \quad \text{when } \mu \text{ is a } p\text{-form}, \tag{8.7b}$$

$$dd\omega = 0, \tag{8.7c}$$

$$da(x^1,\ldots,x^n) = \sum_i dx^i \frac{\partial a}{\partial x^i}. \tag{8.7d}$$

These properties are sufficient to define the action of the exterior derivative on arbitrary $p$-forms:

$$\omega = \sum_{i_1,\ldots,i_p} dx^{i_p} \wedge \cdots \wedge dx^{i_1} a_{i_1 \cdots i_p}(x^1,\ldots,x^n),$$

$$d\omega = \sum_{i_1,\ldots,i_p,l} dx^{i_p} \wedge \cdots \wedge dx^{i_1} \wedge dx^l \frac{\partial}{\partial x^l} a_{i_1 \cdots i_p}(x^1,\ldots,x^n). \tag{8.8}$$

This formalism has the important advantage that equations written in terms of differential forms and exterior derivatives are covariant under local coordinate changes. To see this, we consider coordinate changes in $E^n$:

$$y^r = y^r(x^1,\dots,x^n). \tag{8.9}$$

Every function of $y^1,\dots,y^n$ induces a natural function of $x^1,\dots,x^n$ through the following definition:

$$a(y^1,\dots,y^n) = a(y^1(x),\dots,y^n(x)) = \tilde{a}(x^1,\dots,x^n). \tag{8.10}$$

This definition simply expresses the fact that physical quantities take the same value at the same point, regardless of their labelling in $E^n$. The same should hold true for forms, so we define

$$\begin{aligned}\sum_i \mathrm{d}y^i a_i(y^1,\dots,y^n) &= \sum_{i,j} \mathrm{d}x^j \frac{\partial y^i}{\partial x^j} a_i(y^1(x),\dots,y^n(x)) \\ &= \sum_j \mathrm{d}x^j \tilde{a}_j(x^1,\dots,x^n),\end{aligned} \tag{8.11}$$

for one-forms, and

$$\begin{aligned}&\sum_{i_1,\dots,i_p} \mathrm{d}y^{i_p} \wedge \cdots \wedge \mathrm{d}y^{i_1} a_{i_1\cdots i_p}(y^1,\dots,y^n) \\ &= \sum_{\substack{i_1,\dots,i_p,\\ j_1,\dots,j_p}} \mathrm{d}x^{j_p} \wedge \cdots \wedge \mathrm{d}x^{j_1} \frac{\partial y^{i_p}}{\partial x^{j_1}} \cdots \frac{\mathrm{d}y^{i_1}}{\mathrm{d}x^{j_p}} a_{i_1\cdots i_p}(y^1(x),\dots,y^n(x)) \\ &= \sum_{j_1,\dots,j_p} \mathrm{d}x^{j_p} \wedge \cdots \wedge \mathrm{d}x^{j_1} \tilde{a}_{j_1\cdots j_p}(x^1,\dots,x^n),\end{aligned} \tag{8.12}$$

for $p$-forms. We denote this map of a $y$-basis form into an $x$-basis form by $\phi^*$. The map $\phi^*$ has the following properties:

$$\phi^*(\omega+\eta) = \phi^*\omega + \phi^*\eta, \tag{8.13a}$$

$$\phi^*(\omega \wedge \eta) = (\phi^*\omega) \wedge (\phi^*\eta), \tag{8.13b}$$

$$\mathrm{d}(\phi^*\omega) = \phi^*(\mathrm{d}\omega). \tag{8.13c}$$

The proofs of eqs. (8.13a, b) are obvious. We shall prove (8.13c) by explicit calculation:

$$\omega = \sum \mathrm{d}y^{i_p} \wedge \cdots \wedge \mathrm{d}y^{i_1} a_{i_1 \cdots i_p}(y^1,\ldots,y^n),$$

$$\phi^*\omega = \sum \mathrm{d}x^{j_p} \wedge \cdots \wedge \mathrm{d}x^{j_1} \frac{\partial y^{i_p}}{\partial x^{j_1}} \cdots \frac{\partial y^{i_1}}{\partial x^{j_p}} a_{i_1 \cdots i_p}(y^1(x),\ldots,y^n(x)),$$

$$\begin{aligned} \mathrm{d}(\phi^*\omega) &= \sum \mathrm{d}x^{j_p} \wedge \cdots \wedge \mathrm{d}x^{j_1} \wedge \mathrm{d}x^k \frac{\partial}{\partial x^k} \\ &\quad \times \left( \frac{\partial y^{i_p}}{\partial x^{j_1}} \cdots \frac{\partial y^{i_1}}{\partial x^{j_p}} a_{i_1 \cdots i_p}(y^1(x),\ldots,y^n(x)) \right) \\ &= \sum \mathrm{d}x^{j_p} \wedge \cdots \wedge \mathrm{d}x^{j_1} \wedge \mathrm{d}x^k \frac{\partial y^{i_p}}{\partial x^{j_1}} \cdots \frac{\partial y^{i_1}}{\partial x^{j_p}} \frac{\partial y^r}{\partial x^k} \frac{\partial}{\partial y^r} \\ &\quad \times a_{i_1 \cdots i_p}(y^1(x),\ldots,y^n(x)). \end{aligned} \tag{8.14}$$

The second derivatives of the $y^i$ cancel because of the antisymmetry of the exterior product. This result must be compared with $\phi^*(\mathrm{d}\omega)$:

$$\mathrm{d}\omega = \sum \mathrm{d}y^{i_p} \wedge \cdots \wedge \mathrm{d}y^{i_1} \wedge \mathrm{d}y^r \frac{\partial}{\partial y^r} a_{i_1 \cdots i_p}(y^1,\ldots,y^n),$$

$$\begin{aligned} \phi^*(\mathrm{d}\omega) &= \sum \mathrm{d}x^{j_p} \wedge \cdots \wedge \mathrm{d}x^{j_1} \wedge \mathrm{d}x^k \frac{\partial y^{i_p}}{\partial x^{j_1}} \cdots \frac{\partial y^{i_1}}{\partial x^{j_p}} \frac{\partial y^r}{\partial x^k} \frac{\partial}{\partial y^r} \\ &\quad \times a_{i_1 \cdots i_p}(y^1(x),\ldots,y^n(x)). \end{aligned} \tag{8.15}$$

This is precisely eq. (8.14), so relation (8.13c) is proven.

Relations (8.13a, b, c) show that equations expressed in terms of differential forms and their exterior derivatives are independent of coordinate choice. Note, however, that a change of coordinates induces a change of basis for the differential forms. The basis $\mathrm{d}y^1,\ldots,\mathrm{d}y^n$ in $y$-coordinates becomes $\mathrm{d}x^l(\partial y^1/\partial x^l),\ldots,\mathrm{d}x^l(\partial y^n/\partial x^l)$ in $x$-coordinates.

In applications to physics, local coordinate invariance is not enough. An additional symmetry group must be represented on the fields in the theory. This is a compact Lie group for gauge theories and the Lorentz group for gravity theories. In general, we choose a set of $p$-forms to span a representation space of the symmetry group,

$$(\omega')^l = \omega^k G_k{}^l(x), \tag{8.16}$$

where $l, k = 1,\ldots,L$. We call this group the *structure group*. The matrix $G$

forms an $L$-dimensional representation of the structure group. Note that $G$ is $x$-dependent.

A set of forms which transforms under a linear representation of the structure group is called a *tensor*. Exterior derivatives of tensors are not tensors:

$$\mathrm{d}\sigma'^{I} = \mathrm{d}\sigma^{i} G_{i}^{I} + \sigma^{i} \mathrm{d}G_{i}^{I}. \tag{8.17}$$

We would like to find a derivative which maps $p$-forms into $(p+1)$-forms and tensors into tensors. To do this we must introduce a connection.

*Connections* are Lie algebra valued one-forms with the following transformation law:

$$\varphi' = G^{-1}\varphi G - G^{-1}\mathrm{d}G. \tag{8.18}$$

Here we have suppressed all indices, as we shall do in the remainder of this section. If the structure group is a Lie group, the connection is simply the Yang–Mills vector field,

$$\varphi = \mathrm{d}x^{m} \mathrm{i} A_{m}{}^{a} T^{a}, \tag{8.19}$$

and (8.18) reproduces the well-known transformation law for vector fields.

The transformation law for the connection is chosen so that the *covariant derivative*,

$$\mathcal{D}\omega = \mathrm{d}\omega + \omega \wedge \varphi = \mathrm{d}x^{m} \mathcal{D}_{m}\omega, \tag{8.20}$$

maps $p$-forms into $(p+1)$-forms and tensors into tensors:

$$\begin{aligned} (\mathcal{D}\omega)' &= \mathrm{d}\omega' + \omega' \wedge \varphi' \\ &= \mathrm{d}\omega G + \omega \wedge \mathrm{d}G + \omega G \wedge \left(G^{-1}\varphi G - G^{-1}\mathrm{d}G\right) \\ &= (\mathrm{d}\omega + \omega \wedge \varphi)G = (\mathcal{D}\omega)G. \end{aligned} \tag{8.21}$$

There is precisely one tensor which can be constructed from the connection. It is known as the *curvature tensor*:

$$F = \mathrm{d}\varphi + \varphi \wedge \varphi. \tag{8.22}$$

The curvature tensor is a Lie algebra valued two-form,

$$F = \tfrac{1}{2}\mathrm{d}x^{n} \wedge \mathrm{d}x^{m} \mathrm{i} F_{mn}{}^{a} T^{a}, \tag{8.23}$$

which transforms as follows under the structure group:

$$\begin{aligned} F' &= \mathrm{d}\varphi' + \varphi' \wedge \varphi' \\ &= \mathrm{d}\left(G^{-1}\varphi G - G^{-1}\mathrm{d}G\right) + \left(G^{-1}\varphi G - G^{-1}\mathrm{d}G\right) \wedge \left(G^{-1}\varphi G - G^{-1}\mathrm{d}G\right) \\ &= G^{-1}FG. \end{aligned} \tag{8.24}$$

The suspicious reader will recognize immediately that this is the transformation law for the Yang–Mills field strength:

$$F = \tfrac{1}{2}\,\mathrm{d}x^n \wedge \mathrm{d}x^m\, F_{mn}$$

$$= \tfrac{1}{2}\,\mathrm{d}x^n \wedge \mathrm{d}x^m \left[ \frac{\partial}{\partial x^m} A_n - \frac{\partial}{\partial x^n} A_m - A_m A_n + A_n A_m \right]$$

$$F_{mn} = \partial_m A_n - \partial_n A_m + [A_n, A_m]. \tag{8.25}$$

*Bianchi identities* are derived from the fact that dd = 0. The first set is obtained from the covariant derivative,

$$\mathcal{D}\omega = \mathrm{d}\omega + \omega \wedge \varphi,$$

$$\mathrm{d}\mathcal{D}\omega = \omega \wedge \mathrm{d}\varphi - \mathrm{d}\omega \wedge \varphi$$

$$= \omega \wedge (F - \varphi \wedge \varphi) - (\mathcal{D}\omega - \omega \wedge \varphi) \wedge \varphi,$$

$$\mathcal{D}\mathcal{D}\omega = \omega \wedge F, \tag{8.26}$$

while the second set comes from the curvature tensor:

$$F = \mathrm{d}\varphi + \varphi \wedge \varphi,$$

$$\mathrm{d}F = \varphi \wedge \mathrm{d}\varphi - \mathrm{d}\varphi \wedge \varphi = \varphi \wedge F - F \wedge \varphi,$$

$$\mathcal{D}F = 0. \tag{8.27}$$

These expressions give familiar relations between the coefficient functions:

$$[\mathcal{D}_m, \mathcal{D}_n]\omega = \omega F_{mn},$$

$$\mathcal{D}_m F_{nl} + \mathcal{D}_n F_{lm} + \mathcal{D}_l F_{mn} = 0. \tag{8.28}$$

Bianchi identities play a major role in Yang–Mills theories, supersymmetric Yang–Mills theories, general relativity and supergravity.

## 9. General relativity

General relativity is based upon covariance under general coordinate and local Lorentz transformations. The basic dynamical variable is the vierbein field $E_m{}^a(x)$. The vierbein field defines a local reference frame. Its upper index transforms under the Lorentz group, while its lower index refers to a specific choice of coordinates. To hide this explicit coordinate dependence, we define the vierbein one-form

$$E^a = \mathrm{d}x^m\, E_m{}^a(x). \tag{9.1}$$

The vierbein form transforms as follows under a general coordinate transformation:

$$x^m = x^m(x'),$$

$$\mathrm{d}x^m E_m{}^a(x) = \mathrm{d}x'^n \frac{\partial x^m}{\partial x'^n} E_m{}^a(x(x')). \tag{9.2}$$

This induces a change in the vierbein field:

$$E_n'^a(x') = \frac{\partial x^m}{\partial x'^n} E_m{}^a(x(x')). \tag{9.3}$$

In the infinitesimal case, this becomes

$$x^m = x'^m + \xi^m(x'),$$

$$\delta E_m{}^a(x) = E_m'^a(x) - E_m{}^a(x) = \xi^n \frac{\partial}{\partial x^n} E_m{}^a(x) + \frac{\partial \xi^n}{\partial x^m} E_n{}^a(x). \tag{9.4}$$

Note that the index $a$ is not affected by coordinate transformations. It is reserved for the structure group:

$$E'^a = E^b L_b{}^a(x), \qquad E_m'^a(x) = E_m{}^b(x) L_b{}^a(x). \tag{9.5}$$

Indices transforming under coordinate changes will be denoted by letters from the middle of the alphabet. They will be called Einstein indices. Similarly, indices transforming under the structure group will be taken from the beginning of the alphabet. They will be called Lorentz indices.

The vierbein and its inverse

$$E_b{}^m E_m{}^a = \delta_b{}^a, \qquad E_m{}^a E_a{}^n = \delta_m{}^n, \tag{9.6}$$

connect the two types of indices:

$$v^m = v^a E_a{}^m, \qquad v^a = v^m E_m{}^a. \tag{9.7}$$

Lorentz indices are raised and lowered with the Lorentz metric $\eta_{ab} = (-1,1,1,1)$. This may be used to form a Lorentz invariant symmetric tensor

$$g_{mn} = E_m{}^a \eta_{ab} E_n{}^b. \tag{9.8}$$

We shall identify this tensor with the metric tensor which usually appears in gravity theories.

Covariant derivatives require the introduction of connections. As usual, the connection is a Lie algebra valued one-form:

$$\varphi_{ab} = \mathrm{d}x^m \varphi_{mab}. \tag{9.9}$$

Note that $\varphi_{ab}$ is antisymmetric in its two Lorentz indices.

The covariant derivative of the vierbein is called torsion:

$$\mathcal{D}E^a = \mathrm{d}E^a + E^b\varphi_b{}^a = T^a. \tag{9.10}$$

In terms of coefficient functions, this becomes

$$T^a = \tfrac{1}{2}\mathrm{d}x^m \mathrm{d}x^n T_{nm}{}^a,$$

$$T_{nma} = \partial_n E_{ma} - \partial_m E_{na} - E_n{}^b\varphi_{mba} + E_m{}^b\varphi_{nba}. \tag{9.11}$$

Since torsion is a two-form, $T_{nma}$ is antisymmetric in its Einstein indices:

$$T_{nma} = -T_{mna}. \tag{9.12}$$

These may be converted into Lorentz indices with the inverse vierbein

$$\begin{aligned} T_{bca} &= E_b{}^n E_c{}^m T_{nma} \\ &= E_b{}^n E_c{}^m(\partial_n E_{ma} - \partial_m E_{na}) - \varphi_{cba} + \varphi_{bca}, \end{aligned} \tag{9.13}$$

where

$$\varphi_{cba} = E_c{}^n\varphi_{nba}.$$

The curvature tensor is defined as a Lie algebra valued two-form:

$$R_{ab} = \tfrac{1}{2}\mathrm{d}x^n \mathrm{d}x^m R_{mnab} = \mathrm{d}\varphi_{ab} + \varphi_a{}^c\varphi_{cb},$$

$$R_{mnab} = \partial_m\varphi_{nab} - \partial_n\varphi_{mab} - \varphi_{ma}{}^c\varphi_{ncb} + \varphi_{na}{}^c\varphi_{mcb}. \tag{9.14}$$

It is antisymmetric in two sets of indices:

$$R_{mnab} = -R_{nmab}, \qquad R_{mnab} = -R_{mnba}. \tag{9.15}$$

As usual, we may convert $R_{mnab}$ to Lorentz indices with the inverse vierbein,

$$R_{cdab} = E_c{}^m E_d{}^n R_{mnab}. \tag{9.16}$$

Contracting these indices gives the curvature scalar

$$R_{cdab}\eta^{ac}\eta^{bd} = R_{ab}{}^{ab} = \mathcal{R}. \tag{9.17}$$

Bianchi identities follow from the torsion and the curvature. The torsion gives a Bianchi identity of the first type:

$$\mathcal{D}T = ER. \tag{9.18}$$

In terms of the coefficient functions, this becomes

$$\mathrm{d}x^m \mathrm{d}x^n \mathrm{d}x^l \left( \partial_l T_{nm}{}^a + T_{nm}{}^b \varphi_{lb}{}^a - R_{lnm}{}^a \right) = 0. \tag{9.19}$$

Exploiting the antisymmetry of the three-forms, we find

$$\mathfrak{D}_l T_{nm}{}^a + \mathfrak{D}_n T_{ml}{}^a + \mathfrak{D}_m T_{ln}{}^a = R_{lnm}{}^a + R_{nml}{}^a + R_{mln}{}^a. \tag{9.20}$$

If we had chosen to write the Bianchi identity in terms of quantities with only Lorentz indices, we would have found

$$\begin{aligned}
\mathfrak{D}\left( E^b E^c T_{cb}{}^a \right) &= E^b R_b{}^a \\
&= E^b E^c \mathfrak{D} T_{cb}{}^a + E^b (\mathfrak{D} E^c) T_{cb}{}^a - (\mathfrak{D} E^b) E^c T_{cb}{}^a \\
&= E^b E^c \mathfrak{D} T_{cb}{}^a + E^b T^c T_{cb}{}^a - T^b E^c T_{cb}{}^a \\
&= E^b E^c E^d \left( \mathfrak{D}_d T_{cb}{}^a + \tfrac{1}{2} T_{dc}{}^f T_{fb}{}^a - \tfrac{1}{2} T_{cb}{}^f T_{df}{}^a \right) \\
&= E^b E^c E^d \left( \mathfrak{D}_d T_{cb}{}^a + T_{dc}{}^f T_{fb}{}^a \right).
\end{aligned} \tag{9.21}$$

In components, this becomes

$$\sum_{\substack{b,c,d \\ \text{cyclic perms.}}} \left( \mathfrak{D}_d T_{cb}{}^a + T_{dc}{}^f T_{fb}{}^a - R_{dcb}{}^a \right) = 0. \tag{9.22}$$

This relation differs from eq. (9.20) by terms quadratic in the torsion. Gravity theories are usually formulated with vanishing torsion. When this is the case, eq. (9.22) reproduces the well-known cyclic identities on the curvature tensor.

The curvature tensor gives a Bianchi identity of the second type,

$$\mathfrak{D} R = 0, \tag{9.23}$$

or, in components,

$$\sum_{\substack{l,m,n \\ \text{cyclic perms.}}} \mathfrak{D}_l R_{mna}{}^b = 0. \tag{9.24}$$

Changing to Lorentz indices, we find

$$\sum_{\substack{c,d,f \\ \text{cyclic perms.}}} \left( \mathfrak{D}_f R_{cda}{}^b + T_{fc}{}^k R_{kda}{}^b \right) = 0. \tag{9.25}$$

Again, the torsion terms arise from covariant derivatives of the vierbein form.

In this formulation of general relativity we have treated the vierbein field and the connection as independent variables. It is customary to impose the constraint

$$T_{abc} = 0. \tag{9.26}$$

This allows us to eliminate the degrees of freedom associated with the connection. With this constraint, we can solve eq. (9.13) for $\varphi_{cba}$ by exploiting the antisymmetry $\varphi_{cba} = -\varphi_{cab}$:

$$\varphi_{cba} - \varphi_{bca} = E_b{}^n E_c{}^m (\partial_n E_{ma} - \partial_m E_{na}),$$

$$\varphi_{bac} - \varphi_{abc} = E_a{}^n E_b{}^m (\partial_n E_{mc} - \partial_m E_{nc}),$$

$$\varphi_{acb} - \varphi_{cab} = E_c{}^n E_a{}^m (\partial_n E_{mb} - \partial_m E_{nb}),$$

$$\varphi_{cba} = \tfrac{1}{2} E_b{}^n E_c{}^m (\partial_n E_{ma} - \partial_m E_{na}) + \tfrac{1}{2} E_c{}^n E_a{}^m (\partial_n E_{mb} - \partial_m E_{nb}) - \tfrac{1}{2} E_a{}^n E_b{}^m (\partial_n E_{mc} - \partial_m E_{nc}). \tag{9.27}$$

The constraint (9.26) may also be obtained as an equation of motion. From this point of view, the connection and the vierbein are independent dynamical variables. Variation of the action

$$\mathfrak{L} = \det(E_m{}^a)\mathfrak{R} = \det(E_m{}^a) E^{bn} E^{cr} R_{nrbc} \tag{9.28}$$

with respect to the connection gives eq. (9.26).

To show this explicitly we introduce the following tensors:

$$\tilde{E}_a{}^b = E_a{}^n \delta E_n{}^b = -\delta E_a{}^n E_n{}^b,$$

$$\tilde{\varphi}_{abc} = E_a{}^n \delta \varphi_{nbc}. \tag{9.29}$$

With these tensors, the variation of the scalar

$$(\det E) F_b{}^{adc} R_{cda}{}^b \tag{9.30}$$

takes a manifestly covariant form:

$$\delta\left[(\det E) F_b{}^{adc} R_{cda}{}^b\right]$$

$$= (\det E)\left[F_b{}^{adc}\left(R_{cda}{}^b \tilde{E}_f{}^f - 2\tilde{E}_d{}^f R_{cfa}{}^b - 2T_{cf}{}^f \tilde{\varphi}_{da}{}^b + T_{cd}{}^f \tilde{\varphi}_{fa}{}^b\right)\right.$$

$$\left. - 2\left(\mathfrak{D}_c F_b{}^{adc} \tilde{\varphi}_{da}{}^b\right)\right] + \text{total derivatives}. \tag{9.31}$$

Here $F$ is any antisymmetric Lorentz tensor not subject to variation. If we set $F_b{}^{adc} = -\tfrac{1}{2}(\delta_b{}^c \eta^{ad} - \delta_b{}^d \eta^{ac})$, we find the variation of the Lagrangian (9.28).

## 10. Supergravity

Supergravity theories are obtained by gauging the supersymmetry algebra (5.1). Since this algebra closes into the energy-momentum four-vector $P_m$, four-dimensional translations must also be gauged. This leads to a theory which is covariant under general coordinate transformations in four dimensional space.

The differential operators $Q$ and $\bar{Q}$ of eq. (5.9) may be viewed as the generators of supersymmetry transformations in superspace. These transformations are among the general coordinate transformations of superspace

$$z'^M = z'^M\left(x^m, \theta^\mu, \bar{\theta}_{\dot{\mu}}\right). \tag{10.1}$$

It seems reasonable, therefore, to write supergravity theories in a way which is automatically covariant under superspace transformations.

Differential forms are the natural tools to use. They may be extended to superspace with the following sign changes:

$$\begin{aligned} z^M z^N &= (-1)^{nm} z^N z^M, \\ \mathrm{d}z^M \mathrm{d}z^N &= -(-1)^{nm} \mathrm{d}z^N \mathrm{d}z^M, \\ \mathrm{d}z^M z^N &= (-1)^{nm} z^N \mathrm{d}z^M. \end{aligned} \tag{10.2}$$

Here $n$ and $m$ are functions of $N$ and $M$ which take the values zero or one depending on whether $N$ and $M$ are vector or spinor indices. Keeping this sign factor in mind, we define $p$-forms

$$\Omega = \mathrm{d}z^{M_1} \cdots \mathrm{d}z^{M_p} W_{M_p \cdots M_1}(z), \tag{10.3}$$

and their exterior derivatives

$$\mathrm{d}\Omega = \mathrm{d}z^{M_1} \cdots \mathrm{d}z^{M_p} \mathrm{d}z^N \frac{\partial}{\partial z^N} W_{M_p \cdots M_1}(z). \tag{10.4}$$

The differentials are written to the left of the coefficient functions and the indices are labelled in such a way that there is always an even number of indices between those being summed. With these definitions, the properties (8.7) of the exterior derivative hold exactly, with no additional sign changes.

We now extend the connection form to superspace,

$$\varphi = \mathrm{d}z^M \mathrm{i}\varphi_M{}^a T^a. \tag{10.5}$$

The connection transforms as follows under the structure group:

$$\varphi' = G^{-1}\varphi G - G^{-1}\mathrm{d}G. \tag{10.6}$$

In analogy to Section 8, we use the connection to define covariant derivatives

$$\mathfrak{D}\Omega = \mathrm{d}\Omega + \Omega\varphi, \tag{10.7}$$

and the curvature two-form

$$F = R = \mathrm{d}\varphi + \varphi\varphi. \tag{10.8}$$

We shall treat supergravity as a direct extension to superspace of general relativity as it was introduced in section 9. Our basic dynamical variable will be the vielbein field $E_M{}^A(z)$. Its lower index $M$ transforms under general coordinate transformations,

$$\delta E_M{}^A = \xi^N(z)\partial_N E_M{}^A + \left(\partial_M \xi^N(z)\right)E_N{}^A, \tag{10.9}$$

in accord with eq. (9.4). Its upper index $A$ transforms under the structure group,

$$E'_M{}^A = E_M{}^B L_B{}^A(z). \tag{10.10}$$

We shall retain the Lorentz group as our structure group. Note, however, that $L_B{}^A$ is reducible. In particular, $L_b{}^a$, $L_\beta{}^\alpha$ and $L^{\dot\beta}{}_{\dot\alpha}$ are related through the $\sigma$-matrices:

$$\sigma^a{}_{\alpha\dot\alpha}\sigma^b{}_{\beta\dot\beta}L_{ab} = -2\varepsilon_{\alpha\beta}L_{\dot\alpha\dot\beta} + 2\varepsilon_{\dot\alpha\dot\beta}L_{\alpha\beta}. \tag{10.11}$$

The covariant derivative of the vielbein is called torsion

$$E^A = \mathrm{d}z^M E_M{}^A, \qquad \mathfrak{D}E^A = T^A. \tag{10.12}$$

Torsion is a two-form with several Lorentz-irreducible components. We shall use these components to impose covariant constraint conditions. These constraints reduce the number of independent component fields without restricting their $x$-dependence. In contrast to general relativity, however, it is not possible to obtain the supergravity constraints as Euler–Lagrange equations associated with a given Lagrangian. Instead, they must be imposed by hand:

$$\begin{aligned}
&T_{\underline{\alpha}\underline{\beta}}{}^{\underline{\gamma}} = 0, \qquad T_{\alpha\beta}{}^c = T_{\dot\alpha\dot\beta}{}^c = 0,\\
&T_{\alpha\dot\beta}{}^c = T_{\dot\beta\alpha}{}^c = 2\mathrm{i}\sigma_{\alpha\dot\beta}{}^c,\\
&T_{ab}{}^c = 0, \qquad T_{\underline{\alpha}b}{}^c = T_{a\underline{\beta}}{}^c = 0.
\end{aligned} \tag{10.13}$$

Here $\underline{\alpha}$ denotes either $\alpha$ or $\dot\alpha$.

The constraints (10.13) yield the minimum number of component fields. These are the spin-2 graviton $e_m{}^a(x)$, the spin-3/2 gravitino $\psi_m{}^\alpha(x), \bar{\psi}_{m\dot{\alpha}}(x)$, one complex scalar field $M(x)$ and one real vector field $b_m(x)$. The spin-2 and spin-3/2 fields are to be expected. They couple to the energy-momentum tensor and the spin-3/2 supercurrent. The fields $M$ and $b_m$ are auxiliary fields which equalize the number of bosonic and fermionic degrees of freedom off mass shell.

These fields may be found by solving the Bianchi identities subject to the constraints (10.13). The relevant Bianchi identity is the superspace generalization of eq. (9.22):

$$E^C E^D E^E \left( \mathcal{D}_E T_{DC}{}^A - R_{EDC}{}^A + T_{ED}{}^F T_{FC}{}^A \right) = 0. \tag{10.14}$$

This tells us that all the curvature and torsion components may be expressed in terms of one chiral superfield $R$, one hermitian superfield $G_{\alpha\dot{\alpha}}$ and one chiral superfield $W_{\alpha\beta\gamma}$ symmetric in all indices. For example, we list the following components:

$$\begin{aligned}
T_{\alpha a \dot{\alpha}} &= -\mathrm{i}\sigma_{a\alpha\dot{\alpha}} R^*, \\
T_{\beta a \alpha} &= -\tfrac{1}{8}\mathrm{i}\bar{\sigma}_a{}^{\dot{\varepsilon}\varepsilon}\left( \varepsilon_{\varepsilon\alpha} G_{\beta\dot{\varepsilon}} - 3\varepsilon_{\beta\alpha} G_{\varepsilon\dot{\varepsilon}} - 3\varepsilon_{\beta\varepsilon} G_{\alpha\dot{\varepsilon}} \right), \\
R_{\delta\gamma\beta\alpha} &= 4\left( \varepsilon_{\delta\beta}\varepsilon_{\gamma\alpha} + \varepsilon_{\gamma\beta}\varepsilon_{\delta\alpha} \right) R^*, \\
R_{\delta\dot{\gamma}\beta\alpha} &= -\left( \varepsilon_{\delta\beta} G_{\alpha\dot{\gamma}} + \varepsilon_{\delta\alpha} G_{\beta\dot{\gamma}} \right).
\end{aligned} \tag{10.15}$$

Because of the constraints, the superfields $R$, $G_{\alpha\dot{\alpha}}$ and $W_{\alpha\beta\gamma}$ satisfy certain equations:

$$\begin{aligned}
&\bar{\mathcal{D}}_{\dot{\alpha}} R = 0, \qquad \bar{\mathcal{D}}_{\dot{\varepsilon}} W_{\alpha\beta\gamma} = 0, \\
&G^\dagger_{\alpha\dot{\alpha}} = G_{\alpha\dot{\alpha}}, \qquad \bar{\mathcal{D}}^\alpha G_{\alpha\dot{\gamma}} = \mathcal{D}_{\dot{\gamma}} R^*, \\
&\mathcal{D}^\alpha W_{\alpha\beta\gamma} + \tfrac{1}{2}\mathrm{i}\left( \mathcal{D}_{\gamma\dot{\varepsilon}} G_\beta{}^{\dot{\varepsilon}} + \mathcal{D}_{\beta\dot{\varepsilon}} G_\gamma{}^{\dot{\varepsilon}} \right) = 0.
\end{aligned} \tag{10.16}$$

The $\theta = \bar{\theta} = 0$ components of $R$ and $G$ are identified with the fields $M$ and $b_m$

$$G_a|_{\theta=\bar{\theta}=0} = -\tfrac{1}{3} b_a, \qquad R|_{\theta=\bar{\theta}=0} = -\tfrac{1}{6} M, \tag{10.17}$$

whereas the fields $e_m{}^a$ and $\psi_m{}^\alpha$ are found from the lowest components of the vielbein

$$E_m{}^a|_{\theta=\bar{\theta}=0} = e_m{}^a, \qquad E_m{}^\alpha|_{\theta=\bar{\theta}=0} = \tfrac{1}{2}\psi_m{}^\alpha. \tag{10.18}$$

All other components of $R$, $G$, $W$ and the vielbein are either functions of these fields and their derivatives or may be gauged away.

The action which describes the dynamics of the spin-2 and spin-3/2 fields may be written in the form of an integral over superspace. It is given by

$$S = \int \mathrm{d}^4x\,\mathrm{d}^2\theta\,\mathrm{d}^2\bar{\theta}\,E, \tag{10.19}$$

where $E$ is the superdeterminant of the vielbein $E_M{}^A$.

To make the above procedure more transparent we shall apply it to supersymmetric gauge theories. In this case, the curvature $F$ has the following irreducible components:

$$\begin{aligned}
F_{ba} &= \partial_b\varphi_a - \partial_a\varphi_b - [\varphi_b, \varphi_a],\\
F_{b\alpha} &= \partial_b\varphi_\alpha - D_\alpha\varphi_b - [\varphi_b, \varphi_\alpha],\\
F_{b\dot\alpha} &= \partial_b\varphi_{\dot\alpha} - \bar{D}_{\dot\alpha}\varphi_b - [\varphi_b, \varphi_{\dot\alpha}],\\
F_{\beta\alpha} &= D_\beta\varphi_\alpha + D_\alpha\varphi_\beta - \{\varphi_\beta, \varphi_\alpha\},\\
F_{\dot\beta\dot\alpha} &= \bar{D}_{\dot\beta}\varphi_{\dot\alpha} + \bar{D}_{\dot\alpha}\varphi_{\dot\beta} - \{\varphi_{\dot\beta}, \varphi_{\dot\alpha}\},\\
F_{\beta\dot\alpha} &= D_\beta\varphi_{\dot\alpha} + \bar{D}_{\dot\alpha}\varphi_\beta - \{\varphi_\beta, \varphi_{\dot\alpha}\} + 2\mathrm{i}\sigma^a{}_{\beta\dot\alpha}\varphi_a.
\end{aligned} \tag{10.20}$$

These satisfy the following Bianchi identities:

$$\mathcal{D}_c F_{ba} + \mathcal{D}_b F_{ac} + \mathcal{D}_a F_{cb} = 0, \tag{10.21a}$$

$$\mathcal{D}_\alpha F_{bc} + \mathcal{D}_b F_{c\alpha} + \mathcal{D}_c F_{\alpha b} = 0, \tag{10.21b}$$

$$\bar{\mathcal{D}}_{\dot\alpha} F_{bc} + \mathcal{D}_b F_{c\dot\alpha} + \mathcal{D}_c F_{\dot\alpha b} = 0, \tag{10.21c}$$

$$\mathcal{D}_c F_{\beta\alpha} + \mathcal{D}_\beta F_{\alpha c} - \mathcal{D}_\alpha F_{c\beta} = 0, \tag{10.21d}$$

$$\mathcal{D}_c F_{\dot\beta\dot\alpha} + \bar{\mathcal{D}}_{\dot\beta} F_{\dot\alpha c} - \bar{\mathcal{D}}_{\dot\alpha} F_{c\dot\beta} = 0, \tag{10.21e}$$

$$\mathcal{D}_c F_{\dot\beta\alpha} + \bar{\mathcal{D}}_{\dot\beta} F_{\alpha c} - \mathcal{D}_\alpha F_{c\dot\beta} + 2\mathrm{i}\sigma^a{}_{\alpha\dot\beta} F_{ac} = 0, \tag{10.21f}$$

$$\mathcal{D}_\gamma F_{\beta\alpha} + \mathcal{D}_\beta F_{\alpha\gamma} + \mathcal{D}_\alpha F_{\gamma\beta} = 0, \tag{10.21g}$$

$$\bar{\mathcal{D}}_{\dot\gamma} F_{\beta\alpha} + \mathcal{D}_\beta F_{\alpha\dot\gamma} + \mathcal{D}_\alpha F_{\dot\gamma\beta} + 2\mathrm{i}\sigma^a{}_{\beta\dot\gamma} F_{a\alpha} + 2\mathrm{i}\sigma^a{}_{\alpha\dot\gamma} F_{a\beta} = 0, \tag{10.21h}$$

$$\mathcal{D}_\gamma F_{\dot\beta\dot\alpha} + \bar{\mathcal{D}}_{\dot\beta} F_{\dot\alpha\gamma} + \bar{\mathcal{D}}_{\dot\alpha} F_{\gamma\dot\beta} + 2\mathrm{i}\sigma^a{}_{\gamma\dot\beta} F_{a\dot\alpha} + 2\mathrm{i}\sigma^a{}_{\gamma\dot\alpha} F_{a\dot\beta} = 0, \tag{10.21j}$$

$$\bar{\mathcal{D}}_{\dot\gamma} F_{\dot\beta\dot\alpha} + \bar{\mathcal{D}}_{\dot\beta} F_{\dot\alpha\dot\gamma} + \bar{\mathcal{D}}_{\dot\alpha} F_{\dot\gamma\dot\beta} = 0. \tag{10.21k}$$

The proper constraints are

$$F_{\alpha\beta} = F_{\dot\alpha\dot\beta} = F_{\alpha\dot\beta} = 0. \tag{10.22}$$

We shall solve eq. (10.21) subject to these constraints.

Identities (10.21g, k) are automatically satisfied because of the constraints. Identity (10.21h), however, yields a further restriction on $F$:

$$\sigma^{a}{}_{\alpha\dot\beta}F_{a\beta} + \sigma^{a}{}_{\beta\dot\beta}F_{a\alpha} = 0. \tag{10.23}$$

The vector-spinor $F_{a\alpha}$ has spin-3/2 and spin-1/2 components. Equation (10.23) tells us that the spin-3/2 component vanishes:

$$F_{a\alpha} = \mathrm{i}\sigma_{a\alpha\dot\kappa}\overline{W}^{\dot\kappa}, \qquad \overline{W}^{\dot\kappa} = \tfrac{1}{4}\mathrm{i}\bar\sigma^{a\dot\kappa\alpha}F_{a\alpha}. \tag{10.24}$$

Identity (10.21j) gives a similar result,

$$F_{a\dot\alpha} = \mathrm{i}W^{\beta}\sigma_{a\beta\dot\alpha}, \qquad W^{\alpha} = \tfrac{1}{4}\mathrm{i}F_{a\dot\alpha}\bar\sigma^{a\dot\alpha\alpha}, \tag{10.25}$$

while identity (10.21f) allows us to express $F_{ab}$ in terms of $W$ and $\overline{W}$:

$$\begin{aligned} F_{ab} &= -\tfrac{1}{4}\mathrm{i}\bar\sigma_{a}{}^{\dot\beta\alpha}\left(\overline{\mathcal{D}}_{\dot\beta}F_{\alpha b} + \mathcal{D}_{\alpha}F_{\dot\beta b}\right) \\ &= -\tfrac{1}{4}\left(\overline{\mathcal{D}}\bar\sigma_{a}\sigma_{b}\overline{W} - \mathcal{D}\sigma_{a}\bar\sigma_{b}W\right). \end{aligned} \tag{10.26}$$

Exploiting the antisymmetry of $F_{ab}$, we find

$$\overline{\mathcal{D}}\overline{W} - \mathcal{D}W = 0, \tag{10.27}$$

so

$$F_{ab} = -\tfrac{1}{2}\left(\overline{\mathcal{D}}\bar\sigma_{ab}\overline{W} - \mathcal{D}\sigma_{ab}W\right). \tag{10.28}$$

Identity (10.21e) leads to another restriction on $W$:

$$\left(\sigma^{c}{}_{\beta\dot\alpha}\overline{\mathcal{D}}_{\dot\beta} + \sigma^{c}{}_{\beta\dot\beta}\overline{\mathcal{D}}_{\dot\alpha}\right)W^{\beta} = 0. \tag{10.29}$$

Contracting with $\bar\sigma^{c\sigma\dot\sigma}$ and using eq. (4.11), we have:

$$\left(\delta_{\dot\alpha}{}^{\dot\sigma}\overline{\mathcal{D}}_{\dot\beta} + \delta_{\dot\beta}{}^{\dot\sigma}\overline{\mathcal{D}}_{\dot\alpha}\right)W^{\sigma} = 0. \tag{10.30}$$

Summing over $\dot\alpha$ and $\dot\sigma$ yields:

$$\overline{\mathcal{D}}_{\dot\alpha}W_{\sigma} = 0. \tag{10.31}$$

An analogous result follows from identity (10.21d):

$$\mathcal{D}_{\alpha}\overline{W}_{\dot\sigma} = 0. \tag{10.32}$$

Identities (10.21a, b, c) do not lead to any new results.

Thus we have solved the Bianchi identities (10.21) subject to the constraints (10.22). We found their unique solution to be given by the chiral superfields $W_\sigma$ and $\overline{W}_{\dot\sigma}$, subject to certain constraints:

$$\overline{\mathcal{D}}_{\dot\alpha} W_\sigma = 0, \qquad \mathcal{D}_\alpha \overline{W}_{\dot\sigma} = 0, \qquad \mathcal{D} W = \overline{\mathcal{D}} \overline{W}. \tag{10.33}$$

These are precisely the constraint eqs. (6.8), so $W_\sigma$ has the decomposition (6.9). The corresponding Lagrangian is given in eq. (6.10).

## References

Section 1

S. Coleman and J. Mandula, Phys. Rev. 159 (1967) 1251.

S. Coleman, J. Wess and B. Zumino, Phys. Rev. 177 (1969) 2239.

H.J. Borchers, Nuovo Cimento 15 (1960) 784.

Section 2

E.S. Abers and B.W. Lee, Phys. Rep. 9C (1973) 1.

L.D. Faddeev and A.A. Slavnov, Gauge Fields: Introduction to Quantum Theory (Benjamin/Cummings, Reading, MA, 1980).

R. Balian and J. Zinn–Justin, eds., Methods in Field Theory, Les Houches (1975) Session 28 (North-Holland, Amsterdam, 1976).

Section 3

J.C. Taylor, Gauge Theories of Weak Interactions (Cambridge Univ. Press, Cambridge, 1976).

R. Balian and C.H. Llewellyn-Smith, eds., Weak and Electromagnetic Interactions at High Energies, Les Houches (1976), session 29 (North-Holland, Amsterdam, 1977).

Section 4

E.M. Corson, Introduction to Tensors, Spinors and Relativistic Wave Equations (Blackie, London, 1953).

W. Thirring, Suppl. Nuovo Cimento 14 (1959) 415.

Section 5

P. Fayet and S. Ferrara, Phys. Rep. 32C (1977) 249.

J. Wess and B. Zumino, Nucl. Phys. B70 (1974) 39.

B. Zumino, Nucl. Phys. B89 (1975) 535.

A. Salam and J. Strathdee, Nucl. Phys. B76 (1974) 477.

S. Ferrara, J. Wess and B. Zumino, Phys. Lett. 51B (1974) 239.

J. Wess and B. Zumino, Phys. Lett. 49B (1974) 52.

Section 6

A. Salam and J. Strathdee, Phys. Rev. D11 (1975) 1521.

J. Wess, Acta Phys. Austr., Suppl. XV (1976) 475.

J. Wess and B. Zumino, Nucl. Phys. B78 (1974) 1.

S. Ferrara and B. Zumino, Nucl. Phys. B79 (1974) 413.

Section 7
P. Fayet and J. Iliopoulos, Phys. Lett. 51B (1974) 461.
P. Fayet, Nucl. Phys. B90 (1975) 104.
L. O'Raifeartaigh, Nucl. Phys. B96 (1975) 331.

Section 8
H. Flanders, Differential Forms (Academic Press, New York, 1963).

Section 9
T.W.B. Kibble, J. Math. Phys. 2 (1961) 212.
S. Weinberg, Gravitation and Cosmology (Wiley, New York, 1972).

Section 10
J. Wess and J. Bagger, Supersymmetry and Supergravity (Princeton Univ. Press, 1983).
J. Wess, Supersymmetry–Supergravity, in: Topics in Quantum Field Theory and Gauge Theories, Salamanca, 1977, ed. J.A. de Azcárraga, Springer Lecture Notes in Physics 77 (Springer, Berlin, 1978) p. 81.
R. Grimm, J. Wess and B. Zumino, Nucl. Phys. B152 (1979) 255.
P. van Nieuwenhuizen and D.Z. Freedman, eds., Supergravity, Stony Brook, 1979 (North-Holland, Amsterdam, 1979).

**Seminars related to Course 1**

1. Friedman–Lemaître universes with a cosmological constant; the significance of $\Lambda$ in quantum field theories, by Robert Coquereaux (CNRS, Marseille).

2. The $R_\xi$ gauge in supersymmetric theories, by Burt Ovrut (Inst. for Adv. Stud., Princeton).

3. Duality rotations for interacting fields, by Bruno Zumino (CERN).

4. Particle spectra in extended supersymmetry, by Bruno Zumino (CERN).

COURSE 2

# INTRODUCTION TO PERTURBATIVE QUANTUM CHROMODYNAMICS

C.T.C. SACHRAJDA

*Department of Physics, University of Southampton*
*Southampton, SO9 5NH, England*

*M.K. Gaillard and R. Stora, eds.*
*Les Houches, Session XXXVII, 1981*
*Théories de jauge en physique des hautes énergies/Gauge theories in high energy physics*

# Contents

## Preface

These lectures are intended to provide an introduction to the theoretical ideas needed to calculate the predictions of Quantum ChromoDynamics (QCD) to "hard scattering" processes. We hope that they will provide the student with an overview of the subject and enable him to read recent research papers and more specialized review articles. We have not tried to present a comprehensive review of all the calculations which have been performed in the past ten years or so. Rather we have concentrated on demonstrating why for some quantities we can make reliable predictions while for others we can only speculate or make no predictions at all. Since several other lecturers at this school will make detailed comparison of the QCD predictions with experimental data, we will only comment very briefly on the experimental verification or otherwise of the various predictions derived in the text.

## 1. Introduction

During the past decade Quantum ChromoDynamics (QCD) has emerged as the leading candidate for the theory of strong interactions. With the discovery of the property of "asymptotic freedom" [1,2] came the realization that, at least in certain kinematic régimes, perturbative calculations can be performed in QCD in spite of the fact that it is a theory of strong interactions. These calculations involve the separation of non-calculable (at least in perturbation theory) "long-distance" effects from the calculable "short-distance" contributions, as well as the actual computation of the latter. These lectures are intended to provide an introduction to the subject of perturbative QCD.

QCD is a theory of interactions of quarks and gluons whose Lagrangian density is

$$\mathcal{L} = -\tfrac{1}{4}F^{a}_{\mu\nu}F^{a\mu\nu} + \sum_{f(\text{lavours})} \bar{\psi}_f (i\not{D} - m_f)\psi_f$$
$$+ \text{gauge fixing term} + \text{Fadeev–Popov ghost}, \qquad (1.1)$$

where the field strength tensor $F^a_{\mu\nu}$ is defined by

$$F^a_{\mu\nu} = \partial_\mu A^a_\nu - \partial_\nu A^a_\mu + g f_{abc} A^b_\mu A^c_\nu \tag{1.2}$$

and the covariant derivative $D$ is given by

$$D_\mu = \partial_\mu - \mathrm{i} g T_a A^a_\mu . \tag{1.3}$$

As usual $\psi$ stands for a quark field and $A$ for a gluon field. The $T$'s are the generators of the colour group $SU(3)_c$, with structure constants $f_{abc}$ satisfying

$$[T_a, T_b] = \mathrm{i} f_{abc} T_c . \tag{1.4}$$

The quarks are assumed to transform according to the fundamental representation, i.e. each flavour of quarks is a triplet of the colour group SU(3) [we have suppressed the colour labels on $\bar{\psi}$ and $\psi$ in eq. (1.1)], and the gluons, as always in a non-Abelian gauge theory, transform according to the adjoint representation, so that there are eight gluons.

At this point it seems appropriate to consider briefly some of the reasons which suggest the necessity of the colour quantum number:

(i) *Fermi statistics.* The original motivation for the colour quantum number [3,4] was the problem of quark statistics. Consider for example the $\Delta^{++}$ resonance in the $J_Z = 3/2$ state. We can describe this resonance by

$$|\Delta^{++}, J_Z = 3/2\rangle = |\mathrm{u}\!\uparrow, \mathrm{u}\!\uparrow, \mathrm{u}\!\uparrow\rangle, \tag{1.5}$$

where the upward pointing arrow indicates that the up quark (u) has $z$-component of spin equal to $+1/2$. One expects the ground state of the three quark system to be symmetric under space coordinate exchanges, which leads to the problem that the wavefunction in eq. (1.5) is symmetric under the interchange of two up quarks, in contradiction with Fermi statistics. A possible solution to this problem is to postulate that each quark comes in three "colours", (red, blue and yellow say) and modify the wave function (1.5) to

$$6^{-1/2} \varepsilon^{\mathrm{rby}} |\mathrm{u_r}, \mathrm{u_b}, \mathrm{u_y}\rangle, \tag{1.6}$$

which now is anti-symmetric under the interchange of two quarks.

(ii) *Cancellation of anomalies* [5–7]. In the standard model of weak and electromagnetic interactions, based on the gauge group $SU(2)_L \times U(1)_Y$, the harmful anomalies (i.e. those which would spoil renormalizability) are proportional to

$$\sum_{\text{fermions}} \mathrm{Tr}(YT^aT^a) \propto \sum_{\text{doublets}} Y, \tag{1.7}$$

where $T^a$ here are the generators of weak SU(2) and $Y$ is the weak hypercharge. If, as commonly assumed, with each lepton doublet $(\nu_e, e)_L$ etc. with $Y = -1$, there comes a quark doublet $(u, d_c)_L$ with $Y = 1/3$, then these anomalies will cancel if each flavour of quarks comes in three colours.

(iii) *The process* $\pi^0 \to 2\gamma$ [8]. The amplitude for the process $\pi^0 \to 2\gamma$ is given by the triangle diagram (fig. 1) and leads to the prediction for the width:

$$\Gamma(\pi^0 \to \gamma\gamma) = \frac{1}{64\pi} \frac{m_\pi^3}{f_\pi^2} \left(\frac{\alpha}{\pi}\right)^2 \left(\frac{n}{3}\right)^2, \tag{1.8}$$

where $f_\pi$ is the pion decay constant $\approx 93$ MeV and $n$ is the number of colours. The experimental value is [9]:

$$\Gamma_{\exp}(\pi^0 \to \gamma\gamma) = 7.95 \pm 0.55 \text{ eV}, \tag{1.9}$$

which is in good agreement with eq. (1.8) for $n = 3$. (Indeed the chiral dynamics originally discussed in the context of current algebra is reproduced in QCD [10, 11].)

(iv) *The ratio* $R_{e^+e^-}$. In the parton picture the process $e^+e^- \to$ hadrons is dominated by the channel $e^+e^- \to$ quark + antiquark (fig. 2) so that the ratio $R_{e^+e^-}$, defined by

$$R_{e^+e^-} = \frac{\sigma(e^+e^- \to \text{hadrons})}{\sigma(e^+e^- \to \mu^+\mu^-)}, \tag{1.10}$$

is given by (for three colours):

$$R_{e^+e^-} = 3\sum_i Q_i^2 \ (= 11/3 \text{ for the five flavours u, d, s, c, and b}), \tag{1.11}$$

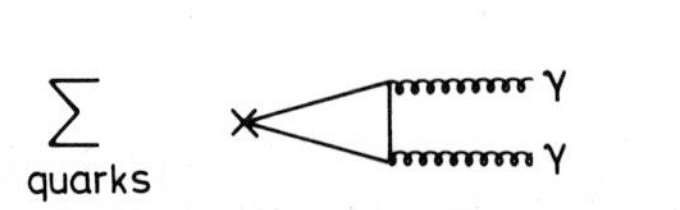


Fig. 1. Amplitude for the decay $\pi^0 \to 2\gamma$.

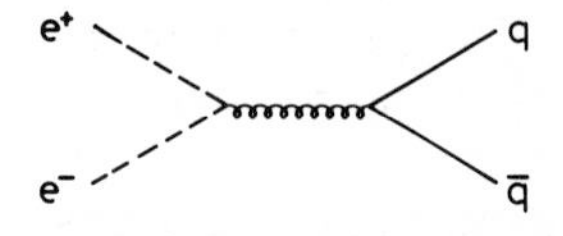


Fig. 2. (Dominant) contribution to the high energy annihilation process, $e^+e^- \to$ hadrons.

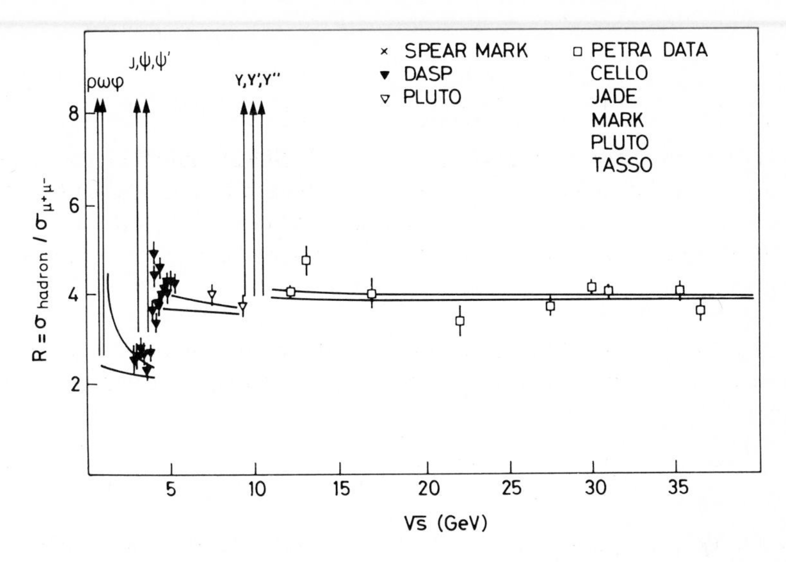


Fig. 3. Recently compiled data for $R = \sigma(e^+e^- \to \text{hadrons})/\sigma(e^+e^- \to \mu^+\mu^-)$.

where $Q_i$ is the charge of the quark with flavour $i$. As can be seen from fig. 3 the data [12] agree well with eq. (1.11) and provide further evidence for the colour quantum number.

(v) *Asymptotic freedom*. QCD is an asymptotically free theory which means that the coupling constant decreases as the scale at which it is defined is increased. This means that most of the parton model "successes" for deep hadronic processes can be approximately reproduced in QCD. This will be discussed in detail below. It is known that only theories which include a non-Abelian gauge symmetry can be asymptotically free [13].

Here we are not going to present the standard procedure for the quantization of non-Abelian theories and the derivation of their Feynman rules*. We just state the Feynman rules for QCD (quantized in a covariant gauge) in fig. 4.

*This is contained in several textbooks on modern field theory [e.g. refs. 14, 15, 16]. For a pedestrian explanation of "Gauge fixing term" and "Fadeev–Popov ghost" of eq. (1.1), see ref. [10].

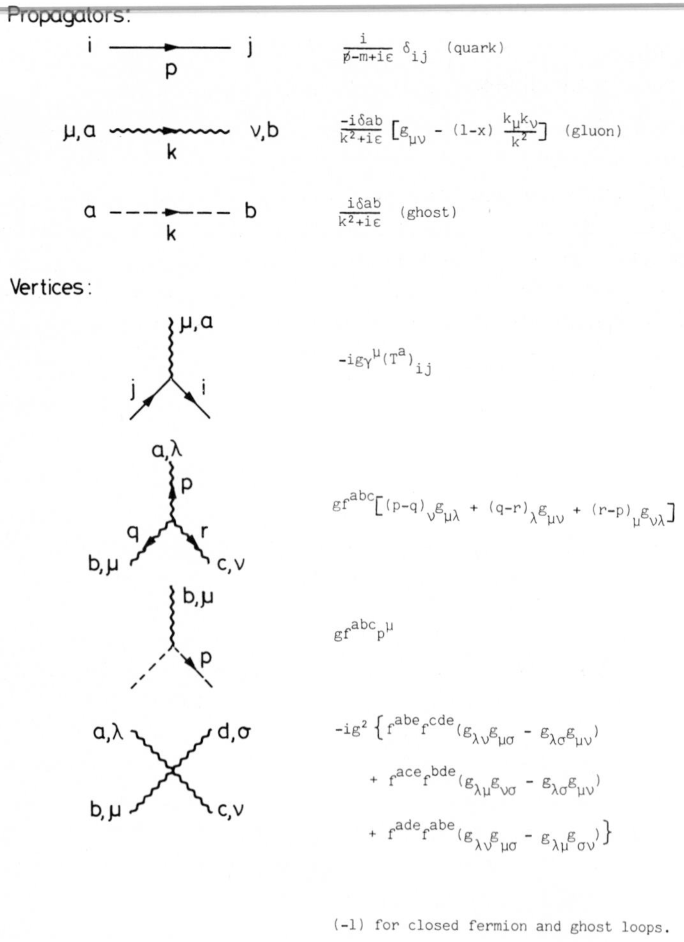

Propagators:

i —— p —— j $\quad \frac{i}{\not{p}-m+i\varepsilon}\,\delta_{ij}$ (quark)

μ,a ~~ k ~~ ν,b $\quad \frac{-i\delta ab}{k^2+i\varepsilon}\left[g_{\mu\nu} - (1-x)\,\frac{k_\mu k_\nu}{k^2}\right]$ (gluon)

a - - - k - - - b $\quad \frac{i\delta ab}{k^2+i\varepsilon}$ (ghost)

Vertices:

(μ,a; j, i) $\quad -ig\gamma^\mu (T^a)_{ij}$

(a,λ, p; b,μ, q; c,ν, r) $\quad gf^{abc}\left[(p-q)_\nu g_{\mu\lambda} + (q-r)_\lambda g_{\mu\nu} + (r-p)_\mu g_{\nu\lambda}\right]$

(b,μ; p) $\quad gf^{abc}p^\mu$

(a,λ; b,μ; c,ν; d,σ)

$$-ig^2\left\{f^{abe}f^{cde}(g_{\lambda\nu}g_{\mu\sigma} - g_{\lambda\sigma}g_{\mu\nu}) + f^{ace}f^{bde}(g_{\lambda\mu}g_{\nu\sigma} - g_{\lambda\sigma}g_{\mu\nu}) + f^{ade}f^{abe}(g_{\lambda\nu}g_{\mu\sigma} - g_{\lambda\mu}g_{\sigma\nu})\right\}$$

(−1) for closed fermion and ghost loops.

Fig. 4. Feynman rules for QCD.

The renormalization constants which we shall need in QCD are:

| | |
|---|---|
| mass renormalization: | $\delta m$ |
| quark wave function renormalization: | $Z_{2\mathrm{f}}$ |
| gluon wave function renormalization: | $Z_{3\mathrm{YM}}$ |
| ghost wave function renormalization: | $\tilde{Z}_3$ |
| one-particle irreducible quark–quark–gluon vertex renormalization: | $Z_{1\mathrm{f}}$ |
| one-particle irreducible three-gluon vertex renormalization: | $Z_{1\mathrm{YM}}$ |

one-particle irreducible ghost–ghost–gluon vertex renormalization: $\tilde{Z}_1$
one-particle irreducible four-gluon vertex renormalization: $Z_5$

In fig. 5 we draw the one-loop diagrams which contribute to the renormalization constants.

Local gauge invariance requires that all the bare vertices of QCD listed in fig. 4 have the same coupling constant $g_0$. If we use a renormalization scheme which preserves this identity between the couplings for the renormalized vertices, then we can write four relations between the renormalized coupling ($g$) and the bare coupling ($g_0$):

$$g = \underset{\text{(a)}}{\frac{Z_{3\mathrm{YM}}^{3/2}}{Z_{1\mathrm{YM}}} g_0} = \underset{\text{(b)}}{\frac{Z_3^{1/2} Z_{2\mathrm{f}}}{Z_{1\mathrm{f}}} g_0} = \underset{\text{(c)}}{\frac{Z_{3\mathrm{YM}}^{1/2} \tilde{Z}_3}{\tilde{Z}_1} g_0} = \underset{\text{(d)}}{\frac{Z_3}{Z_5^{1/2}} g_0}, \tag{1.12}$$

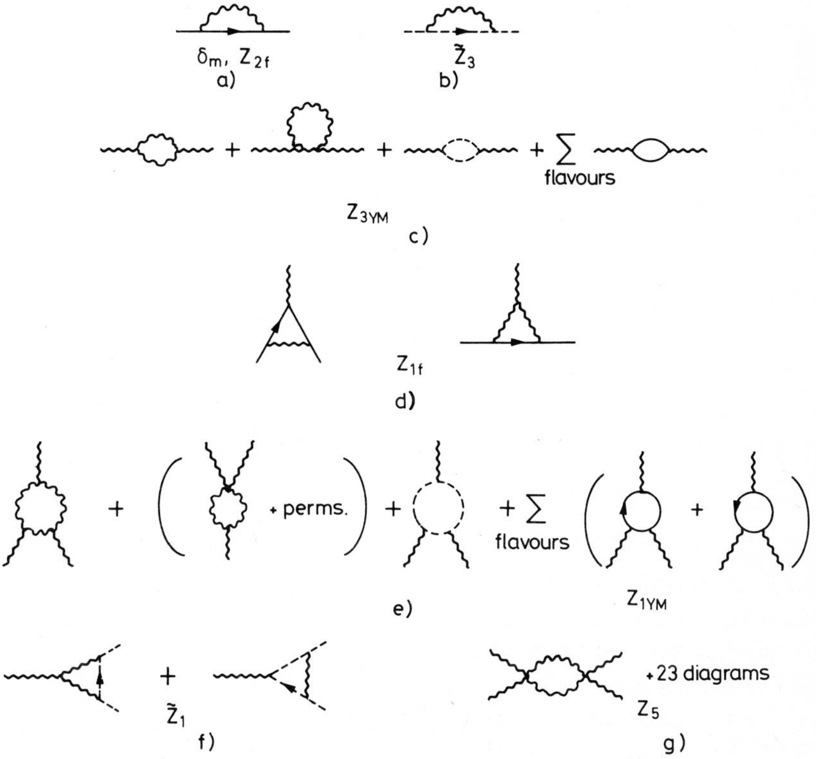

Fig. 5. One-loop contributions to the renormalization constants of QCD.

where we have used (a) the three-gluon vertex, (b) the quark–quark–gluon vertex, (c) the ghost–ghost–gluon vertex and (d) the four-gluon vertex, to define $g$. Thus we have three independent relations among the renormalization constants [17, 18] which in practise can be used either as a check on one's calculations or to avoid calculating the most complicated of these constants.

From the Feynman rules (fig. 4) we see that the computation of QCD Feynman diagrams involves the evaluation of a "group theory weight" for each graph, and for this the fundamental ingredients are the Gell-Mann matrices and the structure constants. The calculation of this SU(3) factor is straightforward, at least for the low order graphs generally computed.

Following Cvitanovic [19, see also 20], let us represent $(T_a)_{ij}$ and $f_{abc}$ as in fig. 6. In fig. 7 we present several group theory factors which occur frequently in QCD. The normalization condition for the generators is

$$\mathrm{Tr}(T_a T_b) = a\delta_{ab} \tag{1.13}$$

with $a = 1/2$ for the Gell-Mann matrices (indeed conventionally one

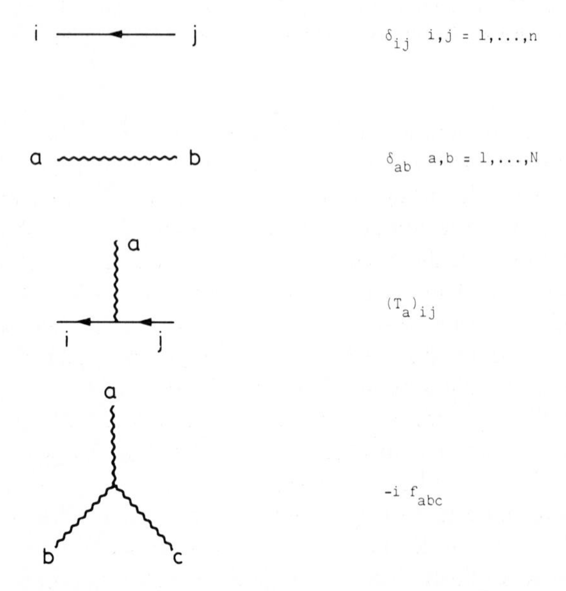


Fig. 6. Diagrammatic definition of the quantities used in evaluating group theory factors in QCD.

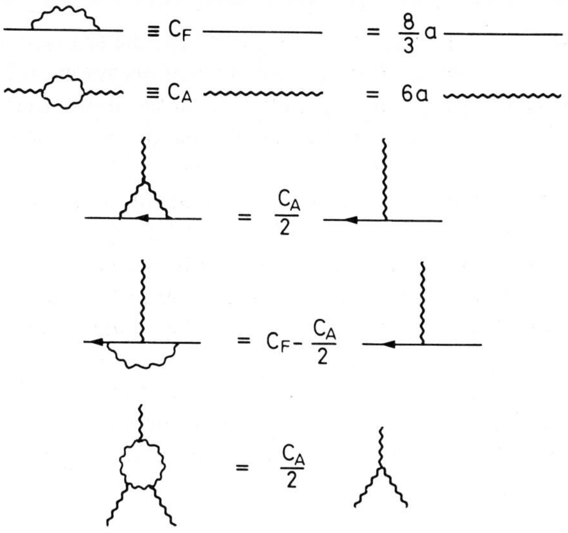

Fig. 7. Some useful identities for the group theory factors.

defines $g$ with the choice $a = 1/2$). The derivation of the results in fig. 7 is left as an exercise, however, in fig. 8 we present several fundamental relations.

The plan for the remainder of these lecture notes is as follows. In the next section we review briefly some of the renormalization group ideas needed later. We start calculating measurable QCD predictions in section 3 where we discuss the application of QCD to $e^+e^-$ annihilation. The problem of infra-red divergences in non-Abelian gauge theories is only partially understood and casts some doubt over various QCD predictions; we discuss some features of this problem in section 4. Section 5 contains a standard discussion of the light-cone techniques used in studying scaling violations in deep inelastic scattering. The same results are obtained using ordinary perturbation theory in section 6 and the predictions are extended to other processes such as massive lepton pair production. Non-asymptotic (logarithmic) corrections can be calculated and are often found to be large; we discuss the reasons and implications for this in section 7. Section 8 contains a presentation of the QCD applications to exclusive processes such as hadronic form factors and fixed-angle scattering. In section 9 we study a few miscellaneous topics and section 10 contains a brief summary.

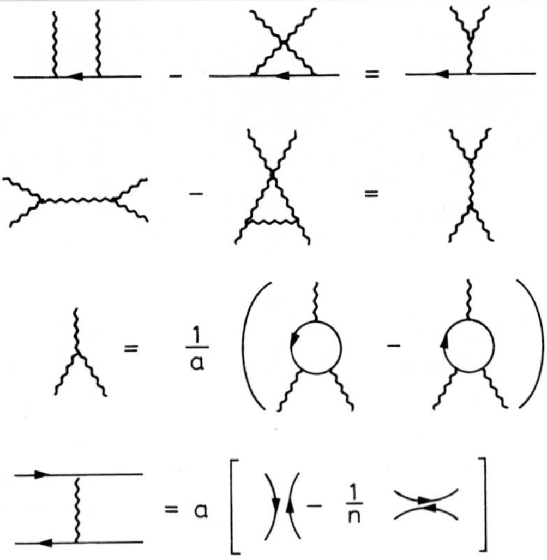

Fig. 8. Fundamental group theory relations used in the derivation of the identities in fig. 7.

## 2. Renormalization group preliminaries*

In Quantum Electrodynamics, the electric charge is normally defined via the three-particle $S$-matrix element coupling the photon to the electron (fig. 9), where the photon and electron are on their mass-shells. The Ward identity $Z_1 = Z_2$ implies that we only have to calculate the vacuum polarization to obtain the relation between the renormalized charge $e$ and the bare charge $e_0$ (the one loop diagram which contributes to the vacuum polarization is shown in fig. 10a). Thus, providing that all the fermions are massive, no infra-red divergences are introduced in renormalization.

In QCD, however, if we try to define the renormalized coupling $g$ in an analogous way, by, e.g., the on-shell coupling of a quark to a gluon (see Fig. 10b), we find that the subtraction procedure introduces infra-red divergences (e.g. the graph of fig. 10b ~ $\log q^2$ for small $q^2$), so that when we try to expand "physical" quantities in terms of powers of this on-shell coupling constant, the expansion coefficients are correspondingly singular. Hence we have to define our renormalized coupling in some other way and of course there are infinitely many ways of doing this. Each of

*For a clear introduction to the use of the renormalization group see ref. [21].

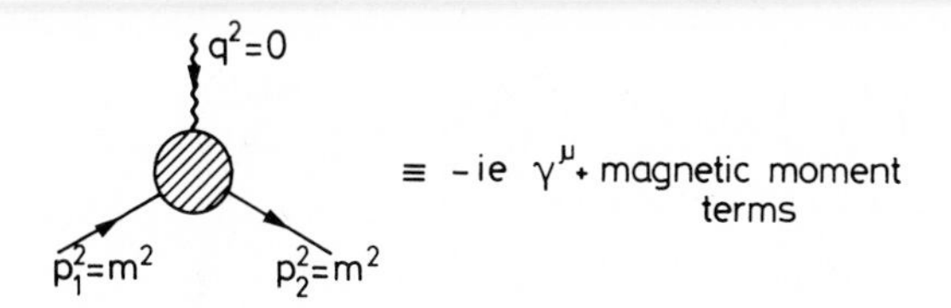


Fig. 9. Definition of (renormalized) electric charge.

these definitions involves the introduction of at least one mass scale. Here we mention some of the commonly used definitions.

(i) *Minimal Subtraction scheme* (*MS*) [22]. The divergent integrals are first regulated using dimensional regularization. For example (after the introduction of Feynman parameters), a typical loop integral

$$I_n = \int \frac{\mathrm{d}^n k}{(2\pi)^n} \frac{1}{(k^2 - V + \mathrm{i}\varepsilon)^N}, \tag{2.1}$$

is evaluated as follows:

$$I_n = \mathrm{i} \frac{(-1)^N \Gamma(N - n/2)}{(4\pi)^{n/2} \Gamma(N)} \frac{1}{(V - \mathrm{i}\varepsilon)^{N - n/2}} \tag{2.2}$$

and the integral over the $(n-1)$-dimensional solid angle is

$$\int \mathrm{d}\Omega_{n-1} = \int_0^{2\pi} \mathrm{d}\theta_1 \int_0^{\pi} \mathrm{d}\theta_2 \sin\theta_2 \ldots \int_0^{\pi} \mathrm{d}\theta_{n-1} (\sin\theta)^{n-2}. \tag{2.3}$$

Fig. 10. Sample one-loop graphs contributing to the definition of electric charge.

The Dirac algebra is also performed in $n$ dimensions, remembering that

$$g_{\mu\nu}g^{\mu\nu} = n. \tag{2.4}$$

The renormalization is performed by absorbing all singular terms as $n \to 4$, and only these terms, into the counterterms. These singular terms are of the form $1/\varepsilon^i$ where $\varepsilon = 4 - n$.

A scale is introduced because in $4 - \varepsilon$ dimensions the coupling constant is dimensional. We can rewrite it in terms of a dimensionless coupling $g$, but then with each factor of $g$ in a perturbation series we will get a factor of $\mu^{\varepsilon/2}$ where $\mu$ is our arbitrary mass scale. For a demonstration that the MS scheme is a consistent renormalization scheme we refer the reader to the original literature [22]. The Taylor–Slavnor identities mentioned in section 1 are respected in the MS scheme.

(ii) *The truncated Minimal Subtraction scheme* ($\overline{MS}$) [23]. Consider an arbitrary Feynman diagram in which there is an insertion of the quark–quark–gluon three-point function (fig. 11). From eq. (2.2) we see that this three-point function can be expanded in terms of the bare coupling $g_0$ in the form

$$g_0 + A g_0^3 \frac{\Gamma(\varepsilon/2)}{(4\pi)^{n/2}}[1 + \mathrm{O}(\varepsilon)] + \mathrm{O}(g_0^5), \tag{2.5}$$

where $A$ is a constant. The dimensionless renormalized coupling constant $g$ is related to the bare coupling $g_0$ by the relation

$$g_0 = \mu^{\varepsilon/2} g Z_g, \tag{2.6}$$

where $Z_g$ is the appropriate product of the renormalization constants

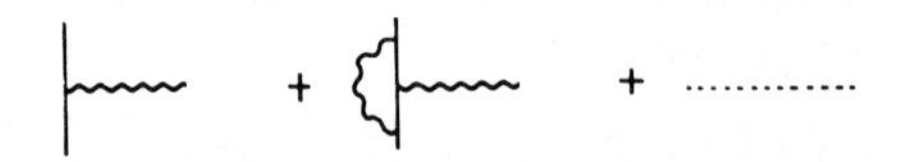


Fig. 11. Auxiliary diagrams for the explanation of $\alpha_{\overline{MS}}$.

defined in section 1. Then in the MS scheme

$$Z_g = 1 - \frac{2A}{16\pi^2\varepsilon} g^2 + \cdots, \tag{2.7}$$

so that we can rewrite eq. (2.5) as

$$g\mu^{\varepsilon/2}\left(1 - \frac{2A}{16\pi^2\varepsilon} g^2\right) + Ag^3\mu^{3\varepsilon/2}\frac{\Gamma(\varepsilon/2)}{(4\pi)^{n/2}}(1 + \cdots). \tag{2.8}$$

This is equal to

$$g + \frac{\beta_0}{16\pi^2\varepsilon} g^3(-\tfrac{1}{2}\varepsilon\gamma + \tfrac{1}{2}\varepsilon\log 4\pi + \varepsilon \log\mu + \cdots) + \cdots, \tag{2.9}$$

where we have written $\beta_0 = 2A$ for a reason that will become clear later in this section. Here $\gamma$ is Euler's constant $\simeq 0.577$. The singularity as $\varepsilon \to 0$ has disappeared as it must if we have done our renormalization correctly, but we notice that in the MS scheme with every factor of $g$ we will get a term

$$\frac{\beta_0 g^3}{32\pi^2}(\log 4\pi - \gamma). \tag{2.10}$$

In practice it is often useful to absorb this term into the definition of the renormalized coupling. Thus for the QCD equivalent of the fine structure constant $\alpha_s = g^2/4\pi$ we define

$$\alpha_{\overline{\text{MS}}} = \alpha_{\text{MS}} + \alpha_{\text{MS}}^2 \frac{\beta_0}{4\pi}(\log 4\pi - \gamma). \tag{2.11}$$

(iii) *Momentum Subtraction scheme* (*MOM*) [24]. It may be argued that the first two schemes are unintuitive in the sense that it is difficult to find a connection between the external momenta in a given process and the scale used to define the renormalized parameters. To this end Celmaster and Gonsalves have defined $g(\mu)$ as the three-gluon coupling (strictly speaking it is the coefficient of the appropriate tensor in fig. 4) at $p^2 = q^2 = r^2 = \mu^2$ in the Landau gauge. Of course there are infinitely many similar choices for the renormalizd coupling, but the above is what is usually referred to as the MOMentum subtraction scheme (MOM). The other renormalization constants are defined similarly by subtractions at $\mu^2$. In terms of $\alpha_{\overline{\text{MS}}}$ one has:

$$\alpha_{\text{MOM}} = \alpha_{\overline{\text{MS}}}(1 + 1.55\beta_0\alpha_{\overline{\text{MS}}}/4\pi + \cdots). \tag{2.12}$$

In the following we shall usually express our results in terms of $\alpha_{\text{MS}}$, $\alpha_{\overline{\text{MS}}}$

or $\alpha_{\text{MOM}}$. (See, however, ref. [25] for an application in which it is convenient to use the dimensional reduction scheme [26].)

### 2.1. *The $\beta$ function*

The $\beta$ function is defined as:

$$\beta(g) \equiv \left.\frac{\partial g(\mu)}{\partial \ln \mu}\right|_{\text{bare parameters fixed}}, \tag{2.13}$$

from which we deduce that in perturbation theory we can write

$$\beta(g) \equiv -\frac{\beta_0}{16\pi^2}g^3 - \frac{\beta_1}{(16\pi^2)^2}g^5 - \frac{\beta_2}{(16\pi^2)^3}g^7 + \cdots, \tag{2.14}$$

which explains why we replaced $2A$ by $\beta_0$ in eq. (2.9). The discovery that $\beta_0 > 0$ [1,2] implies that $g = 0$ is an "ultraviolet-stable fixed point," so that it is possible to have $g \to 0$ as $\mu^2 \to \infty$. This is the property known as asymptotic freedom. Hence if in a given process the "optimal" scale (optimal in the sense that the higher order corrections are reasonably small) at which to define $g$ is large, the "effective coupling" is small and maybe we can sensibly do a perturbative expansion. This will be made more precise in the following chapters.

From eqs. (2.12) and (2.9) we see that the first term in the $\beta$ function, $-\beta_0 g^3/16\pi^2$, is the coefficient of $-1/\varepsilon$ in the graphs of fig. 12. This is easily generalized to the higher order terms [21], so that in the MS scheme:

$$\beta(g) = -\tfrac{1}{2}\varepsilon g + g^2\,\partial a_1/\partial g^2, \tag{2.15}$$

where $a_1$ is the coefficient of $1/\varepsilon$ in $Z_g$. Thus, beyond the one loop order, we need to calculate a non-leading ultraviolet divergence. We leave it as an exercise for the reader to show that for all schemes which can be

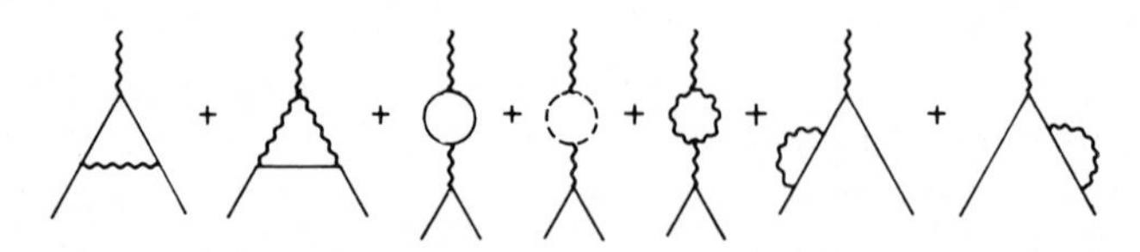


Fig. 12. One-loop diagrams which contribute to the $\beta$ function in QCD.

related by

$$\alpha_1 = \alpha_2(1+\rho\alpha_2 + \cdots), \tag{2.16}$$

where $\rho$ is independent of $\mu$, $\beta_0$ and $\beta_1$ are the same, whereas the higher order terms are scheme dependent [21,27].

So far the first three terms in the $\beta$ function, $\beta_0$ [1,2], $\beta_1$ [28,29], $\beta_2$ [30], have been calculated and they are (in the MS scheme)

$$\beta_0 = 11 - \tfrac{2}{3}f, \tag{2.17}$$

$$\beta_1 = 102 - \tfrac{38}{3}f, \tag{2.18}$$

$$\beta_2 = \tfrac{2857}{2} - \tfrac{5033}{18}f + \tfrac{325}{54}f^2, \tag{2.19}$$

where $f$ is the number of flavours.

### 2.2. The $\Lambda$ parameter

It is often useful to express the results of a QCD calculation in terms of a dimensional parameter $\Lambda$ instead of the dimensionless parameter $g$. $\Lambda$ is defined as

$$\Lambda = \mu \exp - \int^{g(\mu)} \frac{\mathrm{d}g'}{\beta(g')} \tag{2.20}$$

and it is straightforward to check that $\Lambda$ is independent of $\mu$, the explicit $\mu$ dependence being cancelled by the implicit $\mu$ dependence in $g(\mu)$. $\Lambda$ does however depend on the renormalization scheme used to define $g(\mu)$, and also on what is taken for the bottom limit of integration in eq. (2.20). This bottom limit is often implicitly chosen so that

$$\frac{\alpha(\mu^2)}{4\pi} = \frac{1}{\beta_0 \ln \mu^2/\Lambda^2} - \frac{\beta_1}{\beta_0^3}\frac{\ln\ln \mu^2/\Lambda^2}{(\ln \mu^2/\Lambda^2)^2} + \mathrm{O}\left(\frac{1}{(\ln \mu^2/\Lambda^2)^3}\right), \tag{2.21}$$

with no term of the type $1/(\ln \mu^2/\Lambda^2)^2$ on the right hand side of eq. (2.21). Later we shall see how $\Lambda$ is determined.

### 2.3. Running coupling constants and masses

If in a given "hard" process the typical momenta are of order $Q^2$, and we normalize our couplings at $\mu^2$, then we will often get terms in higher order graphs of the form $\ln Q^2/\mu^2$ raised to some power. To avoid such "large logarithms" it is sensible to choose $\mu^2 = \mathrm{O}(Q^2)$, so that the perturbation series is in terms of powers of $g(Q^2)$. We will see later how this works explicitly.

Similarly it is often "sensible" not to define the renormalized mass as the pole in the propagator, but as a running mass $m(Q^2)$ (a mass renormalized at $\mu^2 = Q^2$). The bare ($m_0$) and renormalized ($m$) masses are related by the renormalization constant $Z_m$,

$$m_0 = Z_m m. \tag{2.22}$$

The anomalous dimension of the mass operator ($\gamma_m$) is defined by

$$\gamma_m \equiv -\mu\frac{\partial}{\partial\mu}(\ln Z_m). \tag{2.23}$$

By inspection we see that $\gamma_m$ starts in order $g^2$ (see fig. 5a), specifically

$$\gamma_m = -\gamma_{m_0} g^2/16\pi^2 + \cdots, \tag{2.24}$$

with $\gamma_{m_0} = 8$. From these last two equations we deduce that the dependence of $m(\mu)$ on $\mu$ is given by

$$m(Q^2) = m(\mu^2)\exp\left(\int_{g(\mu^2)}^{g(Q^2)} \mathrm{d}g'\,\gamma_m(g')/\beta(g')\right). \tag{2.25}$$

Asymptotically, as $Q^2$ and $\mu^2 \to \infty$, eq. (2.25) reduces to

$$m(Q^2) = m(\mu^2)\left(\frac{\alpha(Q^2)}{\alpha(\mu^2)}\right)^{\gamma_{m_0}/2\beta_0} \tag{2.26}$$

Thus $m(\mu^2)$ also decreases logarithmically with $\mu^2$.

### 2.4. *Renormalization group equations*

Physical parameters (and bare parameters) $\sigma(g, m, \mu)$ must be independent of $\mu$; the explicit $\mu$ dependence must be cancelled by the implicit $\mu$ dependence in $g$ and $m$ [31,32]. Thus:

$$\mu\frac{\mathrm{d}}{\mathrm{d}\mu}\sigma = 0 \tag{2.27}$$

or in terms of partial derivatives:

$$\left(\mu\frac{\partial}{\partial\mu} + \beta(g)\frac{\partial}{\partial g} + \gamma_m m\frac{\partial}{\partial m}\right)\sigma = 0. \tag{2.28}$$

Equations such as (2.28) are particularly useful when there is only one other scale ($Q^2$, say) in the problem. In this case knowing the $\mu$ dependence of $\sigma$ tells us what the $Q^2$ behaviour is.

Later we shall write renormalization group equations for unphysical Green's functions. It is then useful to define the "anomalous dimension" of the field operator $\gamma_\phi$,

$$\gamma_\phi \equiv \tfrac{1}{2}\mu \frac{\partial}{\partial \mu}(\ln Z_\phi), \tag{2.29}$$

where $Z_\phi$ is the renormalization constant corresponding to the wave function renormalization of the field $\phi$. In the minimal subtraction scheme:

$$\gamma_\phi = -\tfrac{1}{4} g \frac{\partial}{\partial g} Z'_\phi, \tag{2.30}$$

where $Z'_\phi$ is the coefficient of $1/\varepsilon$ in $Z_\phi$.

We will also have to study Green's functions of operators. These will be defined in sections 5 and 9.

## 3. Simple applications of QCD to $e^+e^-$ annihilation

Perhaps the simplest application of QCD is to the ratio $R$ in $e^+e^-$ annihilation, defined by

$$R = \frac{\sigma(e^+e^- \to \text{hadrons})}{\sigma(e^+e^- \to \mu^+\mu^-)}. \tag{3.1}$$

The amplitude for the annihilation of $e^+e^-$ into hadrons is drawn in fig. 13 and we see that it is proportional to the imaginary part of the vacuum polarization. We can calculate $R$ in perturbation theory at least when $Q^2$ is far away from any thresholds (strictly speaking one should calculate a "smeared" cross-section [33]). One finds

$$R = 3\sum_f e_f^2 \left[1 + \frac{\alpha_s(Q^2)}{\pi}\left(1 + r\frac{\alpha_s(Q^2)}{\pi} + \cdots\right)\right], \tag{3.2}$$

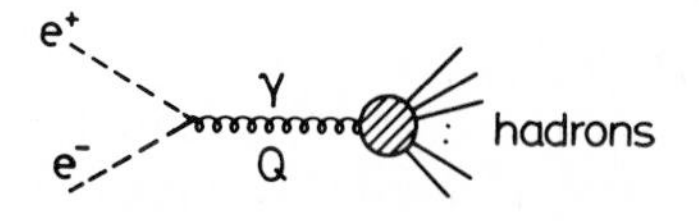


Fig. 13. Schematic representation of $e^+e^- \to$ hadrons.

where [34] for four flavours:

$$r = 5.58\ (\mathrm{MS}), \quad 1.52\ (\overline{\mathrm{MS}}), \quad -1.70\ (\mathrm{MOM}).$$

There is only a mild dependence on the number of flavours.

As will be seen below for $Q^2 \gg m^2$, where $m$ represents typical quark or hadronic masses, one can neglect these masses. Thus it is sensible to choose the renormalization point $\mu^2$ to be equal to $Q^2$. It is for this reason that the expansion in eq. (3.2) is in terms of $\alpha_s(Q^2)$. Following Buras [35], we take $\alpha_{\overline{\mathrm{MS}}}(30\ \mathrm{GeV}^2) \simeq 0.2$ and $\alpha_{\mathrm{MOM}}(30\ \mathrm{GeV}^2) \simeq 0.25$ (from deep inelastic lepton–hadron scattering data). Then the relative sizes of the terms on the right hand side of eq. (32) are

| | 0 loops | | 1 loop | | 2 loops | total |
|---|---|---|---|---|---|---|
| $\overline{\mathrm{MS}}$ scheme: | 1 | : | 0.064 | : | 0.006 | 1.070 |
| MOM scheme: | 1 | : | 0.080 | : | −0.011 | 1.069 |

We conclude that the totals agree very well, and that the corrections are acceptably small so that the perturbative expansion seems to make sense. We see also from fig. 3 that the theoretical prediction agrees with the data.

### 3.1. *First attempts at jet physics*

The data for $R$ are well explained by thinking of the annihilation into hadrons as the process $e^+e^- \to q\bar{q}$. If this picture is correct then there should be two back-to-back jets, with a spread in transverse momentum of about 300 MeV (which is the typical transverse momentum in hadronic interactions). To check this quantitatively the SLAC–LBL group [36] defined the variable "sphericity", $\hat{S}$.

$$\hat{S} \equiv \tfrac{3}{2} \min_{\text{axes}} \left( \sum_i |p^i_\perp|^2 \Big/ \sum_i |\boldsymbol{p}^i|^2 \right). \tag{3.3}$$

The quantity $\Sigma_i |p^i_\perp|^2/\Sigma_i |\boldsymbol{p}^i|^2$ is calculated for a selection of axes and then the value is taken for which it is a minimum. For an event which consists of precisely two back to back jets $\hat{S} = 0$, whereas for an entirely isotropic event $\hat{S} = 1$. The SPEAR data [36] can be summarized as follows: (i) $\mathrm{d}\sigma/\mathrm{d}\hat{S}$ is peaked at low $\hat{S}$. (ii) $\langle \hat{S} \rangle$ decreases with $Q^2$. (iii) The data fit a two-jet model with a Gaussian smearing in $p_\perp$ with $\langle p_\perp \rangle \simeq 315$ MeV. (iv) The angular distribution of the axes which minimize $\hat{S}$ is approximately $1 + \cos^2\theta$ in agreement with spin 1/2 jets.

### 3.2. *Jets in QCD*

The measurement of the sphericity distribution encourages us to believe that indeed the annihilation cross-section is dominated by $e^+e^- \to q\bar{q}$, but we would like to make quantitative predictions. To this end and in order to avoid uncalculable long distance effects we look for processes and quantities at large momentum transfers, which do not depend on hadronic masses (i.e. those which are not singular as we set the quark and gluon masses to zero). We also have to be very careful if we want to probe the small transverse ( < 300 MeV) structure of jets. Thus in order to avoid these long distance effects we have to be aware of all possible sources of $\log m^2$ terms, i.e. we are interested in finding the regions of phase space which yield a divergence as one or more masses vanish. We will now discuss the two types of divergence which occur. These are known as infra-red divergences and mass singularities. Both have been thoroughly studied in field theory. Infra-red divergences [37,38] arise from the presence (in a frame in which the external particles are not at rest) of a soft, real or virtual, massless particle. For example, if we evaluate the diagram of fig. 4 for the process $e^+e^- \to \mu^+\mu^-$ (or $q\bar{q}$), we find that there is a Feynman Integral of the form

$$\int d^4k\left\{(k^2+i\varepsilon)\left[(p_1+k)^2-m^2+i\varepsilon\right]\left[(p_2-k)^2-m^2+i\varepsilon\right]\right\}^{-1}$$

$$\underset{k\to 0}{\sim}\int\frac{d^4k}{k^4}=\infty. \tag{3.4}$$

We see in fig. 14 that when $k$ is soft, three propagators simultaneously come close to their poles. The theorem of Bloch and Nordsieck [37,38] states that in inclusive cross-sections these infra-red divergences cancel. In QED, for example, they cancel between diagrams with real photons and those with virtual photons as in fig. 15. All physical cross-sections are in fact inclusive ones, since in any experiment the energy resolution is not perfect and hence there may be any number of undetected soft

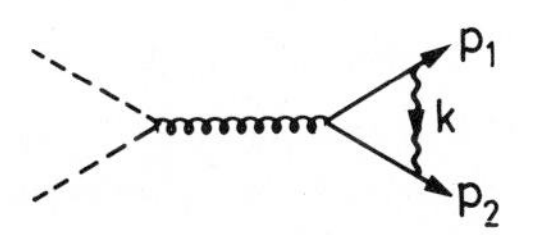


Fig. 14. Example of an infra-red divergent diagram in the process $e^+e^- \to \mu^+\mu^-$ (or $q\bar{q}$).

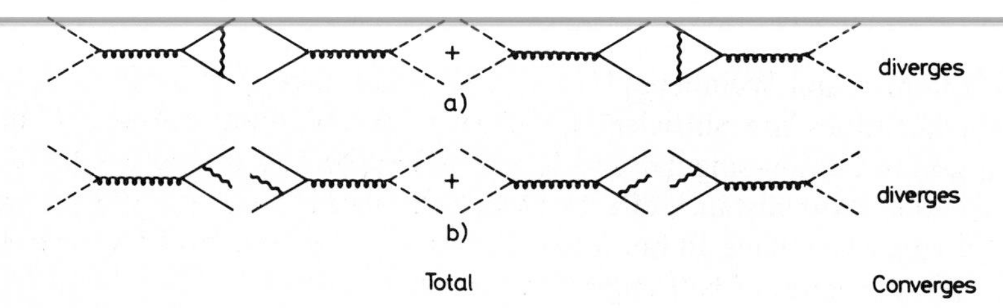


Fig. 15. Example of the Bloch–Nordsieck mechanism for the cancellation of infra-red divergences in the total $e^+e^-$ annihilation cross-section. The sum of a) and b) converges.

photons. Hence, when calculating predictions for measurable cross-sections, we have to sum over states which include these additional photons, and we then arrive at a finite result.

We now come to mass singularities. These occur in theories with coupled massless particles, and are due to the simple kinematical fact that two massless particles (with momenta $k_1, k_2$ say) which are moving parallel to each other, have a combined invariant mass equal to zero

$$k^2 \equiv (k_1 + k_2)^2 = [\omega_1(1,0,0,1) + \omega_2(1,0,0,1)]^2 = 0. \qquad (3.5)$$

For mass singularities there is the theorem of Kinoshita [39] and Lee and Naunberg [40] (KLN) which states that for sufficiently inclusive cross-sections they also cancel. For example, the mass singularities of fig. 15a (in the limit where the mass of the muon is zero) cancel those of fig. 15b, and so for the incluive process $e^+e^- \to \mu^+\mu^- X$ we can set both the $\mu$ and photon masses to zero and still get a finite result. The interpretation of this result is similar to that for infra-red divergences, namely that in any physically measurable process the angular resolution is never perfect, and therefore we should sum over all indistinguishable states, i.e. all states in which there are some (almost) collinear particles. For the exact statement of the KLN theorem we refer the reader to the original papers [39,40]; here we will just note that it assures us that all mass singularities coming from final state undetected particles moving parallel to each other cancel*.

*Later we will see that the mass singularities which come from the regions of phase space in which internal particles have momenta parallel to the momenta of one of the incoming particles do *not* cancel in many processes. (For these to cancel we have to sum over "degenerate" initial states.)

The above properties of infra-red divergences and mass singularities led Sterman and Weinberg [41] to the fact that many physically measurable quantities are sufficiently inclusive for the limit $m \to 0$ (for all masses) to be non-singular and hence these processes are dominated by calculable short distance effects. One such quantity is $f$, the fraction of all events which have all but a fraction $\varepsilon \ll 1$ of their energy in some pair of opposite cones of half angle $\delta$. In QCD,

$$f = 1 - \frac{g^2(Q^2)}{3\pi^2}\left(3\ln\delta + 4\ln\delta\ln 2\varepsilon + \tfrac{1}{3}\pi^2 - \tfrac{7}{4}\right) + \mathrm{O}(g^4, \varepsilon, \delta), \quad (3.6)$$

so that as long as $\varepsilon$ and $\delta$ are not zero, $f$ is finite. In practice one has to be careful about applying eq. (3.6) since it is valid in the region $(\alpha/\pi)\ln\delta\ln\varepsilon \ll 1$ which may be hard to achieve experimentally. We will see later how eq. (3.6) can be improved. To get an idea of how the opening angle of the jet behaves with increasing $Q^2$ we can invert eq. (3.6) to obtain

$$\delta(Q^2, \varepsilon, f) = \left(\frac{\sqrt{Q^2}}{\Lambda}\right)^{(33-2n_f)(1-f)/8(4\log 2\varepsilon + 3)} \times \exp\left[-\left(\tfrac{1}{3}\pi^2 - \tfrac{7}{4}\right)/(4\log 2\varepsilon + 3)\right], \quad (3.7)$$

where $n_f$ is the number of flavours. Writing eq. (3.7) as

$$\delta(Q^2, \varepsilon, f) \sim \left(\sqrt{Q^2}\right)^{-d(f,\varepsilon)}, \quad (3.8)$$

we find for example that for $\varepsilon = 0.1$ and for five flavours, $d(f, \varepsilon) = 0.84$ $(1-f)$. Thus, as should be intuitively expected, jets get narrower with increasing $Q^2$.

For gluons the analogous equation to (3.6) is [42]

$$f_g = 1 - \frac{g^2(Q^2)}{4\pi^2}\left[12\log 2\varepsilon - \left(11 - \tfrac{2}{3}f\right)\right]\log\delta + \cdots, \quad (3.9)$$

which implies that gluon jets should asymptotically be wider than quark jets.

Unfortunately it turns out that the sphericity distribution is not calculable in perturbation theory. To see this, recall that we know from the KLN theorem that in a theory with coupled massless particles we must sum over all "degenerate" states in order to get a cancellation of mass singularities. When calculating the sphericity distribution, however, we weight the amplitude by a function of the square of the momentum so that we weight a state with one massless particle of momentum $p$ differently from that with two massless particles with momentum $xp$ and $(1-x)p$ [since $x^2+(1-x)^2 \neq 1$ in general]. Hence in spite of its phenomenological usefulness, sphericity is not an "infra-red safe" variable.

If we had instead weighted the amplitude by an expression which involves a sum (or sums) over a linear combination of momenta, the mass-singularities would have cancelled out and the corresponding variable would have been calculable. There are many such variables [43,44] used to test QCD, some of which will no doubt be introduced by Wiik [44]. Here we will introduce just one or two of them. Over the past few years a popular variable has been thrust [45], defined by

$$T \equiv \max_{n} \left(\Sigma_i |\boldsymbol{p}_i \cdot \boldsymbol{n}| / \Sigma_i |\boldsymbol{p}_i|\right), \tag{3.10}$$

where $\boldsymbol{n}$ is an arbitrary unit vector and the $\boldsymbol{p}_i$ are the momenta of the final state particles. $T=1$ for an ideal two-jet event and $T=1/2$ for a perfectly spherical event.

Wiik [44, see also 46] will describe in detail the status of the comparison of the theoretical predictions and experimental results. As a brief summary it is probably fair to say that at present values of $Q^2$ the lowest order QCD contribution (gluon bremsstrahlung) to the $(1-T)$ distribution (and many similar quantities) lies below the experimental data, but also that the expected contribution from the long-distance fragmentation process responsible for confinement (which throws out hadrons with a typical transverse momentum of about 300 MeV) is significant, so that models based on QCD can explain the data pretty well with a reasonable value for $\alpha_s(\sim 30\ \mathrm{GeV}^2)$ of about 0.16. Several calculations [47] indicate that the contribution from higher order corrections is large (60–100%), see however [48], but that the shape of the thrust distribution is preserved. In that case maybe $\alpha_s$ is somewhat smaller (0.11?) but we will have more to say about the interpretation of large higher order corrections in the following sections.

Another useful "infra-red safe" variable is the tensor

$$\theta^{ij} = \sum_a \left(p_a^i p_a^j / |\boldsymbol{p}_a|\right) \Big/ \sum_a |\boldsymbol{p}_a|, \tag{3.11}$$

where $i$, $j$ run over 1,2,3 and the sum is over the momenta of the final state particles. This variable involves no optimization over axes, and has the property that for a two-jet event two eigenvalues are zero and for a three-jet event one eigenvalue vanishes. Hence by imposing cuts on the eigenvalues of $\theta^{ij}$ one can select a sample of events biased to contain predominantly those of the desired topology (for example three-jet events).

### 3.3. *Jets in resonance decays*

So far we have been considering the continuum region, far away from any resonances, but it is also possible to make predictions concerning the decays of heavy quarkonia. One obvious prediction is that since a $^3S_1$ state ($J/\psi$, $\Upsilon$ etc) cannot decay via one gluon (colour conservation) or two gluons (charge conjugation invariance) we would expect these resonances to decay primarily into 3 jets. Thus $\langle 1-T \rangle$ and other similar variables should be much larger on resonance than off, and this is indeed the case for the upsilon and provides a useful tool in the search for new particles. For a detailed account of the applications of QCD to jets in resonance decays see section 7.5 of ref. [27].

We will return to look at the more detailed structure of jets in section 9.

## 4. Infra-red divergences in QCD

Many QCD predictions depend on the cancellation (or at least partial cancellation) of infra-red divergences. It is extremely difficult to prove such a cancellation in general and hence this means that many of the QCD applications discussed in these lectures are based on conjectures rather than theorems. We now digress from the main discussion of the lectures to present briefly the status of the cancellation of these infra-red divergences.

(i) *No coloured objects in the initial state.* It has been known for some time [39,40,49] that for totally inclusive cross-sections for processes in which there are no coloured objects in the initial state, (such as the ratio $R$ in $e^+e^-$ annihilation) the infra-red divergences cancel analogously to the Bloch–Nordsieck mehcanism in QED.

(ii) *One coloured object in the initial state.* For processes in which there is just one coloured object in the initial state, (such as a quark scattering of an electromagnetic potential), it has recently been shown [50] that the infra-red divergences cancel, although the proof is fairly complicated.

(iii) *Two coloured objects in the initial state.* For processes with two coloured objects in the initial state the infra-red divergences are known *not* to cancel, (there is nothing analogous to the Bloch–Nordsieck mechanism) even if the colours of the initial quarks or gluons are averaged over [51–53]. An example of such a process is $q\bar{q} \to e^+e^- X$. Clearly if the colours are not averaged over the infra-red divergences cannot cancel, e.g. if we insist that in the process $q\bar{q} \to e^+e^- X$ the initial $q\bar{q}$ system is in a colour singlet state, then the graphs of fig. 15a are infra-red divergent as usual, but the corresponding diagrams with a real gluon (fig. 15b) are zero. Hence in this case the divergence cannot cancel, but this is not surprising since it is not "physically meaningful" to fix the colour of the initial state (since the emission of an arbitrary soft gluon can alter this colour).

We will be interested in the case where the initial colours are indeed averaged over. In this case the infra-red divergences in the $O(g^2)$ graphs (fig. 15) clearly cancel, since up to a factor of 4/3 they are equal to the QED graphs. In two loops they do not cancel. For example for processes with qq or $q\bar{q}$ in the initial state (and with no "Coulomb" divergence) the divergence in the cross-section can be written as

$$\mp \alpha_s^2 C_A C_F \frac{1}{2\varepsilon}\left(\frac{1}{\beta}-1\right)\left(\frac{1}{\beta}\log\frac{1+\beta}{1-\beta}-2\right)\sigma_B, \tag{4.1}$$

where the divergence was regulated by dimensional regularization ($\varepsilon = n - 4$, and $n$ is the number of dimensions in which the integrals are performed). The $-$ $(+)$ sign corresponds to the qq ($q\bar{q}$) initial state and $\beta = |\boldsymbol{q}|/q_0$ in the rest frame of $p$ (see fig. 16). Notice that in the ultra relativistic limit $\beta \to 1$ there is a suppression factor $(1/\beta - 1)$. This will prove very important later although it should be remembered that it has not (yet?) been proved that in higher orders there is never a divergent term without such a suppression factor.

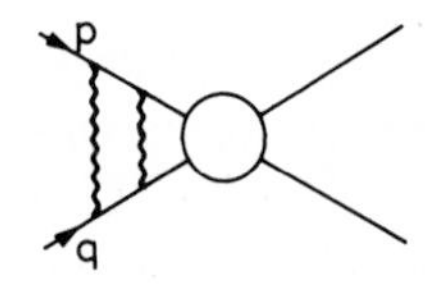


Fig. 16. Example of a two loop infra-red divergent diagram.

## 5. Light-cone techniques and the asymptotic behaviour of deep inelastic structure functions

In this section we review the standard light-cone analysis, using the operator product expansion and renormalization group, as applied to deep inelastic structure functions. In spite of the fact that there are now many excellent reviews on the subject [e.g. refs. [21,54–58], the light-cone expansion is the foundation on which many QCD applications are based, and therefore we felt that we should include at least a brief discussion of the subject here, which we shall now do.

The kinematics of deep inelastic lepton–hadron scattering is defined in fig. 17, and we are interested in the Bjorken limit

$$|q^2| \to \infty, \quad p \cdot q \to \infty, \quad x \equiv -\frac{q^2}{2p \cdot q} \text{ fixed.} \tag{5.1}$$

In the parton model $x$ is found to be the fraction of the target hadron's momentum carried by the "struck" quark, and we shall find that in QCD we can also sensibly interpret it in this way. Since we believe we understand the lepton–photon or lepton–weak-boson coupling, we can concentrate on the absorptive deep-inelastic Compton amplitude $W_{\mu\nu}$,

$$W_{\mu\nu}(p,q) \equiv \frac{1}{4\pi} \int \mathrm{d}^4 y \, \mathrm{e}^{\mathrm{i}q \cdot y} \langle p | \left[ J_\mu^+(y), J_\nu(0) \right] | p \rangle, \tag{5.2}$$

which contains all the strong interaction effects. We define the three structure functions $W_1$, $W_2$ and $W_3$ by

$$\begin{aligned} W_{\mu\nu}(p,q) = & -\left( g_{\mu\nu} - \frac{q_\mu q_\nu}{q^2} \right) W_1(q^2, x) \\ & + \frac{1}{m_\mathrm{H}^2} \left( p_\mu - q_\mu \frac{p \cdot q}{q^2} \right) \left( p_\nu - q_\nu \frac{p \cdot q}{q^2} \right) W_2(q^2, x) \\ & - i \frac{\varepsilon_{\mu\nu\alpha\beta} p^\alpha q^\beta}{2 m_\mathrm{H}^2} W_3(q^2, x), \end{aligned} \tag{5.3}$$

where $m_\mathrm{H}$ is the mass of the hadronic target. We can only have these three structure functions because of the constraints imposed by Lorentz- and CP-invariance and by current conservation. Moreover by parity invariance $W_3 = 0$ for electron and muon scattering. In the parton model

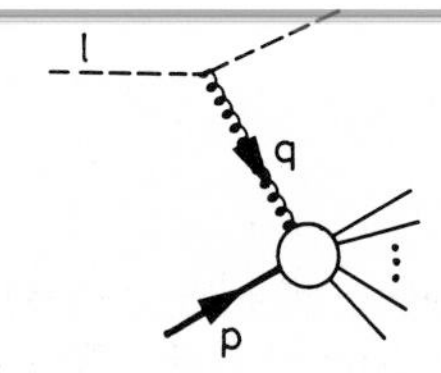


Fig. 17. Schematic representation of deep inelastic lepton–hadron scattering.

the structure functions satisfy Bjorken scaling, i.e.

$$W_1(q^2,x) \underset{q^2\to-\infty}{\to} F_1(x),$$

$$\frac{\nu}{m_{\mathrm{H}}^2} W_2(q^2,x) \underset{q^2\to-\infty}{\to} F_2(x),$$

$$\frac{\nu}{m_{\mathrm{H}}^2} W_3(q^2,x) \underset{q^2\to-\infty}{\to} F_3(x), \tag{5.4}$$

where asymptotically the functions $F_i(x)$ are independent of $q^2$. These functions are just the sum over the quark distributions in the hadron, each quark being weighted by the square of its charge to the weak or electromagnetic current. Thus measuring different structure functions (e.g. $F_1$ and $F_3$ in $\nu$ scattering), and also using different beams gives us considerable information about the quark distribution functions. We will also discuss and use the longitudinal structure function $F_{\mathrm{L}}$ defined by $F_{\mathrm{L}} = F_2 - 2xF_1$.

We now study the structure functions in QCD. The first relevant observation is that the Bjorken limit (5.1) corresponds to the light-cone limit in configuration space (see for example ref. [21]), namely $y^2 \to 0$ in eq. (5.2). Regions far away from the light cone give a rapidly oscillating factor in the integrand of eq. (5.2), and hence a negligible contribution to the integral. We will now calculate the QCD prediction for the violation of Bjorken scaling. There are a number of steps in this calculation:

1). The forward elastic Compton amplitude $T_{\mu\nu}$ is defined by

$$T_{\mu\nu}(p,q) = \mathrm{i}\int \mathrm{d}^4x\, \mathrm{e}^{\mathrm{i}q\cdot x}\langle p|T(J_\mu^+(x)J_\nu(0))|p\rangle, \tag{5.5}$$

so that

$$W_{\mu\nu} = \frac{1}{2\pi} T_{\mu\nu}.$$

We now expand the product of the two currents as an infinite series of local operators and take the Fourier transform obtaining

$$\begin{aligned} & \mathrm{i}\int \mathrm{d}^4x\, \mathrm{e}^{\mathrm{i}q\cdot x} T\left(J_\mu^+(x) J_\nu(0)\right) \\ & = \sum_{N,i} \Big[ -\left( g_{\mu\mu_1} g_{\nu\mu_2} q^2 - g_{\mu_1\mu} q_\nu q_{\mu_2} - g_{\nu\mu_2} q_\mu q_{\mu_1} + g_{\mu\nu} q_{\mu_1} q_{\mu_2} \right) \\ & \qquad \times C_{2,i}^N(q^2, \mu^2, g^2) + \left( g_{\mu\nu} - \frac{q_\mu q_\nu}{q^2} \right) q_{\mu_1} q_{\mu_2} C_{\mathrm{L},i}^N(q^2, \mu^2, g^2) \\ & \qquad - \mathrm{i}\varepsilon_{\mu\nu\alpha\beta} g_{\alpha\mu_1} q_\beta q_{\mu_2} C_{3,i}^N(q^2, \mu^2, g^2) \Big] \\ & \qquad \times q_{\mu_3} \cdots q_{\mu_N} \left( \frac{2}{-q^2} \right)^N O^{\mu_1 \cdots \mu_N}(0), \end{aligned} \tag{5.6}$$

where $i$ labels the different operators with $N$ Lorentz indices. Substituting eq. (5.6) into (5.5) (and working in the Bjorken limit (5.1)) we readily find

$$\begin{aligned} T_{\mu\nu}(q^2, x) = \sum_{N,i} x^{-N} \Bigg[ & \left( g_{\mu\nu} - \frac{q_\mu q_\nu}{q^2} \right) C_{\mathrm{L},i}^N(q^2, \mu^2, g^2) \\ & - \left( g_{\mu\nu} - \frac{p_\mu q_\nu + p_\nu q_\mu}{p\cdot q} + \frac{p_\mu p_\nu}{(p\cdot q)^2} q^2 \right) \\ & \times C_{2,i}^N(q^2, \mu^2, g^2) - \mathrm{i}\varepsilon_{\mu\nu\alpha\beta} \frac{p_\alpha q_\beta}{p\cdot q} \\ & \times C_{3,i}^N(q^2, \mu^2, g^2) \Bigg] \langle p | O_i^N | p \rangle, \end{aligned} \tag{5.7}$$

where

$$\langle p | O_i^{\mu_1 \cdots \mu_N}(0) | p \rangle \equiv p^{\mu_1} \ldots p^{\mu_N} \langle p | O^N | p \rangle. \tag{5.8}$$

The reason why such an expansion is useful is that in the Bjorken limit for each $N$ only three operators contribute to the right-hand side of eqs. (5.6) and (5.7). In this limit we would like the mass dimension of the coefficient functions $C$ to be as large as possible and therefore we want the $\langle p|O^N|p\rangle$ to have as small a mass dimension as possible. Going back to eq. (5.6) this now means that we want $O^{\mu_1\cdots\mu_N}(0)$ to have as low a twist ($\equiv$ dimension − spin) as possible. Thus we look for the lowest twist operators with the correct quantum numbers and we find that for each $N$ there are three twist-two operators

$$\bar{\psi}\gamma^{\mu_1}D^{\mu_2}\dots D^{\mu_N}\psi\tau^k - \text{traces}, \tag{5.9a}$$

$$\bar{\psi}\gamma^{\mu_1}D^{\mu_2}\dots D^{\mu_N}\psi - \text{traces}, \tag{5.9b}$$

$$G^{\mu_1\nu}D^{\mu_2}\dots D^{\mu_N}G^{\mu_N\nu} - \text{traces}, \tag{5.9c}$$

where the covariant derivative $D$ is defined by

$$D^\mu \equiv \partial^\mu - \mathrm{i}gT_aA^{a,\mu}. \tag{1.3}$$

$\tau^k$ is a flavour matrix. Hence the label $i$ in eqs. (5.6) and (5.7) runs over these three operators. A symmetrization over all indices is assumed. Moreover, if we choose a combination of structure function which is a flavour non-singlet, the most common examples being $F_2^{\mathrm{ep}} - F_2^{\mathrm{en}}$, $F_2^{\mu\mathrm{p}} - F_2^{\mu\mathrm{n}}$ or $F_3^{\nu\mathrm{A}}$ where A is an isoscalar target, it is dominated only by the operator (5.9a).

2). We now use a dispersion relation to relate $T_{\mu\nu}$ and its imaginary part $W_{\mu\nu}$. We rewrite the usual dispersion relation in $s \equiv (p+q)^2$ in terms of the variable $x$. The normal threshold cuts in $s$ and $u$ now correspond to a cut in $x$ from $-1$ to 1. The precise form of the dispersion relation depends on which structure function we look at. For scalar currents (in which case there is only one structure function) this relation is

$$\begin{aligned} T(q^2,x) &= \int_{-1}^{1} x'^{s-1}\mathrm{d}x' \frac{W(q^2,x')}{x^{s-1}(x'-x)} + \text{subtractions} \\ &= \sum_{N=s}^{\infty} x^{-N}\int_{-1}^{1} \mathrm{d}x' x'^{N-1} W(q^2,x') + \text{subtractions}, \end{aligned} \tag{5.10}$$

where $s$ is the number of subtractions. Using the crossing properties of the structure functions under $x \to -x$:

$$W_{1,2}^{\mathrm{eH},\mu\mathrm{H}}(q^2,x) = -W_{1,2}^{\mathrm{eH},\mu\mathrm{H}}(q^2,-x) \tag{5.11a}$$

and

$$W_{1,2,3}^{\nu\mathrm{H}}(q^2,x) = -W_{1,2,3}^{\bar{\nu}\mathrm{H}}(q^2,-x), \tag{5.11b}$$

we can rewrite the relations between the $T_i$'s and $W_i$'s in terms of integrals over the physical region $0 < x < 1$,

$$\int_0^1 \mathrm{d}x\, x^{N-2} F_\alpha(x, q^2) = \sum_i C_{\alpha,i}^N(q^2, \mu^2, g^2)\langle p|O_i^N|p\rangle, \quad \alpha = 2, \mathrm{L} \tag{5.12a}$$

and

$$\int_0^1 \mathrm{d}x\, x^{N-1} F_3(x, q^2) = \sum_i C_{3,i}^N(q^2, \mu^2, q^2)\langle p|O_i^N|p\rangle. \tag{5.12b}$$

3). Finally we calculate the $q^2$ behaviour of the Wilson coefficient functions $C^N$, thus calculating the violation of Bjorken scaling in the moments of the structure functions. Since the moments of $\nu W_2$ for example, which we call $M_2^N$, are clearly physically measurable quantities, they must be independent of $\mu$, the renormalization point. Thus

$$\mu \frac{\mathrm{d}}{\mathrm{d}\mu} M_2^N(q^2) = 0. \tag{5.13}$$

As an example let us take the non-singlet combination of structure functions $F_2^{\mathrm{ep}} - F_2^{\mathrm{en}}$, in which case there is only one term in the summation on the right-hand side of eq. (5.12a). Hence the $\mu$ dependence of $C_2^N$ must be compensated for by the $\mu$ dependence of $\langle p|O^N|p\rangle$,

$$\mu \frac{\mathrm{d}}{\mathrm{d}\mu} C_2^N \cdot \langle p|O^N|p\rangle = 0. \tag{5.14}$$

The relevant operator, which is given in eq. (5.9a), is multiplicatively renormalizable with renormalization constant $Z_{O^N}$ defined by

$$O^N \equiv Z_{O^N} O^{N,\mathrm{u}}, \tag{5.15}$$

where the label u signifies the bare (unrenormalized) operator. Equation (5.14) can now be rewritten:

$$\mu \frac{\mathrm{d}}{\mathrm{d}\mu} C_2^N = C_2^N \cdot \mu \frac{\mathrm{d}}{\mathrm{d}\mu}(\ln Z_{O^N}) \equiv C_2^N \cdot \gamma_{O^N}, \tag{5.16}$$

where $\gamma_{O^N}$ is called the anomalous dimension of the operator $O^N$. Hence we get:

$$\left(\mu \frac{\mathrm{d}}{\mathrm{d}\mu} - \gamma_{O^N}\right) C_2^N(q^2, \mu^2, g) = 0, \tag{5.17}$$

which we can rewrite in terms of partial derivatives (working in the

Landau gauge for example, so that the gauge parameter does not get renormalized) as:

$$\left(\mu\frac{\partial}{\partial\mu}+\beta(g)\frac{\partial}{\partial g}-\gamma_{O^N}\right)C_2^N(q^2,\mu^2,g)=0. \tag{5.18}$$

But $C_2^N$ is a dimensionless function which only depends on $q^2$ and $\mu^2$, so it is a function only of the ratio $q^2/\mu^2$. Thus we can make the substitution

$$\mu\frac{\partial}{\partial\mu}\to -q\frac{\partial}{\partial q} \tag{5.19}$$

in eq. (5.18) to obtain:

$$\left(2q^2\frac{\partial}{\partial q^2}-\beta(g)\frac{\partial}{\partial g}+\gamma_{O^N}\right)C_2^N(q^2,\mu^2,g)=0. \tag{5.20}$$

Equation (5.20) gives us the $q^2$ behaviour of the coefficient functions $C_2^N$, and therefore, since the matrix elements of local operators are $q^2$ independent, also the $q^2$ behaviour of the moments of $W_2(x,q^2)$. Its solution is

$$C_2^N(q^2,g,\mu^2)=C_2^N(q^2=\mu^2,g(q^2))\exp-\int_{g(\mu^2)}^{g(q^2)}\frac{\gamma_{O^N}(g')}{\beta(g')}\mathrm{d}g'. \tag{5.21}$$

Expanding $\gamma_{O^N}(g')$ as a power series in $g'$,

$$\Gamma_{O^N}(g')=\frac{g'^2}{16\pi^2}\gamma^0_{O^N}+\mathrm{O}(g'^4), \tag{5.22}$$

we find that the QCD prediction for the asymptotic behaviour of the moments of non-singlet combinations of structure functions is:

$$\frac{M^N(q^2)}{M^N(q_0^2)}=\left[\frac{\alpha(q^2)}{\alpha(q_0^2)}\right]^{\gamma^0_{O^N}/2\beta_0} \tag{5.23}$$

which, combined with eq. (2.21), means that Bjorken scaling (5.4) is violated logarithmically in QCD. Scaling violations of the type predicted in eq. (5.23) seem to agree fairly well with the data [59–61]. Higher order contributions to $\gamma_{O^N}$ and $\beta$ give contributions to the right hand side of eq. (5.23) which are logarithmically suppressed relative to the lowest order term.

Equation (5.21) gives the $q^2$ behaviour of the moments of flavour non-singlet combinations of structure functions. Flavour singlet combinations of structure functions are dominated by the two sets of operators (5.9b) and (5.9c). These mix under renormalization so that the anomalous dimension is now a matrix, and the equation analogous to (5.20) is a matrix equation, which can easily be diagonalized. A compact way to write the solution for moments of singlet combinations of structure functions is

$$M^N(q^2) = A^N\left[\alpha(q^2)\right]^{\gamma_+^N/2\beta_0} + B^N\left[\alpha(q^2)\right]^{\gamma_-^N/2\beta_0}, \tag{5.24}$$

where $\gamma_+^N$ and $\gamma_-^N$ are the two eigenvalues of the anomalous dimension matrix and $A^N$ and $B^N$ are unpredicted constants (reflecting our ignorance of the matrix elements of the operators (5.9b) and (5.9c) between hadronic states).

In general a structure function such as $F_2^{\text{ep}}$ is a combination of singlet and non-singlet terms so that its moments are given by

$$M^N(q^2) = A^N\left[\alpha(q^2)\right]^{\gamma_+^N/2\beta_0} + B^N\left[\alpha(q^2)\right]^{\gamma_-^N/2\beta_0} + C^N\left[\alpha(q^2)\right]^{\gamma_{O^N}^0/2\beta_0}, \tag{5.25}$$

so that for each moment we have three constants $A^N$, $B^N$ and $C^N$ to determine from the data.

We now briefly discuss how one evaluates the ingredients necessary to make quantitative predictions for the violations of Bjorken scaling from eq. (5.21).

(i) *The anomalous dimension* $\gamma_{O^N}$. In the non-singlet case we are interested in the anomalous dimension of the operator (5.9a). To this end we imagine adding a term to the Lagrangian:

$$J_{\mu_1\ldots\mu_N} O^{\mu_1\ldots\mu_N}, \tag{5.26}$$

where $O^{\mu_1\ldots\mu_N}$ is just the operator (5.9a), and the source $J_{\mu_1\ldots\mu_N}$ is conveniently chosen to be [2]:

$$J_{\mu_1\ldots\mu_N} = \Delta_{\mu_1}\Delta_{\mu_2}\ldots\Delta_{\mu_N}, \tag{5.27}$$

with $\Delta^2 = 0$ so that the indices are automatically symmetrized over and the traces are taken out. The Feynman rules for the various vertices in eq. (5.26) are worked out, the two lowest order ones in $g$ are shown in fig. 18, from which the renormalization constants $Z_{O^N}$ and anomalous dimensions $\gamma_{O^N}$ are worked out in the usual way. One finds that in the Minimal

$$\not\Delta(\Delta\cdot k)^{N-1} \qquad -gT_a\Delta_\mu\not\Delta\sum_{j=1}^{N-1}(\Delta\cdot k_1)^{j-1}(\Delta\cdot k_2)^{N-1-j}$$

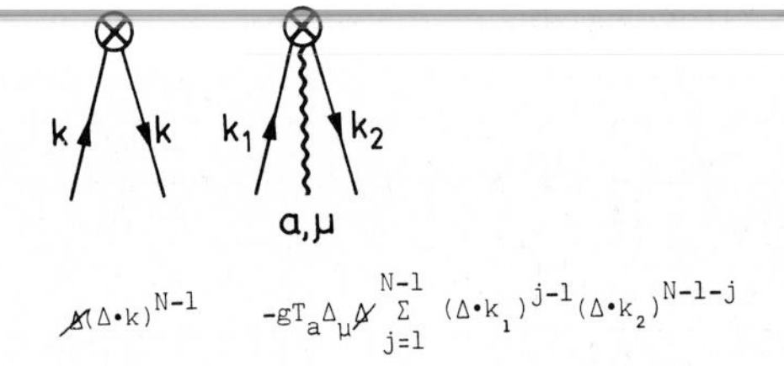

Fig. 18. Feynman rules for some of the twist-two operators which appear in the study of the deep inelastic structure functions.

Subtraction scheme (analogously to the calculation of the $\beta$ function):

$$\gamma_{O^N} = -g^2\frac{\partial Z_1}{\partial g^2}, \tag{5.28}$$

where $Z_1$ is the coefficient of $1/\varepsilon$ in $Z_{O^N}$. The diagrams which have to be calculated in one loop order are drawn in fig. 19a, and their evaluation leads to the result [2]:

$$\gamma^0_{O^N} = C_F\left[2-\frac{4}{N(N+1)}+\sum_{j=0}^{N-2}\frac{8}{j+2}\right]. \tag{5.29}$$

Similarly for the singlet case the diagrams which have to be calculated

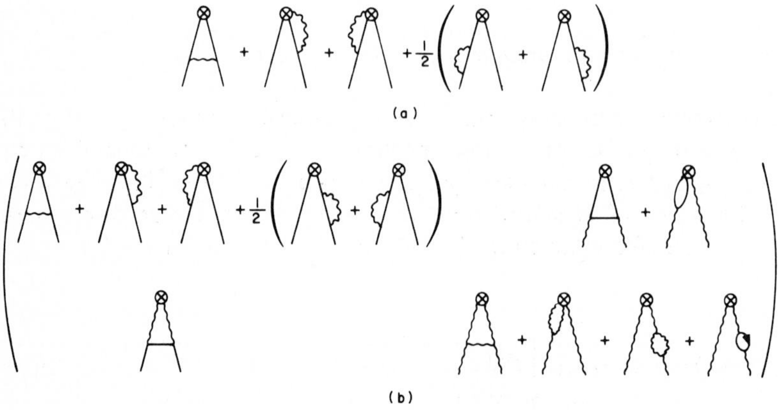

Fig. 19. One-loop diagram contributing to the anomalous dimension of the twist-two (a) non-singlet and (b) singlet operators.

are shown in fig. 19b and the corresponding result is:

$$\begin{pmatrix} \frac{8}{3}\left(1-\dfrac{2}{N(N+1)}+4\displaystyle\sum_{j=2}^{N}\frac{1}{j}\right) & -4f\dfrac{N^2+N+2}{N(N+1)(N+2)} \\ -\dfrac{16}{3}\dfrac{N^2+N+2}{N(N^2-1)} & \frac{4}{3}f+6\left(\frac{1}{3}-\dfrac{4}{(N+1)(N+2)}\right. \\ & \left.-\dfrac{4}{N(N-1)}+4\displaystyle\sum_{j=2}^{N}\frac{1}{j}\right) \end{pmatrix}, \tag{5.30}$$

where $f$ is the number of flavours.

(ii) *The Wilson coefficient function* $C^N(q^2=\mu^2, g(q^2))$. To evaluate the term $C^N(q^2=\mu^2, g(q^2))$ on the right hand side of eq. (5.21), we make use of the important fact that these Wilson coefficient functions are independent of the states between which the operator product expansion (5.6) is sandwiched. We thus evaluate deep inelastic structure functions of a quark in perturbation theory (i.e. expansion in $g(\mu^2)$) directly and also by expanding the right hand side of eq. (5.21) as a power series in $g(\mu^2)$. The only unknown is $C^N(q^2=\mu^2, g(q^2))$ so that by equating the two calculations we can determine it*. To lowest order in $g(q^2)$, $C^N$ is just a constant which can be taken to be one. Similarly one can calculate the coefficient functions for a singlet combination of structure function.

### 5.1. *Probabilistic interpretation of the above results*

An extremely interesting and physically appealing interpretation of the above leading order predictions for deep inelastic structure functions was given by Altarelli and Parisi [62]. We start by considering non-singlet combinations of structure functions, and we will return later to the singlet case. We write eq. (5.23) in terms of the variable $t \equiv \ln q^2/q_0^2$ as

$$M^N(t) = M^N(0)\left[\frac{\alpha(t)}{\alpha(0)}\right]^{\gamma^0_{ON}/2\beta_0}, \tag{5.31}$$

*Note that in lowest order we only have to calculate deep inelastic scattering on a quark at the tree level and therefore to this order the Callan–Gross relation $F_L \approx 0$ is still satisfied.

which satisfies the following equation:

$$\frac{\mathrm{d}}{\mathrm{d}t} M^N(t) = -\frac{\alpha(t)}{8\pi} \gamma^0_{O^N} M^N(t). \tag{5.32}$$

Since the moment of a convolution of two functions is the product of the moments of the functions, we can invert eq. (5.32) if we can find a function whose moments are $\gamma^0_{O^N}$. Thus we define a function $P_{\mathrm{q}\to\mathrm{q}}(z)$, such that

$$\int_0^1 \mathrm{d}z\, z^{N-1} P_{\mathrm{q}\to\mathrm{q}}(z) = -\tfrac{1}{4}\gamma^0_{O^N}. \tag{5.33}$$

Then, inverting eq. (5.32) we get:

$$\frac{\mathrm{d}\tilde{q}}{\mathrm{d}t}(x,t) = \frac{\alpha(t)}{2\pi} \int_x^1 \frac{\mathrm{d}y}{y} \tilde{q}(y,t) P_{\mathrm{q}\to\mathrm{q}}(x/y), \tag{5.34}$$

where $\tilde{q} \equiv q - \bar{q}$, which is a non-singlet combination of structure functions. Equation (5.34) is the "evolution equation" for $\tilde{q}$. $P_{\mathrm{q}\to\mathrm{q}}(x/y)$ is the variation per unit $t$ of the probability density of finding a quark in a quark, with fraction $x/y$ of its momentum. In its infinitesimal form eq. (5.34) is

$$\tilde{q}(x,t) + \mathrm{d}\tilde{q}(x,t) = \int_0^1 \mathrm{d}y \int_0^1 \mathrm{d}z\, \delta(zy - x) \tilde{q}(y,t) \times \left[\delta(z-1) + \frac{\alpha(t)}{2\pi} P_{\mathrm{q}\to\mathrm{q}}(z)\,\mathrm{d}t\right], \tag{5.35}$$

so that to calculate $P_{\mathrm{q}\to\mathrm{q}}(x)$ directly we have to calculate the coefficient of $\log q^2$ in the one-loop diagrams for deep inelastic scattering on a quark. This function must now satisfy eq. (5.33), where the $\gamma^0_{O^N}$ are given in eq. (5.29). The second term in the square brackets is there because, in QCD, quarks can radiate gluons with non-zero energy, see fig. 20a.

We now study the singlet combinations of structure functions. We define $Q^{\mathrm{s}}$ to be the flavour singlet quark distribution:

$$Q^{\mathrm{s}}(x,q^2) = \sum_{i=1}^{f} \left[q^i(x,q^2) + \bar{q}^i(x,q^2)\right] \tag{5.36}$$

and the gluon distribution is denoted by $G(x,q^2)$. We now get a coupled

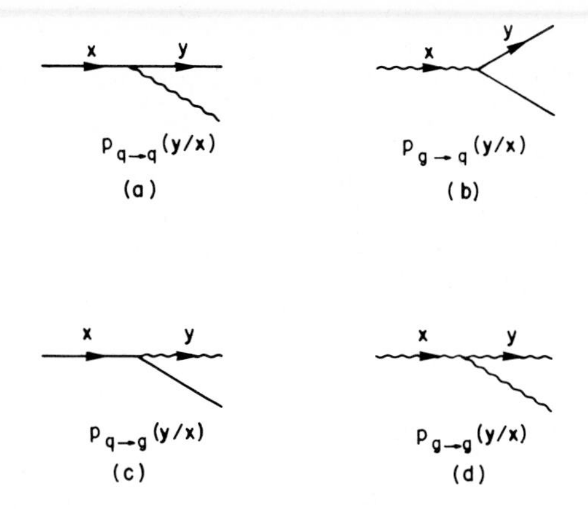


Fig. 20. Vertices which define the Altarelli–Parisi probability functions.

set of evolution equations:

$$\frac{dQ^s}{dt}(x,t) = \frac{\alpha_s(t)}{2\pi}\int_x^1 \frac{dy}{y}\left[Q^s(y,t)P_{q\to q}(x/y) + G(y,t)P_{g\to q}(x/y)\right], \tag{5.37}$$

$$\frac{dG}{dt}(x,t) = \frac{\alpha_s(t)}{2\pi}\int_x^1 \frac{dy}{y}\left[Q^s(y,t)P_{q\to g}(x/y) + G(y,t)P_{g\to g}(x/y)\right]. \tag{5.38}$$

The $Q^s$ distribution varies with $t$ now, not only because of gluon bremsstrahlung as in the non-singlet case, but also because gluons can convert into $q\bar{q}$ pairs (fig. 20b) in a flavour-independent way, so that this variation of $Q^s$ now depends also on the gluon distribution $G$. Similarly the variation of $G$ with $t$ depends on $Q^s$ (because of bremsstrahlung, fig. 20c) and $G$ (because gluons can also radiate gluons, fig. 20d). The moments of $P_{q\to q}$, $P_{g\to q}$, $P_{q\to g}$ and $P_{g\to g}$ are, up to a factor of $-4$ as in eq. (5.33), equal to the entries in the anomalous dimension matrix (5.30).

Thus we have a nice probabilistic interpretation of the violations of Bjorken scaling.

Although the virtual photon or weak bosons do not couple directly to gluons, we can deduce the gluon distribution indirectly from deep inelastic scattering data by using eqs. (5.37) and (5.38), since the gluon distribution does affect the violations of Bjorken scaling. Quantitative results for the gluon distribution will be presented by Para [60].

*5.2. Corrections to the leading order results*

We have calculated the QCD predictions for the asymptotic behaviour of the deep inelastic structure functions. There are numerous sub-asymptotic corrections.

(*a*) *Logarithmic corrections.* These are calculable and will be discussed briefly later. Their calculation involves the evaluation of higher order terms in the anomalous dimensions, the $\beta$ function and the Wilson coefficient functions.

(*b*) *Contributions from higher twist operators.* Although these are suppressed by powers of $q^2$, they may be important, particularly at lower values of $q^2$, and so complicate the comparison of the QCD predictions for scaling violations to the data. It is possible to calculate their anomalous dimensions and coefficient functions [63,64], but the problem is to determine their matrix element between hadronic states. This clearly cannot be done in perturbation theory and is very difficult to do from the data.

(*c*) *Target mass corrections.* Again these are suppressed by powers of $q^2$, they come from taking operators of fixed spin (rather than fixed twist) as a basis for the operator product expansion. These do not get mixed under renormalization. One can take these into account by replacing all the moments in the above discussion by "Nachtmann Moments" [65]; e.g. for $F_3$ the moment is defined as [66]:

$$M_N^{(3)}(q^2) = \int_0^1 \frac{\mathrm{d}x}{x^2} \xi^{N+1} \nu W_3(x, q^2) \times \left[ \frac{1+(N+1)\left(1+4m_{\mathrm{H}}^2 x^2/Q^2\right)^{1/2}}{N+2} \right], \tag{5.39}$$

where

$$\xi \equiv \frac{2x}{1+\left(1+4m_{\mathrm{H}}^2 x^2/Q^2\right)^{1/2}}. \tag{5.40}$$

(*d*) *Quark mass corrections.* Throughout the discussion so far we have neglected the masses of the quarks. Their inclusion will lead to $m_q^2/q^2$ corrections, for attempts to include these effects see refs. [67,68].

## 6. The application of QCD to other hard scattering processes (The factorization of mass singularities)

In this section we will review why we believe that we can make predictions about the asymptotic behaviour of hard scattering processes. The operator product expansion and renormalization group techniques discussed in the preceding lectures are not in general directly applicable, so we have to search for a new approach. The classes of processes which we will consider in this section are those in which there are three mass scales:

(a) a large mass scale $Q^2$,
(b) a small mass scale (of the order of hadronic masses) $p^2$, and
(c) the renormalization scale $\mu^2$.

We will be interested in the limit $Q^2 \to \infty$, $p^2$ fixed; physical cross-sections are, of course, independent of $\mu$. In the relevant process there may be more than one variable which is $O(Q^2)$ (e.g. in deep inelastic scattering there is $q^2$ and $p \cdot q$), but then the ratio of these variables should stay fixed as we take the limit $Q^2 \to \infty$. Similarly, we assume that all hadronic masses are of the same order $p^2$.

We start by reproducing the results of the previous section for the deep inelastic structure functions by using diagrammatic techniques. Later, we will see that the same techniques can be generalized to other hard scattering processes*.

### *6.1. Equivalence of the light-cone result and the summation of leading logarithms in perturbation theory*

In section 5 we have seen how the use of the operator product expansion and the renormalization group enables us to make predictions for the violation of Bjorken scaling in deep inelastic lepton–hadron scattering. In this subsection we will study the same process with a different approach, one which is less rigorous but which, on the other hand, is also applicable to processes which are not light-cone dominated. The basic assumption is that asymptotically hard scattering cross-sections can be written as a convolution of soft hadronic wave functions (which are process-independent) with the cross-section for the "hard subprocess" (which involves only quarks and gluons and can be studied perturbatively). Thus, for example, in the case of deep inelastic scattering we

*See also the "cut-vertex" approach of Mueller [96].

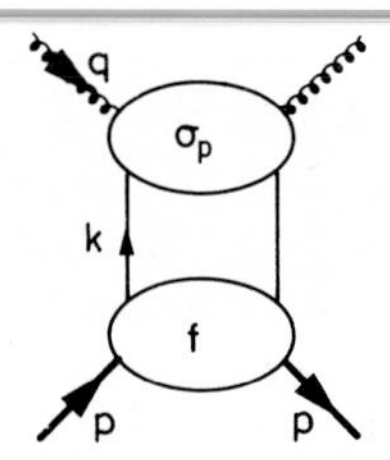


Fig. 21. Schematic representation of the diagrammatic approach to deep inelastic lepton–hadron scattering.

assume that we can write (see fig. 21):

$$F(x,q^2) = \sum_{\text{partons}} \int_x^1 \mathrm{d}\bar{x} \int \mathrm{d}^2\boldsymbol{k}_\perp \, \mathrm{d}k^2 f(\bar{x}, k_\perp, k^2) \times \sigma_{\text{parton}}(x/\bar{x}, k^2, q^2), \tag{6.1}$$

where $f$ is related to the square of the soft wave function and $\sigma_{\text{parton}}$ is the cross-section for deep inelastic scattering with the parton as the target. Within perturbation theory this assumption can be justified (see below). We will now study $\sigma_{\text{parton}}$ in perturbation theory.

When computing $\sigma_{\text{parton}}$ in perturbation theory, we will find terms of the type $g^2 \log Q^2/\mu^2$ and $g^2 \log Q^2/p^2$ ($Q^2$, $p^2$ and $\mu^2$ are defined above). We will see below that with each power of $g^2$ there is at most one logarithm. We would like to demonstrate that the results of the previous section can be obtained by summing the leading logarithms of perturbation theory. For simplicity, we will consider a non-singlet combination of structure functions; the extension to general combinations of structure functions complicates the technical details, but does not alter the essential features.

We take the expression for $q^2$ behaviour of the moment $M^N$ of a structure function:

$$M^N(q^2) = C^N(q^2 = \mu^2, g(q^2)) \exp \int_{g(\mu^2)}^{g(q^2)} \frac{\gamma_{O^N}(g')}{\beta(g')} \mathrm{d}g' \langle p|O^N|p\rangle, \tag{6.2}$$

where e.g. $|p\rangle$ is a quark state of momentum $p$. We now expand the right-hand side of eq. (6.2) in terms of a power series in $\alpha(\mu^2)$, which we can do because we know how to relate $\alpha(Q^2)$ to $\alpha(\mu^2)$ [see eq. (2.13)]. In each order of perturbation theory we keep only the terms which are of the type $g^{2n}(\log X)^n$, where $X$ is a ratio of two of the three mass scales $Q^2$, $p^2$ and $\mu^2$, i.e. we work in the leading logarithm approximation. In this approximation the solution to eq. (2.13) is given by [see eq. (2.21)]:

$$\alpha(q^2) = \alpha(\mu^2)\left(1+\frac{\beta_0}{4\pi}\alpha(\mu^2)\ln q^2/\mu^2\right)^{-1}. \tag{6.3}$$

We now look at the three factors on the right hand side of eq. (6.2) in turn.

1) $C^N$: $C^N$ is a power series in $\alpha(q^2)$ with no logarithms. Therefore in the leading logarithm approximation we need to keep only the first term which is a constant*.

2) The exponential: Here we expand $\gamma_{O^N}$ and $\beta$ and use eq. (6.3) to obtain

$$\exp\left(\int_{g(\mu^2)}^{g(q^2)}\frac{\gamma_{O^N}(g')}{\beta(g')}\mathrm{d}g'\right) \underset{\substack{\text{leading}\\ \text{logarithm}\\ \text{approximation}}}{\approx} \left[1+\frac{\beta_0}{4\pi}\alpha(\mu^2)\ln q^2/\mu^2\right]^{-\gamma^0_{O^N}/2\beta_0}. \tag{6.4}$$

3) $\langle p|O^N|p\rangle$: This can either be calculated directly, or we can use the fact that $M^N$ is $\mu$-independent so that the $\mu$ dependence of eq. (6.4) must be cancelled by that of $\langle p|O^N|p\rangle$. Since $\langle p|O^N|p\rangle$ is independent of $q$, we find

$$\langle p|O^N|p\rangle = \left[1+\frac{\beta_0\alpha(\mu^2)}{4\pi}\ln p^2/\mu^2\right]^{\gamma^0_{O^N}/2\beta_0}. \tag{6.5}$$

Thus in the leading logarithm approximation we find that**

$$M^N(q^2) \sim \left[\frac{1+\frac{\beta_0\alpha(\mu^2)}{4\pi}\ln p^2/\mu^2}{1+\frac{\beta_0\alpha(\mu^2)}{4\pi}\ln q^2/\mu^2}\right]^{d_N}, \tag{6.6}$$

where $d_N = \gamma^0_{O^N}/2\beta_0$. Hence summing leading logarithms and interpret-

*For the moments of $F_L$, the first term in $C^N$ is equal to $g^2$ times a constant.

**If we differentiate the right hand side of eq. (6.6) with respect to $\ln\mu$, we find we get 0 in the leading logarithm approximation, i.e., all terms of the form $g^{2n}(\ln\mu^2)^{n-1}$ cancel.

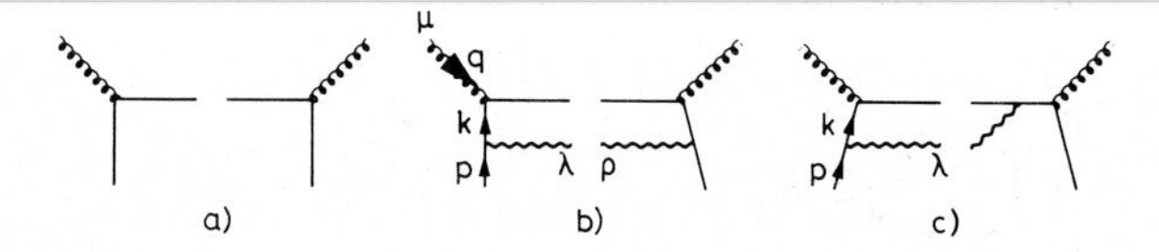


Fig. 22. Some low order diagrams for deep inelastic scattering.

ing eq. (6.6) via eq. (6.3) one is led to the result (5.23), obtained using light cone techniques.

## 6.2. *Reproducing the light-cone results by summing Feynman diagrams*

We will now calculate $\sigma_{\text{parton}}$ (for the non-singlet case) by perturbation theory. Although of course we can calculate it in any gauge, the results look particularly attractive in axial and planar gauges (after some technical problems have been ironed out). Here the presentation will be in the "planar" gauge and follows Dokshitzer et al. [69]*; we start by defining this gauge. Let

$$q' \equiv q + xp, \tag{6.7}$$

so that $q'^2 = 0$. The gluon propagator in this gauge is

$$G_{\mu\nu}(k) \equiv \frac{d_{\mu\nu}(k)}{k^2 + i\varepsilon}, \tag{6.8}$$

where

$$d_{\mu\nu}(k) = g_{\mu\nu} - \frac{k_\mu c_\nu + k_\nu c_\mu}{(k \cdot c)}, \tag{6.9}$$

with

$$c_\mu = Aq'_\mu + Bp_\mu \quad (A \sim B). \tag{6.10}$$

This is not the axial gauge $c \cdot A = 0$ which would have an extra term in $d_{\mu\nu}$ i.e. $k_\mu k_\nu c^2/(c \cdot k)^2$. It can be proved (see the appendices of ref. [69]) that the planar gauge is a ghostless gauge.

The lowest order graph for deep inelastic scattering on a quark is shown in fig. 22a and leads to a contribution to the quark distribution of

*For a pedagogical presentation of the calculation in the Feynman gauge see Llewellyn Smith [70]. For an introduction to some of the calculational techniques see Gribov and Lipatov [70].

$\delta(1-x)$. Consider now the one loop graph of fig. 22b. Its contribution to $W_{\mu\nu}$ is:

$$\tfrac{1}{2}e^2g^2\cdot\tfrac{4}{3}\int\frac{\mathrm{d}^4k}{(2\pi)^4}2\pi\delta^{(+)}\big((k+q)^2\big)2\pi\delta^{(+)}\big((p-k)^2\big)$$

$$\times\mathrm{Tr}\big[\not{p}\gamma_\rho\not{k}\gamma_\nu(\not{k}+\not{q})\gamma_\mu\not{k}\gamma_\lambda\big]d^{\lambda\rho}(p-k)/(k^2+\mathrm{i}\varepsilon)^2, \tag{6.11}$$

where we have neglected all masses, except that in order to regulate the mass singularities we shall keep $p^2$ and take it to be space like. We shall use Sudakov variables $(\alpha,\beta,\boldsymbol{k}_\perp)$ defined by

$$k = \alpha q'_\mu + \beta p'_\mu + \boldsymbol{k}_\perp\,, \tag{6.12}$$

where $q'$ is defined in eq. (6.7),

$$p' = p - p^2q'/s \quad (s\equiv 2p\cdot q), \tag{6.13}$$

and

$$k_\perp\cdot p = k_\perp\cdot q = 0. \tag{6.14}$$

The Jacobian is given by

$$\mathrm{d}^4k = \frac{s}{2}\,\mathrm{d}\alpha\,\mathrm{d}\beta\,\mathrm{d}^2k. \tag{6.15}$$

The two delta functions an now be written as:

$$\begin{aligned}\delta\big((p-k)^2\big) &= \delta\big(-(1-\beta)(\alpha-p^2/s)s-k_\perp^2\big)\\ &= \frac{1}{|1-\beta|s}\delta\big(\alpha-p^2/s+k_\perp^2/(1-\beta)s\big)\end{aligned} \tag{6.16}$$

and

$$\delta\big((k+q)^2\big) = \delta\big((\beta-x)(1+\alpha)s-k_\perp^2\big). \tag{6.17}$$

In the leading logarithm approximation we want to find a term with a factor of $\log q^2/p^2$, and this can only come from the small $k^2$ region (or possibly the small $c\cdot k$ region). In Sudakov variables:

$$k^2 = \alpha\beta s - k_\perp^2 = \beta p^2 - k_\perp^2/(1-\beta), \tag{6.18}$$

where we have used eq. (6.16) to eliminate $\alpha$. By inspection one can see that the dominant region is $\beta$ finite, $k_\perp^2 \ll q^2$ which implies that $\alpha$ is small and that $\beta = x$. Hence the contribution to $W_{\mu\nu}$ can be written as:

$$W_{\mu\nu} = \frac{e^2 g^2}{3(1-x)s} \int \frac{\mathrm{d}^2 k_\perp}{(2\pi)^2} \frac{\mathrm{Tr}\left[\not{p}\gamma_\rho\gamma_\nu(\not{k}+\not{q})\gamma_\mu\not{k}\gamma_\lambda\right] d^{\lambda\rho}(p-k)}{\left[xp^2 - k_\perp^2/(1-x)\right]^2}. \tag{6.19}$$

Evaluating the numerator and keeping only the leading terms in $p \cdot q$ and $q^2$, one readily finds that it is proportional to

$$k^2 \frac{1+x^2}{1-x}, \tag{6.20}$$

which is proportional to $P_{\mathrm{q}\to\mathrm{q}}(x)$ for $x \neq 1$. The $k_\perp$ integration is now

$$\int_0^{\varepsilon Q^2} \frac{\mathrm{d}k_\perp^2}{\left[x(1-x)p^2 - k_\perp^2\right]} = -\log \frac{\varepsilon Q^2}{x(1-x)p^2} \approx -\log Q^2/p^2. \tag{6.21}$$

We notice that $\log x$ and $\log(1-x)$ factors do appear, but since we are working in the leading logarithm approximation and at fixed $x$ we drop them. However we do have to be careful not to choose too large or too small a value of $x$. Combining all the above results one finds that the contribution to the quark distribution from this diagram is

$$e^2 \frac{\alpha_s}{2\pi} \ln \frac{q^2}{p^2} P_{\mathrm{q}\to\mathrm{q}}(x), \tag{6.22}$$

which is the expected result from all the one-loop diagrams (except for the $\delta(x-1)$ terms). This means that the contribution from the diagram of fig. 22c should be zero. To see that this is indeed so we note that in the leading logarithm approximation we are interested in the region where $k$ is parallel to $p$ (by the discussion of section 3) and in particular $k \simeq xp$. Part of the Dirac algebra involves

$$\not{k}\gamma^\lambda\not{p} = x\not{p}\gamma^\lambda\not{p} \simeq 2xp^\lambda\not{p} \simeq \frac{2x}{1-x}(p-k)^\lambda\not{p}. \tag{6.23}$$

But

$$(p-k)^\lambda d_{\lambda\rho}(p-k) = 0$$

and hence we get no logarithm*. Hence for $x \neq 1$ the only diagram which

*Note that in the box diagram fig. 22b we could not make the approximation $k = xp$ in the numerator because there were two singular propagators $1/k^2$. Here we only have one.

contributes to the leading logarithm approximation is the box diagram of fig. 22b.

We notice (from eq. 6.20) that as $x \to 1$ the coefficient of the $\log q^2/p^2$ term is singular. We know from section 4 that these divergences will be cancelled by diagrams with virtual gluons, and this is indeed the case.

Features similar to those in the one-loop diagrams also appear in higher-order graphs. The general feature is that the dominant diagrams in the planar and axial gauges are the generalized ladder diagrams (fig. 23), i.e. ladder diagrams with vertex and self-energy insertions (we denote all these insertions by a cross at the vertex). Gribov and Lipatov [70] have taught us how to evaluate such diagrams and that the dominant region of integration is where

$$p^2 \ll k_1^2 \ll k_2^2 \ll \cdots \ll k_n^2 \ll q^2. \tag{6.24}$$

One of the effects of the vertex and self-energy insertions is that at each vertex we should put $g(t_i)$, where $t_i = k_i^2$ is the largest four-momentum squared at the vertex. This may seem surprising in view of the fact that the momenta in the different legs are not equal. We can choose, however, to associate each self-energy insertion of the fermion lines with the vertex below the insertion, instead of the usual association of half the self-energy insertion with the vertex above and half with the vertex below. In this way we see that we should write $g(t_i)$ at each vertex.

We now outline the calculation of such a set of diagrams. We start by doing the transverse momentum integrals, which we can rewrite in terms

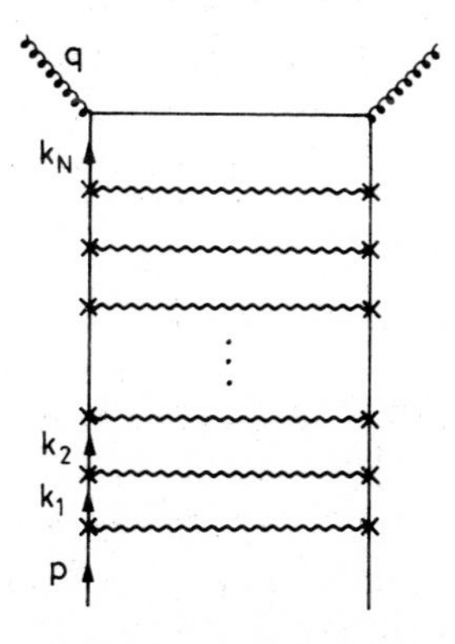


Fig. 23. The dominant structure (in the leading logarithm approximation) for deep inelastic scattering in the Bjorken limit. The ×'s signify that self-energy and vertex insertions must be added.

of the $t_i$. In view of the above discussion we find that for each $t_i$ we have a factor of $\alpha(t_i) \sim (1/2\beta_0)\log t_i/\Lambda^2$, so that

$$I_{\mathrm{T}} = \int_{p^2}^{q^2} \frac{\mathrm{d}t_n}{t_n}\alpha_{\mathrm{s}}(t_n)\ldots\int_{p^2}^{t_3}\frac{\mathrm{d}t_2}{t_2}\alpha_{\mathrm{s}}(t_2)\int_{p^2}^{t_2}\frac{\mathrm{d}t_1}{t_1}\alpha_{\mathrm{s}}(t_1). \tag{6.25}$$

Thus, from the transverse momentum integrals we find a term proportional to*:

$$I_{\mathrm{T}} \propto \frac{1}{n!}(\ln\ln q^2/\Lambda^2)^n\left(\frac{1}{2\beta_0}\right)^n. \tag{6.26}$$

Now we have to calculate the integral over the longitudinal component of momentum. This takes the form

$$\begin{aligned} I_{\mathrm{L}} = {} & \int_x^1 \mathrm{d}x_1 P_{\mathrm{q}\to\mathrm{q}}(x_1)\int_x^{x_1}\frac{\mathrm{d}x_2}{x_1}P_{\mathrm{q}\to\mathrm{q}}\left(\frac{x_2}{x_1}\right)\ldots \\ & \times\int_x^{x_{n-2}}\frac{\mathrm{d}x_{n-1}}{x_{n-2}}P_{\mathrm{q}\to\mathrm{q}}\left(\frac{x_{n-1}}{x_{n-2}}\right) \\ & \times\int\frac{\mathrm{d}x_n}{x_{n-1}}P_{\mathrm{q}\to\mathrm{q}}\left(\frac{x_n}{x_{n-1}}\right)\delta(x_n - x). \end{aligned} \tag{6.27}$$

The fraction of momentum $p$ carried by quark $i$ is $x_i$, and that by quark $i+1$ is $x_{i+1}$, so that the fraction of the momentum of quark $i$ which is carried by quark $i+1$ is $x_{i+1}/x_i$. This is the reason for the arguments in the $P_{\mathrm{q}\to\mathrm{q}}$ functions in eq. (6.27). Since eq. (6.27) is a multiple convolution we take moments.

$$\begin{aligned} \int_0^1 x^{N-1}I_{\mathrm{L}}(x)\,\mathrm{d}x &= \left(\int_0^1 \mathrm{d}x\, P_{\mathrm{q}\to\mathrm{q}}(x)x^{N-1}\right)^n \\ &\propto \left(-\gamma_{O^N}^0\right)^n. \end{aligned} \tag{6.28}$$

Thus we find that for the $N$th moment of the diagram of fig. 23 the asymptotic behaviour is proportional to:

$$\frac{1}{n!}(\ln\ln q^2/\Lambda^2)^n\left(\frac{-\gamma_{O^N}^0}{2\beta_0}\right)^n, \tag{6.29}$$

whence on summing over $n$ we reproduce the standard deep inelastic

*This result is obtained by performing the integrals in eq. (6.25) and keeping the top limit of the integration range. Below we will come back to the terms obtained by keeping one or more of the bottom limits.

result

$$M^N(q^2) \propto (\log q^2/\Lambda^2)^{-\gamma_{ON}^0/2\beta_0}. \tag{6.30}$$

These arguments can also be generalized to the singlet case.

We would like to make a few comments concerning this calculation. First of all, we notice that the corrections to eq. (6.26), obtained by keeping a term which involves at least one of the lower limits, are of the form

$$\frac{1}{(n-1)!}(\ln\ln q^2/\Lambda^2)^{n-1} + \cdots \tag{6.31}$$

and it may therefore seem that the corrections to eq. (6.30) should be suppressed by one or more powers of $\log\log q^2/\Lambda^2$ rather than powers of $\log q^2/\Lambda^2$ which we know to be correct. But this argument is not valid, since if we keep the lower limit in one of the integrals of eq. (6.25), say the $t_i$ integral, then we keep a term from the first $j$ integrals which is $q^2$ independent. If for $j \leqslant i \leqslant n$, we now keep the upper limit in the $t_i$ integrations, we generate a term proportional to

$$\frac{1}{(n-j)!}(\ln\ln q^2/\Lambda^2)^{n-j},$$

whence, summing over $n$ from $j$ to infinity after having performed the integrals over the longitudinal components of momenta, we obtain a new contribution of the form (6.30). This just means that we are in effect including the part of the diagram in which the transverse momenta are finite in the soft wave function $f$ and not in $\sigma_{\text{parton}}$. Of course we are not able to calculate the constant of proportionality in (6.30); this is analogous to not being able to calculate the operator matrix elements.

Another relevant question is: Why are we entitled to substitute the asymptotic form for $\alpha(t_i)$ in eq. (6.25), when we perform the integral from a finite mass $p^2$? This is so because in the leading logarithm approximation, we could equally well have taken the integrals from $Xp^2$, where $X$ is a large fixed number, such that $\alpha(Xp^2)$ is well approximated by its asymptotic form.

Finally, we would like to note that, although the dominant region of phase space is where the transverse momenta $k_i^2 \ll q^2$, nevertheless, the $k_i$ *do* grow with $q^2$. For example, the average transverse momentum of a quark jet is predicted to behave like

$$\langle \boldsymbol{k}_T^2 \rangle \sim \alpha_s(q^2) q^2. \tag{6.32}$$

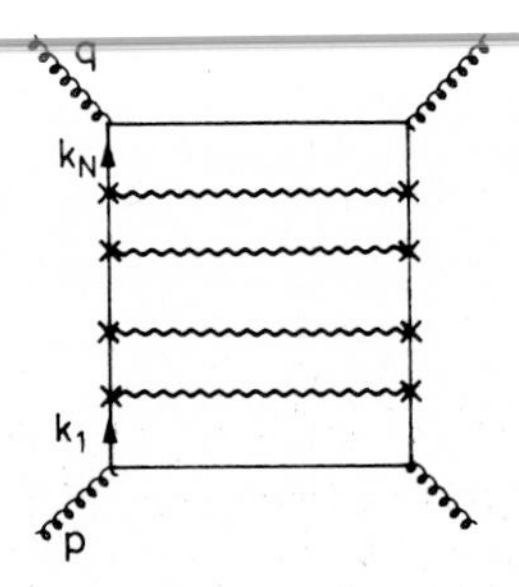


Fig. 24. The dominant structure for deep inelastic scattering on a photon target in the Bjorken limit.

We are now in a position to understand the interesting result of Witten [71] about the structure function of a photon. In the transverse gauges the dominant diagrams are again the generalized ladder diagrams [72, 73] (fig. 24). The only difference is at the bottom of the diagram, where we now have two electromagnetic vertices. These vertices are hard vertices, so that we can have large values of $t_1$, but we have no factor of $\alpha_s(t_1)$. Thus, the transverse momentum integral analogous to eq. (6.25) is now

$$I_T^\gamma = \int_{p^2}^{q^2} \frac{dt_n}{t_n} \alpha_s(t_n) \ldots \int_{p^2}^{t_3} \frac{dt_2}{t_2} \alpha_s(t_2) \int_{p^2}^{t_2} \frac{dt_1}{t_1}, \tag{6.33}$$

which can be evaluated to give

$$I_T^\gamma \propto (\ln Q^2)/(2\beta_0)^n. \tag{6.34}$$

We obtain this result when we keep the upper limit in all the integrals of eq. (6.33). If somewhere we keep a lower limit, we get terms of the form which we encountered in the hadronic case, i.e. eq. (6.26). The $N$th moment of the integral over longitudinal momenta is now equal to (compare with eq. (6.28)):

$$\left(\gamma_{O^N}^0\right)^{n-1} f^N, \tag{6.35}$$

where $f^N$ is the $N$th moment of $P_{\gamma \to q}(x)$, the probability density of finding a quark in a photon, with a fraction $x$ of its momentum. We now sum over $n$, as in the hadronic case, but now we have a geometric series which we sum to find that the $N$th moment of the structure function of a photon is given by:

$$\ln q^2 \frac{f^N}{1+\gamma_0^N/2\beta_0}. \tag{6.36}$$

Witten [71] first obtained this result by using the operator product expansion. In addition to the usual twist-two operators, we now have operators based on $F^{\mu\nu}$, the electromagnetic field-strength tensor. It is these operators which lead to the behaviour (6.36). These predictions will be tested in future $e^+e^-$ experiments at PEP and PETRA.

In a fixed point theory with a small ultraviolet stable fixed point, the structure functions of hadrons are still dominated by the diagrams of fig. 23, but now in the analogous integral to eq. (6.25) we have no $\alpha_s(t_i)$ factors. This means that we can immediately deduce the result; we just make the substitution $\log\log q^2/\Lambda^2 \to \log q^2/\Lambda^2$ and hence the $N$th moment of a structure function in such a theory has a $q^2$ dependence given by

$$(q^2)^{-\gamma^0_{0N}/2\beta_0}. \tag{6.37}$$

Thus, in such theories the violations of Bjorken scaling are given by powers of $q^2$.

### 6.3. Application of QCD to other hard scattering processes

We will now consider the application of QCD to hard scattering processes in which there are two incoming hadrons, for the purposes of illustration we will consider the Drell–Yan process and the inclusive production of particles at large transverse momentum in hadronic collisions.

(*a*) *The Drell–Yan process.* The Drell–Yan process is $h_1h_2 \to \ell^+\ell^- X$ where $h_1$ and $h_2$ are the initial state hadrons, and the lepton pair has an invariant mass squared ($Q^2$) of order $s$. We define the quantity $\tau$ by

$$\tau = Q^2/s \tag{6.38}$$

and the kinematic region we are interested in is $Q^2$, $s \to \infty$, $\tau$ fixed. The explanation of this process proposed by Drell and Yan [74] is that a quark from one of the initial hadrons annihilates an anti-quark from the other, and the resulting massive photon then decays into the observed

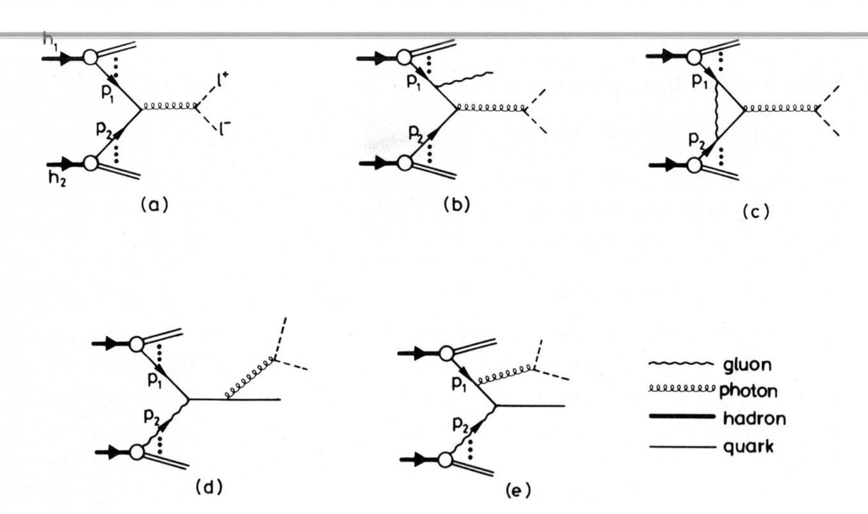


Fig. 25. Sample diagrams which contribute to massive lepton-pair production in hadronic collision.

lepton pair (fig. 25a), so that we can write (in the parton model):

$$\frac{\mathrm{d}\sigma}{\mathrm{d}Q^2}=\frac{4\pi\alpha^2}{3nQ^4}\sum_a e_a^2\int_0^1 \mathrm{d}x_1\,\mathrm{d}x_2\,\delta(x_1x_2-\tau)x_1x_2$$

$$\times\left[q_{1,a}(x_1)\bar{q}_{2,a}(x_2)+q_{2,a}(x_2)\bar{q}_{1,a}(x_1)\right], \qquad (6.39)$$

where $n$ is the number of colours and $q_{i,a}(x)$ $[\bar{q}_{i,a}(x)]$ is the probability density of finding a quark [antiquark] of flavour $a$ in hadron $i$, with fraction $x$ of its longitudinal momentum. We notice that since (i) $q$ and $\bar{q}$ are in principle measurable in deep inelastic lepton–hadron scattering and (ii) the subprocess $q\bar{q}\to\ell^+\ell^-$ is calculable since it only involves QED, eq. (6.39) leads to a prediction for both the $Q^2$ behaviour and normalization of the differential cross-section.

In QCD, however, in addition to the simple diagram of fig. 25a, there are many more contributions; a sample set of low-order graphs is shown in fig. 25.

We now consider the subprocess $q\bar{q}\to\ell^+\ell^-X$, and denote the lowest order contributions to its cross-section (that corresponding to fig. 25a) by $\sigma_0$. When all the radiative corrections of order $g^2$ are calculated (including those of figs. 25b, c) in the leading logarithm approximation, one finds [75] that these corrections are equal to

$$a(\tau)g^2(\ln Q^2/p_1^2 + \ln Q^2/p_2^2). \tag{6.40}$$

It turns out that $a(\tau)$ is a very significant function; it is exactly the same function which appears in the $O(g^2)$ corrections to deep inelastic scattering on a quark. The structure function of a quark is proportional to

$$\delta(x-1) + a(x)g^2 \ln q^2/p^2 + O(g^4 \ln^2 q^2/p^2). \tag{6.41}$$

Thus, to this order, at least, it seems that the *deviations from the naive Drell–Yan picture are intimately related to the violations of Bjorken scaling in deep inelastic scattering*. In other words, the diagrams which spoil the simple Drell–Yan picture (such as those of figs. 25b, c) are related to those which are responsible for the violation of Bjorken scaling (such as those of figs. 26a, b). In particular, to this order we can write a modified Drell–Yan formula:

$$\frac{d\sigma}{dQ^2} = \frac{4\pi\alpha^2}{3nQ^4} \sum_a e_a^2 \int_0^1 dx_1 dx_2 \delta(x_1 x_2 - \tau) x_1 x_2 \times \left[ q_{1,a}(x_1, Q^2) \bar{q}_{2,a}(x_2, Q^2) + q_{2,a}(x_2, Q^2) \bar{q}_{1,a}(x_1, Q^2) \right], \tag{6.42}$$

where the q and $\bar{q}$ distribution functions are the appropriate linear combinations of experimentally determined deep inelastic structure functions.

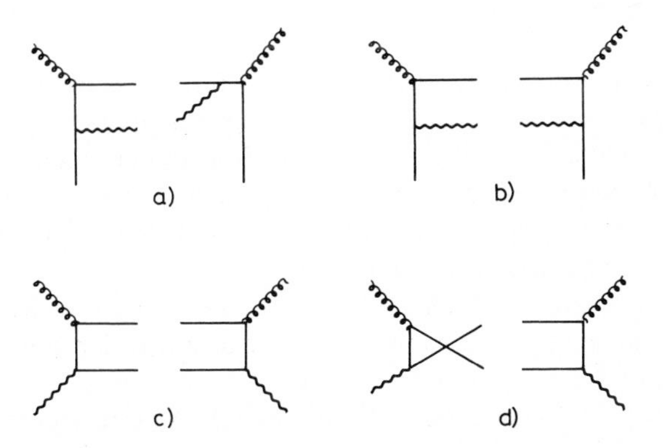


Fig. 26. Sample diagrams contributing non-scaling terms to deep inelastic structure functions.

This result is easy to understand in transverse gauges, because in these gauges it is again the generalized ladder diagrams which dominate. In particular, in the planar gauge $(p_1+\alpha p_2)\cdot A=0$, and in the leading logarithm approximation, the top of the diagram decouples from the bottom, and so we get eq. (6.42) naturally. In other axial gauges we also have to consider the mass singularities in the vertex corrections at the electromagnetic vertex. Equation (6.42) is true to all orders of perturbation theory in the leading logarithm approximation [69,76,77]. The proof is analogous to the derivation of eq. (6.30). Thus the large logarithms which appear in the Drell–Yan process turn out to be the same as those in deep inelastic scattering. So once we have "measured" the sum of these large logarithms in deep inelastic scattering we can use these measurements to make predictions for lepton-pair production.

We now turn our attention to other subprocesses which contribute to massive lepton pair production. As an example, let us consider qg scattering for which some lowest order diagrams are shown in figs. 25d, e. When one calculates the contribution to the cross-section for this subprocess from the diagrams of figs. 25d, e one finds [78] that the result is proportional to

$$g^2[1-2\tau(1-\tau)](\ln Q^2/p_2^2)\sigma_0. \tag{6.43}$$

Again the coefficient of $g^2\sigma_0$ is highly significant. It is the probability of finding an antiquark in a gluon with a fraction $\tau$ of its longitudinal momentum*. When the distribution functions are measured in deep inelastic scattering, they include the effects of gluons through diagrams such as those of figs. 26c, d. Equation (6.43) implies that the contribution to the distribution functions from these diagrams is related to the contribution to the cross-section for lepton pair production from diagrams such as those of figs. 25d, e. Thus, even these diagrams are included in the right-hand side of eq. (6.42). In other words, the dominant contribution from the subprocess $\mathrm{qg}\to\ell^+\ell^-X$ can be expressed as the probability of finding an antiquark in the gluon times the cross-section for this antiquark to annihilate the incident quark.

A similar feature has been found in all other subprocesses which have been studied (e.g. $\mathrm{qq}\to\ell^+\ell^-X$, $\mathrm{gg}\to\ell^+\ell^-X$ [78]). In each case the leading logarithmic terms can be absorbed into the distribution functions

*The moments of $1-2\tau(1-\tau)$ with respect to $\tau$ are thus proportional to one of the off-diagonal elements in the anomalous dimension matrix of the lowest-twist quark and gluon operators.

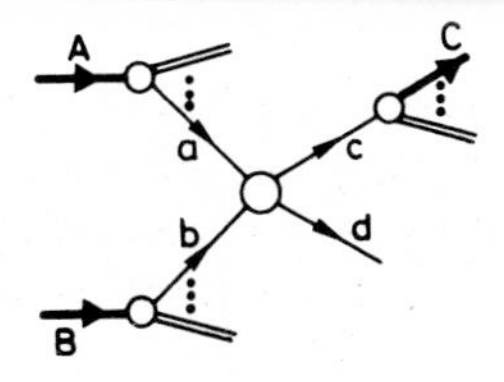


Fig. 27. Schematic representation of the parton model description of the inclusive production of hadrons.

of the incident hadrons, so that the asymptotic behaviour of the cross-section is given by eq. (6.42), where the quark and gluon distribution functions are extracted from linear combination of experimentally measured deep inelastic structure functions.

(*b*) *Inclusive production of particles and jets with large transverse momenta.* Deep inelastic scattering and massive lepton pair production both involve an off-shell photon. We now study a purely strong interaction process, the production of particles and jets with large transverse momenta in hadronic collisions. The mechanism usually believed to be responsible for this process can be summarized by fig. 27, which can be expressed mathematically as

$$E_c \frac{\mathrm{d}\sigma}{\mathrm{d}^3 p_c}(A + B \to C + X)$$

$$= \sum_{a,b,c,d} \int_0^1 \mathrm{d}x_a \int_0^1 \mathrm{d}x_b \frac{\mathrm{d}x_c}{x_c^2} G_{a/A}(x_a) G_{b/B}(x_b) \tilde{G}_{C/c}(x_c)$$

$$\times \delta(s' + t' + u') \frac{s'}{\pi} \frac{\mathrm{d}\sigma^{(\mathrm{B})}}{\mathrm{d}t'}(a + b \to c + d)\Bigg|_{\substack{s' = x_a x_b s \\ t' = (x_a/x_c)t \\ u' = (x_b/x_c)u}}, \tag{6.44}$$

where $G_{a/A}(x_a)$ ($\tilde{G}_{C/c}(x_c)$) is the probability of particle $A$ (constituent $c$) to fragment into constituent $a$ (particle $C$), $a$ ($C$) carrying a fraction $x_a$ ($x_c$) of the longitudinal momentum of $A$ ($c$). To see what, if anything, QCD can teach us about this process, let us study a concrete example: quark ($A$)+quark ($B$) → quark ($C$)+anything, up to order $g^6$ in the cross-section and, as before, in the leading logarithm approximation [79]. For simplicity we start with the case $x < 1$, where $x = (2E_c/\sqrt{s})$, in which case only the inelastic diagrams of fig. 28 contribute. We notice that even in this order there are diagrams with a three-gluon vertex.

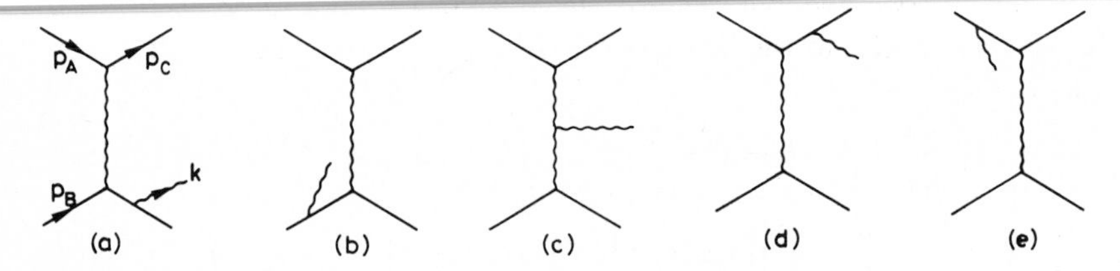


Fig. 28. Lowest order inelastic diagrams which contribute to the processes qq → q$X$.

There are a number of kinematic regions which contribute to the leading logarithmic behaviour. All these contributions can be interpreted in an elegant way. From the region where $k$ is parallel to $p_A$ ($p_B$) we obtain a contribution to $E_c(\mathrm{d}\sigma/\mathrm{d}^3p_c)$ of the form $\log s/p_A^2$ ($\log s/p_B^2$), which can be interpreted as being the convolution of the probability of finding a quark in quark $A$ ($B$), and the Born term for quark–quark elastic fixed angle scattering. In other words, the coefficient of these logarithms is just the function $a$ of eqs. (6.40) and (6.41) (in spite of the presence of the three-gluon vertex). From the region where $k$ is parallel to $p_C$, we obtain a contribution which can be interpreted as the convolution of the Born term for qq elastic fixed angle scattering, and the probability for one of the resulting quarks to fragment into the observed one + anything. The fragmentation function is just that which would be calculated from $e^+e^- \to qX$ and is related to the distribution function (in this order) by the Gribov–Lipatov reciprocity relation [80]. The final dominant contribution comes from the region in which $k$ balances the transverse momentum of $C$, and this can be interpreted as the probability of finding a gluon in quark $B$ with a fraction of the longitudinal momentum of $B$* convoluted with the Born term for gq elastic fixed angle scattering.

At $x = 1$, in addition to the contributions from the diagrams of fig. 28, one must also include the diagrams which contribute to the elastic qq scattering amplitude. The leading $\log s/p^2$ contributions can also be absorbed into the initial distribution and final fragmentation functions. In addition to these $\log s/p^2$ terms, after renormalization there will also be $g^6(\mu^2)\times\log s/\mu^2$ terms. These terms are exactly those which, when combined with the $g^4(\mu^2)$ contribution from the Born term give the first two terms in the expansion of $g^4(s)$, the running coupling constant.

*The moments of this probability are, of course, proportional to an off-diagonal element in the anomalous dimension matrix for the lowest-twist quark and gluon operators.

Thus we conclude that the cross-section for $qq \to qX$ to order $g^6$ in the cross-section and in the leading logarithmic approximation can be written in the form

$$E_c \frac{d\sigma}{d^3 p_c}(A+B\to C+X)$$
$$= \sum_{a,b,c,d} \int_0^1 dx_a dx_b \frac{dx_c}{x_c^2} G_{a/A}(x_a, Q^2)$$
$$\times G_{b/B}(x_b, Q^2) \tilde{G}_{C/c}(x_c, Q^2)$$
$$\times \delta(s'+t'+u') \frac{s'}{\pi} \frac{d\sigma^{(B)}}{dt'}(a+b\to c+d)\Big|_{\substack{s'=x_a x_b s \\ t'=(x_a/x_c)t \\ u'=(x_b/x_c)u}}, \quad (6.45)$$

where $d\sigma^{(B)}/dt'$ is the Born term contribution to the elastic cross-section for qg or qq scattering calculated using the running coupling constant, and $Q^2$ represents a typical large invariant mass squared (we assume $s \sim t \sim u \sim Q^2$). In the leading logarithm approximation we cannot distinguish between $\log s$, $\log t$ and $\log u$. It is interesting to see how the different subprocesses are interrelated: starting with the quark–quark scattering we have found that we also need to consider quark–gluon scattering. If we had started off by looking at radiative corrections to quark–gluon scattering, we would have found that we also need to consider the third subprocess, gluon–gluon scattering. Again, general arguments such as those in ref. [76] ensure the validity of eq. (6.45) to all orders of perturbation theory in the leading logarithm approximation.

Studies, such as the two examples we have discussed above, show that for all hard scattering processes of the type considered here (i.e. those with the three mass scales $Q^2$, $p^2$, and $\mu^2$), the ansatz for calculating the asymptotic behaviour of hard scattering cross-sections as $Q^2 \to \infty$ is:

1). Take the parton model hard scattering formula.

2). Replace the scaling distribution and fragmentation functions by the appropriate non-scaling ones.

3). Keep only the contributions of lowest order in the coupling constant for the hard scattering subprocesses, calculated using the running coupling constant.

Equation (6.42) and (6.45) are specific examples of this ansatz. For other examples see refs. [81,82] and references therein.

We now understand how to calculate the leading asymptotic prediction for these hard scattering processes. In the next section we will see that all the logarithmic corrections are also calculable. We will also mention some problems we have not discussed here, problems which only arise beyond the leading logarithm approximation.

## 7. Higher order corrections

Over the past few years QCD corrections have been calculated to very many processes [82], and we will not try to catalogue them all here. Rather we will make a few general comments and observations.

### 7.1. Higher order corrections to deep inelastic scattering

If we wish to calculate the higher order corrections to deep inelastic scattering, then we can still use the procedure of section 5, but with the $\beta$ function, the anomalous dimensions of the relevant operators and the Wilson coefficient functions calculated more accurately. The $O(\alpha)$ corrections to the asymptotic predictions for the moments of non-singlet combinations of structure functions can be written as

$$M_i^N(q^2) = \text{const.}\left[\alpha(q^2)\right]^{d_N}\left[1+\frac{\alpha(q^2)}{4\pi}a_i^N\right], \tag{7.1}$$

where

$$d_N = \gamma^0_{O^N}/2\beta_0 \tag{7.2}$$

and the results for the $a_i^N$ for $i$ corresponding to $\nu W_2$ and for four flavours in the $\overline{\text{MS}}$ scheme are [83]:

| $N$ | 2 | 4 | 6 | 8 | 10 |
|---|---|---|---|---|---|
| $a_i^N$ | 2.098 | 8.117 | 13.34 | 17.78 | 21.63, |

so that we see that at least for the small moments the corrections are reasonable (they grow logarithmically as $N$ increases).

Once higher order contributions are included the longitudinal structure function is no longer zero, i.e. $F_2 \neq 2xF_1$. To see this consider fig. 29. In the leading logarithm approximation we have seen that the virtualities of the rungs of the ladder are strongly ordered [eq. (6.24)] and that although the propagators were far from their mass shells, the momenta in the loops were always $\ll (q^2)^{1/2}$. This is sufficient to reproduce the parton model argument that $F_2 = 2xF_1$. Now however if we allow the momentum in the top loop (marked with the arrow) to be of order $q^2$ we lose one logarithm [that from the $t_N$ integration in eq. (6.25)] and also the

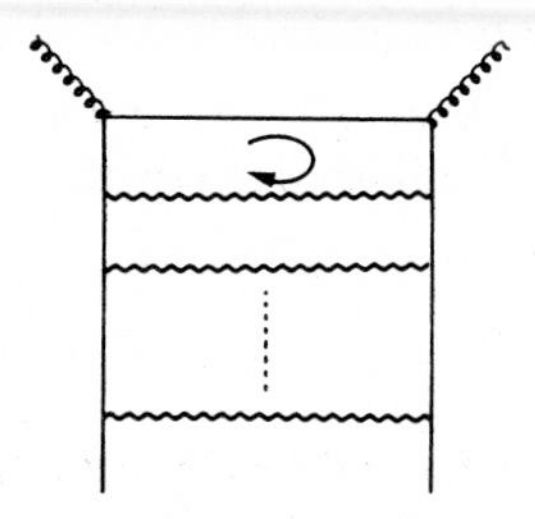

Fig. 29. Auxiliary diagram for the discussion of higher order effects in deep inelastic scattering.

argument that $F_L = 0$. Hence we expect $F_L$ to be only logarithmically suppressed as compared to $F_2$, and this is indeed so. For the non-singlet case, e.g., the Wilson coefficient functions for $F_L$ are:

$$C_L^N = \frac{\alpha(q^2)}{\pi} \frac{4}{3} \frac{1}{N+1}. \tag{7.3}$$

Various parton model sum rules are also violated by terms $O(1/\log q^2)$. All these results are consistent with the data (although the data for $F_L$ are too poor to draw any definitive conclusions).

Another interesting effect of higher order QCD is that the distribution of anti-up quarks is predicted to be different from that of anti-down quarks because of the interference of the graphs of fig. 30a and 30b [84]. This interference is suppressed by a logarithm relative to the $\bar{u}$ and $\bar{d}$ distributions themselves.

## 7.2. *The Sudakov form factor*

We now digress to remind the reader of some features of the Sudakov form factor [85]. If one considers the sum of leading logarithms in the perturbative evaluation of the form factor of the electron at large

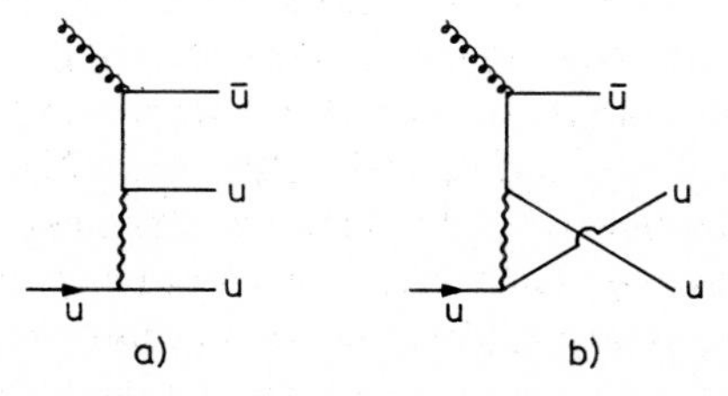


Fig. 30. Diagrams whose interference contributes to $\bar{u}-\bar{d}$.

momentum transfers one finds that:

$$\Gamma \sim \exp\left[-\alpha c\left(\log q^2\right)^2\right], \tag{7.4}$$

where $c$ is a positive constant. Although until recently there were no renormalization group arguments supporting eq. (7.4), the result is in qualitative agreement with intuition, in the sense that we would expect a charged particle to radiate when it is accelerated, and the elastic form factor is a measure of the probability that it does not. Hence we would expect this form factor to be very small at large $q^2$. If we want to write down eq. (7.4) in detail we have to specify whether we give the photon a small mass ($\lambda$) or not, and the result is different in the two cases. In one loop:

(*i*) *Massive photon* (the notation is analogous to that in fig. 14 with $q = p_1 + p_2$):

$$\Gamma_\mu(p_1, p_2) = -\frac{g^2\gamma_\mu}{16\pi^2}\left[\left(\ln q^2/\lambda^2\right)^2 - \left(\ln p_1^2/\lambda^2\right)^2 - \left(\ln p_2^2/\lambda^2\right)^2\right]. \tag{7.5}$$

(*ii*) *Massless photon*:

$$\Gamma_\mu(p_1, p_2) = -\frac{g^2\gamma_\mu}{8\pi^2}\left(\ln q^2/p_1^2\right)\left(\ln q^2/p_2^2\right). \tag{7.6}$$

It is the coefficient of $\gamma_\mu$ which exponentiates. Recently it has been shown [86] that the exponentiation is not spoilt by non-leading logarithms [although since the result falls faster with $q^2$ than any inverse power of $q^2$, it is possible that non-leading powers will spoil the results (7.5) and (7.6)].

In QCD, taking into account renormalization effects in the running coupling constant one finds that the (electromagnetic) form factor of the quark at large momentum transfer in one loop is

$$\begin{aligned}\Gamma_\mu(p, p') = &-\gamma_\mu 2C_F/\left(11 - \tfrac{2}{3}n_f\right)\left[\ln q^2 \ln\ln q^2 - \ln p^2 \ln\ln p^2\right.\\ &\left.- \ln p'^2 \ln\ln p'^2 + \ln(p^2p'^2/q^2)\ln\ln(p^2p'^2/q^2)\right].\end{aligned} \tag{7.7}$$

It is known that this result exponentiates [69] in the leading logarithm approximation and that the non-leading logarithms do not spoil this exponentiation (at least in the massive gluon case) [87]. For a detailed review see ref. [88].

We shall see below that the Sudakov form factor plays an important role in many QCD applications.

### 7.3. Higher order corrections in processes involving parton distributions

To be specific, let us start by considering higher order QCD corrections to the Drell–Yan formula (6.42). Just as there are higher order corrections to the asymptotic prediction for deep inelastic structure functions, so we shall also find that there are logarithmic corrections to eq. (6.42).

Since $F_2 \neq 2xF_1$ after including these corrections, and many parton model sum rules for the structure functions are violated, we have to state precisely what we mean by the quark and gluon distribution functions. Of course there are infinitely many sensible choices.

If we are interested in the next to leading corrections, then although it is no longer true that the relevant graphs in the planar gauge are the ladder diagrams and that the region of integration is (6.24), it is still only a relatively small set of diagrams which contributes. Among these are diagrams where we cross-over two rungs of the ladder (e.g. fig. 31a) but in most cases there is precisely the same contribution to the deep inelastic structure functions and hence these diagrams would not modify the Drell–Yan formula (6.42). In addition there are also diagrams in which we convolute all the one loop graphs contributing to the subprocess parton + parton → lepton pair plus anything, with the ladder graphs (e.g. fig. 31b). These may now involve the subprocess $\mathrm{qg} \to \ell^+ \ell^- X$ (where g stands for a gluon) as well as $\mathrm{q\bar{q}} \to \ell^+ \ell^- X$. Finally we also have to relax the restrictions (6.24) on the relevant region of integration and allow the innermost loop to carry momenta of order $Q^2$. When all this is done one finds [89–91]:

$$\begin{aligned} \frac{\mathrm{d}\sigma^{\mathrm{DY}}}{\mathrm{d}Q^2} = {} & \frac{4\pi\alpha^2}{9Q^2 s} \sum_a e_a^2 \times \int \frac{\mathrm{d}x_1\,\mathrm{d}x_2}{x_1 x_2} \\ & \times \Big\{ \big[q(x_1,Q^2)\bar{q}(x_2,Q^2) + \bar{q}(x_1,Q^2)q(x_2,Q^2)\big] \\ & \times \left[\delta(z-1) + \frac{\alpha_s(Q^2)}{\pi}\big(f_{\mathrm{q,DY}}(z) - 2f_{\mathrm{q},2}(z)\big)\right] \\ & + \big[\big(q(x_1,Q^2)+\bar{q}(x_1,Q^2)\big)g(x_2,Q^2) \\ & + \big(q(x_2,Q^2)+\bar{q}(x_2,Q^2)\big)g(x_1,Q^2)\big] \\ & \times \frac{\alpha_s(Q^2)}{\pi}\big(f_{\mathrm{g,DY}}(z) - f_{\mathrm{g},2}(z)\big)\Big\}, \end{aligned} \tag{7.8}$$

where

$$z = Q^2/x_1 x_2 s \tag{7.9}$$

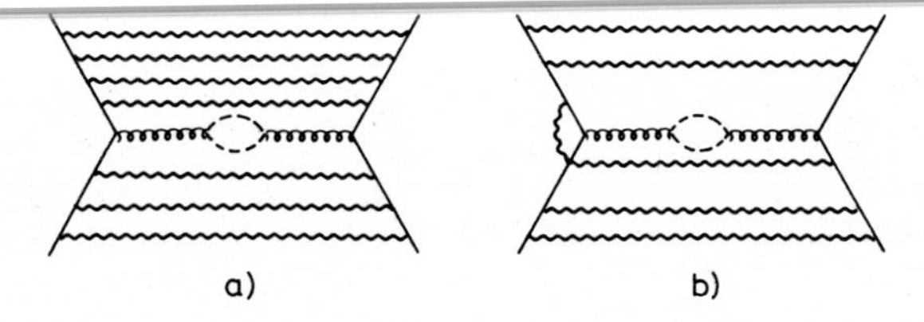


Fig. 31. Auxiliary diagrams for the discussion of higher order terms in massive lepton-pair production.

and we have used $F_2/x$ to define the quark and antiquark distributions. The coefficients of the $O(\alpha_s)$ corrections are

$$f_{q,DY} - 2f_{q,2} = \frac{2}{3}\left[\frac{3}{(1-z)_+} - 6 - 4z + 2(1+z)\left(\frac{\ln(1-z)}{1-z}\right)_+ \right.$$
$$\left. + \left(1 + \tfrac{4}{3}\pi^2\right)\delta(z-1)\right] \tag{7.10}$$

and

$$f_{g,DY} - f_{g,2} = \tfrac{1}{4}\left\{\left[z^2 + (1-z)^2\right]\ln(1-z) + \tfrac{9}{2}z^2 - 5z + \tfrac{3}{2}\right\}. \tag{7.11}$$

The reason for the strange notation $f_{q,DY} - 2f_{q,2}$ etc. is that eq. (7.8) is a relation between the Drell–Yan cross section and the structure functions $F_2$, hence we not only have to calculate the corrections to the lepton-pair cross-section but also to the structure function. Thus we see that we can calculate the $O(\alpha_s)$ correction to the asymptotic predictions (6.42).

Although as $Q^2 \to \infty$, the $O(\alpha_s)$ corrections in eq. (7.8) are indeed negligible, it is easy to see that for all values of $Q^2$ every likely to be reached experimentally they are very large. Consider for example the term*:

$$\frac{\alpha(Q^2)}{\pi}\frac{2}{3}\left(1 + \tfrac{4}{3}\pi^2\right)\delta(z-1) \tag{7.12}$$

in eq. (7.8) (and using eq. (7.10)), which for typical values of $\alpha_s(10\ \text{GeV}^2)$ (0.2 to 0.3 say) gives a correction of the order of 100%. This correction will still be significant ($\sim 10\%$) at $Q^2 \sim 10^{20}\ \text{GeV}^2$, so we really have to try and understand it, it is no good just waiting for the next generation of accelerators. To this order at least we note that the large corrections just

*The term $(\alpha(Q^2)/\pi)\cdot\frac{2}{3}\cdot 2(1+z)[\log(1-z)/(1-z)]_+$ also gives sizeable corrections.

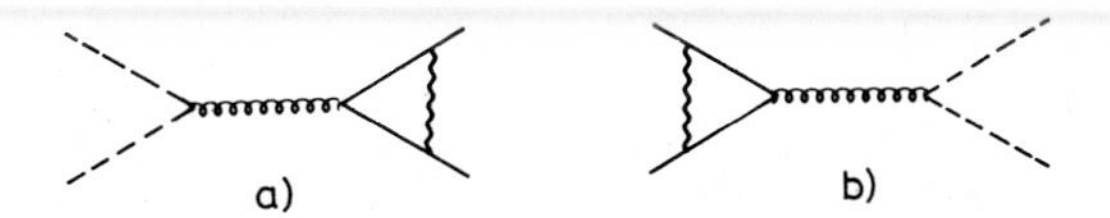


Fig. 32. One-loop diagrams which contribute to the processes (a) $e^+e^- \to q\bar{q}$ and (b) $q\bar{q} \to e^+e^-$.

renormalize the lowest order cross section, they do not change the momentum distribution.

We should not be surprised that the corrections are so large, indeed perhaps we should have expected them to be similarly large in the ratio $R_{e^+e^-}$ (see section 3), but we have seen that this is not the case there. Consider the graph of fig. 32a which contributes to the total annihilation cross section in $e^+e^-$ collisions. For $Q^2 < 0$, we have seen in the discussion on the Sudakov form factor above that this graph has a contribution

$$\sim \frac{\alpha_s}{\pi}\left[\log(-Q^2)\right]^2. \tag{7.13}$$

The argument of the logarithm is $-Q^2$ because for negative $Q^2$ the diagram is purely real. We are of course interested in the contribution for positive $Q^2$ and so we have to continue (7.13) analytically to this region, giving us a contribution to the cross-section

$$\sim \frac{\alpha_s}{\pi}\left[(\log Q^2)^2 - \pi^2\right] + \cdots . \tag{7.14}$$

Thus we see that for timelike regions we expect to have $\alpha_s \cdot \pi[\sim O(1)]$ corrections coming from higher order diagrams. We would like to point out that this analytic continuation is not the only source of $\pi^2$ terms.

The Kinoshita, Lee–Naunberg theorem ensures that all logarithms will cancel in $R$ when contributions from real gluons are also included. It turns out that the $\pi^2$ terms also cancel, and finally the total corrections are small (see eq. (3.2)). For the Drell–Yan process we have essentially the same diagram with a virtual gluon (fig. 32b) and hence from this diagram we get a contribution corresponding to that of eq. (7.14). We have seen in section 6 that the $(\log Q^2)^2$ terms cancel when diagrams with real gluons are considered; $\log Q^2$ terms survive but as we have seen these are the "large logarithms" which appear in deep inelastic scattering, so they are absorbed into the non-scaling parton distribution functions. The $\pi^2$ terms do not cancel!

Clearly we do not expect eq. (7.8) to be the end of the story as far as the large corrections are concerned. In the next order we expect $\alpha_s^2 \cdot \pi^4$ terms and also terms which are more singular as $z \to 1$. Now we know that the $(\log Q^2)^2$ terms in the Sudakov form factor exponentiate and so do the corresponding "$\pi^2$" corrections, so maybe it is sensible to sum these, extracting a factor

$$\exp\left[\frac{2}{3}\frac{\alpha(Q^2)}{\pi}\pi^2\right]. \tag{7.15}$$

Thus the bulk of the term (7.12) has been absorbed into (7.15) but it is not clear that this will be true to all orders. Later we will see that maybe we can also sum the first singular terms as $z \to 1$, with the hope (realized in the one loop case) that the remaining corrections will be small. If this is true then QCD predicts that $d\sigma/dQ^2$ for massive lepton pair production will be about 2.5 times larger than that predicted by the Drell–Yan formula (6.42)*. This seems to be true experimentally.

Analogous factors appear in other "time-like" hard processes, e.g. $e^+e^- \to hX$, $ep \to hX$, $e^+e^- \to h_1h_2X$, where h stands for hadron, and these have also been calculated (see ref. [35] for details and further references).

### 7.4. *Interactions involving "spectator" partons*

In the previous sections (with the exception of section 4), we calculated predictions for hard scattering processes by evaluating low-order Feynman diagrams for quark and gluon subprocesses. We have not considered diagrams such as those of fig. 33, which involve the interactions of "spectator" partons. By assuming that the hadronic wave function $f$ is soft, one can argue that these diagrams are logarithmically suppressed, because the softness of the wave function prevents the loop integral from diverging so that there is no mass singularity logarithm, compared to the leading ones discussed in section 6. But that is all—if we want to calculate the logarithmic corrections we must study these diagrams some more. Moreover, the contributions of these diagrams are gauge dependent. In the case of deep inelastic scattering the light-cone analysis guarantees that eq. (2.56) is correct, irrespective of the contributions of diagrams such as fig. 33a. In a general gauge, the effects of such

*It is ironic that this tends to cancel the famous factor of 1/3 due to colour.

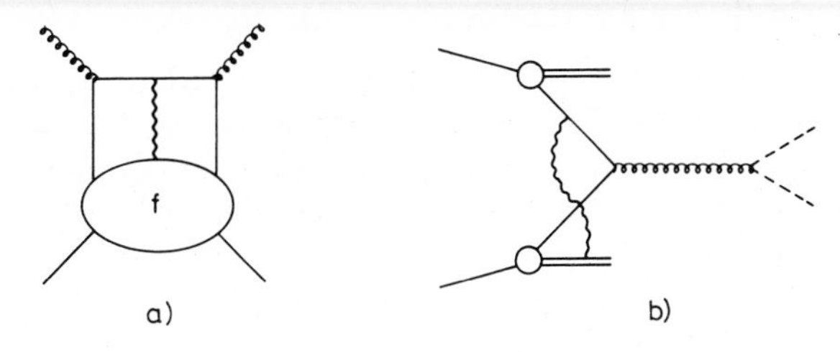


Fig. 33. Some relevant diagrams for (a) deep inelastic scattering and (b) massive lepton-pair production in which "spectator" partons interact.

diagrams for deep inelastic scattering are included in the operator matrix element, and the higher order radiative corrections again generate the series (6.6). However, a priori we have no such guarantee that diagrams involving "spectator" quarks for other hard scattering processes, such as fig. 33b, are negligible [i.e. $O(m^2/Q^2)$]. We shall show in this section that this is nevertheless the case, at least for some of the dominant contributions [92].

For definiteness let us take a simple $\phi^3$ model for the wave function. We take an interaction term of the type $\lambda\phi\chi^+\chi$, where $\phi$ and $\chi$ are scalar fields, with $\phi$ being both electrically neutral and a colour singlet, whereas $\chi$ is a charged and coloured field belonging to the fundamental representation of colour SU(3). Thus $\phi$ is our model for the "hadron" and the $\chi$'s are our "quarks". The quarks interact in the usual way with the coloured gluons, which belong to the adjoint representation. $\lambda$ has dimensions of mass and therefore the $\phi\chi^+\chi$ vertex is soft. We believe that this is the only relevant assumption, the further details of the model (such as whether the quarks have spin 0, as in this case, or spin 1/2) are not relevant. The advantage of working in a specific simple model is that we can obtain explicit answers for the Feynman diagrams. In particular, we can see how gauge invariance is restored between diagrams involving only the "participating" quarks and those which also involve "spectator" quarks.

In order to try and get some insight into the problem, we start by calculating some low-order diagrams. We will work in the Feynman gauge. The lowest order diagram is that of fig. 34a and gives an exactly scaling contribution. Some sample diagrams contributing to the next order are shown in figs. 34b, c, d. The dominant contribution from fig. 34b has a $\log q^2/p^2$ term (analogously to the diagram of fig. 22b), which comes from the usual "collinear" region of integration corresponding to

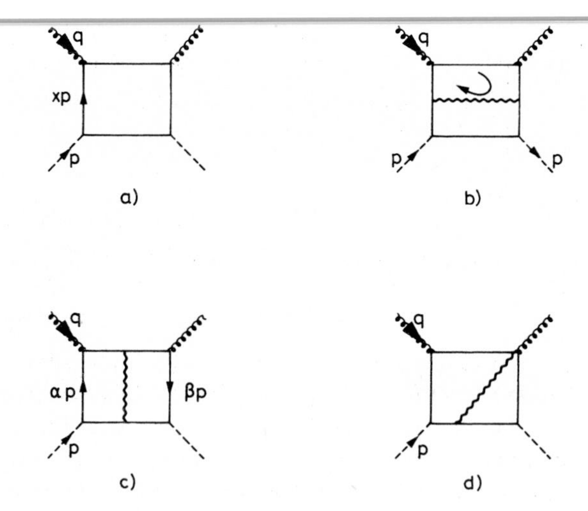


Fig. 34. Some low order diagrams contributing to deep inelastic structure functions.

$p^2 \ll k_T^2 \ll \varepsilon q^2$. The transverse momentum is allowed to get as large as $\varepsilon q$ because in the loop marked with the arrow in fig. 34b there are no soft vertices. This is not the case in figs. 34c, d and we find that for these diagrams we get a scaling contribution, i.e. no logarithms. This is because there is no loop in which the transverse momentum could get large without flowing through a soft vertex. So, in this order of perturbation theory, the contributions from diagrams which include interactions involving "spectator" quarks are suppressed by only one logarithm.

Let us now study the diagram of fig. 34c in some detail. An analysis of the dominant momentum flow in the diagram (using, for example, the techniques of refs. [93, 94]) shows that, as is marked on the diagram, the only large components of momenta in the vertical propagators can be proportional to $p$. Later we shall consider the case where $\alpha$ and $\beta$ are close to $x$, but we start by considering the region where they are different from $x$. In this case the top two horizontal propagators are hard. Comparing this to the lowest order scaling diagram of fig. 34a we now see that we have:

(a) 1 extra hard propagator giving a factor $\sim 1/p \cdot q$,

(b) 1 factor from the numerator, which is $\sim p \cdot q$.

Thus the diagram of fig. 34c gives a scaling contribution. We notice that to obtain the factor of $p \cdot q$ in the numerator, we have had to take a factor

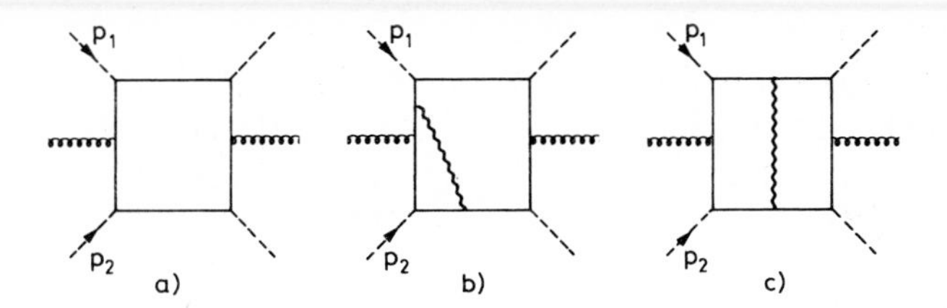


Fig. 35. Some low order diagrams contributing to massive lepton-pair production.

of $p^\lambda$ from the convection current corresponding to the gluon vertex at the bottom of the diagram (the only large momentum available at the bottom is proportional to $p$) and therefore a $q^\lambda$ from the convection current corresponding to the gluon vertex at the top of the diagram.

We now look at the Drell–Yan process $h_1 h_2 \to \ell^+ \ell^- X$, and study some of the diagrams involving interactions of spectator quarks, e.g. those of figs. 35b, c. We leave the diagram of fig. 35c till later, and start by looking at that of fig. 35b, which has many similar features to that of fig. 28c in the deep inelastic case. In the corresponding region of phase space, in the Feynman gauge this diagram also has

($i$) one more hard propagator and hence a factor of $\sim 1/s$,

($ii$) one extra numerator and hence a factor of $\sim s$,

compared to the lowest order diagram, that of fig. 35a. So in the Feynman gauge the diagram of fig. 35b gives a Drell–Yan scaling contribution, and must be considered.

Thus in the Feynman gauge, both the diagrams of figs. 34c and 35b contribute to the next-to-leading logarithms. These diagrams have many similar features and closer examination reveals that they are related by the Drell–Yan formula. Thus, when we write down this formula and insert the experimentally extracted parton distribution functions, we are including the contributions from diagrams such as fig. 34c for these distribution functions, and hence the contributions of diagrams such as fig. 35b for the Drell–Yan cross-section.

The small region of phase space in which $\alpha$ and $\beta$ are very close to $x$ gives a scaling contribution; the size of the phase space is compensated for by the fact that the horizontal propagators are now close to their mass shells. This does not present a problem in this one loop example, the Drell–Yan formula still relates diagram 34c to 35b, but recently Bodwin et al. [97] have claimed that this is not so in higher order graphs. Contributions from this "Glauber" region of phase space spoil the

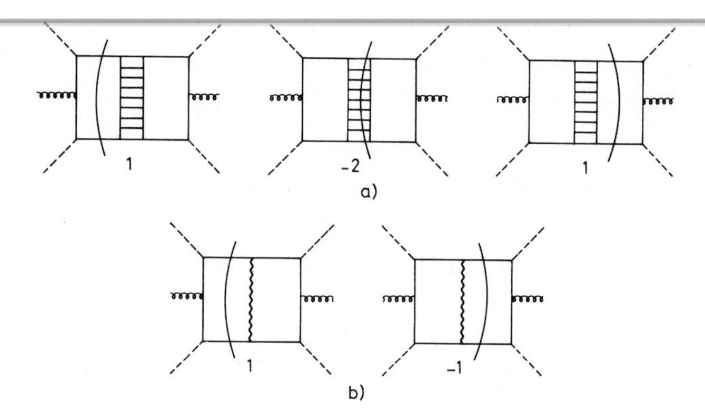


Fig. 36. Example of (a) the cancellation of absorptive corrections to massive lepton-pair production and (b) a similar cancellation of a particular type of gluonic corrections (see text).

factorization*. Mueller [98] however has argued that these non-factorizing effects are likely to be suppressed by the Sudakov form factor. It is clearly of vital importance to know whether this is true or not, so that we can be sure that we can reliably calculate higher order corrections to hard scattering processes.

We are not finished yet; there is a class of diagrams which has not been included, such as that in fig. 35c. This diagram has no analogue in the deep inelastic case. The dominant region of phase space is that in which $k$ is orthogonal to $p_1$ and $p_2$ and is therefore space-like. This means that if we evaluate the contribution to the lepton-pair cross-section from this diagram by taking all the unitarity cuts, we can neglect any cuts through the gluon. This diagram is very reminiscent of that studied by DeTar et al. [95], in which the gluon is replaced by a Pomeron. Their conclusion was that the Pomeron diagram was suppressed by a factor $\sim 1/s$. When evaluating the diagram, since we are dealing with an inclusive cross-section, we have to take care with the i$\varepsilon$'s in the propagators—we are taking the Mueller discontinuity [96]. Having done this, it turns out that there is always an integration which can be shown to be zero by using Cauchy's theorem. In our case we have an extra singularity due to the pole which is also suppressed by a factor of $s$. It is amusing to see how one obtains this answer of zero by taking unitarity cuts. In the Pomeron case, the Pomeron is imaginary and there are three cuts (fig.

*Note added in proof: All potentially non-factorizing contributions in two loop order have recently been shown to cancel [131].

36a) giving the relative contributions $1, -2, 1$. In the gluon case, the gluon is real so there are just two cuts (fig. 36b, remember the cut through the gluon is zero) giving the relative contributions $1, -1$.*

## 8. Application of QCD to exclusive processes

### 8.1. Introduction

All the discussion of the previous sections has concerned inclusive reactions; but in the past two years or so there has been some very interesting progress in the application of QCD to exclusive processes [99], and in this section we will briefly review these developments. Before doing this, however, let us recall how one approached exclusive processes within the context of the parton model.

It was popular, within the parton model, to think of hard exclusive processes in terms of "dimensional counting rules" [100, 101]. For example, for the spin-averaged form factor of a hadron H, these give

$$F_{\mathrm{H}}(t) \sim 1/t^{N-1}, \tag{8.1}$$

where $t = q^2$ and $N$ is the number of fundamental constituents in H. Thus the form factor of a pion is predicted to behave asymptotically as $1/t$, and that of a proton as $1/t^2$, both of which agree with the data, which exist for $|t|$ up to 4 (30) GeV$^2$ for the pion (proton). We can see heuristically that eq. (8.1) is correct by counting the "hard fermion propagators" (fig. 37). We iterate the hadronic wave function, so that the large transverse momentum is routed through the wiggly lines, which represent gluons. In deriving the dimensional counting rules, it is assumed that the four-quark interaction is scale invariant. We see now that in the meson form factor there is at least one hard fermion propagator ($\sim 1/t$), and in the proton case there are at least two such propagators ($\sim 1/t^2$). The generalization to arbitrary $N$ is trivial. For elastic scatter-

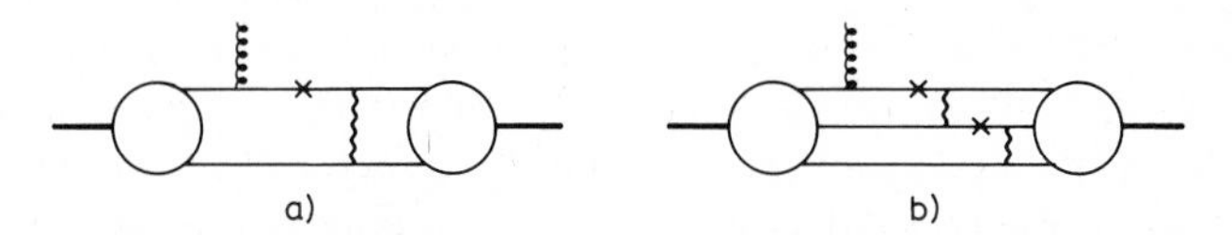


Fig. 37. Diagrams calculated in the derivation of the dimensional counting rules for the form factor of (a) a meson and (b) a baryon.

*Note in proof: See ref. [132] for a more detailed, recent discussion.

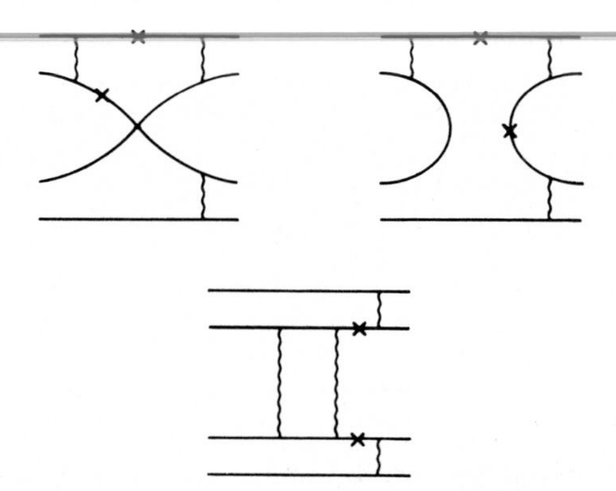

Fig. 38. Diagrams calculated in the derivation of the dimensional counting rules for meson–meson fixed angle scattering.

ing at fixed angle the dimensional counting rules are

$$\frac{\mathrm{d}\sigma}{\mathrm{d}t}(A+B\to C+D) \sim \frac{1}{t^{N-2}} f(\theta), \tag{8.2}$$

where $N$ is the total number of constituents in $A$, $B$, $C$, and $D$, and $\theta$ is the scattering angle. Thus $\mathrm{d}\sigma/\mathrm{d}t$ is predicted asymptotically to behave like $1/t^6$, $1/t^8$, and $1/t^{10}$ for meson–meson, meson–baryon and baryon–baryon scattering, respectively. These predictions seem to be consistent with the experimental data. To see that eq. (8.2) is reasonable we repeat the "hard fermion" counting arguments which we used in the form factor case. For example, in fig. 38 we present some sample diagrams which contribute to meson–meson scattering at large angle. The lines marked with a cross represent hard quark propagators, and we see that in each diagram there are two such propagators. Thus the amplitude $\sim 1/t^2$ and hence $\mathrm{d}\sigma/\mathrm{d}t \sim (1/t^2)(1/t^2)^2 \sim 1/t^6$ which satisfies eq. (8.2). The generalization to higher $N$ is straightforward.

There is one further complication. Landshoff [102] proposed that instead of diagrams such as those in fig. 38, in which there are "hard fermion" propagators, the relevant mechanism for fixed angle scattering may contain no such propagators at all. For example, he suggested that meson–meson scattering may be dominated by two separate scatterings of the two valence quarks through the same angle (fig. 39a). Although there are no hard fermion propagators to suppress the amplitude, there is little phase space for both quarks to scatter through the same angle. Calculation of the diagram of fig. 39 reveals that it contributes to the amplitude $M_{\pi\pi}$ a term

$$M_{\pi\pi} \sim 1/t^{3/2}, \tag{8.3}$$

which leads to:

$$\mathrm{d}\sigma/\mathrm{d}t|_{\pi\pi} \sim 1/t^5, \tag{8.4}$$

which should dominate asymptotically over the dimensional counting contribution which behaves like $1/t^6$. Similarly the analogous "multiple scattering" contribution in the proton–proton case gives a contribution

$$\mathrm{d}\sigma/\mathrm{d}t|_{\mathrm{pp}} \sim 1/t^8, \tag{8.5}$$

which dominates over the dimensional counting contribution which gives $1/t^{10}$. The meson–baryon case is not so clear [103].

In any renormalizable field theory we expect there to be at least some logarithmic modifications to eqs. (8.1)–(8.4). We shall see that this is indeed the case in QCD.

### *8.2. Pion form factor in QCD*

In this section we review the approach and results of Brodsky and Lepage [99, 104] concerning the pion form factor in QCD, which is the most complete discussion of the problem. Other contributions to this problem can be found in refs. [105–109].

Consider the lowest order Feynman diagram which contributes to the pion form factor at large momentum transfers (fig. 40). For the "soft" hadronic wavefunction we take:

$$\gamma_5 \phi_0(x) \not{p}/\sqrt{2}, \tag{8.6}$$

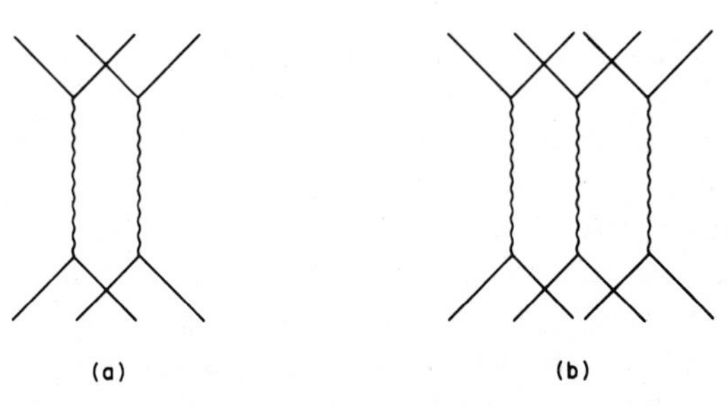


Fig. 39. Some "multiple scattering" diagrams for (a) meson–meson and (b) baryon–baryon fixed angle scattering.

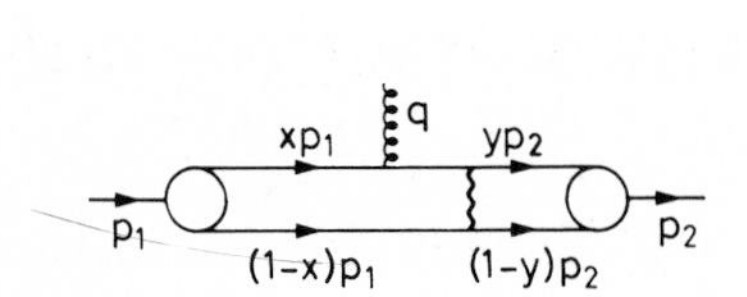


Fig. 40. Lowest order diagram contributing to the pion form factor.

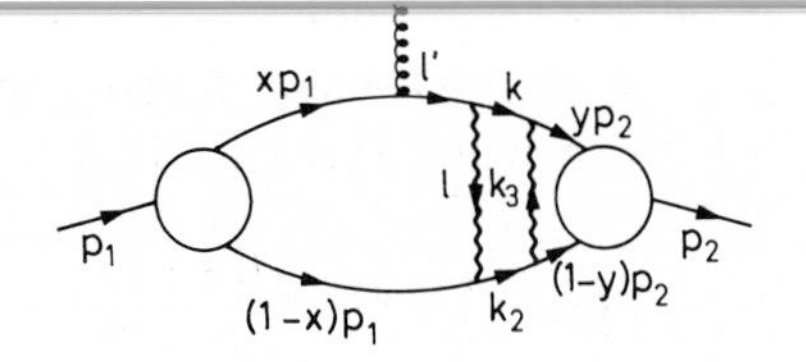


Fig. 41. One-loop correction to the diagram of fig. 40.

where the notation is defined in fig. 40. We leave it as an exercise for the reader to evaluate the graph; its contribution to the pion form factor is:

$$\int \mathrm{d}x\,\mathrm{d}y\,\phi_0^+(y)T(x,y,q^2)\phi_0(x), \tag{8.7}$$

where

$$T = \frac{4g^2C_{\mathrm{F}}}{(1-x)(1-y)q^2}e_1 + \frac{4g^2C_{\mathrm{F}}}{xyq^2}e_2. \tag{8.8}$$

$e_1$ and $e_2$ are the charges of the quarks. We see that eq. (8.8) gives us the dimensional counting result (8.1).

Consider now the higher order graph of fig. 41. We will not evaluate it explicitly, but will give some indication of where the results came from. We introduce Sudakov variables,

$$k = zp_2 + wp_1 + k_\perp, \tag{8.9}$$

which introduces a Jacobian of $-q^2/2$. In terms of the Sudakov variables the singularities of the four propagators are as follows:

$$k^2 + \mathrm{i}\varepsilon = -wzq^2 - k_\perp^2 + \mathrm{i}\varepsilon = 0, \tag{8.10a}$$

$$l^2 + \mathrm{i}\varepsilon = (1-x+w)(1-z)q^2 - k_\perp^2 + \mathrm{i}\varepsilon = 0, \tag{8.10b}$$

$$k_2^2 + \mathrm{i}\varepsilon = w(1-z)q^2 - k_\perp^2 + \mathrm{i}\varepsilon = 0, \tag{8.10c}$$

$$k_3^2 + \mathrm{i}\varepsilon = w(y-z)q^2 - k_\perp^2 + \mathrm{i}\varepsilon = 0, \tag{8.10d}$$

We now do the $w$ integration by closing contours and all the poles in $w$ are on the same side of the contour unless $z$ is in the range from 0 to 1. We also see that we have to treat the $y < z$ case separately from the $y > z$ one. There are many similarities in this problem to the calculation of scaling violations in deep inelastic scattering discussed in section 6. In particular we find mass singularities when $k_3$ is (almost) parallel to $p_2$. In the leading logarithm approximation $w$ is small (in analogy to $\alpha$ being small in eqs. (6.16) and (6.18)), and the coefficient of $\log q^2/p^2$ is (in the light cone gauge $n \cdot A = 0$ with $n^2 = 0$):

$$\frac{z}{y(1-z)}\theta(y-z)\left(1+\frac{1}{y-z}\right)+\frac{1}{1-y}\theta(z-y)\left(1+\frac{1}{z-y}\right), \tag{8.11}$$

where we have not yet performed the $z$ integrations. There is clearly an infra-red divergence as $z \to y$ but this will be cancelled later, so for the moment we leave the $z$ integration undone. We work in the frame in which

$$p_1 = (P, \mathbf{0}, P), \qquad p_2 = \left(P + q_\perp^2/4P, \boldsymbol{q}_\perp, P - q_\perp^2/4P\right),$$

and

$$q = \left(q_\perp^2/4P, \boldsymbol{q}_\perp, -q_\perp^2/4P\right), \tag{8.12}$$

where each vector is written as $l = (l_0, l_\perp, l_z)$ and $P$ is some large number. The vector $n$ which defines the gauge is now written as

$$n = (P, \mathbf{0}, -P). \tag{8.13}$$

This choice of gauge leads to the very welcome feature that in the leading logarithm approximation the dominant graphs are the ladder diagrams (with vertex and self-energy corrections). For a full discussion of this see ref. [104], however many of the arguments are similar to the analogous discussion in section 6. A typical ladder diagram is shown in fig. 42.

At this point let us consider the vertex and self-energy corrections. The $k_\perp$'s and hence $k^2$'s are nested, and the vertex insertions can be absorbed into $\alpha(k_\perp^2)$ at the vertex. The whole of the self-energy is taken with the appropriate vertex (instead of absorbing half the self energy with the vertex on the left and the other half with that on the right). Thus we are left with the two diagrams of fig. (43) which cancel by the Ward identity $Z_1 = Z_2$. Hence if we want to expand everything in terms of

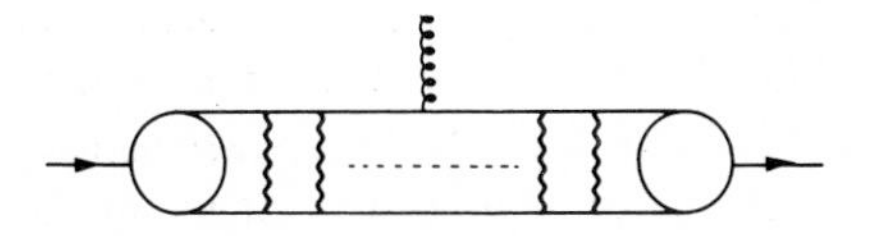

Fig. 42. Dominant structure for the pion form factor. Vertex and self-energy insertions must be included.

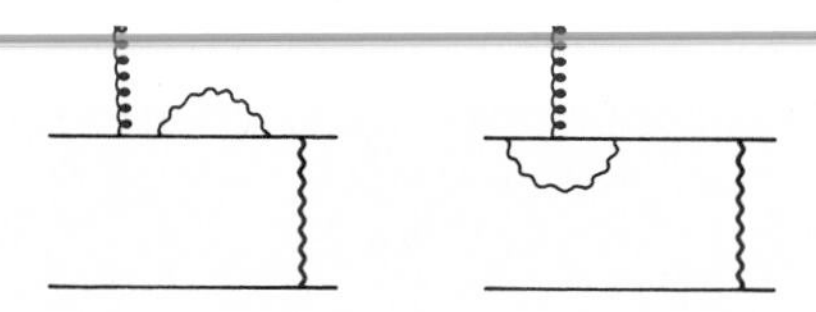

Fig. 43. Auxiliary diagrams for the calculation of the asymptotic behaviour of the pion form factor.

$\alpha_s(q^2)$ we must divide by the square of the self energy. Therefore eq. (8.7) is now modified to

$$F_\pi(q^2) = \int_0^1 dx\, dy\, \phi^+(y, q^2) T_H(x, y, q^2) \phi(x, q^2), \tag{8.14}$$

where

$$T_H(x, y, q^2) = \frac{16\pi C_F}{q^2} \alpha(q^2) \left[ \frac{e_1}{(1-x)(1-y)} + \frac{e_2}{xy} \right] \tag{8.15}$$

and

$$\phi(x_i, Q^2) \equiv (\log Q^2/\Lambda^2)^{-\gamma_F/\beta_0} \int^{Q^2} \frac{dk_\perp^2}{16\pi^2} \psi(x_i, k_\perp). \tag{8.16}$$

In eq. (8.16) $\psi(x_i, k_\perp)$ is the quark–antiquark component of the mesons wave-function. Because of the nesting of the $k_\perp$'s we only have to perform the integral up to $k_\perp^2$ of order $Q^2$ and the factor in front of the integral is precisely the inverse of the sum of the leading logarithms in the quarks self-energy. $\gamma_F$ is the anomalous dimension of the Fermion field (see eq. (2.29)) and in the light cone gauge is equal to

$$\gamma_F = C_F \left( 1 + 4 \int_0^1 dx \frac{x}{1-x} \right). \tag{8.17}$$

We notice that $\gamma_F$ has a divergence as $x \to 1$, but we leave it unintegrated for now.

Since the dominant diagrams are the ladder ones it is rather easy to deduce an integral equation for $\psi$ and hence for $\phi$. The integral equation for $\psi$ is:

$$\frac{\psi(x, q_\perp)}{16\pi^2} = \frac{C_F}{4\pi} \frac{\alpha(q_\perp^2)}{q_\perp^2} \int_0^1 \hat{V}(x, y)\, dy \int^{q_\perp^2} \frac{dl_\perp^2}{16\pi^2} \frac{\psi(y, l_\perp)}{y(1-y)}, \tag{8.18}$$

where

$$\hat{V}(x,y) = 2\left[x(1-y)\theta(y-x)\left(1+\frac{1}{y-x}\right) + y(1-x)\theta(x-y)\left(1+\frac{1}{x-y}\right)\right], \tag{8.19}$$

which can be deduced from the one gluon example worked out above [eq. (8.11)].

Equation (8.18) can easily be rewritten in terms of $\tilde{\phi}$ which is defined by:

$$x(1-x)\tilde{\phi}(x,q^2) \equiv \phi(x,q^2). \tag{8.20}$$

This equation reads

$$x(1-x)q^2\frac{\partial}{\partial q^2}\tilde{\phi}(x,q^2) = C_F\frac{\alpha(q^2)}{4\pi}\int_0^1 dy\,\hat{V}(x,y)\tilde{\phi}(y,q^2) - C_F\frac{\alpha(q^2)}{4\pi}x(1-x)\tilde{\phi}(x,q^2)\left(1+4\int_0^1 dz\frac{z}{1-z}\right). \tag{8.21}$$

We can now see that the divergent terms cancel, for consider these divergent terms for $x < y$; they are proportional to:

$$2\int_x^1 dy\frac{x(1-y)}{y-x}\tilde{\phi}(y,q^2) - 2x(1-x)\tilde{\phi}(x,q^2)\int_0^1 dz\frac{z}{1-z}, \tag{8.22}$$

where we have taken half of the "4" in front of the integral over $z$ here, and the other half will be associated with $x > y$. We now write

$$\int_0^1 dz\frac{z}{1-z} = \frac{1}{1-x}\int_0^{1-x} dz'\frac{z'}{1-x-z'} = \frac{1}{1-x}\int_x^1 dy\frac{1-y}{y-x}, \tag{8.23}$$

so that we can rewrite (8.22) as:

$$2\int_x^1 dy\frac{x(1-y)}{y-x}[\tilde{\phi}(y,q^2)-\tilde{\phi}(x,q^2)] \equiv 2\int_x^1 dy\frac{x(1-y)}{y-x}\Delta\tilde{\phi}(y,q^2), \tag{8.24}$$

which has no divergences. A similar cancellation occurs in the region

$x > y$, so we now rewrite eq. (8.21) as

$$x(1-x)q^2\frac{\partial}{\partial q^2}\tilde{\phi}(x,q^2) = C_F\frac{\alpha(q^2)}{4\pi}\left[\int_0^1 dy\, V(x,y)\tilde{\phi}(y,q^2) - x(1-x)\tilde{\phi}(x,q^2)\right], \tag{8.25}$$

where $V(x, y)$ is equal to $\hat{V}(x, y)$ [eq. (8.19)] but with $1/(y-x)$ [$1/(x-y)$] replaced by $\Delta/(y-x)$ [$\Delta/(x-y)$]. We are now left with the problem of solving eq. (8.25). To this end we change variables to:

$$\bar{x} = 1-2x, \qquad \bar{y} = 1-2y, \tag{8.26}$$

which take values in the range from $-1$ to $+1$. It is a simple matter to check that

$$V(\bar{x},\bar{y}) = V(\bar{y},\bar{x}), \tag{8.27}$$

i.e. $V$ is symmetric in $\bar{x}$ and $\bar{y}$. With a view to diagonalizing the potential we take moments of $V$ and find

$$\begin{aligned}\tfrac{1}{2}\int_{-1}^{1} V(\bar{x},\bar{y})\,\bar{y}^n dy &= \frac{1-\bar{x}^2}{4}\cdot n\text{th order polynomial in } \bar{x}\\ &\equiv \frac{1-\bar{x}^2}{4}\sum_{j=0}^{n}\bar{x}^j U_{jn},\end{aligned} \tag{8.28}$$

where the $U_{jn}$'s are easy to calculate. Since the Gegenbauer polynomials $C_n^{3/2}(x)$ form a complete set of polynomials on the range $-1 \leqslant x \leqslant 1$* we can write:

$$V(\bar{x},\bar{y}) = \frac{(1-\bar{x}^2)(1-\bar{y}^2)}{16}\sum_{n,m} a_{nm}C_n^{3/2}(\bar{x})C_m^{3/2}(y) \tag{8.29}$$

with $a_{nm} = a_{mn}$ in view of eq. (8.27). Then:

$$\begin{aligned}\tfrac{1}{2}\int_{-1}^{1} V(\bar{x},\bar{y})C_n^{3/2}(\bar{y})\,d\bar{y} &= (1-\bar{x}^2)\sum_{m=0}^{\infty} a_{nm}b_n C_m^{3/2}(\bar{x})\\ &= (1-\bar{x}^2)\sum_{j=0}^{n} c_j C_j^{3/2}(\bar{x}),\end{aligned} \tag{8.30}$$

where $b_n$ and $c_j$ are constants and we have used the orthogonality of the

*They are an orthogonal set of polynomials on $-1 \leqslant x \leqslant 1$ with weight function $w(x) = (1-x^2)/4$.

Gegenbauer polynomials and eq. (8.28). Thus:

$$a_{nm} = 0 \quad \text{for } m > n, \tag{8.31}$$

and so by symmetry:

$$a_{nm} = \Gamma_n \delta_{nm}, \tag{8.32}$$

where we leave it as an exercise to calculate $\Gamma_n$. Thus

$$V(\bar{x}, \bar{y}) = \tfrac{1}{16}(1-\bar{x}^2)(1-\bar{y}^2)\sum_n \Gamma_n C_n^{3/2}(\bar{x}) C_n^{3/2}(\bar{y}). \tag{8.33}$$

Having thus diagonalized the potential we can see that anything of the form $f_n(q^2)C_n^{3/2}(x)$ is a solution to eq. (8.25) providing $f_n(q^2)$ is a solution of the following first order differential equation:

$$q^2 \frac{\partial}{\partial q^2} f_n(q^2) + \frac{\alpha(q^2)}{4\pi} C_F f_n(q^2) = \frac{\alpha(q^2) C_F}{4\pi} f_n(q^2) \Gamma_n. \tag{8.34}$$

Equation (8.34) can be solved to give:

$$f_n(q^2) = c_n \left[\alpha(q^2)\right]^{\gamma_n}, \tag{8.35}$$

where

$$\gamma_n = -\frac{C_F}{\beta_0}(\Gamma_n - 1) \tag{8.36}$$

and $c_n$ is the constant of integration. For pions we find

$$\gamma_n = \frac{C_F}{\beta_0}\left(1 + 4\sum_2^{n+1} \frac{1}{k} - \frac{2}{(n+1)(n+2)}\right), \tag{8.37}$$

which is the same anomalous dimension as in deep-inelastic scattering!!* Thus:

$$\tilde{\phi}(\bar{x}, q^2) = \tfrac{1}{4}(1-\bar{x}^2) \sum_{n=0}^{\infty} a_n C_n^{3/2}(\bar{x})\left[\alpha(q^2)\right]^{\gamma_n}, \tag{8.38}$$

where the $a_n$'s are in general uncalculable in perturbation theory. Combining eqs. (8.14), (8.15), (8.20) and (8.38) and noting that the anomalous dimensions $\gamma_n$ are monotonically increasing, we find that the asymptotic

*This will be explained below.

Fig. 44. Decay of the $\pi$-meson.

behaviour of the pion form factor is (note $\gamma_0 = 0$):

$$F_\pi(q^2) \underset{q^2\to\infty}{\to} a_0^2 4\pi C_F \alpha_s(q^2)/q^2. \tag{8.39}$$

Thus we see that the "dimensional counting" result gets modified by a logarithmic factor.

We can go further and determine $a_0$ in terms of the "pion decay constant", $f_\pi$ ($\simeq 93$ MeV) (fig. 44); specifically

$$a_0 = 3f_\pi / \sqrt{n_c}, \tag{8.40}$$

so that

$$F_\pi \underset{q^2\to\infty}{\to} 16\pi\alpha_s(q^2) f_\pi^2/q^2. \tag{8.41}$$

Expression (8.41) is then the asymptotic prediction of QCD for the pion form factor. There is unfortunately insufficient data to check (8.41), but it is a considerable success to have made the theoretical prediction.

The results above can be understood in terms of the operator product expansion and the renormalization group [109]. In the wave functions $\psi$ in fig. 42 the outermost legs are hard, so that we want

$$\langle 0|\psi(z_1)\bar\psi(z_2)|\pi\rangle \tag{8.42}$$

in the limit $z_1^+ = z_2^+$. We expand $\psi\bar\psi$ in terms of local operators (we are still working in the light-cone gauge and in this gauge the path-ordered exponential of the gauge potential just gives a factor 1, so we can neglect it), and again the non-singlet twist-two operators are the dominant ones. There is a slight difference however, since we now have to consider operators of the type

$$\partial_{\mu_{k+1}}\cdots\partial_{\mu_m}\bar\psi\gamma_\mu D_{\mu_1}\cdots D_{\mu_k}\psi. \tag{8.43}$$

In the deep inelastic case we take the forward matrix element of the operators, and in that case it is only the operator with $k = m$ which has a non-zero matrix element. Now, for each value of $m$ we have $m$ operators, and therefore in principle $m$ eigenvalues of an anomalous-dimension matrix. However, the anomalous-dimension matrix is a triangular one; renormalization of the operator with an $m-k$ external derivative gives a linear combination of the operators (8.43) with $m-k, m-k+1,\ldots$ external derivatives. Thus the eigenvalues of this matrix are just the diagonal elements, and these are just the $\gamma_0^N$ for $N<m$. In this way we understand eq. (8.38).

### 8.3. *Form factors of other hadrons*

In the previous section we saw that the QCD prediction for the asymptotic behaviour of the form factor was very similar to the dimensional counting prediction, the only difference being an extra factor of $\alpha_s(Q^2)$ in the QCD case. In general, however, the lowest anomalous dimension is not zero, and there are additional logarithmic modifications to eq. (8.1).

Perhaps the simplest example is the form factor of the vector meson. For the helicity-zero form factor of the $\rho$ meson, for example, the arguments of the previous section go through exactly so that

$$F_\rho(q^2, \lambda_\rho = 0) \sim \alpha(q^2)/q^2. \tag{8.44}$$

In this case the normalization constant can also be determined; it is the same as for the pion (eq. 8.41) with $f_\pi$ replaced by the $\rho$ decay constant $g_\rho$ defined by:

$$\langle 0|\bar{\psi}\gamma_\alpha(\tfrac{1}{2}\tau^+)\psi|\rho^+\rangle = \varepsilon_\alpha g_\rho m_\rho. \tag{8.45}$$

In fact $g_\rho \approx f_\pi$, reflecting the KSFR relation [110], so that $F_\pi \approx F_\rho$ asymptotically. For the transversely polarized $\rho$ (helicity $= \pm 1$) there are two differences:

(i) $T_B$ is suppressed by an additional power of $Q^2$.

(ii) The structure of the potential is different.

The bound state equation can be solved [100] giving

$$F_\rho(q^2, \lambda_\rho = \pm 1) \sim \frac{m^2(q^2)}{q^2}\frac{\alpha_s(q^2)}{q^2}\left[\alpha_s(q^2)\right]^{2C_F/\beta_0}. \tag{8.46}$$

The exponent $2C_F/\beta_0$ is the anomalous dimension of

$$\bar{\psi}\sigma_{\alpha\beta}(\tfrac{1}{2}\tau^+)\psi. \tag{8.47}$$

The subdominant contributions to this form factor have exponents which are just the anomalous dimensions of the tower of operators based on (8.47).

One can carry out a similar analysis for the asymptotic behaviour of the form factors of baryons [111] which are of the form

$$F_{\mathrm{B}}(q^2) = \int_0^1 [\mathrm{d}x][\mathrm{d}y]\phi^+(y, q^2)T_{\mathrm{H}}(x, y, q^2)\phi(y, q^2), \tag{8.48}$$

where

$$[\mathrm{d}x] \equiv \mathrm{d}x_1\mathrm{d}x_2\mathrm{d}x_3\delta\Big(1 - \sum_i x_i\Big). \tag{8.49}$$

The potential is now more complicated and the relevant equations have so far only been solved numerically. For the magnetic form factor of the proton we find

$$G_{\mathrm{M}} \underset{q^2\to\infty}{\to} \text{const.}\frac{\alpha_s^2(q^2)}{q^4}(\ln q^2/\Lambda^2)^{-4/3\beta_0}, \tag{8.50}$$

so that again the "dimensional counting" result is modified by logarithms. Equation (8.50) is consistent with the data, provided that $\Lambda$ is not too large ( $\sim$ 100 MeV).

### 8.4. *End point contributions*

The "dimensional counting" results in the parton model and their QCD analogue came from evaluating the "tail" of the hadronic wavefunction, i.e. the behaviour of the hadronic wavefunction as one (or possibly more) of the constituents are far off their mass-shells. As we have already seen this region of phase space is suppressed by powers of $q^2$, which leads to the power behaviour (modulo QCD logarithms) of the hadronic form factors. There is also another kinematic region which contributes to the large $q^2$ behaviour of the form factor, but which does not involve partons going far off their mass-shells. We call the corresponding contributions "end point contributions".* To illustrate this, let us consider again the form factor of the pion, but this time without iterating the wave function and write

$$F(q^2) = \int_0^1 \mathrm{d}x \int \frac{\mathrm{d}^2k_\perp}{16\pi^3}\psi(x, k_\perp)\psi^*(x, k_\perp + (1-x)q_\perp), \tag{8.51}$$

where we are now working in an infinite momentum frame with $q =$

*For a detailed review see ref. [98].

$(0,\boldsymbol{q}_{\perp},0)$. If $\psi$ is damped for large virtualities (as we have seen it is) then perhaps we have also got to consider the region $x \to 1$, in which there are no hard partons. Just like in the Landshoff mechanism for fixed angle scattering we have to weigh up whether the smallness of the phase space around $x=1$ is sufficiently compensated for by the absence of hard propagators. The contribution from the region around $x=1$ is:

$$F(q^2) = \int_{1-\lambda/q}^{1} \mathrm{d}x\, |\psi(x,\lambda)|^2, \tag{8.52}$$

where $\lambda$ is a small mass cut-off. If the behaviour of the wave function as $x \to 1$ is $(1-x)^{\delta}$, then we see that

$$F(q^2) \sim (\lambda/q)^{1+2\delta}. \tag{8.53}$$

In QCD for the pion's wave functions we find $\delta = 1$ (see e.g. eq. (8.19) and fig. 45a) so that this end point contribution is asymptotically negligible compared to the dimensional counting contribution.

For the proton, eq. (8.52) is replaced by (fig. 45b):

$$F(q^2) \sim \int_{1-\lambda/q}^{1} \mathrm{d}x \int_0^1 (1-x)\, \mathrm{d}\tau\, |\psi|^2, \tag{8.54}$$

which leads to

$$F(q^2) \sim (\lambda/q)^{2+2\delta}. \tag{8.55}$$

Once again we find $\delta = 1$ so that this end point contribution contributes a term $\sim 1/q^4$ to the protons form factor, which is now comparable to the dimensional counting result. Does this mean that we should not take results such as eq. (8.50) seriously? Well it seems that maybe (probably?) these end point contributions are suppressed by the Sudakov form factor [98], since they involve large momentum transfers between on-shell quarks. If this is true (and our understanding of this is as yet incomplete in spite of considerable recent progress), the "modified" dimensional counting results dominate the asymptotic behaviour of hadronic form factors.

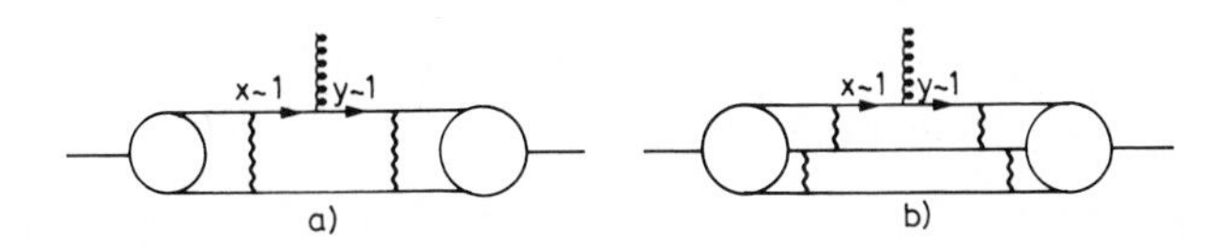


Fig. 45. Diagrams which contain the so called "end-point contributions" to (a) the form factor of a meson and (b) the form factor of a baryon.

### 8.5. *Elastic scattering at fixed angle*

The problem of elastic scattering at fixed angle has not been studied very thoroughly yet in QCD, but certain features seem to have emerged [99]. In particular, in time-ordered perturbation theory in the light-cone gauge the dominant radiative corrections to diagrams such as those of fig. 38, which obey the dimensional counting rules, lead to the ladder-like structure of fig. 46, in which each rung joins together two quarks moving almost parallel to each other. These radiative corrections are identical to those in the form factors so that we can write

$$\frac{\mathrm{d}\sigma}{\mathrm{d}t}(A+B\to C+D)=\frac{\alpha_s^2(t)}{t}F_A(t)F_B(t)F_C(t)F_D(t)f(\theta). \tag{8.56}$$

In principle, the function $f(\theta)$ and the normalization are also calculable, but quite clearly this is a horrendous task. Thus we see that in QCD the diagrams which obeyed the dimensional counting rules are only modified by calculable logarithmic corrections.

What about the Landshoff diagrams, e.g. those of fig. 39, which were the dominant ones in the parton model? When we evaluate the radiative corrections to these diagrams, we find that the leading contributions are exactly those corresponding to the Sudakov form factor; thus each quark–quark scattering amplitude should be multiplied by $F_q^2(t)\times\alpha(t)$ (where $F_q$ is the Sudakov form factor). Since asymptotically the Sudakov form factor falls faster than any power of $t$, the "multiple scattering"

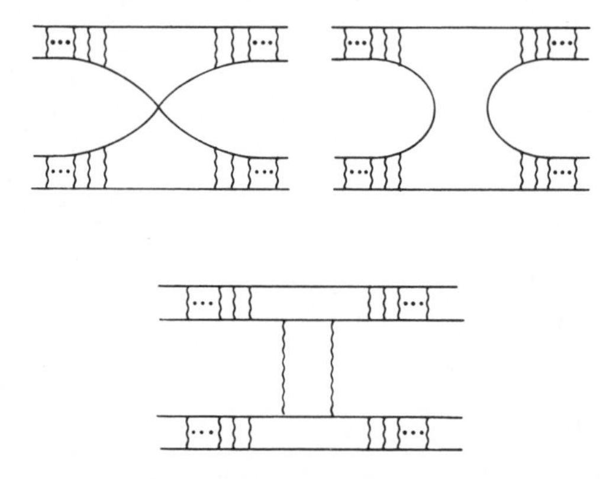

Fig. 46. Dominant gluonic corrections to the diagrams of fig. 38.

diagrams will be suppressed relative to those obeying the dimensional counting rules.

Let us briefly recall why there was no sign of the Sudakov form factor in the pion form factor or in other processes for diagrams obeying the dimensional counting rules (e.g. those of fig. 46). As an example, let us take the pion form factor in the Feynman gauge and consider the diagram of fig. 47a. Clearly, this has a double logarithmic (i.e. $(\log t)^2$) contribution. However, we should also consider the contribution of fig. 47b which also has a $(\log t)^2$ term, which cancels that of fig. 47a. It is crucial that the external particle is a colour singlet; otherwise the colour factors of the two diagrams of figs. 47a, b would not be equal and the cancellation would not take place. This cancellation corresponds to the cancellation of the infrared divergences when we consider radiation from a colour neutral particle. This cancellation occurs in all diagrams which do not have nearly on-shell partons scattering through a finite angle.

Why does such a cancellation not take place in the multiple scattering diagrams? In these diagrams it might seem natural that there should be a cancellation of the $(\log t)^2$ terms within various sets of diagrams, such as the pair in fig. 48. However, this is not the case, since in the diagram of fig. 48b, $k_T$ is bounded above by $\lambda$, as it has to be routed through at least one of the soft hadronic vertices. Notice that this is not the case in fig.

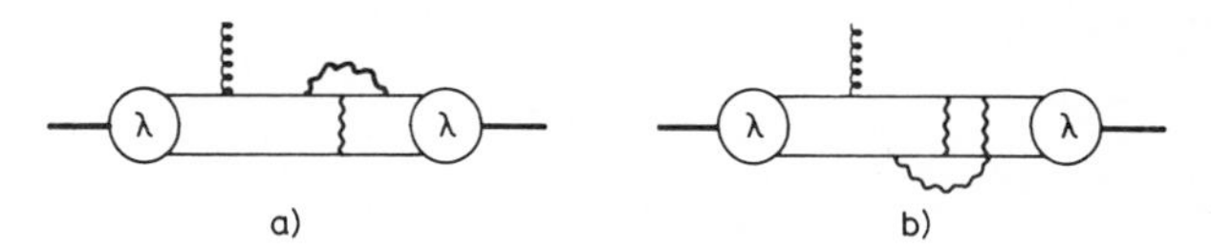


Fig. 47. Example of the cancellation of the "Sudakov double logarithm". Both diagrams a) and b) have such a $(\log t^2)^2$ term; however, their sum does not.

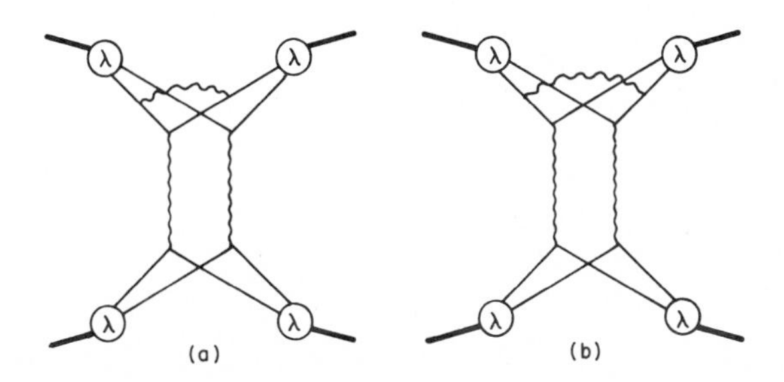


Fig. 48. Examples of radiative corrections to the multiple scattering diagram of fig. 39. Diagram a) has a $(\log t^2)^2$ factor, whereas diagram b) does not.

48a, nor in either of the diagrams of fig. 47. This means that the integral over $k_T$, which in the Sudakov case is

$$\int_{p^2}^{t} \frac{dk_T^2}{k_T^2} \sim \ln t/p^2, \tag{8.57}$$

is independent of $t$ in the case of diagram fig. 48b. Thus we lose a factor of $\log t$ and no cancellation of the $(\log t)^2$ factor occurs. There seems to be no way of avoiding a Sudakov suppression of the multiple scattering diagrams [112].

From the above discussion we see that we are at least beginning to understand elastic scattering processes in QCD.

## 9. Brief review of some selected topics

### 9.1. Processes with two large mass scales

As an example consider the process $h_1h_2 \to \ell^+\ell^- X$, where as usual h and $\ell$ stand for hadron and lepton respectively. We will imagine that both the invariant mass of the lepton pair ($Q$) and its transverse momentum ($Q_\perp$) are measured and are large, in such a way that $Q^2 \sim s$ and $m^2 \ll Q_\perp^2 \ll s$, $Q^2$ where $m$ is a "typical" hadronic mass. There is as yet no renormalization group prediction for this cross-section* and therefore we have to resort to approximations which may be wrong.

One possible approximation is to sum up the leading logarithms in momentum space [69], remembering that there are now two types of large logarithm, $\log s/Q_\perp^2$ and $\log Q_\perp^2/m^2$. Performing the summation one finds

$$\frac{d\sigma}{dQ^2 dQ_\perp^2 dy} = \frac{4\pi\alpha^2}{9} \frac{1}{sQ^2Q_\perp^2} \frac{\partial}{\partial \ln Q_\perp^2} \times \Bigg\{ \sum_{\substack{\text{flavours} \\ f}} e_f^2 \Big[ q_1^f(x_1, Q_\perp^2) q_2^{\bar{f}}(x_2, Q_\perp^2) + q_2^f(x_2, Q_\perp^2) \times q_1^{\bar{f}}(x_1, Q_\perp^2) \Big] T_F^2(Q_\perp^2, Q^2) \Bigg\}, \tag{9.1}$$

*There has however been considerable progress in understanding the related quantity of Energy–Energy correlations in $e^+e^-$ annihilation. See ref. [88] and refs. therein.

where $y$ is the rapidity and $T_F$ is the "remnant" of the Sudakov form factor*

$$T_F = \exp\left[\frac{\alpha_s C_F}{2\pi}\left(\ln Q^2/Q_\perp^2\right)^2\right]. \tag{9.2}$$

There are some technical niceties in this calculation. For instance, even in the planar gauge where ladder diagrams again dominate, the $k$'s are still strongly ordered but all such orderings must now be summed (e.g. $k_{1\perp} \gg k_{2\perp} \gg \cdots$ and $k_{2\perp} \gg k_{1\perp} \gg \cdots$ where the $k$'s are the momenta of the gluons in fig. 31a). Equation (9.1) joins on smoothly at large transverse momentum (i.e. $Q_\perp^2 \sim Q^2$) with the formula which is obtained by using the ansatz at the end of section 6 for processes in which all the large momenta are comparable [113].

Another possible approximation is to sum the leading logarithms in impact parameter space [114], in which case one finds a different result from eq. (9.1). In particular, whereas eq. (9.1) implies that there is a dip at small $Q_\perp^2$, the calculation in impact parameter space (which takes into account the conservation of transverse momentum) suggests that there is no dip. Recent calculations [115–117] in momentum space, but including non-leading logarithmic contributions indicate that the impact parameter calculations are probably correct, i.e. the sum of the leading logarithms is not the leading term. Consider for example the diagram of fig. 31a and the region of phase space in which the transverse momentum of two or more of the gluons is large, but that of the lepton-pair is small. This clearly does not contribute to the leading logarithmic approximation (indeed it is suppressed by three powers of $\log s/Q_\perp^2$), but nevertheless proves to be important. Once the importance of this contribution is recognized, then it becomes clear that there is no dip at small $Q_\perp^2$.

### 9.2. *Probing into jets*

We have seen that often in calculating asymptotic predictions of QCD it is not sufficient to stop in any finite order of perturbation theory, but that one must sum contributions from all orders. Thus in many cases we expect a jet to consist not just of one parton, but a "ladder" of partons with typically a strong ordering of virtualities and transverse momenta.

*Following the discussion in section 7 we can estimate the renormalization effects more carefully in which case one of the logarithms effectively becomes a $\log\log Q^2/Q_\perp^2$.

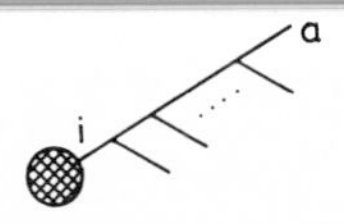


Fig. 49. Auxiliary diagram for the definition of the fragmentation function $D_{a,i}$.

Thus we expect jets to have a rich structure.* We will now illustrate this at the level of partons.

Define the fragmentation function $D_{a,i}(x, Y)$, for finding parton $a$ in parton $i$, with fraction $x$ of its momentum (fig. 49) via some hard scattering process with a typical momentum transfer of order $Q^2$:

$$\sigma^{-1}(\mathrm{d}\sigma/\mathrm{d}x)^{i\to aX} \equiv D_{a,i}(x, Y), \tag{9.3}$$

where

$$Y \equiv \frac{2}{\beta_0}\log\left[\frac{\alpha_s(\mu^2)}{\alpha_s(Q^2)}\right]. \tag{9.4}$$

Following the discussion in sections 5 and 6 we have for the moments of the fragmentation functions:

$$\int_0^1 \mathrm{d}x\, x^N D_{a,i}(x, Q^2) = (e^{A_N Y})_{a,i}, \tag{9.5}$$

where (up to a factor of $-4$) $A_N$ is the anomalous dimension matrix (5.30)**. We represent $D_{a,i}$ by

i a

We can generalize this to multiparticle distributions. For example for two

*For a detailed review see ref. [118].
**In fact $A_{N-1}$ is equal to $-\frac{1}{4}$ of the matrix (5.30).

partons:

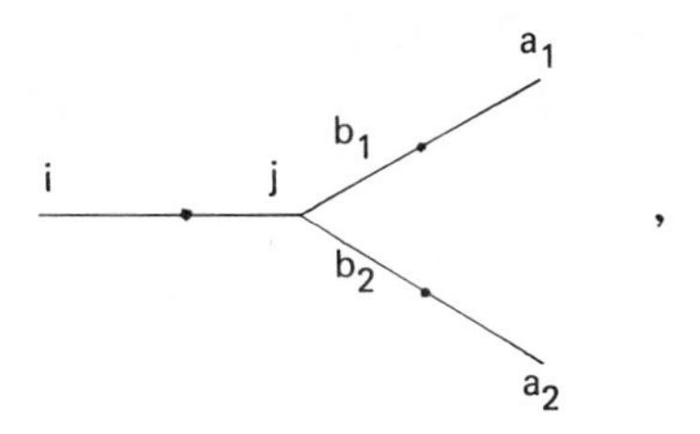


$$\frac{1}{\sigma}\left(\frac{\mathrm{d}\sigma}{\mathrm{d}x_1\,\mathrm{d}x_2}\right)^{i\to a_1a_2X} = \sum_{b_1,b_2,j}\int_0^Y \mathrm{d}y\int \mathrm{d}x\,\mathrm{d}z\,\mathrm{d}w_1\,\mathrm{d}w_2\,D_{a_1,b_1}(w_1,y)$$
$$\times D_{a_2,b_2}(w_2,y)P_{j\to b_1b_2}(z)D_{j,i}(x,Y-y)$$
$$\times\,\delta(x_1 - xzw_1)\delta(x_2 - x(1-z)w_2). \tag{9.6}$$

Equation (9.6) can be generalized to $n$ particle spectra.

The above discussion is not as academic as it might first seem. We can make predictions for measurable (at least in principle) quantities. For example, using

$$\mathrm{d}y = \frac{\mathrm{d}p_\perp^2}{p_\perp^2}\frac{\alpha_s(p_\perp^2)}{2\pi} \tag{9.7}$$

and

$$\sum_a \int \mathrm{d}w\, w D_{a,b}(w,y) = 1 \tag{9.8}$$

(which is the energy momentum sum rule, and is preserved in this leading approximation) we find:

$$\sum_{h_1,h_2} \int dx_1 dx_2 x_1 x_2 \frac{1}{\sigma} \left( \frac{d\sigma}{dx_1 dx_2 dp_\perp^2} \right)^{i \to h_1 h_2 X}$$

$$= \frac{\partial}{\partial p_\perp^2} \sum_b \left[ \frac{\alpha_s(p_\perp^2)}{\alpha_s(Q^2)} \right]^{2A_2/\beta_0}$$

$$= \frac{24}{p_\perp^2 \log 4p_\perp^2/\Lambda^2} \left( A\eta^{0.6086} + B\eta^{1.386} \right), \tag{9.9}$$

where $A = 2.955$ (0.920) and $B = -0.9548$ (3.680) for $i =$ quark (gluon) jet, and $\eta = (\log 2p_\perp/\Lambda)/\log Q/\Lambda$. There is some experimental evidence for eq. (9.9) [118].

### 9.3. *Multiplicity distributions*

At first sight it might seem that multiplicity distributions cannot sensibly be calculated in QCD since the average number of partons is always infinite because of the usual infra-red problem. However it is a tantalizing feature [119–121] that the average number of off-shell partons (with mass $> Q_0^2$) in a process with momentum transfer of order $Q^2$ and in the leading logarithm approximation is*

$$\langle n(Q^2, Q_0^2) \rangle \sim \exp 4(3/\beta_0)^{1/2}\left[ (\log Q^2/\Lambda^2)^{1/2} - (\log Q_0^2/\Lambda^2)^{1/2} \right]. \tag{9.10}$$

We note that all dependence on $Q_0^2$ factors out and therefore maybe eq. (9.10) can be taken as a prediction for the $Q^2$ behaviour of the multiplicity. If this is true then QCD predicts a rather fast increase of the multiplicity distribution with energy, considerably faster than the logarithmic increase we have been used to. However, recent data on the multiplicity distribution in $e^+e^-$ annihilation seems to agree with eq. (9.10). We should be cautious for now in accepting this prediction too easily. It is at present only a leading logarithm prediction for the parton multiplicity, it has as yet no renormalization group stamp of approval, nor is it clear if one can apply it to hadron multiplicities.

*Mueller [122] claims that there is an extra factor of $1/\sqrt{2}$ in front of the logarithms in eq. (9.10).

### 9.4. Summation (?) of leading singularities as $x \to 1$

We have seen that the QCD prediction for the $q^2$ behaviour of the non-singlet combination of structure functions can be written as

$$M^N(q^2) = \text{const.}\left[\alpha_s(q^2)\right]^{d_N}\left[1 + a_N \alpha_s(q^2)/4\pi + \cdots\right], \qquad (9.11)$$

where for large $N$

$$d_N \sim \log N \qquad (9.12)$$

and

$$a_N \sim (\log N)^2. \qquad (9.13)$$

Hence we expect that as $x \to 1$ these predictions become unreliable. There have been several attempts to sum the leading contributions in this region. One can get some intuitive feeling for the answer by noting that from the $\delta$-functions (6.16) and (6.17) (corresponding to the one-loop graph of fig. 22b):

$$k^2 = \beta p^2 - k_\perp^2/(1-x) \qquad (9.14)$$

and that keeping the strong ordering in the virtualities leads to an upper limit of $(1-x)q^2$ rather than $q^2$ for the last integral involving transverse momenta. The net result is that in the Altarelli–Parisi equations $\alpha_s(q^2)$ is replaced by $\alpha_s((1-x)q^2)$.

### 9.5. QCD enhancement factors in weak decays

It seems to be a fact of nature that for strangeness-changing weak decays (e.g. $K \to 2\pi$) the transitions in which the change of isospin $\Delta I$ is equal to 1/2 are enhanced by a factor of about 20 (in amplitude) compared with those in which the change of isospin $\Delta I$ is equal to 3/2. One mechanism which may be at least partially responsible for the surprisingly large difference in rate is the effect of strong interaction radiative corrections [123, 124]. Consider, e.g., the graph of fig. 50b in the Feynman gauge. This is a one-loop correction to the graph of fig. 50a, which is a typical Born graph for a weak interaction process. By inspection one can see that the one-loop graph has not only an extra factor of $g^2$, but also an extra power of $\log M_W^2/\mu^2$, where $\mu$ is a mass scale on the order of the quark masses. This is very reminiscent of the situation in deep inelastic scattering discussed in section 5, and, as there,

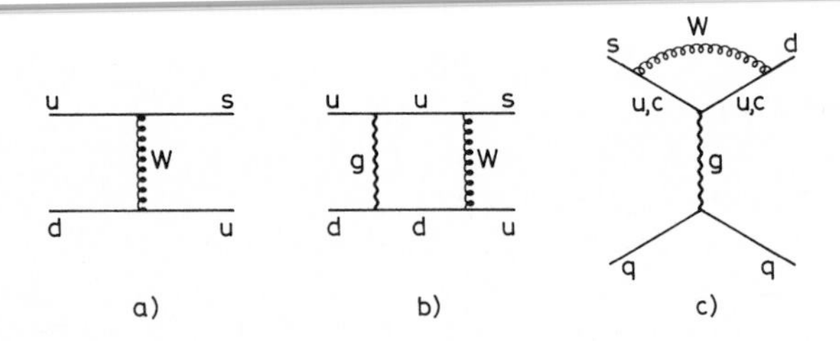


Fig. 50. Some low order graphs which contribute to strangeness-changing weak interactions.

it is possible to sum the leading QCD corrections to the weak processes using an operator product expansion.

The effective Hamiltonian for the strangeness-changing weak interactions is

$$H_{\text{eff}} = g^2 \int \mathrm{d}^4 x\, D_{\mathrm{F}}(x, M_{\mathrm{W}}^2) T\big(j_{\mu_{\mathrm{N}}}^{+}(x)\, j_{\mathrm{S}}^{\mu}(0) + \text{h.c.}\big), \tag{9.15}$$

where $g$ is the weak interaction coupling constant, $D_{\mathrm{F}}$ is the W-meson propagator and $j_{\mu_{\mathrm{N}}}$ and $j_{\mathrm{S}}^{\mu}$ are the strangeness-conserving and strangeness-changing currents. In Euclidean space as $M_{\mathrm{W}}^2$ gets large:

$$D_{\mathrm{F}}(x, M_{\mathrm{W}}^2) \sim \mathrm{e}^{-M_{\mathrm{W}} x}, \tag{9.16}$$

and (provided, as expected, that the Wick rotation can be performed) therefore the integral in eq. (9.15) is dominated by the region of small $x$. Hence we perform the short-distance expansion

$$T\big(j_{\mu_{\mathrm{N}}}^{+}(x)\, j_{\mathrm{S}}^{\mu}(0)\big) = \sum_m F_m(x^2) O_m(0), \tag{9.17}$$

where the $O_m$ are local operators and the dominant behaviour is given by the operators of smallest mass dimension. Since the currents are conserved, the left hand side of eq. (9.17) is independent of $\mu$ so that

$$\left[\mu \frac{\partial}{\partial \mu} + \beta(g) \frac{\partial}{\partial g}\right] \sum_m C_m(M_W^2, \mu^2) O_m = 0, \tag{9.18}$$

where $C_m$ is the integral of $F_m$ over $x$ weighted by the W-meson propagator. Let us assume that we have chosen a basis in which (at least up to the lowest order we are interested in) the relevant operators do not mix under renormalization. Then:

$$\left[\mu\frac{\partial}{\partial\mu}+\beta(g)\frac{\partial}{\partial g}+\gamma_{O^m}(g)\right]C_m\left(M_W^2,\mu^2\right) = 0, \tag{9.19}$$

where the anomalous dimension $\gamma_{O^m}$ includes the derivative of the appropriate wave-function renormalization constants. From eq. (9.19) we can deduce the $M_W^2$ behaviour of $C_m$. We now assume that using SU(6) wave functions or some other technique we can calculate the operator matrix elements renormalized at $\mu^2 \sim O(1\ \mathrm{GeV}^2)$.

The dominant local operators for this process are those of dimension 6, among them is the operator

$$O_1 = \bar{s}_L\gamma_\mu d_L\bar{u}_L\gamma_\mu u_L - \bar{s}_L\gamma_\mu u_L\bar{u}_L\gamma_\mu d_L, \tag{9.20}$$

which is seen to be a purely $\Delta I = 1/2$ operator because of the antisymmetry under $u_L \leftrightarrow d_L$. There are other operators of dimension 6, which in general are a mixture of $\Delta I = 1/2$ and $\Delta I = 3/2$. Thus we find that the effective Hamiltonian can be written as:

$$H_1^{\Delta S=-1} = \frac{G_F}{\sqrt{2}}\cos\theta_c\sin\theta_c\left(\sum C_i O_i + \text{h.c.}\right), \tag{9.21}$$

where

$$C_1 = \left[\alpha_s(\mu^2)/\alpha_s\left(M_W^2\right)\right]^{12/25},$$

$$C_{i\neq 1} = \left[\alpha_s(\mu^2)/\alpha_s\left(M_W^2\right)\right]^{-6/25}. \tag{9.22}$$

Thus if the operator matrix elements are comparable then the $\Delta I = 1/2$ transitions are enhanced by a factor of about 5 in the amplitude (compared to 20 experimentally). It is encouraging to see that the effect is to enhance the $\Delta I = 1/2$ decays relative to the $\Delta I = 3/2$ ones.

A further enhancement comes from the so-called penguin diagrams [125] (fig. 50c). It can be seen that these contribute only to the $\Delta I = 1/2$ processes, and that if $m_u = m_c$ then they would give a zero contribution. Hence they do not have a $\log M_W^2/\mu^2$ term but a $\log m_c^2/\mu^2$ one.

A similar feature occurs in proton decay calculations in Grand Unified Theories, where one is interested in weak, electromagnetic and strong corrections to diagrams such as those of fig. 51 [126–128]. Of course the operators and hence the anomalous dimensions are now different, and an added complication is that the currents are not conserved [129]. This current nonconservation is compensated for by the fact that the different couplings (e.g. $X$uu, $X\mathrm{e}^+\bar{\mathrm{d}}$, etc) are renormalized, and renormalized

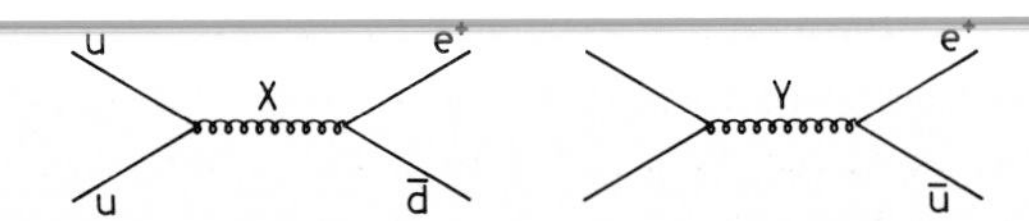


Fig. 51. Some lowest order graphs contributing to baryon-number violating processes in Grand Unified Theories.

differently. We are really interested in these couplings at low energy. They are not gauge invariant even in the leading logarithm approximation; however, as the $\beta$ functions of these couplings are very closely related to the anomalous dimension of the currents, we can write (in the leading logarithm approximation):

$$g_1(M_X) = g_1(\mu)\left[\alpha_s(M_X)/\alpha_s(\mu)\right]^{-\gamma^0_{J_1}/2\beta_0}, \tag{9.21}$$

where $\gamma_{J_1}$ is the anomalous dimension of one of the currents of interest and $g_1$ is the corresponding coupling constant. For simplicity only the strong corrections have been included. Thus we can write the answer in a form analogous to eq. (9.21) where the analogue of $G_F$ is proportional to $\alpha_{GUT} \equiv \alpha(M_X^2)$ [130].

## 10. Conclusions

We have seen that perturbative QCD can be applied to a wide variety of interesting processes. The property of asymptotic freedom means that for these processes we can expand in a power series in $\alpha_s(Q^2)$, which is a small parameter provided that the "appropriate" scale for the process, $Q^2$, is sufficiently large. We also have to ensure that there are no "large logarithms" of the type $\log Q^2/m^2$ (where $Q^2 \gg m^2$) in the coefficients of this expansion. We would now like to highlight some of the points already made in the lectures.

(*a*) *Deep inelastic lepton–hadron scattering and the ratio R in* $e^+e^-$ *annihilation.* The classic QCD predictions, such as those for the ratio $R$ in $e^+e^-$ annihilation and for the scaling violations in deep inelastic lepton–hadron scattering are now well established and, as will be seen in other lectures and seminars at this school, agree at least qualitatively and to a large extent quantitatively with the data. The outstanding theoretical problem here is to determine the size of the "higher twist" contributions.

(*b*) *Jet physics in* $e^+e^-$ *annihilation.* Many theoretical predictions concerning jets in $e^+e^-$ annihilation can be made, and again seem to agree (at least qualitatively) with the data, which certainly require an extension of the two jet picture. Higher order corrections frequently turn out to be large, however we have some "educated guesses" as to the summation of these large terms.

(*c*) *Massive lepton-pair production and related processes.* For hard scattering processes in which there are two incoming hadrons, such as the Drell–Yan process, there are still certain unsolved problems. Although the "collinear singularities" seem to factorize, we still lack a full understanding of the "Glauber singularities" and "infra-red singularities". This is one of the fundamental theoretical problems in perturbative QCD. In addition we have the problem of large higher order corrections already mentioned under (*b*).

(*d*) *Exclusive processes.* The "constituent interchange model" predictions are recovered (modulo logarithmic corrections) in perturbative QCD, providing the generalized Sudakov form factor indeed suppresses the "end point contributions". Considerable progress has recently been made in understanding the significance of the Sudakov form factor, however it is still difficult to carry out detailed phenomenological studies of exclusive processes in QCD.

We are also beginning to make speculative guesses about a number of processes and quantities for which there are as yet no renormalization group justifications at all. It would be very exciting to put some (or all) of these on a firmer theoretical footing.

## References

[1] H.D. Politzer, Phys. Rev. Lett. 30 (1973) 1346.
[2] D.J. Gross and F.A. Wilczek, Phys. Rev. Lett. 30 (1973) 1343.
[3] O.W. Greenberg, Phys. Rev. Lett. 13 (1964) 598.
[4] M. Han and Y. Nambu, Phys. Rev. 139 (1965) 1006.
[5] S.L. Adler, Phys. Rev. 177 (1969) 2426.
[6] J.S. Bell and R. Jackiw, Nuovo Cimento 60A (1969) 47.
[7] D.J. Gross and R. Jackiw, Phys. Rev. D6 (1972) 477.
C.P. Korthals Altes and M. Perrottet, Phys. Lett. 39B (1972) 546.
[8] S.L. Adler and W.A. Bardeen, Phys. Rev. 182 (1969) 1517.
[9] Particle Data Group, Review of Particle Properties, Rev. Mod. Phys. 52 (1980) S1.
[10] E. de Rafael, Quantum Chromodynamics as a Theoretical Framework of the Hadronic Interaction, in: Quantum Chromodynamics, Jaca, Huesca, Spain (1979), eds. J.L. Alonso and R. Tarrach (Springer, Berlin & Heidelberg, 1980) p. 1.
[11] L. Susskind, Coarse Grained Quantum Chromodynamics in: Weak and Electromagnetic Interactions at High Energy, Les Houches (1976), session 29, eds. R. Balian and C.H. Llewellyn Smith (North-Holland, Amsterdam, 1977) p. 209.

[12] B.H. Wiik, Desy preprint DESY 80/129 (1980).
R. Felst, Proc. 1981 Int. Symp. on Lepton and Photon Interactions at High Energies, Bonn (1981), ed. W. Pfeil (Bonn Univ. Phys. Inst., 1981).
[13] S. Coleman and D. Gross, Phys. Rev. Lett. 31 (1973) 1343.
[14] J.C. Taylor, Gauge Theories of Weak Interactions (Cambridge Univ. Press, 1976) and references therein.
[15] C. Itzykson and J.B. Zuber, Quantum Field Theory (McGraw Hill, New York, 1980).
[16] A.A. Slavnov and L.D. Faddeev, Gauge Fields—Introduction to Quantum Theory (Benjamin/Cummins, Reading, MA, 1980).
[17] J.C. Taylor, Nucl. Phys. B33 (1971) 436.
[18] A.A. Slavnov, Sov. J. Part. Nucl. 5 (1975) 303.
[19] P. Cvitanovic, Phys. Rev. D14 (1976) 1536.
[20] J.E. Mandula, Diagrammatic Techniques in Group Theory, Southampton Univ. preprint SHEP 80/81-7 (1981).
[21] D.J. Gross, in: Methods in Field Theory, Les Houches (1975), session 28, eds. R. Balian and J. Zinn-Justin (North-Holland, Amsterdam, 1976).
[22] G.'t Hooft and M. Veltman, Nucl. Phys. B44 (1972) 189.
[23] W.A. Bardeen, A.J. Buras, D.W. Duke and T. Muta, Phys. Rev. D18 (1978) 3998.
[24] W. Celmaster and R.J. Gonsalves, Phys. Rev. Lett. 42 (1979) 1435.
[25] G. Altarelli, G. Curci, G. Martinelli and S. Petrarcha, Nucl. Phys. B187 (1981) 461.
[26] W. Siegel, Phys. Lett. 84B (1979) 193.
[27] J. Ellis and C.T. Sachrajda, Quantum Chromodynamics and its Application, in: Quarks and Leptons, Cargèse (1979), eds. J.L. Basdevant, D. Speiser, J. Weyers, R. Gastmans, M. Jacob and M. Lévy (Plenum, New York, 1980).
[28] W. Caswell, Phys. Rev. Lett. 33 (1974) 224.
[29] D.R.T. Jones, Nucl. Phys. B75 (1974) 531.
[30] O.V. Tarasov, A.A. Vladimirov and A.Yu. Zharkov, Phys. Lett. 93B (1980) 429.
[31] E.C.G. Stueckelberg and A. Peterman, Helv. Phys. Acta 26 (1953) 499.
[32] M. Gell-Mann and F.E. Low, Phys. Rev. 95 (1954) 1300.
[33] E. Poggio, H. Quinn and S. Weinberg, Phys. Rev. D13 (1976) 1958.
[34] M. Dine and J. Sapirstein, Phys. Rev. Lett. 43 (1979) 668.
K.G. Chetyrkin, A.L. Kataev and F.V. Trackov, Phys. Lett. 85B (1979) 277.
[35] A.J. Buras, Phys. Scripta 23 (1981) 863.
[36] G. Hanson et al., Phys. Rev. Lett. 35 (1975) 1609.
[37] F. Bloch and A. Nordsieck, Phys. Rev. 52 (1937) 54.
[38] D.R. Yennie, S.C. Frautschi and H. Suura, Ann. Phys. 13 (1961) 379.
[39] T. Kinoshita, J. Math. Phys. 3 (1962) 650.
[40] T.D. Lee and M. Naunberg, Phys. Rev. 133 (1964) 1549.
[41] G. Sterman and S. Weinberg, Phys. Rev. Lett. 39 (1977) 1436.
[42] M.B. Einhorn and B.J. Weeks, Nucl. Phys. B146 (1978) 445.
K. Shizuya and S.H. Tye, Phys. Rev. Lett. 41 (1978) 787.
[43] A. De Rújula, J. Ellis, E.G. Floratos and M.K. Gaillard, Nucl. Phys. B138 (1978) 387.
[44] B. Wiik, course 8, this volume.
[45] E. Farhi, Phys. Rev. Lett. 39 (1977) 1587.
[46] Proc. 1981 Int. Symp. on Lepton and Photon Interactions at High Energies, Bonn (1981), to be published.
[47] R.K. Ellis, D.A. Ross and A.E. Terrano, Phys. Rev. Lett. 45 (1980) 1226; Nucl. Phys. B178 (1981) 421.
J.A.M. Vermaseren, K.J.F. Gaemers and S.J. Oldham, Nucl. Phys. B187 (1981) 301.
Z. Kunszt, Phys. Lett. 99B (1981) 429; Phys. Lett. 107B (1981) 123.

[48] K. Fabricius, I. Schmitt, G. Schierholz and G. Kramer, Phys. Lett. 97B (1980) 431.
[49] T. Applequist, J. Carazzone, H. Kluberg-Stern and M. Roth, Phys. Rev. Lett. 36 (1976) 761.
[50] S.V. Libby and G. Sterman, Phys. Rev. D19 (1979) 2468.
[51] R. Doria, J. Frenkel and J.C. Taylor, Nucl. Phys. B168 (1980) 93.
[52] C. Di'Lieto, S. Gendron, I.G. Halliday and C.T. Sachrajda, Nucl. Phys. B183 (1981) 223.
[53] A. Andrasi, M. Day, R. Doria, J. Frenkel and J.C. Taylor, Nucl. Phys. B182 (1981) 104.
[54] H.D. Politzer, Phys. Rep. 14C (1974) 129.
[55] A.J. Buras, Rev. Mod. Phys. 52 (1980) 199.
[56] A. Peterman, Phys. Rep. 53C (1979) 157.
[57] E. Reya, Phys. Rep. 69C (1981) 195.
[58] G. Altarelli, Phys. Rep. 81C (1982) 1.
[59] J. Drees, Proc. 1981 Int. Symp. on Lepton and Photon Interactions at High Energies, Bonn (1981), ed. W. Pfeil (Bonn Univ. Phys. Inst., 1981).
[60] A. Para, seminar related to this course.
[61] J.J. Aubert, seminar related to this course.
[62] G. Altarelli and G. Parisi, Nucl. Phys. B126 (1977) 298.
[63] R.L. Jaffe and M. Soldate, Higher Twist in Electroproduction: A systematic QCD Analysis, in: Perturbative Quantum Chromodynamics, Tallahassee (1981), eds. D.W. Duke and J.F. Owens (American Institute of Physics, New York, 1981) p. 60.
S.P. Luttrell, S. Wada and B.R. Webber, Nucl. Phys. B188 (1981) 219.
[64] S. Gottlieb, Nucl. Phys. B139 (1978) 125.
M. Okawa, Nucl. Phys. B172 (1980) 481; B187 (1981) 71.
[65] O. Nachtmann, Nucl. Phys. B63 (1973) 237.
[66] S. Wandzura, Nucl. Phys. B112 (1977) 412.
[67] H. Georgi and H.D. Politzer, Phys. Rev. D9 (1974) 416.
[68] R. Barbieri, J. Ellis, M.K. Gaillard and G.G. Ross, Phys. Lett. B64 (1976) 171; Nucl. Phys. B117 (1976) 50.
[69] Yu.L. Dokshitzer, D.I. Byakonov and S.I. Troyan, Phys. Rep. 58C (1980) 269.
[70] C.H. Llewellyn Smith, Acta Phys. Austr., Suppl. 19 (1978) 331.
V.N. Gribov and I.N. Lipatov, Sov. J. Nucl. Phys. 15 (1972) 438.
[71] E. Witten, Nucl. Phys. B120 (1977) 89.
[72] C.H. Llewellyn Smith, Phys. Lett. 79B (1978) 83.
[73] W.R. Frazer and J.F. Gunion, Phys. Rev. D20 (1979) 147.
[74] S.D. Drell and T.M. Yan, Phys. Rev. Lett. 25 (1970) 316; Ann. Phys. 66 (1971) 578.
[75] H.D. Politzer, Nucl. Phys. B129 (1977) 301.
[76] R.K. Ellis, H. Georgi, M. Machacek, H.D. Politzer and G.G. Ross, Phys. Lett. 78B (1978) 281; Nucl. Phys. B152 (1979) 285.
[77] G. Sterman and S. Libby, Phys. Rev. D18 (1978) 3252, 4737.
[78] C.T. Sachrajda, Phys. Lett. 73B (1978) 185.
[79] C.T. Sachrajda, Phys. Lett. 76B (1979) 100.
[80] V.N. Gribov and I.N. Lipatov, Sov. J. Nucl. Phys. 15 (1972) 675.
[81] C.T. Sachrajda, Physica Scripta 19 (1979) 85.
[82] A.J. Buras, NORDITA preprint 81/43 (1981); Proc. 1981 Int. Symp. on Lepton and Photon Interactions at High Energies, Bonn (1981), ed. W. Pfeil (Bonn Univ. Phys. Inst., 1981).
[83] E.G. Floratos, D.A. Ross and C.T. Sachrajda, Phys. Lett. 80B (1979) 1269.
[84] D.A. Ross and C.T. Sachrajda, Nucl. Phys. B149 (1979) 497.

[85] V. Sudakov, Sov. Phys. JETP 3 (1956) 65.
[86] A.H. Mueller, Phys. Rev. D20 (1979) 2037.
[87] A. Sen, Stony Brook preprint ITP-SB-81-19 (1981).
[88] J.C. Collins and D.E. Soper, Non-Leading Sudakov Effects are Computable in QCD, in: Perturbative Quantum Chromodynamics, Tallahassee (1981), eds. D.W. Duke and J.F. Owens (American Institute of Physics, New York, 1981) p. 41.
[89] G. Altarelli, R.K. Ellis and G. Martinelli, Nucl. Phys. B143 (1978) 521 (Erratum B146 (1978) 544); B157 (1979) 461.
[90] J. Kubar-André and F.E. Paige, Phys. Rev. D19 (1979) 221.
[91] B. Humpert and W.L. Van Neerven, Phys. Lett. 84B (1979) 327; 85B (1979) 293.
[92] C.T. Sachrajda and S. Yankielowicz, unpublished.
[93] N. Nakanishi, Graph Theory and Feynman Integrals (Gordon and Breach, New York, 1971).
[94] I.G. Halliday, J. Huskins and C.T. Sachrajda, Nucl. Phys. B87 (1975) 176.
[95] C.E. DeTar, S.D. Ellis and P.V. Landshoff, Nucl. Phys. B87 (1975) 176.
[95] C.E. DeTar, S.D. Ellis and P.V. Landshoff, Nucl. Phys. B87 (1975) 176.
[96] A.H. Mueller, Phys. Rev. D2 (1970) 2963.
[97] G.T. Bodwin, S.J. Brodsky and G.P. Lepage, SLAC preprint SLAC-PUB-2787 (1981).
[98] A.H. Mueller, Columbia Univ. preprint CU-TP-213 (1981).
[99] S.J. Brodsky and G.P. Lepage, Exclusive Processes in Quantum Chromodynamics, in: Perturbative Quantum Chromodynamics, Tallahassee (1981), eds. D.W. Duke and J.F. Owens (American Institute of Physics, New York, 1981) p. 214.
[100] S.J. Brodsky and G.R. Farrar, Phys. Rev. Lett. 31 (1973) 1153; Phys. Rev. D11 (1975) 1309.
[101] V.A. Matveev, R.M. Muradyan and A.V. Tavkhelidze, Lett. Nuovo Cimento 7 (1973) 719.
[102] P.V. Landshoff, Phys. Rev. D10 1024 (1974).
[103] A. Donnachie and P.V. Landshoff, Z. Phys. C2 55 (1979).
[104] S.J. Brodsky and G.P. Lepage, Phys. Lett. 87B (1979) 359; Phys. Rev. D22 (1980) 2157.
[105] A.V. Efremov and A.V. Radyushkin, Riv. Nuovo Cimento 3 (1980) 1; Phys. Lett. 94B (1980) 245.
[106] A. Duncan and A. Mueller, Phys. Rev. D21 (1980) 1636; Phys. Lett. 98B (1980) 159.
[107] G.R. Farrar and D.R. Jackson, Phys. Rev. Lett. 43 (1979) 246.
[108] G. Parisi, Phys. Lett. 84B (1979) 225.
[109] S.J. Brodsky, Y. Frishman, G.P. Lepage and C.T. Sachrajda, Phys. Lett. 91B (1980) 239.
[110] K. Kawarabayashi and M. Suzuki, Phys. Rev. Lett. 16 255 (1966).
Fayyazuddin and Riazuddin, Phys. Rev. 147 (1966) 1071.
[111] S.J. Brodsky and G.P. Lepage, Phys. Lett. 87B (1979) 359, see also [104].
[112] J. Cornwall and G. Tiktopoulos, Phys. Rev. D13 (1976) 3370; Phys. Rev. D15 (1977) 2937.
[113] S.D. Ellis and W.J. Stirling, Phys. Rev. D23 (1981) 214.
[114] G. Parisi and R. Petronzio, Nucl. Phys. B154 (1979) 427.
[115] S.D. Ellis, N. Fleishon and W.J. Stirling, Univ. of Washington preprint RLO-1388-852 (1981), to be published in Phys. Rev. D.
[116] P.E.L. Rakow and B.R. Webber, Nucl. Phys. B187 (1981) 254.
[117] H.F. Jones and J. Wyndham, Nucl. Phys. B176 (1980) 466.

[118] K. Konishi, CERN preprint TH 2853-CERN (1980).
[119] D. Amati, A. Bassetto, M. Ciafaloni, G. Marchesini and G. Veneziano, Nucl. Phys. B173 (1980) 429.
[120] A. Bassetto, M. Ciafaloni and G. Marchesini, Nucl. Phys. B163 (1980) 477.
[121] W. Furmanski, R. Petranzio and S. Pokorski, Nucl. Phys. B155 (1979) 253.
[122] A.H. Mueller, Columbia Univ. preprint CU-TP-197 (1981).
[123] M.K. Gaillard and B.W. Lee, Phys. Rev. Lett. 33 (1974) 108.
[124] G. Altarelli and L. Maiani, Phys. Lett. 52B (1974) 351.
[125] M.A. Shifman, A.I. Vainshtein and V.I. Zakharov, Nucl. Phys. B120 (1977) 316.
[126] A.J. Buras, J. Ellis, M.K. Gaillard and D.V. Nanópoulos, Nucl. Phys. B135 (1978) 66.
[127] J. Ellis, M.K. Gaillard and D.V. Nanopoulos, Phys. Lett. 88B (1979) 320.
[128] F. Wilczek and A. Zee, Phys. Rev. Lett. 43 (1979) 1571.
[129] M. Daniel and J.A. Penarrocha, Southampton Univ. preprint 80/81-5 (1981).
[130] M. Daniel, J.A. Peñarrocha and C.T. Sachrajda, unpublished.
[131] W.W. Lindsay, D.A. Ross and C.T. Sachrajda, Phys. Lett. 117B (1982) 105; Southampton Univ. preprints SHEP 81/82-6 (1982) and SHEP 82/83-4 (1983), both to be published in Nucl. Phys. B.
[132] P.V. Landshoff and W.J. Stirling, Z. Phys. C14 (1982) 251

**Seminars related to Course 2**

1. Deep inelastic muon–electron scattering, by Jean-Jacques Aubert (LAPP).

2. The infrared problem in QCD, by Christine di Lieto (SLAC).

3. Deep inelastic neutrino scattering, by Adam Para (CERN).

COURSE 3

# PHENOMENOLOGY OF UNIFIED GAUGE THEORIES

John ELLIS

*CERN, Geneva, Switzerland*

*M.K. Gaillard and R. Stora, eds.*
*Les Houches, Session XXXVII, 1981*
*Théories de jauge en physique des hautes énergies / Gauge theories in high energy physics*

## Contents

## Updates in proof

In view of the time that has elapsed since this course was written, the reader may find useful the following updates on the material presented.

*Lecture 1:* There have been considerable amounts of new data on neutral currents, particularly in $e^+e^-$ annihilation [345], all consistent with the Glashow–Salam–Weinberg model.

*Lecture 2:* For a recent review of the phenomenology of the Kobayashi–Maskawa angles, see ref. [346]. The Buras bound on $m_t$ discussed in the text is strengthened by a new measurement of $\eta \to \mu^+\mu^-$ [347] and an improved determination of the $\Delta S = 2$ operator matrix element [348]. It has recently been pointed out that the sign of $\varepsilon'/\varepsilon$ is fixed in the GSW model, and a lower bound of $2 \times 10^{-3}$ has been given for its magnitude [349].

*Lecture 3:* The $W^{\pm}$ have been discovered [350] with masses O(80) GeV. For a recent analysis of neutrino counting in $K^{\pm} \to \pi^{\pm}\bar{\nu}\nu$ decay, see ref. [351].

*Lecture 4:* For a latest attempt at technicolour model-building, see ref. [352].

*Lecture 5:* Many new limits have been published on charged boson production in $e^+e^-$ annihilation. See [345], [353] and references therein.

*Lecture 6:* The topic of Grand Unified Monopoles was neglected in these lectures, and has recently attracted much interest, particularly in connection with the possible catalysis of baryon decay. See the original refs. [354] and [355] for a review.

*Lecture 7:* There has recently been a stringent lower limit on the $p \to e^+\pi^0$ lifetime of about $6 \times 10^{31}$ yr [356] and reports [357] of candidates for $p \to \mu^+ K^0$. The favoured value of $\Lambda_{\overline{MS}}$ no longer appears necessarily to lie in the range used in this lecture, as understanding of theoretical uncertainties in lattice calculations [358] and in data analysis [359] has improved.

*Lecture 8:* For a recent review of upper limits on neutrino oscillation parameters, see ref. [360].

*Lecture 9:* Interest in supersymmetric GUTs and in the phenomenology of supersymmetric particles has increased dramatically. Much interest has centered on theories with spontaneously broken supersymmetry which naturally respect phenomenological constraints [361] on flavour-changing neutral interactions. Satisfactory models with global supersymmetry alone are difficult to construct, and interest currently centers on locally supersymmetric models [362] exploiting the super-Higgs effect [363].

*Lecture 10:* There have recently been negative results [364] on the appearance of dynamical gauge bosons in non-compact analogues of non-linear $\sigma$ models. For an update on the philosophy of this chapter and for a discussion of recent literature, see ref. [365].

## 0. General overview

A schematic picture of the topics of these lectures is shown in fig. 0.1. My brief is to discuss the phenomenological aspects of gauge theories unifying the different fundamental interactions. First we will remind ourselves of the structure of familiar gauge theories including some of the latest analyses of their consequences, and then go on to present some of the more recent and untested ideas going beyond the standard model. In this way two different cardinal aspects of our scientific work may be illustrated. On the one hand there is detailed analysis in the context of a well-established theory, looking for small discrepancies which may illuminate a way forward, and on the other hand there is the speculative leap itself. In either phase, the ultimate arbiter is experiment, and we must be careful not to speculate without a thought for experimental accountability, nor to lose our sense of theoretical direction while threading a maze of detailed experimentation. At the moment the necessary balance between theory and experiment is threatened by the recent

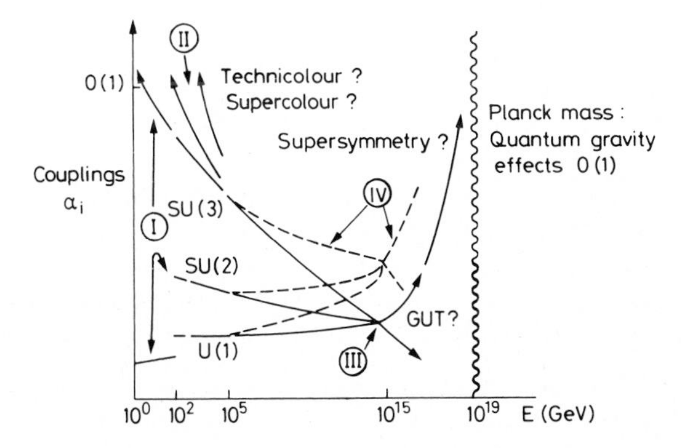


Fig. 0.1. Schematic view of the possible variations of the different gauge coupling constants as functions of Energy. The Roman numerals I, II, III and IV indicate which parts of these lectures treat which interactions.

achievement of consensus on the standard model of strong, weak and electromagnetic interactions. This results in a lack of focus while unbridled theorists develop contradictory ideas with little regard for their experimental testability, and our experimental colleagues push on with elaborate experiments that do not bear upon hot theoretical issues. A major effort will be made in these lectures to maintain as long as possible the essential dialogue between theory and experiment.

Part I of these lectures treats the standard Glashow–Weinberg–Salam model [1] of weak and electromagnetic interactions, discussing in turn its basic structure and weak neutral currents, charged currents, mixing angles and *CP* violation, and the phenomenology of weak vector and Higgs bosons. Part II of the lectures discusses the structure of theories of dynamical symmetry breaking [2] such as technicolour, phenomenological consequences, frustrations and alternatives. The third part of these lectures offers the standard menu of grand unified theories (GUTs) [3] of the strong, weak and electromagnetic interactions, including an hors d'oeuvre of constraints on the parameters of the standard model, a main course of baryon number violating processes, and desserts which violate lepton number and *CP*. The fourth and final part goes through different attempts [4] to remedy the inadequacies of previous theories by invoking supersymmetry and reaching out towards gravitation. Figure 0.1 indicates how these different theoretical lines fit together in the domain of coupling constants as functions of energy. It remains to be seen how many of them will be confirmed by experiment.

# I. The Standard $SU(2) \times U(1)$ Model

## 1. Structure and neutral currents

### 1.1. How to make a model and win a prize

Nowadays the name of the theoretical game is to build, work out and test gauge theories. You can play a subsidiary rôle in this game by working on somebody else's theory, or you can make up your own theory. To see how the latter is done, this lecture starts with instructions for a do-it-yourself gauge theory [5] construction kit, whose steps are illustrated by the prize-winning Glashow–Weinberg–Salam model [1]. Later on we will see how other hopeful players fare.

(a) Choose a group. The Standard Model choice was SU(2)×U(1), pioneered by Glashow.

(b) Choose representations for fermions. Since gauge interactions conserve helicity, we can and should make independent choices for left-handed and right-handed fermions $f_{L,R} = \frac{1}{2}(1 \mp \gamma_5)f$. The Standard Model choice [1] was to place $f_L$ in doublets of SU(2), with the $f_R$ in SU(2) singlets. The U(1) hypercharges were then chosen so as to fit the observed electromagnetic charges.

(c) Choose Higgs representations, or else construct a realistic model of dynamical symmetry breaking [2], which no one has yet succeeded in doing. The Standard Model choice was a single complex SU(2) doublet $\phi \equiv \begin{pmatrix} \phi^+ \\ \phi^0 \end{pmatrix}$ of elementary Higgs fields, together with the antiparticles $\phi^\dagger \equiv \begin{pmatrix} \bar{\phi}^0 \\ -\phi^- \end{pmatrix}$.

(d) Determine the pattern of symmetry breaking. This involves choosing a potential $V(\phi)$ for elementary Higgses which when minimized leaves intact the observed exact gauge symmetries: SU(2)×U(1) → U(1)$_{em}$ in the Standard Model [1]. More work is needed in theories of dynamical symmetry breaking [2] to determine the form of the effective potential and the pattern of vacuum alignment.

The final steps are then

(e) Publish and await experimental confirmation. If this is forthcoming, go to (f). If not, go direct to (g).

(f) Go to Stockholm, and if you are ambitious

(g) Return to (a).

We will now see in more detail how this programme is carried out in the Standard SU(2)×U(1) Model [1]. The pure gauge boson part of the Lagrangian is just

$$\mathcal{L}_G = -\tfrac{1}{4}\left(G_{\mu\nu a}G^{\mu\nu}{}_a\right) - \tfrac{1}{4}\left(F_{\mu\nu}F^{\mu\nu}\right), \tag{1.1}$$

where

$$G_{\mu\nu a} \equiv \partial_\mu W_{\nu a} - \partial_\nu W_{\mu a} + g\varepsilon_{abc}W_{\mu b}W_{\nu c} \tag{1.2a}$$

and

$$F_{\mu\nu} \equiv \partial_\mu B_\nu - \partial_\nu B_\mu, \tag{1.2b}$$

respectively, are the field strength tensors of the SU(2) gauge fields $W_{\mu a}$ and the U(1) gauge field $B_\mu$. When inserted into (1.1) the bilinear terms in $G_{\mu\nu a}$ (1.2a) generate the trilinear and quadrilinear boson interactions characteristic of non-Abelian gauge theories. In addition to (1.1), one should in principle allow [6] for a possible $P$- and $CP$-violating SU(2) gauge term

$$\delta \mathcal{L} = + \frac{g^2 \theta_2}{32\pi^2} G_{\mu\nu a} \tilde{G}^{\mu\nu}{}_{a}, \tag{1.3}$$

where

$$\tilde{G}_{\mu\nu a} \equiv \tfrac{1}{2} \varepsilon_{\mu\nu\alpha\beta} G^{\alpha\beta}{}_{a}.$$

Because it is a total divergence, effects of $\delta \mathcal{L}$ (1.3) only show up when non-perturbative phenomena are involved. These are negligible for weak SU(2) (see however lecture 7) but the analogous term for strong SU(3) QCD is potentially significant (see lectures 2, 3 and 8).

The fermion kinetic terms in the Standard Model Lagrangian take the form

$$\mathcal{L}_{\mathrm{f}} = -\sum_{\mathrm{f}} \Big[ \bar{f}_{\mathrm{L}} \gamma^{\mu} \Big( \partial_{\mu} + \mathrm{i} g \frac{\tau_a}{2} W_{\mu a} + \mathrm{i} g' Y_{\mathrm{L}} B_{\mu} \Big) f_{\mathrm{L}} + \bar{f}_{\mathrm{R}} \gamma^{\mu} \big( \partial_{\mu} + \mathrm{i} g' Y_{\mathrm{R}} B_{\mu} \big) f_{\mathrm{R}} \Big], \tag{1.4}$$

where the quantities in round brackets are just the covariant derivatives of the fermion fields which yield the trilinear fermion–fermion gauge boson vertices. Note that there is no $W_{\mu a}$ interaction for the $\mathrm{f_R}$: this reflects the choice (b) above of singlet SU(2) representations for right-handed fermion fields. The overall scale of the Abelian U(1) hypercharge coupling $g'$ and the individual fermion hypercharges $Y_{\mathrm{L,R}}$ in (1.4) are a priori arbitrary: they will be fixed later so as to get the right electromagnetic charges for the fermions.

The kinetic term for the Higgs fields (assuming they exist, see however [2] lectures 4 and 5) takes the form:

$$\mathcal{L}_{\phi} = -\left| \Big( \partial_{\mu} + \mathrm{i} g \frac{\tau_a}{2} W_{\mu a} + \mathrm{i} \frac{g'}{2} B_{\mu} \Big) \phi \right|^2. \tag{1.5}$$

Notice again the $W_{\mu a}$ interaction term reflecting the choice (c) of an SU(2) doublet of Higgs fields. When $\langle 0|\phi|0\rangle \neq 0$ this interaction will give $W_{\mu} - W^{\mu}$ terms and a direct $W-\phi$ coupling, thereby giving masses to the $W$ bosons. Notice that in (1.5) we have fixed the overall scale of the U(1) coupling $g'$ and the different hypercharges by specifying $Y_{\phi} = \frac{1}{2}$.

The fermions and the Higgs fields interact through generalized Yukawa couplings

$$\mathcal{L}_{\mathrm{Y}} = -\sum_{\mathrm{f,f'}} \left[ H_{\mathrm{ff'}}(\bar{f}_{\mathrm{L}} \cdot \phi) f'_{\mathrm{R}} + H^{*}_{\mathrm{ff'}} \bar{f}'_{\mathrm{R}} (\phi^{\dagger} \cdot f_{\mathrm{L}}) \right], \tag{1.6}$$

where $H_{\mathrm{ff'}}$ is a general coupling matrix in the space of different fermion species (generations or families), which will give rise to the fermion masses as soon as $\langle 0|\phi|0\rangle \neq 0$. The couplings (1.6) will also give the generalized Cabibbo mixing angles of the Kobayashi–Maskawa [7] matrix, as we will see in section 1.3. At this point we appreciate one of the reasons for choosing SU(2) doublet Higgs fields: given the choices (b) of fermion representations, we need $I = 1/2$ Higgs to give non-zero fermion masses.

Finally we come to the Higgs self-interactions, i.e. the potential

$$\mathcal{L}_{V} = -V(\phi), \quad V(\phi) = -\mu^2(\phi^{\dagger}\phi) + \tfrac{1}{2}\lambda(\phi^{\dagger}\phi)^2, \tag{1.7}$$

which is the most general SU(2)-invariant and renormalizable (dimension $\leqslant 4$) form. In order to have a potential bounded below we must have $\lambda > 0$ and we also need $\mu^2 > 0$ if we want to get spontaneous symmetry breaking as illustrated in fig. 1.1. When $\langle 0|\phi|0\rangle \neq 0$ the potential (1.7) will give trilinear Higgs self-interactions as well as the physical Higgs mass and quadrilinear couplings.

### *1.2. Analysis of the Standard Model Lagrangian*

The full Lagrangian

$$\mathcal{L}_{\mathrm{WS}} = \mathcal{L}_{\mathrm{G}} + \delta\mathcal{L} + \mathcal{L}_{\mathrm{f}} + \mathcal{L}_{\phi} + \mathcal{L}_{\mathrm{Y}} + \mathcal{L}_{V} \tag{1.8}$$

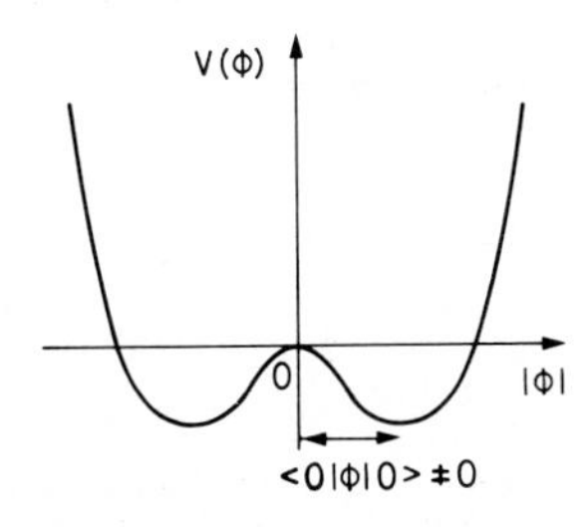


Fig. 1.1. The Higgs potential for $\mu^2$ and $\lambda > 0$, showing how spontaneous symmetry breaking occurs.

is the most general SU(2)-invariant and renormalizable one we can construct given the set of fields we chose. To analyze its content we first minimize the Higgs potential $V(\phi) = -\mathcal{L}_V$. As a convention we can choose to write the Higgs vacuum expectation value of fig. 1.1 in the form

$$\langle 0|\phi|0\rangle = \langle 0|(\phi^\dagger)^{\mathrm{T}}|0\rangle = \frac{v}{\sqrt{2}}\begin{pmatrix}0\\1\end{pmatrix}, \tag{1.9}$$

with $v$ determined by the potential parameters:

$$v = \sqrt{2\mu^2/\lambda}\,. \tag{1.10}$$

If we now substitute

$$\phi = \langle 0|\phi|0\rangle + \hat{\phi} \tag{1.11}$$

into the Higgs kinetic term $\mathcal{L}_\phi$ we get quadratic terms

$$\tfrac{1}{4}\left( -g^2\frac{v^2}{2} W_{\mu a}W^{\mu}{}_{a} + gg'v^2 B_\mu W^{\mu}{}_{3} - g'^2\frac{v^2}{2} B_\mu B^\mu \right), \tag{1.12}$$

and $W_\mu - \partial^\mu\hat{\phi}$ mixing terms such as

$$W_\mu^{\pm}\,\partial^\mu\phi^{\mp}\,(gv/2), \tag{1.13}$$

where

$$W_\mu^{\pm} \equiv (W_{\mu 1} \pm \mathrm{i}W_{\mu 2})/\sqrt{2}\,.$$

The quadratic terms (1.12) give masses to the $W^\pm$ and a combination of neutral vector bosons. This can also be seen from the mixing terms (1.13) which can be iterated as seen diagrammatically in fig. 1.2 to give a longitudinal piece to the $W^\pm$ propagator of the form

$$\begin{aligned}&\frac{1}{q^2} + \frac{1}{q^2}(\mathrm{i}q_\mu gv/2)\frac{1}{q^2}(-\mathrm{i}q_\mu gv/2)\frac{1}{q^2} + \left(\frac{1}{q^2}\right)^5 (q^2g^2v^2/4)^2 + \cdots \\ &= \frac{1}{q^2(1-g^2v^2/4q^2)} = \frac{1}{q^2 - m_{\mathrm{W}^\pm}^2},\end{aligned} \tag{1.14}$$

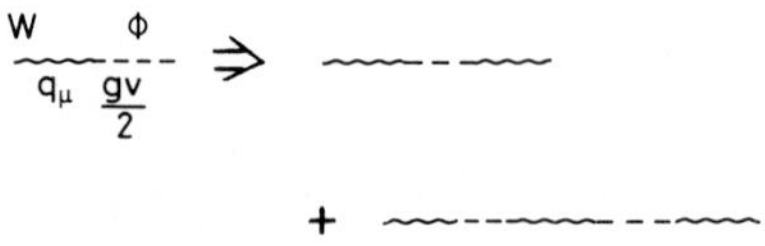

Fig. 1.2. The W–Higgs coupling which gives rise to the W mass when it is iterated.

where

$$m_{W^\pm} = gv/2. \tag{1.15}$$

The analysis of the neutral bosons proceeds similarly except for the minor complication of $W_\mu^3 - B_\mu$ mixing seen in (1.12). We can diagonalize these mass terms by introducing the mixed field

$$Z_\mu = (g'B_\mu - gW_{\mu 3})/\sqrt{g^2 + g'^2}, \tag{1.16}$$

which has the well-defined mass

$$m_Z = \tfrac{1}{2}\sqrt{g^2 + g'^2}\, v. \tag{1.17}$$

The orthogonal combination

$$A_\mu \equiv (g'W_{\mu 3} + gB_\mu)/\sqrt{g^2 + g'^2} \tag{1.18}$$

clearly acquires no mass from (1.12) and is therefore a candidate for the photon field. It may seem at first mysterious that out of the chaos of spontaneous symmetry breaking there should emerge one massless field, but this is not in fact so. The masslessness of $A_\mu$ corresponds to the unbroken Abelian U(1) symmetry of local phase rotations which leave unchanged the direction (1.9) of spontaneous symmetry breaking.

The weak isospin currents of the SU(2) group are

$$J_a^\mu \equiv \sum_{\text{SU(2) doublets}} \bar{f}_L \gamma^\mu \frac{\tau_a}{2} f_L, \quad a = 1,2,3, \tag{1.19}$$

and the $W^\pm$ couple to the weak charged current combinations

$$J_\pm^\mu \equiv \tfrac{1}{2}(J_1^\mu \pm i J_2^\mu) \tag{1.20}$$

through the interaction

$$\mathcal{L}_{\text{int}}^{\text{CC}} \equiv \frac{g}{\sqrt{2}}(W_\mu^+ J_-^\mu + W_\mu^- J_+^\mu). \tag{1.21}$$

Now that we have masses (1.15) for the $W^\pm$ bosons, it is clear that their exchanges at low energies give rise via fig. 1.3 to an effective four-fermion interaction

$$\tfrac{1}{4}\mathcal{L}_{\text{eff}} \equiv \frac{G_F}{\sqrt{2}}(J_\mu^+ J_-^\mu) \tag{1.22}$$

and that the magnitude of the Fermi constant $G_F$ is now fixed by

$$G_F/\sqrt{2} = g^2/8m_W^2. \tag{1.23}$$

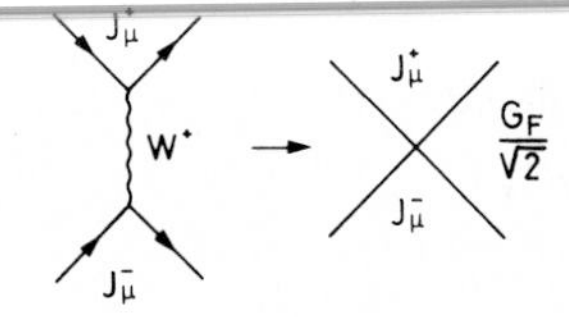


Fig. 1.3. The exchange of a heavy W gives rise to an effective four-fermion interaction at low energies.

In view of eq. (1.15) we can now fix the Higgs vacuum expectation value

$$v = \left(\sqrt{2}\, G_{\mathrm{F}}\right)^{-1/2}, \tag{1.24}$$

but notice that only the ratio (1.10) of the parameters $\mu^2$ and $\lambda$ is now known, and not their individual values.

It is easy to check using (1.4) and (1.18) that the photon $A_\mu$ couples to the combination $I_3 + Y$ of SU(2) generators and hypercharges, which means that we should choose the hypercharges

$$Y_{\mathrm{L}} = Q_{\mathrm{em}} - I_3, \qquad Y_{\mathrm{R}} = Q_{\mathrm{em}}, \tag{1.25}$$

if we are to get the right electromagnetic charges for all the fermions. Notice at this stage that there is no explanation why the electromagnetic current

$$J^{\mu}_{\mathrm{em}} \equiv \sum_{\ell \equiv \mathrm{e}, \mu, \tau} \bar{\ell}\gamma^{\mu}(-1)\ell + \sum_{p \equiv u, c, t} \bar{p}\gamma^{\mu}\left(\tfrac{2}{3}\right)p + \sum_{n \equiv d, s, b} \bar{n}\gamma^{\mu}\left(-\tfrac{1}{3}\right)n \tag{1.26}$$

should have turned out vector-like, let alone why the electromagnetic charges of different fermions should be commensurate (i.e. have ratios which are rational numbers) as is found experimentally. The magnitude of the unit of electromagnetic charge $e$ is easily checked to be

$$e = gg'/\sqrt{g^2 + g'^2}\,. \tag{1.27}$$

The usual parametrization of the mixing (1.16), (1.18) and of the electric charge unit (1.27) is in terms of the weak mixing angle $\theta_{\mathrm{W}}$ introduced by Glashow,

$$\cos\theta_{\mathrm{W}} \equiv \frac{g}{\sqrt{g^2 + g'^2}}, \qquad \sin\theta_{\mathrm{W}} \equiv \frac{g'}{\sqrt{g^2 + g'^2}}, \qquad e = g\sin\theta_{\mathrm{W}}, \tag{1.28}$$

in terms of which the neutral boson $Z_\mu^0$ couples to the combination of isospin (1.19) and electromagnetic (1.26) currents

$$J_\mu^0 \equiv J_\mu^3 - \sin^2\theta_W J_\mu^{em} \tag{1.29}$$

with a coupling

$$\mathcal{L}_{int}^{NC} = \sqrt{g^2 + g'^2}\, Z^\mu J_\mu^0, \quad \sqrt{g^2 + g'^2} = g/\cos\theta_W \tag{1.30}$$

Therefore massive $Z^0$ exchange gives an effective neutral current interaction

$$\tfrac{1}{4}\mathcal{L}_{eff} = \tfrac{1}{2}\sqrt{2}\, G_F \rho \left( J_\mu^0 J_0^\mu \right) \tag{1.31}$$

with the strength parameter $\rho = 1$ because $m_W = m_Z \cos\theta_W$. More general Higgs representations give different values of $\rho$: in general [8] we have

$$\rho = \frac{m_W^2}{m_Z^2 \cos^2\theta_W} = \frac{\Sigma_\phi \left( I^2 - I_3^2 + I \right) \langle 0 | \phi_{I,I_3} | 0 \rangle^2}{2\Sigma_\phi \langle 0 | \phi_{I,I_3} | 0 \rangle^2 I_3^2}. \tag{1.32}$$

Since the experimental value of $\rho$ is very close to 1, this is another piece of support for the $I = 1/2$ Higgs assignment made earlier.

Before turning to more details of the fermion masses, mixing and charged currents in the next section, it is worthwhile to gather together some properties of the intermediate bosons of the Standard Model [1]. The gauge bosons have masses

$$m_{W^\pm} = \frac{\left( \pi\alpha/\sqrt{2}\, G_F \right)^{1/2}}{\sin\theta_W} = \frac{37.4\,\text{GeV}}{\sin\theta_W},$$

$$m_Z = \frac{\left( \pi\alpha/\sqrt{2}\, G_F \right)^{1/2}}{\sin\theta_W \cos\theta_W} = \frac{m_W}{\cos\theta_W}, \tag{1.33}$$

which are about 80 and 90 GeV, respectively, for the experimentally preferred (section 1.4) range of $\sin^2\theta_W$. For more details of $W^\pm$ and $Z^0$ phenomenology, see lecture 3. In addition to the $W^\pm$ and $Z^0$ the theory contains one physical Higgs boson. The combinations $(\phi^+, (i/\sqrt{2})(\phi^0 - \bar{\phi}^0), \phi^-)$ of the original complex doublet of Higgs fields have been "eaten" by the $W^\pm$ and $Z^0$ to become their longitudinal polarization states and give them masses. Counting physical degrees of freedom, out of the original four components of the complex Higgs doublet therefore one physical Higgs particle survives [9], the combination

$$H \equiv \frac{\phi^0 + \bar{\phi}^0}{\sqrt{2}} - v. \tag{1.34}$$

Its mass is almost completely arbitrary, since

$$m_H = \sqrt{2}\,\mu \tag{1.35}$$

and we saw earlier that only the ratio $v$ (1.10) is as yet determined (1.24), and not yet the value of $\mu$ (see however lecture 3). The couplings of the physical Higgs H are on the other hand completely determined, as we shall start seeing in the next section.

### 1.3. Fermion masses and mixing

We will now analyze [10] the quark Yukawa interaction terms $\mathcal{L}_Y$ (1.6). It suffices to consider the couplings of the neutral Higgs $(1/\sqrt{2})(\phi^0 + \bar{\phi}^0) = H + v$, which are directly related to the fermion masses and mixing matrix:

$$\mathcal{L}_Y \ni -\frac{H+v}{v}\left(\bar{P}_L H_P P_R + \bar{N}_L H_N N_R + \text{h.c.}\right). \tag{1.36}$$

The first term in eq. (1.36) gives masses to the charge $+2/3$ quarks denoted generically by P, and the second term to the charge $-1/3$ quarks denoted generically by N, where $(P_L, N_L)$ are partners in SU(2) doublets: the "interaction" basis. Notice that the general $H_{ff'}$ matrix in $\mathcal{L}_Y$ (1.6) has been separated into two independent matrices $H_{P,N}/v$ in the space of quark generations (or families). We can diagonalize the bilinear mass terms in (1.36) by suitable unitary transformations $U$ and $V$ on the different helicities of P and N quarks:

$$\begin{aligned} P_L &= U_P p_L, & N_L &= U_N n_L, \\ P_R &= V_P p_R, & N_R &= V_N n_R, \end{aligned} \tag{1.37}$$

to the basis of mass eigenstates denoted generically by p and n. The transformations (1.38) turn the interactions (1.36) into

$$-\frac{H+v}{v}\left[\bar{p}_L\left(U_P^\dagger H_P V_P\right)p_R + \bar{n}_L\left(U_N^\dagger H_N V_N\right)n_R + \text{h.c.}\right], \tag{1.38}$$

and we see that the diagonalized mass matrices are

$$m_p \equiv U_P^\dagger H_P V_P, \qquad m_n \equiv U_N^\dagger H_N V_N. \tag{1.39}$$

It is immediately apparent from eq. (1.38) that the couplings $H_{P,N}/v$ (1.36) of the physical Higgs are diagonalized simultaneously with the quark masses (1.39) and that

$$g_{\mathrm{H f \bar{f}}} \equiv U^{\dagger}_{\mathrm{P,N}} \frac{H_{\mathrm{P,N}}}{v} V_{\mathrm{P,N}} = \frac{m_{\mathrm{f}}}{v} = \left(2^{1/4}\sqrt{G_{\mathrm{F}}}\right) m_{\mathrm{f}} = \frac{g m_{\mathrm{f}}}{2 m_{\mathrm{W}}}, \tag{1.40}$$

where use has been made of the relations (1.23) and (1.24).

It is clear that the neutral currents, which were flavour-diagonal in the interaction basis (P, N), remain flavour-diagonal in the mass basis (p, n), because

$$\begin{aligned} \bar{P}_{\mathrm{L}}\gamma_{\mu}P_{\mathrm{L}} &= \bar{p}_{\mathrm{L}}U^{\dagger}_{\mathrm{P}}\gamma_{\mu}U_{\mathrm{P}}p_{\mathrm{L}} = \bar{p}_{\mathrm{L}}\gamma_{\mu}p_{\mathrm{L}}, \\ \bar{N}_{\mathrm{L}}\gamma_{\mu}N_{\mathrm{L}} &= \bar{n}_{\mathrm{L}}U^{\dagger}_{\mathrm{N}}\gamma_{\mu}U_{\mathrm{N}}n_{\mathrm{L}} = \bar{n}_{\mathrm{L}}\gamma_{\mu}n_{\mathrm{L}}, \end{aligned} \tag{1.41}$$

since the matrices $U_{\mathrm{P,N}}$ are unitary in flavour space. However, the charged currents clearly look different in the mass basis:

$$W^{+}_{\mu}\bar{P}_{\mathrm{L}}\gamma^{\mu}N_{\mathrm{L}} = W^{+}_{\mu}\bar{p}_{\mathrm{L}}\gamma^{\mu}\left(U^{\dagger}_{\mathrm{P}}U_{\mathrm{N}}\right)n_{\mathrm{L}}, \tag{1.42}$$

so that the $W^{\pm}$ couplings are given [10] by a generalized Cabibbo mixing matrix

$$U \equiv U^{\dagger}_{\mathrm{P}}U_{\mathrm{N}}. \tag{1.43}$$

If there are $N_{\mathrm{G}}$ generations, *a priori* the matrix $U$ contains $N_{\mathrm{G}}^2$ parameters. However, $(2N_{\mathrm{G}}-1)$ of these are the relative phases between the fields of different quark flavours which are not in principle observable. We can therefore make phase rotations on the quark flavours to remove these parameters from $U$, which therefore contains $N_{\mathrm{G}}^2-(2N_{\mathrm{G}}-1)=(N_{\mathrm{G}}-1)^2$ true observable parameters. To see how this general rule works in some specific cases, consider the following cases

$N_G = 1$. $(N_{\mathrm{G}}-1)^2=0$ parameter: just $\binom{u}{d}$ coupling, no mixing.

$N_G = 2$. $(N_{\mathrm{G}}-1)^2=1$ parameter: that of the Cabibbo matrix

$$\begin{pmatrix} \cos\theta_{\mathrm{C}} & \sin\theta_{\mathrm{C}} \\ -\sin\theta_{\mathrm{C}} & \cos\theta_{\mathrm{C}} \end{pmatrix}. \tag{1.44}$$

$N_G = 3$. $(N_{\mathrm{G}}-1)^2=4$ parameters: three orthogonal and one distinctively "unitary" complex phase [7], and we can use the phase freedoms to write

$$U = \begin{pmatrix} c_1 & -s_1c_3 & -s_1s_3 \\ s_1c_2 & c_1c_2c_3-s_2s_3\mathrm{e}^{\mathrm{i}\delta} & c_1c_2s_3+s_2c_3\mathrm{e}^{\mathrm{i}\delta} \\ s_1s_2 & c_1s_2c_3+c_2s_3\mathrm{e}^{\mathrm{i}\delta} & c_1s_2s_3-c_2c_3\mathrm{e}^{\mathrm{i}\delta} \end{pmatrix}, \tag{1.45}$$

where $c_i$ $(s_i) \equiv \cos\theta_i$ $(\sin\theta_i)$ $(i = 1,2,3)$ and the phase $\delta$ parametrizes *CP* violation. Kobayashi and Maskawa [7] were the first to point out that this source of *CP* violation is present for the first time when there are six quarks. In lecture 2 we will return to a phenomenological discussion of the angles and phase in the Kobayashi–Maskawa (KM) matrix (1.45).

So far we have only talked about quarks, not leptons. If we assume that the neutrinos are massless, then even after diagonalizing the charged lepton masses, we can still make a unitary rotation among the indistinguishable neutrino fields to render the couplings of the $W^{\pm}$ diagonal, with specific neutrinos $\nu_e, \nu_\mu, \nu_\tau, \ldots$ associated with specific charged leptons e, $\mu$, $\tau$,.... (An analogous freedom in the quark sector would remove the KM phase $\delta$ and hence *CP* violation if two quarks of the same charge were degenerate in mass.) If the neutrinos have masses which are unequal, then there is a KM mixing matrix (1.45) for leptons too, with the possibility of *CP* violation etc. in the lepton sector (see lecture 8).

### *1.4. Neutral current phenomenology*

In this section we will be testing the form of the effective Lagrangian

$$\tfrac{1}{4}\mathcal{L}_{\text{eff}} = \frac{G_F}{\sqrt{2}}\left(J_\mu^+ J_-^\mu + \rho J_\mu^0 J_0^\mu\right). \tag{1.46}$$

The questions we should be asking are:

–do the neutral interactions factorize into a current–current form as in (1.46)?

–do the neutral currents $J_\mu^0$ take the specific form of eq. (1.29) for either the first and/or subsequent generations of fermions?

–if so, what is the value of $\sin^2\theta_W$?

–and what is the magnitude of the strength parameter $\rho$ in (1.46)?

For reasons of space, the question of factorization [11,12] will not be discussed in detailed here, though we will make some remarks about it. We assume for the moment that the neutral interactions are products of combinations of vector and axial currents. It is possible [13] to parametrize them in a form sufficiently general to include all models based on the group SU(2)×U(1). Writing a form for the neutral current $J_\mu^0$ which is more general than (1.29):

$$J_\mu^0 = \sum_{\text{fermions}} \left(g_L^f \bar{f}_L \gamma^\mu f_L + g_R^f \bar{f}_R \gamma^\mu f_R\right), \tag{1.47}$$

we can have

$$\begin{aligned} g_{\mathrm{L}}^{\mathrm{e}}(=g_{\mathrm{L}}^{\mu}=g_{\mathrm{L}}^{\tau}?) &= -\tfrac{1}{2}+x, & g_{\mathrm{R}}^{\mathrm{e}}(=g_{\mathrm{R}}^{\mu}=g_{\mathrm{R}}^{\tau}?) &= I_{\mathrm{R}}^{3}+x,\\ g_{\mathrm{L}}^{\mathrm{u}}(=g_{\mathrm{L}}^{\mathrm{c}}=g_{\mathrm{L}}^{\mathrm{t}}?) &= \tfrac{1}{2}-\tfrac{2}{3}x, & g_{\mathrm{R}}^{\mathrm{u}}(=g_{\mathrm{R}}^{\mathrm{c}}=g_{\mathrm{R}}^{\mathrm{t}}?) &= I_{\mathrm{R}}^{3}-\tfrac{2}{3}x,\\ g_{\mathrm{L}}^{\mathrm{d}}(=g_{\mathrm{L}}^{\mathrm{s}}=g_{\mathrm{L}}^{\mathrm{b}}?) &= -\tfrac{1}{2}+\tfrac{1}{3}x, & g_{\mathrm{R}}^{\mathrm{d}}(=g_{\mathrm{R}}^{\mathrm{s}}=g_{\mathrm{R}}^{\mathrm{b}}?) &= I_{\mathrm{R}}^{3}+\tfrac{1}{3}x, \end{aligned} \tag{1.48}$$

where we use the notation $x \equiv \sin^2\theta_{\mathrm{W}}$. It is sometimes convenient to use the vector and axial couplings

$$g_{\mathrm{V}}^{\mathrm{f}} = g_{\mathrm{L}}^{\mathrm{f}} + g_{\mathrm{R}}^{\mathrm{f}}, \qquad g_{\mathrm{A}}^{\mathrm{f}} \equiv -g_{\mathrm{L}}^{\mathrm{f}} + g_{\mathrm{R}}^{\mathrm{f}}. \tag{1.49}$$

Of course, in the Standard Model [1] described in previous sections $I_{\mathrm{R}}^{3}=0$ for all fermions, so that for example

$$g_{\mathrm{V}}^{\mathrm{e}} = -\tfrac{1}{2}+2\sin^2\theta_{\mathrm{W}}, \qquad g_{\mathrm{A}}^{\mathrm{e}} = \tfrac{1}{2}. \tag{1.50}$$

Even if we restrict ourselves to $J_{\mu}^{\mathrm{em}}$ and left-handed currents alone, it is possible to have an extra term in $\mathcal{L}_{\mathrm{eff}}$:

$$\tfrac{1}{4}\mathcal{L}_{\mathrm{eff}} = \frac{G_{\mathrm{F}}}{\sqrt{2}}\left(cJ_{\mu}^{\mathrm{em}}J_{\mathrm{em}}^{\mu}\right), \tag{1.51}$$

if there are more than one neutral vector bosons [14]. Later on we will see limits on this factorization violating term, as well as experimental determinations of the neutral current coupling $g_{\mathrm{L,R}}^{\mathrm{f}}$ and values for $\sin^2\theta_{\mathrm{W}}$ and $\rho$.

Table 1.1 lists major classes of neutral current experiments and their principal impact on determinations of the neutral current parameters, and we will just go through a few recent highlights. Most of the statistical weight in the determination of neutral current parameters comes from the big deep inelastic $\nu$N and $\bar{\nu}$N scattering experiments [15, 16]. It has long been clear that the neutral currents are predominantly left-handed, but [15] the existence of a small right-handed piece (coming from the mixture with $J_{\mu}^{\mathrm{em}}$?) has also been established, as shown in fig. 1.4 [16]. Deep inelastic scattering experiments are subject to some uncertainties (the validity of the quark–parton model with or without QCD scaling violations for what value of $\Lambda_{\mathrm{QCD}}$, and what about higher twist effects [17]?). In principle purely leptonic processes are much cleaner, but it is difficult to get a large number of events with small background. Figure 1.5 shows a recent result [18] from a counter experiment: clearly the problem here is background subtraction rather than statistics. It has

Table 1.1
Neutral current experiments

| Experiments | Principal impacts |
| --- | --- |
| Deep inelastic $\nu$N, $\bar{\nu}$N | High statistics measurements of $(g_L^u)^2+(g_L^d)^2$, $(g_R^u)^2+(g_R^d)^2$ |
| Deep inelastic $\nu$p, $\nu$n | Separation of $(g_L^u)^2/(g_L^d)^2$ $[(g_R^u)^2/(g_R^d)^2]$ |
| $\nu N \to \nu \pi X$<br>$\nu p$, $\bar{\nu} p$ elastic<br>$\nu p, n \to \nu p, n\pi$ | determine isospin properties of neutral currents—rule out theories which are mainly isoscalar |
| $\nu_\mu e \to \nu_\mu e$<br>$\bar{\nu}_\mu e \to \bar{\nu}_\mu e$<br>$\bar{\nu}_e e \to \bar{\nu}_e e$ | leptonic processes very clean in principle, but problems with statistics and backgrounds |
| $e^+e^- \to \mu^+\mu^-$ | forward–backward asymmetry seen, rules out ambiguity in leptonic neutral currents, tests for factorization |
| $e^+e^- \to$ hadrons | no effect seen, no information on neutral currents of heavy quarks |
| eD $\to$ eX interference | confirms V, A structure of neutral currents, good measure of $\sin^2\theta_W$ if $\rho = 1$ |
| atomic physics parity violation | determines electronic wave function at centre of atom, and little else |

become possible to use $e^+e^- \to \ell^+\ell^-$ data to remove [19] an ambiguity in the determination of leptonic neutral currents, and pin $g_V^e$ and $g_A^e$ down to a neighbourhood of the Standard Model values as shown [20] in fig. 1.6. This same class of experiments can be used to test factorization, and gives [20] an upper bound on the parameter $c$ of eq. (1.51):

$$c < 0.03 \quad \text{(PETRA)}. \tag{1.52}$$

Experiments to date determine very well the neutral currents of the first generation of fermions (e, u, d), as can be seen [13] in table 1.2. It is clear from this table that the possibilities $I_R^3 = \pm 1/2$ are totally excluded for (e, u, d).

Some information is now becoming available about the neutral current couplings of the heavier fermions, but this is so far rather sketchy, as can be seen from table 1.2. Information to date comes from three sources. The CHARM collaboration [21] has deduced from its $\nu$N and

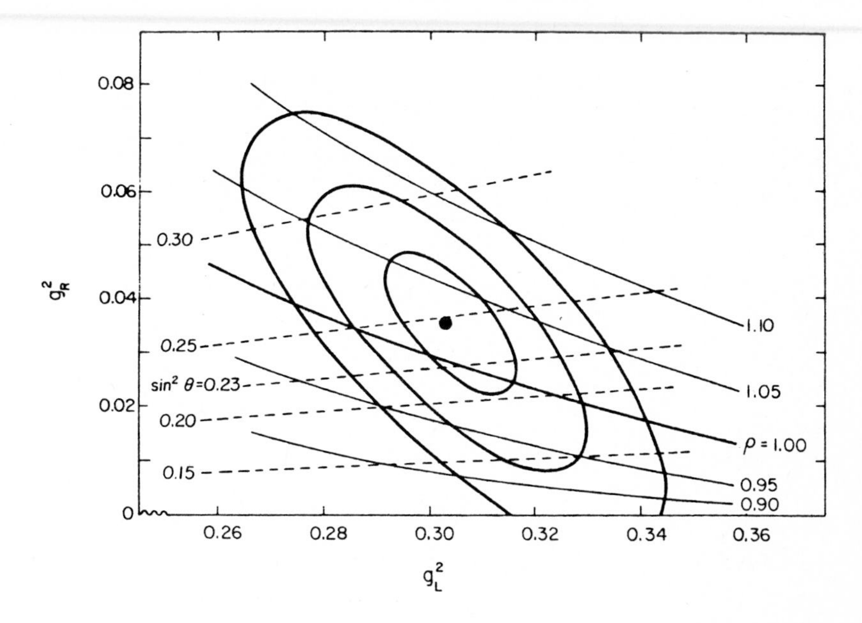


Fig. 1.4. Evidence for a non-zero right-handed component in the neutral current couplings of the up and down quarks [16].

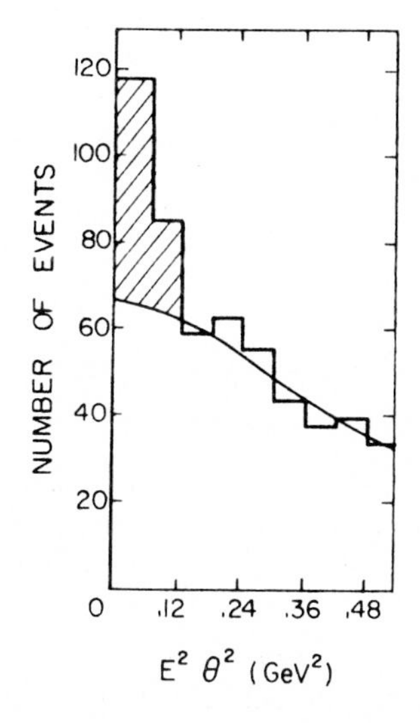


Fig. 1.5. Experimental results on antineutrino–electron scattering [18] which illustrate the background one has to subtract.

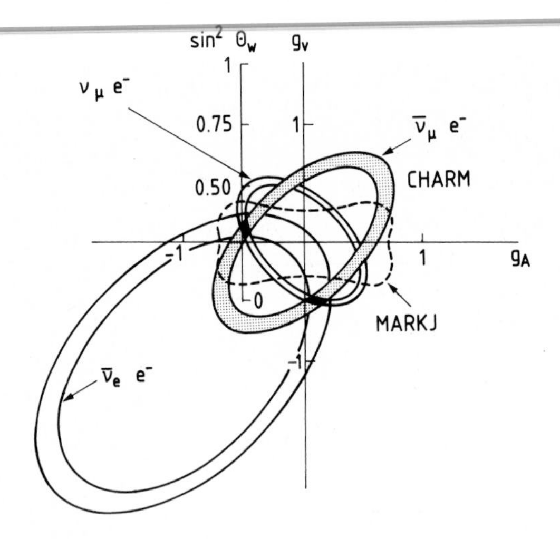


Fig. 1.6. A compilation [20] of information on the neutral current couplings of the electron, coming from $\nu$e scattering and $e^+e^-$ annihilation experiments.

Table 1.2
Neutral current parameters [13]

| Parameter | Experimental value | Standard model | With $\sin^2\theta_W = 1/4$ |
|---|---|---|---|
| $g_L^u$ | $0.339 \pm 0.033$ | $-\frac{1}{2} - \frac{2}{3}\sin^2\theta_W$ | 0.333 |
| $g_L^d$ | $-0.424 \pm 0.026$ | $-\frac{1}{2} + \frac{1}{3}\sin^2\theta_W$ | $-0.417$ |
| $g_R^u$ | $-0.179 \pm 0.019$ | $-\frac{2}{3}\sin^2\theta_W$ | $-0.167$ |
| $g_R^d$ | $-0.016 \pm 0.058$ | $\frac{1}{3}\sin^2\theta_W$ | 0.083 |
| $g_V^e$ | $0.043 \pm 0.063$ | $-\frac{1}{2} + 2\sin^2\theta_W$ | 0 |
| $g_A^e$ | $-0.545 \pm 0.056$ | $-\frac{1}{2}$ | $-\frac{1}{2}$ |
| $\dfrac{(g_L^s)^2 + (g_R^s)^2}{(g_L^d)^2 + (g_R^d)^2}$ | $1.39 \pm 0.43$ | 1 | 1 |
| $\dfrac{(g_V^c)^2 + (g_A^c)^2}{(g_V^u)^2 + (g_A^u)^2}$ | $1.6 \pm 0.5$ | 1 | 1 |
| $g_A^\mu$ | $0.72 \pm 0.21$[a] | 0.5 | 0.5 |

[a]Assuming $g_A^e = 1/2$.

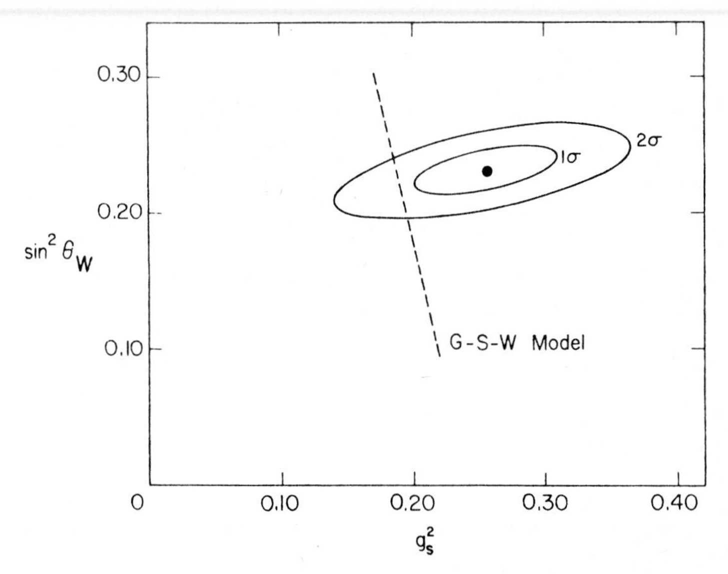


Fig. 1.7. Determination [21] of the neutral current coupling of the strange quark.

$\bar{\nu}$N data the neutral current coupling of the s quark, taking its density in the nucleon sea from $\overset{(-)}{\nu} \to \mu^-\mu^+$ events and the simple Glashow–Iliopoulos–Maiani [22] (GIM) charm model (more about this in lecture 2). They deduce [21] from the results of fig. 1.7 that

$$\frac{(g_L^s)^2+(g_R^s)^2}{(g_L^d)^2+(g_R^d)^2} = 1.39 \pm 0.43, \tag{1.53}$$

which can be compared with the values expected if

$$I_R^3(s) = \begin{cases} -\frac{1}{2} & \Rightarrow 1.92, \\ +\frac{1}{2} & \Rightarrow 2.83. \end{cases} \tag{1.54}$$

The first of these unorthodox assignments cannot yet be excluded, while the second more unorthodox one does seem to be impossible. A second piece of information comes from the observation by the CDHS collaboration [23] of the reaction $\nu N \to \nu N$ ($J/\psi \to \mu^+\mu^-$), which is believed to proceed by diffractive $Z^0$–hadron scattering as in fig. 1.8. By compar-

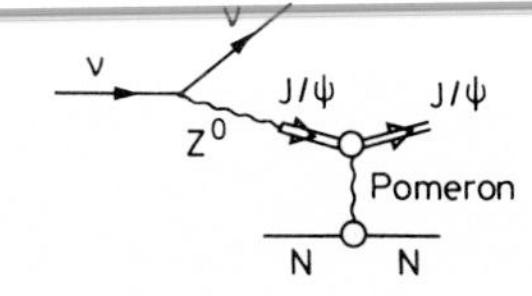


Fig. 1.8. Sketch of the diffractive mechanism believed to dominate J/ψ production by neutral currents.

ing with the reaction $\mu N \to \mu N$ ($J/\psi \to \mu^+\mu^-$), they deduce

$$\sigma/\sigma_{\text{Standard Model}} = 1.6 \pm 0.5, \tag{1.55}$$

as seen in fig. 1.9. This should be a direct measure of $(g_V^c)^2 + (g_A^c)^2$ and should be compared with

$$\frac{(g_V^c)^2 + (g_A^c)^2}{(g_V^c)^2 + (g_A^c)^2|_{\text{Standard Model}}} = \begin{cases} \frac{8}{5} \\ 4 \end{cases} \text{if } I_R^3(c) = \begin{cases} +\frac{1}{2} \\ -\frac{1}{2} \end{cases}. \tag{1.56}$$

The first, less unorthodox, one of these non-standard alternatives cannot yet be ruled out, whereas the second one can. A third piece of information on heavier fermion couplings comes from the reactions $e^+e^- \to \mu^+\mu^-$ and $\tau^+\tau^-$ [19,20,24]. From measurements [24] of the forward–backward asymmetry in $e^+e^- \to \mu^+\mu^-$ (see lecture 3 for definitions and more detailed formulae), which is proportional to $g_A^e g_A^\mu$, and assuming that $g_A^e = 1/2$ (reasonable in view of table 1.2) one can deduce

$$g_A^\mu = 0.72 \pm 0.21, \tag{1.57}$$

whereas one would expect

$$I_R^3(\mu, \tau) = \begin{cases} -\frac{1}{2} \\ +\frac{1}{2} \end{cases} \Rightarrow g_A^{\mu,\tau} = \begin{cases} 0 \\ 1 \end{cases}. \tag{1.58}$$

The first of these alternatives is excluded for the $\mu$ but not the $\tau$, and $I_R^3 = +1/2$ is allowed for both the $\mu$ and $\tau$. Not much information so far! A final measurement, which is sometimes [25] believed to give us information about the neutral current couplings of heavy quarks, is that of

$$\frac{\sigma_{\text{tot}}(e^+e^- \to \text{hadrons})}{\sigma_{\text{tot}}(e^+e^- \to \gamma^* \to \text{hadrons})} = 1 - \sum_f \frac{8s}{Q_f} g_V^e g_V^f \frac{G_F}{8\sqrt{2}\,\pi\alpha} + \cdots, \tag{1.59}$$

at low energies $\sqrt{s} \ll m_Z$. No deviation from the single photon exchange cross section is to be seen, which is to be expected since $g_V^e \approx 0$ (cf. table

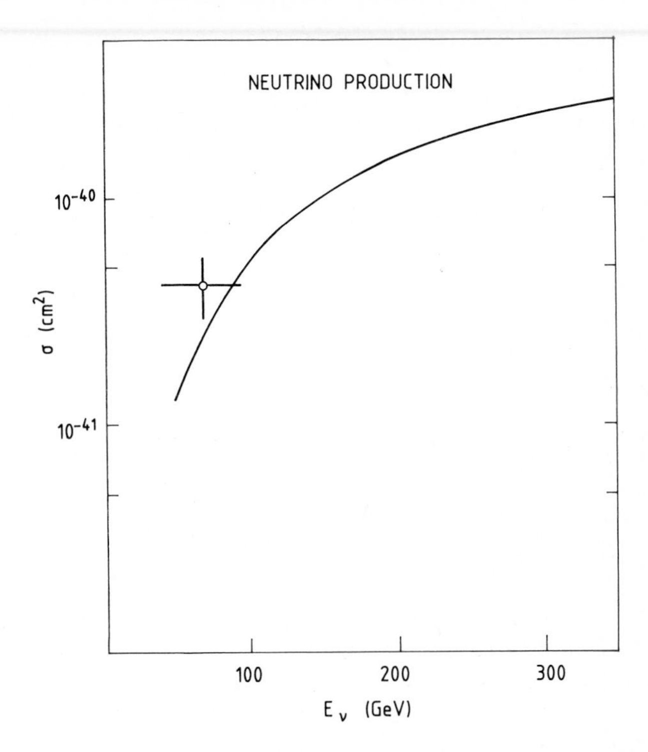


Fig. 1.9. Data [23] on the production of the $J/\psi$ by neutral currents, compared with a theoretical curve whose absolute normalization matches muon production data if the charm quark has the conventional neutral current couplings.

1.2) and would be exactly zero if $\sin^2\theta_W = 1/4$. No matter what the values of the $g_V^f$ in eq. (1.55), it would be essentially impossible to get an observable effect on $\sigma_{tot}(e^+e^- \to \text{hadrons})$. These experiments check once more that $g_V^e \approx 0$.

In view of the general success of the Standard Model in fitting the data shown in table 1.2, it is appropriate to perform a detailed analysis [13, 26] in the framework of the Standard Model [1] and determine the parameters $\rho$ and $\sin^2\theta_W$. This is tricky because the values of these two parameters are strongly correlated in many fits, and one loses information if one chooses uncorrelated subsets of the experimental information. Furthermore, the precision of the experimental data and of predictions (see lecture 6) from Grand Unified Theories (GUTs) are now so high that one must take into account radiative corrections, some of which are

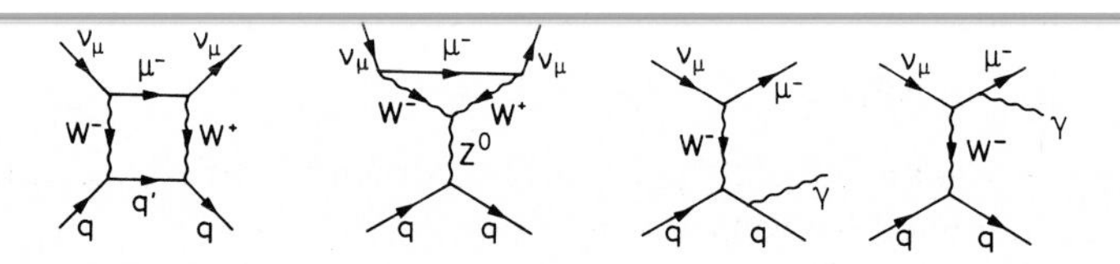


Fig. 1.10. Some radiative corrections to neutrino–quark scattering.

shown in fig. 1.10. These affect both $\rho$ and $\sin^2\theta_W$, rendering both parameters process dependent. As an example, it has been computed [27,28] that in the Standard Model with one $I=1/2$ Higgs field the effective $\rho$ parameter in $\nu_\mu$N scattering at an average momentum transfer $\langle Q^2\rangle$ should be

$$\rho^2 = 1 - \frac{\alpha}{\pi}\left[\ln\left(\frac{m_Z^2}{-2\langle Q^2\rangle}\right) + 2 - \frac{3}{8x^2}\ln(1-x)\right.$$
$$+ \frac{7}{8x} - \frac{1}{2(1-x)^2 x}\left(\frac{5}{2} - \frac{15x}{4} - \frac{x^2}{5} + \frac{14x^3}{9}\right)$$
$$\left. - \frac{3}{8x}h\left(\frac{\ln(1-x/h)}{1-x-h} + \frac{1}{1-x}\frac{\ln h}{1-h}\right) - \frac{3m_t^2}{8xm_W^2}\right] + O(\alpha^2), \tag{1.60}$$

where $x=\sin^2\theta_W$ as before, and $h \equiv m_H^2/m_Z^2$. If we set $\sin^2\theta_W = 0.23$, $m_H = m_Z$, $m_t = 20$ GeV and $\langle Q^2\rangle = -20$ GeV$^2$, then we get [27]

$$\rho^2 = 0.983, \quad \text{cf. experiment: } \rho^2 = 0.998 \pm 0.050. \tag{1.61}$$

While we do not yet see the radiative correction to $\rho = 1$, it is clear that this $I=1/2$ Higgs value is strongly favoured.

As for $\sin^2\theta_W$, a naive average of $\nu$N and $\bar{\nu}$N scattering data would give [13,26]

$$x = \sin^2\theta_W = 0.227 \pm 0.010 \quad \text{if } \rho = 1. \tag{1.62}$$

But we have just seen that $\rho$ should not equal 1, and if we put in the theoretical value of $\rho$ (1.61) we find [27]

$$\delta x = 0.49(\rho^2 - 1) = -0.0084. \tag{1.63}$$

There are other small corrections [27] due to the fact that the radiative corrections are different for u and d quarks, destroying the isoscalarity of target nuclei:

$$\delta x = -0.0007. \tag{1.64}$$

A more substantial effect [27] is that of real photon radiative corrections, when the observed hadronic energy $E_{\mathrm{H}}$ is cut off, e.g.

$$\delta x = -0.0035 \quad \text{for } E_H > 10 \text{ GeV}. \tag{1.65}$$

A "final" radiatively corrected value of an effective $\sin^2\theta_{\mathrm{W}}$ extracted from $\nu$N and $\bar{\nu}$N scattering data is [27, 29]

$$\sin^2\theta_{\mathrm{W}}\left(\nu_{\mu\mathrm{N}}, \langle Q^2\rangle = -20 \text{ GeV}^2\right)$$
$$= 0.216 \pm 0.010(\text{expt.}) \pm 0.004(\text{theor.}). \tag{1.66}$$

For comparison with other classes of experiments it is convenient to relate the value of this effective $\sin^2\theta_{\mathrm{W}}$ parameter to some useful "universal" definitions of $\sin^2\theta_{\mathrm{W}}$. Examples are the parameter appearing in a modified minimal subtraction ($\overline{\mathrm{MS}}$) renormalization prescription:

$$\sin^2\hat{\theta}_{\mathrm{W}}(m_{\mathrm{W}}) = 0.215 \pm 0.010 \pm 0.004, \tag{1.67}$$

or the parameter related to the physical (radiatively corrected) masses of the $\mathrm{W}^{\pm}$ and $\mathrm{Z}^0$:

$$\sin^2\theta_{\mathrm{W}}^{m} \equiv 1 - m_{\mathrm{W}^{\pm}}^2/m_{\mathrm{Z}}^2 = 0.217 \pm 0.010 \pm 0.004, \tag{1.68}$$

from deep inelastic $\nu$N, $\bar{\nu}$N scattering data. Combining with a similar analysis of eD interference data [30] gives the "best" world average [27, 29]

$$\sin^2\hat{\theta}_{\mathrm{W}}(m_{\mathrm{W}}) = 0.215 \pm 0.012, \tag{1.69}$$

which is to be compared with the GUT prediction of lecture 6.

## 2. Charged current angles and phases

### 2.1. *Are the charged currents $V - A$?*

In most of these lectures it is assumed that the weak interactions are well described by the conventional Glashow–Weinberg–Salam model [1] with the isospin assignments for fermions described in section 1.1, in which case the charged currents should be purely left-handed. What is the evidence for this behaviour?

We start with the simplest fermions to analyse, namely the leptons. We can parametrize [31] the charged leptonic currents by the following parameters in addition to the traditional Fermi coupling constant $G_F$:

$$\lambda_e \equiv -g_A^e/g_V^e, \qquad \lambda_\mu \equiv -g_A^\mu/g_V^\mu, \qquad \lambda_\tau \equiv -g_A^\tau/g_V^\tau, \tag{2.1a}$$

$$\kappa_\mu \equiv g_V^\mu/g_V^e, \qquad \kappa_\tau \equiv g_V^\tau/g_V^e. \tag{2.1b}$$

Clearly the charged currents are V−A if $\lambda_e=\lambda_\mu=\lambda_\tau=1$, and conventional weak universality implies $\kappa_\mu=\kappa_\tau=1$. Constraints on the parameters (2.1) come from several experiments. Several parameters are measurable in $\mu\to e\bar{\nu}\nu$ *decay*: among them we might mention [32]

$$\rho_\mu \equiv \frac{3}{8}+\frac{3}{2}\frac{\lambda_e\lambda_\mu}{(1+\lambda_e^2)(1+\lambda_\mu^2)}:$$

$$\frac{d\Gamma}{dx}(\mu\to e\bar{\nu}\nu) = \frac{G_F^2}{16\pi^3}m_\mu^5x^2\left[1-x+\tfrac{2}{3}\rho_\mu\left(\tfrac{4}{3}x-1\right)\right], \tag{2.2}$$

where $x=2E_e/m_\mu$, which becomes 3/4 for pure V−A charged currents, and

$$\Gamma_{tot}(\mu\to e\bar{\nu}\nu) = \frac{G_F^2m_\mu^5}{192\pi^3}\underset{\substack{\uparrow\\ \text{(phase space)}}}{F(m_e^2/m_\mu^2)}\left[(g_V^e)^2+(g_A^e)^2\right]\left[(g_V^\mu)^2+(g_A^\mu)^2\right]. \tag{2.3}$$

Another useful observable bearing on $\lambda_e$, $\lambda_\mu$ and $\kappa_\mu$ is

$$\frac{\Gamma(\tau\to e\bar{\nu}\nu)}{\Gamma(\tau\to\mu\bar{\nu}\nu)} = \underset{\substack{\uparrow\\ \text{(phase space)}}}{1.028}\ \frac{1+\lambda_e^2}{\kappa_\mu^2(1+\lambda_\mu^2)}, \tag{2.4}$$

which has the experimental value [33] of $0.950\pm0.086$.

Surprisingly, perhaps, the same combination of parameters is observable in purely leptonic $\pi\to e\bar{\nu},\mu\bar{\nu}(\equiv\pi_{\ell2})$ and $K\to e\bar{\nu},\mu\bar{\nu}(\equiv K_{\ell2})$ decays:

$$R_P \equiv \frac{\Gamma(P_{e2})}{\Gamma(P_{\mu2})} = \frac{1+\lambda_e^2}{\kappa_\mu^2(1+\lambda_\mu^2)}\times\begin{cases}1.139\times10^{-4} & \text{for } \pi,\\ 2.474\times10^{-5} & \text{for K.}\end{cases} \tag{2.5}$$

After radiative corrections [34], experiments determine the prefactor to be $1.023\pm0.019$ in $\pi_{\ell2}$ decays and $0.978\pm0.044$ in $K_{\ell2}$ decays. Clearly the best determination of $(1+\lambda_e^2)/\kappa_\mu^2(1+\lambda_\mu^2)$ comes from $\pi$ decays, though the information from $\tau$ decays [33] is not outclassed. Another useful constraint comes from the observation [35] that inverse $\mu$ decay proceeds at essentially the expected rate:

$$S \equiv \frac{\sigma(\nu_\mu + e^- \to \mu^- + \nu_e)}{\sigma_{V-A}} = 0.98 \pm 0.18. \tag{2.6}$$

Putting together all the experimental results on (2.2) to (2.6) and other related parameters, the Helsinki group [31] finds an overall best fit with

$$\lambda_e = 1.11^{+0.07}_{-0.27}, \qquad \lambda_\mu = 1.00 \pm 0.10, \qquad \kappa_\mu = 1.05^{+0.07}_{-0.15}. \tag{2.7}$$

Let us henceforward assume that $\lambda_e = \lambda_\mu = \kappa_\mu = 1$ and ask what are the constraints on the charged current couplings of the $\tau$. Two pieces of information are now available, on the spectral shape parameter [36] in $\tau \to e\nu\bar{\nu}$ or $\mu\bar{\nu}\nu$ decay:

$$\rho_\tau = \frac{3}{8} + \frac{3}{4}\frac{\lambda_\tau}{1+\lambda_\tau^2}, \tag{2.8}$$

and more recently [37] on the total $\tau$ lifetime

$$\frac{\tau(\tau)}{\tau_{V-A}} = \frac{1}{\kappa_\tau^2(1+\lambda_\tau^2)}, \tag{2.9}$$

if we assume pure $V-A$ couplings for the light quarks as well as the e and the $\mu$. The experimental value [36] of $0.72 \pm 0.15$ for the quantity (2.8) certainly seems to exclude a pure $V+A$ coupling $\lambda_\tau = -1$, but does not tell us much more:

$$0.28 < \lambda_\tau < 3.57. \tag{2.10}$$

If we assume that $\lambda_\tau = 1$ then the experimental ratio [37]

$$\frac{\tau(\tau)}{\tau_{V-A}} = \frac{(4.9 \pm 1.8)\times 10^{-13}\,\mathrm{s}}{(2.8 \pm 0.2)\times 10^{-13}\,\mathrm{s}} \tag{2.11}$$

tells us that

$$\kappa_\tau = 0.8 \pm 0.2. \tag{2.12}$$

We see from the comparison between (2.7) and (2.12) that there is clearly some way to go in the determination of $\tau$ charged currents, though it is most probably a conventional universal $V-A$ lepton.

After this exhaustive study of leptonic charged currents we will not dwell at length on the charged currents of the light u, d and s quarks. They are known to be mostly V – A on the basis of a multitude of decay and deep inelastic neutrino scattering measurements. In the latter experiments, the conventional expectations

$$\frac{d\sigma}{dy}(\nu_\mu q \to \mu^- + q') \text{ flat}, \qquad \frac{d\sigma}{dy}(\bar{\nu}_\mu q \to \mu^+ q') \propto (1-y)^2, \tag{2.13}$$

where $y \equiv 1 - E_\mu / E_{\nu_\mu}$ are consequences of the left-handed charged current couplings of the quarks, and are well checked at high neutrino energies [38]. Information about the handedness of the charged current couplings of the c quark can be extracted from recent data [23, 39] on the $y$ distributions in opposite-sign dimuon events $\nu(\bar{\nu}) + \mathrm{N} \to \mu^- \mu^+ + \mathrm{X}$. The observed $y$ distributions are consistent with being flat, and if one parametrizes them as

$$\frac{d\sigma}{dy} \propto (1-\alpha) + \alpha(1-y)^2, \tag{2.14}$$

then the measured value is

$$\nu : \alpha < 0.1, \qquad \bar{\nu} : \alpha < 0.25, \tag{2.15}$$

which in turn means that

$$(g_\mathrm{R}^\mathrm{c})^2 / (g_\mathrm{L}^\mathrm{c})^2 < 0.1. \tag{2.16}$$

Unfortunately, the constraint (2.16) does not yet translate into a severe constraint on the parameter $\lambda_\mathrm{c}$ defined analogously to eq. (2.1a): roughly

$$\tfrac{1}{2} < \lambda_c \equiv -g_\mathrm{A}^\mathrm{c} / g_\mathrm{V}^\mathrm{c} < 2, \tag{2.17}$$

though a systematic analysis of the data has not yet been completed by the experimentalists themselves. It would be interesting and useful to follow through the analysis chain (2.14) to (2.17) separately for the charged current couplings of d to c ($\lambda_\mathrm{c}^\mathrm{d}$) and of s to c ($\lambda_\mathrm{c}^\mathrm{s}$). The process $\bar{\nu} + \mathrm{N} \to \mu^+ \mu^- + \mathrm{X}$ mainly gives us the parameter $\lambda_\mathrm{c}^\mathrm{s}$, whereas $\nu + \mathrm{N} \to \mu^- \mu^+ + \mathrm{X}$ at moderate to large $x$ gives us information on the parameter $\lambda_\mathrm{c}^\mathrm{d}$. They could in principle be different, and once upon a time [40] there were interesting and valid weak interaction models with $\lambda_\mathrm{c}^\mathrm{s} \neq \lambda_\mathrm{c}^\mathrm{d}$. Nowadays of course we all believe that

$$\lambda_\mathrm{c}^\mathrm{s} = \lambda_\mathrm{c}^\mathrm{d} = 1, \tag{2.18}$$

but this should be checked experimentally. For the rest of this lecture we will assume the validity of (2.18), and indeed that all charged currents are left-handed.

### 2.2. *Constraints on the generalized Cabibbo angles*

If all the charged currents between three quark generations are indeed left-handed, then as described in section 1.3 they can be parametrized in terms of a four-parameter unitary coupling matrix [7]

$$U \equiv \begin{pmatrix} U_{\mathrm{ud}} & U_{\mathrm{us}} & U_{\mathrm{ub}} \\ U_{\mathrm{cd}} & U_{\mathrm{cs}} & U_{\mathrm{cb}} \\ U_{\mathrm{td}} & U_{\mathrm{ts}} & U_{\mathrm{tb}} \end{pmatrix} \tag{2.19a}$$

$$\equiv \begin{pmatrix} c_1 & -s_1c_3 & -s_1s_3 \\ s_1c_2 & c_1c_2c_3 - s_2s_3\mathrm{e}^{\mathrm{i}\delta} & c_1c_2s_3 + s_2c_3\mathrm{e}^{\mathrm{i}\delta} \\ s_1s_2 & c_1s_2c_3 + c_2s_3\mathrm{e}^{\mathrm{i}\delta} & c_1s_2s_3 - c_2c_3\mathrm{e}^{\mathrm{i}\delta} \end{pmatrix}, \tag{2.19b}$$

where $c_i(s_i) = \cos\theta_i$ $(\sin\theta_i)$, $i = 1, 2, 3$. Constraints on the parameters $U_{\mathrm{qq'}}$ or $(\theta_i, \delta)$ come from the following sources. The ratio between $0^+ \to 0^+$ nuclear $\beta$-decay rates and $\mu$ decay indicates [10,41–43] a slight deviation from weak universality so that

$$|U_{\mathrm{ud}}| = 0.9737 \pm 0.0025. \tag{2.20}$$

The number (2.20) comes after making weak radiative corrections [44] to the ratio of decay rates which are O(2)% and hence by no means negligible. An overall fit [42,43] to hyperon decays yields

$$|U_{\mathrm{us}}| = 0.219 \pm 0.003. \tag{2.21}$$

Combining the constraints (2.20) and (2.21) one finds that

$$|U_{\mathrm{ud}}|^2 + |U_{\mathrm{us}}|^2 = 0.996 \pm 0.004, \tag{2.22}$$

and hence that the "leakage" $|U_{\mathrm{ub}}|$ of the u quark's charged weak coupling to the b quark should be [10,42,43]

$$|U_{\mathrm{ub}}| = 0.06 \pm 0.06. \tag{2.23}$$

In terms of the Kobayashi–Maskawa [7] parameters (2.19b) it follows that

$$|c_1| = 0.9737 \pm 0.0025, \qquad |s_3| = 0.28^{+0.21}_{-0.28}, \tag{2.24}$$

so that $s_3^2$ is constrained to be considerably smaller than $c_3^2$, and could be of the order of the Cabibbo suppression factor:

$$s_3^2 \approx \sin^2\theta_C = s_1^2?, \tag{2.25}$$

though this is by no means directly established.

Another constraint comes from the rates of neutrino-induced dilepton events $\overset{(-)}{\nu} + \mathrm{N} \to (\mu\mathrm{e} \text{ or } \mu\mu) + \mathrm{X}$ coming from charmed particle production and decay. They can either be produced off d quarks or antiquarks $\propto |U_{cd}|^2$ or off s quarks or antiquarks $\propto |U_{cs}|^2$. Comparing the rates of neutrino- and antineutrino-induced $\mu$e events, some professionals [45] found

$$0.19 < |U_{cd}| < 0.34, \tag{2.26a}$$

whereas my own amateur analysis of the CDHS collaboration's $\mu\mu$ events [39] suggests

$$0.17 < |U_{cd}| < 0.23. \tag{2.26b}$$

The values (2.26) can be extracted without knowing the amount of sea s quarks in the nucleon. If one turns around the neutral current analysis [21] of section 1.4 by assuming universality between the neutral current couplings of the s and d quarks and equal integrated densities for the sea s and $\bar{\mathrm{s}}$, one can deduce the absolute normalization of the s quarks in the nucleon, and hence determine

$$|U_{cs}| = 0.85^{+0.16}_{-0.11}. \tag{2.27}$$

An alternative source of information on $|U_{cs}|$ is charm decays [45]. If one approximates from experiment [46]

$$\tau(\mathrm{D}^+ \to \text{all}) \sim 1 \times 10^{-12}\ \mathrm{s}, \qquad \frac{\Gamma(\mathrm{D}^+ \to \mathrm{K}^0\mathrm{e}^+\nu)}{\Gamma(\mathrm{D}^+ \to \text{all})} \sim 10\%, \tag{2.28}$$

and uses the rate for $\mathrm{D}^+ \to \mathrm{K}^0\mathrm{e}^+\nu$ calculated using naïve SU(4) symmetry, one gets [45]

$$|U_{cs}| = 0.66 \pm 0.33. \tag{2.29}$$

Neither of the determinations (2.27), (2.29) is above theoretical suspicion, but at least comparing them with the estimates (2.26) of $|U_{cd}|$ supports the general contention that the naïve Glashow–Illiopoulos–Maiani [22] pattern of Cabibbo-suppressed and favoured couplings is experimentally favoured. It remains to be seen whether $|U_{cs}|$, though large, is perhaps somewhat less than unity.

Two constraints on charged weak couplings are now forthcoming from measurements of the decays of b-flavoured hadrons. There is an upper limit of $5\times 10^{-12}$ s [47] on their mean lifetime, which, when compared with the total decay rate [48]

$$\Gamma(\mathrm{b}\to \mathrm{q}\bar{\mathrm{q}}'\mathrm{q}'') = \frac{\overset{\substack{\left(\text{perturbative QCD enhancement}\right)\\ \uparrow}}{\mathrm{O}(2)}\times \overset{\substack{\left(\text{colour factor}\right)\\ \uparrow}}{3}}{192\pi^3} G_{\mathrm{F}} m_{\mathrm{b}}^5 \underset{\substack{\uparrow\\ \text{(phase space)}}}{F(m_{\mathrm{q}}^2/m_{\mathrm{b}}^2)} |U_{\mathrm{bq}}|^2 |U_{\mathrm{q'q''}}|^2, \tag{2.30}$$

tells us that

$$\left(|U_{\mathrm{cb}}|^2+2.5|U_{\mathrm{ub}}|^2\right)^{1/2} > 0.02, \tag{2.31}$$

which is not very surprising news. More interesting are recent measurements [49] of the shape of the lepton spectrum in $\mathrm{B}\to \mathrm{e}\nu\mathrm{X}$ decays, which are well fitted by $\mathrm{B}\to(\mathrm{D}$ or $\mathrm{D}^*)\mathrm{e}\nu$ and badly by $\mathrm{B}\to(\pi$ or $\rho)\mathrm{e}\nu$—consistently with the original suggestion [48] that b quarks should couple more to c quarks than to u quarks. Computing the ratio of these decay rates using naïve SU(4) and taking account of the phase space differences, one deduces

$$|U_{\mathrm{ub}}|^2/|U_{\mathrm{cb}}|^2 < 0.1. \tag{2.32}$$

The constraints enumerated in this section still allow a value of $s_2^2$ which is relatively large compared with the known value of $s_1^2$. Rather than quote numerical limits on $\theta_i$ I prefer to quote [50] a "representative" Cabibbo–Kobayashi–Maskawa [7] mixing matrix which is plausible and more or less consistent with the constraints (2.20), (2.21), (2.23), (2.26), (2.27), (2.29), (2.31), (2.32):

$$U = \begin{pmatrix} 0.97 & 0.22 & 0.068 \\ -0.22 & 0.85-0.66\times 10^{-3}\mathrm{i} & -0.48+2.1\times 10^{-3}\mathrm{i} \\ -0.046 & 0.48+3.2\times 10^{-3}\mathrm{i} & -0.88-1.0\times 10^{-3}\mathrm{i} \end{pmatrix}. \tag{2.33}$$

The precise form of this matrix is only intended to be indicative. The values of the imaginary parts in the matrix elements of $U$ (2.33) come from an analysis of $CP$ violation in the $\mathrm{K}^0$–$\overline{\mathrm{K}}^0$ system (see section 2.4), in which a specific value of $m_{\mathrm{t}}=15$ GeV has been assumed [50]. In the next section we will address the question of the $\mathrm{K}^0$ system and the mass of the top quark.

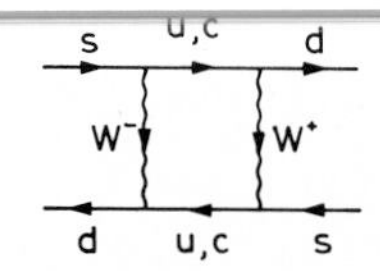


Fig. 2.1. Box diagram contributing to $K^0$–$\overline{K}^0$ mixing, which exhibits a GIM [22] cancellation.

## 2.3. *The $K^0$ system and $m_t$*

A showpiece of suppression of flavour-changing neutral interactions in the original four-quark GIM model [22] was the successful estimation of the $K^0$–$\overline{K}^0$ mass mixing parameter $\Delta m_K$, using the box diagram of fig. 2.1 to generate an effective $|\Delta S| = 2$ operator $[\bar{s}\gamma_\mu(1-\gamma_5)d]^2$. This calculation was used by Gaillard and Lee [51] to estimate the mass of the charmed quark. The calculation falls into two parts, a reliable weak perturbation theory computation of the coefficient of the $|\Delta S| = 2$ operator, and an unreliable strong coupling QCD computation of the operator's matrix element between $K^0$ and $\overline{K}^0$, which introduces an inherent ambiguity into the calculation. The result is [52]

$$\frac{\Delta m_K}{m_K} = \frac{0.42}{R}\frac{G_F}{\sqrt{2}}\frac{2}{3}f_K^2\frac{\alpha}{4\pi}\frac{1}{\sin^4\theta_W}F\left(m_c^2/m_W^2, m_t^2/m_W^2; \theta_i, \delta\right), \tag{2.34}$$

where $F(x, y; \theta_i, \delta)$ is known explicitly [52], and all the matrix element uncertainty resides in the parameter $R$. Gaillard and Lee [51] estimated the order of magnitude of the matrix element by making the vacuum insertion approximation

$$\begin{aligned}&\langle\overline{K}^0|\left[\bar{s}\gamma_\mu(1-\gamma_5)d\right]^2|K^0\rangle\\ &\quad\approx \langle\overline{K}^0|\bar{s}\gamma_\mu(1-\gamma_5)d|0\rangle\langle 0|\bar{s}\gamma_\mu(1-\gamma_5)d|K^0\rangle,\end{aligned} \tag{2.35}$$

corresponding to $R = 0.42$ in eq. (2.34). Unfortunately, it is known that an intermediate $\pi^0$ would contribute with a similar magnitude but with the opposite sign. An attempt [53] has been made to make a more complete computation using the MIT bag model, which yields $R = 1$. Taking from experiment $\Delta m_K/m_K \approx 0.7 \times 10^{-14}$, neglecting the as then

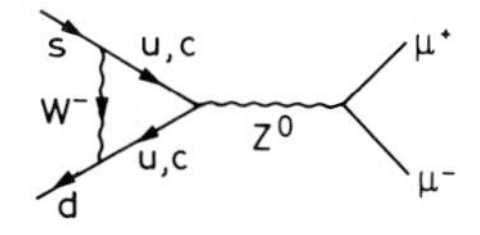


Fig. 2.2. Diagram contributing to the flavour-changing neutral decay $K^0 \to \mu^+\mu^-$, which also exhibits a GIM [22] cancellation.

unimagined t quark and putting in the known value of the Cabibbo angle $\theta_C$, the Gaillard and Lee [51] calculation with $R = 0.42$ gave $m_c \lesssim 2$ GeV. The question now arises whether our knowledge of $m_c(\sim 1.5$ GeV) and the Kobayashi–Maskawa [7] angles, combined with the bag model result [53] $R = 1$, can give us some restriction on $m_t$.

The answer is: not without extra information coming, for example, from the short-distance contribution $\hat{\Gamma}(K_L^0 \to \mu^+\mu^-)$ to the $K_L^0 \to \mu^+\mu^-$ decay illustrated in fig. 2.2. Shrock and Voloshin [54] pointed out that this process could give useful constraints on very heavy t quarks, and Buras [52] has recently published a combined analysis of $K^0$–$\overline{K}^0$ mixing and the $K_L^0 \to \mu^+\mu^-$ decay. Removing the long-distance contribution $K_L^0 \to \gamma\gamma \to \mu^+\mu^-$ illustrated in fig. 2.3, Buras finds, in a crude approximation to the ratio $\Delta m_K / \hat{\Gamma}(K_L^0 \to \mu^+\mu^-)$, that

$$\left(\text{monotonically increasing function of } m_t^2/m_W^2\right) \lesssim \frac{1.63}{R} F_{QCD} s_1^2 c_3^2, \tag{2.36}$$

where $F_{QCD} \approx 0.77$ is a fudge factor calculable in QCD perturbation theory. It is apparent from the approximate eq. (2.36) that an upper bound on $m_t$ can be obtained in this way, and this remains true if the complete expression for $\Delta m / \hat{\Gamma}(K_L^0 \to \mu^+\mu^-)$ is used [52]. The final bound on $m_t$ as a function of $R$, under the assumption that $m_c = 1.5$ GeV and putting in the perturbative QCD factor $F_{QCD} \approx 0.77$, is shown as the top solid line in fig. 2.4. The solid and dashed lines show the increase in the bound if one neglects $F_{QCD}$, and the reduction if one takes $m_c = 1.2$ GeV. It seems from the dashed lines in fig. 2.4 that one should expect

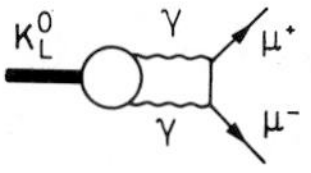

Fig. 2.3. Contribution to $K^0 \to \mu^+\mu^-$ from a $\gamma\gamma$ intermediate state.

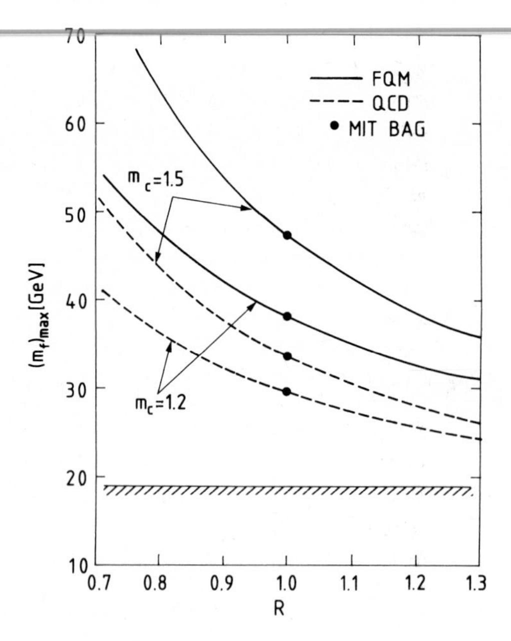


Fig. 2.4. Upper bound on the top quark mass coming from a consideration [52] of $K^0-\bar{K}^0$ mixing and $K^0 \to \mu^+\mu^-$ decay.

[52]

$$m_t < 33 \text{ GeV } (?), \tag{2.37}$$

if $R = 1$ as in the bag model.

However, this analysis is subject to objections, one of which [55] is to the procedure of subtracting the long-distance contribution to $K_L^0 \to \mu^+\mu^-$. It has been observed that the decay $\eta \to \mu^+\mu^-$ should be very analogous to the long-distance contribution to $K_L^0 \to \mu^+\mu^-$, and that [56]

$$\left.\frac{\Gamma(\eta \to \mu^+\mu^-)}{\Gamma(\eta \to \gamma\gamma)}\right|_{exp} = (5.9 \pm 2.2) \times 10^{-5} \gg \frac{\Gamma(\eta \to \gamma\gamma \to \mu^+\mu^-)}{\Gamma(\eta \to \gamma\gamma)}$$
$$= 1.1 \times 10^{-5}. \tag{2.38}$$

Perhaps there is another large long-distance contribution to $K_L^0 \to \mu^+\mu^-$ besides that shown in fig. 2.3, due to the dispersive off-shell photon process [55] $K_L^0 \to \gamma^*\gamma^* \to \mu^+\mu^-$. If so, substantial interference between the long- and short-distance contributions to $K_L^0 \to \mu^+\mu^-$ is possible, the previous extraction of $\hat{\Gamma}(K_L^0 \to \mu^+\mu^-)$ is no longer reliable, and no limit on $m_t$ can be obtained. I personally find this possibility somewhat contrived: it would be useful to have a reliable theoretical calculation of $K_L^0 \to \gamma^*\gamma^* \to \mu^+\mu^-$ and also experimental confirmation of the apparent fact (2.38) [56]. My guess is that when the dust settles it will turn out that there is a non-trivial theoretical upper bound [52] on $m_t$, and that the experimental value of $m_t$ will respect it. Though this may be pure coincidence!

### 2.4. *CP violation in the $K^0$ system*

Having discussed the charged weak mixing angles $\theta_i$ up to now, we turn to *CP* violation and the phase $\delta$. We start with the only *CP* violation actually observed, namely that in the $K^0$ system. Phenomenologically, it seems that this *CP* violation arises principally in the $K^0 - \overline{K}^0$ mass matrix, and we will start by setting up the general formalism [57]. The mass matrix for the mixing of two neutral mesons can be written as

$$\begin{pmatrix} M_{11} - \frac{1}{2}\mathrm{i}\Gamma_{11} & M_{12} - \frac{1}{2}\mathrm{i}\Gamma_{12} \\ M_{21} - \frac{1}{2}\mathrm{i}\Gamma_{21} & M_{22} - \frac{1}{2}\mathrm{i}\Gamma_{22} \end{pmatrix}, \tag{2.39}$$

where the $\Gamma_{ij}$ are the absorptive parts,

$$\Gamma_{ij} \equiv 2\pi \sum_X \langle K_j | \mathcal{H}_{wk} | X \rangle \langle X | \mathcal{H}_{wk} | K_i \rangle, \tag{2.40}$$

and $M_{ij}$ the dispersive parts of the mass matrix. *CPT* invariance ensures that

$$M_{11} = M_{22}, \qquad \Gamma_{11} = \Gamma_{22}, \tag{2.41}$$

while the hermiticity of the Lagrangian ensures

$$M_{ij} = M_{ji}^*, \qquad \Gamma_{ji} = \Gamma_{ij}^*. \tag{2.42}$$

If *CP* invariance were valid, it would imply that

$$M_{ij} \stackrel{?}{=} M_{ji}, \qquad \Gamma_{ji} \stackrel{?}{=} \Gamma_{ij}, \tag{2.43}$$

but this is not valid experimentally. Applying the conditions (2.41), (2.42), the mass matrix (2.39) becomes

$$\begin{pmatrix} M - \frac{1}{2}\mathrm{i}\Gamma & M_{12} - \frac{1}{2}\mathrm{i}\Gamma_{12} \\ M_{12}^* - \frac{1}{2}\mathrm{i}\Gamma_{12}^* & M - \frac{1}{2}\mathrm{i}\Gamma \end{pmatrix}. \tag{2.44}$$

It is simple to diagonalize the matrix (2.44), finding the eigenstates

$$K_{S,L} \equiv \left[2(1+|\varepsilon|^2)\right]^{-1/2}\left[(1+\varepsilon)K^0 \pm (1-\varepsilon)\bar{K}^0\right] \tag{2.45}$$

with masses

$$m_{S,L} = M \pm \mathrm{Re}\left[\left(M_{12} - \tfrac{1}{2}\mathrm{i}\Gamma_{12}\right)\left(M_{12}^* - \tfrac{1}{2}\mathrm{i}\Gamma_{12}^*\right)\right]^{1/2}. \tag{2.46}$$

and lifetimes

$$\Gamma_{S,L} = \Gamma \pm 2\,\mathrm{Im}\left[\left(M_{12} - \tfrac{1}{2}\mathrm{i}\Gamma_{12}\right)\left(M_{12}^* - \tfrac{1}{2}\mathrm{i}\Gamma_{12}^*\right)\right]^{1/2}, \tag{2.47}$$

where the $CP$ violating mixing parameter

$$\varepsilon = \frac{-\mathrm{Re}\,M_{12} + \tfrac{1}{2}\mathrm{i}\,\mathrm{Re}\,\Gamma_{12} + \left[\left(M_{12} - \tfrac{1}{2}\mathrm{i}\Gamma_{12}\right)\left(M_{12}^* - \tfrac{1}{2}\mathrm{i}\Gamma_{12}^*\right)\right]^{1/2}}{\mathrm{i}\,\mathrm{Im}\,M_{12} + \tfrac{1}{2}\,\mathrm{Im}\,\Gamma_{12}}. \tag{2.48}$$

If $\varepsilon$ is small, it can be written approximately as

$$\varepsilon \approx \frac{\tfrac{1}{2}\,\mathrm{Im}\,\Gamma_{12} + \mathrm{i}\,\mathrm{Im}\,M_{12}}{\tfrac{1}{2}\mathrm{i}\,\Delta\Gamma - \Delta m}, \tag{2.49}$$

where

$$\Delta m \equiv m_S - m_L, \qquad \Delta\Gamma \equiv \Gamma_S - \Gamma_L. \tag{2.50}$$

Experimentally, it turns out that $\tfrac{1}{2}\Delta\Gamma_K \sim \Delta m_K$ while one expects $\mathrm{Im}\,\Gamma_{12} \ll \mathrm{Im}\,M_{12}$. In this case, the phase of $\varepsilon$ should be of order $\pi/4$, which is the case experimentally.

In addition to this $CP$ violation in the mass matrix, there can also be intrinsic $CP$ violating phases in individual decay amplitudes. These are phases left over after removal of the final state interactions, e.g. in $K^0 \to 2\pi$ decays:

$$\eta_{+-} \equiv \frac{\langle \pi^+\pi^- | \mathcal{H}_{wk} | K_L \rangle}{\langle \pi^+\pi^- | \mathcal{H}_{wk} | K_S \rangle} \approx \varepsilon + \varepsilon',$$

$$\eta_{00} \equiv \frac{\langle \pi^0\pi^0 | \mathcal{H}_{wk} | K_L \rangle}{\langle \pi^0\pi^0 | \mathcal{H}_{wk} | K_S \rangle} \approx \varepsilon - 2\varepsilon', \tag{2.51}$$

where

$$\varepsilon' \equiv \sqrt{2}\,\mathrm{i}\exp\mathrm{i}(\delta_2 - \delta_0)\frac{\mathrm{Im}\,A_2}{A_0}, \tag{2.52}$$

with $A_I$ the decay amplitudes [57] into final states of isospin $I$, the $\delta_I$ the corresponding $\pi\pi$ phase shifts, and where we have used the convention that the dominant amplitude $A_0$ is real. The relative phase between the decay amplitudes measured by (2.52) is directly related to phase differences between the $\Delta I = 1/2$ and $\Delta I = 3/2$ pieces of $\mathcal{H}_{\text{wk}}$.

We now turn to calculations [10,58–61] of $\varepsilon$ and $\varepsilon'$ using the six-quark Kobayashi–Maskawa (K–M) mechanism for *CP* violation introduced in section 1.3, using the choice of phase convention manifested in eq. (1.42). We use [10] the box diagrams of fig. 2.1 to estimate

$$\varepsilon_m \equiv \left|\frac{\operatorname{Im} M_{12}}{\Delta m}\right| \times \frac{2c_1 s_2 c_2 s_3 c_3 \sin\delta \left(\cos 2\theta_2 \dfrac{m_t^2}{m_t^2 - m_c^2} \ln \dfrac{m_t^2}{m_c^2} + s_2^2 \dfrac{m_t^2}{m_c^2} - c_2^2\right)}{\text{complicated expression found in ref. [59]}} \tag{2.53}$$

if $m_t \ll m_W$, which takes the approximate value

$$\varepsilon_m \approx 18 s_2 s_3 c_2 \sin\delta \tag{2.54}$$

if $m_c = 1.5$ GeV, $m_t = 30$ GeV, and $\theta_2$ and $\theta_3$ are small. A non-zero K–M phase $\delta$ therefore gives *CP* violation in the $K^0$–$\overline{K}^0$ mass matrix. However, it also gives a non-zero phase in the $I = 0$ amplitude $A_0$, called $2\xi_{2\pi}$, which we must remove if we are to change from the K–M convention (1.42) to the phase convention [57] used in writing eq. (2.52). When this is done, we find that

$$\varepsilon = \varepsilon_m + 2\xi_{2\pi}, \qquad \arg A_2 = -2\xi_{2\pi}, \tag{2.55}$$

and the magnitude of the intrinsic *CP* violating parameter of eqs. (2.51) and (2.52) is

$$|\varepsilon'| = \frac{1}{\sqrt{2}}\left|\frac{\operatorname{Im} A_2}{A_0}\right| \approx \tfrac{1}{10}|\xi_{2\pi}|. \tag{2.56}$$

The second, approximate, equality in formula (2.56) comes from the experimental ratio of $|A_2|/|A_0|$, known from the ratio $\Gamma(K_S \to \pi^+\pi^-)/\Gamma(K_S \to \pi^0\pi^0)$. To calculate $|\varepsilon'/\varepsilon|$ we need a dynamical model for $A_2$ and $A_0$, so as to estimate a value for $|\xi_{2\pi}|$. One approach [50,62] to this problem is to look at the simplest possible quark diagrams which can contribute to $K \to 2\pi$ decays, shown in fig. 2.5. It is clear that the diagrams shown in figs. 2.5a, b, c and d give decays through both $I = 0$ and $I = 2$ amplitudes, whereas the "Penguin" diagrams [48] of figs. 2.5 e

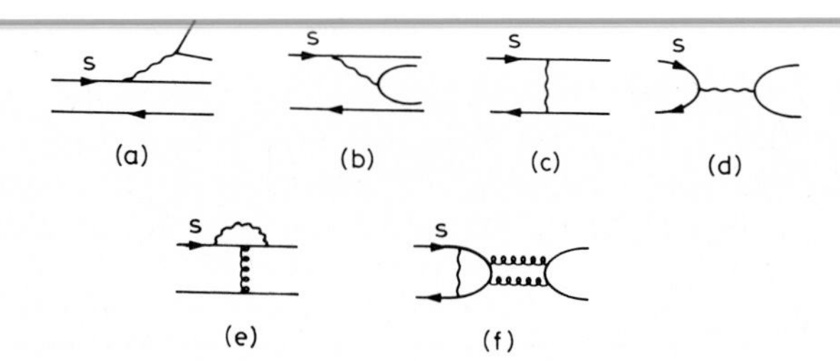


Fig. 2.5. Topologically distinct diagrams contributing to K decays.

and f only contribute to $A_0$. Explicitly one finds

$$A(\mathrm{K}^0 \to \pi^+\pi^-) = U_{\mathrm{us}}U^*_{\mathrm{ud}}(\mathrm{a}+\mathrm{c}+\mathrm{e}+2\mathrm{f}) + U_{\mathrm{cs}}U^*_{\mathrm{cd}}(\mathrm{e}+2\mathrm{f}),$$
$$A(\mathrm{K}^0 \to \pi^0\pi^0) = \tfrac{1}{2}U_{\mathrm{us}}U^*_{\mathrm{ud}}(\mathrm{b}+\mathrm{c}+\mathrm{e}) + U_{\mathrm{cs}}U^*_{\mathrm{cd}}(\mathrm{e}), \tag{2.57}$$

and it is easy to see that thanks to the presence of the Penguin diagrams, it is possible that $\arg(A_2) \neq \arg(A_0)$, implying that $|\varepsilon'|$ (2.52) is non-zero. A similar conclusion can be reached using the more formal method of operator analysis [59–61]. The most general $\Delta S = 1$ piece of $\mathcal{H}_{\mathrm{wk}}$ can be written as

$$\mathcal{H}^{\Delta S=1}_{\mathrm{wk}} = -\frac{G_{\mathrm{F}}}{\sqrt{2}} s_1 c_1 c_3 \sum_{i=1}^{5} e_i O_i, \tag{2.58}$$

where the $e_i$ are coefficients calculable using perturbative QCD, since they involve gluonic corrections [63] to the weak diagrams in which weakly interacting quarks are treated as free (cf. fig. 2.6 for some examples), and the $O_i$ are a complete linearly independent set of $\Delta S = 1$ operators:

$$O_1 \equiv (\bar{s}^i_{\mathrm{L}}\gamma^\mu d^i_{\mathrm{L}})(\bar{u}^j_{\mathrm{L}}\gamma_\mu u^j_{\mathrm{L}}), \qquad O_2 \equiv (\bar{s}^i_{\mathrm{L}}\gamma_\mu d^j_{\mathrm{L}})(\bar{u}^j_{\mathrm{L}}\gamma^\mu u^i_{\mathrm{L}}),$$
$$O_3 \equiv (\bar{s}^i_{\mathrm{L}}\gamma^\mu d^i_{\mathrm{L}})[(\bar{u}^j_{\mathrm{L}}\gamma_\mu u^j_{\mathrm{L}}) + (\bar{d}^j_{\mathrm{L}}\gamma_\mu d^j_{\mathrm{L}}) + (\bar{s}^j_{\mathrm{L}}\gamma_\mu s^j_{\mathrm{L}})],$$
$$O_4 \equiv (\bar{s}^i_{\mathrm{L}}\gamma^\mu d^i_{\mathrm{L}})[(\bar{u}^j_{\mathrm{R}}\gamma_\mu u^j_{\mathrm{R}}) + (\bar{d}^j_{\mathrm{R}}\gamma_\mu d^j_{\mathrm{R}}) + (\bar{s}^j_{\mathrm{R}}\gamma_\mu s^j_{\mathrm{R}})],$$
$$O_5 \equiv (\bar{s}^i_{\mathrm{L}}\gamma^\mu d^j_{\mathrm{L}})[(\bar{u}^j_{\mathrm{R}}\gamma_\mu u^i_{\mathrm{R}}) + (\bar{d}^j_{\mathrm{R}}\gamma_\mu d^i_{\mathrm{R}}) + (\bar{s}^j_{\mathrm{R}}\gamma_\mu s^i_{\mathrm{R}})]. \tag{2.59}$$

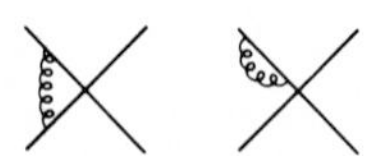

Fig. 2.6. Gluonic radiative corrections to the four-fermion operator in non-leptonic decays.

The coefficients $e_1$ and $e_2$ are likely to be quite large since the operators $O_1$ and $O_2$ contribute to the expansion (2.58) even in the absence of gluonic corrections. The other operators need gluons, and specifically operators $O_4$ and $O_5$ get contributions from the Penguin diagrams of figs 2.5e and f. One then finds [59–61] that

$$\xi_{2\pi} = p \operatorname{Im} e_5 / \operatorname{Re} e_5, \tag{2.60}$$

where $p$ is the proportion of the $A_0$ amplitude due to Penguin diagrams. The ratio of the real and imaginary parts of $e_5$ appearing in (2.60) is determined by the K–M factors and a perturbative QCD factor about whose magnitude opinions [59–61] differ, but which is probably O(1). Thus we find

$$\xi_{2\pi} = \mathrm{O}(1) \times p s_2 c_2 s_3 \sin\delta, \tag{2.61}$$

and inserting eqs. (2.54) and (2.61) into the ratio (2.56) we eventually get

$$|\varepsilon'/\varepsilon| = \mathrm{O}(\tfrac{1}{200})\,p. \tag{2.62}$$

It is thus clear that one expects $|\varepsilon'/\varepsilon| \lesssim \mathrm{O}(1/200)$ in the K–M model, to be compared with the experimental limit [64] of $|\varepsilon'/\varepsilon| < 1/50$. Opinions differ whether the Penguins make a dominant contribution to $A_0$ ($p \approx 1$) or a very small contribution ($p \ll 1$). In the former case a non-zero value of $\varepsilon'$ might be detectable in a forthcoming experiment [65] at FNAL, in the latter case our experimental colleagues will have to work harder to measure this second possible manifestation of *CP* violation.

Before leaving this section it should be noted that models with multiple Weinberg–Salam Higgs doublets [66] contain other mechanisms for *CP* violation, which consistently predict [67] values of $|\varepsilon'/\varepsilon|$ larger than the K–M model (2.62). Indeed, they seem to conflict with the experimental upper bound, and should therefore perhaps be discarded as models for the bulk of *CP* violation in the $K^0$ system. (They also have problems [68] with the $\theta$ vacuum parameter discussed in section 2.6.)

### 2.5. *CP violation in other weak decays?*

What other *CP*-violating quantities might be observable in the future? As far as the decays of strange particles are concerned, there are possible differences [62] in the partial decay rates of particles and antiparticles, e.g.

$$\begin{aligned} &\Gamma(K^+ \to \pi^+\pi^+\pi^-) \neq \Gamma(K^- \to \pi^-\pi^-\pi^+), \\ &\Gamma(\Lambda \to \pi^- p) \neq \Gamma(\bar{\Lambda} \to \pi^+ \bar{p}), \\ &\Sigma \text{ decays}, \ldots . \end{aligned} \tag{2.63}$$

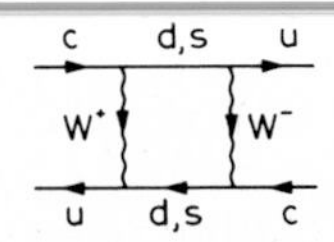


Fig. 2.7. Box diagram contributing to $D^0-\overline{D}^0$ mixing, which exhibits yet another GIM [22] cancellation.

It would perhaps be more interesting to see $CP$ violation in charm or bottom decays. Unfortunately, one expects [69] relatively little $D^0-\overline{D}^0$ mixing, compared with $K^0-\overline{K}^0$ mixing, since the box diagram of fig. 2.7 gives

$$
\begin{aligned}
&M_{12}^{\mathrm{D}} \propto m_s^2 \quad \left(\text{cf. } M_{12}^{\mathrm{K}} \propto m_c^2\right),\\
&\Gamma(\mathrm{D}\to \text{all}) \propto \cos^2\theta_{\mathrm{C}} \quad \left(\text{cf. } \Gamma(\mathrm{K}\to\text{all}) \propto \sin^2\theta_{\mathrm{C}}\right),
\end{aligned}
\tag{2.64}
$$

so that D particles tend to decay before they can mix. Thus one is unlikely to get a repeat of the rich $K^0-\overline{K}^0$ system discussed in the previous section. One could again look for differences [62] in partial decay rates, e.g. for

$$
\begin{aligned}
&\mathrm{D}^{\pm} \to \left\{\begin{matrix}\overline{\mathrm{K}}^0\mathrm{K}^+\\ \mathrm{K}^0\mathrm{K}^-\end{matrix}\right., \eta\pi^{\pm}, \eta\pi^{\pm}(\mathrm{X}: S=0),\ldots,\\
&\mathrm{F}^{\pm} \to \left\{\begin{matrix}\mathrm{K}^0\pi^{\pm}\\ \overline{\mathrm{K}}^0\pi^{\pm}\end{matrix}\right., \mathrm{K}^{\pm}\pi^0, \mathrm{K}^{\pm}\eta,\ldots,\\
&\left.\begin{matrix}\mathrm{D}^0\\ \overline{\mathrm{D}}^0\end{matrix}\right\} \to \mathrm{K}^-\mathrm{K}^+, \pi^+\pi^-, \pi^0\pi^0,\ldots .
\end{aligned}
\tag{2.65}
$$

Unfortunately, all the interesting decay modes are Cabibbo-disfavoured, and any decay asymmetries would probably be very small. Ugh!

It was at one time hoped [48] that bottom decays could provide a useful laboratory for studies [70] of $CP$ violation, but this now seems [71] unlikely. There could indeed be large mixing in the neutral $\mathrm{B}^0(\equiv \mathrm{b}\bar{\mathrm{d}})-\overline{\mathrm{B}}^0(\equiv \bar{\mathrm{b}}\mathrm{d})$ system from box diagrams:

$$
M_{12}^{\mathrm{B}} \approx \frac{G_{\mathrm{F}}^2 f_{\mathrm{B}}^2 m_{\mathrm{B}}^2}{12\pi}\left(\xi_{\mathrm{t}}^2\left(m_{\mathrm{t}}^2 + \tfrac{1}{3}m_{\mathrm{b}}^2 + \tfrac{3}{4}m_{\mathrm{b}}^2 \ln\left(m_{\mathrm{t}}^2/m_{\mathrm{b}}^2\right)\right) + \xi_{\mathrm{c}}^2 \mathrm{O}\left(m_{\mathrm{c}}^2\right)\right),
\tag{2.66}
$$

where

$$\xi_t \equiv s_1 s_2\left(c_1 s_2 s_3 - c_2 c_3 e^{i\delta}\right), \qquad \xi_c \equiv s_1 c_2\left(c_1 c_2 c_3 + s_2 c_3 e^{i\delta}\right), \tag{2.67}$$

and $f_B$ is the B meson decay constant analogous to $f_\pi$ or $f_K$. Neglecting the $m_b^2/m_t^2$ and $m_c^2/m_t^2$ in (2.66) we get [71]

$$M_{12}^B \approx \frac{G_F^2 f_B^2 m_B^2}{12\pi} \xi_t^2 m_t^2, \tag{2.68}$$

which should be compared to the total B meson decay rate (2.30). We see that, depending on $m_t$, one could easily have

$$\Delta m_B/\Gamma_B \gtrsim O(1/10), \tag{2.69}$$

and hence observable mixing $\propto (\Delta m_B/\Gamma_B)^2 \gtrsim O(1)\%$. One is therefore encouraged [48,70] to look for manifestations of this mixing in (e.g.) like-sign dilepton events due to

$$e^+e^- \to \Upsilon''' \to B^0\overline{B}^0 \begin{cases} \nearrow B^0 B^0 \begin{array}{l} \to \ell^- + X \\ \to \ell^- + X \end{array} \\[2ex] \searrow \overline{B}^0 \overline{B}^0 \begin{array}{l} \to \ell^+ + X \\ \to \ell^+ + X \end{array} \end{cases} . \tag{2.70}$$

One should be careful to select dilepton events from the primary semi-leptonic decays of B particles, and avoid contamination from (e.g.) secondary $B^0 \to D \to \ell^+ + X$, $\overline{B}^0 \to \overline{D} \to \ell^- + X$ decays. One possible signal for *CP* violation would then be an asymmetry in the number of dilepton events of opposite charges: $\#(\ell^-\ell^-) \neq \#(\ell^+\ell^+)$, which would measure the *CP* violating mass mixing parameter for the B system [cf. eq. (2.46)]:

$$A = \frac{\#(++) - \#(--)}{\#(++) + \#(--)} \approx -4\,\mathrm{Re}\,\varepsilon_B \quad \text{if } |\varepsilon| \text{ small.} \tag{2.71}$$

Unfortunately, one expects $M_{12}^B$ and $\Gamma_{12}^B$ to have similar phases [71,72], so that $\varepsilon_B$ is almost purely imaginary [cf. (2.48)]. Specifically, one calculates [72]

$$\Gamma_{12}^B \approx \frac{-G_F^2 f_B^2 m_B^3}{6\pi} \left( \tfrac{3}{4}\xi_t^2 + 2\xi_c \xi_t \frac{m_c^2}{m_b^2} + \cdots \right), \tag{2.72}$$

so that the *CP* violating effect (2.71) goes away as $m_c^2/m_b^2 \to 0$. This is to

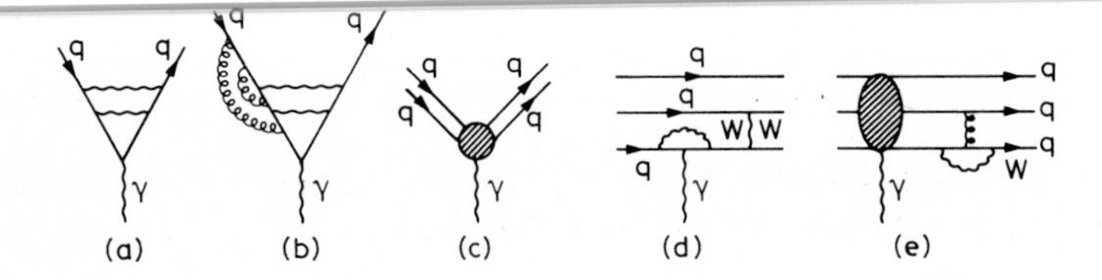


Fig. 2.8. Diagrams contributing to the neutron electric dipole moment in the Kobayashi–Maskawa [7] model of *CP* violation.

be expected, since one needs distinct masses for *all* six quarks if the analysis of section 1.3 is to be useful, and a *CP* violating phase $\delta$ is to be observable. More detailed computations [72] confirm the heuristic discussion given here: it is possible to find values of the $(\theta_i, \delta)$ such that $\mathrm{Re}\,\varepsilon_{\mathrm{B}}$ is large (or $\mathrm{B}^0-\overline{\mathrm{B}}^0$ mixing is substantial) but not both at the same time. Ugh! again! It seems we must again look elsewhere for a *CP*-violating observable.

## 2.6. *The neutron electric dipole moment*

Experimentally [73], this *CP* violating quantity $d_{\mathrm{n}}$ is bounded by

$$|d_{\mathrm{n}}| \lesssim 6 \times 10^{-25}\, e\,\mathrm{cm}, \tag{2.73}$$

corresponding to a limit on the distortion of the neutron's charge distribution comparable to a bump on the surface of a spherical Earth which is less than 1/10 mm high! In the K–M model [7], no neutron electric dipole moment can arise in second order of weak perturbation theory, because the K–M matrix factor is just $UU^*$, which is real [10]. One gets a complex K–M coupling factor in fourth order of weak perturbation theory with two $\mathrm{W}^{\pm}$ or unphysical $\mathrm{H}^{\pm}$ exchanges (cf. the one-quark diagram in fig. 2.8a). However, there is still no *CP* violation [74] if such a diagram is not dressed by at least two QCD gluons (cf. fig. 2.8b), and it might even cancel them. One then has an upper bound [68] on a weak perturbative contribution to $d_{\mathrm{n}}$:

$$\left|\frac{d_{\mathrm{n}}}{e}\right|_{1\mathrm{q}} \lesssim m_{\mathrm{u,d}} \left(\frac{\alpha_{\mathrm{s}}}{\pi}\right)^2 \left(\frac{\alpha}{\pi}\right)^2 s_1^2 s_2 s_3 \sin\delta \, \frac{m_{\mathrm{s,c}}^2}{m_{\mathrm{W}}^2} f\left(m_{\mathrm{t}}^2/m_{\mathrm{b}}^2\right), \tag{2.74}$$

where most of the factors are self-explanatory, except perhaps the external factor of $m_{\mathrm{u}}$ or $m_{\mathrm{d}}$, which reflects the helicity-flip nature of the dipole moment operator. Putting numbers into (2.74) one gets a bound

$$|d_{\mathrm{n}}/e|_{1\mathrm{q}} \lesssim \mathrm{O}(10^{-30\pm1})\ \mathrm{cm}. \tag{2.75a}$$

Turning to two-quark operators like that in fig. 2.8*c*, it has recently been argued [75] that Penguin-less diagrams such as fig. 2.8*d* only give

$$|d_{\mathrm{n}}/e|_{2\mathrm{q},\not\mathrm{P}} \sim \mathrm{O}(10^{-34})\ \mathrm{cm}, \tag{2.75b}$$

whereas Penguin diagrams such as fig. 2.8*e* give [76]

$$|d_{\mathrm{n}}/e|_{2\mathrm{q},\mathrm{P}} \sim \mathrm{O}(10^{-30\pm1})\ \mathrm{cm}, \tag{2.75c}$$

This therefore stands as the best current estimate of the K–M model contribution to the neutron electric dipole moment:

$$|d_{\mathrm{n}}/e|_{\mathrm{K-M}} \approx \mathrm{O}(10^{-30\pm1})\ \mathrm{cm}. \tag{2.76}$$

Happily for our experimental colleagues, there is another possible contribution to $d_{\mathrm{n}}$ which may make it larger, and perhaps observable. Remember that a study [6] of non-perturbative phenomena in QCD requires the introduction of a new parameter $\theta_{\mathrm{QCD}}$, characterizing the way in which the contributions of topologically distinct configurations of the gluon fields are added together in the functional integral

$$\int \mathfrak{D}[G_\mu] \to \sum_{n=-\infty}^{\infty} \mathrm{e}^{in\theta_{\mathrm{QCD}}} \int \mathfrak{D}[G_\mu]_n, \tag{2.77}$$

where the index $n$ is the Pontryagin index of the gauge field configuration [77]. The effect of the parameter $\theta_{\mathrm{QCD}}$ in eq. (2.77) can be summarized by introducing a new parameter into $\mathfrak{L}_{\mathrm{QCD}}$:

$$\mathfrak{L}_{\mathrm{QCD}} \to \mathfrak{L}_{\mathrm{QCD}} + \frac{\theta_{\mathrm{QCD}}}{32\pi^2} F^i_{\mu\nu}\tilde{F}^{i\mu\nu}, \quad \tilde{F}^i_{\mu\nu} \equiv \tfrac{1}{2}\varepsilon_{\mu\nu\alpha\beta}F^{i\alpha\beta}, \tag{2.78}$$

where $F_{\mu\nu}$ is the QCD field strength [cf. eq. (1.3)]. This new parameter clearly violates parity [because of the $\varepsilon$ symbol in (2.78)] and *CP* (because the colour singlet product of two gauge fields is *C*-even). Although the additional term in (2.78) is a total divergence, it can have a non-zero contribution to the action $A = \int \mathrm{d}^4x\, \mathfrak{L}$ when non-perturbative field configurations [78] are taken into account.

We can estimate the contribution that $\theta_{\mathrm{QCD}}$ makes to $d_{\mathrm{n}}$ by making an axial U(1) transformation related to the current

$$A_\mu \equiv \sum_{\text{quarks q}} \bar{q}\gamma^\mu\gamma_5 q. \tag{2.79}$$

In addition to its usual pieces

$$\partial^\mu A_\mu \ni \mathrm{i}\sum_{\mathrm{q}} m_{\mathrm{q}} \bar{q}\gamma_5 q, \tag{2.80}$$

proportional to the quark masses, the divergence of $A_\mu$ (2.79) also contains an anomalous QCD gluon piece

$$\partial^\mu A_\mu \ni \frac{g^2}{16\pi^2} F^i_{\mu\nu} \tilde{F}^{i\mu\nu}. \tag{2.81}$$

When we make an axial transformation through an angle $\sigma$,

$$\theta_{\mathrm{QCD}} \to \theta_{QCD} - 2\sigma. \tag{2.82}$$

We can use (2.82) to rotate away $\theta_{\mathrm{QCD}}$ at the expense of altering via (2.80) the quark-mass-dependent chiral symmetry breaking part of the QCD Lagrangian: in the case of three flavours

$$\delta \mathcal{L}_{\mathrm{CSB}} = - m_{\mathrm{u}} \bar{u} u - m_{\mathrm{d}} \bar{d} d - m_{\mathrm{s}} \bar{s} s$$

$$\delta \mathcal{L}_{\mathrm{CSB}} + \delta \mathcal{L}_{\mathrm{CP}} = \delta \mathcal{L}_{\mathrm{CSB}} + \mathrm{i}\theta_{\mathrm{QCD}} \frac{m_{\mathrm{u}} m_{\mathrm{d}} m_{\mathrm{s}}}{m_{\mathrm{u}} m_{\mathrm{d}} + m_{\mathrm{d}} m_{\mathrm{s}} + m_{\mathrm{u}} m_{\mathrm{s}}} (\bar{u}\gamma_5 u + \bar{d}\gamma_5 d + \bar{s}\gamma_5 s). \tag{2.83}$$

The effect of this modification to $\delta \mathcal{L}_{\mathrm{CSB}}$ can be estimated using normal chiral perturbation theory on the object

$$T\langle \mathrm{n} | \mathrm{i} \int \mathrm{d}^4 x \, \delta \mathcal{L}_{\mathrm{CP}}(x) J_\mu^{\mathrm{em}}(0) | \mathrm{n} \rangle. \tag{2.84}$$

Two calculations have been made of the $d_{\mathrm{n}}$ given by non-zero $\theta_{\mathrm{QCD}}$. Inserting baryon resonances, calculated using the MIT bag model, as intermediate states in (2.84), Baluni [79] finds

$$d_{\mathrm{n}}/e = 2.7 \times 10^{-16} \theta_{\mathrm{QCD}} \text{ cm}, \tag{2.85a}$$

while Crewther et al. [80] use a slightly different technique to get

$$(d_{\mathrm{n}}/e) = 3.6 \times 10^{-16} \theta_{\mathrm{QCD}} \text{ cm}. \tag{2.85b}$$

Comparing the experimental bound (2.73) with the calculations (2.85) it seems reasonable to conclude that

$$|\theta_{\mathrm{QCD}}| \lesssim 2 \times 10^{-9}. \tag{2.86}$$

How and why is this so? Once one couples together the K–M model of *CP* violation and QCD, it is no longer natural that $\theta_{\mathrm{QCD}}$ be zero or small.

Indeed the $\theta_{QCD}$ parameter is infinitely renormalized [68] by weak perturbation theory in the K–M model, as we shall see in lecture 3. We will have to look beyond QCD and the minimal Weinberg–Salam model for an explanation why $\theta_{QCD}$ obeys the bound (2.86), and we will return to this point in lectures 3 and 8.

## 3. Weak bosons

The weak bosons fall into two categories: the gauge vector bosons: $W^{\pm}$, $Z^0$, etc., which have relatively well-defined properties, and the scalar bosons: Higgs, axions, etc., whose properties are somewhat more vague [81]. The first section deals with the intrinsic properties of the $W^{\pm}$ and $Z^0$, while the subsequent section deals with their production in high energy collisions. The rest of the lecture is devoted to elementary Higgses [9] and axions [82], while composite Higgses are the province of lectures 4 and 5.

### *3.1. Masses and widths of the $W^{\pm}$ and $Z^0$*

We start this section with an overview of the leading-order predictions [81] of the Standard Model for the $W^{\pm}$ and $Z^0$, and then go into some aspects of higher-order corrections and rare decay modes. It is trivial to compute the $W^{\pm} \to e^{\pm}\nu$ partial decay width as a function of the $W^{\pm}$ mass:

$$\Gamma(\mathrm{W} \to \mathrm{e}\nu) \approx G_{\mathrm{F}} m_{\mathrm{W}}^3 / 6\pi\sqrt{2}\,, \tag{3.1}$$

and the branching ratios into fermion–antifermion channels:

$$\Gamma(\mathrm{W} \to \mathrm{e}\nu) : \Gamma(\mathrm{W} \to \mu\nu) : \Gamma(\mathrm{W} \to \tau\nu) : \Gamma(\mathrm{W} \to \bar{\mathrm{q}}\mathrm{q}') \approx 1 : 1 : 1 : 3 |U_{\mathrm{qq}'}|^2\,, \tag{3.2}$$

where the factor of 3 is due to colour, and $U_{qq'}$ is the relevant entry in the K–M matrix (1.45). We have neglected phase space corrections in writing (3.2): they will be negligible except possibly for decays involving the t, or heavier quarks if they exist. It is clear from (3.2) that one expects the branching ratios

$$B(\mathrm{W} \to \mathrm{e}\nu) = B(\mathrm{W} \to \mu\nu) \approx \frac{1}{4N_{\mathrm{G}}}\,, \tag{3.3}$$

where $N_G$ is the number of generations lighter than the $W^{\pm}$. Believing that there are at least three generations and that the t quark is lighter than the $W^{\pm}$ we deduce that

$$B(W \to e\nu) \lesssim 1/12, \tag{3.4}$$

a fact to be borne in mind when thinking about detection of the $W^{\pm}$. The leading-order prediction for $m_{W^{\pm}}$ in the Standard Model has already been given in formula (1.30):

$$M_W = \frac{\left(\pi\alpha/\sqrt{2}\, G_F\right)^{1/2}}{\sin\theta_W} = \frac{37.38\ \text{GeV}}{\sin\theta_W} \approx 84\ \text{GeV} \quad \text{if } \sin^2\theta_W \approx 0.20. \tag{3.5}$$

Substituting into eq. (3.1) one finds a decay width

$$\Gamma(W \to e\nu) \approx 260\ \text{MeV}, \qquad \Gamma(W \to \text{all}) \approx N_G\ \text{GeV}. \tag{3.6}$$

Correspondingly we find for the $Z^0$

$$\Gamma(Z^0 \to \nu\bar{\nu}) = N_\nu G_F m_Z^3/24\pi\sqrt{2}, \tag{3.7}$$

where $N_\nu$ is the number of neutrinos, assumed all to be much lighter than $m_Z/2$. Generalizing to other fermion–antifermion pairs one has

$$\Gamma(Z^0 \to \text{all}) \approx \frac{G_F m_Z^3}{12\pi\sqrt{2}}\left(\sum_\ell\left[\left(g_V^\ell\right)^2 + \left(g_A^\ell\right)^2\right] + 3\sum_q\left[\left(g_V^q\right)^2 + \left(g_A^q\right)^2\right]\right), \tag{3.8}$$

which in the Standard Model becomes

$$\begin{aligned}\Gamma(Z^0 \to \text{all}) = \frac{G_F m_Z^3}{24\pi\sqrt{2}}\Big\{&\left[1 + \left(1 - 4\sin^2\theta_W\right)^2\right]N_{\ell^-} + 2N_\nu \\ &+ 3\left[1 + \left(1 - \tfrac{8}{3}\sin^2\theta_W\right)^2\right]N_{2/3} \\ &+ 3\left[1 + \left(1 - \tfrac{4}{3}\sin^2\theta_W\right)^2\right]N_{-1/3}\Big\}.\end{aligned} \tag{3.9}$$

If we again take for orientation $\sin^2\theta_W = 0.20$ we deduce the following ratios of partial decay rates:

$$\begin{aligned}&\Gamma(Z \to \nu\bar{\nu}) : \Gamma(Z \to \ell\bar{\ell}) : \Gamma(Z \to u\bar{u}) : \Gamma(Z \to d\bar{d}) \\ &\qquad = 2 : 1.04 : 3.63 : 4.67,\end{aligned} \tag{3.10}$$

and it is clear that the leptonic branching ratios

$$B(\mathrm{Z} \to \mathrm{e}^+\mathrm{e}^-) = B(\mathrm{Z} \to \mu^+\mu^-) \approx \frac{1}{11 N_\mathrm{q}} \lesssim 3\%. \tag{3.11}$$

The leading-order prediction for $m_\mathrm{Z}$ has also been given in formula (1.30):

$$m_\mathrm{Z} = \frac{\left(\pi\alpha/\sqrt{2}\,G_\mathrm{F}\right)^{1/2}}{\sin\theta_\mathrm{W}\cos\theta_\mathrm{W}} = \frac{m_\mathrm{W}}{\cos\theta_\mathrm{W}} \approx 94\ \mathrm{GeV} \quad \text{for } \sin^2\theta_\mathrm{W} \approx 0.20, \tag{3.12}$$

implying via eq. (3.9) that

$$\Gamma(\mathrm{Z} \to \mathrm{e}^+\mathrm{e}^-) \approx 90\ \mathrm{MeV}, \qquad \Gamma(\mathrm{Z} \to \mathrm{all}) \approx N_\mathrm{G}\ \mathrm{GeV}. \tag{3.13}$$

So much for the leading-order predictions: what about the higher-order corrections? There is an important effect [27,83] of weak radiative corrections on the masses of the $\mathrm{W}^\pm$ and $\mathrm{Z}^0$. If we use the effective $\sin^2\theta_\mathrm{W}$ (1.62) relevant to deep inelastic $\nu_\mu$ scattering at an average $Q^2 \approx 20\ \mathrm{GeV}^2$, then the radiative corrections of $\mathrm{O}(\alpha)$ in

$$m_\mathrm{W} = \frac{37.38\ \mathrm{GeV}}{\sin\theta_\mathrm{W}\left(\nu_\mu\mathrm{N}, \langle Q^2\rangle \approx 20\ \mathrm{GeV}^2\right)} + \mathrm{O}(\alpha),$$

$$m_\mathrm{Z} = \frac{m_\mathrm{W}}{\cos\theta_\mathrm{W}\left(\nu_\mu\mathrm{N}, \langle Q^2\rangle \approx 20\ \mathrm{GeV}^2\right)} + \mathrm{O}(\alpha), \tag{3.14}$$

are quite substantial. Already mentioned in (1.64) was a possible definition [27] of a renormalized $\sin^2\theta_\mathrm{W}$ parameter which is more directly related to the $\mathrm{W}^\pm$ and $\mathrm{Z}^0$ masses:

$$\sin^2\theta_\mathrm{W} \equiv 1 - m^2_{\mathrm{W}^\pm}/m^2_{\mathrm{Z}^0}. \tag{3.15}$$

This definition of $\sin^2\theta_\mathrm{W}$ is related to an effective low energy parameter at $\langle Q\rangle \ll m_\mathrm{f}$ as follows:

$$\sin^2\theta_\mathrm{W} = \sin^2\theta_\mathrm{W}(\mathrm{low}\,Q^2)\left[1 + \frac{2}{3}\frac{\alpha(m_\mathrm{W})}{\pi}\left(\sum_\mathrm{f} Q_\mathrm{f}^2 \ln\frac{m_\mathrm{W}}{m_\mathrm{f}} + C\right)\right], \tag{3.16}$$

where the potentially large logarithm comes from the vacuum polarization diagrams of fig. 3.1, which change the effective fine structure constant, and the non-logarithmic constant $C$ is also not negligible.

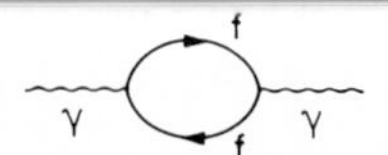


Fig. 3.1. A typical vacuum polarization diagram which changes the effective value of the fine structure constant between $Q^2 = 0$ and $O(m_W^2)$.

Marciano and Sirlin [7] find that

$$\sin^2\theta_W = \left(1.074^{+0.004}_{-0.005}\right)\sin^2\theta_W\left(\text{low } Q^2\right), \tag{3.17}$$

and the related effect that the predicted masses of the $W^{\pm}$ and $Z^0$ are increased by O(4)%:

$$m_W = \frac{38.64^{+0.07}_{-0.09}\ \text{GeV}}{\sin\theta_W}, \qquad m_Z = \frac{m_W}{\cos\theta_W}. \tag{3.18}$$

Alternatively, in terms of the modified minimal subtraction scheme definition of $\sin^2\hat{\theta}_W(m_W)$ they get

$$m_W = \frac{38.5^{+0.07}_{-0.09}\ \text{GeV}}{\sin^2\hat{\theta}_W(m_W)}. \tag{3.19}$$

Inserting into (3.18) or (3.19) the experimental value (1.63) of $\sin^2\theta_W$ deduced [27] from present experiments [$\sin^2\hat{\theta}_W(m_W) = 0.215 \pm 0.012$] they finally estimate

$$m_{W^{\pm}} = 83.0 \pm 2.4\ \text{GeV}, \qquad m_{Z^0} = 93.8 \pm 2.0\ \text{GeV}. \tag{3.20}$$

These predictions are somewhat higher than what one would deduce from the first-order formulae (3.5), (3.12) and the uncorrected experimental value of $\sin^2\theta_W \approx 0.227$ (1.58). This is the reason why I used a somewhat out-of-line value of $\sin^2\theta_W = 0.20$ in the mass formulae (3.5) and (3.12), but it should be emphasized that using this value in the width formula (3.9) to get the values (3.10) is not justified in the absence of a complete calculation of radiative corrections to $Z^0$ decay rates.

Let us now focus in more detail on $Z^0$ decays, as they can in principle be studied in great detail in $e^+e^- \to Z^0$ factories [84, 85]. QCD radiative corrections to the rates for decays into $q\bar{q}$ pairs are not negligible: a factor

$$1 + \alpha_s/\pi + \cdots \approx 1.04 \quad \text{for } \alpha_s(m_Z) = 0.13. \tag{3.21}$$

Finite mass corrections for the decay $Z^0 \to t\bar{t}$ are also non-negligible [86]:

$$\Gamma(Z \to t\bar{t}) = \frac{G_F m_Z^3}{4\sqrt{2}\,\pi}\left(1 - \frac{4m_t^2}{m_Z^2}\right)^{1/2} \times \left(1 - \tfrac{8}{3}x + \tfrac{32}{9}x^2 - \frac{m_t^2}{m_Z^2}\left(1 + \tfrac{16}{3}x - \tfrac{64}{9}x^2\right)\right), \qquad (3.22)$$

where $x \equiv \sin^2\theta_W$ as in section 1.4. The corrections (3.22) are already substantial for $m_t \approx 20$ GeV: the experimental lower bound $\Gamma(Z^0 \to t\bar{t})$ is reduced by 23%. Including the effects (3.21) and (3.22) gives us finally

$$\Gamma(Z(93.8\ \text{GeV}) \to \text{all}) \approx 3.06\ \text{GeV}, \qquad (3.23)$$

if there are just three generations.

Of great phenomenological interest is the decay rate into neutrinos:

$$\Gamma(Z(93.8\ \text{GeV}) \to \nu\bar{\nu}) \approx 0.181\ \text{GeV} \times N_\nu, \qquad (3.24)$$

which gives a substantial change in $\Gamma(Z^0 \to \text{all})$:

$$\Delta\Gamma_Z/\Gamma_Z \approx +6\%\ \text{per neutrino}. \qquad (3.25)$$

Precision measurements of the $Z^0$ (width of the peak, height of the peak, rate of radiative correction events $e^+e^- \to Z^0 + \gamma, Z^0 \to \nu\bar{\nu}$) can therefore be used to count the total number of neutrinos [84, 85]. Cosmologists [87] assure us, more or less convincingly, that the validity of cosmological nucleosynthesis calculations requires $N_\nu \lesssim 3$ or at most 4. The best present direct limit on $N_\nu$ from particle physics seems to come from $K^+$ decay:

$$B(K^+ \to \pi^+\nu\bar{\nu}) < 6 \times 10^{-7} \Rightarrow N_\nu \lesssim 10^5. \qquad (3.26)$$

(This result is a correction of that given in ref. [81], taking account of possible suppressions in loop diagrams if charged leptons have masses $\gtrsim m_W$, and of the fact that ref. [51] already included two neutrinos in their estimated branching ratio for $K^+ \to \pi^+\nu\bar{\nu}$.) One could in principle get good limits from heavy quarkonium decays: for example [81, 88],

$$\frac{\Gamma({}^3S_1(t\bar{t}) \to Z^0 \to \nu\bar{\nu})}{\Gamma({}^3S_1(t\bar{t}) \to \gamma \to e^+e^-)} \approx 2 \times 10^{-9}[m_{t\bar{t}}(\text{GeV})]^4 \times N_\nu. \qquad (3.27)$$

Unfortunately the $J/\psi$ is too light: the experimental bound

$$\frac{\Gamma(J/\psi \to \nu\bar{\nu})}{\Gamma(J/\psi \to e^+e^-)} \lesssim \frac{1}{10} \Rightarrow N_\nu \lesssim 5\times 10^5, \tag{3.28}$$

while one could do rather better with the $\Upsilon$: if one could establish

$$\frac{\Gamma(\Upsilon \to \nu\bar{\nu})}{\Gamma(\Upsilon \to e^+e^-)} < 1 \text{ ? then } N_\nu \lesssim 5000 \text{ ?} \tag{3.29}$$

In all this analysis of the total $Z^0$ decay width we have assumed that there are no other important decay modes of the $Z^0$ apart from $f\bar{f}$. Other decay modes have been calculated [89,90], but none seem very large; for example

$$B(Z^0 \to W^+ + e^- + \nu) \approx 10^{-8}, \qquad B(Z^0 \to W^\pm + X) \approx 2\times 10^{-7}, \tag{3.30}$$

while decay modes into Higgs particles are relatively small [91,92], as discussed in section 3.3. The time is now ripe to move on to the production and detection of the $Z^0$ and the $W^\pm$.

### 3.2. *Production of* $W^\pm$ *and* $Z^0$

The cleanest ways of producing the $W^\pm$ and $Z^0$ are in $e^+e^-$ collisions, as at the proposed LEP, SLC and CESR II accelerators [84,85]. In discussing the cross sections it is convenient to normalize them relative to the naïve point-like cross section:

$$\sigma_{pt} \equiv \sigma(e^+e^- \to \gamma^* \to \mu^+\mu^-) = 4\pi\alpha^2/3s, \tag{3.31}$$

which is exactly $10^{-2}$ nb or $10^{-35}$ cm$^2$ at a centre-of-mass energy $\sqrt{s} \equiv Q = E_{cm} = 2E_{beam} = 94$ GeV, the expected mass of the $Z^0$. At the peak of the $Z^0$ one naïvely has an event rate [84]

$$\frac{\sigma(e^+e^- \to Z^0 \to \text{all})}{\sigma_{pt}} = \frac{9}{\alpha^2} B(Z^0 \to e^+e^-) \approx 5000, \tag{3.32}$$

which corresponds to five events per second if the luminosity for collisions is $10^{32}$ cm$^{-2}$ s$^{-1}$. This rate is in fact reduced by a factor of 2 by radiative corrections, and furthermore the luminosity expected is somewhat lower, so one finishes up at LEP with the data collection rates shown in fig. 3.2. These are nonetheless very encouraging, and one happily contemplates experiments with up to $10^7$ $Z^0$ decays observed.

The first of the classic experiments around the $Z^0$ peak is of course the measurement of the total cross sections for fermion–antifermion pair production, described in terms of

$$\begin{aligned} R_f &\equiv \frac{\sigma(e^+e^- \to f\bar{f})}{\sigma_{pt}} \\ &= Q_f^2 - \frac{2s\rho Q_f g_V^e g_V^f}{(s/m_Z^2 - 1)^2 + \Gamma_Z^2/(s - m_Z^2)} \\ &\quad + \frac{s^2\rho^2\left[(g_V^e)^2 + (g_A^e)^2\right]\left[(g_V^f)^2 + (g_A^f)^2\right]}{(s/m_Z^2 - 1)^2 + \Gamma_Z^2/m_Z^2}, \end{aligned} \tag{3.33}$$

where we have introduced the notation

$$\rho \equiv \frac{G_F}{8\sqrt{2}\,\pi\alpha} \approx \frac{0.39}{m_Z^2} \quad \text{if } m_Z = 94 \text{ GeV}. \tag{3.34}$$

Notice that finite-width effects have been included in (3.33), which are not important when the centre-of-mass energy $\sqrt{s}$ differs greatly from $m_Z$. As an example of the general formula (3.33), consider the particular case of $e^+e^- \to \mu^+\mu^-$, for which

$$R_\mu = 1 + 2v^2\chi + (v^2 + a^2)^2\chi^2, \tag{3.35}$$

where we have assumed lepton universality:

$$g_V^e = g_V^\mu \equiv v, \qquad g_A^e = g_A^\mu \equiv a \tag{3.36}$$

($v$ is not to be confused with the Higgs vacuum expectation value) and have denoted

$$(\rho m_Z^2)s/(s - m_Z^2) \equiv \chi \tag{3.37}$$

and neglected finite width effects. It is easy to check that $R_\mu$ has a minimum at

$$\frac{s}{s - m_Z^2} = \frac{1}{m_Z^2\rho} \frac{v^2}{(v^2 + a^2)^2}, \tag{3.38}$$

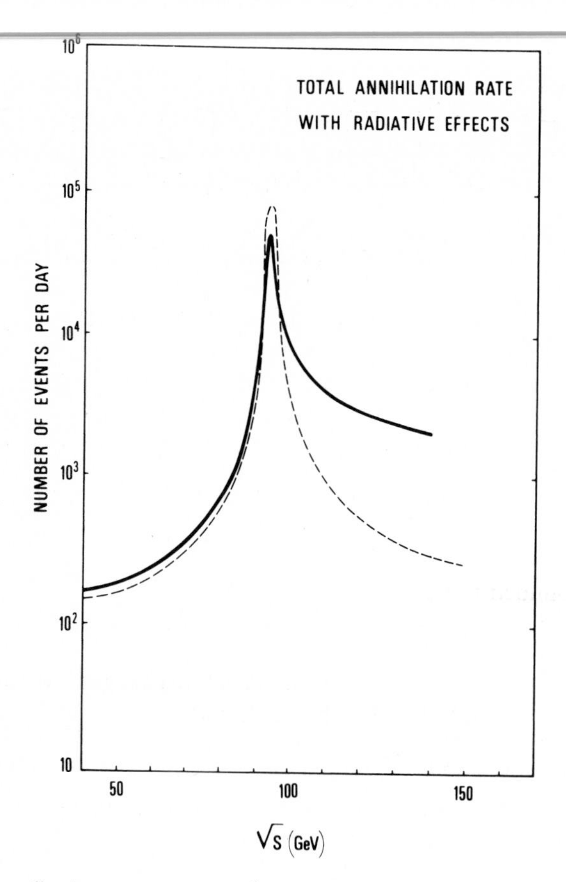


Fig. 3.2. Data collection rates near the $Z^0$ peak expected [84] at LEP. The dashed (solid) lines are before (after) including radiative corrections.

which is at $\sqrt{s} = 29$ GeV in the Standard Model with $\sin^2\theta_W = 0.20$. This is comfortably within the PEP–PETRA energy range, which sounds exciting; unfortunately for $\sin^2\theta_W = 0.20$

$$R_\mu^{\min} = 0.9985, \tag{3.39}$$

which conceivable experiments cannot distinguish from unity!

The second classic $Z^0$ experiment is the measurement of a forward–backward asymmetry. The angular distribution for $e^+e^- \to \bar{f}f$

takes the form

$$\frac{d\sigma}{d\cos\theta}(e^+e^- \to f\bar{f}) = \frac{\pi\alpha^2}{2s}\Big\{ Q_f^2(1+\cos^2\theta) - 2Q_f\chi\big[g_V^e g_V^f(1+\cos^2\theta)+2g_A^e g_A^f\cos\theta\big] + \chi^2\Big\{\big[(g_V^e)^2+(g_A^e)^2\big]\big[(g_V^f)^2+(g_A^f)^2\big] \times(1+\cos^2\theta)+8g_V^e g_V^f g_A^e g_A^f\cos\theta\Big\}\Big\}, \quad (3.40)$$

which exhibits $\cos\theta$ terms proportional to the product of a pair of axial current couplings. We can define a forward–backward asymmetry

$$A_f \equiv \frac{\int_0^1 d\cos\theta\left(d\sigma^{f\bar{f}}/d\cos\theta\right) - \int_{-1}^0 d\cos\theta\left(d\sigma^{f\bar{f}}/d\cos\theta\right)}{\int_{-1}^1 d\cos\theta\left(d\sigma^{f\bar{f}}/d\cos\theta\right)}, \quad (3.41)$$

which is bounded to lie in the range

$$|A_f| \lesssim 3/4 \quad (3.42)$$

if one is restricted to direct-channel $0, \pm 1$ exchanges. At low energies, eqs. (3.40) and (3.41) yield for the asymmetry

$$A_f = -\tfrac{3}{2}\chi\left(g_A^e g_A^f/Q_f\right), \quad (3.43)$$

which is about 10% for $e^+e^- \to \mu^+\mu^-$ or $\tau^+\tau^-$ at the highest PEP–PETRA centre-of-mass energies $\sqrt{s} \approx 40$ GeV. The minimum of the $e^+e^- \to \mu^+\mu^-$ asymmetry is at

$$\frac{s}{m_Z^2} = \frac{1}{1+\left(\rho m_Z^2\right)\left(a^2+3v^2\right)}: \qquad A_\mu^{\min} = -\frac{3}{4}\,\frac{1}{1+2v^2/a^2}, \quad (3.44)$$

which is $-0.69$ at $\sqrt{s} = 78$ GeV for $\sin^2\theta_W = 0.20$. Then there is a maximum asymmetry at

$$\frac{s}{m_Z^2} = \frac{1}{1-\left(\rho m_Z^2\right)\left(a^2-v^2\right)}: \qquad A_\mu^{\max} = +3/4, \quad (3.45)$$

which is at $\sqrt{s} = 118$ GeV for $\sin^2\theta = 0.20$. Fig. 3.3 shows the asymmetry for $e^+e^- \to \mu^+\mu^-$ as a function of energy for both the Standard Model [1] and also a non-standard model based on the group SU(2)×SU(2)×U(1). It shows clearly that, while measurements [19, 20, 24] of the asymmetry at

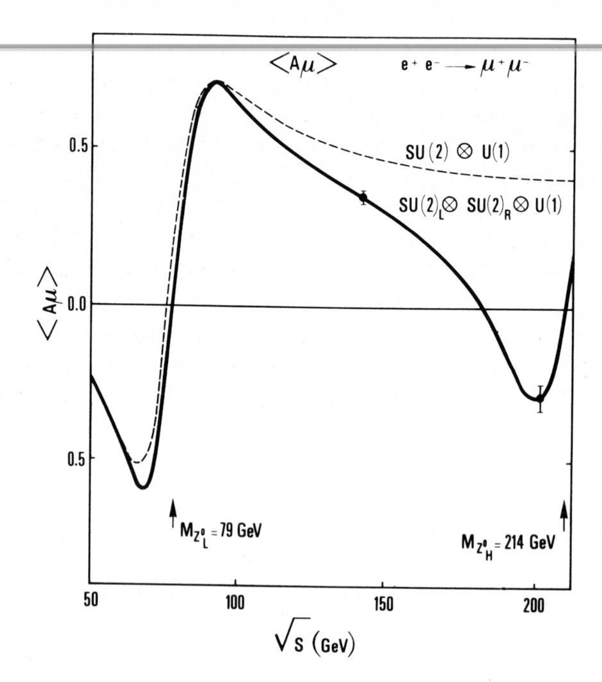


Fig. 3.3. The forward–backward asymmetry in $e^+e^- \to \mu^+\mu^-$ as a function of the centre-of-mass energy [84], showing the sensitivity to different weak interaction models.

low energies [cf. eq. (3.43)] are very interesting, they are limited in their capacity to distinguish between different weak interaction models.

A final set of measurements concerns polarization- and helicity-dependent cross sections. The total cross sections depend on the $e^+e^-$ helicity —in fact they vanish in the single gauge boson exchange approximation if the $e^+e^-$ are not spinning parallel (i.e. with opposite helicities). Hence it is useful to have polarized beams available. Also, the average helicity of a final state fermion depends on the centre-of-mass energy and angle at which it is measured. The easiest helicity to measure is likely to be that of the $\tau$ (via its $\pi\nu$ or e, $\mu\bar{\nu}\nu$ decays) which takes the following value in the Standard Model when emitted forward:

$$\langle H_\tau \rangle(\cos\theta = +1) = \frac{4\chi a v\left[1+\chi(a^2+v^2)\right]}{1+2\chi(v^2+a^2)+\chi^2\left[(a^2+v^2)^2+4a^2v^2\right]}. \tag{3.46}$$

This measurement is clearly sensitive to the sign of the axial coupling $a$ (we have again assumed lepton universality), which was not determined by the forward–backward asymmetry, which only fixes $|a|$. This sign information is also obtainable from polarized beam experiments: it is a matter of taste which type of experiment one prefers to do.

Turning now to the $W^{\pm}$, we have seen previously in eq. (3.30) that the $Z^0$ is not a copious source of single $W^{\pm}$, and the same is true everywhere below the $e^+e^- \to W^+W^-$ reaction threshold. The reaction $e^+e^- \to W^+W^-$ is a showcase [89,93] of gauge theories, exhibiting the crucial gauge theory cancellations between the direct channel $\gamma$ and $Z^0$ exchange diagrams and the crossed channel $\nu$ exchange shown in fig. 3.4, which are vital for the renormalizability of the theory. The total cross section [89,93] is

$$\sigma(e^+e^- \to W^+W^-) = \frac{\pi\alpha^2\beta}{2\sin^4\theta_W s}\left((1+2w+2w^2)\frac{L}{\beta} - \frac{5}{4}\right.$$

$$+\frac{m_Z^2(1-2\sin^2\theta_W)}{s-m_Z^2}\left[2w^2\left(1+\frac{2}{w}\right)\frac{L}{\beta} - \frac{1}{12w} - \frac{5}{3} - w\right]$$

$$\left.+\frac{m_Z^2(8\sin^4\theta_W - 4\sin^2\theta_W + 1)}{48(s-m_Z^2)^2}\frac{\beta^2}{w^2}(1+20w+12w^2)\right), \qquad (3.47)$$

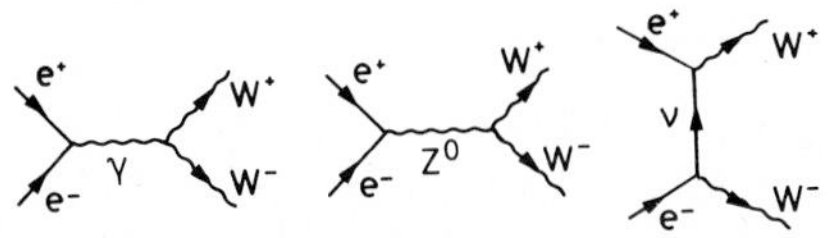

Fig. 3.4. Diagrams contributing to the reaction $e^+e^- \to W^+W^-$ which exhibit the famed gauge theory cancellations [89,93,94].

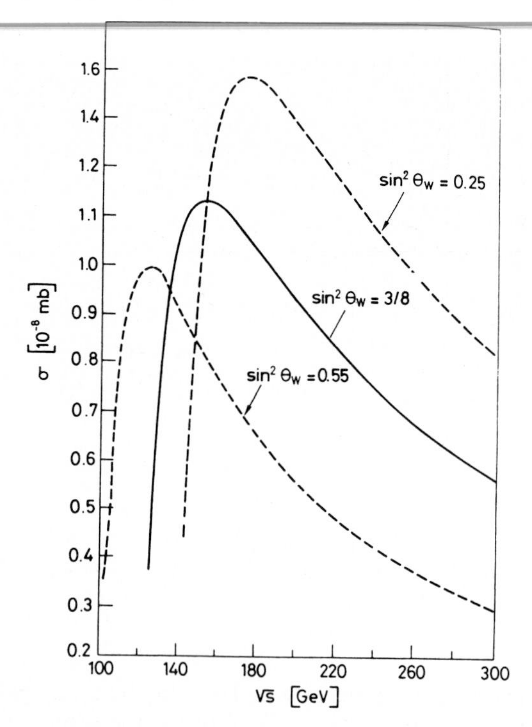


Fig. 3.5. The $e^+e^- \to W^+W^-$ cross section exhibits a clear peak about 50 GeV above threshold, and then falls as the gauge theory cancellations start to bite [84].

where

$$w \equiv m_W^2/s, \qquad \beta \equiv \sqrt{1-4w}\,, \qquad L \equiv \ln\left|\frac{1+\beta}{1-\beta}\right|. \tag{3.48}$$

We see in fig. 3.5 that this cross section exhibits a well-defined peak at about 40 GeV above threshold, and then falls with increasing $s$. This fall is a manifestation of the gauge theory cancellations [89] which are shown in fig. 3.6. The cross section is very sensitive to the polarization of the $e^+e^-$ beams, as is shown in fig. 3.7. It is also sensitive to the precise form of the three-boson vertex appearing in the graph of fig. 3.4. Shown in fig. 3.8 are the effects [94] of deviations from the gauge theory predictions of the magnetic moment and quadrupole moment of the $W^{\pm}$. Finally, fig. 3.9 shows the sensitivity [95] of the cross section, via radiative correc-

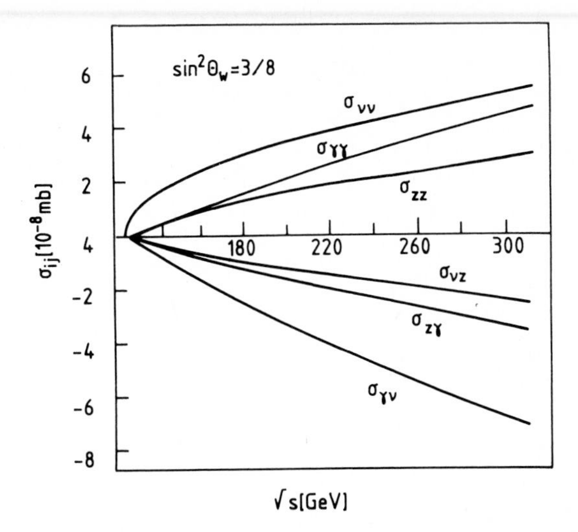


Fig. 3.6. Comparison [89] of different contributions to the $e^+e^- \to W^+W^-$ cross-section, showing how the interferences between the different diagrams in fig. 3.4 bring about the gauge theory cancellation.

tions, to the mass of the Higgs boson. This may be a way of detecting the indirect effects of the Higgs boson even if it is too heavy to be produced directly (see section 3.3). This discussion of the reaction $e^+e^- \to W^+W^-$ has perhaps convinced you that this reaction is an important testing ground for gauge theories.

I will not say very much about the production of $W^\pm$ and $Z^0$ in hadronic collisions, as other people are talking at this School about these processes, and there exist several good reviews [96] on the subject. Suffice it to say that naïve calculations using the simple Drell–Yan annihilation model give cross sections

$$\sigma(\bar{p}p \to W^- \text{ or } W^+ + X) \sim 2 \times 10^{-33}\ \text{cm}^2,$$
$$\sigma(\bar{p}p \to Z^0 + X) \sim 8 \times 10^{-34}\ \text{cm}^2, \tag{3.49}$$

at the CERN SPS $p\bar{p}$ collider energy $\sqrt{s} = 540$ GeV. QCD corrections are unlikely to change these cross sections by more than a factor of 2, though they should spread out the $p_T$ distributions of the produced bosons and their decay leptons [97]. Detection will be easiest through their leptonic decays, and when one bears in mind the branching ratios (3.4) and (3.11),

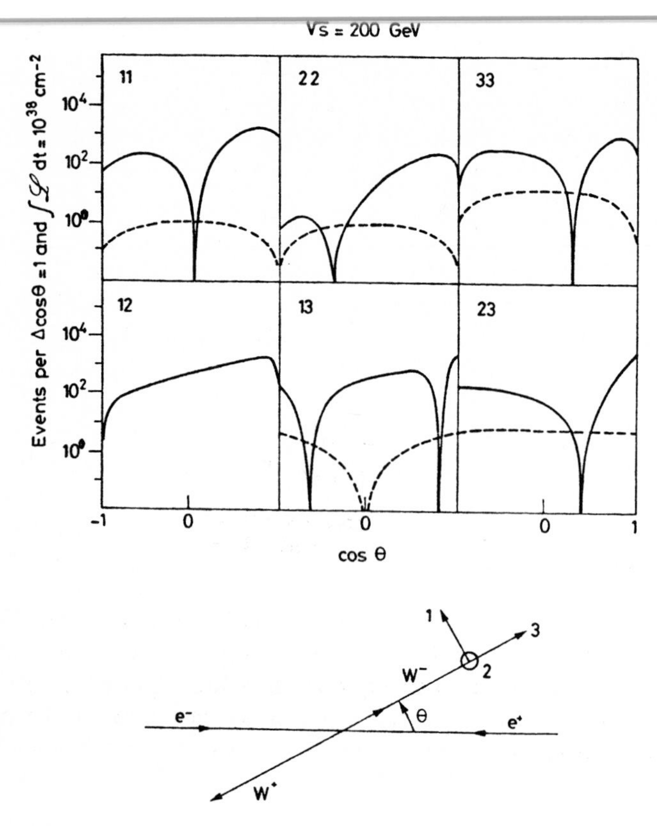


Fig. 3.7. The sensitivity [84] of the $e^+e^- \to W^+W^-$ cross-section to the polarization of the initial $e^\pm$. The solid curves are for $e_L^- e_R^+$, the dashed curves for $e_R^- e_L^+$. The notation for different W helicity states is shown in the bottom part of the figure.

it seems that a luminosity $\geqslant 10^{29}\ \mathrm{cm}^{-2}\ \mathrm{s}^{-1}$ will be necessary. Let us hope that the CERN $p\bar{p}$ collider speedily acquires such a high luminosity.

Finally I mention ep collisions [98], where the $W^\pm$ and $Z^0$ are expected to be produced with relatively low cross sections $\approx 10^{-36}$ or $10^{-37}\ \mathrm{cm}^2$ at presently accessible energies. However, in contrast to hadron–hadron collisions, ep collisions have the virtue that up to 1/2 of the $W^\pm$ or $Z^0$ production events should be very clean, with only a lepton and a proton accompanying the final state boson. Let us hope that a high energy ep collider gets built somewhere, some day.

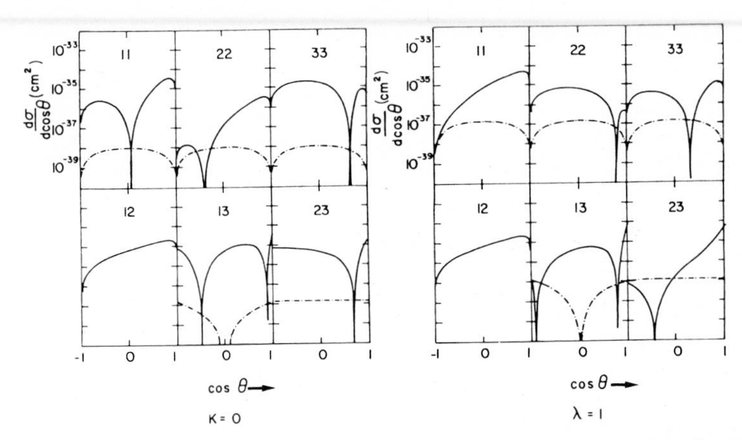


Fig. 3.8. The effects of modifying the gauge theory predictions for the dipole and quadrupole moments on the cross-sections for producing $W^+W^-$ in different helicity states [94].

### 3.3. *Higgs bosons*

These are Cinderellas of gauge theories, which people often wish were not necessary and did not exist, and in lectures 4 and 5 we will study scenarios [2] in which elementary Higgses are replaced by composite scalar fields. However, for the moment we just look at the minimal Weinberg–Salam model of section 1.2, in which there is one physical

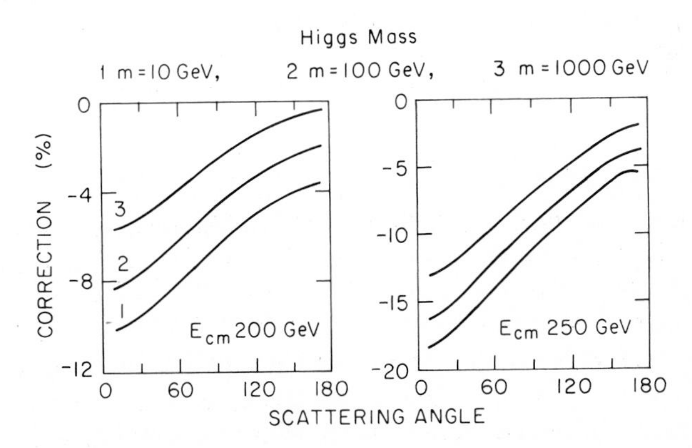


Fig. 3.9. The sensitivity [95] of the $e^+e^- \to W^+W^-$ cross-section to radiative corrections which depend on the mass of the Higgs boson.

Higgs boson H [9]. The couplings of this particle are completely fixed:

$$g_{\bar{f}fH} = m_f/v = (2^{1/4}G_F^{1/2})m_f, \tag{3.50a}$$

$$g_{W^+W^-H} = (2^{1/4}G_F^{1/2})2m_W^2, \qquad g_{Z^0Z^0H} = (2^{1/4}G_F^{1/2})2m_Z^2, \tag{3.50b}$$

from which we see that the H likes to couple to the heaviest particles available. For example, in its decays

$$\Gamma\left(H \to \begin{Bmatrix} \ell\bar{\ell} \\ q\bar{q} \end{Bmatrix}\right) = \begin{Bmatrix} 1 \\ 3 \end{Bmatrix} \times \frac{G_F m_f^2 m_H}{4\sqrt{2}\,\pi}\left(1 - \frac{4m_f^2}{m_H^2}\right)^{3/2}, \tag{3.51}$$

so that if $2m_b \ll m_H < 2m_t$ one expects its most important decay modes to be

$$\Gamma(H \to \bar{b}b) : \Gamma(H \to \bar{c}c) : \Gamma(H \to \bar{\tau}\tau) : \Gamma(H \to \bar{s}s) : \Gamma(H \to \mu^+\mu^-)$$
$$\approx 3m_b^2 : 3m_c^2 : m_\tau^2 : 3m_s^2 : m_\mu^2, \tag{3.52}$$

so that the branching ratio into $\mu^+\mu^-$ would be $O(10^{-4})$, and that into $e^+e^-$ another factor of $4 \times 10^4$ smaller. In contrast to its coupling, the mass of the Higgs H is essentially arbitrary; recall from section 1.2 that

$$m_H = \sqrt{2}\,\mu, \tag{3.53}$$

where

$$\mu^2 = \lambda(2\sqrt{2}\,G_F)^{-1}. \tag{3.54}$$

To get large $m_H$, all we need is a large $\mu$ (3.53) and hence a large $\lambda$ (3.54), i.e. strong coupling.

There are some phenomenological constraints on light Higgses [9]. The facts that they have not been seen in nuclear transitions $Z^* \to Z + H(\to e^+e^-)$, nor have shown up via anomalies in muonic atom spectra due to an extra Higgs exchange potential, nor have made an anomalous contribution to low energy neutron–nucleus scattering, tell [99] us that

$$m_H \gtrsim (15 \text{ to } 20)\ \text{MeV}. \tag{3.55}$$

Unfortunately, high energy physics experiments do not seem to give us any stronger bound on $m_H$; for example, we know that in K decay for some range of Higgs masses larger than (3.55):

$$B(K^+ \to \pi^+ + H) < O(10^{-7}), \tag{3.56}$$

but there is considerable uncertainty in the theoretical estimation [9] of

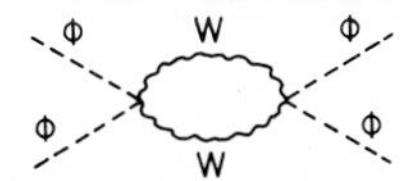


Fig. 3.10. Radiative correction to the effective Higgs potential.

the branching ratio which is also $O(10^{-7})$. There are also debilitating uncertainties in the estimation of $\eta' \to H+X$ decays [99].

There is, however, a theoretical argument pointing to a more stringent lower bound on $m_H$, which comes from considering radiative corrections to the Higgs potential, and cosmology. What happens to the Higgs potential when $\mu^2 \to 0$? The potential then becomes flat at the origin and eq. (3.53) says that the Higgs boson mass goes to zero. But in this limit radiative corrections to the Higgs potential, such as those in fig. 3.10, become important and generate [100] an absolute minimum away from the origin $|\phi| = 0$ as shown in fig. 3.11a. They have the form

$$\delta V(\phi) = \frac{\sqrt{2}\,G_F}{16\pi^2}\left(|\phi|^2\right)^2 \ln\left(|\phi|^2/M^2\right)\left(3\sum_V m_V^4 - 4\sum_f m_f^4\right), \qquad (3.57)$$

which is clearly negative for small $|\phi|$ unless some fermion f is very heavy. Notice that $\delta V(\phi)$ (3.57) has a scale size $M$ introduced as a renormalization scale parameter. It also appears in the specification of a renormalized $\lambda(M)$ in such a way that the full potential $V(\phi) + \delta V(\phi) + \cdots$ is independent of the renormalization scale. Because of (3.57) one expects radiative corrections to (3.53), and for $\mu^2 = 0$ one in fact gets

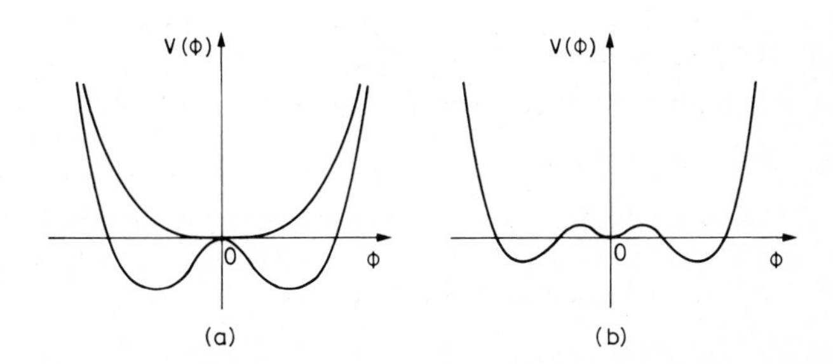


Fig. 3.11. The effective Higgs potential with radiative corrections included (a) for the case when $\mu^2 = 0$, and (b) for the case when $\mu^2$ is slightly less than zero.

[100]

$$m_{\rm H}^2 = \frac{3\alpha^2}{8\sqrt{2}\,G_{\rm F}}\left[\frac{2+\sec^4\theta_{\rm W}}{\sin^4\theta_{\rm W}} - {\rm O}(m_{\rm f}/m_{\rm W})^4\right], \tag{3.58}$$

which gives $m_{\rm H} = 10.4$ GeV for $\sin^2\theta_{\rm W} = 0.20$ [101]. If one takes into account two-loop corrections [102] one finishes up with $m_{\rm H} = 10.6$ GeV for the experimentally preferred value of $\sin^2\hat{\theta}_{\rm W}(m_{\rm W}) = 0.215$.

This value is not yet a lower bound on $m_{\rm H}$ since one could take $\mu^2$ slightly negative and still keep the absolute minimum of the potential at $|\phi| \neq 0$, as indicated in fig. 3.11b. Going to the limit of negative $\mu^2$ with this property one gets [103]

$$m_{\rm H} \gtrsim 7.3 \text{ GeV}. \tag{3.59}$$

This would be the final bound on $m_{\rm H}$ if it were not for cosmology. Initially the Universe was at high temperature $T$ and the mean value of the Higgs field was presumably zero:

$$\langle 0|\phi(T > T_{\rm c})|0\rangle = 0. \tag{3.60}$$

As the Universe cooled, the probability per unit time that it made a transition to $\langle 0|\phi|0\rangle \neq 0$ depends on the parameter $\mu^2$ in the Higgs potential. It turns out that if $\mu^2 < 0$, the rate of transition to the asymmetric vacuum is too slow to be cosmologically acceptable, and it seems [104] that the $\mu^2 = 0$ theory is almost the limit of acceptability, so that finally

$$m_{\rm H} \gtrsim 10.6 \text{ GeV}. \tag{3.61}$$

What about an upper limit on $m_{\rm H}$? The theory becomes strongly interacting [105] if $m_{\rm H} > 1$ TeV, and this is just what happens in the Technicolour models of lectures 4 and 5. If one wishes to keep the coupling $\lambda$ small all the way up to the grand unification mass (see lectures 6 and 7) then one finds [106] that

$$m_{\rm H} \lesssim 200 \text{ GeV}, \tag{3.62}$$

but this is not on such a solid basis as the lower bounds (3.55) or (3.61).

How does one produce Higgs bosons? One promising reaction is heavy $^3{\rm S}_1$ quarkonium ${\rm O} \to {\rm H} + \gamma$, where Wilczek [107] has calculated the graphs of fig. 3.12 to get a decay rate

$$\frac{\Gamma({\rm O} \to {\rm H} + \gamma)}{\Gamma({\rm O} \to \gamma \to {\rm e}^+{\rm e}^-)} \approx \frac{G_{\rm F} m_{\rm O}^2}{4\sqrt{2}\,\pi\alpha}\left(1 - \frac{m_{\rm H}^2}{m_{\rm O}^2}\right). \tag{3.63}$$

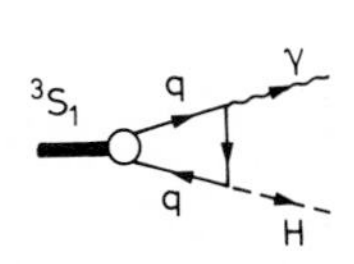


Fig. 3.12. Graph giving $^3S_1$ quarkonium $O \to H + \gamma$ decay [107].

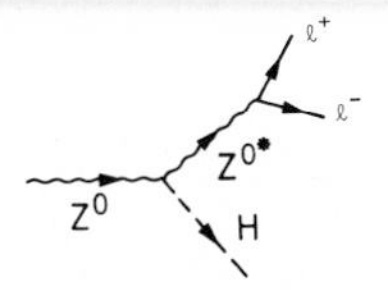


Fig. 3.13. Graph giving $Z^0 \to H + (\ell^+ \ell^-)$ decay [91].

This ratio is $\gtrsim 10\%$ for $m_O \geqslant 40$ GeV as seems to be the case for toponium, so that one expects a branching ratio

$$B(O(t\bar{t}) \to H + \gamma) \gtrsim 1\%, \tag{3.64}$$

unless an unexpected and hence even more exciting decay mode turns out to drown $O \to$ three gluon decays. Another promising reaction [91] is $Z^0 \to H + \ell^+ \ell^-$, which goes through the graph of fig. 3.13 with a branching ratio of order $10^{-4}$ (for $m_H = 10$ GeV) to $10^{-6}$ (for $m_H = 50$ GeV) as shown in fig. 3.14. Also shown in this figure is the branching ratio [92] for $Z^0 \to H + \gamma$, which is of order $10^{-6}$ for 10 GeV $\leqslant m_H \leqslant 50$ GeV. It seems likely that the former decay would have a more distinctive signature as well as a possibly larger rate. Another interesting reaction in which the $Z^0$ is used to catalyze Higgs production is $e^+e^- \to Z^0 + H$ [108]. The cross section is

$$\sigma(e^+e^- \to Z^{0*} \to Z^0 + H)$$
$$= \frac{\pi\alpha^2}{24}\left(\frac{2K}{\sqrt{s}}\right)\frac{K^2 + 3m_Z^2}{\left(s - m_Z^2\right)^2}\,\frac{1 - 4\sin^2\theta_W + 8\sin^4\theta_W}{\sin^4\theta_W\left(1 - \sin^2\theta_W\right)^2}, \tag{3.65}$$

where $K$ is the centre-of-mass momentum of the outgoing particles, which yields

$$\sigma/\sigma_{pt} \gtrsim 0.1 \quad \text{for } m_H < 100 \text{ GeV} \quad \text{at } \sqrt{s} = 200 \text{ GeV}. \tag{3.66}$$

This is the most powerful way to look for the Higgs, and may be the only feasible way [9] if $m_H \geqslant O(50)$ GeV.

Finally let us recall hadron + hadron $\to H + X$. At accessible centre-of-mass energies $\sqrt{s}$, this reaction is expected [109] to have a cross section $\lesssim 10^{-35}$ cm$^2$ for $m_H \gtrsim 10$ GeV. Furthermore, there is no clean decay signature to be picked out from all the hadronic junk, since $B(H \to \mu^+\mu^-) \sim 10^{-3}$ or $10^{-4}$. It therefore seems that $e^+e^-$ collisions afford the best chances to produce and detect the Higgs boson.

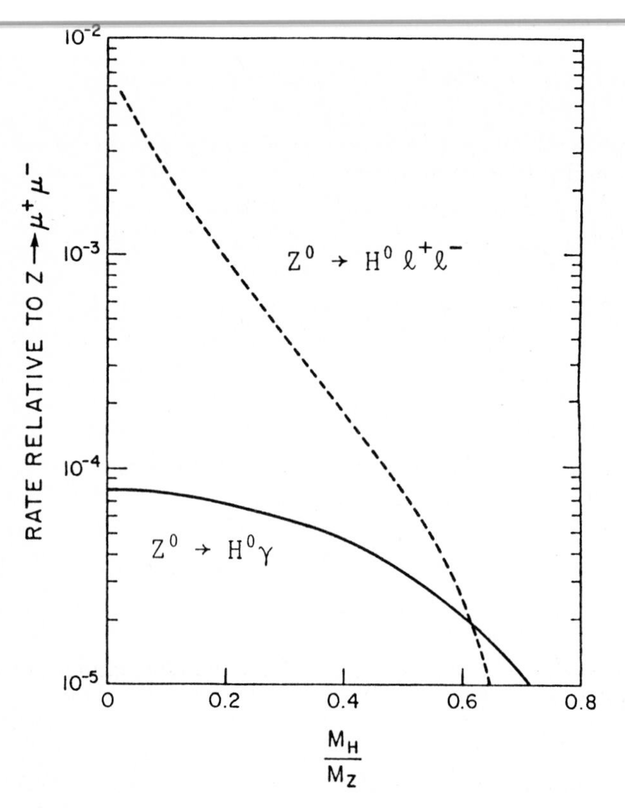


Fig. 3.14. Branching ratios for $Z^0 \to H + (\ell^+ \ell^-)$ and $Z^0 \to H + \gamma$ [92].

### 3.4. *Axions*

So far we have concentrated on the minimal Higgs system in the Weinberg–Salam model, though infinite complications are possible. There is one class of models which is rather strongly motivated, namely those which have an extra symmetry capable of removing the strong interaction *CP* violation due to the $\theta_{\mathrm{QCD}}$ parameter discussed in section 2.6. Recall that the characterization of non-perturbative phenomena required us to introduce the parameter:

$$\mathcal{L}_{\mathrm{QCD}} \to \mathcal{L}_{\mathrm{QCD}} + \theta_{\mathrm{QCD}} \frac{g^2}{32\pi^2} F^i_{\mu\nu} \tilde{F}^{i\mu\nu} . \tag{3.67}$$

Recall also that due to the anomaly

$$\partial^\mu A_\mu = \frac{g^2}{16\pi^2} F^i_{\mu\nu} \tilde{F}^{i\mu\nu} \tag{3.68}$$

in the U(1) axial current $\bar{\psi}\gamma_\mu\gamma_5\psi$, we change $\theta_{\mathrm{QCD}} \to \theta_{\mathrm{QCD}} - 2\sigma$ when we make a chiral rotation through an angle $\sigma$. If we have a theory in which the perturbative Lagrangian possesses such a U(1) chiral symmetry then all values of $\theta_{\mathrm{QCD}}$ are equivalent, and presumably conserve $CP$ as would have been the case if $\theta_{\mathrm{QCD}} = 0$. As an example of such a theory, consider [110] the toy model

$$\begin{aligned}\mathcal{L} &= -\tfrac{1}{4}F_{\mu\nu}F^{\mu\nu} + \mathrm{i}\bar{\psi}\not{D}\psi + \bar{\psi}\left[G\phi(1+\gamma_5) + G^*\phi^*(1-\gamma_5)\psi\right] \\ &\quad + \frac{\mathrm{i}\theta g^2}{32\pi^2} F_{\mu\nu}\tilde{F}^{\mu\nu} - |D_\mu\phi|^2 - \mu^2|\phi|^2 + \lambda|\phi|^4.\end{aligned} \tag{3.69}$$

This theory is invariant under the axial U(1) transformations

$$\psi \to \exp(\mathrm{i}\sigma\gamma_5)\psi, \qquad \phi \to \exp(-2\mathrm{i}\sigma)\phi, \tag{3.70}$$

whose axial anomaly causes a change

$$\delta\mathcal{L} = -2\sigma \frac{g^2}{32\pi^2} F_{\mu\nu}\tilde{F}^{\mu\nu}, \tag{3.71}$$

with changes $\theta$ to $\theta - 2\sigma$. All values of $\theta$ are therefore equivalent, and will conserve $CP$ if

$$\alpha \equiv \arg(\mathrm{e}^{\mathrm{i}\theta} G\langle\phi\rangle) = 0, \tag{3.72}$$

i.e., the fermion mass $m_\psi$ is real when $\theta = 0$. It can be argued [110] that this is the case for this model, and hence that the theory behaves just as if $\theta = 0$.

Does such a U(1) symmetry exist in the minimal Weinberg–Salam model? The only possibly non-invariant piece in $\mathcal{L}_{\mathrm{WS}}$ (1.8) is the Yukawa term

$$\mathcal{L}_{\mathrm{Y}} = -\frac{H_{\mathrm{N}}}{v}(\bar{F}_{\mathrm{L}}\cdot\phi)N_{\mathrm{R}} - \frac{H_P}{v}(\bar{F}_{\mathrm{L}}\cdot\phi^\dagger)P_{\mathrm{R}}, \tag{3.73}$$

where the expression has been rewritten to exhibit explicitly the two different terms which give masses (1.33) to the charge 2/3 quarks P and the charge $-1/3$ quarks N. If we try to make chiral transformations analogous to (3.70) on the left-handed fields $F_{\mathrm{L}}$ ($\sigma_{\mathrm{L}}$), on the $P_{\mathrm{R}}$ fields ($\sigma_{\mathrm{P}}$), $N_{\mathrm{R}}$ fields ($\sigma_{\mathrm{N}}$) and $\phi$ ($\sigma$), under which the interactions (3.73) are invariant, then we find from the first and second terms, respectively, that

$$-\sigma_{\mathrm{L}} + \sigma + \sigma_{\mathrm{N}} = 0 = -\sigma_{\mathrm{L}} - \sigma + \sigma_{\mathrm{P}}. \tag{3.74}$$

Thus $\mathcal{L}_Y$ is invariant if and only if

$$\sigma = \sigma_L - \sigma_N \quad and \quad \sigma_N + \sigma_P = 2\sigma_L. \tag{3.75}$$

The latter condition implies that there is no net axial component in this set of phase transformations, and hence we have no anomaly (3.68) in the transformation which can be used to change $\theta_{QCD}$ and set it to zero. Minimal Weinberg–Salam theories with different values of $\theta_{QCD}$ are inequivalent. As we discussed in section 2.6, the electric dipole moment of the neutron seems to be telling us (2.83) that we live in a world with $|\theta_{QCD}| < 2 \times 10^{-9}$.

One could perhaps imagine that by some unknown and mysterious "relaxation" mechanism [111] the bare value of $\theta_{QCD}$ was set to zero. But then higher-order weak and electromagnetic interactions [68] renormalize $\theta_{QCD}$. They renormalize the Higgs coupling matrices $H_P, H_N$ (1.33), (3.73), necessitating a renormalization of the unitary matrices $U_{P,N}$ and $V_{P,N}$ (1.34) used to diagonalize the quark mass matrices. This renormalization will in general contain a net axial U(1) piece

$$\arg\det\left(V_N^\dagger U_N\right) + \arg\det\left(V_P^\dagger U_P\right), \tag{3.76}$$

and when we rotate the quark fields through this amount we change $\theta_{QCD}$ by the same amount. The first renormalization of $\theta_{QCD}$ in the K–M model occurs in fourth order (cf. fig. 3.15) and gives [68]

$$\delta\theta_{QCD} = O(10^{-16}), \tag{3.77}$$

comfortably below our phenomenological bound (2.83) on $|\theta_{QCD}|$. However, since we are talking about the renormalization of dimensionless coefficients $H_P$ and $H_N$ of dimension-four terms in the renormalizable field theory $\mathcal{L}_{WS}$, we must expect to find a logarithmic divergence in $\delta\theta$ sooner or later. The first one turns up in 14th order (cf. fig. 3.16) and gives [68]

$$\delta\theta_{QCD} \sim \left(\frac{\alpha}{\pi}\right)^7 \left(\frac{m_q}{m_W}\right)^{12} (\text{angles}) \ln(\Lambda/m_W), \tag{3.78}$$

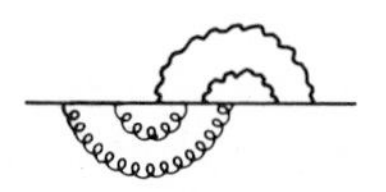

Fig. 3.15. Finite contribution to $\theta$ renormalization in the Kobayashi–Maskawa model [7] coming from a 4th order weak diagram with strong radiative corrections [68].

where $\Lambda$ is a cut-off parameter. If we take $\Lambda$ to be $\lesssim$ the Planck mass $m_{\rm P}$ —surely some new post-Weinberg–Salam physics must appear before then—we estimate [68] that the infinite piece of

$$\delta\theta_{\rm QCD} \lesssim {\rm O}(10^{-32}), \tag{3.79}$$

which is comfortably small. Thus, although the minimal Weinberg–Salam model by itself does not automatically have $\theta_{\rm QCD} = 0$, the observed value of $\theta_{\rm QCD}$ would be small enough to be phenomenologically acceptable if we could embed QCD and the Weinberg–Salam model into a more complete (grand unified?) theory with its $\theta = 0$.

However, it is attractive to try to construct a non-minimal Weinberg–Salam model with $\theta_{\rm QCD} = 0$ automatically, and this can be done [110] with two Higgs doublets $\phi_1$ and $\phi_2$. Consider the Yukawa interactions

$$\mathcal{L}_{\rm Y}^{\rm PQ} = -\frac{H_{\rm N}}{v}(\bar{F}_{\rm L}\cdot\phi_1)N_{\rm R} - \frac{H_{\rm P}}{v}(\bar{F}_{\rm L}\cdot\phi_2^{\dagger})P_{\rm R}, \tag{3.80}$$

and now try to make independent phase transformations $(\sigma_{\rm L}, \sigma_{\rm P}, \sigma_{\rm N}, \sigma_1, \sigma_2)$ analogous to (3.74). We find that invariance requires

$$-\sigma_{\rm L} + \sigma_1 + \sigma_{\rm N} = 0 \quad \textit{and} \quad -\sigma_{\rm L} - \sigma_2 + \sigma_{\rm P} = 0 \tag{3.81}$$

and we can easily get a net axial U(1) transformation $\sigma_{\rm N} + \sigma_{\rm P} - 2\sigma_{\rm L} \neq 0$ if $\sigma_1 \neq \sigma_2$. Hence this theory with two Higgs doublets should automatically have $\theta_{\rm QCD} = 0$.

As it has two Higgs doublets, the Higgs particle spectrum is more complicated than in the minimal model. In particular, the two complex doublets $\phi_1$ and $\phi_2$ give us four neutral fields, one of which is massless and eaten by the $\rm Z^0$, two of which are massive and analogous to the H of section 3.3, and one of which—the axion a [112]—is the (pseudo-) Goldstone boson of the $\rm U_A(1)$ symmetry. We say "pseudo" because the $\rm U_A(1)$ symmetry is broken by non-perturbative QCD effects, such as the condensation in the vacuum of $\bar{\rm q}$q pairs: $\langle 0|\bar{q}_{\rm L}q_{\rm R}|0\rangle \neq 0$, implying that

$$m_{\rm a}^2 \propto m_{\rm q}\langle 0|\bar{q}_{\rm L}q_{\rm R}|0\rangle \propto \Lambda_{\rm QCD}^3. \tag{3.82}$$

Fig. 3.16. A generic 14th order diagram making an infinite contribution to $\theta$ renormalization [68].

(The non-perturbative vacuum is just the lowest energy state of the system, and need not be annihilated by the quark operator: $q|0\rangle \neq 0$, unlike the perturbative vacuum.) More precisely, the axion has been shown to have the following properties in the minimal Peccei–Quinn [110] model (3.80): its mass is

$$m_{\mathrm{a}} = \frac{N_{\mathrm{G}} m_{\pi} f_{\pi}}{(m_{\mathrm{u}}+m_{\mathrm{d}})^{1/2}} \frac{m_{\mathrm{u}} m_{\mathrm{d}} m_{\mathrm{s}}}{m_{\mathrm{u}} m_{\mathrm{d}} + m_{\mathrm{d}} m_{\mathrm{s}} + m_{\mathrm{s}} m_{\mathrm{u}}} \frac{2^{1/4} G_{\mathrm{F}}^{1/2}}{\sin 2\alpha}, \tag{3.83}$$

where $\tan\alpha \equiv \langle 0|\phi_1|0\rangle / \langle 0|\phi_2|0\rangle$, which gives numerically

$$m_{\mathrm{a}} = \frac{23 \times N_{\mathrm{G}}}{\sin 2\alpha} \text{ keV}. \tag{3.84}$$

The couplings of the axion to quarks are well defined [112]:

$$\mathcal{L}_{\mathrm{aq}} = \mathrm{i}(2^{1/4} G_{\mathrm{F}}^{1/2}) a (m_{\mathrm{p}} \bar{p} \gamma_5 p \tan\alpha + m_{\mathrm{n}} \bar{n} \gamma_5 n \cot\alpha), \tag{3.85}$$

where we have again used (p, n) [eq. (1.34)] to denote the mass eigenstates of charge $+2/3$ and charge $-1/3$ quarks, respectively. The couplings of the axion to charged leptons have a twofold ambiguity:

$$\mathcal{L}_{\mathrm{a}\ell} = \mathrm{i}(2^{1/4} G_{\mathrm{F}}^{1/2}) a m_{\ell} \bar{l} \gamma_5 l (\tan\alpha \text{ or } \cot\alpha). \tag{3.86}$$

The couplings to light quarks (u, d, s) at small momentum transfers are best parametrized in terms of direct couplings between the axion, the $\pi$ and the $\eta$, which can be derived by diagonalizing the pseudoscalar meson mass matrix:

$$\xi_{\pi} \equiv \xi \left( \frac{3m_{\mathrm{d}} - m_{\mathrm{u}}}{m_{\mathrm{d}} + m_{\mathrm{u}}} \tan\alpha - \frac{3m_{\mathrm{u}} - m_{\mathrm{d}}}{m_{\mathrm{u}} + m_{\mathrm{d}}} \cot\alpha \right), \tag{3.87a}$$

$$\xi_{\eta} \equiv \xi (3^{1/2} \tan\alpha + 3^{-1/2} \cot\alpha), \tag{3.87b}$$

where

$$\xi \equiv 2^{1/4} G_{\mathrm{F}}^{1/2} f_{\pi} \approx 1.9 \times 10^{-4}. \tag{3.88}$$

Can the axion exist? Studies [113] of heat transport in stars suggest that

$$m_{\mathrm{a}} > 200 \text{ keV}, \tag{3.89}$$

but this is consistent with the mass (3.84) for suitable values of $\alpha$. A more serious objection [114] to the axion comes from beam dump experiments

[115] at CERN which looked for the production reaction $p+N \to a+X$, $a+N \to X$, and did not find them:

$$\sigma(p+p \to a+X)\sigma(a+p \to X)|_{expt} \lesssim O(10^{-69})\ cm^4, \tag{3.90}$$

whereas one can argue [114] that there is a theoretical "lower bound" on the product (3.90) coming from the $a-\eta$ coupling $\xi_\eta$ (3.87b), which gives

$$\sigma(p+p \to a+X)\sigma(a+p \to X)|_{theory} \gtrsim O(10^{-67})\ cm^4. \tag{3.91}$$

Similar negative results come from other beam dump experiments, notably an electron beam dump at SLAC [115]. Also, there are negative results on axion production in nuclear reactors, followed by subsequent secondary interaction or decay into $\gamma\gamma$ [116].

However, there is a recent claim [117] of a positive result from a beam dump experiment at SIN looking at

$$p(600\ MeV) + N \to a + X, \quad a \to \gamma\gamma\ decay. \tag{3.92}$$

The rate at which events are seen seems compatible with the original axion model, but the statistical significance of the claimed effect is not completely convincing. In particular, the interpretation of the observed events as decays depends on comparing two data sets with and without a heavy iron absorber in front of the decay region: both sets should show a similar peak of events coming from the direction of the target. The statistical significance of the effect is reduced when one has to divide the events into two such subclasses.

Since these experimental results were published [117], there have been four more negative results. Some theorists [118] have re-estimated the branching ratio for $K^+ \to \pi^+ + a$ and found that it is generically much larger than the experimental upper limit of $1.4 \times 10^{-6}$, unless one invokes a conspiratorial cancellation between different contributing diagrams. There have been two negative searches [119] for $Z^* \to Z + a$ decays at a level one hundred or more lower than the theoretically predicted [120] rate. Finally, there has been an unsuccessful search [121] for $p+N \to a+X$, $a+N \to \mu^+\mu^- + X$ by the CDHS collaboration.

However, the axion is not dead yet. It is possible [120] to construct axion models which evade the latest experimental setbacks, and the Aachen group [117] claims new evidence of axion production at the Jülich nuclear reactor. Nevertheless my gut feeling is that the original axion of Peccei and Quinn [110], Weinberg and Wilczek [112] probably does not exist. In lecture 8 we will meet a variant [122, 123] of the idea which is not excluded.

# II. Dynamical symmetry breaking

## 4. Technicolour theories

### 4.1. *Motivations*

Figure 4.1a shows the ascetic GUT view of the fundamental interactions, according to which nothing very much happens in physics between the Weinberg–Salam [1] energy scale of $10^2$ GeV, and the energy scale of $10^{15}$ GeV or so at which GUTters [3] believe that grand unification takes place. In the great desert between these energy scales the low energy couplings just evolve logarithmically until they finally come together. A basic hypothesis of this picture is that the presently known "elementary" fields—gauge bosons, quarks and leptons, Higgs fields—all remain point-like down to distance scales of order $10^{-28}$ cm or less. This is clearly an enormous extrapolation of our present knowledge which is only that quarks and leptons are point-like down to distances of order $10^{-16}$ cm.

Many theorists reject the ascetic view of fig. 4.1a as overly simplistic and guilty of hubris in the grandiose nature of the extrapolations involved. They propose [2] the less boring picture of fig. 4.1b in which there is plenty of action in the form of new strong interactions at energy scales of 1 to 100 TeV. A common theoretical thread in their approaches is the belief that there are by now so many "elementary" fields that they

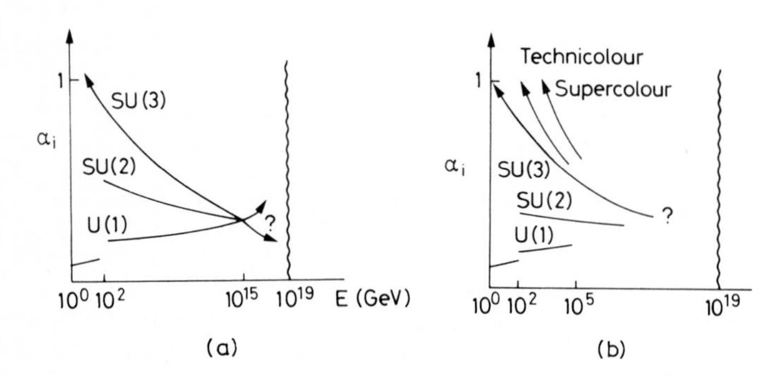


Fig. 4.1. (a) The ascetic view of gauge theories proposed in GUTs [3], contrasted with (b) the richer prospects offered by theories of dynamical symmetry breaking [2].

must in fact presumably all be composite on some sufficiently small distance scale. What should this scale be? In the case of scalar (Higgs) fields there are some technical problems [124], which motivate the idea that they are "precociously" composite. The reason is that if one calculates quantum corrections to the masses of elementary scalars through, for example, the diagrams of fig. 4.2, one finds that they are naïvely quadratically divergent:

$$|\delta m_S^2| \sim \Lambda^2, \tag{4.1}$$

where $\Lambda$ is an ultraviolet cut-off. (Since these divergences can be renormalized away, it is not clear how much we should worry about them: see the later discussion.) On the other hand, we saw in the last lecture that the physical Weinberg–Salam Higgs should have a mass less than 1 TeV if its self-couplings are not to be strong. It is "natural" to want [124] a theory in which the corrections to the mass are smaller than the mass itself:

$$|\delta m_S^2| < m_S^2 < 1\ \text{TeV}^2, \tag{4.2}$$

in which case one should regulate the divergence (4.1) at an energy scale $<1$ TeV. Two ways of doing this have been proposed, one of which postulates an extra (super-)symmetry [125] which forces cancellations between the different diagrams in fig. 4.2. This approach will be discussed in lectures 9 and 10. The other approach [2] "dissolves" the diagrams by making Higgs fields composite, and this alternative is what we will discuss in this and the next lecture. A general overview of theoretical ideas and problems in this approach is shown in fig. 4.3.

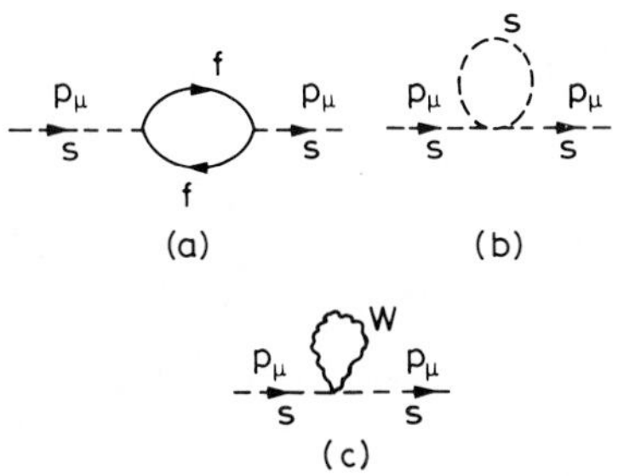

Fig. 4.2. Diagrams which yield quadratically divergent contributions to the mass$^2$ of a scalar boson S.

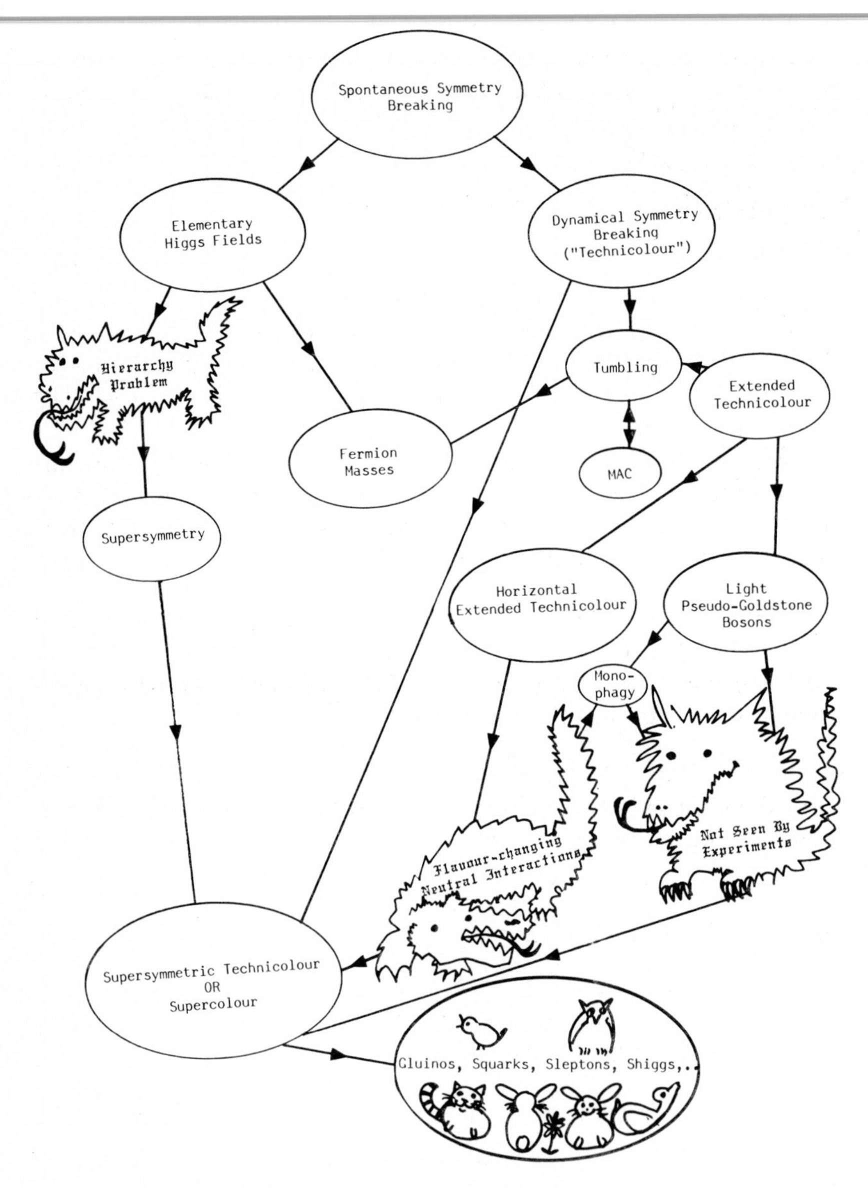


Fig. 4.3. A general overview of the ideas and implications of theories [2] of dynamical symmetry breaking. I thank Tatiana Fabergé for the artwork.

Some remarks are perhaps in order about the use above of the term "precociously composite" for the scalar fields. If one looks at radiative corrections to fermion masses one finds that they are only logarithmically divergent [124]:

$$\delta m_{\mathrm{f}} \sim m_{\mathrm{f}}(\alpha/\pi)\ln(\Lambda/m_{\mathrm{f}}), \tag{4.3}$$

and hence become comparable with $m_{\mathrm{f}}$ only when

$$\ln(\Lambda/m_{\mathrm{f}}) \sim (\alpha/\pi)^{-1}, \tag{4.4}$$

or perhaps when $\Lambda \sim 10^{19}$ GeV, the Planck mass? This is then a "natural" scale to look for compositeness of fermion fields, as will be explored in lecture 10. The question we will study now is whether scalar fields might be composite on a distance scale much larger than that on which fermions are composite.

Before exploring this scenario in more detail, let us briefly recall how elementary Higgs bosons should work—and why they may not. We recall from section 1.1 that in the minimal Weinberg–Salam model [1] we have a complex doublet of Higgs fields:

$$\phi \equiv \begin{pmatrix} \phi^+ \\ \phi^0 \end{pmatrix}, \quad \phi^\dagger = \begin{pmatrix} \bar{\phi}^0 \\ -\phi^- \end{pmatrix}, \tag{4.5}$$

and that spontaneous symmetry breakdown arises with vacuum expectation values for the neutral components:

$$\langle 0|\phi^0|0\rangle = v/\sqrt{2} = \langle 0|\bar{\phi}^0|0\rangle \neq 0. \tag{4.6}$$

The three gauge bosons $W^\pm$ and $Z^0$ then acquire masses by eating the massless Higgs fields

$$\left(\phi^+, (\mathrm{i}/\sqrt{2})(\phi^0 - \bar{\phi}^0), \phi^-\right), \tag{4.7}$$

thanks to the coupling of fig. 4.4, which, when iterated as in eq. (1.14):

$$\frac{1}{q^2} + \left(\frac{1}{q^2}\right)^2\left(\frac{g^2v^2}{4}\right) + \left(\frac{1}{q^2}\right)^3\left(\frac{g^2v^2}{4}\right)^2 + \cdots, \tag{4.8}$$

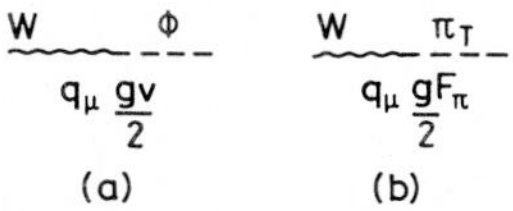

Fig. 4.4. (a) The direct W–Higgs coupling of lecture 1 can be replaced by (b) the W–technipion coupling to give the $W^\pm$ masses.

gives a mass to the $W^{\pm}$ bosons:

$$m_{W^{\pm}} = gv/2. \tag{4.9}$$

It has already been glibly asserted [124] that the problem with this picture is that there are radiative corrections to the masses of the Higgs fields which are quadratically divergent [eq. (4.1)]: let us just verify this and look at the divergences more carefully. The fermion loop in fig. 4.2a gives a correction

$$g_{\mathrm{Sff}}^2 \int \frac{\mathrm{d}^4 k}{(2\pi)^4} \frac{1}{\not{k} + \not{p}/2 - m_{\mathrm{f}}} \frac{1}{-\not{k} + \not{p}/2 - m_{\mathrm{f}}} \tag{4.10}$$

to the scalar propagator. At large $k$ the integral in (4.10) clearly diverges quadratically:

$$g_{\mathrm{Sff}}^2 \int \frac{\mathrm{d}^4 k}{(2\pi)^4} \left(\frac{1}{\not{k}}\right)^2, \tag{4.11}$$

but this divergence is independent of the external momentum scale $p$. The same is also true of the boson loops in fig. 4.2b, c: the complete integrals

$$(\lambda, g^2) \int \frac{\mathrm{d}^4 k}{(2\pi)^4} \frac{1}{k^2 - m_{\mathrm{S}}^2} \tag{4.12}$$

are actually completely independent of $p$. In the dimensional regularization scheme [126], the quadratic divergences in (4.11) and (4.12) show up as poles at $d$ ( = space–time dimensionality) = 2, and hence do not have to be renormalized, unlike the usual poles at $d = 4$. This corresponds to the observation in momentum space that because the quadratic divergences (4.11), (4.12) are momentum independent, they do not alter the effective scalar mass as a function of its momentum. This is to be contrasted with the residual logarithmic divergence in (4.10), or the corresponding logarithmic divergences in fermion propagators, which can be summed using the renormalization group [127] to obtain scale-dependent masses [128] varying logarithmically:

$$m_{\mathrm{S}}(p) \text{ or } m_{\mathrm{f}}(p) \approx (\ln p/\mu)^{\mathrm{power}} \tag{4.13}$$

Since even the divergences of fig. 4.2 can be summed in this way, it is to some extent a matter of taste how much one worries about the "naturalness" problem (4.2) of elementary scalars. However, when one tries to embed the Weinberg–Salam model [1] in a larger context, the problem recurs. For example in GUTs [3] in the normal way one gets

$$\delta m_{\text{light Higgs}} \sim m^2_{\text{superheavy}} \tag{4.14}$$

from diagrams of the type of fig. 4.5—the famous hierarchy problem [129] tackled in lecture 9. Also, it has been argued [130] that when one takes into account quantum gravity, there are contributions to

$$\delta m^2_{\text{Higgs}} \sim m^2_{\text{Planck}}\,. \tag{4.15}$$

Of course all these big contributions (4.11), (4.12), (4.14) and (4.15) to the scalar (mass)$^2$ could just be wished away or ignored, but this would leave a bad taste in the mouth (cf. the monster on the left of fig. 4.3), which is the reason why we will now investigate an alternative.

### *4.2. The basic technicolour idea*

We postulate [2] a new set of exactly symmetric, asymptotically free gauge interactions based on a group such as $SU(N)_{TC}$ or $SO(N)_{TC}$, which become strong on a momentum scale $\Lambda_{TC} \lesssim 1$ TeV. We further assume [2] that there are some massless technifermions which feel these technicolour forces, whose left-handed and right-handed components transform differently under the weak SU(2)×U(1) group. The simplest possibility is to give them the same SU(2)×U(1) transformation properties as ordinary fermions, e.g. left-handed doublets and right-handed singlets in the fundamental representation of the technicolour group:

$$\begin{pmatrix} U_L \\ D_L \end{pmatrix}_{1,2,\ldots,N}, \quad (U_R)_{1,2,\ldots,N}, \quad (D_R)_{1,2,\ldots,N}\,. \tag{4.16}$$

Such a theory possesses the global chiral flavour symmetry

$$SU(2)_L \times SU(2)_R \times U(1)_V \times U(1)_A \tag{4.17}$$

if non-technicolour interactions and non-perturbative technicolour effects are ignored. These latter are believed to break the axial $U(1)_A$ symmetry

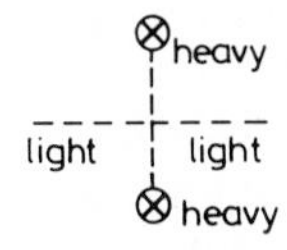


Fig. 4.5. There is a contribution to the mass$^2$ of a "light" Higgs coming from its quartic interactions with "heavy" Higgses with large vacuum expectation values in GUTs [3].

just as we believe happens in QCD, while the $U(1)_V$ is just the technibaryon number. It is thought that non-perturbative technicolour effects would also break the chiral symmetry

$$SU(2)_L \times SU(2)_R \rightarrow SU(2)_{L+R} \tag{4.18}$$

through the formation of condensates of technifermions

$$\begin{aligned}\langle 0|\bar{U}_L U_R|0\rangle &= \langle 0|\bar{U}_R U_L|0\rangle = \langle 0|\bar{D}_L D_R|0\rangle \\ &= \langle 0|\bar{D}_R D_L|0\rangle = O(\Lambda^3_{TC}).\end{aligned} \tag{4.19}$$

This is again by analogy with QCD [77] where it is believed that the light quarks u, d condense in the vacuum:

$$\langle 0|\bar{u}_L u_R|0\rangle = \langle 0|\bar{d}_L d_R|0\rangle = \cdots. \tag{4.20}$$

The breaking (4.18) of the global chiral symmetry must be accompanied by the appearance of massless Goldstone bosons carrying the quantum numbers of the spontaneously broken generators (they are pseudoscalars only with respect to the technicolour interactions):

$$\pi_T^+ \equiv \bar{D}\gamma_5 U, \qquad \pi_T^0 \equiv \frac{1}{\sqrt{2}}\left(\bar{U}\gamma_5 U - \bar{D}\gamma_5 D\right), \qquad \pi_T^- \equiv \bar{U}\gamma_5 D. \tag{4.21}$$

These are very analogous to the usual pions of QCD which in the absence of non-strong interactions would have masses

$$m_\pi^2 = O(m_{u,d}\Lambda_{QCD}), \tag{4.22}$$

and hence would also be massless if the u and d quarks were strictly massless. The $(\pi_T^\pm, \pi_T^0)$ (4.21) can be eaten by the $W^\pm$ and $Z^0$ and give them masses, in just the same way as the previous massless Higgs fields (4.7). In the same way that the conventional $\pi^\pm$ couple to the $W^\pm$ through the pion decay constant $f_\pi$, so also the $\pi_T$ would couple through the analogous technipion decay constant $F_\pi$ as in fig. 4.4b, which is formally equivalent to the usual Higgs–$W^\pm$ coupling of fig. 4.4a. This coupling is then iterated as in (4.8) to give the $W^\pm$ masses:

$$m_{W^\pm} = gF_\pi/2. \tag{4.23}$$

The structure so obtained is summarized in the Higgs–Technicolour translation dictionary shown in table 4.1. A few remarks about table 4.1 are perhaps in order. Just as there is believed to be a scalar isoscalar meson $\varepsilon$ made out of u and d quarks (though it has never been conclusively identified), so the corresponding TC analogue $\varepsilon_T$ should

Table 4.1
Higgs–Technicolour dictionary

| Higgs | Technicolour |
|---|---|
| Higgs vev:<br>$\langle 0 \vert \phi \vert 0 \rangle = v/\sqrt{2} \neq 0$ | Technifermion condensate:<br>$\langle 0 \vert \bar{F}_R F_L \vert 0 \rangle = O(\Lambda^3_{TC}) \neq 0$ |
| massless eaten Higgses:<br>$(\phi^+, (1/\sqrt{2}\,i)(\phi^0 - \bar{\phi}^0), \phi^-)$ | Goldstone bosons:<br>$(\pi_T^+, \pi_T^0, \pi_T^-)$ |
| Higgs–W coupling:<br>$iq_\mu g v/2$<br>$m_W = gv/2$ | Technipion decay coupling:<br>$iq_\mu g F\pi/2$<br>$m_W = gF_\pi/2$ |
| physical Higgs H | Techni-analogue of QCD:<br>$\varepsilon$(~~500~~, ~~600~~, ~~700~~, ~~1000~~, ~~1200~~, ~~1300~~,..?) meson |
| — | Techni-$\eta$ and other technihadrons |
| — | new strong interactions on scale of 1 TeV |
| fermion masses $m_f = g_{Hf\bar{f}} \langle 0 \vert \phi \vert 0 \rangle$<br>from ff–H diagrams | ? ⇒ extended technicolour |

exist [131] as a replacement for the physical Higgs discussed in section 3.3. The technicolour theory also contains many other strongly interacting particles with masses of order 1 TeV, such as the techni-$\eta$, which is expected [132] to acquire a mass from the same sort of $U_A(1)$-breaking non-perturbative effects which give the $\eta$ its mass. This and the other technihadrons have no analogues in the simple Higgs scheme. However, there is one important feature of the Higgs model for which there is as yet no analogue in the technicolour framework, namely the Yukawa couplings (1.6) which give masses to the "elementary" fermions. Whereas the technicolour scheme outlined so far is rather economical and elegant, remedying this defect gets us into considerable trouble, as we will see in the next section and on the right side of fig. 4.3.

### 4.3. *Fermion masses and extended technicolour*

If we want to find a techni-analogue of the usual Higgs mechanism for giving masses to fermions, fig. 4.6a, yielding

$$g_{H\bar{f}f} \bar{f} f \phi: \quad \langle 0 | \phi | 0 \rangle = v \quad \Rightarrow m_f = g_{H\bar{f}f} v, \tag{4.24}$$

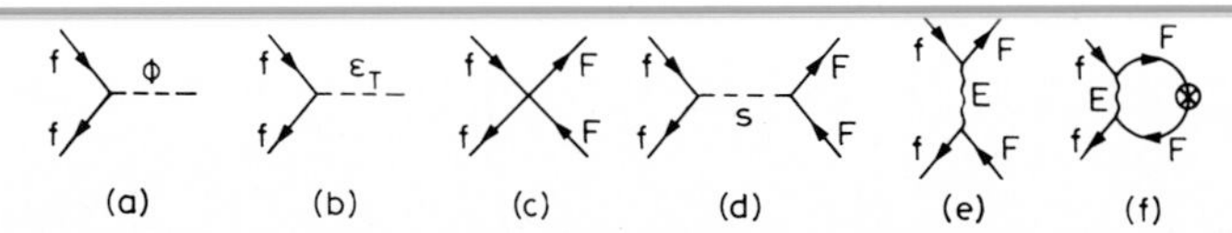


Fig. 4.6. The generation of fermion masses (a) by a coupling to a Higgs, (b) by coupling to a technimeson, which (c) requires a four-fermion coupling, which could in turn be generated either (d) by a scalar boson exchange [133, 134], or (e) by a crossed-channel vector boson exchange [135] which (f) gives rise to a mass through the technifermion condensate.

our first guess might be to replace fig. 4.6a by a coupling of conventional fermions to the composite "Higgs" as illustrated in fig. 4.6b:

$$g_\varepsilon \bar{f} f \varepsilon_{\mathrm{T}}: \quad \varepsilon_{\mathrm{T}} \equiv \frac{1}{\sqrt{2}}(\bar{U}U + \bar{D}D). \tag{4.25}$$

Unfortunately, since the $\varepsilon_{\mathrm{T}}$ is a composite state (4.25), this involves introducing a four-fermion coupling as illustrated in fig. 4.6c:

$$g_\varepsilon(\bar{f}f)(\bar{F}F), \tag{4.26}$$

such a coupling $g_\varepsilon$ must have mass dimensions:

$$[g_\varepsilon] = [\mathfrak{L}] - 4[f] = 4 - 6 = 2, \tag{4.27}$$

which conflicts with the renormalizability requirement that couplings have zero dimensionality, $[g] = 0$. However, we already know full well that such an apparently non-renormalizable four-fermion interaction (4.26) can be generated at low energies by the exchange of a heavy boson, either a spin-zero boson in the direct channel (fig. 4.6d), or a spin-one boson in the crossed channel (fig. 4.6e). The former possibility is developed in the supercolour [133] or supersymmetric technicolour [134] theories discussed at the end of lecture 5. Let us for the moment study the possible existence [135] of heavy vector bosons E (fig. 4.6e), which have a directly analogous rôle to that the $\mathrm{W}^\pm$ and $\mathrm{Z}^0$ play in generating the Fermi four-fermion weak interaction (1.43).

It is apparent from fig. 4.6e that these bosons must possess technicolour indices, and they are often called extended technicolour (ETC) bosons. Their exchange generates an effective coupling

$$g_\varepsilon = g_{\mathrm{E}}^2 / m_{\mathrm{E}}^2, \tag{4.28}$$

and hence fermion masses. This mechanism is illustrated diagrammatically in fig. 4.6f: the condensate $\langle 0|\bar{F}F|0\rangle \neq 0$ (4.19) of technifermions generates a fermion mass

$$m_{\mathrm{f}} = \left(g_{\mathrm{E}}^2/m_{\mathrm{E}}^2\right)\langle 0|\bar{F}F|0\rangle = \mathrm{O}\left(\Lambda_{\mathrm{TC}}^3\right)g_{\mathrm{E}}^2/m_{\mathrm{E}}^2, \tag{4.29}$$

at which point a new complication rears its ugly head. The known fermions have masses which vary over a wide range:

$$m_{\mathrm{e}} \sim \tfrac{1}{2}\ \mathrm{MeV} \quad \text{to} \quad m_{\mathrm{t}} \gtrsim 20\ \mathrm{GeV}, \tag{4.30}$$

with hierarchical ratios between the fermions in different generations. The idea of grouping fermions into generations is discussed in more detail in lecture 6: see for example fig. 6.1. Looking at the formula (4.29), it appears then that we need several very different values of $g_{\mathrm{E}}^2/m_{\mathrm{E}}^2$. Since the E bosons extend technicolour, we expect $g_{\mathrm{E}}^2$ to be of order unity, and it is difficult to see how to get the necessary factors (4.30) anywhere except from the denominators $m_{\mathrm{E}}^2$ in (4.29). Thus we are led to a scenario with several different mass scales for the different ETC bosons $\mathrm{E}_{\mathrm{f}}$:

$$m_{\mathrm{f}} = \left(g_{\mathrm{E}}^2/m_{\mathrm{E}_{\mathrm{f}}}^2\right)\langle 0|\bar{F}F|0\rangle \quad \Rightarrow m_{\mathrm{E}_{\mathrm{f}}}^2 = \mathrm{O}\left(\Lambda_{\mathrm{TC}}^3\right)/m_{\mathrm{f}}. \tag{4.31}$$

Equation (4.31) is probably an oversimplification, since more than one $\mathrm{E}_{\mathrm{f}}$ boson can contribute to the mass of any given fermion, and we must allow for generalized Cabibbo–Kobayashi–Maskawa [7] mixing, etc. Nevertheless, it serves as a useful guide to the likely spectrum of $\mathrm{E}_{\mathrm{f}}$ masses [136], as shown in table 4.2.

It is apparent that the structure of ETC interactions must be quite complicated, with a plethora of mass scales whose origins we will discuss later. The general belief [135] is that the ETC generators should commute with the weak SU(2)×U(1) generators $[G_{\mathrm{ETC}}, \mathrm{SU}(2)\times\mathrm{U}(1)] = 0$, because otherwise one would have to unify them in a larger simple group:

$$G \supset G_{\mathrm{ETC}}, \mathrm{SU}(2)\times\mathrm{U}(1), \tag{4.32}$$

Table 4.2
Spectra of fermion and ETC boson masses [136]

| Quark | u | d | s | c | b | t |
|---|---|---|---|---|---|---|
| Assumed mass | 6 Mev | 10 MeV | 200 MeV | 1.5 GeV | 5 GeV | 20 GeV |
| $m_{\mathrm{E}_{\mathrm{q}}}^2$ | 2200 $\mathrm{TeV}^2$ | 1200 $\mathrm{TeV}^2$ | 60 $\mathrm{TeV}^2$ | 8 $\mathrm{TeV}^2$ | $2\frac{1}{2}$ $\mathrm{TeV}^2$ | $\frac{1}{2}$ $\mathrm{TeV}^2$ |

with a unique coupling constant. It is difficult to see how to unify

$$g_{\mathrm{E}} = \mathrm{O}(1) \gg g_{\mathrm{weak}} \tag{4.33}$$

without introducing a hierarchy of mass scales à la GUT, which is anathema to the philosophy of dynamical symmetry breaking [2]. The same argument suggests that $G_{\mathrm{ETC}}$ should also commute with strong SU(3), since this interaction is also weak in the TeV region:

$$\alpha_{\mathrm{s}} \equiv g_{\mathrm{s}}^2/4\pi \ll 1. \tag{4.34}$$

We are therefore led to a theory [137, 138] with at least one complete technigeneration of fermions and the basic ETC scenario exhibited in fig. 4.7. One imagines that the three or more conventional fermion generations are coupled to the technigenerations by a sequence of ETC bosons with progressively smaller masses (cf. table 4.2) corresponding to a sequential breakdown:

$$G_{\mathrm{ETC}} \underset{10^3\,\mathrm{TeV}^2}{\rightarrow} G_{\mathrm{E'TC}} \underset{30\,\mathrm{TeV}^2}{\rightarrow} G_{\mathrm{E''TC}} \underset{1\,\mathrm{TeV}^2}{\rightarrow} G_{\mathrm{TC}}. \tag{4.35}$$

An important point to note is the presence [135, 136] of so-called "horizontal ETC" (HETC) generators and bosons in the one-technigeneration model. The existence of generators $\bar{\mathrm{f}}_i\mathrm{F}$ and $\bar{\mathrm{F}}\mathrm{f}_j$, where $\mathrm{f}_i$ and $\mathrm{f}_j$ are conventional fermions of the same charge from generations $i$ and $j$, and F is a technifermion, necessarily implies the existence of generators and bosons

$$\left[\bar{f}_i F, \bar{F} f_j\right] = \bar{f}_i f_j, \tag{4.36}$$

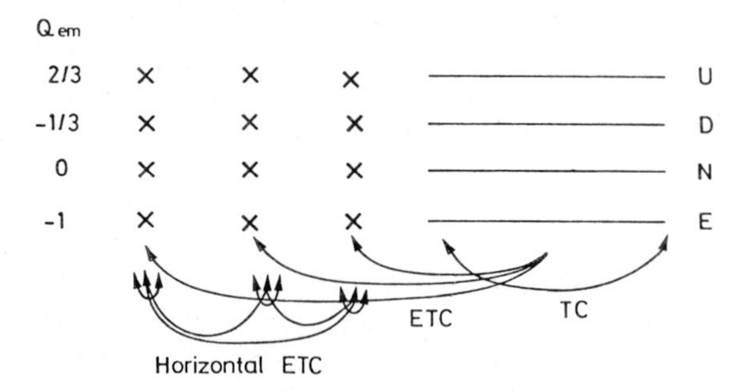


Fig. 4.7. The multiplet structure and interactions in an extended technicolour model [135] containing one technigeneration, showing the action of Extended TechniColour (ETC) and Horizontal Extended TechniColour (HETC) interactions.

which couple directly the fermions of ordinary generations. Some of them will be essentially flavour-diagonal ($i=j$), while others are intrinsically flavour-changing ($i \neq j$). One would expect [136] them generically to have masses comparable to the heavier of the ETC bosons $E_{f_i}, E_{f_j}$, for example

$$\begin{aligned} m^2_{\bar{f}_1 f_1} &= \text{Clebsch–Gordan coefficient} \times m^2_{\bar{f}_1 f_2} \\ &= \text{Clebsch–Gordan coefficient} \times m^2_{E_{f_1}}, \end{aligned} \tag{4.37}$$

with the only corrections to the relations (4.37) being of order of the ratio of the E′TC and ETC mass scales (4.35). While the above picture is complicated by generalized Cabibbo mixing, it seems clear that the "horizontal" ETC bosons will give rise to flavour-changing neutral interactions (see the corresponding monster in fig. 4.3). The only way to avoid their existence is to postulate a theory with multiple technigenerations $F_a$:

$$\left[\bar{f}_i F_a, \bar{F}_b f_j\right] = 0 \quad (i \neq j, a \neq b), \tag{4.38}$$

but this scenario runs into more problems [139] concerned with light spin-zero bosons [137, 138] (cf. lecture 5) and breaking the degeneracy between the different generations. Let us persevere for the moment with the basic ETC scenario of fig. 4.7.

### 4.4. Origin of the ETC masses

Where do the widely varying mass scales of table 4.2 or eq. (4.35) come from? We want to generate them by spontaneous symmetry breaking, and this should be of dynamical origin [140] according to the philosophy of this lecture. In order to break down the ETC group as in (4.35) we need condensates $\bar{F}_R F_L$ of fermions which are not singlets of ETC. Thus we need fermion representations $R$ that are complex with respect to $G_{ETC}$:

$$R(F_R) \neq R(F_L), \tag{4.39}$$

though they are real with respect to the TC subgroup. Then given (4.39) condensates $\langle 0|\bar{F}_R F_L|0\rangle \neq 0$ will necessarily break $G_{ETC}$. To get all the masses of table 4.2 we need several successive stages of dynamical symmetry breaking, called "tumbling" [140].

We need a criterion for deciding the sequence of symmetry breaking, and the only candidate [140] on the market is the heuristic "Most

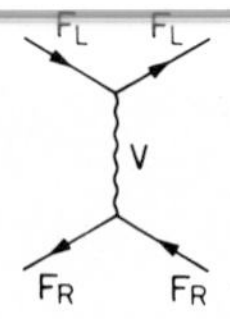


Fig. 4.8. The diagram whose algebraic properties are supposed to determine the channel in which fermions condense, according to the Most Attractive Channel (MAC) hypothesis [140].

Attractive Channel" (MAC) hypothesis. According to this, one looks at the one gauge boson exchange potential (fig. 4.8) between an $F_R$ and an $F_L$ (with group Casimirs $C_R$ and $C_L$, respectively) and analyses it in different direct channel representations (with group Casimirs $C_D$). The effective Coulomb potential is proportional to

$$(g_E^2/4\pi)(C_D - C_R - C_L), \tag{4.40}$$

and the idea is that the dynamics will choose the channel with the maximum value of $(C_R + C_L - C_D)$ to condense first in the vacuum: $\langle 0|\bar{F}_R F_L|0\rangle \neq 0$ at an energy scale such that the quantity (4.40) is of order unity:

$$g_E^2(\Lambda_{ETC})/4\pi = O(1)/(C_R + C_L - C_D). \tag{4.41}$$

After the first stage of symmetry breaking one performs a similar MAC analysis of the residual unbroken theory, and repeats the same cycle as many times as possible, as illustrated in fig. 4.9. One stops at the stage where the MAC is a singlet of the residual gauge group, which therefore remains unbroken and is to be identified with $G_{TC}$. In this tumbling/MAC scenario all mass scales $(\Lambda_{TC}, m_{ETC}, m_f, \ldots)$ would be determined dynamically in a rather elegant way, but there are unfortunately some objections to this programme [140].

We have seen that condensation in some channel (e.g. the MAC) generates some non-zero $m_{ETC}$. The exchange potential due to these bosons will therefore be screened. Should they or should they not [141] be included in the Coulomb potential (4.40)? It is possible [141] to set up models in which the MAC changes when the screened exchanges are excluded from the computation. In such a case, is it the original MAC that condenses first, or the revised one, or do they both break at essentially the same energy scale? In fact, it is quite difficult in realistic models to get a large ratio (cf. table 4.2) between successive E$^n$TC

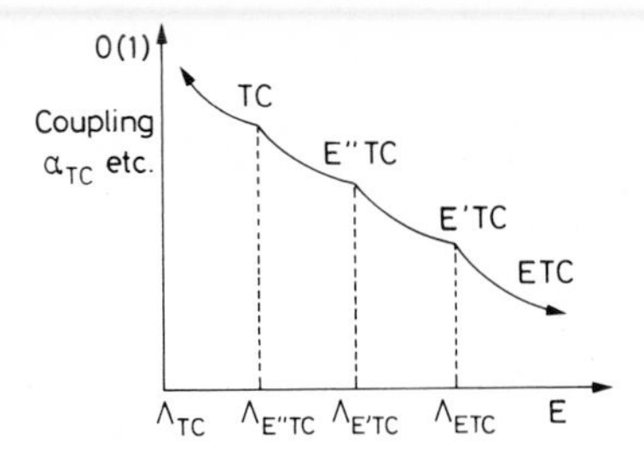


Fig. 4.9. The evolution of the effective coupling with energy, showing how [140] an ETC group can be broken in successive stages ("tumbling") down to an exact TC subgroup.

breaking scales: the coupling constant in fig. 4.9 typically evolves too rapidly from

$$\mathrm{E}_1\colon\ \frac{g^2(\mathrm{E}_1)}{4\pi}\left(C_{\mathrm{R}_1}+C_{\mathrm{L}_1}-C_{\mathrm{D}_1}\right)=\mathrm{O}(1)$$

$$\text{to}\ \ \mathrm{E}_2\colon\ \frac{g^2(\mathrm{E}_2)}{4\pi}\left(C_{\mathrm{R}_2}+C_{\mathrm{L}_2}-C_{\mathrm{D}_2}\right)=\mathrm{O}(1). \tag{4.42}$$

It is also not obvious why it is only the scalar channel $\bar{F}_{\mathrm{R}}F_{\mathrm{L}}$ which can condense. Why not [142] a vector channel such as $\bar{F}_{\mathrm{R}}\gamma_{\mu}F_{\mathrm{R}}$ or $\bar{F}_{\mathrm{L}}\gamma_{\mu}F_{\mathrm{L}}$, which always contains a group singlet, and so should be favoured according to the MAC criterion? A condensate $\langle 0|\bar{F}_{\mathrm{L\ or\ R}}\gamma_{\mu}F_{\mathrm{L\ or\ R}}|0\rangle$ would of course violate Lorentz invariance [143], but how do we know this cannot happen? (Or maybe has happened: perhaps at very short distances there are more than four space–time dimensions?) Some of these objections to the MAC criterion have been studied [144] in two-dimensional models, where the one-boson exchange potential actually confines, so that the quantity (4.40) is a priori more likely to be relevant than in four dimensions. It was found [144] that condensation in the MAC often reduces the global flavour group, implying the existence of Goldstone bosons which are of course forbidden [145] in two dimensions. Therefore the MAC criterion must fail in these cases. Even if this problem can be avoided, sometimes the MAC criterion leads to the necessity to satisfy anomaly conditions with fermions of the wrong handedness: left- and right-fermions in two dimensions are completely distinct and obey different equations of motion. These failures of the MAC criterion are discouraging, though they occur for reasons specific to two dimensions and the applicability of these conclusions to four dimensions is doubtful. We do not in fact need to know the origin of the ETC mass scales in the phenomenological analysis of lecture 5, so let us ignore the MAC controversy from here onwards.

### 4.5. *Pseudo-Goldstone bosons*

The type of ETC theory shown in fig. 4.7 predicts the existence of many light spin-zero bosons [137, 138]. They are the "pseudo"-Goldstone bosons corresponding to the breakdown of the large chiral flavour symmetry group

$$\mathrm{SU}(8)_{\mathrm{L}} \times \mathrm{SU}(8)_{\mathrm{R}} \rightarrow \mathrm{SU}(8)_{\mathrm{L+R}} \tag{4.43a}$$

if the technifermion representation is complex (expected for $\mathrm{SU}(N)_{\mathrm{TC}}$), or

$$\mathrm{SU}(16) \rightarrow \mathrm{SO}(16) \tag{4.43b}$$

if the technifermion representation is real (true in general for $\mathrm{SO(N)}_{\mathrm{TC}}$). These bosons will be pseudoscalar with respect to the TC interactions, but not necessarily for the other interactions. In the complex fermion case (4.43a) the bosons are

$$\mathrm{P}_{\bar{a}b} \equiv \bar{\mathrm{F}}_a \gamma_5 \mathrm{F}_b \quad (a, b = 1, \ldots, 8), \qquad \mathrm{F}_a = (\mathrm{U}_{\mathrm{R,Y,B}}, \mathrm{D}_{\mathrm{R,Y,B}}, \mathrm{N}, \mathrm{E}), \tag{4.44}$$

and analogous to the conventional $\bar{\mathrm{q}}\mathrm{q}$ pseudoscalars $(\pi, \mathrm{K}, \mathrm{D}, \eta, \ldots)$ of QCD, with the exception that the latter can have masses because the quarks are not massless. We might expect on the basis of (4.44) 64 light bosons, but one of them is expected to acquire a mass $\mathrm{O}(\Lambda_{\mathrm{TC}})$ from the $\mathrm{U_A}(1)$ anomaly, leaving us with 63 corresponding to the chiral symmetry breakdown (4.43a) [137, 138]. Of these, three are truly massless and eaten by the $\mathrm{W}^{\pm}$ and $\mathrm{Z}^0$ to give them masses. We are therefore left with 60 "pseudo-Goldstone bosons" (PGBs), so-called because they are not strictly massless since the flavour symmetries to which they correspond are broken by perturbations on the TC interactions due for example to ETC and $\mathrm{SU(3)} \times \mathrm{SU(2)} \times \mathrm{U(1)}$ interactions. An analogous situation arises in QCD, where one has $\mathrm{SU(2)_L} \times \mathrm{SU(2)_R} \rightarrow \mathrm{SU(2)_{L+R}}$ if the u and d quarks are massless, but the $\mathrm{SU(2)_{L+R}}$ symmetry is broken by the U(1) of electromagnetism, so that one expects

$$m^2_{\pi^+} - m^2_{\pi^0} \neq 0. \tag{4.45}$$

The spectrum of light PGBs in the complex fermion case (4.43a) is shown

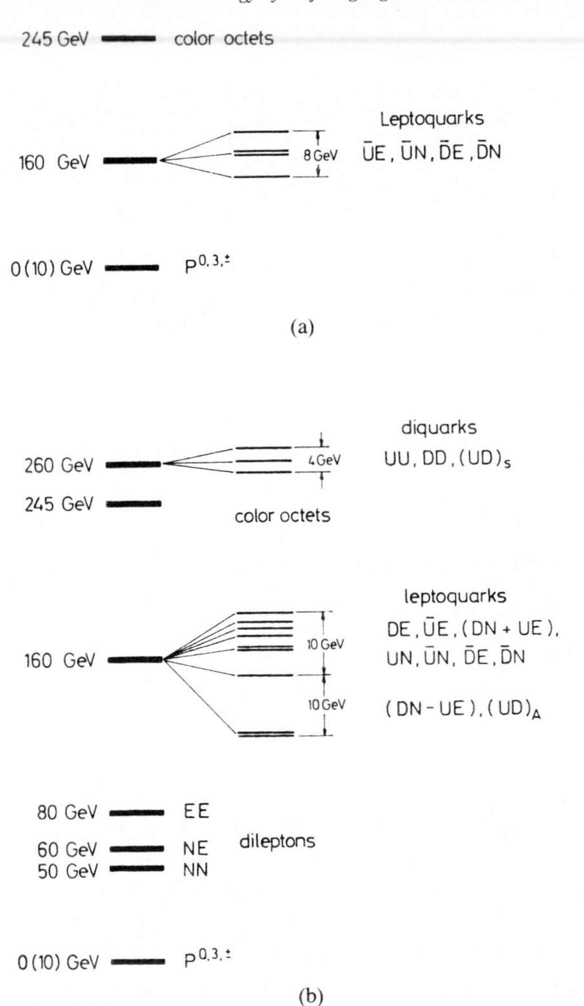


Fig. 4.10. The spectra of light pseudoscalar bosons expected in a one-technigeneration model based (a) on an $SU(N)_{TC}$ group, and (b) on an $SO(N)_{TC}$ group [137, 138].

in fig. 4.10a, while the richer spectrum expected in the real fermion case (4.43b) is shown in fig. 4.10b (see also fig. 4.3). The technifermion contents of the PGBs in fig. 4.10a are shown in table 4.3.

The figures 4.10 include estimates of the PGB masses, which we will now review briefly. It has been shown [146] that in a single-technigeneration model the ETC interactions do not give masses to the colour singlets

Table 4.3
PGBs in models with complex fermion representations

$$P^0 \equiv (\bar{U}\gamma_5 U + \bar{D}\gamma_5 D) - 3(\bar{N}\gamma_5 N + \bar{E}\gamma_5 E)$$
$$P^3 \equiv (\bar{U}\gamma_5 U - \bar{D}\gamma_5 D) - 3(\bar{N}\gamma_5 N - \bar{E}\gamma_5 E)$$
$$P^{\pm} \equiv \begin{cases} \bar{D}\gamma_5 U - 3\bar{E}\gamma_5 N \\ \bar{U}\gamma_5 D - 3\bar{N}\gamma_5 E \end{cases}$$
$$P_8 \equiv \bar{Q}\gamma_5 \lambda Q$$
$$P_{\bar{L}Q} \equiv \bar{E}\gamma_5 U, \text{ etc.}$$

$P^{0,3,\pm}$ or the colour octets $P_8^{0,3,\pm}$, though they may give masses to the leptoquark PGBs $P_{\bar{L}Q}$. Colour SU(3) interactions give [137, 138]

$$m^2 = O(\alpha_s)\Lambda_{TC}^2 \tag{4.46}$$

to all the colour nonsinglets, and are estimated to give

$$m_{P_8} \approx 250 \text{ GeV}, \qquad m_{P_{\bar{L}Q}} \approx 160 \text{ GeV}. \tag{4.47}$$

Weak SU(2)×U(1) interactions give masses much smaller [135, 137, 138] than those (4.46) donated by colour SU(3), and are most significant for the light colour singlets $P^{\pm}$:

$$m_{P^{\pm}} \approx \begin{cases} 5 \text{ to } 8 \text{ GeV for complex fermions,} \\ 8 \text{ to } 14 \text{ GeV for real fermions [147].} \end{cases} \tag{4.48}$$

So far there are no masses for the electromagnetically neutral colour singlets $P^{0,3}$ of table 4.3. Giving them masses requires extra interactions [135, 148] which couple leptons to quarks directly. The simplest possibility is Pati–Salam (PS) SU(4) interactions [149]:

$$SU(4)_{\text{Pati-Salam}} \left\{ \begin{pmatrix} u_R \\ u_Y \\ u_B \\ \cdots \\ \nu \end{pmatrix} \begin{pmatrix} d_R \\ d_Y \\ d_B \\ \cdots \\ e^- \end{pmatrix} \right. \quad \text{SU(3) colour.} \tag{4.49}$$

The exchange of PS gauge bosons can contribute to $K_L^0 \to \mu e$ decay as indicated in fig. 4.11. To be compatible with the experimental limit [150]

$$\Gamma(K_L^0 \to \mu e)/\Gamma(K^+ \to \mu^+ \nu) < 6 \times 10^{-10}, \tag{4.50}$$

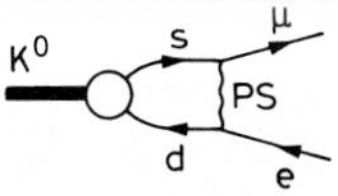

Fig. 4.11. Contribution to $K_L^0 \to \mu e$ decay from the exchange of a Pati–Salam vector boson in the crossed channel [148].

we need [136] to have

$$m_{PS}/g_{PS} \gtrsim 300 \text{ TeV}. \tag{4.51}$$

At such high energies, ETC is presumably a good symmetry (cf. table 4.2) and it is reasonable to assume that $SU(4)_{PS}$ commutes with ETC. The $P^{0,3}$ acquire masses from PS interactions:

$$m_{P^{0,3}} = O(\Lambda_{TC}^2)/m_{PS}, \tag{4.52}$$

which can be no larger than [148, 151, 152]

$$m_{P^{0,3}} \lesssim 1\tfrac{1}{2} \text{ GeV } (\times 2^{0\pm 1} \text{ theoretical uncertainty}), \tag{4.53}$$

if we apply the bound (4.51). The $P^{0,3}$ must therefore be relatively light.

The next lecture contains a detailed discussion of technicolour phenomenology in general and of PGB phenomenology in particular. To close this lecture setting up the theory, let us just note that generically one expects the couplings to fermions of technicolour PGBs to be of comparable magnitude to those of conventional Higgs bosons:

$$g_{P\bar{f}f} \sim \frac{g_E^2}{m_E^2}\Lambda_{TC}^2 \sim \frac{m_f}{\Lambda_{TC}} \sim \frac{g m_f}{m_W}, \tag{4.54}$$

cf. eq. (3.50a). The magnitudes of the expected couplings (4.54) and the lightness of the expected masses (4.53) will turn out to restrict considerably the choice of possible ETC theories.

## 5. Technicolour: phenomenology and alternatives

### *5.1. Flavour-changing neutral interactions*

In the previous lecture we set up the basic machinery of TC and ETC theories [2, 135]; now it is time to see whether they are phenomenologically acceptable. We will see that ETC theories run into phenomenological problems with flavour-changing neutral interactions and with the proliferation of light PGBs exposed in fig. 4.10. It is not clear that these challenges can be overcome, and some theorists are therefore starting to explore alternative models of dynamical symmetry breaking, which incorporate supersymmetry [133, 134]. We will come to these theories at the end of this lecture, but first we look at the phenomenological aspects of ETC theories [135].

The first set of problems concerns flavour-changing neutral interactions [136], which are strongly suppressed in the standard Glashow–Weinberg–Salam model [1] as a result of some well-understood mechanisms [153]. In principle, neutral gauge boson exchanges could have flavour-changing ($\Delta F \neq 0$) components, but these amplitudes are suppressed by the Glashow–Iliopoulos–Maiani [22] (GIM) mechanism to

$$A(\Delta F \neq 0)/A(\Delta F = 0) = \mathrm{O}(G_{\mathrm{F}} m_{\mathrm{f}}^2). \tag{5.1}$$

Whenever (A): *all quarks of the same charge have the same values of the other weak quantum numbers* [153] (e.g. $I$ and $I_3$ in the case of an SU(2) group, with possible generalizations to larger groups [154]), then there are no $\Delta F \neq 0$ amplitudes at the tree level, and loop diagrams (cf. figs 5.1a and b) are subject to strong cancellations and finish up proportional to the differences of quark (mass)$^2$ as in eq. (5.1). In lecture 2 we saw in eq. (2.34) an explicit example of how effective a $\Delta F \neq 0$ suppression can be. The exchanges of neutral Higgs bosons H also conserve flavour to a very good approximation in the Glashow–Weinberg–Salam model [1]:

$$g_{\mathrm{H}\bar{\mathrm{f}}\mathrm{f}}(\Delta F \neq 0) = \mathrm{O}(G_{\mathrm{F}} m_{\mathrm{f}}^3/m_{\mathrm{W}}). \tag{5.2}$$

The suppression (5.2) occurs in all models where (B): *all fermions of the same charge get their masses from couplings to the same Higgs field* [153]. It is very easy to see from eq. (1.38) how this comes about: since the

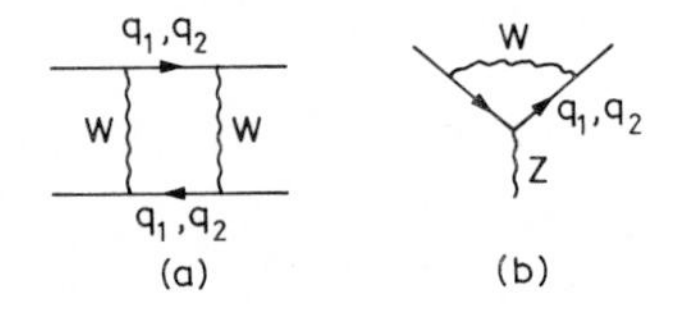


Fig. 5.1. Two classes of diagrams which are subject to GIM [22] cancellations in theories satisfying the conditions of ref. [153] (cf. figs. 2.1 and 2.2).

physical Higgs couplings and the fermion masses are proportional,

$$\frac{H+v}{v}\bar{f}fg_{\mathrm{H\bar{f}f}}, \tag{5.3}$$

it is clear that the $g_{\mathrm{H\bar{f}f}}$ are automatically diagonalized along with the fermion masses. The minimal single-Higgs model [9] of lecture 1 is an example of this phenomenon (5.2), but so also is the two-Higgs axion model [110] of section 3.4, since there also all fermions of the same charge get their masses from the same Higgs field as shown in (3.80).

If we now look at ETC theories in the light of condition (A) above, then we see an immediate problem because in the basic ETC scenario of fig. 4.7 the different fermions of the same charge *necessarily* occupy inequivalent positions in the same representation of ETC. Indeed, we saw in section 4.3 that in this scenario there *necessarily* exist HETC bosons (4.36), which change flavour with essentially unit coupling strength. Our only hope is that these bosons are sufficiently massive that their exchanges are too suppressed to be significant. It is easy [136] to check this hope because the masses of these HETC bosons are simply related (4.37) to the masses of the ETC bosons determined previously in table 4.2 by fitting the fermion mass spectrum. As for condition (B) above, it will be clear that we will need an analogous condition to suppress the $\Delta F \neq 0$ couplings of the light PGBs (4.54), and such a condition ("monophagy") will be discussed [155] in section 5.2.

After this preamble, let us get down to some specifics of HETC gauge boson exchanges. In view of eq. (4.37) and table 4.2 we expect [136] to find

$$\begin{aligned} m^2_{\mathrm{HETC}_{\mathrm{d}\leftrightarrow\mathrm{d}}} &\approx m^2_{\mathrm{HETC}_{\mathrm{d}\leftrightarrow\mathrm{s}}} \approx m^2_{\mathrm{E_d}} \approx 1200\ \mathrm{TeV}^2, \\ m^2_{\mathrm{HETC}_{\mathrm{s}\leftrightarrow\mathrm{s}}} &\approx m^2_{\mathrm{E_s}} \approx 60\ \mathrm{TeV}^2, \end{aligned} \tag{5.4}$$

and

$$\begin{aligned} m^2_{\mathrm{HETC}_{\mathrm{u}\leftrightarrow\mathrm{u}}} &\approx m^2_{\mathrm{HETC}_{\mathrm{u}\leftrightarrow\mathrm{c}}} \approx m^2_{\mathrm{E_u}} \approx 2200\ \mathrm{TeV}^2, \\ m^2_{\mathrm{HETC}_{\mathrm{c}\leftrightarrow\mathrm{c}}} &\approx m^2_{\mathrm{E_c}} \approx 8\ \mathrm{TeV}^2. \end{aligned} \tag{5.5}$$

By comparison, we expect on the basis of fig. 5.2 to find

$$\mathcal{L}^{\mathrm{eff}}_{(\bar{\mathrm{s}}\mathrm{d})^2} = \left\{\left(\bar{s}_{\mathrm{L}}\gamma_\mu d_{\mathrm{L}}\right)^2, \left(\bar{s}_{\mathrm{L}}\gamma_\mu d_{\mathrm{L}}\right)\left(\bar{s}_{\mathrm{R}}\gamma_\mu d_{\mathrm{R}}\right), \left(\bar{s}_{\mathrm{R}}\gamma_\mu d_{\mathrm{R}}\right)^2\right\}\left(\frac{\theta^2}{m^2_{\mathrm{E_d}}}, \frac{\theta^2}{m^2_{\mathrm{E_s}}}\right), \tag{5.6}$$

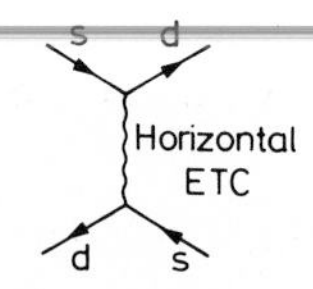


Fig. 5.2. Example of the exchange of an HETC boson giving rise to a flavour-changing neutral interaction, in this case $K^0-\overline{K}^0$ mixing.

where we have put in generalized Cabibbo angle factors $\theta$ to take account of the mixing between the generations, which means, e.g., that a d $\leftrightarrow$ d boson probably has an $O(\sin\theta)$ d $\leftrightarrow$ s coupling. Assuming $\theta = O(\theta_C)$ and comparing (5.6) with the experimental value of the coefficient of the $\mathcal{L}^{\text{eff}}_{(\bar{s}d)^2}$ operator deduced from $K^0-\overline{K}^0$ mixing (cf. lecture 2), we deduce [136] that we need

$$m^2_{E_d}, m^2_{E_s} \gtrsim 5\times 10^4\ \text{TeV}^2, \tag{5.7}$$

which clearly conflicts with eq. (5.4) by up to three orders of magnitude. This discrepancy is accentuated if we consider the *imaginary part of the $\Delta S = 2$ transition operator*. There is no obvious reason in ETC theories why it should be much less than the real part, but the experimental value of the coefficient of the imaginary part of $\mathcal{L}^{\text{eff}}_{(\bar{s}d)^2}$ is $O(10^{-15})$ GeV$^{-2}$, so that the discrepancy (5.7) increases to six orders of magnitude. A similar analysis can be carried out for the $\Delta C = 2$ operator. The experimental upper limit [156] on $D^0-\overline{D}^0$ mixing tells us that the coefficient of the $\mathcal{L}^{\text{eff}}_{(\bar{c}u)^2}$ operator $\lesssim 3\times 10^{-11}$ GeV$^{-2}$, which requires [136]

$$m^2_{E_u}, m^2_{E_c} \gtrsim 800\ \text{TeV}^2. \tag{5.8}$$

Comparing with eq. (5.5) we again find a discrepancy, this time by two orders of magnitude (see the associated monster in fig. 4.3.).

There are several other cases [136] where HETC exchanges are only marginally consistent with the experimental limits on $\Delta F \neq 0$ processes. HETC exchanges can contribute to $K^0_L \to \mu e$ decay as shown in fig. 5.3, and the experimental upper limit (4.50) [150] on this decay translates into a lower bound on $m^2_{E_s}$ which conflicts with (5.4) by about an order of magnitude [136]. In view of all the uncertainties and approximations involved, perhaps this discrepancy need not disturb us too much. However, it does suggest that the decay $K^0_L \to \mu e$ might be detectable in a slightly improved experiment. Turning to the decay $K^+ \to \pi^+ e\mu$, the

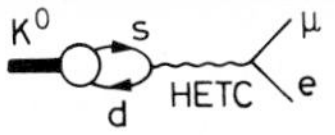

Fig. 5.3. HETC exchange contribution to $K_L^0 \to \mu e$ decay.

present experimental limit of

$$\Gamma(K^+ \to \pi^+ e\mu)/\Gamma(K^+ \to \pi^0 \nu \mu^+) < 1.5 \times 10^{-7} \tag{5.9}$$

translates into a lower bound on $m^2_{E_s}$ which is close to the deduced value (5.5), suggesting that the decay should occur within two orders of magnitude of the present experimental limit (5.9). A similar conclusion applies to the decay $\mu \to e\bar{e}e$.

So far we have concentrated on tree diagram HETC exchanges such as those of figs. 5.2 and 5.3. What about higher-order diagrams such as the box of fig. 5.4? It is unlikely that these could be larger than the tree diagram HETC exchanges, and they may even be smaller as the loop integration typically introduces a factor of order $1/128\pi^2 \approx 10^{-3}$. This factor could of course be overwhelmed by large gauge couplings $g_E^4$ in the numerator, but let us postpone worrying about such higher-order diagrams until the problems of HETC exchange are solved. One other question concerns *CP* violation: it has been estimated [157] that in an ETC theory one should expect to get

$$\theta_{QCD} \sim 10^{-9}, \tag{5.10}$$

corresponding, according to the estimates (2.82), to a neutron electric dipole moment $d_n \sim 3 \times 10^{-25}$ $e$ cm [79, 80], within a factor of two of the present experimental bound (2.73) [73]. It may even be [136] that an ETC theory would give $\theta_{QCD}$ an order of magnitude or two larger than the estimate (5.10). In any case, it seems reasonable in the context of ETC theories to expect a non-zero electric dipole moment to show up very soon.

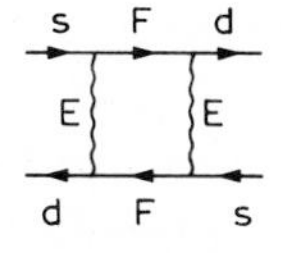


Fig. 5.4. Box diagram with internal technifermions which is not subject to a GIM cancellation in a single technigeneration model.

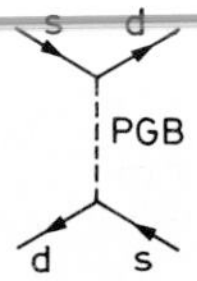

Fig. 5.5. Exchange of a neutral PGB which may contribute to $K^0-\bar{K}^0$ mixing.

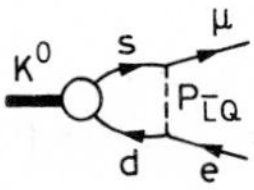

Fig. 5.6. Contribution to $K_L^0 \to \mu e$ decay from leptoquark PGB exchange.

To finish off this section, let us quantify our earlier comments on the $\Delta F \neq 0$ transitions mediated by light PGBs, which we suppose to have couplings

$$g_{P\bar{f}_1 f_2} = (g m_f / m_W)\theta_{12}, \tag{5.11}$$

where we have borrowed the generic estimate (4.54) and put in a generalized mixing angle factor $\theta_{12}$. Comparing fig. 5.5 with the experimental upper limit on the coefficient of $\mathcal{L}^{\text{eff}}_{(\bar{s}d)^2}$ of $\frac{1}{2} \times 10^{-12}\ \text{GeV}^{-2}$ coming from $K^0-\bar{K}^0$ mixing we deduce [136] that the masses $m_P$ of the $P^{0,3}$ must obey

$$m_P \gtrsim 6000\theta_{ds}\ \text{GeV}, \tag{5.12}$$

with a factor 30 more if we worry about the imaginary part of the $\Delta S = 2$ operator. Correspondingly from the $D^0-\bar{D}^0$ system we deduce [136]

$$m_P \gtrsim 2000\theta_{uc}\ \text{GeV}, \tag{5.13}$$

to be compared with the upper limit (4.53) on $m_{P^{0,3}}$ of about 3 GeV [148, 151, 152]. Conversely, if we take this limiting value for $m_{P^{0,3}}$ we deduce

$$\theta_{ds} \lesssim 1/2000, \qquad \theta_{uc} \lesssim 1/700, \tag{5.14}$$

so that these mixing angles must be much smaller than the Cabibbo angle. We must also consider a contribution to the $K_L^0 \to \mu e$ decay from "leptoquark" PGB $P_{\bar{L}Q}$ exchange in the crossed channel (fig. 5.6), which should not exhibit any mixing angle suppression, as the $\mu$ and the s quark are expected to be in the same generation, as are the e and the d quark. Taking the upper limit (4.50) again we find

$$m_{P_{\bar{L}Q}} \gtrsim 150\ \text{GeV}, \tag{5.15}$$

which is just consistent with the theoretical expectation (4.47). This is another suggestion from ETC theories, that the decay $K_L^0 \to \mu e$ should be seen soon: it would be crucial for ETC theories to push further the search for this decay.

It is clear from the analysis of this section that some tricks are needed to suppress the $\Delta F \neq 0$ amplitudes due to HETC and PGB exchanges, and the next section is devoted to one such trick.

### 5.2. *Monophagy*

We choose to attack first the problems (5.13), (5.14) of flavour-changing PGB couplings by seeking an analogue to the corresponding condition (B) of section 5.1, which suppressed the $\Delta F \neq 0$ couplings of Higgs bosons. The obvious analogue is that all fermions of the same charge get their masses from the same condensate of technifermions, so that for example

$$m_{\mathrm{u,c,t}} = \left(g_{\mathrm{E}}^2 / m_{\mathrm{E_{u,c,t}}}^2\right)\langle 0|\bar{F}F|0\rangle, \tag{5.16}$$

where the $m_{\mathrm{E_{u,c,t}}}^2$ are flavour-dependent but the $\langle 0|\bar{F}F|0\rangle$ is universal. It is then easy to show [155] that the dominant couplings (4.54) of the neutral PGBs $\mathrm{P}^{0,3}$, $\mathrm{P}_8^{0,3}$ conserve flavour. This is because the masses of the fermions come from couplings

$$\left(\bar{f}_{\mathrm{L}}\gamma_\mu F_{\mathrm{L}}\right)\left(g_{\mathrm{E}}^2/m_{\mathrm{E}}^2\right)\left(\bar{F}_{\mathrm{R}}\gamma_\mu f_{\mathrm{R}}\right) + \mathrm{h.c.}, \tag{5.17}$$

when they are Fierz-transformed to get

$$-2\left(\bar{F}_{\mathrm{R}}F_{\mathrm{L}}\right)\left(\bar{f}_{\mathrm{L}}(\Gamma_1 + \mathrm{i}\Gamma_2)f_{\mathrm{R}}\right) + \mathrm{h.c.} \tag{5.18}$$

The interaction (5.18) can be divided into a symmetric part

$$-\left(\bar{F}_{\mathrm{R}}F_{\mathrm{L}} + \bar{F}_{\mathrm{L}}F_{\mathrm{R}}\right)\left(\bar{f}\Gamma_1 f + \mathrm{i}\bar{f}\Gamma_2\gamma_5 f\right), \tag{5.19}$$

which gives fermion masses $m_{\mathrm{f}}$ via $\langle 0|\bar{F}F|0\rangle \neq 0$, and an antisymmetric part

$$\left(\bar{F}_{\mathrm{R}}F_{\mathrm{L}} - \bar{F}_{\mathrm{L}}F_{\mathrm{R}}\right)\left(\mathrm{i}\bar{f}\Gamma_2 f + \bar{f}\Gamma_1\gamma_5 f\right), \tag{5.20}$$

which gives the PGB–$\bar{\mathrm{f}}$f couplings. Simple algebra shows [155] that (5.19) and (5.20) are diagonalized simultaneously if only one condensate of technifermions contributes to the $\bar{\mathrm{f}}$f mass matrix $m_{\mathrm{f}}$. It is also a remarkable fact, which we will not prove here [155], that the dominant couplings (4.52) of the $\mathrm{P}^{0,3}$ and $\mathrm{P}_8^{0,3}$ are determined, and the charged coulour

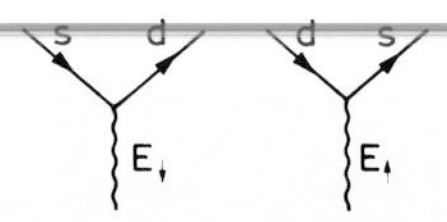


Fig. 5.7. The basic idea of the schizon mechanism [158, 159] for suppressing $\Delta F \neq 0$ interactions: the $E_\downarrow$ and $E_\uparrow$ are distinct generation raising and lowering operators which do not mix.

singlet PGB $P^\pm$ couplings are fixed in terms of the Kobayashi–Maskawa [7] matrix $U_{KM}$, to within a finite ambiguity, which is removed in the single technigeneration model of fig. 5.7 if $[G_{ETC}, SU(3)\times SU(2)\times U(1)] = 0$ as was argued in section 4.3.

However, it should be noted [152] that there are necessarily subdominant couplings of the $P^{0,3}$ which violate flavour conservation, due to effective four-fermion interactions of the type

$$(\bar{f}_L\gamma_\mu F_L)(g_E^2/m_E^2)(\bar{F}_L\gamma_\mu f_L)$$

$$\overset{\text{Fierz}}{\rightarrow} (\bar{f}_L\gamma_\mu f_L)(g_E^2/m_E^2)(\bar{F}_L\gamma_\mu F_L) + (L \leftrightarrow R). \tag{5.21}$$

The $(\bar{F}_L\gamma_\mu F_L)$ factor in (5.21) couples to the PGBs with a magnitude $\mathrm{i}p_\mu F_\pi$, so that the coupling (5.21) becomes

$$\mathrm{i}p_\mu(\bar{f}_L\gamma_\mu f_L)(g_E^2/m_E^2)F_\pi P, \tag{5.22}$$

and the equation of motion of the fermion f can be used to write the resulting PGB–$\bar{\mathrm{f}}$f coupling as

$$m_f(g_E^2/m_E^2)F_\pi = O(m_f^2/\Lambda_{TC}^2) = O(g^2 m_f^2/m_W^2). \tag{5.23}$$

If one is to have non-trivial Cabibbo angles, the couplings (5.21) or (5.23) cannot be diagonalized simultaneously for both charge $+2/3$ and for charge $-1/3$ quarks [152]. We just note here that the couplings (5.23) are sufficiently small to be consistent with the constraints (5.12), (5.13) if $m_{P^{0,3}} \gtrsim O(500)$ MeV, and will return later to other phenomenological implications of these couplings. In the meantime, let us return to the dominant couplings generated by (5.20).

If indeed $[G_{ETC}, SU(3)\times SU(2)\times U(1)] = 0$ so that the conventional quarks $(u, c, t)_{R,Y,B} \overset{ETC}{\rightarrow} U_{R,Y,B}$ and similarly for the other quarks and leptons, then the PGB couplings are [155]

$$\frac{P^{0,3}}{F_\pi}\left[\sqrt{1/3}\,(\bar{p}m_{\rm p}\gamma_5 p \pm \bar{n}m_{\rm n}\gamma_5 n) \mp \sqrt{3}\,(\bar{l}m_\ell\gamma_5 l)\right], \tag{5.24a}$$

$$\frac{P^+}{F_\pi}\left[\sqrt{2/3}\,\bar{p}\left(U_{\rm KM}m_{\rm n}\frac{1+\gamma_5}{2} - m_{\rm p}U_{\rm KM}\frac{1-\gamma_5}{2}\right)n - \sqrt{6}\left(\bar{\nu}m_\ell\frac{1+\gamma_5}{2}l\right)\right], \tag{5.24b}$$

$$\frac{P^{0,3}_{8\alpha}}{F_\pi}\sqrt{2}\left[\bar{p}\,(m_{\rm p}\gamma_5\lambda_\alpha)\,p \pm \bar{n}\,(m_{\rm n}\gamma_5\lambda_\alpha)\,n\right], \tag{5.24c}$$

where $p, n$ denote generically the mass eigenstates of charge $+2/3$ and charge $-1/3$ quarks, the $m_{\rm p,n}$ and $m_\ell$ are the diagonalized mass matrices of the quarks and the charged leptons, respectively, and $U_{\rm KM}$ is the Kobayashi–Maskawa [7] matrix (1.42). The couplings of the leptoquark PGBs $\rm P_{\bar{L}Q}$ are not so constrained, cf. for example

$$\frac{P_{\bar{\rm N}{\rm U}}}{F_\pi}\left[\bar{p}\mathfrak{M}^{\ell{\rm q}}\frac{1-\gamma_5}{2}\nu\right], \tag{5.24d}$$

where $\mathfrak{M}^{\ell{\rm q}}$ is a general matrix with the dimension of mass which is not constrained by knowledge of $U_{\rm KM}$ or of the mass matrices $m_{{\rm p,n},\ell}$, though one might naïvely expect it to be roughly diagonal in generation space ($\rm e \leftrightarrow u$, $\mu \leftrightarrow \rm c$, $\tau \leftrightarrow \rm t$). The reader is directed to ref. [155] for details of the PGB couplings in another, less plausible, monophagic model in which $[G_{\rm ETC}, {\rm SU(3)}] \neq 0$: a typical feature is that the factors of $\sqrt{3}$ in eqs. (5.24) get somewhat shuffled around. Finally, it should not be forgotten that to the couplings (5.24) must be added [152] rather more model-dependent $O(g^2m_{\rm f}^2/m_{\rm W}^2)$ couplings (5.23) of the type discussed in the previous paragraph.

At this point we have made no effort to solve the flavour-changing HETC exchange problems of section 5.1, but it seems worthwhile to construct an explicit monophagic model and see whether the problems are as serious as advertized. There is a little model [158] based on the group

$$\mathrm{SU(3)\times SU(2)\times U(1)}\times\left[\mathrm{SU}(N+3)_{\rm TC}\to \mathrm{SO}(N)_{\rm TC}\right], \tag{5.25}$$

in which we assign the doublet $\rm (U, D)_L$ and the singlet $\rm U_R$ to fundamental $\boldsymbol{N+3}$'s of $\mathrm{SU}(N+3)$, and the singlet $\rm D_R$ to an $\overline{\boldsymbol{N+3}}$ of $\mathrm{SU}(N+3)$. In this case we get independent mass matrices for the charge $+2/3$

quarks P and the charge $-1/3$ quarks N:

$$m_{\mathrm{P}} = \tfrac{1}{2}g^2\Lambda^3_{\mathrm{TC}}\lambda = m_{\mathrm{P}}^{\dagger} \text{ (hermitean)},$$

$$m_{\mathrm{N}} = \tfrac{1}{2}g^2\Lambda^3_{\mathrm{TC}}\lambda' = m_{\mathrm{N}}^{\mathrm{T}} \text{ (symmetric)}. \tag{5.26}$$

It is then possible to impose a pattern of $\mathrm{SU}(N+3)$ breaking down to $\mathrm{SO}(N)$ which gives quasi-realistic mass matrices and Cabibbo angles. For example, in a toy $\mathrm{SU}(N+2)$ model with two generations, diagonalization of (5.26) to the p, n basis gives

$$m_{\mathrm{c}} = 1.5 \text{ GeV}, \qquad m_{\mathrm{s}} = 117 \text{ MeV}, \qquad m_{\mathrm{u}} = 28 \text{ MeV},$$

$$m_{\mathrm{d}} = 75 \text{ MeV}, \qquad \theta_{\mathrm{C}} = 13°, \tag{5.27}$$

and thanks to monophagy the PGB couplings are predominantly flavour-diagonal as described earlier (5.24). What about the HETC exchanges? First the bad news:

$$\mathcal{L}^{\mathrm{eff}}_{(\bar{\mathrm{s}}\mathrm{d})^2} = \mathrm{O}(10^{-9}) \text{ GeV}^{-2}(\bar{s}d)^2, \tag{5.28}$$

which violates the experimental constraint by three orders of magnitude, as expected from (5.7), and then the good news:

$$\mathcal{L}^{eff}_{(\bar{\mathrm{c}}\mathrm{u})^2} = \mathrm{O}(10^{-11}) \text{ GeV}^{-2}(\bar{c}u)^2, \tag{5.29}$$

which does not conflict with experiment, unlike what we expected on the basis of (5.8). The reason for this unexpected suppression by $\mathrm{O}(10^2)$ can be traced to a relic of a techni-analogue to the old "schizon" mechanism proposed [159] before the discovery of charm for the suppression of $\Delta F \neq 0$ neutral currents. The principle is illustrated in fig. 5.7: there are generation-"raising" and "lowering" bosons $\mathrm{E}_{\uparrow}$ and $\mathrm{E}_{\downarrow}$, which have relatively little mixing. This is possible in an $\mathrm{SU}(N)_{\mathrm{ETC}}$ theory, where the two generators are independent, but not in an $\mathrm{SO}(N)_{\mathrm{ETC}}$ theory, where they are combined into a single generator. It remains to be seen whether a cleverer monophagic model can be constructed in which the residual $\Delta F \neq 0$ problem (5.28) is solved. The authors of ref. [158] were unable to do it, but were not able to prove it impossible: it is left as an exercise for the reader. Let us assume optimistically that the flavour-changing interaction can eventually be overcome, and move on to another aspect of technicolour phenomenology.

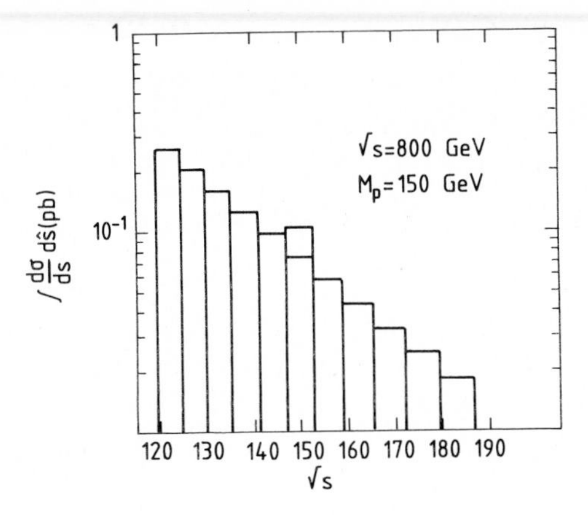


Fig. 5.8. The production of a $P_8$ boson in hadron–hadron collisions, compared with the background continuum production of $\bar{t}t$ pairs and binned according to a reasonable mass resolution [161].

## 5.3. *PGB phenomenology*

We start with the heaviest of the PGBs in fig. 4.10a, namely the colour octet PGBs $P_8$. The neutral $P_8^0$ can be produced by gluon–gluon fusion in hadron–hadron collisions [151, 160, 161], though with a rather small cross section. Its dominant decay is expected [155] to be into $\bar{t}t$ pairs, providing the rather indistinct experimental signature [161] shown in fig. 5.8. For further details you are referred to the lectures of Gordy Kane at this School (Seminar 3) [151].

The colour triplets $P_{\bar{L}Q}$ can be produced [162] in high energy ep collisions as in fig. 5.9, and should have a distinctive decay signature into $t\bar{\tau}$ (assuming, as is not strictly warranted by eq. (5.24d), that the

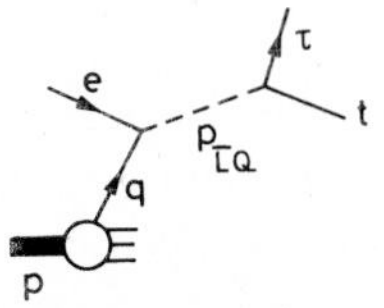

Fig. 5.9. Possible mechanism for producing leptoquark PGBs in electron–proton collisions. The heavy quark q is best obtained by splitting a gluon [162].

couplings to conventional fermions are roughly diagonal in generation, and increase in magnitude with the fermion masses). However, it is seen from fig. 5.10 that the cross section is not very encouraging, even at HERA. An indirect way to look for the $P_{\bar{L}Q}$ would be to search for $K_L^0 \to \mu e$ decay mediated by fig. 5.6. Here a negative result would be significant, but the interpretation of a positive result would be ambiguous, since there are other mechanisms (figs. 4.11, 5.3) for this decay in ETC theories. It is difficult to imagine that *all* these mechanisms could be strongly suppressed in such theories, and it therefore would be very interesting to push further the experimental limit on this decay.

We now turn to the charged colour singlets $P^{\pm}$, which should have masses [cf. (4.48)] [137, 138, 147]

$$m_{P^{\pm}} \approx \begin{cases} 5 \text{ to } 8 \text{ GeV for complex fermions,} \\ 8 \text{ to } 14 \text{ GeV for real fermions.} \end{cases} \tag{5.30}$$

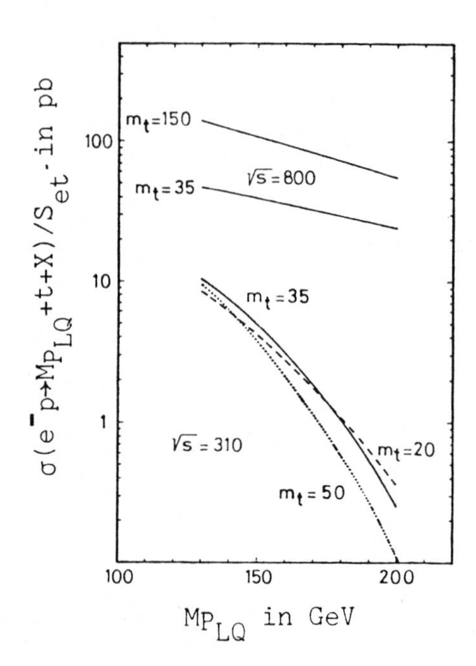


Fig. 5.10. Production rate of leptoquark PGBs, taken from ref. [162]. $S_{et}$ is the modulus squared of a mixing angle factor.

They should be pair-produced in $e^+e^-$ annihilation with a cross section

$$R \equiv \frac{\sigma(e^+e^- \to \gamma^* \to P^+P^-)}{\sigma(e^+e^- \to \gamma^* \to \mu^+\mu^-)} = \tfrac{1}{4}\beta^3, \quad \beta \equiv \left(1 - 4m_{P^\pm}^2/s\right)^{1/2}, \tag{5.31}$$

and are expected, on the basis of eq. (5.24b), to have dominant decays [155]

$$\Gamma(b\bar{c}) : \Gamma(s\bar{c}) : \Gamma(\tau\bar{\nu}) \approx 10|U_{bc}|^2(\approx 1?) : 1 : 3. \tag{5.32}$$

Unfortunately, particles with a cross section (5.31), masses between 5 and 14 GeV, and decay branching ratio into $\tau\bar{\nu}$ between 5 and 95%, seem to be ruled out by recent PETRA data [47, 163], as shown in fig. 5.11. It seems necessary, therefore, either to find an extra contribution to the $P^\pm$ masses invalidating (5.30), or else to find another model for its couplings which makes the $\tau\bar{\nu}$ branching ratio compatible with the constraints of

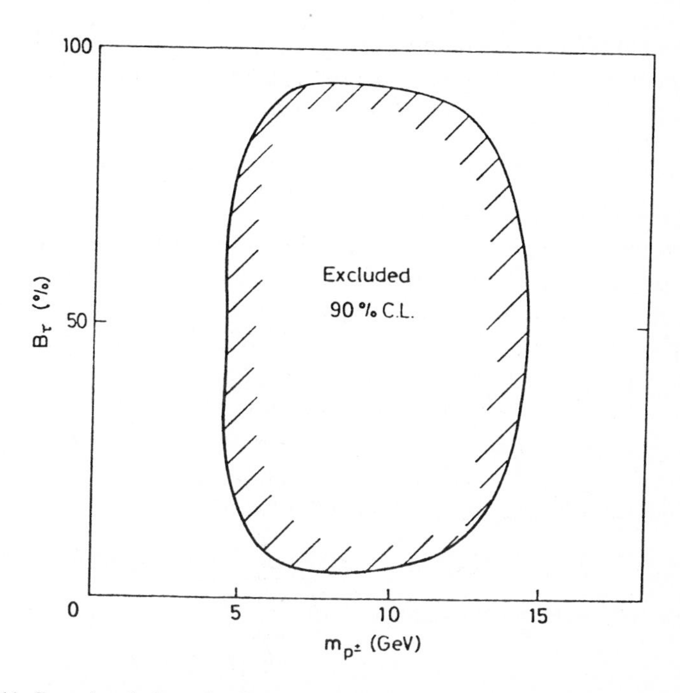


Fig. 5.11. Domain of charged colour-neutral PGB masses and leptonic branching ratios excluded by a recent PETRA experiment [163].

fig. 5.11 if one wants to avoid the monster in fig. 4.3. If the $P^{\pm}$ have masses less than the top quark, as expected from (5.30), then they would be among the dominant decay products of t-flavoured hadrons [164]:

$$\frac{\Gamma(t \to P^+ + b)}{\Gamma(t \to b q \bar{q})} = O(10^3 \text{ to } 10^4)(20\ \text{GeV}/m_t)^2, \tag{5.33}$$

and would even dominate the decays of toponium [164]:

$$\Gamma(\zeta(\bar{t}t) \to b + P^+ + \bar{t} + X) = O(5)(m_t/20\ \text{GeV})^3 \ggg \Gamma(\zeta \to 3 \text{ gluons}),$$
$$\llcorner\!\!\rightarrow P^- + X \tag{5.34}$$

i.e. the $\zeta$ falls apart before its constituent quarks find each other and annihilate. The decays (5.33), (5.34) would give events in the $t\bar{t}$ continuum or in toponium final states which could have relatively little hadronic energy:

$$e^+e^- \to \tau + \bar{\nu} + \bar{\tau} + \nu + X, \tag{5.35}$$

which could affect the efficiency of scanning for the $t\bar{t}$ threshold.

We turn finally to the neutral colour singlets $P^{0,3}$, which should have masses [cf. (4.53)] [148, 151, 152]

$$m_{P^{0,3}} \lesssim 1\tfrac{1}{2}\ \text{GeV}\ (\times 2^{0 \pm 1} \text{ theoretical uncertainty}). \tag{5.36}$$

The bound (5.36) would of course be strengthened by an improved upper limit on the $K^0_L \to \mu e$ decay rate, implying a stronger lower limit on the PS boson mass (4.51) and hence reducing $m_{P^{0,3}}$ (4.52). As was remarked in section 5.2, interactions of the type (5.21) give $\Delta F \neq 0$ interactions of the type (5.23) *either* to charge 2/3 quarks (class P) *or* to charge $-1/3$ quarks (type N) *or* to both [152]. In either case P or case N, the fact that $P^{0,3}$ have not been seen in $K^{\pm}$ decays means that

$$m_{P^{0,3}} \gtrsim 350\ \text{MeV}. \tag{5.37}$$

In case P we would expect [152] branching ratios

$$B(D \to P^{0,3} + X) \approx 1\%, \qquad B(B \to P^{0,3} + X) \approx 0.1\%, \tag{5.38P}$$

whereas in case N we would expect [152]

$$B(D \to P^{0,3} + X) \approx O(10^{-10}), \qquad B(B \to P^{0,3} + X) \approx 100\%. \tag{5.38N}$$

In view of (5.24a) we expect the decay branching ratio

$$\frac{\Gamma(\mathrm{P}^3 \to \mu^+\mu^-)}{\Gamma(\mathrm{P}^3 \to \text{hadrons})} \approx 3\frac{m_\mu^2}{m_\mathrm{s}^2} \gtrsim 10\%, \tag{5.39}$$

which is incompatible with (5.38N) and the experimental upper limit [165] on

$$B(\mathrm{B} \to (\mu^+\mu^-) + X) \lesssim 0.2\%, \tag{5.40}$$

coming from CESR, for all the range (5.36) of expected masses. Hence models of class N appear excluded. However, models of class P are not yet excluded: probably the predictions (5.38P) cannot be ruled out, and even if $\mathrm{D} \not\to \mathrm{P}^{0,3} + \mathrm{X}$ at the expected level, it could just be that $m_{\mathrm{P}^{0,3}} \gtrsim 1500$ MeV [cf. (5.36)]. While they may appear slightly ill, the $\mathrm{P}^{0,3}$ are not dead yet. For reference and comparison with the Higgs boson H of section 3.3, let us just note that [155, 166, 167]

$$B(\mathrm{Z}^0 \to \mathrm{P}^{0,3} + \gamma) \approx 10^{-7} \text{ to } 10^{-9},$$

$$\frac{\sigma(\mathrm{e}^+\mathrm{e}^- \to \mathrm{Z}^0 + \mathrm{P}^{0,3})}{\sigma(\mathrm{e}^+\mathrm{e}^- \to \mu^+\mu^-)} \lesssim 10^{-2} \quad \text{for } \sqrt{s} \lesssim 250 \text{ GeV}, \tag{5.41}$$

both of which are considerably less than the corresponding predictions for the H, enabling a distinction to be drawn between the $\mathrm{P}^{0,3}$ and the H.

### 5.4. *Supercolour and supersymmetric technicolour*

We have seen in sections 5.1 and 5.3 that ETC theories run into severe problems with flavour-changing neutral interactions and with PGB phenomenology. It is not clear that either of these monsters is fatal to the ideas [2, 135] of lecture 4, but it does start to seem worthwhile to explore alternatives. It was already mentioned in section 4.1 that an alternative philosophy concerning the quadratic divergences of fig. 4.3 was to make them cancel by supersymmetry. This may either be incorporated [133, 134] into the framework of dynamical symmetry breaking treated in this lecture and the previous one, or else into the GUTs of lectures 6, 7 and 8. We defer discussion of the latter possibility to lecture 9, and here concentrate on the combination of supersymmetry and technicolour ideas called supercolour [133] or supersymmetric technicolour [134] (see also fig. 4.3).

It is proposed that one consider the following set of supersymmetric gauge interactions:

$$\mathrm{SU}(3)_c \times \mathrm{SU}(2)_L \times \mathrm{U}(1) \times \mathrm{SU}(M)_{SC} \times (\mathrm{SU}(N)_{TC}) \times (N=1 \text{ supersymmetry}). \tag{5.42}$$

The supersymmetry (susy) gives a direct connection between bosons and fermions, e.g. in number (associating Majorana fermions to gauge bosons, spin-zero squarks $\tilde{q}$ and sleptons $\tilde{\ell}$ to conventional quarks and leptons, fermion shiggs $\tilde{H}$ to the Higgs fields) and in imposing mass formulae such as

$$m_{\tilde{B}}^2 = m_{\tilde{F}}^2 \tag{5.43}$$

for supersymmetric partners B and F. These latter are experimentally unacceptable [47, 168] (any smuon $\tilde{\mu}$ must have mass $\gtrsim 15$ GeV, any selectron $\tilde{e}$ must have mass $\gtrsim 16$ GeV, etc.) and we must therefore break supersymmetry. It is proposed [133, 134] to achieve this dynamically by condensates generated by the $\mathrm{SU}(M)_{SC}$ interactions of (5.42):

$$\langle 0|f_s f_s|0\rangle = O(\Lambda_{SC}^3), \qquad \langle 0|S_s S_s|0\rangle = O(\Lambda_{SC}^2), \qquad \text{etc.} \tag{5.44}$$

for fermions $f_s$ and spin-zero particles $S_s$ which feel the supercolour force. It is not clear that such dynamical breaking of supersymmetry is in fact possible [169], but let us assume it is, as is necessary to avoid mass degeneracy between susy partners. Then the unseen supersymmetric partners of known particles acquire large masses, for example the Wino $\tilde{W}$ and the Bino $\tilde{B}$, susy partners of the SU(2) gauge bosons W and the U(1) boson B, respectively, get masses [134] from fig. 5.12 of order

$$m_{\tilde{W}} \sim 4\ \text{TeV}, \quad m_{\tilde{B}} \sim 1\ \text{TeV} \quad \text{if } \Lambda_{SC} \sim 10\ \text{TeV}. \tag{5.45}$$

Similarly the squarks and sleptons acquire masses

$$m_{\tilde{q},\tilde{\ell}} \approx \text{a few hundred GeV}, \tag{5.46}$$

and the Higgs boson (which must sit in a different chiral supermultiplet from the conventional leptons if it is to have non-zero Yukawa couplings

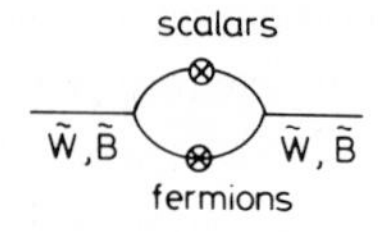


Fig. 5.12. Diagram whereby fermion and scalar condensates (5.44) can give rise to masses for the $\tilde{W}$ and $\tilde{B}$ in a theory of supersymmetric technicolour [133, 134].

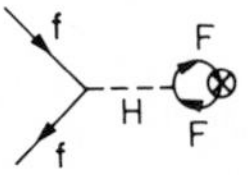

Fig. 5.13. Diagram giving masses to fermions in a theory of supersymmetric technicolour [133, 134]: cf. fig. 4.6d.

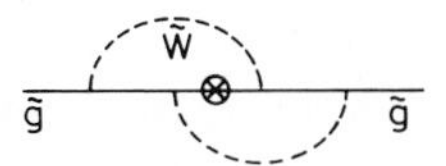


Fig. 5.14. Diagram giving rise to a gluino mass in a theory of supersymmetric technicolour [133, 134].

to them) also gets a mass

$$m_{\mathrm{H}} \approx \text{a few hundred GeV}. \tag{5.47}$$

The Higgs boson is protected from rising higher by the supersymmetry, which prevents it from acquiring a quadratically divergent mass renormalization. In addition to the Higgs fields, there are believed to be ordinary technicolour interactions, whose condensates

$$\langle 0|f_{\mathrm{T}}f_{\mathrm{T}}|0\rangle = \mathrm{O}\left(\Lambda^{3}_{\mathrm{TC}}\right), \qquad \langle 0|S_{\mathrm{T}}S_{\mathrm{T}}|0\rangle = \mathrm{O}\left(\Lambda^{2}_{\mathrm{TC}}\right), \qquad \text{etc.} \tag{5.48}$$

break $\mathrm{SU(2)_L \times U(1)}$. They are believed to give most of the $\mathrm{W}^{\pm}$ and $\mathrm{Z}^{0}$ masses, $\approx 100$ GeV, while the Higgs vacuum expectation value gives

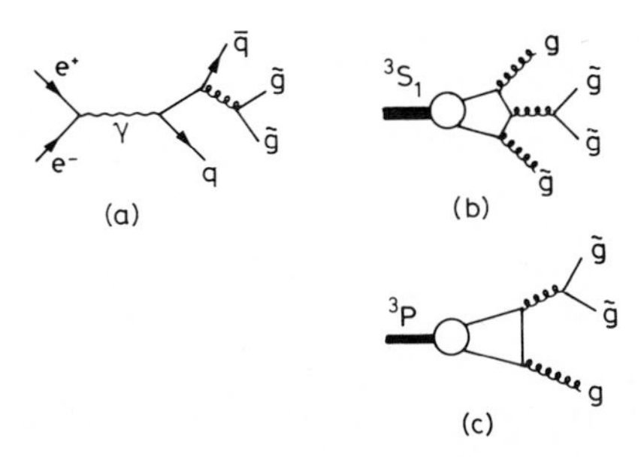


Fig. 5.15. Possible mechanisms [170] for gluino pair production (a) in the $e^{+}e^{-}$ continuum, (b) in $^{3}\mathrm{S}_{1}$ quarkonium decay, and (c) in $^{3}\mathrm{P}$ quarkonium decay.

somewhat smaller masses to the $W^\pm$ and $Z^0$, and to the conventional fermions through diagrams like that of fig. 5.13 (cf. fig. 4.6d). It has been found [134] that theories without a technicolour factor as in (5.42), or with the supercolour and technicolour interactions identified, tend to suffer from unwanted axions and/or Goldstone bosons. By avoiding the ETC interactions because of the mechanism of fig. 5.13 for giving masses to fermions, these theories seem to avoid the flavour-changing neutral interaction and PGB phenomenology problems of sections 5.1 and 5.3.

While these theories are not yet very highly developed, it is interesting that there appear to be some relatively light susy partner particles (see the cuddly new animals in fig. 4.3). For example, the gluino $\tilde{g}$ (spin-1/2 susy partner of the gluon g) gets a mass through the higher-order diagram of

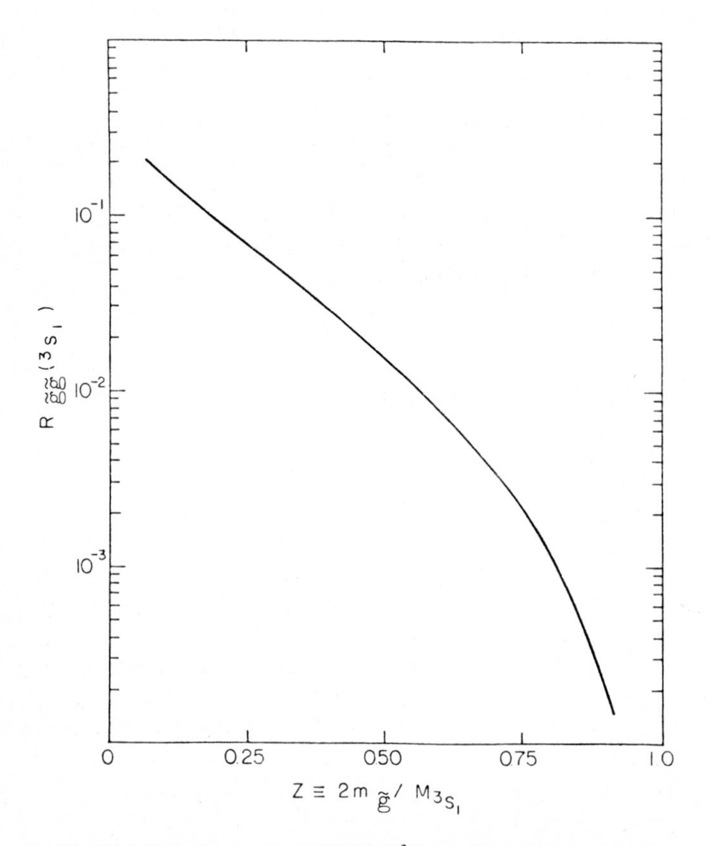


Fig. 5.16. Relative decay rate [170] of $^3S_1$ quarkonium into gluinos.

fig. 5.14 of order

$$m_{\tilde{g}} = O(5)\ \text{GeV}. \tag{5.49}$$

It should be relatively long-lived:

$$\tau_{\tilde{g}} = O(10^{-10})\ \text{s}, \tag{5.50}$$

and hence have a clear production signature. Good places [170] to look for it might be peculiar multiple jet events in either the $e^+e^-$ continuum (fig. 5.15a) or else in heavy quarkonium decays (figs. 5.15b, c). As shown in fig. 5.16, the rates of production in the latter decays are not negligible [170], and if it could be shown that a particle living as long as (5.50) could only be produced in $\Upsilon$ decays with a branching ratio $\lesssim 1\%$, it would mean that $m_{\tilde{g}} \gtrsim 3$ GeV. The predictions (5.49), (5.50) should be easily testable in toponium decays.

Another light susy partner (this time of the Higgs boson) should be the shiggs $\tilde{H}$ expected [134] to have a mass O(30) GeV, which would enable it to be pair-produced at LEP, the SLC or CESR II.

While still at a rather embryonic stage, the models of this section, which combine supersymmetry and technicolour [133, 134], have a rather complicated structure (5.42) and a correspondingly rich phenomenology (5.45), (5.46), (5.47), (5.49), (5.50). We will see in lecture 9 whether the alternative susy models, which abandon dynamical symmetry breaking, look any more elegant and appealing.

# *III. Grand Unification*

## 6. Introduction to Grand Unified Theories

### *6.1. The philosophy of Grand Unification*

It is no longer controversial that our Standard SU(3)×SU(2)×U(1) Model works well at present energies ( < 100 GeV) and is probably correct in the sense of being a valid low-energy approximation to the full dynamics. However, the Standard Model is clearly inadequate in many respects. For example, it is not completely unified but contains three different gauge coupling constants for the different SU(3), SU(2) and U(1) factors: $g_3 \neq g[\equiv g_2$ for the SU(2) factor$] \neq g'$ [for the U(1) factor]. Furthermore, no explanations are provided for many fundamental facts and/or quantities. Why are the electron and proton charge commensurate, with [171]:

$$|Q_e|/|Q_p| = +O(10^{-20}) \tag{6.1}$$

(the charge quantization problem)? Why are the fermion masses and weak mixing angles the way they are, etc., etc.? Even if one exempts the U(1) hypercharge assignments of the fundamental fermions, the Standard Model contains at least twenty arbitrary parameters: $g_3$, $g_2$ and $g_1$; two non-perturbative vacuum parameters [6] $\theta_3$ and $\theta_2$; six quark masses; three generalized Cabibbo angles $\theta_i$ and one Kobayashi–Maskawa [7] phase $\delta$; $m_H$ and $m_W$; three charged lepton masses. Everyone agrees that we should seek a theory with fewer parameters. Theories of dynamical symmetry breaking [2] like those discussed in lectures 4 and 5 seek ultimately to explain all parameters in terms of gauge coupling constants. However, that line of attack seems to have got rather bogged down, and we need an alternative philosophy.

In our search for empirical indications how to proceed, let us take a closer look at the "generation" or "family" structure of fundamental fermions illustrated in fig. 6.1. Plotted along the horizontal axis are the SU(3) colour properties of the fermions, while the vertical axis represents the weak SU(2)×U(1) properties. The third is the "generation" axis. The known fermions seem to occur in sets of 15 helicity states whose SU(3)×SU(2) transformation properties are

$$\begin{array}{llll} (3,2) + 2(\bar{3},1) & + (1,2) & + (1,1) \\ \left[(\mathrm{u},\mathrm{d})_\mathrm{L} + \bar{\mathrm{u}}_\mathrm{L} + \bar{\mathrm{d}}_\mathrm{L}\right. & + (\nu_\mathrm{e},\mathrm{e}^-)_\mathrm{L} & \left. + \mathrm{e}^+_\mathrm{L}\right], \end{array} \tag{6.2}$$

with the first generation examples given below in square parentheses. The masses of the fermions in each generation are qualitatively similar, e.g.

$$\begin{aligned} m_{\mathrm{u,d}} &\sim 0.01\ \mathrm{GeV}, & m_\mathrm{e} &\sim 0.0005\ \mathrm{GeV}, \\ m_{\mathrm{s,c}} &\sim 1\ \mathrm{GeV}, & m_\mu &\sim 0.1\ \mathrm{GeV}, \\ m_{\mathrm{b,t}} &\sim 10\ \mathrm{GeV}, & m_\tau &\sim 1.8\ \mathrm{GeV}, \end{aligned} \tag{6.3}$$

and rather closer together than the mass splittings between generations. The known interactions occur mainly between fermions in the same generation. This is strictly true for the photon and gluon at zero momentum transfer, and is approximately true [22,51] of the neutral currents mediated by the $Z^0$. Even though the charged weak interactions do exhibit the generalized Cabibbo mixing between different generations, this seems to be relatively small, i.e. the $\theta_q^2$ are $\ll 1$. The correlations in masses and weak mixings between the different particles conventionally assigned to the same generation have been shown to be statistically significant [172], and are not just the products of theorists' over-creative imaginations. It is natural to seek to unify all the interactions between the different members of the same generation, which involves introducing interactions coupling quarks to leptons. For the moment (see however lecture 10) we set aside the problem of explaining the total number of generations.

The strategy usually adopted [3] is to embed the Standard Model interactions into a simple group [173, 174]:

$$G \underset{10^{15}\,\mathrm{GeV}}{\rightarrow} \mathrm{SU}(3)\times\mathrm{SU}(2)\times\mathrm{U}(1) \underset{10^{2}\,\mathrm{GeV}}{\rightarrow} \mathrm{SU}(3)\times\mathrm{U}(1), \tag{6.4}$$

where we have indicated the energy scales at which $G$ is successively broken down. The scale of $10^2$ GeV $\approx m_{\mathrm{W,Z}}$ is already familiar to us, the scale of $10^{15}$ GeV will be motivated shortly [175]. Because of the simple group structure $G$, we expect to explain charge quantization (6.1) as the consequence of a Grand Unified Clebsch–Gordan coefficient [173]. Because quarks and leptons are in the same multiplet we may hope to get some explanations [173, 176] for some of the quark-to-lepton mass ratios in eq. (6.3). Because of the new direct quark-to-lepton interactions we may also hope for some exciting predictions for baryon and lepton number violating processes such as proton decay [173–177] and neutrino masses.

However, there is a major hurdle to be jumped on the route to Grand Unification, namely the gross inequality

$$g_3 \gg g_2, g' \tag{6.5}$$

between the strong and weak couplings observed at present energies, whereas we would expect $g_3 = g_2$ in a unified theory, with $g'$ to be related to their common value by a Clebsch–Gordan coefficient. A possible solution [175] to this problem is provided by the renormalization group [178], which tells us that the effective couplings vary with energy, and specifically by asymptotic freedom [179] according to which the strong coupling $\alpha_{\mathrm{s}}$ or $\alpha_3$ decreases as the energy $Q$ increases:

$$\alpha_3(Q) \approx \frac{12\pi}{(33-2N_{\mathrm{q}})\ln Q^2/\Lambda^2_{\mathrm{QCD}}}. \tag{6.6}$$

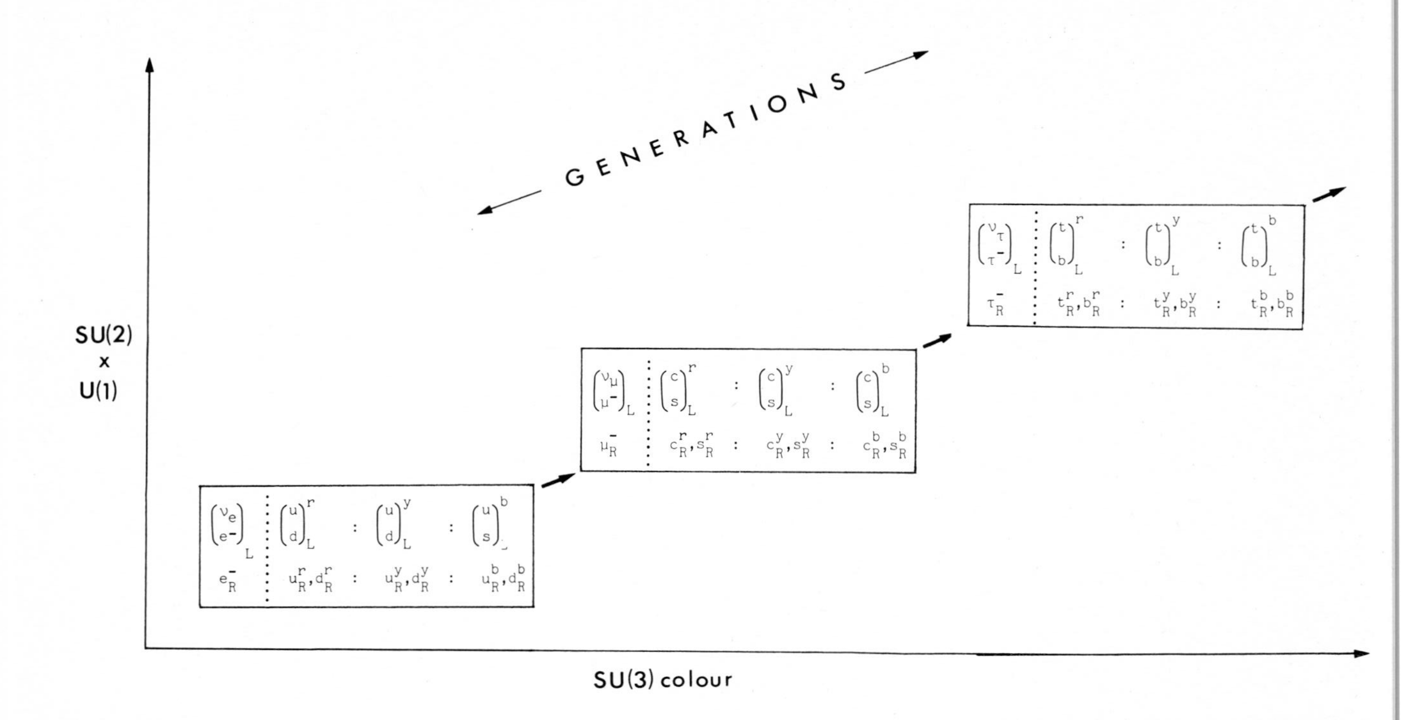


Fig. 6.1. The apparent “generation” or “family” structure of fundamental fermions. The horizontal axis corresponds to SU(3) colour properties, the vertical axis to SU(2)×U(1) representation contents.

In (6.6), $N_q$ is the number of quarks with masses much less than $Q$, $\Lambda_{QCD}$ is a parameter describing the overall mass scale of the strong interactions which is probably between 0.1 and 1 GeV [180], and we approximated the full renormalization group equations by their one-loop terms in obtaining the quoted solution for $\alpha_3(Q)$. Of course $\alpha_2$ and $\alpha_1$ also evolve with $Q$, but not as rapidly as the strong coupling $\alpha_3(Q)$ in eq. (6.6). If $\alpha_3$ decreases, one suspects that there may be some energy at which $\alpha_3(Q) = \alpha_2(Q)$, as illustrated in fig. 6.2. However, because of the relatively slow logarithmic variation in eq. (6.6) it is clear that this Grand Unification scale will be rather large. Indeed, calculations [175–177], [181–189] with $\Lambda_{QCD} \sim$ (0.1 to 1 GeV) and $\alpha = 1/137$ to set the weak interaction scale yield unification at about $10^{15}$ GeV, the scale mentioned in eq. (6.4). While this scale may seem vertiginously high to particle physicists, it is in fact considerably less than the Planck scale of $10^{19}$ GeV, which is the energy at which quantum gravity effects,

$$G_N E^2 \approx O(1), \tag{6.7}$$

become of order unity. It is therefore a self-consistent first step to neglect gravitation in building a Grand Unified Theory, though we will later (in lecture 10) go on to consider the unification with gravity [4]. For the moment, armed with figs. 6.1 and 6.2, we look for a suitable Grand Unifying group $G$ (6.4).

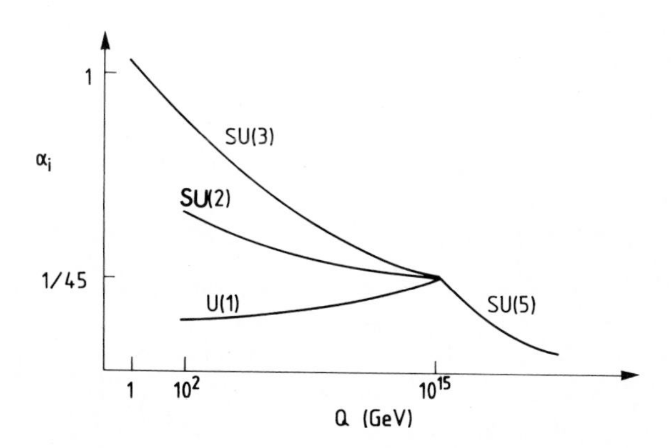


Fig. 6.2. A sketch of the manner in which the SU(3), SU(2) and U(1) coupling constants are supposed to come together in a GUT.

### 6.2. Models for Grand Unification

Before starting our Odyssey through simple Lie groups in search of a GUT, let us first spend a moment on a technical point concerning particle helicities. Since gauge vertices conserve helicity: $\bar{f}_{L,R}\gamma_\mu f_{L,R}$, it is natural to work consistently with fermions of a specified helicity (say left) and classify them into representations of the GUT group $G$. Right-handed fermions can be replaced by the corresponding left-handed antifermions: $f_R \to \bar{f}_L$. This was already done in eq. (6.2), where we chose to write down the SU(3)×SU(2) transformation properties of $\bar{u}_L$ and $\bar{d}_L$ rather than $u_R$ and $d_R$. We can make conventional four-component spinors out of two of our two-component left-handed fermion fields:

$$\psi = \begin{pmatrix} \psi_1 \\ i\sigma_2(\psi_2)^* \end{pmatrix}. \tag{6.8}$$

These will be "Dirac" (complex) fermions if $\psi_1 \neq \psi_2$, and "Majorana" (real) fermions if $\psi_1 = \psi_2$.

How large a group [173] must we use for Grand Unification? Since the strong SU(3) group has rank 2, and the weak SU(2) and U(1) groups have rank 1 apiece, our GUT group must have rank $\geqslant 4$. In order to have a single gauge coupling constant, it should either be simple, or else be the product of identical simple factors whose couplings constants can be set equal by a discrete symmetry. A complete catalogue of candidate groups of rank 4 is:

$$[\mathrm{SU}(2)]^4, [\mathrm{O}(5)]^2, [\mathrm{SU}(3)]^2, [\mathrm{G}_2]^2, \mathrm{O}(8), \mathrm{O}(9), \mathrm{Sp}(8), \mathrm{F}_4 \text{ and } \mathrm{SU}(5). \tag{6.9}$$

The first two of these do not contain SU(3) factors. If we wanted to use the group $[\mathrm{SU}(3)]^2$ we would have to put the quarks in a representation of a weak SU(3) group. It would contain electromagnetic U(1) as a generator, and hence we would need $\Sigma Q_q = 0$, where the sum is over all quarks q. This does not seem very natural: we would need to discard the generation structure of fig. 6.1, e.g. by putting (u, d, s) into a triplet representation. All the other groups except SU(5) are excluded because they only have real representations, and the representation (6.2) is clearly complex. It will turn out that the representation (6.2) can indeed be comfortably [173] accommodated within a (reducible) representation of SU(5), as we will see shortly.

The minimal SU(5) version of the breaking sequence (6.4) proceeds [190] via

$$\mathrm{SU}(5) \underset{\mathbf{24}\text{ of Higgs}}{\rightarrow} \mathrm{SU}(3)\times\mathrm{SU}(2)\times\mathrm{U}(1) \underset{\mathbf{5}\text{ of Higgs}}{\rightarrow} \mathrm{SU}(3)\times\mathrm{U}(1). \quad (6.10)$$

The SU(5) theory contains 24 gauge bosons, 12 of which are the well-known $\gamma$, $\mathrm{W}^{\pm}$ $\mathrm{Z}^0$ and $g_{1,2,\ldots,8}$. The other 12 fall into the simplest possible SU(3)×SU(2) representation which is non-trivial for both factors, namely (3,2) and its complex conjugate $(\bar{3},2)$, which are denoted by $\mathrm{X}_{\mathrm{R,Y,B}}(Q=+4/3)$ and $\mathrm{Y}_{\mathrm{R,Y,B}}(Q=+1/3)$ and their antiparticles $\bar{\mathrm{X}}$, $\bar{\mathrm{Y}}$. These bosons will be superheavy and mediate baryon decay in this model. The generations of $N_{\mathrm{G}}$ fermions are each assigned [173] to *reducible* and anomaly-free $\bar{\mathbf{5}}+\mathbf{10}$ representations of SU(5). If we identify colour SU(3) with the first 3 out of the 5 SU(5) indices, and SU(2) with the last 2, it is clear that

$$\left.\begin{aligned} \bar{\mathbf{5}} &= (\bar{3},1)+(1,2) \\ \mathbf{10} &= (3,2)+(\bar{3},1)+(1,1) \end{aligned}\right\} \text{ of } \mathrm{SU}(3)\times\mathrm{SU}(2), \quad (6.11)$$

which together coincide with our requirement (6.2). The only ambiguity we must resolve is which of the two $(\bar{3},1)$ in eq. (6.11) is to be identified with which of $\bar{\mathrm{d}}_{\mathrm{L}}$ and $\bar{\mathrm{u}}_{\mathrm{L}}$. Since the electromagnetic charge is to be a generator of SU(5), the sum of the $Q_{\mathrm{em}}$ over any representation must be zero. Applying this requirement to the $\bar{\mathbf{5}}$ in eq. (6.11) we see that each member of the $(\bar{3},1)$ must have $Q_{\mathrm{em}}=+1/3$, so that we must identify it with the $\bar{\mathrm{d}}_{\mathrm{L}}$. The full assignments to the representations (6.11) are

$$\bar{\mathbf{5}} = \begin{pmatrix} d_{\mathrm{r}}^{\mathrm{c}} \\ d_{\mathrm{y}}^{\mathrm{c}} \\ d_{\mathrm{b}}^{\mathrm{c}} \\ \hline e^- \\ \nu_{\mathrm{e}} \end{pmatrix}_{\mathrm{L}}, \quad \mathbf{10} = \frac{1}{\sqrt{2}} \left(\begin{array}{ccc|cc} 0 & u_{\mathrm{b}}^{\mathrm{c}} & -u_{\mathrm{y}}^{\mathrm{c}} & -u_{\mathrm{r}} & -d_{\mathrm{r}} \\ -u_{\mathrm{b}}^{\mathrm{c}} & 0 & u_{\mathrm{r}}^{\mathrm{c}} & -u_{\mathrm{y}} & -d_{\mathrm{y}} \\ u_{\mathrm{y}}^{\mathrm{c}} & -u_{\mathrm{r}}^{\mathrm{c}} & 0 & -u_{\mathrm{b}} & -d_{\mathrm{b}} \\ \hline u_{\mathrm{r}} & u_{\mathrm{y}} & u_{\mathrm{b}} & 0 & -e^+ \\ d_{\mathrm{r}} & d_{\mathrm{y}} & d_{\mathrm{b}} & e^+ & 0 \end{array}\right)_{\mathrm{L}}. \quad (6.12)$$

In writing out the particle content of the irreducible representations we have used the charge-conjugate spinors $f^{\mathrm{c}} \equiv C\bar{f}^{\mathrm{T}}$ where $C$ is the charge conjugation operator. Note that there is no room in the representations (6.12) for a $\nu_{\mathrm{L}}^{\mathrm{c}}$, which would correspond to a right-handed neutrino field [191].

Equation (6.10) has already introduced the minimal Higgs structure for the SU(5) theory. The adjoint **24** can be represented as a traceless $5\times5$ matrix $\phi$, and the vacuum expectation value,

$$\langle 0|\phi|0\rangle = \left(\begin{array}{ccc|cc} 1 & 0 & 0 & 0 & 0 \\ 0 & 1 & 0 & 0 & 0 \\ 0 & 0 & 1 & 0 & 0 \\ \hline 0 & 0 & 0 & -3/2 & 0 \\ 0 & 0 & 0 & 0 & -3/2 \end{array}\right) \times \left(v = \mathrm{O}(10^{15})\ \mathrm{GeV}\right), \tag{6.13}$$

gives the desired pattern of breakdown to SU(3)×SU(2)×U(1), and masses O($10^{15}$ GeV) to the X and Y bosons which couple the first three to the last two indices. [Looking at the vacuum expectation value (6.13) it appears to have a U(3) symmetry on the first two indices and a U(2) symmetry on the last two indices. But the representation is traceless, which forces us to remove a combination of the two U(1) factors, getting S(U(3)×U(2)) which is locally equivalent to SU(3)×SU(2)×U(1).] It is easy to figure out [177] the action of the X and Y bosons on the fermion representations (6.12):

$$\frac{1}{\sqrt{2}} g X_{i\mu}\left(-\bar{d}_{i_{\mathrm{R}}}\gamma^{\mu}e^{+}_{\mathrm{R}} + \varepsilon_{ijk}\bar{u}^{\mathrm{c}}_{\mathrm{k_L}}\gamma^{\mu}u_{j_{\mathrm{L}}} + \bar{d}_{i_{\mathrm{L}}}\gamma^{\mu}e^{+}_{\mathrm{L}}\right) + \mathrm{h.c.},$$

$$\frac{1}{\sqrt{2}} g Y_{i\mu}\left(\bar{d}_{i_{\mathrm{R}}}\gamma^{\mu}\nu^{\mathrm{c}}_{e_{\mathrm{R}}} + \varepsilon_{ijk}\bar{u}^{\mathrm{c}}_{\mathrm{k_L}}\gamma^{\mu}d_{j_{\mathrm{L}}} - \bar{u}_{i_{\mathrm{L}}}\gamma^{\mu}e^{+}_{\mathrm{L}}\right) + \mathrm{h.c.} \tag{6.14}$$

It is evident from (6.14) that the X and Y bosons will mediate baryon number $B$-violating transitions such as $\mathrm{u}+\mathrm{u}\to\bar{\mathrm{d}}+\mathrm{e}^{+}$, $\mathrm{u}+\mathrm{d}\to\bar{\mathrm{u}}+\mathrm{e}^{+}$, $\mathrm{u}+\mathrm{d}\to\bar{\mathrm{d}}+\bar{\nu}$, as shown in fig. 6.3. Their structure will be discussed in more detail in lecture 7. The SU(2)×U(1) symmetry can be broken down to electromagnetic U(1) by a **5** of Higgs $H$ whose vacuum expectation value points in one of the last two directions (say the fifth):

$$\langle 0|H|0\rangle = \begin{pmatrix} 0 \\ 0 \\ 0 \\ 0 \\ 1 \end{pmatrix} \times \left(v_0 = \mathrm{O}(10^{2})\ \mathrm{GeV}\right). \tag{6.15}$$

The Higgs vacuum expectation value gives masses to the $\mathrm{W}^{\pm}$ and the $\mathrm{Z}^{0}$, and also the fermions (6.12) because it appears in the products

$$\bar{\mathbf{5}}\times\mathbf{10} = \mathbf{5}+\overline{\mathbf{45}} \quad \text{and} \quad \mathbf{10}\times\mathbf{10} = \bar{\mathbf{5}}+\mathbf{45}+\mathbf{50}. \tag{6.16}$$

We will return to the problem of fermion masses in section 6.4.

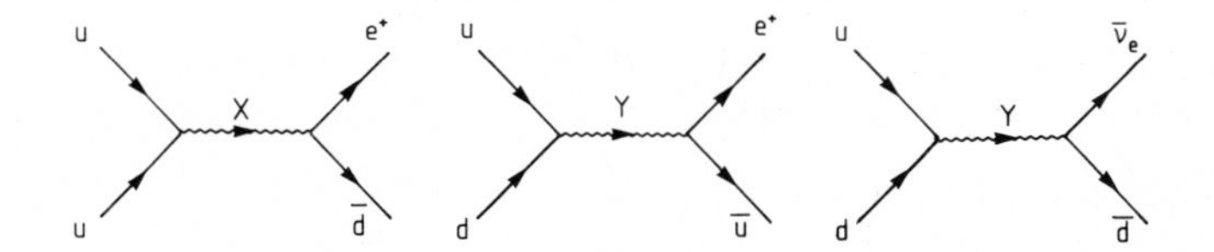


Fig. 6.3. Lowest order X and Y exchanges give a $B$ and $L$ violating, but $B-L$ conserving interaction in minimal SU(5).

Among the elementary properties of the SU(5) model which should be noted is its explanation of the charge quantization mystery. We just saw from the tracelessness of $Q_{\text{em}}$ in the $\bar{\mathbf{5}}$ representation that:

$$3Q_{\bar{d}} + Q_{e^-} = 0 \Rightarrow Q_{\bar{d}} = +1/3 \tag{6.17}$$

and hence $Q_d = -1/3$ and since $Q_u = Q_{\bar{d}} + 1 = 2/3$ it follows that

$$Q_p = 2Q_u + Q_d = +1, \tag{6.18}$$

so that we have derived (6.1). Another important aspect of the model is the determination of the normalization of the U(1) charge operator and the gauge coupling constant $g'$ introduced in the Standard Model of lecture 1, and hence a determination [173, 175] of $\sin^2\theta_W$. Let us write

$$Q_{\text{em}} = I_3 + cI_0 \tag{6.19}$$

as a linear superposition of two similarly normalized generators of SU(5), one an isotriplet and one an isosinglet with a coefficient to be determined. If we identify the conventional U(1) hypercharge

$$Y = cI_0, \tag{6.20}$$

then the relation between the associated U(1) gauge couplings is

$$g' = \frac{1}{c} g_1, \tag{6.21}$$

where $g_1 = g_2(\equiv g) = g_3$ in the SU(5) symmetry limit. We deduce from eq. (6.19) that:

$$\sum_{\text{rep.}} Q_{\text{em}}^2 = (1+c^2) \sum_{\text{rep.}} I_3^2, \tag{6.22}$$

and putting in the fermions of a single generation for which $\Sigma T_3^2 = 2$ and $\Sigma Q_{\text{em}}^2 = 16/3$, we deduce the correct relation between the conventional U(1) coupling $g'$ of the standard model, and the $g_1$ of SU(5):

$$C^2 = 5/3 \Rightarrow g_1 = \sqrt{5/3}\, g', \tag{6.23}$$

where $g_1 = g_2(\equiv g) = g_3$ in the SU(5) symmetry limit. We have the corresponding prediction [173] for (see eq. (1.28)):

$$\sin^2\theta_W = e^2/g^2 = g'^2/(g^2+g'^2) = 1/(1+c^2) = 3/8, \qquad (6.24)$$

which applies in the symmetry limit at energies $\gg 10^{15}$ GeV. We will see in section 6.3 how it gets renormalized [175] at the present low energies.

Let us now go on to consider possible GUTs of rank 5. Again we write [176] a list of all groups which are simple or the products of identical simple factors:

$$[\mathrm{SU}(2)]^5, \mathrm{SO}(11), \mathrm{Sp}(10), \mathrm{SU}(6) \text{ and } \mathrm{SO}(10). \qquad (6.25)$$

Of these, $[\mathrm{SU}(2)]^5$ contains no SU(3) factor and the next two only have real representations [not complex like (6.2)] and do not in any case have 15- or 16-dimensional representations. There is a 15-dimensional representation of SU(6) but it has the unsuitable [cf. eq. (6.2)] SU(3)×SU(2) decomposition

$$\mathbf{15} = (3,2)+(3,1)+(\bar{3},1)+(1,2)+(1,1), \qquad (6.26)$$

and we are left with SO(10) as a candidate group [176, 192].

Many patterns of breaking SO(10) down to a lower symmetry are possible [193], e.g.

$$\mathrm{SO}(10) \begin{array}{l} \nearrow \mathrm{SU}(5) \to \mathrm{SU}(3)\times\mathrm{SU}(2)\times\mathrm{U}(1) \searrow \\ \searrow \mathrm{SU}(4)\times\mathrm{SU}(2)\times\mathrm{SU}(2) \to \cdots \nearrow \end{array} \mathrm{SU}(3)\times\mathrm{U}(1), \qquad (6.27)$$

but even if SU(5) is not present as an intermediate energy symmetry, it is often convenient to use it as a tool in decomposing SO(10) representations. The gauge bosons of SO(10) form a

$$\mathbf{45} = \mathbf{24}+\mathbf{10}+\overline{\mathbf{10}}+\mathbf{1} \text{ of } \mathrm{SU}(5), \qquad (6.28)$$

including extra superheavies which provide new ways for baryons to decay beyond those provided in SU(5). The fermion generations can be accommodated in irreducible

$$\mathbf{16} = \mathbf{10}+\bar{\mathbf{5}}+\mathbf{1} \qquad (6.29)$$

representations of SO(10). By comparison with the minimal SU(5) assignments (6.11), (6.12) we see that (6.29) contains an extra singlet of SU(5), which therefore is a colourless isosinglet with $Q_{em}=0$. It is therefore an ideal candidate to be a $\nu_L^c$ or $\nu_R$, and we will analyse in lecture 8 how such a field can avoid being detected at low energies but may play an important rôle in generating neutrino masses. The minimal set of Higgs fields one can get away with in SO(10) is that which realizes the following breaking pattern [193]:

$$\mathrm{SO}(10) \underset{\mathbf{16}}{\rightarrow} \mathrm{SU}(5) \underset{\mathbf{45}}{\rightarrow} \mathrm{SU}(3) \times \mathrm{SU}(2) \times \mathrm{U}(1) \underset{\mathbf{10}}{\rightarrow} \mathrm{SU}(3) \times \mathrm{U}(1), \tag{6.30}$$

and we will return later to its implications for fermion masses.

Many other groups have been proposed for Grand Unification, including $E_6$ [194], SU(7) [195], SU(8) [4], SU(9) [196], SO(14) or SO(18) [197], etc., but my feeling is that the simplest SU(5) [173] and SO(10) [176, 192, 193] models exhibit most of the interesting features. Consequently, mainly these groups will be used in these lectures to explore the predictions of GUTs.

### 6.3. *Predictions for low energy couplings*

A simple Grand Unified Theory such as SU(5) contains two important parameters describing its gauge interactions: the scale $\approx m_X$ at which the symmetry is first broken, and the value $g(m_X)$ of the gauge coupling at that energy. Since the low-energy Standard Model contains 3 gauge couplings $g_3$, $g_2$ and $g_1 = (5/3)^{1/2} g'$ (6.23), it follows that there must be some prediction interrelating them. To obtain this prediction we use the renormalization group [178] expressions [198] for the variation of the gauge couplings, assuming that the only particles at energies $Q \ll m_X$ are conventional fermion generations:

$$\frac{1}{\alpha_i(Q)} = \frac{1}{\alpha_i(m_X)} + \frac{1}{12\pi}[-4N_G + 11i] \ln m_X^2/Q^2 \quad \text{for } i = 2,3, \tag{6.31a}$$

$$\frac{1}{\alpha_1(Q)} \left( = \frac{12\pi}{5g'^2(Q)} \right) = \frac{1}{\alpha_1(m_X)} + \frac{1}{12\pi}[-4N_G] \ln m_X^2/Q^2. \tag{6.31b}$$

Using the equalities $\alpha_1(m_X) = \alpha_2(m_X) = \alpha_3(m_X)$ and the expressions (6.23), (6.31), we can deduce that for $Q \lesssim m_X$:

$$\frac{1}{\alpha_3(Q)} - \frac{1}{\alpha_2(Q)} = \frac{11}{12\pi} \ln Q^2/m_X^2, \tag{6.32}$$

and that

$$\frac{8}{3} \frac{1}{\alpha_3(Q)} - \frac{1}{\alpha(Q)} = \frac{11}{2\pi} \ln Q^2/m_X^2, \tag{6.33}$$

the right-hand sides of which both vanish at $Q = m_X$, as they should. Before we go on to precision tests of (6.32) and (6.33), it is amusing to note one simple qualitative prediction [199]. It is apparent from eq. (6.33) that [185]:

$$(m_X/\Lambda_{QCD}) = \exp[O(1)/\alpha + \cdots], \tag{6.34}$$

showing that the value of $m_X$, given a value of $\Lambda_{QCD}$, is relatively sensitive to the value of the fine structure constant $\alpha$. If one takes the known scale of the strong interactions $0.1\ \text{GeV} < \Lambda_{QCD} < 1\ \text{GeV}$ and demands that $m_X < 10^{19}$ GeV (so that our GUT does not have to include gravity) and $m_X > 10^{14}$ GeV (so that baryons do not decay too quickly), then it turns out [199] that the fine structure constant must lie in the range

$$\tfrac{1}{120} > \alpha > \tfrac{1}{170}\,. \tag{6.35}$$

While it is desirable to have a more profound derivation of the magnitude of the fine structure constant [4], we are impressed that the experimental value of $1/137$ fits neatly into the "Grand Unifiable" range (6.35), and are thereby encouraged to move on to more precise tests of the Grand Unification philosophy outlined in section 6.1.

It is simple to re-express eqs. (6.32), (6.33) as a prediction for the weak mixing parameter $\sin^2\theta_W$, which is related to the ratio of SU(2) and U(1) coupling constants:

$$\sin^2\theta_W(Q < m_X) = \frac{\frac{3}{5}g_1^2(Q)}{g_2^2(Q) + \frac{3}{5}g_1^2(Q)} \to \frac{g'^2}{g^2 + g'^2} \tag{6.36}$$

and $\sin^2\theta_W = 3/8$ in the symmetry limit (6.24) at $Q \gg m_X$. Including the leading logarithmic variations (6.31) of the gauge couplings in (6.36) we obtain [175]:

$$\sin^2\theta_W(Q^2) = \tfrac{3}{8}\left[1 - \frac{\alpha}{4\pi}\left(\frac{110}{9}\right)\ln m_X^2/Q^2\right] + \text{higher order corrections.} \tag{6.37}$$

This yields [176, 177]:

$$\sin^2\theta_W(Q \approx m_W) = 0.20 \pm 0.01 \tag{6.38}$$

if we put in the value of $\ln Q^2/m_X^2$ deduced from knowing $\Lambda_{QCD}$ and the low energy fine structure constant $\alpha$. We remember that the experimental value [13, 26, 27, 29] (1.66) of $\sin^2\theta_W$ is quite close to the region (6.38), and it is worthwhile to calculate the next corrections to eq. (6.37) to see how they affect the agreement. Calculating these higher order corrections in the same modified minimal subtraction ($\overline{\text{MS}}$) renormalization prescription which was used in the phenomenological extraction of (1.69),

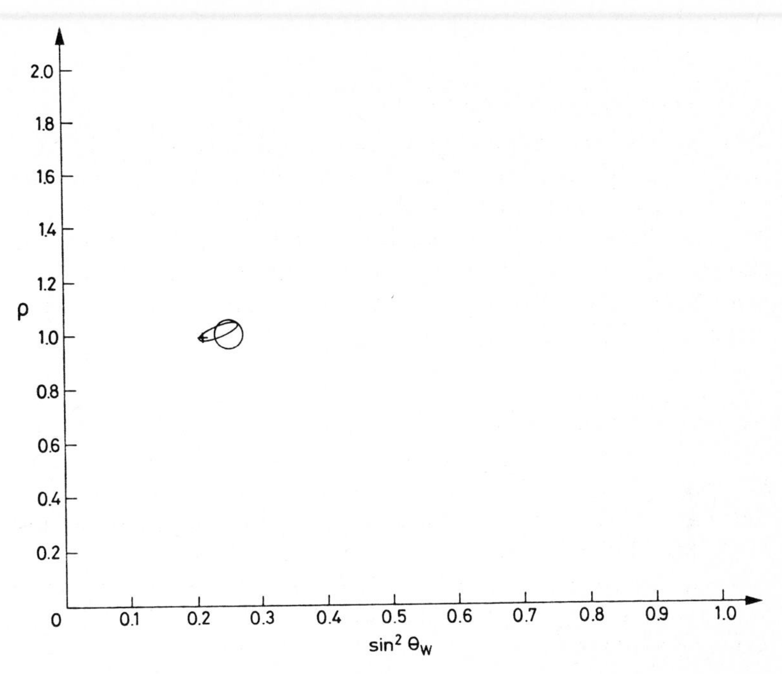


Fig. 6.4. An overview of the experimental determination of neutral current parameters [13,26]. The ellipse and circle are two different ways of determining the values of $\rho$ and $\sin^2\theta_W$, the cross marks the prediction of minimal SU(5) [200].

Marciano and Sirlin [200] find:

$$\sin^2\hat{\theta}_W(m_W)|_{SU(5)} = 0.208 + 0.004(N_H - 1) + 0.006 \ln 400 \text{ MeV}/\Lambda_{\overline{MS}} \tag{6.39}$$

in the minimal SU(5) Grand Unified Theory. In eq. (6.39), $N_H$ denotes the number of light Higgs doublets with masses $\ll m_X$ which is generally taken to be one (*pace* section 3.4). For $\Lambda_{\overline{MS}}$, the QCD parameter in the $\overline{MS}$ scheme, we find:

$$\Lambda_{\overline{MS}} = (100\text{–}200) \text{ MeV} \tag{6.40}$$

in most recent analyses of experimental data [201]. Taking $N_H = 1$ and the range (6.40) for $\Lambda_{\overline{MS}}$ we get

$$\sin^2\hat{\theta}_W(m_W) = 0.214 \pm 0.002, \tag{6.41}$$

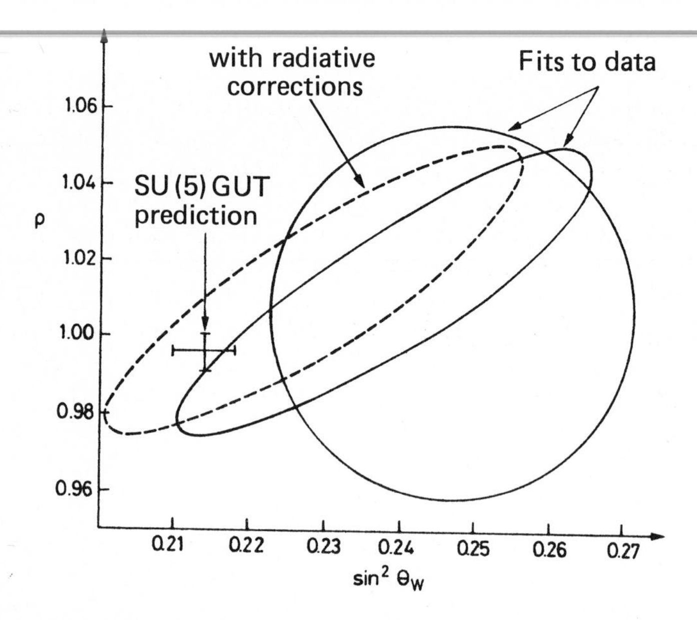


Fig. 6.5. A close-up of the action in fig. 6.4, showing the effect of incorporating radiative corrections in the analysis of the experimental data [27,29].

whose agreement [27,29,200] with the experimental value is almost too good to be true! Despite the heroic achievements of our experimental colleagues in determining $\sin^2\theta_W$ and the neutral current strength parameter $\rho$ (1.46) shown in fig. 6.4, they have not yet succeeded in disproving the minimal GUT prediction (6.39). [Note in passing that most GUTs automatically make the usual prediction (1.60), (1.61) for $\rho$, since they break SU(2) predominantly through a weak isodoublet as in (6.15).] Figure 6.5 shows a close-up of the action in comparing theory with experiment. The solid lines are the results of two different ways of fitting the experimental data [26], with and without correlations between the parameters $\rho$ and $\sin^2\theta_W$. The dashed ellipse is the result of shifting the experimental determination by the amounts (1.63–65) mandated by the theory [27,29] of the radiative corrections. Clearly the SU(5) GUT prediction is still alive and well, and it will be exciting to see if it survives further improvements in the determination of neutral current parameters, for example when the $Z^0$ and $W^\pm$ are discovered and their properties measured.

## 6.4. *Predictions for fermion masses*

In Section 6.1 we mentioned as one motivation for Grand Unification the apparent clustering (6.3) of fermion masses within a generation, and expressed the hope that we might be able to explain some of the quark-to-lepton mass ratios now that there are quarks and leptons in the same GUT multiplet [e.g. (6.12) in the case of SU(5)]. We saw in section 6.2 that in the minimal SU(5) the **5** representation can give rise (6.16) to some fermion masses. Since there are two independent couplings $\bar{\mathbf{5}}\times\mathbf{10}\to\mathbf{5}$ and $\mathbf{10}\times\mathbf{10}\to\bar{\mathbf{5}}$, and three different fermion masses per generation $(m_{+2/3}, m_{-1/3}, m_{-1})$ even if the neutrinos are massless, we might expect to get one new mass formula per generation. We would expect even more mass relations in the minimal SO(10) model since there is just coupling of the **10** of Higgs to the fermions:

$$\mathbf{16}\times\mathbf{16} = \mathbf{10}+\mathbf{120}+\mathbf{126}. \tag{6.42}$$

Let us first look for the relation expected in the SU(5) model if the **45** of Higgs is absent. The possible fermion–fermion Higgs coupling terms can be written as

$$\mathcal{L}_{\mathrm{Y}} = \frac{1}{\sqrt{2}}(\chi^{\dagger})^{\alpha\beta}\gamma_0\mathfrak{M}_1\left[H_\alpha\psi_\beta - H_\beta\psi_\alpha\right] - \tfrac{1}{4}\varepsilon^{\alpha\beta\gamma\delta\varepsilon}\chi_{\alpha\beta}\mathfrak{M}_2 H_\gamma\chi_{\delta\varepsilon}, \tag{6.43}$$

where the **5** fermions are written as $\psi_\alpha$, the **10** as $\chi_{\alpha\beta}$, both of which are to be regarded as vectors in generation space, while the $\mathfrak{M}_{1,2}$ are matrices in generation space. We can always diagonalize $\mathfrak{M}_1$ by rotations:

$$\psi' = U_1\psi, \qquad \chi' = U_2\chi \tag{6.44}$$

in generation space. Then the **5** Higgs vacuum expectation value (6.15) generates in the first term of (6.43) a coupling

$$(\chi^{\dagger})^{5\beta}\psi_\beta, \quad \beta = 1,2,3,4. \tag{6.45}$$

Checking with the explicit representations (6.12) we see that eq. (6.45) implies [175]:

$$m_{\mathrm{d}_{\mathrm{R,Y,B}}} = m_{\mathrm{e}}, \quad \text{and similarly } m_{\mathrm{s}} = m_\mu, m_{\mathrm{b}} = m_\tau, \tag{6.46}$$

reflecting the residual SU(4) symmetry of the couplings when the $H$ takes the vacuum expectation value (6.15). The predictions (6.46) are also obtained in the SO(10) model if only the **10** of Higgs is present from

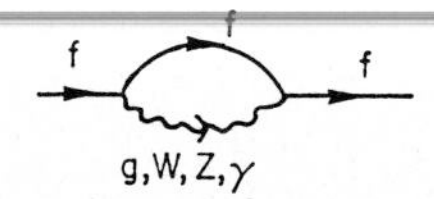


Fig. 6.6. Dominant contribution to the renormalization [176, 177] of the quark-to-lepton mass ratio in GUTs.

among the right-hand side of eq. (6.42). So we have our advertized mass formulae (6.46), but they do not look very successful phenomenologically. Nor should they, because they only apply in the symmetry limit, to effective quark and lepton mass parameters measured on an energy scale $Q \gg m_X$. To relate them to the observed fermion masses we must compute [176, 177] the renormalization of (6.46) using a $Q$-dependent mass parameter such as $m_f(Q)$ [202]:

$$S_f^{-1}(Q) \equiv \not{Q} - m_f(Q) \tag{6.47}$$

in the inverse fermion propagator where the wave function renormalization has been made off-shell at a momentum $Q$ [203]. The $Q$-dependence of $m_f(Q)$ (6.47) is controlled by the renormalization group [178] via the anomalous dimension of the $\bar{\psi}\psi$ operator, and we find [176, 177] that the dominant contribution comes from virtual gluon diagrams as in fig. 6.6. They have the general effect shown in fig. 6.7, for example

$$\left[\frac{m_b(Q)}{m_\tau(Q)}\right] = \left[\frac{\alpha_3(m_Q)}{\alpha_3(m_X)}\right]^{4/(11-4N_q/3)} [1 + O(20)\% \text{ corrections}], \tag{6.48}$$

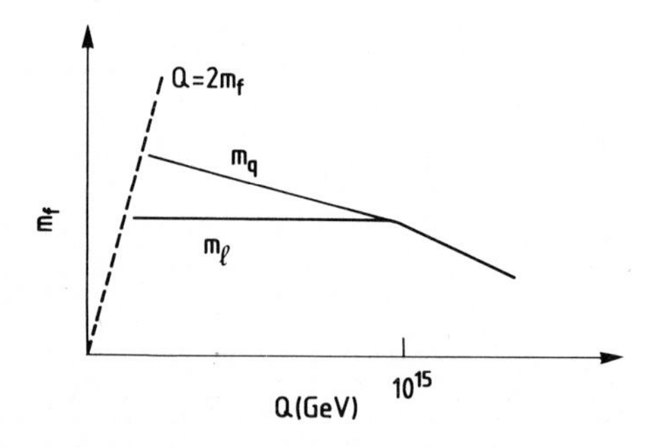


Fig. 6.7. A sketch of the renormalization of the effective quark and lepton mass parameters (6.47) as a function of momentum/energy scale.

where the corrections come from renormalization due to U(1) gauge interactions, higher order SU(3) corrections, finite-mass effects, etc. If we use (6.48) to calculate the effective quark mass [202] $m_q(Q)$ at the threshold $Q = 2m_q$ as indicated in fig. 6.7, then we find [176, 177]:

$$m_b = (5\text{–}5\tfrac{1}{2})\ \text{GeV} \tag{6.49}$$

if there are only 3 generations—in agreement with experiment! Note however the sensitivity of the renormalization (6.48) to the number of generations: the successful prediction (6.49) is lost if there are 4 or more low-mass generations [204] or even if the top quark mass is too high ($>155$ GeV) [205] as seen in fig. 6.8. A similar analysis [177] of the strange quark mass yields:

$$m_s = O(1/2)\ \text{GeV}, \tag{6.50}$$

with rather larger uncertainties because $Q = 2m_s = O(1)$ GeV is well into the strong coupling régime of QCD.

There is some debate whether the prediction (6.50) is experimentally correct: while everyone accepts that the constituent mass of the strange quark is about 1/2 GeV, it has been argued [206] that the current algebra

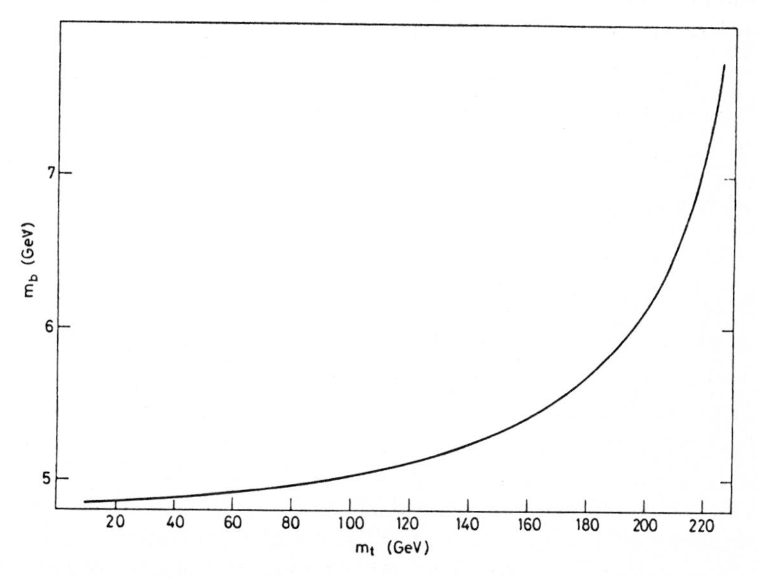


Fig. 6.8. The results of a recent recalculation [205] of $m_b$, assuming six quark flavours and demonstrating the sensitivity to the top quark mass.

(short distance) mass, which is what we are computing here, should rather be $\approx 150$ MeV. The argument for this value is not very solid [206], as it depends on identifying $m_\Lambda - m_N \approx 150$ MeV with the difference between the *constituent* masses of the s and d quarks, *and* then identifying this with the difference in their *current algebra* masses:

$$m_\Lambda - m_N \stackrel{?}{=} m_s - m_d|_{\text{constituent}} \stackrel{?}{=} m_s - m_d|_{\text{current algebra}} \tag{6.51}$$

*and* then using the commonly accepted ratios of current algebra masses:

$$m_d/m_s \approx 1/20 \tag{6.52}$$

to infer that the current algebra $m_s \approx 150$ MeV. It is not at all clear that the second step in eq. (6.51) is valid: the constituent quark in a hadron is stuck in a non-perturbative bag and we do not know how its mass is related to the current algebra mass we may know how to calculate. For example, perhaps

$$m_q|_{\text{constituent}} = \left[m_q^2|_{\text{current algebra}} + X^2_{\text{non-pert.}}\right]^{1/2} ? \tag{6.53}$$

with $X_{\text{non-pert.}} \approx 300$ MeV, in which case:

$$m_s|_{\text{constituent}} = 500 \text{ MeV} \Rightarrow m_s|_{\text{current algebra}} = 400 \text{ MeV}, \tag{6.54}$$

which is not in bad disagreement with the minimal SU(5) or SO(10) prediction (6.50). QCD sum rules [207] are compatible with either the low or the high mass possibility for $m_s$, while recent lattice calculations [208] seem to prefer $m_s$ small.

There is one prediction [173, 176, 177] of the minimal SU(5) or SO(10) theories which *must* be incorrect, namely that

$$m_d/m_s|_{\text{current algebra}} \stackrel{?}{=} m_e/m_\mu \,(\approx 1/200), \tag{6.55}$$

in clear conflict with the generally accepted value (6.52) based on the analysis of meson masses. There are several possible "fixes" for this problem. One is to postulate [209] that Higgses in the **45** of SU(5) or the **126** of SO(10) contribute to the quark mass matrix, in which case it is easy to get (6.52). However, while it is still possible in such models to retain the successful prediction (6.49), it loses some of its natural lustre. An alternative "fix" is to suppose that there is some small extra contribution of the order of a few MeV to the fermion mass matrix, which does not respect the SU(4) relations (6.46), originating perhaps from radiative corrections [210], or perhaps even from quantum gravity [211].

Despite these quantitative problems at the lower end of the fermion spectrum, the qualitative successes of the predictions (6.49), (6.50) vindicate our naïve associations of quarks and leptons into the generations of fig. 6.1. To date they are the only pieces of evidence that quarks and leptons are related in this way. So far we have made no remarks about the masses of the charge $+2/3$ quarks, and in the SU(5) model they are determined by the $\mathfrak{M}_2$ matrix of eq. (6.43) and are completely unrelated to the other fermion masses. If there is a simple Higgs structure in larger GUTs such as SO(10) or $E_6$ one can make the prediction [212] analogous to (6.55) that

$$m_t/m_c = m_\tau/m_\mu. \tag{6.56}$$

Putting in the appropriate renormalization group corrections, this formula predicts that

$$m_t \approx 20 \text{ GeV}, \tag{6.57}$$

tantalizingly close to the present experimental limit. Perhaps we will soon have another successful GUT prediction in low-energy physics.

## 7. Baryon number violation

### 7.1. *Qualitative introduction*

We have already seen in lecture 6 that in general GUTs one expects baryon number $B$ to be violated due to the introduction of direct quark–lepton and quark–antiquark transitions. This should not shock us: while formerly sacred, $B$ conservation is no longer, thanks to the dogma of the gauge age which asserts that:

> "The only exact symmetries are gauge symmetries"

of which examples are electromagnetic U(1), colour SU(3) and perhaps technicolour if it exists. Baryon and lepton numbers are not protected by such an exact gauge symmetry, and it is therefore to be expected that they will be violated at some level. In fact, two mechanisms for $\Delta B \neq 0$ processes are already known. One involves non-perturbative gravitational phenomena [130, 213], where it is known from the "no-hair" theorem that the only conserved quantum numbers that can rigorously be associated with a black hole are those corresponding to long-range (gauge) fields:

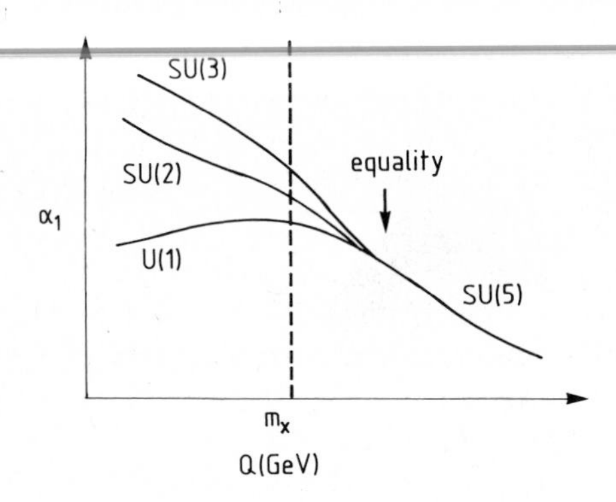


Fig. 7.1. In a momentum space renormalization scheme the coupling constants in a GUT do not come together exactly at $Q = m_X$.

mass, angular momentum and electromagnetic charge. One can therefore imagine a black hole catalyzing a $\Delta B \neq 0$ process:

$$\mathrm{p} + \mathrm{BH}(m, J, Q) \to \mathrm{BH}(m', J + \tfrac{1}{2}, Q + 1) \to \mathrm{BH}(m, J, Q) + \mathrm{e}^+. \tag{7.1}$$

Obviously such a process would be unconscionably slow if one needed an astrophysical black hole. However, Zeldovich [213] has pointed out that one could get an effective $\Delta B \neq 0$ interaction from virtual black holes with masses of the order of the Planck mass of $10^{19}$ GeV, which could cause baryons to decay with a lifetime

$$\tau_{\mathrm{B}} \approx (10^{45}\text{–}10^{50})\ \mathrm{yr}. \tag{7.2}$$

Furthermore, 't Hooft [214] has pointed out that instantons of the Glashow–Weinberg—Salam theory [1] can cause $\Delta B = N_{\mathrm{G}}$ reactions which would allow for example

$$\mathrm{N} + \mathrm{N} \to \overline{\mathrm{N}} + \text{leptons}, \tag{7.3}$$

but the rate for this is expected to be even slower than the black hole process (7.2). Fortunately one expects baryons to decay much faster in GUTs.

In minimal SU(5) one has the X and Y interactions shown in fig. 7.1 which give rise to an effective four-fermion interaction at low energies $\ll m_{\mathrm{X,Y}}$. Neglecting the grand unified analogue of generalized Cabibbo mixing, to which we return in section 7.4, the effective Lagrangian takes the form [177]

$$\frac{1}{4}\mathcal{L}_{\rm GU} = \frac{1}{\sqrt{2}} G_{\rm GU}\Big[\big(\varepsilon_{ijk}\bar{u}^{\rm c}_{k_{\rm L}}\gamma_\mu u_{j_{\rm L}}\big)\big(2\bar{e}^+_{\rm L}\gamma^\mu d_{i_{\rm L}} + \bar{e}^+_{\rm R}\gamma^\mu d_{i_{\rm R}}\big)$$

$$-\big(\varepsilon_{ijk}\bar{u}^{\rm c}_{k_{\rm L}}\gamma_\mu d_{j_{\rm L}}\big)\big(\bar{\nu}^{\rm c}_{\rm e}\gamma^\mu d_{i_{\rm R}}\big) + {\rm h.c.}\Big], \tag{7.4}$$

where the effective coupling $G_{\rm GU}$ is very analogous to the conventional Fermi coupling $G_{\rm F}$ mediated by $\mathrm{W}^{\pm}$ exchange:

$$\frac{G_{\rm GU}}{\sqrt{2}} = \frac{g^2}{8m_{\rm X}^2} = \frac{g^2}{8m_{\rm Y}^2}, \quad \text{cf.} \quad \frac{G_{\rm F}}{\sqrt{2}} = \frac{g^2}{8m_{\rm W}^2}. \tag{7.5}$$

We would get from the interaction (7.4) a decay amplitude

$$A \propto 1/m_{\rm X}^2 \tag{7.6}$$

and thence a decay rate

$$\Gamma \propto |A|^2 \propto 1/m_{\rm X}^4, \tag{7.7}$$

so that the baryon lifetime $\tau_{\rm B}$ is $\propto m_{\rm X}^4$. But $\tau_{\rm B}$ has the mass-energy dimension $-1$, so dimensional analysis tells us that it must be scaled by some inverse mass to the fifth power, say the nucleon mass:

$$\tau_{\rm B} = C\big(m_{\rm X}^4/m_{\rm N}^5\big), \tag{7.8}$$

where $C$ is a model-dependent coefficient to be calculated.

Looking at the minimal SU(5) results (7.4 to 7.8) several questions leap to mind. Does one expect the $\Delta B \neq 0$ interaction in general GUTs to resemble that (7.4) in minimal SU(5)? In particular, should one always anticipate a four-fermion interaction, and is $B - L$ always conserved as in (7.4)? These points will be addressed in section 7.2. If one has a general form (7.4), so that the baryon lifetime is proportional to $m_{\rm X}^4$ (7.8), we need to know the value of $m_{\rm X}$ as precisely as possible, a point discussed in section 7.3. In order to be sure that baryons are not trying to decay into t quarks or $\tau$ leptons we need to study the generalized Cabibbo mixing structure we neglected in writing down the effective interaction (7.4), which is the subject of section 7.4. Next we need to know the value of the coefficient $C$ in (7.8), and hence determine the total decay rate and branching ratios (section 7.5). Finally we should ask whether there are any other possible manifestations of $\Delta B \neq 0$ interactions, and neutron–antineutron oscillations are discussed in section 7.6.

*7.2. Form of the effective Lagrangian for baryon decay*

Since the Lagrangian has mass dimension 4, any operator appearing as an effective Lagrangian which has dimension $d > 4$ must have a coefficient of order $M^{4-d}$. We will be generating such effective interactions from the exchanges of heavy particles, so we will evidently concentrate on the operators of lowest dimension $d$. Since SU(3)×SU(2)×U(1) is a good low-energy symmetry, we should classify [215,216] all possible interactions using invariance under this group. A simple tool for this task has been provided by Weinberg [215] in the guise of $F$-parity. This quantity has the following assignments:

$$F = +1: \mathrm{q}, \ell,$$

$$F = -1: \bar{\mathrm{q}}, \bar{\ell}, \text{ gauge bosons, Higgses, derivatives } \partial_\mu, \tag{7.9}$$

which are arrived at in the following way. Lorentz invariance can be implemented by classifying particles in representations $(a, b)$ (where $2a$ and $2b$ are integers) and writing down interactions in which $a$ and $b$ are separately conserved. This implies in particular that $(-1)^{2a}$ and $(-1)^{2b}$ must be multiplicatively conserved. On the other hand, weak SU(2) invariance requires that $(-1)^{2T}$ be conserved. Weinberg [215] defines $F$ parity as the combination of these conserved quantities:

$$F = (-1)^{2a+2I}. \tag{7.10}$$

Once one has assured $F$ conservation, it is easy to implement invariance under the full SU(3)×SU(2)×U(1) group. Using the definition (7.10) one easily arrives at the assignments (7.9):

$$\left.\begin{array}{l} \mathrm{f_L}: a = I = \tfrac{1}{2}, b = 0 \\ \mathrm{f_R}: a = I = 0, b = \tfrac{1}{2} \end{array}\right\} F = +1, \qquad \left.\begin{array}{l} \bar{\mathrm{f}}_\mathrm{L}: a = 0, b = I = \tfrac{1}{2} \\ \bar{\mathrm{f}}_\mathrm{R}: a = \tfrac{1}{2}, b = I = 0 \end{array}\right\} F = -1,$$

$$\left.\begin{array}{l} \mathrm{W, Z, g}, \gamma, \partial_\mu: a = b = \tfrac{1}{2}, I = 0 \text{ or } 1 \\ H: a = b = 0, I = \tfrac{1}{2} \end{array}\right\} F = -1. \tag{7.11}$$

We are now in a position to look for the $\Delta B \neq 0$ operators of lowest possible dimension:

$$d = 6: \quad \begin{array}{lll} \mathrm{qqq}\ell & \text{has} & F = +1: \text{allowed}, \\ \mathrm{qqq}\bar{\ell} & \text{has} & F = -1: \text{disallowed}. \end{array} \tag{7.12}$$

One can construct [215,216] several different $d = 6$ operators of the allowed form using the first generation of fermions:

$$\left(\varepsilon_{ijk}\bar{u}^{c}_{k_{L}}\gamma_{\mu}u_{jL}\right)\left(\bar{e}^{+}_{L}\gamma^{\mu}d_{i_{L}}\right), \tag{7.13a}$$

$$\left(\varepsilon_{ijk}\bar{u}^{c}_{k_{L}}\gamma_{\mu}\right)\left(u_{j_{L}}\bar{e}^{+}_{R}+d_{j_{L}}\bar{\nu}^{c}_{e_{R}}\right)\gamma^{\mu}d_{i_{R}}, \tag{7.13b}$$

and operators with similar particle content which cannot be generated by the exchanges of vector bosons. As examples, the minimal SU(5) effective Lagrangian (7.4) contains 2×(7.13a) and 1×(7.13b), while SO(10) with its extra gauge bosons (6.28) can yield a more general [217] combination of (7.13a and b). There are some common properties of the allowed qqq$\ell$ operators (7.12) which are quite striking. The first is that $B$ and $L$ are violated in such a way that $\Delta B=\Delta L$, so that

$$\mathrm{p,n} \to (\ell^{+},\bar{\nu})+X, \not\to (\ell^{-},\nu)+X. \tag{7.14}$$

$B-L$ conservation is in fact an *exact* global symmetry of minimal SU(5), but only an *accidental*, approximate feature of more general models such as SO(10). Another striking feature of (7.12) is that

$$\Delta S/\Delta B \leqslant 0 \Rightarrow \mathrm{p,n} \not\to \mathrm{K}^{-}+X. \tag{7.15}$$

We now turn to the next lowest dimension operators.

$$d=7: \quad \begin{array}{lll} \mathrm{qqq}\bar{\ell}\mathrm{B} & \text{has} & F=+1:\text{ allowed,} \\ \mathrm{qqq}\ell\mathrm{B} & \text{has} & F=-1:\text{ disallowed.} \end{array} \tag{7.16}$$

In eq. (7.16), B stands for any gauge boson, Higgs boson or space–time derivative. The allowed interaction qqq$\bar{\ell}$B has $\Delta B=-\Delta L$, in contrast to (7.12). If our GUT only has one large mass scale $m_{\mathrm{X}}$, then we would expect the coefficient of (7.16) to be suppressed relative to that of (7.12) by a factor of $(m_{\mathrm{W}}$ or $\Lambda_{\mathrm{QCD}})/m_{\mathrm{X}}$. Thus we expect $B-L$ to be a good symmetry of baryon decays unless there is some intermediate mass scale in between $m_{\mathrm{W}}$ and $m_{\mathrm{X}}$. The prediction (7.14) is therefore a very general test of the GUT philosophy which planners of experiments should try to test.

It is often asked whether baryons can decay into 3 leptons [218]. To do this requires an operator of dimension as least as large as

$$d=9: \quad \mathrm{qqq}\ell\ell\ell \text{ has } F=+1:\text{ allowed,} \tag{7.17}$$

which requires a very low mass scale $M \approx 10^{5}$ GeV to scale its coefficient if it is to yield a baryon lifetime close to the present experimental limit. The decay of a baryon into 3 antileptons requires

$$\mathrm{d}=10: \quad \mathrm{qqq}\bar{\ell}\bar{\ell}\bar{\ell}\mathrm{B} \text{ has } F=+1:\text{ allowed} \tag{7.18}$$

and hence an even lower "large" mass scale if it is to be competitive.

In the next few sections we will focus on the preferred form (7.12) of $\Delta B \neq 0$, $B - L$ conserving interaction and study it in detail. In section 7.6 we will change gear to consider neutron–antineutron oscillations [219, 220] which can only be mediated by higher dimensional operators.

### 7.3. *Determination of the Grand Unification mass*

The basic idea of computing the unification mass $m_X$ was given in section 6.3. In this section we mention a few of the subtleties that go into a more precise calculation, but refer you to the original literature [175–177, 181–189] for all the details. Roughly speaking, the SU(3), SU(2) and suitably (6.21) normalized U(1) couplings are all equal at energies $Q \gg m_X$, and vary at lower energies according to eq. (6.31), so that for example the SU(3) and SU(2) couplings approach each other according to (6.32). We then deduced from the rate of approach that

$$(m_X/\Lambda_{QCD}) = \exp\{O(1)/\alpha + \cdots\}. \tag{7.19}$$

Our job in this section is to determine precisely the O(1) terms in (7.19) and the leading dotted terms which are

$$\exp(\cdots) = \exp[O(1)\times\ln\alpha + O(1) + \cdots] \tag{7.20}$$

as can easily be deduced [180] from the solutions to the two-loop renormalization group equations for the coupling constants $\alpha_i$.

If we assume the conventional generation structure for all particles with masses $\ll m_X$, there is nevertheless one small modification to be made to the leading approximation (6.32) to the rate of approach of $\alpha_1$ and $\alpha_2$. This is due to the inclusion of $N_H$ light Higgs doublets:

$$\frac{1}{\alpha_3(Q)} - \frac{1}{\alpha_2(Q)} = -\left(\frac{11 + N_H/2}{12\pi}\right)\ln(m_X^2/Q^2) + \cdots. \tag{7.21}$$

Setting $N_H$ equal to the minimal number of one reduces the Grand Unification mass by a factor of O(2) compared with what one would estimate on the basis of (6.32). This now determines correctly the O(1) coefficient in the exponent of eq. (7.19).

To determine the O(1) coefficients in eq. (7.20) we need first of all to choose a consistent renormalization prescription for the coupling constants. This may either be the modified minimal subtraction [183, 185–187, 221] or the momentum space scheme [181, 182, 184, 185, 188], both giving similar results. The former is technically slightly more convenient, the latter is more physical and will be used as a guide

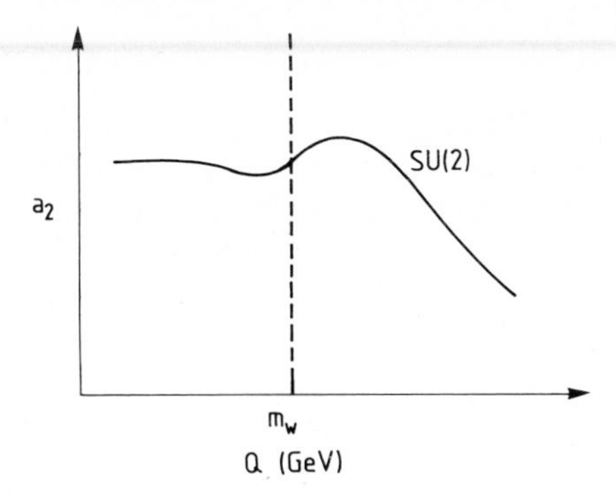


Fig. 7.2. The SU(2) coupling constant in momentum space evolves in a non-trivial manner close to the W threshold.

to the discussion here. The vague term $\Lambda_{\text{QCD}}$ in formula (7.19) should therefore be interpreted as $\Lambda_{\text{MOM}}$ [180]. Several effects contribute to the expression (7.20). One is using [182, 183] the two-loop approximation to the renormalization group equations between the weak interaction threshold $O(m_W)$ and the Grand Unification threshold $O(m_X)$: their inclusion has the effect of decreasing the estimate of $m_X$ by O(4). Another effect is that of taking into account carefully the Grand Unification threshold [181]. The couplings defined in momentum space do not come together strictly at the point $Q = m_X$, but rather approach each other gradually as indicated in fig. 7.1. Looked at from the low-energy side, the momentum scale at which the coupling constants seem to come together is actually rather larger than $m_X$, meaning that our previous naïve estimate of $m_X$ was too high by a factor of two or three. An analogous phenomenon [181] occurs at the weak interaction threshold. One might naïvely use the full SU(2) renormalization group equation for $\alpha_2$ immediately one passes $Q = m_W$, in a sort of $\theta$-function approximation [221]. However, the true evolution of $\alpha_2$ in momentum space looks rather complicated, as indicated in fig. 7.2, and a better place to start the SU(2) evolution is somewhat above $Q = m_W$. This means that $\alpha_2$ at $Q > m_W$ is somewhat larger than has been naïvely thought, and hence that the naïve estimate of $m_X$ is further reduced by a factor of order two or three. However, the most significant reduction in $m_X$ comes from a very banal source [182, 183], namely the variation in the fine structure constant between the Thompson limit ($Q = 0$) where it is measured to be $1/137$, and the region of $Q \approx m_W$ where it is embedded into the SU(2)×

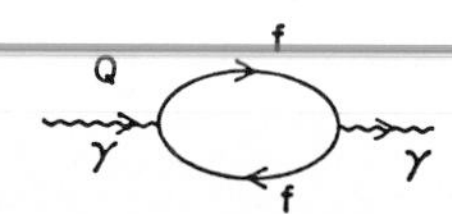


Fig. 7.3. Dominant contribution to the renormalization of the effective value of $\alpha$ between $Q=0$ and $Q=O(m_W)$.

U(1) weak interaction theory. The dominant renormalization of $\alpha$ in this range comes from the vacuum polarization graphs of fig. 7.3, which change $\alpha^{-1}$ by about nine:

$$1/\alpha|_{Q\approx m_W} = 1/\alpha|_{Q=0} - (9 \pm 1/2) = 128 \pm 1/2, \tag{7.22}$$

where the main uncertainty [185] comes from not knowing how to treat exactly the strong corrections to the vacuum polarization fig. 7.3 for the light quarks. This change again pushes up the weak coupling constants at large $Q$, and hence decreases the estimate of $m_X$ by almost an order of magnitude [182, 183]. As a result of all these corrections, the current best estimate [184–188, 200] of $m_X$ is two orders of magnitude smaller than early estimates. The principal residual uncertainty is in the value of the parameter $\Lambda_{QCD}$, which must either be extracted from perturbative QCD analyses of experimental data [180, 201] or else from non-perturbative calculations [222]. Values are most often quoted for $\Lambda_{\overline{MS}}$ which is related to $\Lambda_{MOM}$ by a known multiplicative factor. The end result is the following estimate [184–188, 200]:

$$(m_X/\Lambda_{\overline{MS}}) = (1\text{–}2)\times 10^{15} \tag{7.23}$$

in minimal SU(5) and also GUTs which reduce to it at energies $Q \lesssim m_X$. The $\Lambda_{\overline{MS}}$ parameter in equation (7.23) is that corresponding to four operational flavours $m_q \ll Q$ in two-loop QCD calculations. The remaining uncertainty in (7.23) reflects possible higher order effects [the dots in eq. (7.20)], the possible effects of varying $m_H$, uncertainty in the precise value (7.22) of $\alpha(m_W)$ to use, etc....

Grand Unified Theories with a symmetry breaking pattern which does not take them through SU(5) at any stage, e.g. the second pattern of SO(10) breaking in (6.27), may well have values of the Grand Unification mass very different [193] from (7.23). Supersymmetric Grand Unified Theories may also yield large deviations from (7.23), as we shall see in lecture 9.

## 7.4. Mixing angles in baryon decay

Now that we have the form (7.13) of the dominant $B$-violating interaction in simple GUTs, and know (7.23) the Grand Unification mass to expect in such theories, we should address the question of generalized Cabibbo mixing which was neglected in writing down (7.13) and (7.4). In particular we hope to exclude the possibility that baryons would want to try to decay into heavier quarks or leptons, which would obviously greatly suppress their decay rates! To analyse this problem in the minimal SU(5) model we must return [223] to the fermion–fermion Higgs couplings (6.43)

$$\mathcal{L}_Y = \frac{1}{\sqrt{2}}(\chi^\dagger)^{\alpha\beta}\gamma_0 \mathcal{M}_1 [H_\alpha \psi_\beta - H_\beta \psi_\alpha] - \tfrac{1}{4}\varepsilon^{\alpha\beta\gamma\delta\varepsilon}\chi_{\alpha\beta}\mathcal{M}_2 H_\gamma \chi_{\delta\varepsilon}, \tag{7.24}$$

written in such a basis that the matrix $\mathcal{M}_1$ is diagonal in generation space (cf. section 6.4). Because of Fermi statistics, the matrix $\mathcal{M}_2$ must be symmetric in generation space, and can therefore be diagonalized by a transformation $U_3$ on the **10** representations of fermions $\chi$:

$$\mathcal{M}_2 = U_3^{\mathrm{T}} \mathcal{M}_2^{\mathrm{D}} U_3. \tag{7.25}$$

If we want, we can remove the phase factors $e^{i\phi_{ij}}$ from the first row and column of $U_3$, thus putting it in the standard form (1.42) of the Kobayashi–Maskawa matrix, by making further transformations $U_4$ and $U_5$:

$$U_3 = U_5 U U_4; \quad U_4 = \begin{pmatrix} e^{i\phi_{11}} & & & O \\ & e^{i\phi_{21}} & & \\ O & & \ddots & e^{i\phi_{n1}} \end{pmatrix},$$

$$U_5 = e^{-i\phi_{11}} \begin{pmatrix} e^{i\phi_{11}} & & & O \\ & e^{i\phi_{12}} & & \\ O & & \ddots & e^{i\phi_{1n}} \end{pmatrix}. \tag{7.26}$$

If we absorb the matrix $U_4$ into the definition of the $\chi$, and make the corresponding transformation on the $\psi$ so as to keep $\mathcal{M}_1$ real and diagonal, the second term of eq. (7.24) becomes:

$$\mathcal{L}_Y \ni -\tfrac{1}{4}\varepsilon^{\alpha\beta\gamma\delta\varepsilon}\chi_{\alpha\beta} U^{\mathrm{T}} U_5^2 \mathcal{M}_2^{\mathrm{D}} U H_\gamma \chi_{\delta\varepsilon}. \tag{7.27}$$

The charge $+2/3$ quark mass matrix comes from the $1 \leqslant \alpha, \beta, \delta, \varepsilon \lesssim 4$

pieces of this interaction. We can diagonalize it by going to a new basis,

$$U\chi_{\alpha\beta} \equiv \chi'_{\alpha\beta}; \quad 1 \leqslant \alpha, \beta \leqslant 4, \tag{7.28}$$

for the top left-hand 4×4 submatrix of the SU(5) matrix $\chi$. The mass matrix is still not real, a problem we can cure by making extra phase transformations,

$$U_5^2\chi'_{\alpha\beta} = \chi''_{\alpha\beta}; \quad 1 \leqslant \alpha, \beta \leqslant 3, \tag{7.29}$$

on the top left-hand 3×3 submatrix of the SU(5) matrix $\chi'$.

It is clear from the assignments (6.12) of quarks to the matrix $\chi$ that the rotation $U$ (7.28) rotates the left-handed charge $+2/3$ quarks relative to the left-handed charge $-1/3$ quarks, and can therefore be identified with the Kobayashi–Maskawa [7] matrix (1.45). The phase rotations (7.29) modify the relative phases of left-handed antiquarks relative to the charge $+2/3$ quarks, and hence do not show up in the conventional SU(2) weak interactions. On the other hand they will show up [177,224,223] in interactions involving the X and Y bosons, which couple together the first three and the last two indices of SU(5).

The most important implication of this analysis is the close connection that exists between generalized Cabibbo angles and Grand Unified mixing angles. This means in particular that baryon decay cannot [223] be "Cabibbo-rotated away". If one incorporates mixing between the first two generations into the SU(5) effective interaction (7.4) one finds

$$\begin{aligned}\tfrac{1}{4}\mathcal{L}_{\mathrm{GU}} &= \mathrm{e}^{\mathrm{i}\phi}\frac{G_{\mathrm{GU}}}{\sqrt{2}} \\ &\times\Big[\big(\varepsilon_{ijk}\bar{u}^{\mathrm{c}}_{k_{\mathrm{L}}}\gamma_{\mu}u_{j_{\mathrm{L}}}\big)\Big\{\big[\big(1+\cos^2\theta_{\mathrm{c}}\big)\bar{\mathrm{e}}^{+}_{\mathrm{L}}+\sin\theta_{\mathrm{c}}\cos\theta_{\mathrm{c}}\bar{\mu}^{+}_{\mathrm{L}}\big]\gamma^{\mu}d_{i_{\mathrm{L}}}+\bar{\mathrm{e}}^{+}_{\mathrm{R}}\gamma^{\mu}d_{i_{\mathrm{R}}} \\ &\qquad+\big[\big(1+\sin^2\theta_{\mathrm{c}}\big)\bar{\mu}^{+}_{\mathrm{L}}+\sin\theta_{\mathrm{c}}\cos\theta_{\mathrm{c}}\bar{\mathrm{e}}^{+}_{\mathrm{L}}\big]\gamma^{\mu}s_{i_{\mathrm{L}}}+\bar{\mu}^{+}_{\mathrm{R}}\gamma^{\mu}s_{i_{\mathrm{R}}}\Big\} \\ &-\big[\varepsilon_{ijk}\bar{u}^{\mathrm{c}}_{k_{\mathrm{L}}}\gamma_{\mu}\big(d_{j_{\mathrm{L}}}\cos\theta_{\mathrm{c}}+s_{j_{\mathrm{L}}}\sin\theta_{\mathrm{c}}\big]\big[\nu^{\mathrm{c}}_{\mathrm{e_R}}\gamma^{\mu}d_{i_{\mathrm{R}}}+\bar{\nu}^{\mathrm{c}}_{\mu_{\mathrm{R}}}\gamma^{\mu}s_{i_{\mathrm{R}}}\big]\Big]+\mathrm{h.c.}\end{aligned} \tag{7.30}$$

Notice the occurrence of the new phase parameter, which does not of course affect baryon decay rates. It is evident from (7.30) that one has Cabibbo-favoured decay modes (into $\mathrm{e}^+$ + non-strange, $\mu^+$ + strange) and disfavoured modes ($\mathrm{e}^+$ + strange, $\mu^+$ + non-strange), whose ratios can be predicted [223] quantitatively:

$$\frac{\Gamma(\mathrm{N}\to\mu^+ + \text{non-strange})}{\Gamma(\mathrm{N}\to\mathrm{e}^+ + \text{non-strange})} = \frac{\sin^2\theta_c\cos^2\theta_c}{(1+\cos^2\theta_c)^2+1},$$

$$\frac{\Gamma(\mathrm{N}\to\mathrm{e}^+ + \text{strange})}{\Gamma(\mathrm{N}\to\mu^+ + \text{strange})} = \frac{\sin^2\theta_c\cos^2\theta_c}{(1+\sin^2\theta_c)^2+1}. \tag{7.31}$$

If baryon decay is ever found, it will be very interesting and important to check the predictions (7.31), as they go right to the guts of our GUTs. Baryon decay may be our only window opening directly onto $10^{15}$ GeV physics. The predictions (7.30, 7.31) clearly rest very strongly on our assumed assignments (fig. 6.1) of fermions to different generations, the only evidence for which comes indirectly from the patchy successes of the renormalized mass predictions (6.49), (6.50).

The crucial element in deriving (7.30) was the symmetry of the mass matrix $\mathcal{M}_2$. It would no longer be valid if there was an antisymmetric contribution to $\mathcal{M}_2$ as happens in some Grand Unified models [185, 225, 226]. However, minimal versions of SO(10) which only have a **10** of Higgs (6.42) contributing to the fermion mass matrix would have a similar form of effective interaction. But SO(10) models have more freedom than SU(5) in the ratio of the two types (7.13a, b) of interaction term, and one can hope [217] to distinguish between SO(10) and SU(5) models by detailed measurements of baryon decay branching ratios.

### *7.5. Estimation of baryon decay rates*

Armed with the full effective Lagrangian (7.30), with its normalization fixed by (7.23), we are now in a position to calculate the total decay rates of protons and bound neutrons, as well as individual branching ratios. Our basic problem is to calculate the blob in fig. 7.4. This calculation can be factored into two parts, one of which is the reliable computation of short distance enhancement effects, and the other is the unreliable

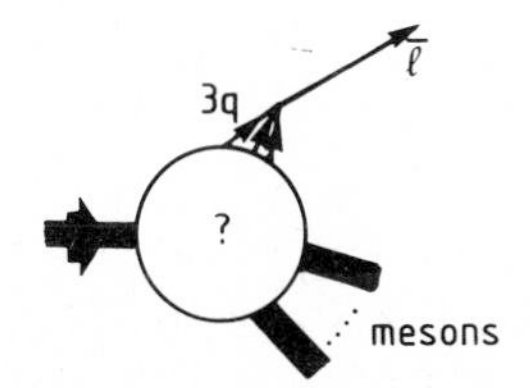


Fig. 7.4. The strong interaction problem in baryon decay.

estimation of hadronic matrix elements of operators renormalized at large distances.

The basic qqqℓ interaction takes place over a distance scale $\Delta x \sim 1/m_X \sim 10^{-28}$ cm, whereas the conventional nonleptonic decay technology of hadronic physics which we will borrow for the second part of the calculation can in principle be used to estimate the matrix elements of operators renormalized at $\Delta x \sim 1/1$ GeV $\sim 10^{-14}$ cm. To get from the short scale to the long one we must calculate the exchanges at distances $10^{-28}$ cm $< \Delta x < 10^{-14}$ cm of gluons, $W^{\pm}$, $Z^0$, photons, etc. as shown in fig. 7.5. This can be done [185,216,223] using the renormalization group [178] and the anomalous dimensions of the qqqℓ operators. (It was recently argued [227] that one should also take account of the anomalous dimensions of the currents to which the X and Y bosons couple. However this is not necessary, as their effects are cancelled [228] by the scale dependence of the effective coupling constant (7.5) appearing in the effective interaction, and all such complications in fact vanish in the Landau gauge.) The leading order enhancement in amplitude due to gluon exchange is [185]

$$A_3 = \left[\frac{\alpha_3(1\ \text{GeV})}{\alpha_5(m_X)}\right]^{2/(11-4N_G/3)}, \tag{7.32a}$$

while that from $W^{\pm}$, $Z^0$ and $\gamma$ exchange is [216,223]:

$$A_{21} = \left[\frac{\alpha_2(100\ \text{GeV})}{\alpha_5(m_X)}\right]^{27/(86-4N_G)} \begin{cases} \times\left[\dfrac{\alpha_1(100\ \text{GeV})}{\alpha_5(m_X)}\right]^{-69/(6+20N_G)} \\ \text{for operator (7.13a)} \\ \times\left[\dfrac{\alpha_1(100\ \text{GeV})}{\alpha_5(m_X)}\right]^{-33/(6+20N_G)} \\ \text{for operator (7.13b)} \end{cases} \tag{7.32b}$$

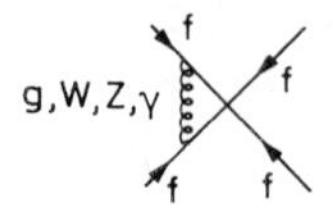


Fig. 7.5. The effective four-fermion operator responsible for baryon decay gets renormalized by g, W, Z and γ exchanges at short distances $1/m_X < \Delta x < 1/1$ GeV.

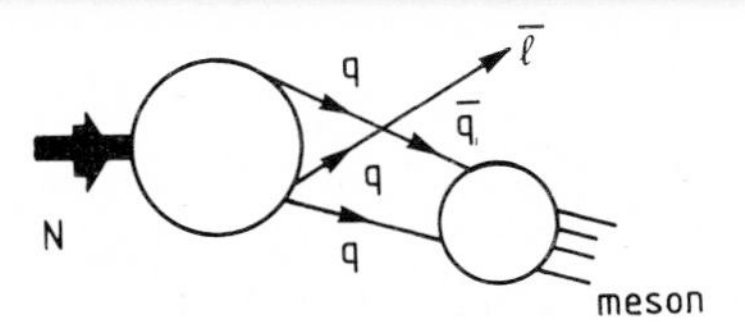


Fig. 7.6. The usual model for baryon decay into an antilepton and a meson.

The overall amplitude enhancement factor is about $3\frac{1}{2}$ to 4, resulting in a decrease in the calculated decay rate by O(15).

The trickier part of the calculation is that of the hadronic matrix elements. One believes that the dominant contribution to the blob of fig. 7.4 comes from the quark–quark annihilation diagram of fig. 7.6. One then calculates the overlap of two quarks in the initial nucleon and the probability that the produced antiquark will combine with the spectator quark into a given meson system using either naïve old-fashioned non-relativistic SU(6) [177, 184, 229–231]:

$$|p_\uparrow\rangle = |u_\uparrow(u_\uparrow d_\downarrow - u_\downarrow d_\uparrow)\rangle, \quad \text{etc.} \tag{7.33}$$

or else the ultra-relativistic kinematics of the MIT bag model [232–234]. Even different authors using the same calculational scheme get widely differing answers, varying over

$$\tau_{\mathrm{p,n}} = (0.6\text{–}25)\times 10^{30}\ \mathrm{yr}\times\left(m_{\mathrm{X}}/5\times 10^{14}\ \mathrm{GeV}\right)^4. \tag{7.34}$$

If one takes $\Lambda_{\overline{\mathrm{MS}}}$ from the latest phenomenological analyses [201]:

$$\Lambda_{\overline{\mathrm{MS}}} = (100\text{–}200)\ \mathrm{MeV}, \tag{7.35}$$

one obtains from (7.23) the estimate

$$m_{\mathrm{X}} = (1\text{–}4)\times 10^{14}\ \mathrm{GeV} \tag{7.36}$$

in minimal SU(5). Substituting into the estimated range (7.34) one finishes up with

$$\tau_{\mathrm{p,n}} = 1\times 10^{27}\text{–}1\times 10^{31}\ \mathrm{yr} \tag{7.37}$$

(lower than the range quoted in [185] because of the reduction [180, 201] in the preferred value of $\Lambda_{\overline{\mathrm{MS}}}$). Since the present experimental limit [235] on the baryon lifetime is

$$\tau_{\mathrm{p,n}} \gtrsim (1\text{–}2)\times 10^{30}\ \mathrm{yr}, \tag{7.38}$$

and experiments [236] now coming into operation should reach a sensitiv-

Table 7.1
Nucleon decay branching ratios[a] in minimal SU(5)

| Decay mode | Non-relativistic model | Preferred "recoil" model | Relativistic model |
|---|---|---|---|
| $p \to e^+\omega$ | 21 | 25 | 26 |
| $e^+\rho^0$ | 2 | 7 | 11 |
| $e^+\pi^0$ | 36 | 40 | 38 |
| $e^+\eta$ | 7 | 2 | 0 |
| $\bar{\nu}\rho^+$ | 1 | 3 | 4 |
| $\bar{\nu}\pi^+$ | 14 | 16 | 15 |
| $\mu^+K^0$ | 18 | 8 | 5 |
| $\nu_\mu K^+$ | 0 | 0 | 1 |
| $n \to \bar{\nu}\omega$ | 5 | 5 | 5 |
| $\bar{\nu}\rho^0$ | 1 | 1 | 2 |
| $\bar{\nu}\pi^0$ | 8 | 7 | 7 |
| $\bar{\nu}\eta$ | 2 | 0 | 0 |
| $e^+\rho^-$ | 6 | 12 | 19 |
| $e^+\pi^-$ | 79 | 72 | 68 |
| $\bar{\nu}_\mu K^0$ | 1 | 3 | 1 |

[a]Adapted from ref. [237].

ity to $\tau_{p,n} \approx 10^{33}$ yr, we can hope that the prediction (7.37) will soon be confirmed or refuted.

The same methods of fig. 7.6 and either non-relativistic or ultra-relativistic kinematics can be used [230–234, 237] to estimate branching ratios into individual mesonic final states. Some typical results for minimal SU(5) are shown in table 7.1, together with the values expected in a preferred intermediate model [237]. In addition, estimates [232, 238] have been made of three-body decay modes such as $p \to e^+(\pi\pi)_{\text{S-wave}}$, etc.: typically they are smaller than the quasi-two-body modes of table 7.1. We see from this table that decays into relatively easily detectable modes such as $p \to e^+\pi^0$ or $n \to e^+\pi^-$ are expected to have relatively large branching ratios. This is good news for our experimental colleagues, and we wish them luck with their tests of the prediction (7.37).

### *7.6. Neutron–antineutron oscillations?*

There has been some interest recently [220] in this phenomenon [219], which would require a $\Delta B = 2$, $\Delta L = 0$ transition. It can only be obtained from an operator containing at least six quarks, and the simplest possibility is for dimension

$$d = 9: \qquad \text{qqqqqq has } F = +1: \text{ allowed}, \tag{7.39}$$

according to the general analysis of section 7.2. Since this operator has dimension $d = 9$, its coefficient $G_{\mathrm{n\bar{n}}}$ must have the dimension $M^{-5}$. One can make an order of magnitude analysis of n–n̄ oscillations as follows: consider the mass matrix

$$(n, \bar{n})\begin{pmatrix} m_{\mathrm{n}} + \Delta E/2 & \delta m \\ \delta m & m_{\mathrm{n}} - \Delta E/2 \end{pmatrix}\begin{pmatrix} n \\ \bar{n} \end{pmatrix}, \tag{7.40}$$

where we have neglected any possible $CP$ violation and included via the terms $\pm\Delta E/2$ a possible difference in the neutron and antineutron energy levels due to the environment in which they are embedded. Possible sources of this "environmental" pollution are stray magnetic fields or interactions inside a nucleus. One can estimate the order of magnitude of

$$\delta m = [C = \mathrm{O}(1)?] \times m_{\mathrm{n}}^6/M^5 \tag{7.41}$$

in the same way as we did for the baryon lifetime (7.8). It seems that the experimental sensitivity [239] for oscillation times $t$ is:

$$t \lesssim \mathrm{O}(10^8)\ \mathrm{s}. \tag{7.42}$$

Comparing this with eq. (7.41) for $\delta m$, and setting $\delta m \sim t^{-1}$, we see that such experiments (7.42) may be sensitive to

$$M \approx 10^6\ \mathrm{GeV}, \tag{7.43}$$

but probably to rather smaller $M$ in many cases, since any simple model (see, e.g., fig. 7.7) will tend to have powers of small coupling constants in the denominator of (7.41), so that the uncalculated coefficient $C$ may be rather less than unity. Corresponding as it does to an oasis in the desert, an energy scale such as (7.43) is not found in minimal GUTs, though it could be present [240] in a non-minimal GUT such as SO(10) broken through the second chain of (6.27).

In fact, no $\Delta B = 2$ term of the type (7.41) could ever arise in the minimal SU(5) model, because it conserves the combination $B - L$. One can set up [240] non-minimal SU(5) models with a **15** of Higgses in which $\delta m \neq 0$, as shown in fig. 7.7a, but they give very small values. The

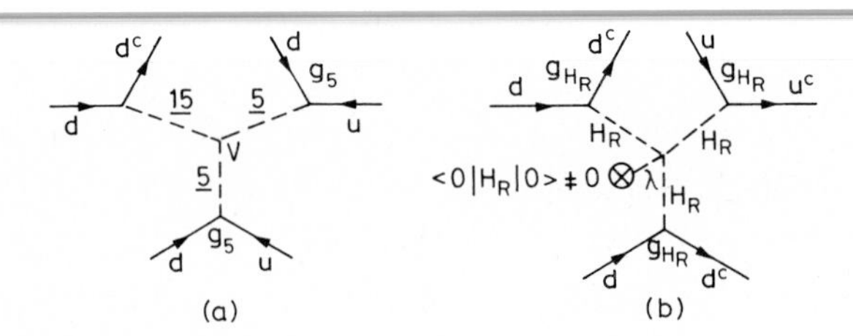


Fig. 7.7. Diagrams which could yield n–n̄ oscillations (a) in a non-minimal SU(5) model, and (b) in an SO(10) model [220].

amplitude from fig. 7.7a:

$$A \sim \frac{V g_{15} g_5^2}{m_{15}^2 m_5^2}, \tag{7.44}$$

where the suffices on the couplings $g$ and masses $m$ indicate the dimensionality of the corrresponding Higgs representations. From the lower limit (7.38) on the lifetime of the proton we know that

$$\frac{g_5^2}{m_5^2} \lesssim 10^{-30}\ \mathrm{GeV}^{-2}, \tag{7.45}$$

and plausible values for the other parameters in (7.44) eventually give [240] unobservable oscillation times

$$t \gtrsim 10^{20}\ \mathrm{yr!} \tag{7.46}$$

One can do better in non-minimal SO(10) broken down in the manner

$$\begin{aligned} \mathrm{SO}(10) &\underset{\mathbf{54}}{\rightarrow} \mathrm{SU}(2)_\mathrm{L}\mathrm{SU}(2)_\mathrm{R} \times \mathrm{SU}(4) \\ &\underset{\mathbf{45}}{\rightarrow} \mathrm{SU}(2)_\mathrm{L} \times \mathrm{SU}(2)_\mathrm{R} \times \mathrm{U}(1)_{B-L} \times \mathrm{SU}(3)_C \\ &\underset{\mathbf{126}}{\rightarrow} \mathrm{SU}(2)_\mathrm{L} \times \mathrm{U}(1) \times \mathrm{SU}(3)_C \underset{\mathbf{10}}{\rightarrow} \mathrm{SU}(3)_C \times \mathrm{U}(1)_\mathrm{em}, \end{aligned} \tag{7.47}$$

in which case the $\Delta B = 2$ transition is associated with the vacuum expectation value of the **126** of Higgses $H_\mathrm{R}$. Fig. 7.7b gives:

$$A \sim \lambda g_{H_\mathrm{R}}^3 \langle 0 | H_\mathrm{R} | 0 \rangle / m_{H_\mathrm{R}}^6, \tag{7.48}$$

and if

$$\lambda = \mathrm{O}(\alpha^2), \quad g_{H_\mathrm{R}} = \mathrm{O}(\alpha), \quad \langle 0|H_\mathrm{R}|0\rangle = \mathrm{O}(10^3\ \mathrm{GeV}),$$

$$m_{H_\mathrm{R}} = \mathrm{O}(10^4\ \mathrm{GeV}), \tag{7.49}$$

it has been estimated [240] that:

$$t \approx (10^5\text{–}10^6)\ \mathrm{s}. \tag{7.50}$$

Let us close with some remarks about the phenomenology of n–$\bar{\mathrm{n}}$ oscillations. If they exist, one should also observe other $\Delta B = 2$ transitions such as $\mathrm{N}+\mathrm{N} \to$ pions in nuclei. The rate for such interactions is bounded below by the same experiments that establish the limit (7.38) on the baryon lifetime. One can then estimate [241] that:

$$t \sim 1/\delta m \approx [\Gamma(\mathrm{N}+\mathrm{N} \to \text{pions}) \times \mathrm{O}(1)\ \mathrm{GeV}]^{-1/2} \gtrsim (10^6\text{–}10^7)\ \mathrm{s}, \tag{7.51}$$

so that there is a relatively narrow window between this and the accessible bound (7.42). To reach this sensitivity one must be very careful about the environmental pollution $\Delta E$ of eq. (7.40). As an example, if we suppose that the energy difference between a neutron and a neutron in a nucleus might be $\Delta E = \mathrm{O}(10)$ MeV, this is clearly catastrophically larger than the $\delta m \approx 10^{-30}$ GeV corresponding to $t \approx 10^8$ s, rendering oscillations inside a nucleus unobservable. In an external magnetic field $B$, $\Delta E = 2\mu_\mathrm{n} B$ where $\mu_\mathrm{n}$ is the neutron's magnetic moment, and for the Earth's magnetic field of O(1) Gauss one finds:

$$\Delta E = O(10^{-11})\ \mathrm{eV}, \tag{7.52}$$

which is still too large. However, if one shields the Earth's magnetic field by a factor $\mathrm{O}(10^{-3})$ so that

$$\mu_\mathrm{n} B \tau \ll 1, \tag{7.53}$$

where $\tau$ is the typical flight time of a neutron, of order $10^{-1}$ or $10^{-2}$ s in practicable experiments, then the environmental pollution no longer kills the n–$\bar{\mathrm{n}}$ oscillations and one can expect

$$\frac{\#\ \text{antineutrons}}{\#\ \text{neutrons}} = \tfrac{1}{4}(\delta m \tau)^2. \tag{7.54}$$

Several experiments [239] are now being set up to look for this phenomenon, though it should be emphasized that it is considerably more speculative than the search for baryon decay, which occurs in minimal GUTs with many less uncertainties as to the rate (7.37).

Actually, it has even been speculated [242] that the indirect consequences of n–n̄ oscillations may already have been detected in cosmic rays. There is an excess [243] of cosmic ray antiprotons, particularly in a recent experiment at low energies $E \approx 130$–$320$ MeV, as shown in fig. 7.8. Conventional calculations [244] of p̄ production at low energies by primary matter cosmic rays are much too low. It has been suggested that cosmic ray antiprotons may be stochastically decelerated by stray galactic magnetic fields, but the calculated effect seems too small. It might be that there are primary antimatter cosmic rays coming from regions of the Universe containing antimatter, but then one would have expected to see cosmic ray $\overline{^4\mathrm{He}}$ nuclei with a flux higher than the present experimental limit [243]. It would help to have n–n̄ oscillations so that one could imagine the sequence of events

$$\mathrm{p} + X \to \mathrm{n} + X', \quad \mathrm{n} \to \bar{\mathrm{n}}, \quad \bar{\mathrm{n}} \to \bar{\mathrm{p}} + \mathrm{e}^+ + \nu, \tag{7.55}$$

but one would need [242] an oscillation time $t \approx 10^4$ s to have the desired magnitude of effect, and this seems unlikely in view of the indirect phenomenological bound (7.51) [245]. There is probably a much more mundane explanation for the p̄ excess in fig. 7.8.

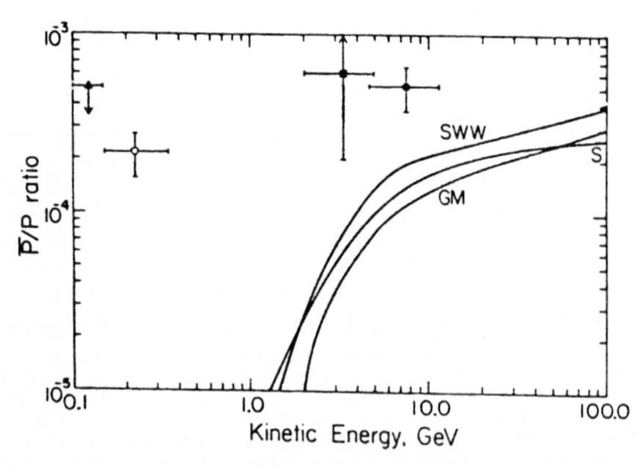


Fig. 7.8. Recent observations [243] of antiprotons in cosmic rays, compared with various theoretical predictions based on production by primary matter cosmic rays.

## 8. Neutrino masses and *CP* violation

### 8.1. Neutrino masses in GUTs

Neutrinos and their masses are clearly different from other fermions. We know that the neutrino masses are much smaller than those of the corresponding leptons and quarks:

$$m_{\nu_e/m_e} \lesssim O(10^{-4}), \qquad m_{\nu\mu}/m_{\mu} \lesssim 1/200, \qquad m_{\nu_\tau}/m_{\tau} < 1/10. \tag{8.1}$$

The most sensitive experiments on the neutrino masses come from the endpoint of tritium $\rightarrow {}^3\text{He}+e^- + \bar{\nu}_e$ decay [246], the $\pi \rightarrow \mu\nu_\mu$ decay [247], and the shape of the spectrum in $\tau \rightarrow \mu\nu\bar{\nu}$ decay [248]. A Russian group has recently reported [249] a positive result of a non-zero $\bar{\nu}_e$ mass from observations of tritium decay:

$$14 \text{ eV} < m_{\bar{\nu}_e} < 46 \text{ eV}, \tag{8.2}$$

and it will be exciting to see if this result is confirmed by other experiments, a point we will return to in section 8.3.

Another peculiarity of the neutrino is that only left-handed neutrinos $\nu_L$ have ever been observed. As we saw in section 2.1, all data on weak charged currents are consistent with their being entirely left-handed, which means that only $\nu_L$ beams are available for neutral current studies. In contrast, the quarks and charged leptons are known to have both right- and left-handed helicity states. This enables them to acquire a "Dirac" mass $m^D$ of the type discussed in section 1.3, through couplings to Higgs fields of the type:

$$g_{H\bar{f}f}\langle 0|H^{\Delta I=1/2}_{\Delta L=0}|0\rangle \bar{f}_R f_L \Rightarrow m_f^D = g_{H\bar{f}f}\langle 0|H^{\Delta I=1/2}_{\Delta L=0}|0\rangle. \tag{8.3}$$

Whether or not the right-handed neutrino $\nu_R$ exists, we have a problem to solve. If a $\nu_R$ exists, why are the neutrino masses (8.1) so small? If a $\nu_R$ does not exist, can neutrinos acquire any mass at all?

Let us answer the second question first—in the affirmative. We reminded ourselves in section 6.2 that the antiparticle of a right-handed fermion is a left handed fermion, hence we could in general imagine replacing $\bar{f}_R$ in (8.3) by $\overline{f^c_L}$. The only difference between $\overline{f^c_L}$ and $f_L$ is in a charge conjugation matrix $C$: hereafter we will lazily write $\overline{f^c_L}$ as $f_L$ as we are only interested in the internal symmetry properties. Replacing $\overline{f_R}$ by $f_L$ could not give masses to the quarks or charged leptons, because $q_L q_L$

and $l_L l_L$ have non-trivial transformation properties under the exact gauge symmetry group SU(3)×U(1), and hence are not legal mass terms. However, since the neutrino has zero electric charge and is colourless, a Majorana mass term of the type

$$m^M \nu_L^T C \nu_L = m^M \overline{\nu_L^c} \nu_L = m^{M} {}'' \nu_L \nu_L'' \tag{8.4}$$

is quite legal *if we allow ourselves to violate lepton number conservation*, and can generate an interaction with weak isospin $\Delta I = 1$. The minimal Glashow–Weinberg–Salam model [1] of lecture 1 has lepton number conservation built into it, and only has $I = 1/2$ Higgs fields. However, as we saw in lectures 6 and 7, GUTs in general violate $L$ conservation and have Higgs fields with $I \neq 1/2$, for example in the **24** of minimal SU(5). Hence we might anticipate finding non-zero Majorana masses $m^M$ (8.4) in GUTs.

There exist several possible mechanisms for generating such Majorana masses in GUTs. The simplest possibility would just be

$$\tilde{g}_{H\bar{\nu}\nu} \langle 0 | H_{\substack{\Delta I = 1 \\ \Delta L = 2}} | 0 \rangle'' \nu_L \nu_L'' \Rightarrow m^M = \tilde{g}_{H\bar{f}f} \langle 0 | H_{\substack{\Delta I = 1 \\ \Delta L = 2}} | 0 \rangle. \tag{8.5}$$

This possibility does not arise in minimal SU(5) where the only Higgs representation with an $I = 1$ component is the **24** $\phi$ which does not couple to pairs of $\bar{\mathbf{5}} + \mathbf{10}$ fermions [cf. eq. (6.16)]. However, this possibility does exist in SO(10) models [193] containing a **126** of Higgs fields, which contain an $I = 1$ piece and can couple to pairs of fermions in the **16**, as seen in eq. (6.42). Another mechanism for generating a Majorana mass (8.4) is via a two-Higgs–two-fermion interaction [215,250] such as:

$$\frac{1}{M} \left( H_{I=1/2} \nu_L \right) \left( M_{I=1/2} \nu_L \right), \tag{8.6}$$

where we have indicated explicitly the inverse power of a mass $M$ which enters since the interaction term has dimension $d = 5$. The interaction (8.6) is formally unrenormalizable and so normally disallowed in gauge theory. However we know that gravity is not renormalizable in the usual sense, and quantum effects are of O(1) at the Planck mass, so one could imagine a term like (8.6) with $M \approx m_P \approx 10^{19}$ GeV perhaps emerging from a theory of quantum gravity [250]. It could perhaps be generated by virtual black holes which have no reason to conserve the global quantum number $L$, just as they had no reason to conserve baryon number $B$, as discussed in section 7.1! Somewhat more conventionally, an interaction (8.6) could be generated by the exchange of a conventional superheavy

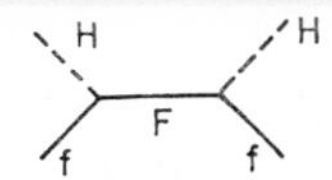


Fig. 8.1. Diagram which may contribute to an effective $(Hf)^2$ interaction (8.6).

fermion of mass $M$ as illustrated in fig. 8.1. This would give

$$m^{\mathrm{M}} = \mathrm{O}\big(\langle 0|H_{I=1/2}|0\rangle^2/M\big), \tag{8.7}$$

and putting in $\langle 0|\mathrm{H}_{I=1/2}|0\rangle = \mathrm{O}(m_{\mathrm{W}})$, $M = \mathrm{O}(m_{\mathrm{X}})(\times 10^{\pm 4}?)$ we get:

$$m^{\mathrm{M}} = \mathrm{O}\big(m_{\mathrm{W}}^2/m_{\mathrm{X}}\big) \ll m_{\mathrm{q},\ell}, \tag{8.8}$$

consistent with the limits (8.1). In point of fact, one would expect the same order of magnitude (8.8) from the alternative source (8.5). The reason for this is that specific models such as minimal SU(5), as well as general arguments, suggest [251,252] that for Higgses of arbitrary weak isospin $I$:

$$\langle 0|\mathrm{H}_I|0\rangle = \mathrm{O}(m_{\mathrm{W}}/m_{\mathrm{X}})^{2I} m_{\mathrm{X}}. \tag{8.9}$$

Specific SU(5) examples are the $I = 0$ component of the **24** of Higgs with a vacuum expectation value (6.13) of order $m_{\mathrm{X}}$, and the $i = 1/2$ component of the **5** of Higgs with vacuum expectation value (6.15) of order $m_{\mathrm{W}}$. In lecture 9 we will meet another example in the $I = 1$ component of the **24** of Higgs which will turn out [177] to have vacuum expectation value $\mathrm{O}(m_{\mathrm{W}}^2/m_{\mathrm{X}})$.

We have crossed a Rubicon in drawing fig. 8.1. We have introduced a new neutral fermion field: why could it not be the long-lost $\nu_{\mathrm{R}}$? and then why could we not have a simple "Dirac" mass term (8.3)?

$$g_{\mathrm{H}\bar{\nu}\nu}\langle 0|H_{I=1/2}|0\rangle \bar{\nu}_{\mathrm{R}}\nu_{\mathrm{L}} \Rightarrow m^{\mathrm{D}} = g_{\mathrm{H}\bar{\nu}\nu}\langle 0|H_{I=1/2}|0\rangle. \tag{8.10}$$

The simplest way to permit (8.10) is for the $\nu_{\mathrm{R}}$ to have $I = 0$. In this case it would be simplest to assign it to a singlet of SU(5), and such an object is found for example in the **16** of fermions used in SO(10) models (6.29). But if such a term (8.10) exists, why is it not the case that

$$m^{\mathrm{D}} = \mathrm{O}(m_{\mathrm{W}}), \text{ or at least } \mathrm{O}(m_{\mathrm{q}}, m_{\ell})?. \tag{8.11}$$

The answer is provided by the observation [253,254,251] that, once we have $\nu_{\mathrm{R}}$ as well as $\nu_{\mathrm{L}}$, we have to consider the most general form of mass matrix whose internal symmetries are shown explicitly below:

$$(\nu_L, \bar{\nu}_R)\begin{pmatrix} m^M & m^D \\ m^D & M^M \end{pmatrix}\begin{pmatrix} \nu_L \\ \bar{\nu}_R \end{pmatrix}. \tag{8.12}$$

We have already met the top left and off-diagonal elements of (8.12): they are expected to be of order $O(m_W^2/m_X)$ [251] (8.8) and $O(m_W)$ (8.11), respectively. The new element in (8.12) is in the bottom right-hand corner. If the $\nu_R$ is a singlet of SU(5) as we hypothesized above, then a Majorana $\bar{\nu}_R\bar{\nu}_R$ mass term is SU(5) invariant and hence can be as large as it likes, and quite possibly much larger than $m_X$. If we now diagonalize the mass matrix (8.12) we find two mass eigenstates:

$$\nu_m \equiv \nu_L + O(m_W/m_X)\bar{\nu}_R : m_{\nu_m} = O(m_W^2/m_X), \tag{8.13a}$$

$$N \equiv \bar{\nu}_R + O(m_W/m_X)\nu_L : m_N = O(m_X). \tag{8.13b}$$

where we should allow several orders of magnitude latitude in the estimates (8.13). For example, perhaps

$$m_{\nu_m} = O\left(\frac{m_\ell^2 \text{ or } m_q^2}{m_X}\right)? \tag{8.13a'}$$

or perhaps:

$$m_N = O\left(m_X \times 10^{\pm 4}\right) = O\left(m_p \text{ or } 10^{11} \text{ GeV}\right)? \tag{8.13b'}$$

In writing (8.12) and (8.13) the generation degree of freedom has been neglected. However, it should be borne in mind that there is no particular reason why the mass eigenstates $\nu_m$ should be the same as the weak eigenstates $\nu_e, \nu_\mu, \nu_\tau, \ldots$, and we will return to the implications of this fact in section 8.2.

So far we have made a general analysis punctuated by illustrations in different GUTs. Let us now look systematically at what happens in some popular models. As observed before, in *minimal SU(5)* there is only a $\nu_L$ (so we do not need the matrix (8.12)) and $B - L$ is strictly conserved so that one would naïvely expect $m_\nu = 0$. But the point has already been made that it is only a global symmetry that protects the neutrino mass, and we may expect these to be broken at the Planck mass even if not before. Therefore we could anticipate [250] finding an interaction of the type (8.6) with $M \approx m_P$ and hence a Majorana mass (8.7) of order

$$m_\nu = O\left(m_W^2/m_P\right) = O(10^{-5}) \text{ eV}, \tag{8.14}$$

which might turn out to be a sort of lower bound on the neutrino mass, as we shall see. In *non-minimal SU(5)* models we can easily introduce $\nu_R$ as singlets of SU(5), introduce new Higgses such as a **15** of SU(5), and violate $B-L$ conservation. Since we already have a **5** of Higgs, and

$$\mathbf{5}\times\mathbf{1} = \mathbf{5}, \tag{8.15}$$

we can easily generate an $m^{\mathrm{D}}$ for (8.12). We can get an $m^{\mathrm{M}}$ from a **15** of Higgs, since

$$\mathbf{5}\times\mathbf{5} = \mathbf{10}+\mathbf{15}, \tag{8.16}$$

and the **15** contains a weak isotriplet of Higgses whose vacuum expectation value is naturally (8.9) of order $(m_{\mathrm{W}}^2/m_{\mathrm{X}})$. We can get an $M^{\mathrm{M}}$ term from an SU(5) invariant interaction, and hence go down the standard route (8.12), (8.13), (8.13′) and easily get [255]:

$$m_\nu = \mathrm{O}(1)\ \mathrm{eV}. \tag{8.17}$$

Even the *minimal SO(10)* model [176, 192] has a $\nu_R$, and $B-L$ is necessarily violated, since it is a broken generator:

$$\begin{aligned}\mathrm{SO}(10) \supset\ & \mathrm{SU}(4)_{\text{Pati-Salam}} \times \mathrm{SU}(2)_{\mathrm{L}} \times \mathrm{SU}(2)_{\mathrm{R}} \\ & \hookrightarrow \supset \mathrm{U}(1)_{B-L}\end{aligned} \tag{8.18}$$

[see also eq. (7.47)]. Once again we can get contributions to all entries in the neutrino mass matrix (8.12), using the SO(10) Higgs representations shown below [193, 251]:

$$(\nu_{\mathrm{L}}, \bar{\nu}_{\mathrm{R}})\begin{pmatrix} \mathbf{126}_{15} & \mathbf{10}_5 \\ \mathbf{10}_{\bar{5}} & \mathbf{126}_1 \end{pmatrix}\begin{pmatrix} \nu_{\mathrm{L}} \\ \bar{\nu}_{\mathrm{R}} \end{pmatrix}, \tag{8.19}$$

where we have indicated by subscripts the SU(5) representation content of the contributing Higgs fields. At this point you might object that minimal SO(10) (6.30) actually does not contain an explicit **126** of Higgses. However, it has been pointed out [256] that even in this case one can get an "effective" **126** from higher order diagrams as in fig. 8.2. In

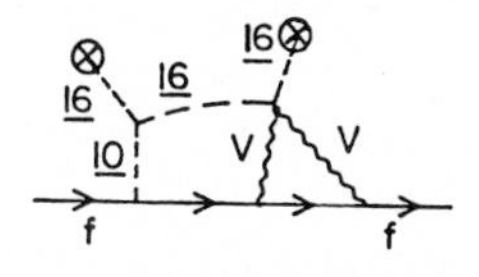


Fig. 8.2. Diagram generating an effective **126** of Higgs fields in the minimal SO(10) model [256].

this case one gets:

$$M^{\mathrm{M}} = \mathrm{O}(\alpha/\pi)^2 m_{\mathrm{X}} \tag{8.20}$$

and the off-diagonal elements $m^{\mathrm{D}}$ are naturally of order $m_\ell$ or $m_{\mathrm{q}}$, so that

$$m_{\nu_{\mathrm{m}}} \approx \frac{m_\ell^2 \text{ or } m_{\mathrm{q}}^2}{\mathrm{O}(\alpha/\pi)^2 m_{\mathrm{X}}} \lesssim \mathrm{O}(10)\ \mathrm{eV}, \tag{8.21}$$

which sets an upper bound to our range of neutrino masses expected in GUTs.

It is possible to get masses larger than (8.21) at the expense [240] of introducing an intermediate energy scale between $m_{\mathrm{W}}$ and $m_{\mathrm{X}}$, for example the scheme (7.47) discussed in connection with $\mathrm{n}$–$\bar{\mathrm{n}}$ oscillations. And it is of course possible to get neutrino masses without invoking GUTs at all. Hence neutrino masses do not constitute such a crucial and specific test of GUT ideas as does baryon decay. However, GUTs suggest the following general prejudices which are experimentally interesting:

$$\text{a range:} \quad \mathrm{O}(10^{-5})\ \mathrm{eV} \lesssim m_\nu \lesssim \mathrm{O}(10)\ \mathrm{eV}, \tag{8.22a}$$

$$\text{a hierarchy:} \quad m_{\nu_1} : m_{\nu_2} : m_{\nu_3} \approx m_{\mathrm{e,u}}^{1 \text{ or } 2} : m_{\mu,\mathrm{c}}^{1 \text{ or } 2} : m_{\tau,\mathrm{t}}^{1 \text{ or } 2}, \tag{8.22b}$$

$$\text{mixing:} \quad \nu_1, \nu_2, \nu_3 \neq \nu_{\mathrm{e}}, \nu_\mu, \nu_\tau, \tag{8.22c}$$

whose implications we shall now investigate.

### 8.2. *Phenomenology of neutrino oscillations*

For an introduction and a review of this subject we refer the reader to [257] and [258], respectively.

The conventional charged currents of neutrinos,

$$J_\mu = \bar{e}\gamma_\mu(1-\gamma_5)\nu_{\mathrm{e}} + \bar{\mu}\gamma_\mu(1-\gamma_5)\nu_\mu + \bar{\tau}\gamma_\mu(1-\gamma_5)\nu_\tau, \tag{8.23}$$

automatically produce, either in decays or scattering, weak interaction eigen-neutrinos $\nu_\ell$. These are in general related to mass eigen-neutrinos $\nu_{\mathrm{m}}$ by a unitary mixing matrix,

$$\nu_\ell = U_{\ell \mathrm{m}} \nu_{\mathrm{m}}, \tag{8.24}$$

which will normally generate oscillations. Let us consider a beam which is initially of pure $\nu_{\mathrm{e}}$, and has a well-defined momentum $\boldsymbol{p}$. (We could have considered a well-defined energy $E$, or a more general wave-packet, but the bottom line (8.28) would be the same [259].) As time passes and the beam propagates [258], the different mass eigen-neutrinos propagate with slightly different phase factors:

$$|\nu_e\rangle \to U_{e1}|\nu_1\rangle e^{i(\boldsymbol{p}\cdot\boldsymbol{x}-E_1t)} + U_{e2}|\nu_2\rangle e^{i(\boldsymbol{p}\cdot\boldsymbol{x}-E_2t)} + \cdots, \tag{8.25}$$

where

$$E_i = |\boldsymbol{p}| + \frac{m_i^2}{2|\boldsymbol{p}|} + \cdots. \tag{8.26}$$

For simplicity, we will henceforward truncate equation (8.25) at $|\nu_2\rangle$, and assume

$$E_i \approx |\boldsymbol{p}| \equiv E \gg m_i \tag{8.27}$$

so that the neutrinos propagate with essentially the speed of light and we can neglect the dots in eq. (8.26). The different phase factors in (8.25) mean that an initial $\nu_e$ eventually becomes a superposition of $\nu_e$, $\nu_\mu$ and $\nu_\tau$. For example, the persistence probability,

$$P(\nu_e \to \nu_e) \approx \left|\sum_{i,j} U_{ei}\exp\left[\frac{i\left(m_i^2 - m_j^2\right)R}{2E}\right]U_{ej}^*\right|^2, \tag{8.28}$$

where $R$ is the distance of propagation, is clearly less than unity in general. It is convenient to define oscillation lengths

$$L_{ij} \equiv 4\pi E/|m_i^2 - m_j^2|, \tag{8.29}$$

which in convenient units become:

$$L_{ij}\left\{\begin{matrix}\mathrm{m}\\ \mathrm{km}\end{matrix}\right\} = 2.5\times E\left\{\begin{matrix}\mathrm{MeV}\\ \mathrm{GeV}\end{matrix}\right\}/|m_i^2 - m_j^2|(\mathrm{eV}^2). \tag{8.30}$$

In the simple case of two neutrino species there is just one mixing angle $\theta$:

$$U = \begin{pmatrix} \cos\theta & \sin\theta \\ -\sin\theta & \cos\theta \end{pmatrix}, \tag{8.31}$$

and the transition probability is:

$$P(\nu_1 \to \nu_2) = \tfrac{1}{2}\sin^2 2\theta\left(1 - \cos\frac{2\pi R}{L_{12}}\right). \tag{8.32}$$

The above analysis applies to the free propagation of neutrinos, but neutrinos are frequently passing through matter, with which they may interact. This can affect the oscillation phenomenology [260]. In fact, the different weak eigen-neutrinos $\nu_\ell$ all have the same neutral current interactions and hence the same forward elastic scattering cross-sections, in which case the relative amplitudes of the different terms in the wave (8.25) are unaffected. Well, almost. We should recall that the $\nu_e$ and $\bar{\nu}_e$ have additional interactions with the electrons due to charged currents, which means that:

$$\sigma\left(\overset{(-)}{\nu}_e e^- \to \overset{(-)}{\nu}_e e^-\right) \neq \sigma\left(\overset{(-)}{\nu}_{\mu,\tau} e^- \to \overset{(-)}{\nu}_{\mu,\tau} e^-\right). \tag{8.33}$$

This means that the $\nu_e$ and $\bar{\nu}_e$ weak eigen-neutrinos have slightly different interaction energies (and hence "effective masses") than do the other $\nu_\ell$. It is convenient to express this using the Schrödinger equation

$$\mathrm{i}\frac{\mathrm{d}}{\mathrm{d}t}|\nu_i\rangle = E_i|\nu_i\rangle \mp \sum_j \sqrt{2}\, G_F N_e U_{ei} U_{je}^* |\nu_j\rangle, \tag{8.34}$$

where the relative sign in (8.34) changes between $\nu_e$ and $\bar{\nu}_e$, and $N_e$ is the density of electrons in the matter being traversed.

If we restrict [261] ourselves to two neutrino species as above, we can introduce, in addition to the usual oscillation length (8.29), (8.30), a matter oscillation length

$$L_M \equiv \frac{2\pi}{\sqrt{2}\, G_F N_e}, \tag{8.35}$$

which takes the value

$$L_M = 1.8 \times 10^7 \times (2\text{–}5)\ \mathrm{m}, \tag{8.36}$$

where the precise value depends on the material. The effect of the extra term (8.34) in the propagation of the neutrino is to change the effective mixing angle

$$\theta \to \theta_{\mathrm{eff}} : \tan 2\theta_{\mathrm{eff}} = \sin 2\theta / (\cos 2\theta - L/L_M), \tag{8.37}$$

and the effective (mass)$^2$ difference:

$$\delta m^2 \to \delta^2_{m_{\mathrm{eff}}} = \delta m^2 \left[1 - 2(L/L_M)\cos 2\theta + (L/L_M)^2\right]^{1/2}. \tag{8.38}$$

Because of the very large length $L_M$ (8.36), these matter effects are negligible except for neutrinos which cross most of the Earth, and even then only for

$$E\ (\mathrm{MeV}) \gtrsim 10^5 |\delta m^2\ (\mathrm{eV}^2)|. \tag{8.39}$$

Thus they are potentially important for cosmic ray neutrino oscillation

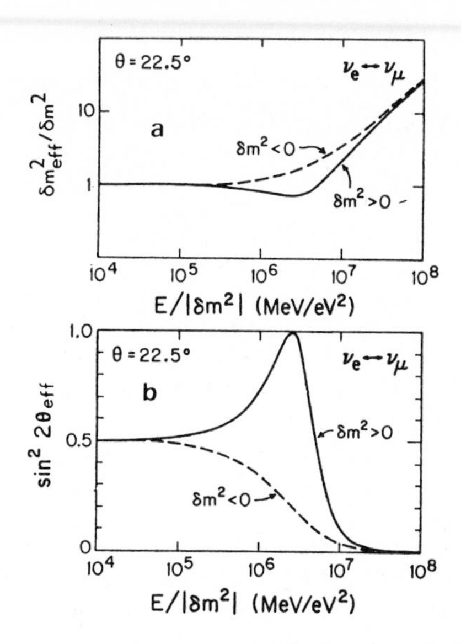


Fig. 8.3. The effects of matter on neutrino oscillations [260]: (a) the effect on $\delta m^2$, and (b) the effect on $\sin^2 2\theta$ [261].

experiments, for which

$$E \approx 1\ \text{GeV}(=10^3\ \text{MeV}), \quad \delta m^2 \gtrsim 10^{-3}\ \text{eV}^2. \tag{8.40}$$

Figure 8.3a shows [261] $\delta m^2_{\text{eff}}/\delta m^2$ for various different values of $\theta$ as a function of $E/\delta m^2$, and exhibits significant effects when the condition (8.39) applies. Figure 8.3b shows [261] $\sin^2 2\theta_{\text{eff}}$ as a function of $E/\delta m^2$ for two different values of the original mixing angle $\theta$ (8.31). These figures show that care will need to be taken in the interpretation of cosmic ray neutrino oscillation experiments.

## 8.3. *Experiments on neutrino masses and oscillations*

Let us start with the most direct experimental information about neutrino masses and proceed to the most indirect. We have already mentioned the experiment [249] claiming positive evidence for a neutrino mass in the range (8.2). This result was announced some time ago and has not been shot down: it should be taken seriously. Confirmation is of the highest priority, but the ITEP group has spent many years perfecting its sophisticated technique, and it will not be easy for another group to

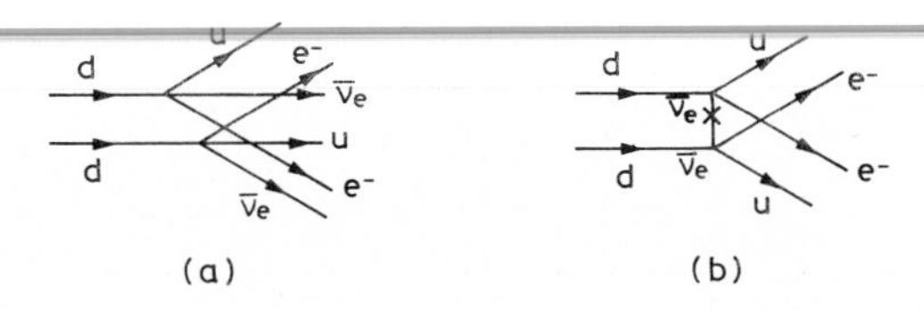


Fig. 8.4. Mechanisms for (a) $(\beta\beta)_{2\nu}$ decay, and (b) $(\beta\beta)_{0\nu}$ decay via a Majorana neutrino mass.

do as well as them, or better, in the near future. Unless someone comes up with a better place than the tritium end-point for measuring the electron neutrino mass. De Rújula [262] has proposed using the end-point of the photon spectrum in internal bremsstrahlung electron capture (IBEC):

$$e^- + (Z, A) \rightarrow (Z-1, A) + \gamma + \nu_e. \tag{8.41}$$

It may be that this process can eventually become competitive with the tritium end-point, but it will not be an easy or short road to travel.

An indirect way to look for a Majorana neutrino mass is neutrinoless double $\beta$ decay $(\beta\beta)_{0\nu}$. There are various geochemical and laboratory indications for double $\beta$ decay, but the only sure example [263] seems to be of conventional two-neutrino double $\beta$ decay $(\beta\beta)_{2\nu}$ (fig. 8.4a). If the neutrino has a Majorana mass, fig. 8.4b shows how $(\beta\beta)_{0\nu}$ could occur. There are various discrepancies [264] between measured or inferred double $\beta$ decay rates and calculations of the conventional $(\beta\beta)_{2\nu}$, and several authors [265–267] have discussed the possibility that $(\beta\beta)_{0\nu}$ may be the culprit. Unfortunately the calculations of such decay rates are notoriously uncertain, and the best approach may be to compare [266, 267] two different, but closely related, processes where some of the uncertainties may cancel out. An example of this is the ratio of half-lives of $^{128}$Te and $^{130}$Te, which seems [266] to disagree with the $(\beta\beta)_{2\nu}$ calculations. It is possible to get agreement if one invokes $(\beta\beta)_{0\nu}$ and the average Majorana mass in the loop of fig. 8.4b is [266]:

$$\langle m_\nu \rangle_e \equiv \sum_j m^{\mathrm{M}}_{\nu_j} |U_{ej}|^2 \approx 34 \text{ eV}. \tag{8.42}$$

The agreement with (8.2) is striking, but remember that these $\beta\beta$ discrepancies have been around a lot longer than the result (8.2), and did not excite much interest until recently because of the considerable

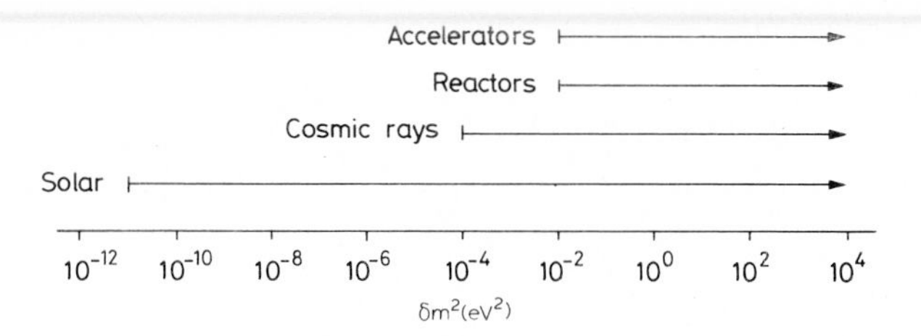


Fig. 8.5. The ranges of $\delta m^2$ to which different classes of neutrino experiments [275] are sensitive.

uncertainties involved. (See ref. [267] for inferred upper limits on $\langle m_\nu \rangle_e$ which are more stringent than (8.42), and a possible way of reconciling these limits with the result (8.2) of ref. [249]).

For the moment, most of our indirect information about neutrino masses and mixing is likely to come from neutrino oscillation experiments. Figure 8.5 shows qualitatively the ranges of neutrino mass (difference)$^2$ to which different classes of experiment are sensitive. Clearly the most sensitive are solar neutrino experiments [268]. So far chlorine 37 has been used [269] as a target for the reaction

$$\nu_e + {}^{37}\mathrm{Cl} \to {}^{37}\mathrm{Ar} + e^-, \tag{8.43}$$

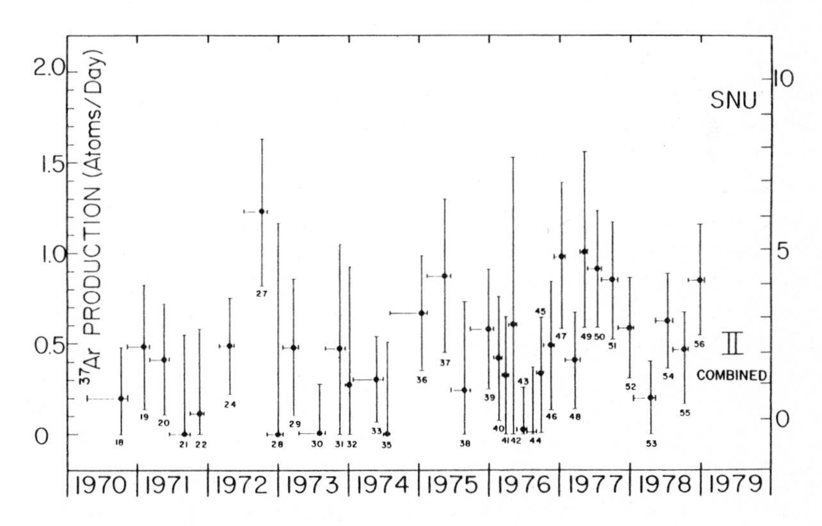


Fig. 8.6. The observed rate [269] of solar neutrino events.

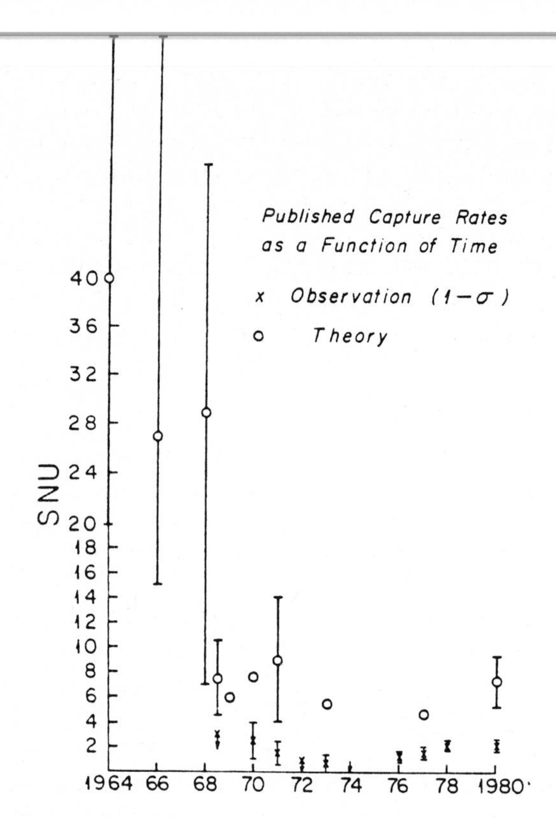


Fig. 8.7. A historical comparison [269] of the theoretically expected and experimentally observed rates of solar neutrino events.

which unfortunately has a rather high threshold in $E_\nu$ (0.814 MeV), which means that most of the solar neutrinos have too low an energy to excite (8.43). The observed rate (fig. 8.6) of events corresponds to a flux of

$$2.2 \pm 0.4 \text{ SNU} \tag{8.44a}$$

(a SNU is defined to be $10^{-36}$ captures per target atom per second), whereas the expected flux has been quoted as

$$7.5 \pm 1.5 \text{ SNU}, \tag{8.44b}$$

a discrepancy [268] by a factor of about $3\frac{1}{2}$. However, there is a general feeling that the error in (8.44b) is under-estimated. Figure 8.7 shows the evolution of the published values [268] of both the experimental (8.44a) and theoretical (8.44b) capture rates. They both vary as a function of time, but always maintain a discrepancy. The expected flux (8.44b) depends on knowing very well an obscure side reaction chain involving $^3\mathrm{He}+{}^4\mathrm{He}$ and $^7\mathrm{Be}+\mathrm{p}$ collisions, whose interaction rates are subject to some uncertainty [270], and which do not contribute significantly to the thermodynamics of the Sun. Furthermore, the calculated rate is very sensitive to the temperature at the core of the Sun, and the discrepancy would vanish if this were reduced by 10%, perhaps as a result of internal convection (for indirect limits on this possibility from measurements of solar oscillations, see [271]). It is planned to use soon a gallium target and look for:

$$\nu_e + {}^{71}\mathrm{Ga} \to {}^{71}\mathrm{Ge} + e^-, \tag{8.45}$$

transitions which have a much lower threshold (0.236 MeV) and are therefore sensitive to neutrinos produced by the vital heat-producing reactions in the Sum. The flux of these can be calculated much more reliably [270], so the motto with solar neutrinos seems to be: wait and see!

Several cosmic ray neutrino experiments have set out to measure the $\nu_\mu$ flux and some experiments have reported discrepancies with theory:

$$R_\mu = \frac{\#\nu_\mu \text{ observed}}{\#\ \nu_\mu \text{ expected}} = \begin{cases} 0.62 \pm 0.17 & [272], \\ \sim 1/2 & [273], \end{cases} \tag{8.46a}$$

but the statistics so far have been rather meagre. Recently a Russian experiment [274] has reported a less interesting result

$$R_\mu = 0.99 \pm 0.23, \tag{8.46b}$$

but all three experiments (8.46) are obviously consistent within the errors. One can expect a breakthrough soon in this class of measurements as the big new underground baryon decay experiments [236] come into operation, so once more: wait and see!

Finally we have accelerator and reactor neutrino experiments, of which a great many have recently reported results [275], and a great many more are planned. Their results are typically complicated constraints on the allowed range of the $\delta m^2$ and $\theta$ parameters as illustrated in fig. 8.8. To get a feeling for the sensitivity of each experiment it is often convenient to quote limits on

$$\begin{aligned} &\delta m^2 \quad \text{as } \sin^2 2\theta \to 1, \\ &\sin^2 2\theta \quad \text{as } \delta m^2 \to \infty, \end{aligned} \tag{8.47}$$

Table 8.1
Limits on neutrino oscillation parameters[a]

| Type of oscillation | $\delta m^2$ (eV$^2$) | $\sin^2 2\theta$ |
|---|---|---|
| $\nu_\mu \to \nu_e$ | < 0.7 | < 0.002 |
| $\nu_\mu \to \nu_\tau$ | < 2.8 | < 0.027 |
| $\nu_e \to \nu_\tau$ | < 8 | < 0.6 |
| $\bar{\nu}_\mu \to \bar{\nu}_e$ | < 0.9 | < 0.2 |
| $\bar{\nu}_\mu \to \bar{\nu}_\tau$ | < 6.3 | < 0.9 |
| $\nu_e \to$ all | < 2.5 | < 0.07 |
| $\bar{\nu}_e \to$ all | < 0.15 | < 0.32 |
| $\nu_\mu \to$ all | < 25 | < 0.1 |

[a]Extracted from refs. [268,275].

corresponding to the limits indicated by solid arrows in fig. 8.8. Table 8.1 sets out the best [275] extant limits on $\delta m^2$ and $\sin^2 2\theta$ (8.47) for various different varieties of $\nu \to \nu'$ (or $\bar{\nu} \to \bar{\nu}'$) transitions. Note that most analyses ignore the possible complications of three-way mixing, which may be an error! Note that there is *no generally accepted evidence* for neutrino oscillations. There is a *prima facie* contradiction (fig. 8.9) between last year's claim [276] of a positive signal and a recent negative result [277], though they may perhaps be reconciled [278] as suggested by the crosses in fig. 8.9. It is striking that the $\delta m^2$ limits in table 8.1 are all much smaller than the Russian [249] (mass)$^2$ (8.2). Therefore unless all the neutrinos have almost identical masses [conflicting with our prejudice

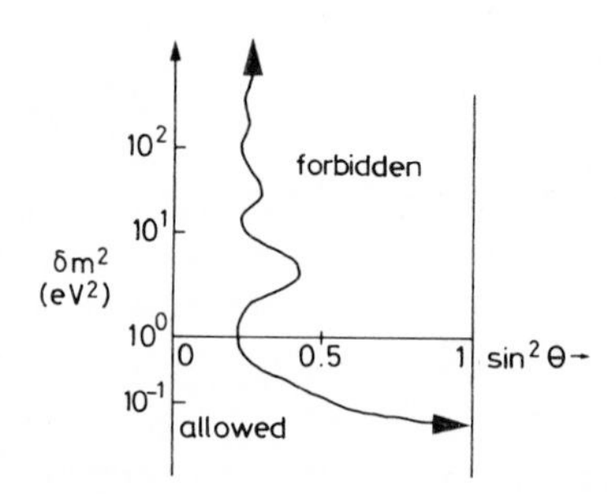


Fig. 8.8. Sketch of the domains of $\delta m^2$ and $\sin^2\theta$ allowed and forbidden by a typical neutrino oscillation experiment.

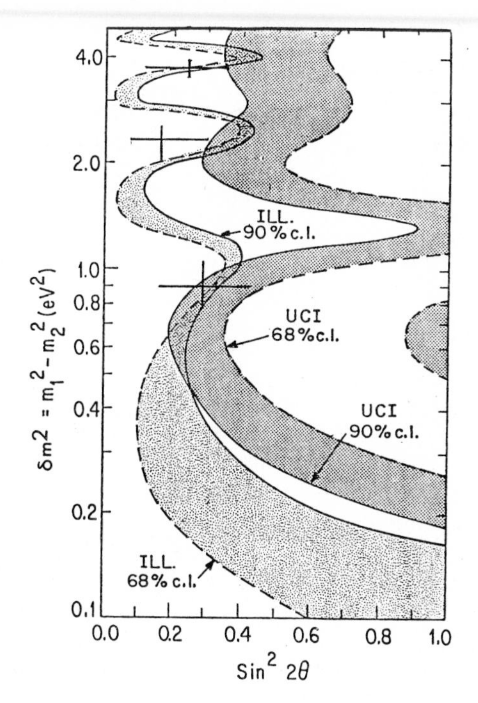


Fig. 8.9. Comparison [278] between two different reactor experiments [276, 277] on electron neutrino oscillations. The crosses indicate possible regions of compatibility.

(8.22b)] so that

$$|m - m_{\text{Russian}}| \approx \delta m^2/2m_{\text{Russian}} \lesssim \mathrm{O}(10^{-1}\text{–}10^{-2})\ \text{eV}, \tag{8.48}$$

one has to accept the upper limits on $\sin^2 2\theta$ in table 8.1. Some of these bounds are rather smaller than $\sin^2 2\theta_c$, but this may not be a fundamental problem if

$$\theta_{ij}^2 \sim m_{\ell_i}/m_{\ell_j} \quad (\text{e.g. } m_e/m_\mu \approx 1/200), \tag{8.49}$$

analogously to the favourite formula

$$\theta_c^2 \sim m_d/m_s \quad (\text{i.e. } \approx 1/20). \tag{8.50}$$

Then one could perhaps have $|m - m_{\text{Russian}}|$ larger than the limit (8.48). Naïvely (8.22b) we expected the electron neutrino to be the lightest, and the others to be hierarchically larger. This is clearly acceptable to laboratory experiments if we accept the bounds of table 8.1 on $\sin^2 2\theta$.

But then cosmologists would object, because the value (8.2) is close to the limit that they can accept [279]:

$$\sum_{\nu_i} m_{\nu_i} < \mathrm{O}(50)\ \mathrm{eV}, \tag{8.51}$$

and is indeed on the hairy edge of closing the Universe. Curiouser and curiouser, to quote Alice. In view of the forthcoming crop of new experimental results, let us once more wait and see!

### 8.4. *CP violation in GUTs*

[Do not worry—this topic has nothing directly to do with the material in previous sections of this lecture—this just happened to a convenient place in the programme to cover this material!]

It is clear that GUTs must contain some *CP*-violating parameters, since they include the weak interactions which violate *CP* [perhaps through the Kobayashi–Maskawa [7] phase $\delta$ (1.42)?] and may well contain extra phase parameters, as we saw (7.29), (7.30) in the discussion of section 7.4. Furthermore, some *CP* violation must be present at high energies and temperatures if one is to explain [219,224,280,281] the baryon number of the Universe as reviewed by Mike Turner at this School (seminar 2). Furthermore, any GUT will also contain a non-perturbative vacuum parameter $\theta_{\mathrm{GUT}}$ analogous to the $\theta$ parameter of QCD (and in principle the Glashow–Weinberg–Salam model [1]) discussed in sections 2.6 and 3.4:

$$\mathcal{L}_{\mathrm{GUT}} \to \mathcal{L}_{\mathrm{GUT}} + \frac{\theta_{\mathrm{GUT}}}{32\pi^2} g^2 F^{\mathrm{a}}_{\mu\nu} \tilde{F}^{\mu\nu}_{\mathrm{a}}. \tag{8.52}$$

At low energies when the GUT is broken down to SU(3)×SU(2)×U(1), $\theta_{\mathrm{GUT}}$ will give rise to two parameters $\theta_3$ and $\theta_2$ for the non-Abelian SU(3) and SU(2) subgroups. In simple models $\theta_{\mathrm{GUT}}$ is generally non-zero and gets renormalized by CP violation elsewhere in the GUT, for example in complex couplings of the Higgses to the fermions. We already discussed the formalism [68] for this renormalization in section 3.4. It will in general be logarithmically divergent, and minimal SU(5) acquires its first such infinity [224] from the 8th order graph of fig. 8.10, appreciably sooner than the 14th order divergence [68] of the Weinberg–Salam model exhibited in fig. 3.16. Just as we argued (3.79) that the divergence in the Weinberg–Salam model was not large enough to worry about, so also we can argue [68] that the divergence of fig. 8.10 need not conflict with the

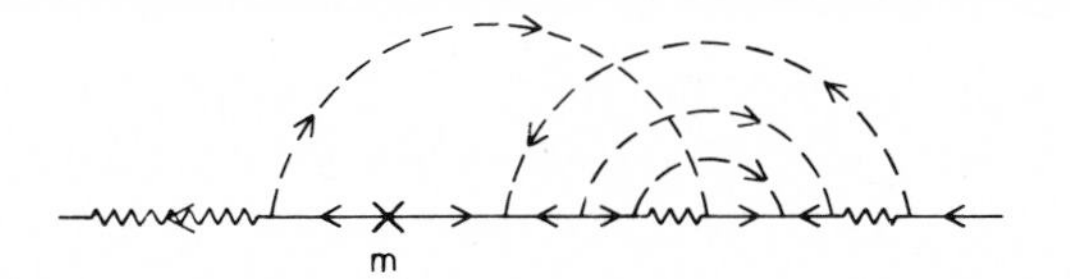


Fig. 8.10. Example of an 8th order diagram contributing to infinite $\theta$ renormalization in a minimal SU(5) GUT [68]. The straight lines are **10**'s and the zigzag lines **5**'s of SU(5) fermions, while the dashed lines are **5**'s of SU(5) Higgses.

experimental bound (2.86):

$$\delta\theta = O(\alpha/\pi)^4 (m_q/m_W)^8 (\text{mixing angles}) \ln(\Lambda/m_X) \tag{8.53}$$

is much smaller than $10^{-9}$ if $\Lambda \lesssim m_P$.

However, the minimal SU(5) model is inadequate to get enough baryons in the Universe [224, 282, 283]. Most of the primordial baryon–antibaryon asymmetry, and hence the present baryon-to-photon ratio, is believed to come from the decays of heavy Higgses, and the lowest order *B*-, *C*- and *CP*-violating diagram [229] is 8th order (fig. 8.11) and closely related [284] to the $\theta$ renormalization diagram of fig. 8.10. It is estimated [285] to yield a net baryon-to-photon ratio

$$(n_B/n_\gamma) \lesssim O(10^{-15})(m_t/m_W), \tag{8.54}$$

whereas astrophysical limits on deuterium and $^3$He abundances [286] require

$$(n_B/n_\gamma) \gtrsim 3 \times 10^{-10}. \tag{8.55}$$

It seems that we need a GUT with more *CP* violation and we start to wonder whether this also yields a $\delta\theta_{GUT}$ larger than (8.53). Indeed, it is possible to argue [284] that the resemblance between figs. 8.10 and 8.11 may carry over to more complicated GUTs.

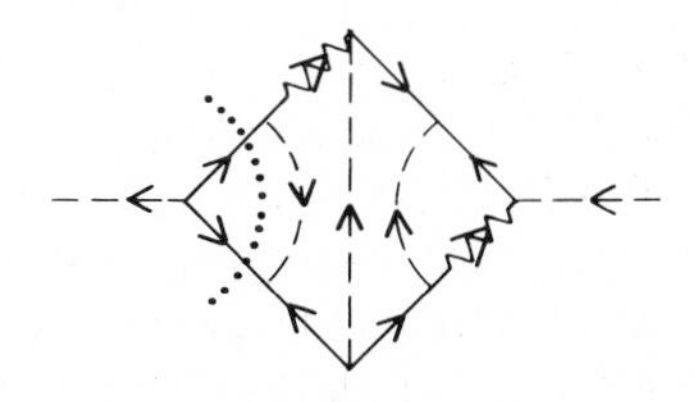

Fig. 8.11. Example of an 8th order diagram contributing to the cosmological baryon asymmetry in a minimal SU(5) GUT [224].

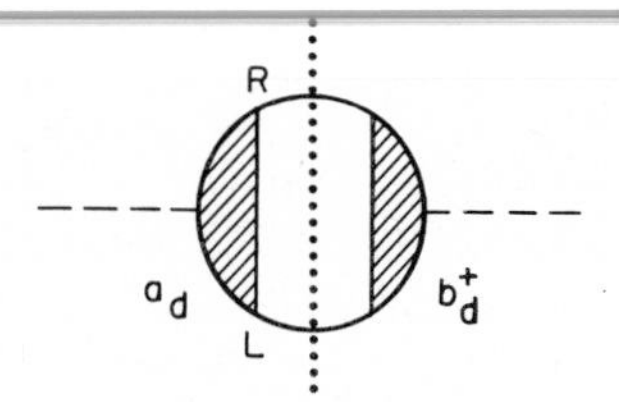


Fig. 8.12. Generic structure of a contribution to the baryon asymmetry from the decays of a d Higgs.

We saw in section 3.4 that $\theta$ renormalization is caused by the necessity to make chiral rotations on the quarks as a result of renormalization of the quark mass matrix. This in turn corresponds to renormalization of the quark–Higgs coupling matrix, and we have learnt that $(n_{\mathrm{B}}/n_{\gamma})$ is usually generated by $CP$ violation in the couplings and decays of heavy Higgs bosons. This connection between $\theta$ renormalization and baryon generation enables us [284, 287, 288] to "derive" a cosmological "lower bound" on the neutron electric dipole moment $d_{\mathrm{n}}$, since we saw in section 2.6 that there is a contribution [77, 79] from $\theta$ to $d_{\mathrm{n}}$:

$$\Delta(d_{\mathrm{n}}/\mathrm{e}) \approx 3\times 10^{-16}\theta \text{ cm}. \tag{8.56}$$

To "derive" the "bound" we start by considering the generic diagram of fig. 8.12 which contributes to baryon generation:

$$\left(\frac{n_{\mathrm{B}}}{n_{\gamma}}\right) \approx \frac{1}{10}\,\frac{\mathrm{Im\,Tr}\left(a_{\mathrm{d}}b_{\mathrm{d}}^{\dagger}\right)}{\mathrm{Tr}\left(H_{\mathrm{d}}H_{\mathrm{d}}^{\dagger}\right)}, \tag{8.57}$$

where $a_{\mathrm{d}}$ and $b_{\mathrm{d}}^{\dagger}$ are contributions to the couplings of incoming and outgoing decaying Higgs (d Higgs), and $H_{\mathrm{d}}$ is the full coupling matrix of the d Higgs. The renormalization of $\theta$ is given by the generic diagram of fig. 8.13a, where $M$ is the quark mass matrix. This is equivalent to the mass generating Higgs (m Higgs) coupling of fig. 8.13b:

$$\delta\theta \approx \mathrm{Im\,Tr}\left(M^{-1}\ \ {}_{\mathrm{L}}\!\!\frown\!\!{}_{\mathrm{R}}\right) \approx \mathrm{Im\,Tr}\left(H_{\mathrm{m}}^{-1}\ \ {}_{\mathrm{L}}\!\!\frown\!\!{}_{\mathrm{R}}\ \text{(m Higgs)}\right), \tag{8.58}$$

where $H_{\mathrm{m}}$ is the full coupling matrix of the m Higgs. There is in general a contribution to fig. 8.13b from diagrams of the form of fig. 8.14, which are got from fig. 8.12 by a little bit of cutting and pasting. We find from fig. 8.14 a contribution to $\delta\theta$:

$$\delta\theta \gtrsim \mathrm{Im\,Tr}\left(a_{\mathrm{d}}b_{\mathrm{d}}^{\dagger}\right)|U_{\mathrm{dm}}|^{2}, \tag{8.59}$$

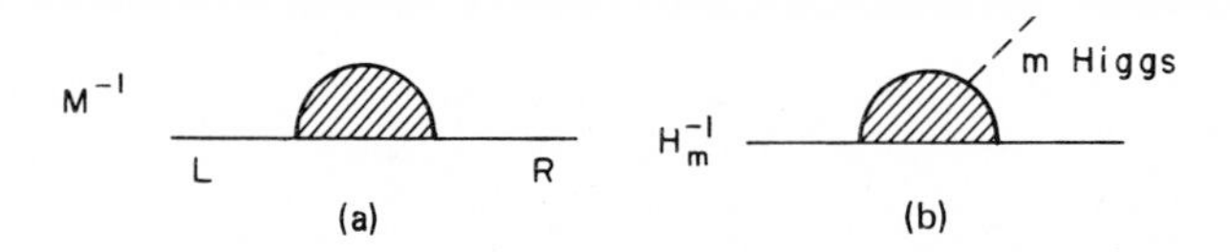


Fig. 8.13. (a) Generic structure of contribution to $\theta$ renormalization, and (b) generic structure of contribution to renormalization of the m Higgs coupling.

where $U_{\mathrm{dm}}$ is the mixing between the m Higgs and the d Higgs, which are not in general pure partners in a GUT multiplet. We now have all the elements to "derive" the "bound" on $d_{\mathrm{n}}$. From (8.56) we have

$$(d_{\mathrm{n}}/e) \gtrsim \Delta(d_{\mathrm{n}}/e) \approx 3\times 10^{-16}\theta, \tag{8.60}$$

and if we accept that $\theta \gtrsim \delta\theta$ (8.58) we get from equation (8.59) neglecting the factor $|U_{\mathrm{dm}}|^2$:

$$\left(\frac{d_{\mathrm{n}}}{e}\right) \gtrsim 3\times 10^{-16}\,\mathrm{Im\,Tr}\left(a_{\mathrm{d}}b_{\mathrm{d}}^{\dagger}\right). \tag{8.61}$$

If we now use (8.57) and take

$$\mathrm{Tr}\left(H_{\mathrm{d}}H_{\mathrm{d}}^{\dagger}\right) \approx \sqrt{2}\,\mathrm{G_F}\left(m_{\mathrm{t}}^2/m_{\mathrm{W}}^2\right) \gtrsim 6\times 10^{-4}, \tag{8.62}$$

we can then deduce from (8.61) that

$$(d_{\mathrm{n}}/e) \gtrsim 2\times 10^{-18}\left(n_{\mathrm{B}}/n_{\gamma}\right), \tag{8.63}$$

and from the cosmological limit (8.55) we finally get [284]

$$\left(\frac{d_{\mathrm{n}}}{e}\right) \gtrsim 6\times 10^{-28}\ \mathrm{cm}. \tag{8.64}$$

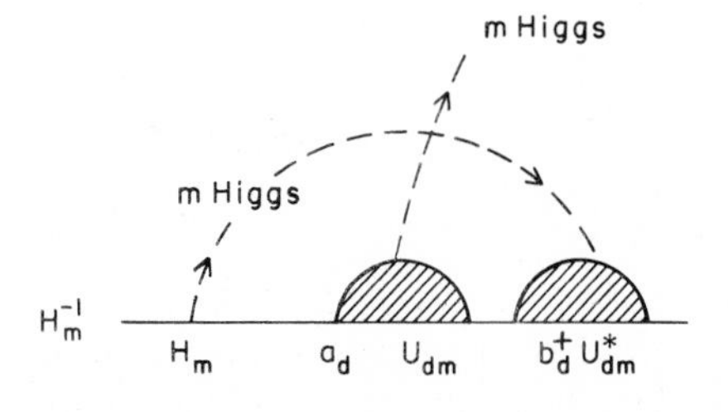


Fig. 8.14. Contribution to the m Higgs coupling of Fig. 813b which is related to the baryon asymmetry generating diagram of Fig. 8.12 [284].

This "bound" is considerably larger than the Kobayashi–Maskawa estimate (2.76) of $10^{-30 \pm 1}$ cm, and about three orders of magnitude below the present experimental limit [73] (2.73) of $6 \times 10^{-25}$ cm. Experiments are now in progress to improve this limit by about 2 orders of magnitude, and it would indeed be exciting if they found an effect, as it would indicate that the Kobayashi–Maskawa phase is not the sole source of *CP*-violation.

The above argument clearly contains more holes than a sieve [288]: we have been totally cavalier about numerical factors and neglected the possible occurrence of cancellations between other diagrams and the ones we have picked out. There is in fact a class of models in which $\theta$ is fated to vanish automatically, namely those with a Peccei–Quinn $U(1)_{PQ}$ symmetry. We saw in section 3.4 that incorporating [110] such a symmetry into the Weinberg–Salam model [1] involves introducing an axion [112] which has probably been excluded. However, it is possible [289] to construct GUTs with a $U(1)_{PQ}$ symmetry, and they have a modified axion which seems phenomenologically successful. In such a theory the $U(1)_{PQ}$ symmetry is spontaneously broken by a Higgs field with a large vacuum expectation value $V$, in which case the previous mass formula (3.83) becomes

$$m_a = O(f_\pi m_\pi / V), \tag{8.65}$$

and the previous couplings (3.85), (3.86) become

$$g_{a\bar{f}f} = O(m_f / V). \tag{8.66}$$

Particles with the properties (8.65), (8.66) give too much heat transport in stars, unless [113]:

$$V > O(10^9)\ \text{GeV}, \tag{8.67}$$

which suppresses their couplings sufficiently for them to conduct a negligible amount of heat. The previous limit (3.89) originated with a Boltzmann cut-off $e^{-m/T}$ for strongly interacting but heavy axions. The limit (8.67) is clearly consistent with the GUT expectation that $V = O(10^{15})$ GeV. The simplest GUT realizing [290] the $U(1)_{PQ}$ symmetry is based on SU(5) with two **5**s of Higgs $H_1$, $H_2$ and a complex **24** $\Phi$. The fermion couplings are:

$$f_{\bar{5}} f_{10} \overline{H}_1, \quad f_{10} f_{10} H_2, \tag{8.68}$$

and the Higgs self-couplings are mostly invariant under arbitrary phase transformations except for

$$\bar{H}_1\Phi^2H_2, \quad (\bar{H}_1\cdot H_2)\mathrm{Tr}(\Phi^2), \tag{8.69}$$

which restrict us to the desired $U(1)_{\mathrm{PQ}}$ phase symmetry. In this type of GUT the axion mass (8.65) becomes

$$m_{\mathrm{a}} \approx f_\pi m_\pi/10^{15}\ \mathrm{GeV} \approx 10^{-8}\ \mathrm{eV}, \tag{8.70}$$

and its couplings

$$g_{\mathrm{a}\bar{\mathrm{f}}\mathrm{f}} \approx m_{\mathrm{f}}/10^{15}\ \mathrm{GeV} \approx 10^{-15} \tag{8.71}$$

The Compton wavelength corresponding to (8.70) is in the range of tens of metres, but the pseudoscalar form of the axion couplings (3.85), (3.86) prevents us from experiencing unwelcome long-range forces [289–291]. The couplings (8.71) are so small that the Grand Unified axion is unobservable in conventional laboratory experiments.

The only "observable" consequence of such a Grand Unified axion theory is that $\theta$ is guaranteed to be very small: we estimate [292]

$$\theta \lesssim \mathrm{O}(10^{-17})? \tag{8.72}$$

In this case the $\theta$ contribution (8.56) to the neutron electric dipole moment is negligible and we expect $d_{\mathrm{n}}$ to be given essentially by the Kobayashi–Maskawa estimate (2.76). If $d_{\mathrm{n}}$ were seen at the level of our "bound" (8.64) it probably would mean that there is no Grand Unified axion [292]. Conversely, if the experimental limit on $d_{\mathrm{n}}$ were pushed below the "bound" (8.64) a most natural explanation would be that a Grand Unified axion exists, but probably we will never see it!

# *IV. Superunification*

## 9. The hierarchy problem and supersymmetric GUTs

### *9.1. Preamble*

GUTs contain many different mass scales whose ratios are very different from unity:

$$m_{\nu_i} \ll m_{\mathrm{e}} < m_{\mathrm{u,d}} < m_\mu, \Lambda_{\mathrm{QCD}}, m_{\mathrm{s}} < m_\tau, m_{\mathrm{c}} < m_{\mathrm{b}}, m_{\mathrm{t}}$$
$$\lesssim m_{\mathrm{W}} \lll m_{\mathrm{X}}? \ll m_{\mathrm{P}}, \tag{9.1}$$

where $m_P$ is the Planck mass, or in numbers:

$$\begin{aligned} \mathrm{O}(10^{-9})\ \mathrm{GeV?} \ll \mathrm{O}(10^{-3})\ \mathrm{GeV} &< \mathrm{O}(10^{-2})\ \mathrm{GeV} \\ &< \mathrm{O}(10^{-1})\ \mathrm{GeV} < \mathrm{O}(1)\ \mathrm{GeV} < \mathrm{O}(10)\ \mathrm{GeV} \\ &\lesssim \mathrm{O}(10^{2})\ \mathrm{GeV} \lll \mathrm{O}(10^{15})\ \mathrm{GeV?} \\ &\ll \mathrm{O}(10^{19})\ \mathrm{GeV}. \end{aligned} \tag{9.2}$$

Some of these ratios of mass scales are understood within the GUTs that spawned them. For example, we have already seen in lecture 6 how the basic equation

$$\frac{1}{\alpha_3(Q)} - \frac{1}{\alpha_2(Q)} = \frac{-(11+N_{\mathrm{H}/2})}{12\pi} \ln(m_{\mathrm{X}}^2/Q^2), \tag{9.3}$$

leads to the implication [175] that the strong interactions become strong on a scale $\Lambda_{\mathrm{QCD}}$ given by:

$$(\Lambda/m_{\mathrm{X}}) = \exp[-\mathrm{O}(1)/\alpha_{\mathrm{em}}] \ll 1. \tag{9.4}$$

Furthermore, we have seen in lecture 8 how the small ratio $m_{\mathrm{W}}/m_{\mathrm{X}}$ can lead [251], [253], [254] to very small neutrino masses:

$$m_\nu \simeq \mathrm{O}(m_{\mathrm{W}}^2/m_{\mathrm{X}}) = \mathrm{O}(m_{\mathrm{W}}/m_{\mathrm{X}}) m_{\mathrm{q},\ell}. \tag{9.5}$$

However, most of the ratios (9.1) are still not understood. In lectures 4 and 5 we looked at some attempts to understand the hierarchies of conventional fermion masses in (9.1) in theories of dynamical symmetry breaking [2]. Unfortunately, such extended technicolour theories [135] seem to have run into a phenomenological dead end. The most severe problem [129] in (9.1) is the famous "Hierarchy Problem":

$$\text{why is} \quad m_{\mathrm{W}}/m_{\mathrm{X}} \, [=\mathrm{O}(10^{-12})?] \lll 1\,? \tag{9.6}$$

But we should not forget the other [291] hierarchy problem:

$$\text{why is} \quad m_{\mathrm{X}}/m_{\mathrm{P}} \, [=\mathrm{O}(10^{-4}\text{–}10^{-5})?] \ll 1\,? \tag{9.7}$$

Most of this lecture will concern the problem (9.6). It will be posed in section 9.2 and possible solutions [169,292–294] using supersymmetry will be discussed in section 9.4. Section 9.3 will discuss attempts [100,101,295] to use radiative corrections to break gauge symmetries and understand the other [291] hierarchy problem (9.7).

### 9.2. *The "Hierarchy Problem"*

Rather than make a very general formulation of this problem [129], it is most convenient to illustrate it with the simplest SU(5) GUT [177], which will amply reveal the magnitude of the problem. As discussed in section 6.2, we need two Higgs representations to realize the desired pattern of symmetry breaking: a **24** $\phi$ to break SU(5) $\to$ SU(3)$\times$SU(2)$\times$U(1) and a **5** $H$ to break SU(2)$\times$U(1) $\to$ U(1). These should have vacuum expectation values

$$\langle 0|\phi|0\rangle = v \left(\begin{array}{ccc|cc} 1 & 0 & 0 & 0 & 0 \\ 0 & 1 & 0 & 0 & 0 \\ 0 & 0 & 1 & 0 & 0 \\ \hline 0 & 0 & 0 & -3/2 & 0 \\ 0 & 0 & 0 & 0 & -3/2 \end{array}\right)$$

$$m_X^2 = m_Y^2 = \tfrac{25}{8} g_5^2 v^2 \tag{9.8}$$

and

$$\langle 0|H|0\rangle = v_0 \begin{pmatrix} 0 \\ 0 \\ 0 \\ 0 \\ 1 \end{pmatrix} \qquad m_W^2 = m_Z^2 \cos^2\theta_W = \tfrac{1}{4} g_2^2 v_0^2. \tag{9.9}$$

We will try to arrange these vacuum expectation values by setting up in steps the full SU(5) Higgs potential. We start [177] with the potential for $\phi$ alone:

$$V(\phi) = -\mu^2 \mathrm{Tr}(\phi^2) + \tfrac{1}{4} a \left[\mathrm{Tr}(\phi^2)\right]^2 + \tfrac{1}{2} b \,\mathrm{Tr}(\phi^4), \tag{9.10}$$

where we have imposed for the sake of simplicity an unnecessary discrete symmetry: $\phi \leftrightarrow -\phi$. The potential (9.10) will give the desired $\langle 0|\phi|0\rangle$ (9.8) if

$$\mu^2, b > 0; \qquad a > -\tfrac{7}{15} b, \tag{9.11}$$

with the magnitude of the vacuum expectation value given by

$$\mu^2 = \tfrac{15}{2} a v^2 + \tfrac{7}{2} b v^2. \tag{9.12}$$

Turning to the most general form of the $H$ potential,

$$V(H) = -\tfrac{1}{2} \nu^2 |H|^2 + \tfrac{1}{4} \lambda \left(|H|^2\right)^2, \tag{9.13}$$

we get $\langle 0|H|0\rangle$ of the desired form (9.9) if:

$$\nu^2, \lambda > 0 \text{ (positivity)}, \tag{9.14}$$

with the magnitude of the vacuum expectation value given by

$$\nu^2 = \tfrac{1}{2}\lambda v_0^2. \tag{9.15}$$

At this point it seems possible to arrange an arbitrary ratio for $v_0/v$ and hence for $m_W/m_X$, just by adding together $V(\phi)$ (9.10) and $V(H)$ (9.13), but there are two things [177] wrong with this naïve procedure.

One difficulty is that there is a massless triplet of Higgs fields left uneaten by the gauge bosons:

$$\mathcal{H}_i \equiv H_i + \frac{\sqrt{2}}{9}\frac{v_0}{v}\phi_i + \cdots, \quad i = 1,2,3. \tag{9.16}$$

These could mediate baryon decay at a rate [177]

$$\frac{\Gamma\left(\mathrm{p,n} \xrightarrow{\mathcal{H}} \mathrm{e}^+ + X\right)}{\Gamma\left(\mathrm{p,n} \underset{\mathrm{X,Y}}{\rightarrow} \mathrm{e}^+ + X\right)} = \mathrm{O}\left(\frac{m_\mathrm{f}^2}{m_\mathrm{W}^2}\frac{m_\mathrm{X}^2}{m_\mathcal{H}^2}\right), \tag{9.17}$$

which is less than unity only if $m_\mathcal{H} \gtrsim 10^{11}$ GeV or so. The second difficulty is that the potential $V = V(\phi) + V(H)$ by itself is not renormalizable [296]. This is because radiative corrections like those in fig. 9.1 will in general give rise to couplings between the $H$ and $\phi$ fields. To make the theory renormalizable [296] we should include all possible dimension $d = 4$ terms respecting the $\phi \leftrightarrow -\phi$ discrete symmetry:

$$V(\phi, H) = \alpha|H|^2\mathrm{Tr}(\phi^2) + \beta\bar{H}\phi^2 H, \tag{9.18}$$

as well as the terms $V(\phi)$ (9.10) and $V(H)$ (9.13). Adding in the terms (9.18) can even solve the other problem, as it gives [177] to the boson $\mathcal{H}$ a mass

$$m_\mathcal{H}^2 = -\tfrac{5}{2}\beta v^2 + \mathrm{O}(v_0^2). \tag{9.19}$$

However, the terms (9.18) raise a new problem.

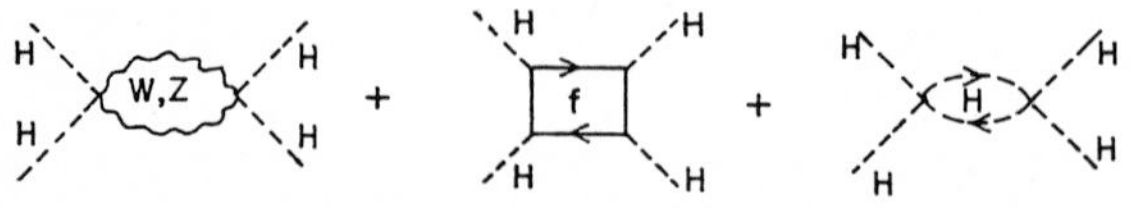

Fig. 9.1. Some typical radiative corrections to the effective Higgs potential.

The conditions (9.12), (9.15) for the Higgs vacuum expectation values are now modified [177] to become

$$\text{for } \langle 0|\phi|0\rangle: \quad \mu^2 = \tfrac{15}{2}av^2 + \tfrac{7}{2}bv^2 + \alpha v_0^2 + \tfrac{9}{30}\beta v_0^2 \tag{9.12$'$}$$

and

$$\text{for } \langle 0|H|0\rangle: \quad \nu^2 = \tfrac{1}{2}\lambda v_0^2 + 15\alpha v^2 + \left(\tfrac{9}{2} - 3\varepsilon\right)\beta v^2. \tag{9.15$'$}$$

The revised condition (9.12′) for $v$ is not much changed, but the form of $\langle 0|\phi|0\rangle$ is now subtly different:

$$\langle 0|\phi|0\rangle = v\begin{pmatrix} 1 & 0 & 0 & 0 & 0 \\ 0 & 1 & 0 & 0 & 0 \\ 0 & 0 & 1 & 0 & 0 \\ 0 & 0 & 0 & -3/2-\varepsilon/2 & 0 \\ 0 & 0 & 0 & 0 & -3/2+\varepsilon/2 \end{pmatrix}$$

$$\varepsilon \approx \frac{3}{20}\frac{\beta v_0^2}{bv^2}. \tag{9.20}$$

The modified vacuum expectation value (9.20) contains a very small isospin $I=1$ piece whose value $O(v_0^2/v) = O(m_W^2/m_X)$ is consistent with the general order of magnitude estimate [252] (8.9). It is too small to be detectable by a deviation from unity of the ratio $\rho = m_W^2/m_Z^2\cos^2\theta_W$ (1.32), but is inevitable when we couple the $\phi$ and $H$ fields together (9.18). The new problem comes from the $O(v^2)$ terms in the modified equation (9.15′) for $v_0$. They change the character of this equation as they introduce new terms which are individually much larger than $v_0^2$. They imply that the natural order of magnitude of $v_0$ is in fact $O(v)$. What is happening here is that $\langle 0|\phi|0\rangle$ and the $H^2\phi^2$ interaction terms (9.18) combine as in fig. 9.2 to give a large effective $(\text{mass})^2$

$$\delta m^2 = O(\alpha, \beta)v^2 \tag{9.21}$$

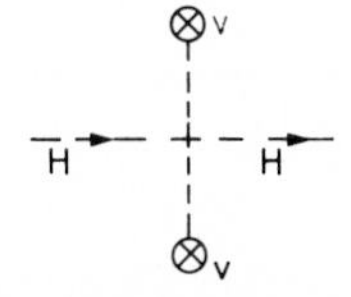


Fig. 9.2. Contribution to the effective mass$^2$ of light Higgses from their interactions with Higgs fields having large vacuum expectation values.

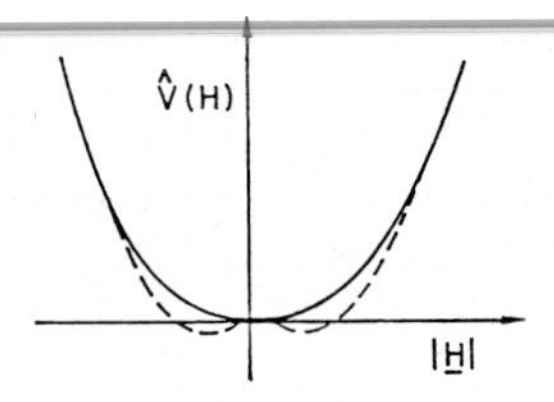


Fig. 9.3. The effect of the radiative corrections on the Higgs potential at small values of $|H|$.

to the Weinberg–Salam Higgs field. Our only remedy to keep $v_0$ small is to hope for a bizarre cancellation such that

$$\nu^2 - \left(15\alpha + \tfrac{9}{2}\beta\right)v^2 = \mathrm{O}\left(v_0^2\right) = \mathrm{O}(10^{-24})v^2. \tag{9.22}$$

How and why could such a 24-decimal point cancellation come about? Even if we imposed it on the potential at the tree level, it would be messed up by radiative corrections such as those in fig. 9.1, which would normally yield

$$\delta\alpha, \delta\beta = \mathrm{O}\left(g^4\right) \gg \mathrm{O}(10^{-24}). \tag{9.23}$$

To enforce (9.22) we need this conspiracy to work to $\mathrm{O}(\alpha_{\mathrm{GUT}}^{12})$!

This then is the "Hierarchy Problem," which comes in two parts:—one must adjust the parameters of the Lagrangian to many decimal places, and—this adjustment should survive radiative corrections.

### 9.3. *Symmetry breaking by radiative corrections?*

Looking at equation (9.22) it is natural to suspect that such a small quantity might actually be zero. Also, perhaps it would be easier [100, 101, 295] to find a theory in which such a combination was zero, rather than one in which it was merely "small". If we impose

$$\nu^2 - \left(15\alpha + \tfrac{9}{2}\beta\right)v^2 = 0, \tag{9.24}$$

then the Weinberg–Salam Higgs field $\hat{H}$ becomes effectively massless. We can write the potential $\hat{V}$ for the light Higgs doublet in the form

$$\hat{V}(\hat{H}) = \tfrac{1}{4}\lambda\left(|\hat{H}|^2\right)^2, \tag{9.25}$$

which is illustrated as the solid line in fig. 9.3. Since $\hat{V}$ is positive everywhere except at the origin, it appears naïvely as if no further

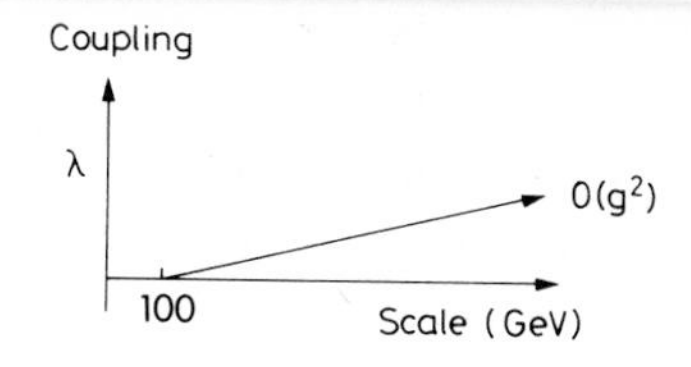


Fig. 9.4. The variation (9.27) of the coupling $\lambda$ in the effective Higgs potential as a function of the Higgs scale.

symmetry breaking is possible. However, in this case radiative corrections come to the rescue. They have the functional form

$$\delta V_{\text{rad}} \approx \text{O}(\lambda^2, \lambda g^2, g^4)(|\hat{H}|^2)^2 \ln(|\hat{H}|^2/M^2), \tag{9.26}$$

where $M$ is a renormalization parameter for the scale of the Higgs field, which must also be incorporated into $\lambda$ (9.25) so that the full potential $\hat{V} + \delta V_{\text{rad}}$ is independent of the choice of renormalization scale. The overall shape of $\hat{V} + \delta V_{\text{rad}}$ is shown as the dashed line in fig. 9.3. Although the $\text{O}(g^4)$ coefficient in $\delta V_{\text{rad}}$ (9.26) is usually much smaller than $\lambda$, which we might expect to be $\text{O}(g^2)$, $\lambda$ is now a function of $M$, and for some values of $M$ it may become small enough for the radiative corrections (9.26) to be significant. We can compute the variation in $\lambda(M)$, corresponding to the variation of the potential with the scale of the Higgs field, by using a renormalization group equation

$$M\frac{\mathrm{d}\lambda(M)}{\mathrm{d}M} = a\lambda^2 + b\lambda g^2 + cg^4 + \cdots, \tag{9.27}$$

whose coefficients at the one loop order are determined by the leading terms in the radiative correction $\delta V_{\text{rad}}$ (9.26). In this order the renormalization group equations for the effective Higgs potential couplings are identical with those of coupling constants in momentum space [100]. In this setting we are familiar with the asymptotic unfreedom of scalar couplings which means that they increase at large momentum scales, and decrease at smaller ones. This suggests the idea [295] that even if $\lambda$ is relatively large, $\text{O}(g^2)$ (?), at large scales $M = \text{O}(m_X)$, it will become small, $\text{O}(g^4)$, at some renormalization scale

$$M_0 = m_X \exp[-(\text{O}(1)/g^2)], \tag{9.28}$$

as illustrated in fig. 9.4. At this scale $M_0$ the radiative corrections (9.26) are no longer negligible compared with the tree potential (9.25). Clearly (9.26) is not positive at small values of $|\hat{H}|$, and hence the dips below zero in the dashed curve of fig. 9.3. These dips mean that a vacuum expectation value develops for $\hat{H}$, and the symmetry is spontaneously broken on a scale

$$\langle 0|\hat{H}|0\rangle = v_0 = \mathrm{O}(M_0) = m_{\mathrm{X}} \exp[-(\mathrm{O}(1)/g^2)]. \tag{9.29}$$

This looks like the sort of hierarchy we have been looking for. If this is the way the hierarchy problem is solved, then the Higgs boson of the Weinberg–Salam model has the special "radiative correction" value [101, 102]:

$$m_{\mathrm{H}} = 10.6\ \mathrm{GeV} \quad \text{for } \sin^2\hat{\theta}_{\mathrm{W}}(m_{\mathrm{W}}) = 0.215, \tag{9.30}$$

which we discussed in section 3.3. The picture of fig. 9.4 and the hierarchy formula (9.29) are all very well, but they beg the central question: why and how does the light field have an effectively zero mass?

Leaving this question on one side for the moment, it may be interesting to mention one possible scenario [291] for understanding why $m_{\mathrm{X}}/m_{\mathrm{P}} = \mathrm{O}(10^{-4}) \ll 1$, which is illustrated in fig. 9.5. It is supposed that (A) all Higgs couplings start out $\mathrm{O}(g^2)$ at a scale $M = \mathrm{O}(m_{\mathrm{P}})$. Then (B) the adjoint Higgs $\phi^4$ couplings decrease faster with decreasing $M$ than do the $H^4$ couplings, basically because of the larger non-Abelian charge of the adjoint fields. (This point has been checked [291] numerically in an SU(5) example.) In the region (C) the $\phi^4$ coupling (actually a particular

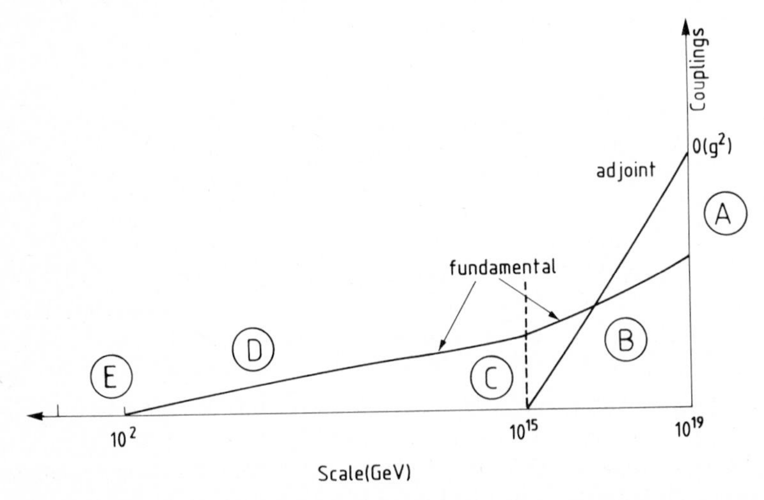


Fig. 9.5. A possible scenario [291] for arranging a hierarchy of hierarchies in a GUT.

combination in the SU(5) model studied) becomes of order $g^4$, and the radiative corrections can drive spontaneous symmetry breaking. Explicit calculations show that this can very naturally occur at a scale of order $10^{-4}$ to $10^{-5}$ times the Planck mass. At lower scales (D) the $\hat{H}^4$ coupling can continue to decrease, but does so slower than it did when $M > m_X$, because particles which have had masses $O(m_X)$ generated by spontaneous symmetry breaking have dropped out of their renormalization group equations. *A fortiori* the rate of charge of the $\hat{H}^4$ coupling is much slower than that of the $\phi^4$ coupling in region (B), and hence the region (D) can cover a much longer logarithmic range of scales. Finally, around (E) the $\hat{H}^4$ coupling finally becomes $O(g^4)$ and radiative corrections drive $\langle 0|\hat{H}|0\rangle \neq 0$ as described previously.

In this scenario we not only explain why $(m_X/m_P) = O(10^{-4})$ but also why

$$m_W/m_X \ll m_X/M_P \ll 1, \tag{9.31}$$

the so-called "hierarchy of hierarchies". However, the problem why the effective Higgs masses are zero still arises—twice in fact, once at the Planck mass scale and a second time in the region $M < m_X$.

### 9.4. *Can supersymmetry help?*

We now have two major difficulties with Higgs fields. Much was made in lecture 4 of the occurrence of quadratic divergences

$$\delta m_S^2 \sim \Lambda^2_{\text{cut off}} \tag{9.32}$$

in the radiative corrections to Higgs boson masses. Earlier sections of this lecture have stressed the problems arising when one tries to arrange a hierarchy of vacuum expectation values for Higgs fields, and emphasized that imposing one requires that the effective masses of certain Higgses be either very small or zero. This requirement meshes rather badly with the divergences (9.32), though we should perhaps retain some of the scepticism of section 4.1 whether these quadratic divergences are really such a serious problem. Even if we forget about these divergences, imposing small effective masses after a first phase of spontaneous symmetry breaking requires a bizarre conspiracy (9.22) between coupling constants which must be re-adjusted in each order of perturbation theory.

We saw in lectures 4 and 5 that theories of dynamical symmetry breaking [2] have not yet advanced to a practicable stage, and now is the time to see if supersymmetry can help. Veltman [297] has pointed out that the one-loop quadratic divergences (9.32) in the Weinberg–Salam model cancel if one imposes the sum rule relation

$$\sum_f m_f^2/m_W^2 = 3/2 + 3/2\cos^2\theta_W + \tfrac{3}{4}m_H^2/m_W^2. \tag{9.33}$$

If one takes $\theta_W$ and the light fermion masses from experiment, puts in the theoretically determined value of $m_W$ and assumes that $0 \lesssim m_H \lesssim m_W$, one can use (9.33) to make a prediction [297] for the mass of the top quark if $\sin^2\theta_W = 0.215$:

$$m_t \approx (90\text{–}99)\ \text{GeV}. \tag{9.34}$$

(There are assumed to be only six quarks in all.) Veltman [297] goes on to observe that higher loop quadratic divergences can be cancelled among diagrams involving superheavy particles as well as the observable light ones, and is happy to leave them to fend for themselves. He concludes that the relation (9.33) and the prediction (9.34) would be "suggestive of supersymmetry if true". Are they in fact true in supersymmetric theories?

We will look at theories in which the intermediate energy symmetry is:

$$[G \supset \mathrm{SU}(3)\times\mathrm{SU}(2)\times\mathrm{U}(1)]\times[N=1\ \text{supersymmetry}] \tag{9.35}$$

and ask whether it is ever possible to get a quadratic divergence. The answer [133] is yes whenever $G$ contains a U(1) factor for which one can generate the corresponding Fayet–Iliopoulos [298] $D$-term with a quadratically divergent coefficient. Low order computations revealed [133] that such a quadratic divergence did not arise in the Weinberg–Salam model at the one- or two-loop level. It was then realized [169] that this cancellation worked to all orders in theories where the U(1) factor was eventually embedded into a simple group at some higher energy scale. The reason [299] is that the potential quadratic divergences in all orders of perturbation theory are proportional to

$$\mathrm{Tr}(Y_i)\times\Lambda^2_{\text{cut off}}, \tag{9.36}$$

where the trace over the hypercharges includes all particles in the theory. It is well known that $\mathrm{Tr}(Y_i)=0$ in the Weinberg–Salam model [1], and this enables it to be embedded in a GUT with hypercharge a generator. In a conventional GUT the fermions with masses $\ll m_X$ form complete GUT multiplets [e.g. $\bar{\mathbf{5}}+\mathbf{10}$ of SU(5)] and hence have $\mathrm{Tr}(Y_i)=0$ and there is no quadratic divergence (9.32). However, if GUT multiplets get split so that the light fermions form incomplete multiplets with $\mathrm{Tr}(Y_i)\neq 0$ (which does happen to the **5** of Higgs fields in the minimal SU(5) model), then one could generate an effective Fayet–Iliopoulos [298] D-term with coefficient $\mathrm{O}(m_{\mathrm{X}}^2)$ replacing the quadratic divergence (9.32). In this case Veltman's [297] sum rule (9.33) would be replaced by

$$\sum_{\text{bosons}} m_{\text{boson}}^2 - \sum_{\text{fermions}} m_{\text{fermion}}^2 = \operatorname*{Tr}_{\substack{\text{low}\\ \text{mass}}} (Y_i) \times [\Lambda = \mathrm{O}(m_{\mathrm{X}})]^2 \times (\text{coefficient}). \qquad (9.37)$$

It is therefore correct that the confirmation of the prediction (9.34) would be very suggestive of supersymmetry if true. However, one could imagine supersymmetric models where it was not true, and one should not give up on supersymmetry if the top quark mass does not lie in the range (9.34)!

To continue our discussion how supersymmetry can help us with our theoretical problems, let us now turn to the concrete analysis of some specific supersymmetric GUT models [292, 293].

### 9.5. *Supersymmetric GUTs*

The general scheme of this section is that one has a GUT with an $N = 1$ supersymmetry that persists down to "low" energies $\mathrm{O}(10^3)$ GeV:

$$G \times [N = 1 \text{ supersymmetry}] \rightarrow \mathrm{SU}(3) \times \mathrm{SU}(2) \times \mathrm{U}(1) \times [N = 1 \text{ supersymmetry}]. \qquad (9.38)$$

Clearly supersymmetry is not exact at zero energy: there must be a "supergap" and we suppose it to be $\lesssim \mathrm{O}(10^3)$ GeV, and due to soft (dimension $d < 3$) supersymmetry breaking terms in the Lagrangian which protect the Higgs mass [292–294, 300]. It has recently been demonstrated that in certain cases [301] the quadratic divergences still cancel if one breaks supersymmetry with interactions of dimension $d_2 < 2$ (e.g. $\phi$ or $\phi^2$) but they may reoccur if the breaking term, though "soft", has dimension $d = 3$ (e.g. $\phi^3$ or $\bar{\psi}\psi$). Recall, however, that it is not clear how much one should worry about these famous quadratic divergences. What will be important is to avoid disturbing too much the relation (9.44) which guarantees the lightness of the Weinberg–Salam Higgs doublets.

Before getting into specific models, there is one striking consequence of the scenario (9.38), namely that the Grand Unification mass $m_{\mathrm{X}}$ has a tendency to increase [302]. The reason is that in contrast to the assumption we made in part III of these lectures, the low energy spectrum now contains incomplete GUT matter multiplets. We recall that GUTs have always had incomplete GUT gauge multiplets at low energies (the low energy spectrum contains the $W^{\pm}$, $Z^0$, g and $\gamma$, but not the X and Y). Supersymmetry means that the superpartners of these bosons, the Majorana fermions $\tilde{W}$, $\tilde{Z}$, $\tilde{g}$ and $\tilde{\gamma}$, will have masses differing from those of the W, Z, g and $\gamma$ by $O(10^3)$ GeV. The incompleteness of the GUT multiplets of Majorana fermions means that the low energy coupling constants approach each other more slowly than we thought (6.32) before [175–177]. The rates of variation of $\alpha_3$, $\alpha_2$ and $\alpha_1$ are changed [302] to:

$$\frac{1}{\alpha_3(Q)} = \frac{1}{\alpha_3(m_X)} + \frac{1}{12\pi}\ln\left(Q^2/m_X^2\right) \times \left[(27-4N_G) \text{ not } (33-4N_G)\right], \tag{9.39a}$$

$$\frac{1}{\alpha_2(Q)} = \frac{1}{\alpha_2(m_X)} + \frac{1}{12\pi}\ln\left(Q^2/m_X^2\right) \times \left[(18-4N_G) \text{ not } (22-4N_G)\right], \tag{9.39b}$$

$$\frac{1}{\alpha_1(Q)} = \frac{1}{\alpha_1(m_X)} + \frac{1}{12\pi}\ln\left(Q^2/m_X^2\right) \times (-4N_G). \tag{9.39c}$$

Taking the difference between (9.39a) and (9.39b) we get

$$\frac{1}{\alpha_3(Q)} - \frac{1}{\alpha_2(Q)} = \frac{9}{12\pi}\ln\left(Q^2/m_X^2\right), \tag{9.40}$$

and the comparison with (6.32) suggests that

$$\left[\ln(m_X/m_W)\right]_{\text{supersymmetry}} = \tfrac{11}{9}\left[\ln(m_X/m_W)\right]_{\text{lecture 6}}. \tag{9.41}$$

We could therefore expect our precious estimate (7.23) of $m_X$ to be increased by two orders of magnitude. There is however a further subtlety [303, 304]. The more careful analysis of section 7.3 revealed that incomplete GUT multiplets of Higgs fields at low energies also alter (7.21) the rate of approach of the coupling constants:

$$\frac{1}{\alpha_3(Q)} - \frac{1}{\alpha_2(Q)} = \left(\frac{11 + N_H/2}{12\pi}\right)\ln\left(Q^2/m_X^2\right), \tag{9.42a}$$

with the effect of decreasing the naïve estimate of $m_X$. In supersymmetric GUTs we have to contend not only with the Higgses, but also with their chiral fermion superpartners, so that [303, 304]:

$$\left.\frac{1}{\alpha_3(Q)} - \frac{1}{\alpha_2(Q)}\right|_{\text{susy}} = \left(\frac{9 + 3N_H/2}{12\pi}\right)\ln\left(Q^2/m_X^2\right). \tag{9.42b}$$

Table 9.1
Susy GUT calculations of $m_X$ and $\sin^2\theta_W$

| Number of Higgs supermultiplets | $m_X$ (GeV) | $\sin^2\hat{\theta}_W(m_W)$ |
|---|---|---|
| 0 | $7.5\times10^{17}$ | 0.206 |
| 2 | $1.7\times10^{16}$ | 0.233 |
| 4 | $8.0\times10^{14}$ | 0.256 |
| 6 | $6.2\times10^{12}$ | 0.275 |

The quoted results [304] are for three conventional generations and $\Lambda_{\overline{MS}}$ (four flavours) $\approx 300$ MeV. The results for $m_X$ scale approximately $\propto \Lambda_{MS}$, while $\Delta[\sin^2\hat{\theta}_W(m_W)] \approx 0.005 \ln(0.3\text{ GeV}/\Lambda_{\overline{MS}})$.

Furthermore, in order to be free of anomalies and give masses to the conventional quarks, one must include at least two Higgs supermultiplets [292, 293, 303, 304]. With sufficiently many Higgses, $m_X$ can be reduced again to the range discussed in lecture 7. Table 9.1 shows the results of calculations of $m_X$ and $\sin^2\theta_W$ in susy GUTs with different numbers of Higgs supermultiplets $N_H$. Our experimental colleagues may be getting nervous about baryon decay in supersymmetric GUTs, but they should not abandon hope, as we will see later.

To focus better our discussion, let us now look at the simplest supersymmetric GUT based on the SU(5) group [292]. Particles will now be grouped in an adjoint gauge supermultiplet (containing the vector bosons and their Majorana superpartners) and several left-handed chiral supermultiplets (containing one left-handed spin 1/2 particle and its antiparticle, a pseudoscalar and a scalar) which we will denote by their most familiar representatives. We will need **5** and $\bar{\mathbf{5}}$ Higgs supermultiplets $H_\alpha$, $H'^\alpha$ and a **24** Higgs supermultiplet $\phi^\alpha_\beta$, which will also contain various light and superheavy shiggses. We will also have three $\bar{\mathbf{5}}+\mathbf{10}$ generations of fermion supermultiplets $\psi^\alpha$ and $\chi_{\alpha\beta}$ which will also contain spin-zero superpartners (squarks and sleptons). Note that because of chirality selection rules [292], one cannot use the scalars in the fermion supermultiplets to give masses to the quarks and leptons, and that one in fact needs two Higgs supermultiplets, one for the $m_{u,c,t}$ masses and one for the $m_{d,s,b}$ masses. This structure is reminiscent of the doubling found necessary to realize a $U(1)_{PQ}$ symmetry at the GUT level in section 8.4, and the first Grand Unified axion ideas [122] indeed emerged from studies [134] of supersymmetric theories.

The interactions in a supersymmetric theory comprise the normal interactions of the gauge supermultiplet and a set of interactions between chiral supermultiplets characterized [305] by a superpotential $W$ which is a cubic function of the available chiral supermultiplets $\Phi$. The usual Higgs potential $V$ is obtained from $W(\Phi)$ by

$$V = W_a(\Phi)\overline{W}^a(\Phi) + \tfrac{1}{4}K_\alpha K^\alpha, \tag{9.43}$$

where

$$W_a(\Phi) \equiv \partial W(\Phi)/\partial\Phi_a, \qquad K_\alpha \equiv g\,\mathrm{Tr}(\bar{\Phi}T_\alpha\Phi), \tag{9.44}$$

and $T_\alpha$ is a conventional gauge generator. In the case of the minimal SU(5) theory [292], the superpotential can take the form

$$\begin{aligned} W(\Phi) = {} & \lambda_1\left(\tfrac{1}{3}\phi^\alpha{}_\beta\phi^\beta{}_\gamma\phi_\alpha{}^\gamma + \tfrac{1}{2}m\phi^\alpha{}_\beta\phi^\beta{}_\alpha\right) \\ & + \lambda_2 H'^\alpha\left(\phi^\beta{}_\alpha + 3m'\delta^\beta_\alpha\right)H_\beta \\ & + \mathcal{M}_1 H'^\alpha\chi_{\alpha\beta}\psi^\beta + \mathcal{M}_2\varepsilon^{\alpha\beta\gamma\delta\varepsilon}H_\alpha\chi_{\beta\gamma}\chi_{\delta\varepsilon}, \end{aligned} \tag{9.45}$$

where $\mathcal{M}_{1,2}$ are matrices in generation space analogous to the previous matrices in eq. (6.43). It has been shown [292] that the full potential $V$ (9.43) derived from the superpotential (9.45) has three degenerate minima [306] corresponding to different breaking patterns for SU(5):

$$\begin{aligned} & \mathrm{SU}(5) \to \mathrm{SU}(5)\ (\text{no breaking!})\ \text{or} \\ & \mathrm{SU}(4)\times\mathrm{U}(1)\ \text{or}\ \mathrm{SU}(3)\times\mathrm{SU}(2)\times\mathrm{U}(1). \end{aligned} \tag{9.46}$$

The desired (and possible) vacuum expectation value for $\phi^\alpha_\beta$ is

$$\langle 0|\phi^\alpha_\beta|0\rangle = 2m\left(\begin{array}{ccc|cc} 1 & 0 & 0 & 0 & 0 \\ 0 & 1 & 0 & 0 & 0 \\ 0 & 0 & 1 & 0 & 0 \\ \hline 0 & 0 & 0 & -3/2 & 0 \\ 0 & 0 & 0 & 0 & -3/2 \end{array}\right), \tag{9.47}$$

with

$$\langle 0|H|0\rangle = \langle 0|H'|0\rangle = \langle 0|\psi|0\rangle = \langle 0|\chi|0\rangle = 0, \tag{9.48}$$

at this first stage of symmetry breaking, giving an intermediate energy symmetry SU(3)×SU(2)×U(1)×($N=1$ supersymmetry). If in (9.45) we now set

$$m' = m = \mathrm{O}(m_\mathrm{X}), \tag{9.49}$$

then it can be checked that the SU(2) doublet Higgses among $H$ and $H'$

have zero (mass)$^2$, just as we wanted in sections 9.1 and 9.2. There our problem was renormalization of this desirable situation: what happens in supersymmetric GUTs?

NOTHING

There is a remarkable no-renormalization theorem first proved by Iliopoulos, Wess and Zumino [307] to the effect that in globally $N=1$ supersymmetric theories the only renormalizations necessary are wave function renormalizations—there are no separate vertex renormalization factors. Looking at the superpotential (9.45), this means that there is no renormalization of the ratio of $m/m'$: if we set it to unity (9.49), it stays that way. The line of argument is that the renormalization of $m$ is in the first term of (9.45) given by $Z_\phi^3/Z_\phi^2 = Z_\phi$, while the renormalization of $m'$ in the second term of (9.45) is given by $Z_\phi Z_{H'} Z_H / Z_{H'} Z_H = Z_\phi$ also. Note that it is not just a question of the masslessness of scalars being guaranteed by supersymmetry and the chiral symmetry of the associated fermions. In these theories there is no symmetry reason why they should not acquire masses, but the renormalization turns out to vanish. (In some sense, supersymmetric theories are not renormalizable [296] in the usual way, as one does not necessarily have to include in the Lagrangian all possible supersymmetric interaction terms of dimension $d \leqslant 4$.) This no-renormalization theorem means that light doublet Higgses are technically "natural" in that they avoid large radiative corrections.

To get a realistic mass spectrum from a supersymmetric theory, we must include [292, 293] some explicit (soft) supersymmetry breaking so as to have a non-zero supergap. In particular, we must stick in (mass)$^2$ terms for the squarks and sleptons, masses for the shiggses, masses for the Majorana partners of the gauge bosons, and (mass)$^2$ parameters to get the desired pattern of symmetry breaking. Points to watch in this last area include the necessity to break the degeneracy between the different breaking patterns (9.46) so as to favour SU(3)×SU(2)×U(1) as an intermediate energy symmetry [306], and avoidance of a light axion. It is apparent from this catalogue that this type of theory must have a large number of "small" (i.e. $\ll m_{\mathrm{P}}$) mass parameters. This makes the theory ugly, but it is nevertheless "natural" as mentioned earlier. It solves the second part of the "Hierarchy Problem" [129] as formulated at the end of section 9.2.

Ultimately one might hope to reduce some of the arbitrariness by breaking supersymmetry dynamically [134, 169] instead of explicitly, as was discussed in section 5.4. However, a dynamically broken supersymmetric GUT synthesizing the ideas of that section and this has not been worked out in detail. Another trick one might like to work is to "derive" the equality (9.49) rather than impose it in a *deus ex machina* style. This is the other half of the "Hierarchy Problem" as formulated at the end of section 9.2. Attempts have been made to do this using larger GUT groups than SU(5), and examples have been given based on the groups SU(7), SO(10) and $E_6$ [308]. However, no elegant and satisfying model of this type has yet been found (for example the SU(7) model only has 2 conventional generations while neutrinos are too heavy [309] in the $E_6$ model), and further work in this direction would also be useful.

Putting aside the question of an elegant way of imposing the hierarchy condition (9.49), one may ask [310] whether the explicit SU(5) GUT of this section is in fact phenomenologically acceptable. The three main successes of conventional GUTs are $\sin^2\theta_W$, $m_b/m_\tau$, and a nucleon lifetime (barely) long enough to be compatible with experiment. We see from table 9.1 that a minimal SU(5) susy GUT with $N_H = 2$ has [304] an uncomfortably high value of $\sin^2\theta_W$:

$$\sin^2\hat{\theta}_W(m_W) \approx 0.236 \tag{9.50}$$

if $\Lambda_{\overline{MS}} \approx 140$ MeV as suggested by lattice QCD calculations [180, 208]. On the other hand, $m_b/m_\tau$ is essentially unchanged [304, 310] in susy GUTs, contrary to some claims in the literature [302, 303]. If we assume the susy breaking threshold is at $m_W$:

$$\frac{m_b/m_\tau|_{\text{susy}}}{m_b/m_\tau|_{\text{minimal SU(5)}}} \approx \frac{\left[\alpha_3(m_W)/\alpha_3(m_X)|_{\text{susy}}\right]^{8/9}}{\left[\alpha_3(m_W)/\alpha_3(m_X)|_{\text{minimal SU(5)}}\right]^{4/7}} = 1.0, \tag{9.51}$$

if one puts in

$$\alpha_3(m_W) = 0.12, \qquad \alpha_s(m_X)|_{\text{susy}} = 1/23$$
$$\text{and} \quad \alpha_s(m_X)|_{\text{minimal SU(5)}} = 1/41$$

as suggested by detailed calculations [304]. The end result (9.51) is due to a bizarre conspiracy between changes in the $\beta$-function for the SU(3) group and in the anomalous dimension of the quark mass operator. Turning to the nucleon lifetime, one might guess [302] from the second row of table 9.1 that the lifetime in a minimal susy GUT would be much larger than in a minimal SU(5) GUT where $m_X$ is much smaller [$m_X \approx 5 \times 10^{14}$ GeV for $\Lambda_{\overline{MS}} = 300$ MeV (4 flavours)]. However, Weinberg and Sakai and Yanagida [309] have pointed out that in the minimal SU(5) susy GUT of Dimopoulos and Georgi [292], the one-loop diagrams

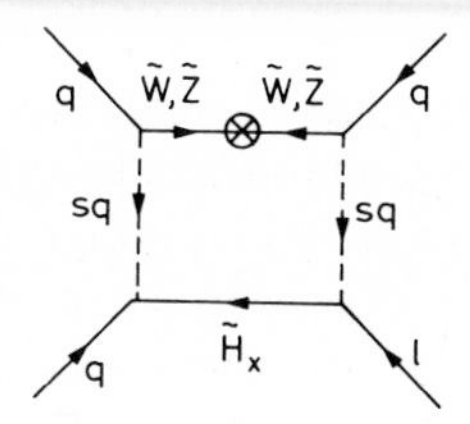


Fig. 9.6. One of the diagrams [295] contributing to baryon decay in a minimal SU(5) supersymmetric GUT [292, 293].

illustrated in fig. 9.6 give a nucleon decay rate much faster than the usual X boson exchange:

$$A \propto \underset{(\text{for } m_{\tilde{\mathrm{W}}} \gg m_{\mathrm{sq}})}{\frac{1}{m_{\tilde{\mathrm{W}}} m_{\tilde{\mathrm{H}}_{\mathrm{X}}}}} \quad \text{or} \quad \underset{(\text{for } m_{\tilde{\mathrm{W}}} \ll m_{\mathrm{sq}})}{\frac{m_{\tilde{\mathrm{W}}}}{m_{\mathrm{sq}}^2 m_{\tilde{\mathrm{H}}_{\mathrm{X}}}}}. \tag{9.52}$$

Putting in the important numerical factors and taking

$$m_{\tilde{\mathrm{H}}_{\mathrm{X}}} \approx m_{\mathrm{X}} \approx 10^{16}\ \mathrm{GeV},$$

$$m_{\mathrm{sq}} \ll m_{\tilde{\mathrm{W}}} \approx m_{\mathrm{W}} \approx 10^{2}\ \mathrm{GeV}, \tag{9.53}$$

we find [310] that fig. 9.6 actually can give a nucleon lifetime $\tau \gtrsim 10^{30}$ yr, comparable with the previous minimal SU(5) estimate and (barely) compatible with the present experimental limit. However, the errors in the susy GUT estimate of the nucleon lifetime are considerably greater than in the conventional GUT case, as a result of the spread of $m_{\mathrm{X}}$ values in table 8.1 and of our inability to estimate $m_{\tilde{\mathrm{W}}}$, $m_{\mathrm{sq}}$ and $m_{\tilde{\mathrm{H}}_{\mathrm{X}}}$ directly. As far as nucleon decay modes are concerned, we find [310] the following hierarchy of decay modes in the simplest SU(5) susy GUT [292, 293]:

$$\begin{aligned}
&\Gamma(\mathrm{N} \to \bar{\nu}_{\tau}\mathrm{K}) \\
&\quad \overset{?}{\gg} \Gamma(\mathrm{N} \to \bar{\nu}_{\mu}\mathrm{K}) \overset{?}{\approx} \Gamma(\mathrm{N} \to \bar{\nu}_{\tau}\pi) \\
&\quad \overset{?}{\gg} \Gamma(\mathrm{N} \to \bar{\nu}_{\mu}\pi, \mu^{+}\mathrm{K}, \bar{\nu}_{\mathrm{e}}\mathrm{K}) \\
&\quad \gg \Gamma(\mathrm{N} \to \mu^{+}\pi, \bar{\nu}_{\mathrm{e}}\pi) \\
&\quad \gg \Gamma(\mathrm{N} \to \mathrm{e}^{+}\mathrm{K}) \\
&\quad \gg \Gamma(\mathrm{N} \to \mathrm{e}^{+}\pi).
\end{aligned} \tag{9.54}$$

There are two other problems with the present formulations of susy GUTs which are of a less phenomenological nature. One is that mass terms for the fermions in chiral supermultiplets, which were originally included in the model of Dimopoulos and Georgi [292], in fact give rise to quadratic divergences in perturbation theory [310]. "Softness" in the traditional sense ($d < 4$ terms in the Lagrangian) is not equivalent to "softness" in the susy sense of an absence of quadratic divergences. Finally, since the Dimopoulos–Georgi model [292] has degenerate SU(5), SU(4)×U(1) and SU(3)×SU(2)×U(1) invariant vacua (9.46), it may have cosmological problems [306] with the generation of baryon number. In view of these difficulties and the possibly uncomfortable value of $\sin^2\theta_W$ (9.50), it seems that susy GUTs still require some development.

## 10. Unification in extended supergravity

### *10.1. Motivations*

The present GUTs [3] described in Part III of these lectures are clearly incomplete and unsatisfactory in many respects. For example, they do not explain or predict the number of generations, they have a hierarchy problem with elementary scalars, etc., etc. They have many parameters: even the minimal SU(5) GUT has 23: one gauge coupling $g_5$, one non-perturbative vacuum parameter $\theta_5$, nine parameters to characterize the Higgs potential, six quark and lepton masses and six generalized Cabibbo–Kobayashi–Maskawa mixing angles and phases. Eventually, we would like to have fewer parameters in our physical theory—perhaps 0, 1 or 3, depending on one's religious persuasion!

On the other hand, supersymmetry [125] is so beautiful that it must surely be true. It may even be useful [169], [292–294] to solve the hierarchy problem [129] as discussed in lecture 9. Let us assume we need *supersymmetry*. So far the theories we have discussed in lectures 5 and 9 have only invoked the structure

$$G \times [N=1 \text{ supersymmetry}], \tag{10.1}$$

which is not really true superunification in that the supersymmetry generators do not carry internal quantum numbers. Therefore it is desirable to go to *extended supersymmetry* [311] with $N>1$ spinorial charges. Lastly, our grandiosely named "Grand Unified Theories" have not yet included gravity. It is known that this can only be done in a supersymmetric theory if one makes the supersymmetry *local* rather than global, forming a *supergravity* theory [312].

Putting together the italicized items in the previous paragraph we are led to consider extended supergravity theories and how one can use them as a framework for unification. For reasons of space, this lecture does not contain a detailed description of such attempts: the interested reader is referred to ref. [4] for more information.

### *10.2. Extended supergravity theories*

There are very few extended supergravities [313], and they are essentially characterized by the number $N$ of supersymmetry generators: $N=2$, 3, 4, 5, 6, 7 or 8. It is in fact generally believed that the $N=7$ and 8 theories are equivalent, as their spectra are identical (see table 10.1) and also their internal symmetries and interactions as far as they are known. The only ambiguity among supergravity theories is the possibility of gauging an internal SO(8) symmetry using the helicity $\pm 1$ states in the spectrum of table 10.1. The gauging of the $N \leqslant 4$ theories has long been known to be possible, while the gauging of $N=5$ [314], 6, 7 and 8 [315] now also seems to be possible. Gauging necessarily involves the introduction [314,315] of a cosmological constant

$$\Lambda_{\text{cosmo}} = \mathrm{O}(g^2) m_{\mathrm{P}}^4. \tag{10.2}$$

It is not clear whether this is a virtue or an embarrassment: clearly $\Lambda_{\text{cosmo}}$ must be very small, corresponding to an energy density $< 10^{-47}$ GeV$^4$, in the Universe viewed on a large scale. However, Hawking and others

Table 10.1
Multiplets in extended supergravities

| Helicity | $N=2$ | 3 | 4 | 5 | 6 | 7=8 |
|---|---|---|---|---|---|---|
| 2 | 1 | 1 | 1 | 1 | 1 | 1 = 1 |
| 3/2 | 2 | 3 | 4 | 5 | 6 | 7 + 1 = 8 |
| 1 | 1 | 3 | 6 | 10 | 15 + 1 | 21 + 7 = 28 |
| 1/2 | — | 1 | 4 | 10 + 1 | 20 + 6 | 35 + 21 = 56 |
| 0 | — | — | 1 + 1 | 5 + 5 | 15 + 15 | 35 + 35 = 70 |
| −1/2 | — | 1 | 4 | 1 + 10 | 6 + 20 | 21 + 35 = 56 |
| −1 | 1 | 3 | 6 | 10 | 1 + 15 | 7 + 21 = 28 |
| −3/2 | 2 | 3 | 4 | 5 | 6 | 1 + 7 = 8 |
| −2 | 1 | 1 | 1 | 1 | 1 | 1 = 1 |

[130,316] have argued that if $\Lambda_{\mathrm{cosmo}}$ is zero on a large distance scale, it must necessarily be large $O(m_P^4)$ in space–time viewed on a Planck distance scale $O(1/M_P)$.

The full spectra of the states appearing in extended supergravity theories are exhibited in table 10.1. Notice how the *TCP* conjugate multiplets gradually overlap with increasing $N$ until we get to $N=7$, where they conspire to reproduce the spectrum of the $N=8$ theory. Notice that in each theory the number of $\pm 3/2$ helicity states corresponds to the number of supersymmetry generators, while the helicity $\pm 1$ states have the right number $N(N-1)/2$ to gauge an internal SO($N$) symmetry (plus 1 in the $N=6$ theory to gauge SO(6)×SO(2) [315]).

Can one use the supermultiplets of table 10.1 directly for unification? The answer is no [317] for several reasons. One is that the group

$$\mathrm{SO}(8) \not\supset \mathrm{SU}(3)\times\mathrm{SU}(2)\times\mathrm{U}(1), \tag{10.3}$$

which means that there are not enough spin 1 fields to accommodate all the gauge bosons of the Standard Model, let alone a GUT. In fact the closest SO(8) comes to a useful subgroup is

$$\mathrm{SO}(8) \supset \mathrm{SU}(3)\times\mathrm{U}(1)\times\mathrm{U}(1), \tag{10.4}$$

from which we see that while the photon and the $Z^0$ could conceivably be extracted from table 10.1, the $W^{\pm}$ could not and would have to be identified as composite fields [318]. Setting aside any lingering doubts about the desirability of gauging the $N=8$ theory, there is the problem that SO(8) only admits real representations. Thus the fermion spectrum is real, which does not correspond to observation (6.2) and it is not clear why the spin 1/2 particles do not immediately acquire large masses of order $m_P$ [319]. Furthermore [317] the fermion spectrum does not contain candidates for such well-known particles as the $\mu$, $\tau$, and b: presumably they would also have to be interpreted as composite states.

Hawking and collaborators have pointed out technical problems with elementary spin 0 [130] and spin 1/2 fields [316]. They have argued [130] that when quantum gravity effects are taken into account, elementary spin 0 fields would acquire masses $O(m_P)$ from propagation through space–time foam as illustrated in fig. 10.1a. They conclude that any low ($\ll m_P$) mass scalar fields must be composite. More recently they have argued [316] that in supergravity theories elementary spin 1/2 particles would acquire masses $O(m_P)$ from propagating through super space–time foam as indicated in fig. 10.1b. In this case, presumably any low ($\ll m_P$) mass spin 1/2 fermion must also be composite.

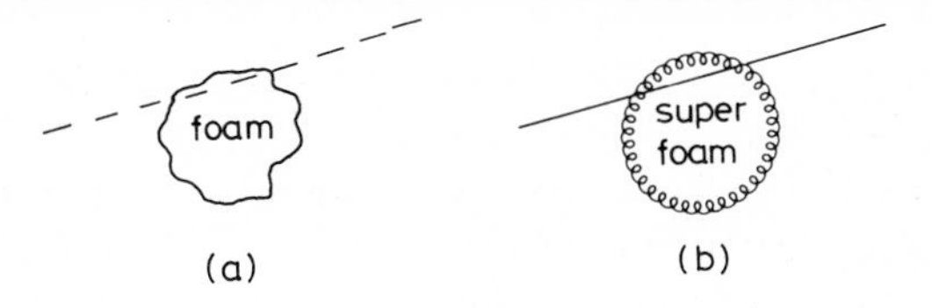


Fig. 10.1. Elementary particles propagating through space–time foam may acquire large masses $O(m_P)$ [213, 316].

In view of all these suggestions that spin 1, 1/2 and 0 states should be composite in a supergravity theory, it seems worthwhile to delve somewhat further below the surface of these theories.

## 10.3. Concealed symmetries of extended supergravities

It has been observed that one can write these theories in a form invariant under a larger global "parent" group, which is a non-compact form of $E_7$ (often denoted $E_{7(7)}$) in the case of the $N=8$ theory [320]. This formulation incorporates a local (gauge) symmetry, SU(8) in the case of the $N=8$ theory, which enables one to reduce the number of fields down to the physical degrees of freedom shown in table 10.1. For example in the $N=8$ theory:

$$\underset{E_7}{133 \text{ scalars}} - \underset{SU(8)}{63 \text{ scalars}} = \underset{N=8}{70 \text{ scalars}}\,. \tag{10.5}$$

In each case the number of scalar fields is equal to the dimensionality of the relevant coset space, $E_7/SU(8)$ in the case of $N=8$, and the theory has the structure of a nonlinear σ-model. (Note, however, that the $E_{7(7)}$ symmetry is lost in the gauged version [315] of the $N=8$ theory.)

In the $N=8$ case, the scalars can be represented by a $56\times 56$ matrix $S$ subject to many constraints which reduce the number of parameters to 133, and which plays the rôle of a vielbein field [320]:

$$S = \begin{pmatrix} U_{[AB]}{}^{[MN]} & V_{[AB][MN]} \\ \overline{V}^{[AB][MN]} & \overline{U}^{[AB]}{}_{[MN]} \end{pmatrix}, \tag{10.6}$$

whose left indices $A, B$ sit in SU(8) and whose right indices $M, N$ sit in $E_7$. As usual, the symbols [ ] denote antisymmetrization, so that the $U, V$ in (10.6) are $28\times 28$ submatrices. The SU(8) gauge transformations are generated [320] by a vector connection which is a composite of the elementary fields of the theory, with a scalar part

$$(\partial_\mu S)S^{-1} = \begin{pmatrix} 2Q_{\mu[A}^{[C}\delta_{B]}^{D]} & P_{\mu[ABCD]} \\ \bar{P}_\mu^{[ABCD]} & 2\bar{Q}_{\mu[A}^{[C}\delta_{B]}^{D]} \end{pmatrix}, \tag{10.7}$$

where the $E_7$ indices $M$, $N$ have been contracted. The matrix $Q_{\mu A}^{C}$ is an SU(8) adjoint vector field associated with the gauge transformations. [In the cases of $N = 4, 5, 6$, the relevant local symmetry is $U(N)$.]

Cremmer and Julia pointed out in their original work [320] a possible analogy with $CP^{N-1}$ models [321]. These contain $N$ complex scalar fields $Z_i$ subject to the constraint

$$\sum_{i=1}^{N} |Z_i|^2 = 1. \tag{10.8}$$

One can write down an invariant Lagrangian consistent with this constraint in the form

$$\mathcal{L} = -\sum_{i=1}^{N} (\partial_\mu - iv_\mu) Z_i^* (\partial_\mu + iv_\mu) Z_i, \tag{10.9}$$

where it exhibits a local U(1) invariance under

$$Z_i(x) \to e^{ia(x)} Z_i(n), \qquad v_\mu(x) \to v_\mu(x) - \partial_\mu a(x). \tag{10.10}$$

Since there is no kinetic term for the $v_\mu$ field in (10.9), one can easily use the equations of motion to eliminate $v_\mu(x)$, obtaining

$$v_\mu(x) = \tfrac{1}{2} i \sum_{i=1}^{N} Z_i^*(x) \overleftrightarrow{\partial}_\mu Z_i(x), \tag{10.11}$$

which is somewhat reminiscent of $Q_{\mu A}^{C}$ in equation (10.7). In the two-dimensional version of the $CP^{N-1}$ model it has been shown [321] that quantum effects generate a pole at momentum $k^2 = 0$ in the $v_\mu$ propagator, i.e. the kinetic term for $v_\mu$ lacking in eq. (10.9) is now generated. Thus the gauge field (10.11) becomes in some sense dynamical, and gives rise to a long-range potential which confines the $Z_i$ fields in this two-dimensional model. A similar phenomenon occurs in a supersymmetric $CP^{N-1}$ model in two dimensions, with the fermion superpartner of the $v_\mu$ field also becoming dynamical. More recently, similar phenomena have also been found to occur in a supersymmetric non-linear σ-model in three dimensions [322, 323]. There is a dynamical gauge boson (now with one non-trivial helicity state) and a dynamical superpartner corresponding to the unbroken $N = 1$ supersymmetry [323].

It was conjectured by Cremmer and Julia [320] that a similar phenomenon might occur in the four-dimensional extended supergravity theories, with the composite $Q_{\mu B}^{A}$ field becoming a dynamical, physical gauge boson. The natural extension [324] of this conjecture is that the superpartners of $Q_{\mu B}^{A}$ *also* become dynamical, physical particles. All known "elementary" particles with the possible exception of the graviton would therefore be composites made out of "preons" taken from the basic supermultiplets of table 10.1.

If we want to use the $Q_{\mu B}^{A}$ as gauge field and its superpartners for other particles, we need to know what is the supermultiplet containing $Q_{\mu B}^{A}$. We proceed [324, 325] by analogy with the known supercurrent multiplets in the known $N=1$ case, where it is a doublet of helicities

$$(3/2, 1), \tag{10.12}$$

and the $N=2$ case, where it is the set

$$(2\times 3/2, 4\times 1, 2\times 1/2). \tag{10.13}$$

We note that in each of the cases (10.12), (10.13) the number of helicity 1 states is just that needed to gauge a $U(N)$ internal symmetry. Supported by other arguments as well, we suggest [324, 325] that in general the $Q_{\mu B}^{A}$ sit in a supercurrent multiplet:

$$\left((3/2)^{A}, (1)_{B}^{A}, (1/2)_{[BC]}^{A}, (0)_{[BCD]}^{A}, \ldots, ((3-N)/2)^{A}\right). \tag{10.14}$$

The lower indices are totally antisymmetrized a maximum of $N$ times, which explains the formula for the minimum helicity in the supermultiplet (10.14). In the $N=8$ case the supercurrent multiplet (10.14) has the form shown in table 10.2. Shown in parentheses are the SU(8) multiplets obtained by tracing the upper index in (10.14) with one of the lower ones. In the helicity $-1$ case this trace field would be an SU(8) singlet $Q_{\mu A}^{A}$ which does not play any special symmetry rôle in the analysis of Cremmer and Julia [320] [cf. eq. (10.7)]. It is tempting to conjecture [325] that this field and the analogous trace fields with other helicities also have no dynamical rôle, and do not appear as physical particles in the low energy spectrum, though this assumption is not shared by all authors

Table 10.2
The $N=8$ supercurrent multiplet

| Helicity | $-3/2$ | $-1$ | $-1/2$ | $0$ | $+1/2$ | $+1$ | $+3/2$ | $+2$ | $+5/2$ |
|---|---|---|---|---|---|---|---|---|---|
| SU(8) content | $\bar{\mathbf{8}}$ | **63** (+**1**) | **216** (+**8**) | **420** (+**28**) | **504** (+**56**) | $\overline{\mathbf{378}}$ (+**70**) | $\overline{\mathbf{168}}$ (+$\overline{\mathbf{56}}$) | $\overline{\mathbf{36}}$ (+$\overline{\mathbf{28}}$) | $\bar{\mathbf{8}}$ |

[326,327]. The problems of chirality that we encounter later in this lecture occur whether or not one neglects the trace representations.

### 10.4. Attempts at Superunification

In furtherance of our programme [324] of making all "elementary" particles, except possibly the graviton, composites of supergravitational preons, we must now ask whether all the observed "elementary" particles can be identified with states in the supercurrent multiplet. We encounter here a major problem of chirality, in that GUTs certainly want complex (chiral) representations of fermions with inequivalent representations for left- and right-handed fermions:

$$R(f_{\mathrm{L}}) \neq R(f_{\mathrm{R}}), \tag{10.15}$$

so that the combination $\bar{f}_{\mathrm{R}} f_{\mathrm{L}}$ is not a GUT singlet, and hence cannot acquire a GUT-invariant mass $\gg m_{\mathrm{X}}$. Unfortunately the supercurrent multiplet (10.14) is too chiral [324]. Ultimately we want to give some mass to all the "elementary" fermions except possibly the neutrino, and this is only possible if they form together a vector-like real representation of the exact low-energy gauge group SU(3)×U(1). Unfortunately the fermions in the supermultiplet of equation (10.14) or table 10.2 are not real with respect to this subgroup of SU(8). For example the helicity 5/2 fermions of table 10.2 contain a $\bar{\mathbf{3}}$ of SU(3) embedded in the naïve way, while the *TCP* conjugate helicity $-5/2$ fermions sit in a **3** of this SU(3). (One cannot [324] avoid this problem by embedding SU(3) into SU(8) in a more devious way.) Similar problems exist also for fermions with smaller helicities. Another problem of the supercurrent multiplet concerns anomalies. One is not too sure what the anomaly calculation means for a helicity 5/2 fermion, since, while algebraic formulae do exist [328], no consistent local interacting field theory is known for such high-spin particles, and there are physical arguments [329] that only spin 1/2 fermions can contribute to anomalies. It seems unlikely that any prescription for a helicity 5/2 "anomaly" could cancel the anomalies coming from lower helicities [324]:

$$\begin{array}{lcccc} \text{Helicity:} & 3/2 & 1/2 & -1/2 & -3/2 \\ \mathrm{SU}(8)\ \text{anomaly:} & -3 & +3 & -75 & -165. \end{array} \tag{10.16}$$

We therefore conclude that the supercurrent multiplet as it stands is not satisfactory: we must either discard some of its helicity states, or add some more supermultiplets, or both.

It is generally believed [330, 4] that if one is not prepared to throw away *ad hoc* any helicity states, one would need an infinity of supermultiplets to give conventional masses to all the undesired fermions while retaining a few with masses $\ll m_{\mathrm{P}}$. We will return [4] to this possibility later, but pursue in the meantime the alternative of throwing away some of the helicity states in table 10.1.

We appeal [325] to Veltman's "theorem" [297] and demand that the only particles surviving with masses $\ll m_{\mathrm{P}}$ are those forming a (maximal) renormalizable subtheory. Renormalizability means that there can be no states of $|\text{helicity}| > 1$, that the only states of $|\text{helicity}| = 1$ are gauge bosons, and that the $|\text{helicity}| = 1/2$ fermions must be anomaly-free with respect to SU(8) or whatever gauge subgroup we believe to be relevant at energies $\ll m_{\mathrm{P}}$. By maximality we mean [325] that whenever possible we will look for the largest number of $|\text{helicity}| = 1/2$ states satisfying the anomaly-freedom constraint for the largest plausible low-energy gauge group. In addition to conventional renormalizable interactions, our scenario would also include possible non-renormalizable interactions scaled by inverse powers of the Planck mass determined by naïve dimensionality: $m_{\mathrm{P}}^{4-d}$ for an interaction of dimension $d$. These could include interactions fixing up the ratio $m_{\mathrm{d}}/m_{\mathrm{e}}$ (section 6.4) [211], causing baryon decay (section 7.1) [213] and giving neutrino masses (section 8.1) [250].

Later on we will discuss possible alternative scenarios for the disposal of the unwanted helicity states in the supermultiplet of table 10.2 which we discard [325] using Veltman's "theorem" [297]. For the moment we just look for candidate superGUTs respecting the criteria outlined above. It seems that the only possible scenario [324, 325] is to embed an SU(5) GUT into the $N = 8$ supergravity because of the following two Theorems.

*Theorem 1.* The only extended supergravity whose supercurrent supermultiplet is large enough to include all particles in a GUT is $N = 8$.

The proof requires enumerating the SU(5) transformation properties of all the states in the $N = 6$ supercurrent supermultiplet and observing that they do not form a rich enough spectrum. It has helicity $1/2$ fermions in the

$$\begin{aligned}&\underset{\text{of SU(6)}}{\mathbf{84}+\mathbf{70}(+\mathbf{6}+\mathbf{20})}\\&= \underset{\text{of SU(5)}}{\mathbf{45}+\mathbf{40}+\mathbf{24}+\mathbf{15}+2\times\mathbf{10}+2\times\mathbf{5}(+\mathbf{10}+\overline{\mathbf{10}}+\mathbf{5}+\mathbf{1})}\,.\end{aligned} \tag{10.17}$$

which does not contain three generations. On the other hand the SU(8) supercurrent multiplet is amply sufficient, as we will see shortly.

*Theorem 2*. The only plausible GUT group contained in the $N=8$ theory is SU(5).

The proof of this assertion consists of first observing that the trendiest alternative GUT groups to SU(5), namely SO(10) [193] and $E_6$ [194], are not subgroups of SU(8). However, one could in principle try to construct a GUT based on SU(6) or SU(7) [195]. However, the only anomaly-free subsets of the $N=8$ supercurrent multiplet which are vector-like with respect to SU(3)×U(1) are also vector-like with respect to the full SU(6) or SU(7) group, and hence have unacceptable fermion spectra.

There is in fact another empirical suggestion [325] that SU(8) may break directly down to SU(5) at the Planck mass. The SU(8) supermultiplet of table 10.1 contains a **420** of scalars, and if any component of this multiplet acquires a vacuum expectation value SU(6) and SU(7) are broken. The maximal simple factor which can be left as an exact symmetry is SU(5) (suitable for a GUT) but there may be an additional SU(2) factor (suitable for a generation group?) [326].

Let us now look more carefully at the helicity $-1/2$ fermions in the $N=8$ supercurrent multiplet and see what SU(5) representations they contain in a decomposition with respect to SU(5)×SU(3) (the SU(3) factor is just for classification and does not necessarily play any dynamical rôle):

$$\begin{aligned}\mathbf{216} &= (\overline{\mathbf{45}},\mathbf{1}) + (\mathbf{24},\mathbf{3}) + (\mathbf{10},\bar{\mathbf{3}}) + (\bar{\mathbf{5}},\bar{\mathbf{3}}) + (\mathbf{5},\mathbf{8}) \\ &\quad + (\mathbf{5},\mathbf{1}) + (\mathbf{1},\bar{\mathbf{6}}) + (\mathbf{1},\mathbf{3}) + (\bar{\mathbf{5}},\mathbf{1}) + (\mathbf{1},\bar{\mathbf{3}}), \qquad (10.18a)\\ \overline{\mathbf{504}} &= (\mathbf{45},\mathbf{3}) + (\mathbf{40},\bar{\mathbf{3}}) + (\mathbf{24},\mathbf{1}) + (\mathbf{15},\mathbf{1}) + (\mathbf{10},\mathbf{1}) \\ &\quad + (\mathbf{10},\mathbf{8}) + (\overline{\mathbf{10}},\mathbf{6}) + (\overline{\mathbf{10}},\bar{\mathbf{3}}) + (\mathbf{5},\mathbf{3}) + (\bar{\mathbf{5}},\bar{\mathbf{3}}), \qquad (10.18b)\end{aligned}$$

while the **420** of scalars decomposes to:

$$\begin{aligned}\mathbf{420} &= (\mathbf{24},\bar{\mathbf{3}}) + (\mathbf{5},\mathbf{3}) + (\mathbf{5},\bar{\mathbf{6}}) + (\mathbf{45},\mathbf{3}) + (\overline{\mathbf{40}},\mathbf{1}) \\ &\quad + (\overline{\mathbf{10}},\bar{\mathbf{3}}) + (\bar{\mathbf{5}},\mathbf{1}) + (\mathbf{10},\mathbf{1}) + (\mathbf{10},\mathbf{8}) + (\mathbf{1},\bar{\mathbf{3}}). \qquad (10.19)\end{aligned}$$

It is evident that the fermion representations (10.18) contain many $\bar{\mathbf{5}}$ and **10** of fermions suitable for GUTting, while the scalars (10.19) contain many **24**, **5** and **45** representations suitable for breaking the SU(5) symmetry in the desired way (6.10) and giving masses (6.16) to the $\bar{\mathbf{5}}+\mathbf{10}$ fermions. The fermion representations (10.18) also contain many surplus fermions which we can pair off [324, 325] into real combinations which can acquire very large masses $O(m_X - m_P)$. The maximal chiral, SU(5) anomaly-free representation of fermions obtainable from (10.18) is:

$$(\mathbf{45}+\overline{\mathbf{45}}) + \mathbf{424} + 9(\mathbf{10}+\overline{\mathbf{10}}) + 3(\mathbf{5}+\bar{\mathbf{5}}) + \mathbf{91} + 3(\bar{\mathbf{5}}+\mathbf{10}). \qquad (10.20)$$

Notice that in addition to the vector-like sets we also obtain three generations of chiral SU(5) fermions. This number of generations agrees with our phenomenological prejudices [177, 204, 205, 87]. We know since the discovery of the bottom quark that we need at least three generations, and to get $m_b/m_\tau$ correct (6.49) in minimal GUTs we can tolerate at most three generations [204, 205]. This number gets some support from cosmology, since the rate of cosmological $^4$He synthesis is too high unless there are only three or at the most four generations [87].

There is one other amusing observation [324, 331] to be made about the candidate superGUT (10.20). There is so much vector-like stuff with masses $\gtrsim m_X$ that above all thresholds the renormalization group function is:

$$\beta_{SU(5)} = +147\tfrac{1}{2} \qquad (10.21)$$

in a normalization convention where the conventional $N_G$ generations would give

$$\beta_{SU(5)} = -55 + 4N_G = -43 \quad \text{if } N_G = 3. \qquad (10.22)$$

The positivity of (10.21) means that $g_5$ increases when all the superheavy

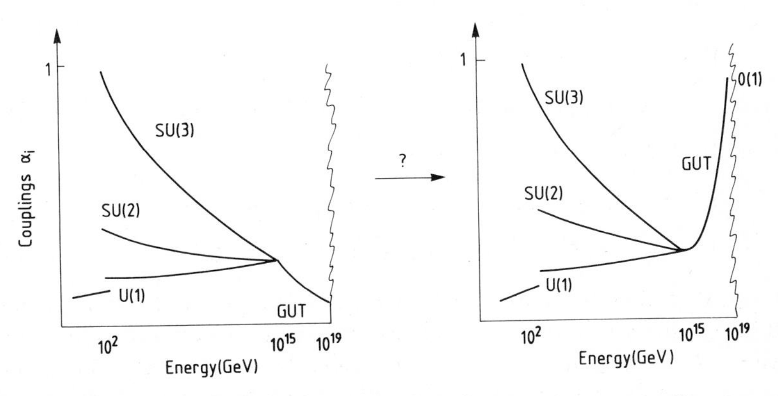


Fig. 10.2. Sketches of the possible evolution [324, 331] of the gauge coupling constant between the Grand Unification mass and the Planck mass.

thresholds are passed, and may easily become U(1) at $m_{\mathrm{P}}$ as illustrated in fig. 10.2. This trick only requires

$$\langle \beta_{\mathrm{SU(5)}} \rangle \approx +70 \tag{10.23}$$

in the energy range between $m_{\mathrm{X}}$ and $m_{\mathrm{P}}$. There is therefore no need for nonperturbative supergravity to generate a small coupling constant $\alpha_{\mathrm{GUT}}$ at the Planck scale. Looked at another way, we have a "natural" explanation why the fine structure constant $\alpha$ is so small: even if $\alpha_{\mathrm{GUT}}$ starts as large as unity at the Planck mass, it gets driven down to a small value at the Grand Unification point, and the observed $\alpha$ is then little different from this value.

Thus we have a candidate superGUT (10.20) with some attractive features, but we have swept an enormous number of problems under the carpet, some of which we address [4] in the next section.

### *10.5. What about the unwanted helicity states?*

How do we get rid of the unwanted helicity states with helicities $1/2$ and up in the supercurrent multiplet of table 10.1? And how do the dynamics select which (if any) renormalizable subtheory out of all the possibilities presented by the representations (10.18)? We have nothing useful to say in response to the second question, but it is possible to imagine three scenarios answering the first question, which we will now discuss. Perhaps

A. the unwanted helicity states were never bound in the first place. Or

B. some of them are present in the physical spectrum as massless states but we have not yet observed them. Or

C. all the unwanted states exist, but have found "partner" helicity states to "eat" and become massive.

Our preference [4] is for the third of these options, but we will first discuss the other two.

Did the unwanted helicity states in fact have to bind? If they had done so they would have made the theory unrenormalizable, and it could well be that the incipient singularities in the relevant bound-state equation might well have prevented binding from occurring. In fact, we know of no theorem proving that one must necessarily have a bound state pole recognizable in every composite channel. (Remember our old friend the scalar isoscalar $\varepsilon$ meson? Every year or two a candidate state with higher mass is reported, which seems to evaporate the year after. But no one gives up on the quark model because this particular $\bar{q}q$ composite cannot be found.) There is indeed a theorem [332] that one cannot have massless states (bound or elementary) with helicity $|\lambda| > 1$ if the underlying theory contains a Lorentz-covariant energy–momentum tensor, or $|\lambda| > 1/2$ if the theory contains a Lorentz-covariant conserved current. Unfortunately, this promising-sounding theorem is not applicable to most cases of interest including our own because physical theories generally do not satisfy the assumptions of Lorentz covariance. Gravity theories are only Lorentz-covariant up to a general coordinate transformation, and gauge theories are only Lorentz-covariant up to a gauge transformation. These weaker forms of Lorentz covariance are all that apply in our supergravity theory too, and so it seems that the assumptions of this powerful theorem are too stringent for our purposes [333]. This is just as well, as the theorem [332] would also have forbidden us from extracting the massless dynamical gauge fields that we want for our low energy effective gauge theory!

We now turn to the possibility that some of the unwanted helicity states may in fact be present in the physical spectrum, but not yet observed. This possibility may seem bizarre, but there are some general theorems [334] in the kinematical singularities in scattering amplitudes for massless particles which indicate that the coupling constants of such states must necessarily be dimensional [335]. A dimensionless helicity amplitude $F$ for $\lambda_1 + \lambda_2 \to \lambda_3 + \lambda_4$ has the singularity structure [334]

$$F = \hat{F}(\sqrt{s})^{|\lambda_1+\lambda_2+\lambda_3+\lambda_4|}(\sqrt{-t})^{|\lambda_1-\lambda_2-\lambda_3+\lambda_4|}(\sqrt{-u})^{|\lambda_1-\lambda_2+\lambda_3-\lambda_4|}, \tag{10.24}$$

where $\hat{F}$ contains just dynamical singularities. It is simple to check [4] by considering scattering amplitudes for $|\lambda| \geqslant 1$ that the dynamical poles in $\hat{F}$ will in general have dimensional coefficients. In our case these can only involve inverse powers of the Planck mass and the interactions of these particles are negligible at low energies. This type of argument is familiar for particles of helicity $|\lambda| \geqslant 2$ (cf. the graviton). It is known how to give helicity $|\lambda| = 3/2$ dimensionless gauge interactions, but this necessarily involves making them massive [336]. The only known consistent theory of dimensionless interactions for particles of helicity $|\lambda| = 1$ is a gauge theory [337], and this cannot accommodate the non-adjoint vector states appearing in table 10.2. We regret that we cannot see how one could extend this line of argument to the unwanted states of helicity $|\lambda| = 1/2$.

Massless particles with interactions scaled by inverse powers of the Planck mass are undetectable in laboratory experiments unless they can

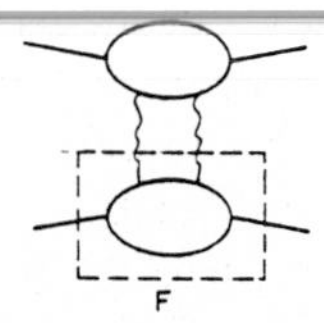


Fig. 10.3. Candidate diagram for the generation of a long-range potential by the exchange of a pair of spin 3/2 particles, which is in fact suppressed by the kinematic factors (10.23) —see ref. [4].

interact coherently with large bodies of matter, which is only possible if they have helicity $|\lambda| = 2$ [338]. (One might worry about the exchange of a pair of fermions as in fig. 10.3, however one can argue [4] that this diagram does not give a long-range potential.) Extra massless particles could in principle show up indirectly by changing the rate of expansion of the early Universe during primordial nucleosynthesis and altering the predicted $^4$He abundance [339]. Such massless states speed up the expansion rate during nucleosynthesis by a factor $1 + \delta$ where

$$\delta = \left( \frac{7}{86} \sum_{\substack{\text{fermions} \\ f}} N_f + \frac{8}{86} \sum_{\substack{\text{bosons} \\ b}} N_b \right) \Big/ \left( \frac{43}{4N_\mathrm{d}} \right)^{4/3}, \tag{10.25}$$

where $N_f$ and $N_b$ are the number of putative massless fermions and bosons respectively, and $N_\mathrm{d}$ is the total number of helicity states thermally excited at the temperature when the unseen particles decoupled, probably $\mathrm{O}(10^{19}\ \mathrm{GeV})$. The upper limit on $\delta$ inferred from the success of nucleosynthesis calculations is about 0.15. It then turns out [4] that if one includes in $N_\mathrm{d}$ all the states in our candidate superGUT (10.20) then many of the unwanted helicity states could be massless and present in the Universe without disturbing the primordial nucleosynthesis calculations.

Finally we turn to the possibility that the unwanted helicity states may have found partners and acquired a large mass. To play this trick we need extra helicity states to "eat" the unwanted ones, and it is generally believed [330] that an infinite set of supermultiplets is necessary if one tries to remove all the unwanted states in the original supermultiplet. Where could such additional supermultiplets come from? Going back to the original two-dimensional $\mathrm{CP}^{N-1}$ models, it has been shown [340] that the final physical bound state spectrum actually contains unitary representations of the "parent" global $\mathrm{SU}(N)$ symmetry. There are also indications [341] of an infinite spectrum in two-dimensional models with a non-compact "parent" symmetry group $SO(N, 1)$, redolent of a unitary infinite-dimensional representation of the non-compact group. There also seem [322] to be unitary representations of the "parent" symmetry group in three-dimensional nonlinear $\sigma$-models.

We therefore conjecture [4] that in the $N = 8$ supergravity theory the non-compact form of $E_7$ may play a rôle as a symmetry of the physical spectrum, which would contain infinite-dimensional $E_7$ representations built up from basic SU(8) supermultiplets. The structure of such representations seems to be

$$\{\lambda, R_\lambda\} \otimes \sum_{n=0}^{\infty} \{[\phi_{[ABCD]}]^n\}, \tag{10.26}$$

where the first factor is an SU(8) supermultiplet with maximal helicity $\lambda$ in the representation $R_\lambda$, and the infinite sum is of symmetrized $n$-fold products of preon scalars in the **70** representation of table 10.1. A set of SU(8) representations like (10.26) is sufficiently rich to "eat" all the unwanted helicity states provided SU(8) is broken down to SU(6) or a smaller subgroup—which we expect to be the case. However, there is no indication why it should be the preferred finite subset of helicity states which should be left massless, and not some other less desirable subset (or even none at all!). However, this approach [4] at least demonstrates that it is possible to dispose of all the unwanted helicity states in a way which does not conflict with general ideas about conservation of the number of helicity states. One should, however, remember that the $E_7$ symmetry is absent from the gauged version [315] of the $N = 8$ theory.

### *10.6. Open problems in Superunification*

There are certainly plenty of these! Starting with the supersymmetric theories in earlier lectures, there is still the fundamental problem whether non-perturbative breaking of supersymmetry is in fact possible [169], and then the phenomenological question whether a viable supercolour [133, 134] model can be found. Turning to supersymmetric GUTs [292–294], one would like to understand in a more fundamental way the origin of the hierarchy condition of massless scalars, rather than just imposing it *ad hoc*. Next, one would like to understand how large the soft supersymmetry breaking terms can be without destroying the successful non-renormalization of the hierarchy condition. If possible, one would like to avoid the inelegant proliferation [292–294] of supersymmetry breaking parameters on a scale of 1 TeV, which may bring us back to the question of dynamical symmetry breaking.

As far as the ideas of this lecture are concerned, it is clear that one must understand the significance of the gauging [315] of $N=8$ extended supergravity. If one wants to use composite states of the theory, one must learn how to do non-perturbative dynamical studies like those now being done in gauge theories [180, 208], with a view to learning how to dispose of unwanted helicity states, and seeing whether and how the dynamics select which renormalizable subtheory. Some attempts [4, 342] have been made to connect up with the supersymmetric GUTs of lecture 9, but this point should be pursued, with a view to seeing if one can extract a sensible $N=1$ supersymmetric GUT which also solves the $\theta$ vacuum problem if possible (cf. section 8.4). The key to understanding the physical spectrum of extended supergravity theories may lie with understanding better the rôle and unitary representations of the "parent" non-compact symmetry groups such as $E_{7(7)}$ and more studies of these more mathematical problems [343] would be useful.

The stakes are high. It is not unconceivable that in $N=8$ extended supergravity we may have the ultimate physical theory [344]—maybe all we have to do is solve it.

## Acknowledgements

It is a great pleasure to thank participants in the School for their enthusiasm, interest and incisive questioning during the presentation of these lectures. I would particularly like to thank Robert Coquereaux and Graciela Gelmini for going carefully through the manuscript, setting it straight on many important points and influencing substantially its final form. It is also a pleasure to thank Raymond Stora for his support and Nicole Berger for her heroic struggle to make sense of my handwriting. Thanks also go to my collaborators for the many happy hours we have spent working together on topics discussed here. Finally, very special thanks go to Mary K. Gaillard for the pleasure of a fruitful collaboration shared over many years, which is reflected in these lectures, as well as for the invitation to such an enjoyable School.

## References

[1] S.L. Glashow, Nucl. Phys. 22 (1961) 579.
S. Weinberg, Phys. Rev. Lett. 19 (1967) 1264.
A. Salam, Proc. 8th Nobel Symp., Stockholm 1968, ed. N. Svartholm (Almqvist and Wiksells, Stockholm, 1968) p. 367.

[2] J. Schwinger, Phys. Rev. 125 (1962) 397; 128 (1969) 2425.
R. Jackiw and D. Johnson, Phys. Rev. D8 (1973) 2386.
M.A.B. Bég and A. Sirlin, Ann. Rev. Nucl. Sci. 24 (1974) 379.
S. Weinberg, Phys. Rev. D13 (1976) 974; D19 (1979) 1277.
L. Susskind, Phys. Rev. D20 (1979) 2619.
A recent review containing more complete references is:
E. Farhi and L. Susskind, Phys. Rep. 74C (1981) 277.

[3] For reviews and references, see:
J. Ellis, Gauge Theories and Experiments at High Energies, eds. K.C. Bowler and D.G. Sutherland (Scottish Univ. Summer School in Physics, Edinburgh, 1981) p. 201.
P. Langacker, Phys. Rep. 72C (1981) 185.

[4] For a recent review and references, see:
J. Ellis, M.K. Gaillard and B. Zumino, Acta Phys. Pol. B13 (1982) 253.

[5] For reviews of gauge theories, see:
E.S. Abers and B.W. Lee, Phys. Rep. 9C (1973) 1.
J.C. Taylor, Gauge Theories of Weak Interactions (Cambridge Univ. Press, 1976).

[6] G. 't Hooft, Phys. Rev. Lett. 37 (1976) 8; Phys. Rev. D14 (1976) 3432.
R. Jackiw and C. Rebbi, Phys. Rev. Lett. 37 (1976) 172.
C.G. Callan, R.F. Dashen and D.J. Gross, Phys. Lett. 63B (1976) 334.

[7] M. Kobayashi and T. Maskawa, Prog. Theor. Phys. 49 (1973) 652.

[8] D.A. Ross and M. Veltman, Nucl. Phys. B95 (1975) 135.

[9] For reviews of Higgs properties, see:
J. Ellis, M.K. Gaillard and D.V. Nanopoulos, Nucl. Phys. B106 (1976) 292.
M.K. Gaillard, Comm. Nucl. Part. Phys. 8 (1978) 31.
ECFA/LEP Exotic Particles working group: G. Barbiellini et al., DESY preprint 79/27 (1979).

[10] J. Ellis, M.K. Gaillard and D.V. Nanopoulos, Nucl. Phys. B109 (1976) 213.

[11] J.D. Bjorken, Phys. Rev. D19 (1979) 335, raises even more fundamental questions.

[12] P.Q. Hung and J.J. Sakurai, Nucl. Phys. B143 (1978) 81.

[13] J.E. Kim, P. Langacker, M. Levine and H.H. Williams, Rev. Mod. Phys. 53 (1981) 211.

[14] D. Schildknecht, Bielefeld Univ. preprint BI-TP 81/12 (1981) and references therein.

[15] M. Holder et al., Phys. Lett. 72B (1977) 254.
CDHS collaboration: C. Geweniger, Proc. Int. Conf. Neutrino 79, Bergen, eds. A. Haatuft and C. Jarlskog (Bergen, 1979) p. 392.

[16] CHARM collaboration: M. Jonker et al., Phys. Lett. 99B (1981) 265.

[17] M. Glück and E. Reya, Phys. Rev. Lett. 47 (1981) 1104.

[18] CHARM collaboration: M. Jonker et al., Phys. Lett. 105B (1981) 242.

[19] Mark J collaboration: D.P. Barber et al., Phys. Rev. Lett. 46 (1981) 1663.

[20] M. Pohl, Proc. XVIth Rencontre de Moriond, ed. J. Tran Thanh Van (Editions Frontières, Gif-sur-Yvette, 1981) vol. 1, p. 161.

[21] CHARM collaboration: M. Jonker et al., Phys. Lett. 102B (1981) 67.

[22] S.L. Glashow, J. Iliopoulos and L. Maiani, Phys. Rev. D2 (1970) 1285.

[23] CDHS collaboration: J. Knobloch, Proc. Neutrino 81, Hawaii, eds. R.J. Cence, E. Ma and A. Roberts (HEP group, Hawaii, 1981) vol. 1, p. 323.

[24] JADE collaboration: W. Bartel et al., DESY preprint 81-072 (1981).

[25] JADE collaboration: W. Bartel et al., Phys. Lett. 101B (1981) 361.

[26] I. Liede and M. Roos, Nucl. Phys. B167 (1980) 397.

[27] W. Marciano and A. Sirlin, Phys. Rev. D22 (1980) 2695.

A. Sirlin and W. Marciano, Nucl. Phys. B189 (1981) 442. See also:
S. Dawson, J.S. Hagelin and L. Hall, Phys. Rev. D23 (1981) 2666.
[28] M.B. Einhorn, D.R.T. Jones and M. Veltman, Nucl. Phys. B191 (1981) 146.
[29] Results almost identical to those of ref. [27] have been obtained by:
C.H. Llewellyn Smith and J.F. Wheater, Phys. Lett. 105B (1981) 486.
[30] C.Y. Prescott et al., Phys. Lett. 77B (1978) 347; 84B (1979) 524.
[31] K. Enqvist, K. Mursula, J. Maalampi and M. Roos, Helsinki Univ. preprint HU-TFT-81-18 (1981).
[32] L. Michel, Proc. Roy. Soc. 63A (1950) 514.
[33] See for example:
R. Brandelik et al., Phys. Lett. 73B (1978) 109.
[34] T.A. Goldman and W.J. Wilson, Phys. Rev. D15 (1977) 709.
[35] N. Armenise et al., Phys. Lett. 84B (1979) 137.
CHARM collaboration: M. Jonker et al., Phys. Lett. 93B (1980) 203.
[36] W. Bacino et al., Phys. Rev. Lett. 42 (1979) 749.
[37] Mark II collaboration: J. Hollebeek, Proc. 1981 Int. Symp. on Lepton and Photon Interactions at High Energies, Bonn, ed. W. Pfeil (Univ. of Bonn, 1981) p. 1.
[38] M. Holder et al., Phys. Rev. Lett. 39 (1977) 433.
[39] J.V. Allaby, Proc. Neutrino 81, Hawaii, eds. R.J. Cence, E. Ma and A. Roberts (HEP group, Hawaii, 1981) vol. 1, p. 349.
[40] See for example:
A. De Rújula, in: Weak and Electromagnetic Interactions at High Energy, Les Houches (1976), session 29, eds. R. Balian and C.H. Llewellyn Smith (North-Holland, Amsterdam, 1977) p. 569.
[41] L. Ryder, Phys. Rep. 34C (1977) 55.
[42] V. Barger, W.F. Long and S. Pakvasa, Phys. Rev. Lett. 42 (1979) 1585.
[43] R.E. Shrock, S.B. Treiman and L.-L. Wang, Phys. Rev. Lett. 42 (1979) 1589.
[44] A. Sirlin, Rev. Mod. Phys. 50 (1978) 573; Phys. Rev. D22 (1980) 971.
[45] S. Pakvasa, S.F. Tuan and J.J. Sakurai, Phys. Rev. D23 (1981) 2799.
[46] M. Aguilar-Benitez et al., CERN preprint EP-81-131 (1981).
[47] JADE collaboration: J. Bürger, Proc. 1981 Int. Symp. on Lepton and Photon Interactions at High Energies, Bonn, ed. W. Pfeil (Univ. of Bonn, 1981) p. 115.
[48] J. Ellis, M.K. Gaillard, D.V. Nanopoulos and S. Rudaz, Nucl. Phys. B131 (1977) 285.
[49] L.J. Spencer et al., Phys. Rev. Lett. 47 (1981) 771.
[50] L.-L. Chau Wang, Proc. Cornell $Z^0$ Theory Workshop, eds. M.E. Peskin and S.-H.H. Tye, CLNS 81-485 (1981) p. 25.
[51] M.K. Gaillard and B.W. Lee, Phys. Rev. D10 (1974) 897.
[52] A.J. Buras, Phys. Rev. Lett. 46 (1981) 1354.
[53] R.E. Shrock and S.B. Treiman, Phys. Rev. D19 (1979) 2148.
[54] R.E. Shrock and M. Voloshin, Phys. Lett. 87B (1979) 375.
[55] V. Barger, W.F. Long, E. Ma and A. Pramudita, Phys. Rev. D24 (1981) 1410.
[56] B. Hyams et al., Phys. Lett. 29B (1969) 128.
[57] For a phenomenological review and early references, see:
K. Kleinknecht, Proc. 17th Int. Conf. on High Energy Physics, London (1974), ed. J.R. Smith (Rutherford Lab., Chilton, Didcot, 1974) p. III-23.
[58] L. Maiani, Phys. Lett. 62B (1976) 183.
S. Pakvasa and H. Sugawara, Phys. Rev. D14 (1976) 305.
[59] F. Gilman and M.B. Wise, Phys. Lett. 83B (1979) 83; Phys. Rev. D20 (1979) 2392.
[60] B. Guberina and R.D. Peccei, Nucl. Phys. B163 (1980) 254.

C.T. Hill and G.G. Ross, Phys. Lett. 94B (1980) 234.
[61] R.D.C. Miller and B.H.J. McKellar, Los Alamos preprint LA-UR-82-1855 (1982).
[62] L.-L. Chau Wang and F. Wilczek, Phys. Rev. Lett. 43 (1979) 816.
M. Bander, D. Silverman and A. Soni, Phys. Rev. Lett. 43 (1979) 242.
A.B. Carter and A.I. Sanda, Phys. Rev. Lett. 45 (1980) 952; Phys. Rev. D23 (1981) 1567.
[63] M.K. Gaillard and B.W. Lee, Phys. Rev. Lett. 33 (1974) 108.
G. Altarelli and L. Maiani, Phys. Lett. 52B (1974) 351.
[64] Particle Data Group, Rev. Mod. Phys. 52 (1980) S1.
[65] R. Bernstein et al., FNAL proposal 617 (1979).
[66] S. Weinberg, Phys. Rev. Lett. 37 (1976) 657.
D.W. McKay, Phys. Rev. D13 (1976) 645.
[67] A.I. Sanda, Phys. Rev. D23 (1981) 2647.
N.G. Deshpande, Phys. Rev. D23 (1981) 2654.
J.G. Körner and D. McKay, DESY preprint 81/034 (1981).
J.F. Donoghue, J.S. Hagelin and B.R. Holstein, Phys. Rev. D25 (1982) 195.
[68] J. Ellis and M.K. Gaillard, Nucl. Phys. B150 (1979) 141.
[69] A. Pais and S.B. Treiman, Phys. Rev. D12 (1975) 2744.
L.B. Okun, V.I. Zakharov and B.M. Pontecorvo, Lett. Nuovo Cim. 13 (1975) 218.
[70] A. Ali and Z.Z. Aydin, Nucl. Phys. B148 (1979) 165.
[71] J.S. Hagelin, Phys. Rev. D20 (1979) 2893.
E. Ma, W.A. Simmons and S.F. Tuan, Phys. Rev. D20 (1979) 2888.
V. Barger, W.F. Long and S. Pakvasa, Phys. Rev. D21 (1980) 174.
[72] J.S. Hagelin and M.B. Wise, Nucl. Phys. B189 (1981) 87.
J.S. Hagelin, Nucl. Phys. B193 (1981) 123.
E. Franco, M. Lusignoli and A. Pugliese, Nucl. Phys. B194 (1982) 403.
[73] I.S. Altarev et al., Phys. Lett. 102B (1981) 13; Zh. Eksp. Teor. Fiz. Pis'ma 29 (1979) 794.
W.B. Dress et al., Phys. Rev. D15 (1977) 9.
N.F. Ramsey, Phys. Rep. 43 (1978) 409.
[74] E.P. Shabalin, Sov. J. Nucl. Phys. 28 (1978) 75; 31 (1980) 864.
[75] M.B. Gavela et al., Phys. Lett. 109B (1982) 215.
For previous work, see:
B.F. Morel, Nucl. Phys. B157 (1979) 23.
E.P. Shabalin, Sov. J. Nucl. Phys. 32 (1980) 228.
D.V. Nanopoulos, A. Yildiz and P.H. Cox, Phys. Lett. 87B (1979) 53; Ann. Phys. 127 (1980) 126.
[76] M.B. Gavela et al., LAPP preprint TH-41 (1981).
[77] For reviews, see:
R.J. Crewther, Acta Phys. Austriaca, Suppl. XIV (1978) 47.
[78] A. Belavin, A. Polyakov, A. Schwartz and Y. Tyupkin, Phys. Lett. 59B (1979) 85.
[79] V. Baluni, Phys. Rev. D19 (1979) 2227.
[80] R.J. Crewther, P. Di Vecchia, G. Veneziano and E. Witten, Phys. Lett. 88B (1979) 123 (Erratum 91B (1980) 487).
[81] For a previous review on the properties of $W^{\pm}$, $Z^0$ and Higgs, see:
J. Ellis, Proc. 1978 SLAC Summer Inst. on Particle Physics, ed. M. Zipf (SLAC-215, 1978) p. 69.
[82] For a review of the axion and its properties, see:
R. Peccei, Proc. Neutrino 81, Hawaii, eds. R.J. Cence, E. Ma and A. Roberts (HEP

group, Hawaii, 1981) vol. 1, p. 149.
M. Veltman, Phys. Lett. 91B (1980) 95.
[83] M.A. Green and M. Veltman, Nucl. Phys. B169 (1979) 137 (Erratum B175 (1980) 547).
F. Antonelli and L. Maiani, Nucl. Phys. B186 (1981) 269.
[84] L. Camilleri et al., CERN report 76-18 (1976), especially section 2.
Proc. of the LEP Summer Study, Les Houches 1978—CERN report 79-01 (1979).
[85] M.E. Peskin and S.-H.H. Tye, eds., Proc. Cornell $Z^0$ Theory Workshop, CLNS 81-485 (1981).
[86] W.J. Marciano and Z. Parsa, Proc. Cornell $Z^0$ Theory Workshop, eds. M.E. Peskin and S.-H.H. Tye, CLNS 81-485 (1981) p. 127.
[87] G. Steigman, Proc. Neutrino 81, Hawaii, eds. R.J. Cence, E. Ma and A. Roberts (HEP group, Hawaii, 1981) vol. 2, p. 271.
[88] For a previous estimate, see:
J. Rich and D.R. Winn, Phys. Rev. D14 (1976) 1283.
[89] W. Alles, C. Boyer and A.J. Buras, Nucl. Phys. B119 (1977) 125.
[90] D. Albert, W.J. Marciano, D. Wyler and Z. Parsa, Nucl. Phys. B166 (1980) 460.
[91] J.D. Bjorken, Proc. 1976 SLAC Summer Inst. on Particle Physics (SLAC-198, 1977) p. 1.
[92] R.N. Cahn, M.S. Chanowitz and N. Fleishon, Phys. Lett. 82B (1979) 113.
[93] O.P. Sushkov, V.V. Flambaum and I.B. Khriplovich, Sov. J. Nucl. Phys. 20 (1975) 537.
[94] K.J.F. Gaemers and G.J. Gounaris, Z. Phys. C1 (1979) 259.
[95] M. Lemoine and M. Veltman, Nucl. Phys. B164 (1980) 445.
[96] For reviews, see:
C. Quigg, Rev. Mod. Phys. 49 (1977) 297.
R.F. Peierls, T.L. Treiman and L.-L. Wang, Phys. Rev. D16 (1977) 1397.
[97] For a review, see:
P. Aurenche and J. Lindfors, LAPP preprint TH-36 (1981).
[98] J. Ellis, B.H. Wiik and K. Hübner, eds., CHEEP, An ep facility in the SPS, CERN report 78-02 (1978).
ECFA study on the proton–electron storage ring project HERA-ECFA 80/42, DESY HERA 80/01 (1980).
CHEER feasibility study (Carleton Univ., Ottawa, 1980).
[99] D. Kohler, B.A. Watson and J.A. Becker, Phys. Rev. Lett. 33 (1974) 1628.
R. Barbieri and T.E.O. Ericson, Phys. Lett. 57B (1975) 270.
R.I. Dzhelyadin et al., Phys. Lett. 105B (1981) 239.
[100] S. Coleman and E. Weinberg, Phys. Rev. D7 (1973) 1888.
[101] J. Ellis, M.K. Gaillard, D.V. Nanopoulos and C.T. Sachrajda, Phys. Lett. 83B (1979) 339.
[102] K.T. Mahanthappa and M.A. Sher, Phys. Rev. D22 (1980) 1711.
[103] S. Weinberg, Phys. Rev. Lett. 36 (1976) 294.
A.D. Linde, Zh. Eksp. Teor. Fiz. Pis'ma 23 (1976) 73.
[104] E. Witten, Nucl. Phys. B177 (1981) 477.
A. Guth and E. Weinberg, Phys. Rev. Lett. 45 (1980) 1131.
[105] M. Veltman, Acta Phys. Pol. B8 (1977) 475.
C. Vayonakis, Lett. Nuovo Com. 17 (1976) 383; New threshold of weak interactions Athens Univ. preprint (1978), unpublished.
B.W. Lee, C. Quigg and H.B. Thacker, Phys. Rev. D16 (1977) 1519.

[106] L. Maiani, G. Parisi and R. Petronzio, Nucl. Phys. B136 (1978) 115.
N. Cabibbo, L. Maiani, G. Parisi and R. Petronzio, Nucl. Phys. B158 (1979) 295.
[107] F.A. Wilczek, Phys. Rev. Lett. 39 (1977) 1304.
J. Ellis, M.K. Gaillard and D.V. Nanopoulos, Nucl. Phys. B106 (1976) 292.
[108] B.W. Lee, C. Quigg and H.B. Thacker, Phys. Rev. D16 (1977) 1519.
[109] H. Georgi, S.L. Glashow, M. Machacek and D.V. Nanopoulos, Phys. Rev. Lett. 40 (1978) 692.
[110] R.D. Peccei and H.R. Quinn, Phys. Rev. Lett. 38 (1977) 1440; Phys. Rev. D16 (1977) 1791.
[111] S. Weinberg, Phys. Lett. 82B (1979) 387.
[112] S. Weinberg, Phys. Rev. Lett. 40 (1978) 223.
F.A. Wilczek, Phys. Rev. Lett. 40 (1978) 279.
[113] D.A. Dicus, E.W. Kolb, V.L. Teplitz and R.V. Wagoner, Phys. Rev. D18 (1978) 1829; D22 (1980) 839.
K. Sato and H. Sato, Prog. Theor. Phys. 54 (1975) 1564.
[114] J. Ellis and M.K. Gaillard, Phys. Lett. 74B (1978) 374.
See also E. Bellotti, E. Fiorini and L. Zanotti, Phys. Lett. 76B (1978) 223.
[115] H. Faissner et al., Phys. Lett. 60B (1976) 401.
P. Alibran et al., Phys. Lett. 74B (1978) 134.
T. Hansl et al., Phys. Lett. 74B (1978) 139.
P.C. Bosetti et al., Phys. Lett. 74B (1978) 143.
P. Coteus et al., Phys. Rev. Lett. 42 (1979) 1438.
P.F. Jacques et al., Phys. Rev. D21 (1980) 1206.
J.M. Losecco et al., Phys. Lett. 102B (1981) 209.
P. Fritze et al., Phys. Lett. 96B (1980) 427.
M. Jonker et al., Phys. Lett. 96B (1980) 435.
[116] T.W. Donnelly et al., Phys. Rev. D18 (1978) 1607.
[117] H. Faissner et al., Phys. Lett. 103B (1981) 234.
For updates, see:
H. Faissner, Proc. Neutrino 81, Hawaii, eds. R.J. Cence, E. Ma and A. Roberts (HEP group, Hawaii, 1981) vol. 1, p. 159; Proc. 1981 Int. Symp. on Lepton and Photon Interactions at High Energies, Bonn, ed. W. Pfeil (Univ. of Bonn, 1981) p. 797.
[118] J.-M. Frère, M.B. Gavela and J.A.M. Vermaseren, Phys. Lett. 103B (1981) 129.
[119] A. Zehnder, Phys. Lett. 104B (1981) 494.
P. Lehmann (1981) private communication.
[120] A. Barroso and N.C. Mukhopadhyay, SIN preprint PR-81-03 (1981), but see also Phys. Lett. 106B (1981) 91.
[121] CDHS collaboration: F. Eisele and B. Renk (1981) private communication.
[122] M. Dine, W. Fischler and M. Srednicki, Phys. Lett. 104B (1981) 199.
[123] M.B. Wise, H. Georgi and S.L. Glashow, Phys. Rev. Lett. 47 (1981) 402.
[124] K. Wilson, private communication cited by L. Susskind, Phys. Rev. D20 (1979) 2619.
G. 't Hooft, in: Recent developments in Gauge Theories, eds. G. 't Hooft et al., (Plenum, New York, 1980).
[125] Y.A. Gol'fand and E.P. Likhtman, Zh. Eksp. Teor. Fiz. Pis'ma 13 (1971) 323.
D. Volkov and V.P. Akulov, Phys. Lett. 46B (1973) 109.
J. Wess and B. Zumino, Nucl. Phys. B70 (1974) 39.
[126] G. 't Hooft and M. Veltman, Nucl. Phys. B44 (1972) 189; Diagrammar, CERN report 73-9 (1973).

[127] E.C.G. Stueckelberg and A. Peterman, Helv. Phys. Acta 26 (1953) 499.
M. Gell-Mann and F.E. Low, Phys. Rev. 95 (1954) 1300.
[128] H. Georgi and H.D. Politzer, Phys. Rev. D14 (1976) 1829.
[129] E. Gildener and S. Weinberg, Phys. Rev. D13 (1976) 3333.
E. Gildener, Phys. Rev. D14 (1976) 1667.
[130] S.W. Hawking, D.N. Page and C.N. Pope, Phys. Lett. 86B (1979) 175; Nucl. Phys. B170 (FS1) (1980) 283.
[131] B. Ward, SLAC preprint SLAC-PUB-2618 (1980).
[132] P. Di Vecchia and G. Veneziano, Phys. Lett. 95B (1980) 247.
[133] S. Dimopoulos and S. Raby, Nucl. Phys. B192 (1981) 353.
[134] M. Dine, W. Fischler and M. Srednicki, Nucl. Phys. B189 (1981) 575.
[135] S. Dimopoulos and L. Susskind, Nucl. Phys. B155 (1979) 237.
E. Eichten and K. Lane, Phys. Lett. 90B (1980) 125.
[136] S. Dimopoulos and J. Ellis, Nucl. Phys. B182 (1981) 505.
[137] S. Dimopoulos, Nucl. Phys. B168 (1980) 69.
[138] M.E. Peskin, Nucl. Phys. B175 (1980) 197.
J. Preskill, Nucl. Phys. B177 (1981) 21.
[139] For a recent review, see:
J. Ellis, Proc. 1981 SLAC Summer Inst. on Particle Physics, ed. A. Mosher (SLAC-245, 1981) p. 621.
[140] S. Raby, S. Dimopoulos and L. Susskind, Nucl. Phys. B169 (1980) 373.
[141] H. Georgi, L. Hall and M.B. Wise, Phys. Lett. 102B (1981) 315.
[142] F.A. Bais and J.-M. Frère, Phys. Lett. 98B (1981) 431.
[143] J.D. Bjorken, Ann. Phys. (NY) 24 (1963) 174.
[144] T. Banks, Y. Frishman and S. Yankielowicz, Nucl. Phys. B191 (1981) 493.
[145] S. Coleman, Comm. Math. Phys. 31 (1973) 259.
[146] P. Binétruy, S. Chadha and P. Sikivie, Phys. Lett. 107B (1981) 425; Nucl. Phys. B207 (1982) 505.
[147] S. Chadha and M.E. Peskin, Nucl. Phys. B185 (1981) 61; B187 (1981) 541.
[148] S. Dimopoulos, S. Raby and P. Sikivie, Nucl. Phys. B182 (1981) 77.
E. Eichten and K. Lane, Phys. Lett. 90B (1980) 125.
[149] J.C. Pati and A. Salam, Phys. Rev. Lett. 31 (1973) 661; Phys. Rev. D8 (1973) 1240; Phys. Rev. D10 (1974) 275.
[150] A.R. Clark et al., Phys. Rev. Lett. 26 (1970) 1667.
[151] G.L. Kane, Univ. of Michigan preprint UM HE 81-56 (1981); seminar 3, this volume.
[152] J. Ellis, D.V. Nanopoulos and P. Sikivie, Phys. Lett. 101B (1981) 387.
[153] S.L. Glashow and S. Weinberg, Phys. Rev. D15 (1977) 1958.
E.A. Paschos, Phys. Rev. D15 (1977) 1966.
[154] M.S. Chanowitz, J. Ellis and M.K. Gaillard, Nucl. Phys. B128 (1977) 506.
[155] J. Ellis, M.K. Gaillard, D.V. Nanopoulos and P. Sikivie, Nucl. Phys. B182 (1981) 529.
[156] G.H. Trilling, Phys. Rep. 75C (1981) 57, and references therein.
[157] E. Eichten, K. Lane and J.P. Preskill, Phys. Rev. Lett. 45 (1980) 225.
[158] J. Ellis and P. Sikivie, Phys. Lett. 104B (1981) 141.
[159] C.H. Llewellyn Smith, Proc. Int. Symp. on Electron and Photon Interactions at High Energies, Bonn 1973, eds. H. Rollnik and V. Pfeil (North-Holland, Amsterdam, 1974) p. 449.
[160] F. Hayot and O. Napoly, Z. Phys. C7 (1981) 229.
For the pair production of coloured technicolour particles, see:
J.A. Grifols and A. Méndez, Phys. Rev. D26 (1982) 324.

[161] G. Girardi, P. Mery and P. Sorba, Nucl. Phys. B195 (1982) 410.
[162] S. Rudaz and J. Vermaseren, CERN preprint TH-2961 (1980), unpublished.
[163] TASSO collaboration: M. Althoff et al., Phys. Lett. 122B (1983) 95.
[164] ECFA/LEP Exotic Particles Working Group: G. Barbiellini et al., DESY preprint 81/064 (1981).
A. Ali, talk at Orbis Scientiae 1981, DESY preprint 81/032 (1981).
[165] K. Chadwick et al., Phys. Rev. Lett. 46 (1981) 88.
A. Silverman, Proc. 1981 Int. Symp. on Lepton and Photon Interactions at High Energies, Bonn, ed. W. Pfeil (Univ. of Bonn, 1981) p. 138.
[166] A. Ali and M.A.B. Bég, Phys. Lett. 103B (1981) 376.
For other recent studies on technicolour phenomenology in $e^+e^-$ annihilation, see:
G.J. Gounaris and A. Nicolaidis, Phys. Lett. 102B (1981) 144.
J.A. Grifols, Phys. Lett. 102B (1981) 277.
A. Ali, H.B. Newman and R.Y. Zhu, Nucl. Phys. B191 (1981) 93.
[167] L. Arnellos, W.J. Marciano and Z. Parsa, Nucl. Phys. B196 (1982) 365.
[168] J. Branson et al., Phys. Rev. Lett. 45 (1980) 1904.
[169] E. Witten, Nucl. Phys. B188 (1981) 513.
M.V. Voloshin and V.I. Zakharov, Zh. Eksp. Teor. Fiz. 34 (1981) 508.
[170] B.A. Campbell, J. Ellis and S. Rudaz, CERN preprint TH-3184 (1981).
[171] If this were not so, the electromagnetic forces between large clumps of matter would be larger than the gravitational forces, which is manifestly not the case.
[172] C. Froggatt and H.B. Nielsen, Nucl. Phys. B164 (1980) 114.
[173] H. Georgi and S.L. Glashow, Phys. Rev. Lett. 32 (1974) 438.
[174] See, however, for an alternative philosophy which is currently out of favour: J.C. Pati and A. Salam, Phys. Rev. Lett. 31 (1973) 661; Phys. Rev. D8 (1973) 1240; D10 (1974) 275.
[175] H. Georgi, H.R. Quinn and S. Weinberg, Phys. Rev. Lett. 33 (1974) 451.
[176] M.S. Chanowitz, J. Ellis and M.K. Gaillard, Nucl. Phys. B128 (1977) 506, obtain a rather more precise value of $\sin^2\theta_W$ than does ref. [175], and give the first approximate calculation of mass renormalization in GUTs.
[177] A.J. Buras, J. Ellis, M.K. Gaillard and D.V. Nanopoulos, Nucl. Phys. B135 (1978) 66.
[178] For a review and references, see:
A. Peterman, Phys. Rep. 53C (1979) 157.
[179] H. Politzer, Phys. Rev. Lett. 30 (1973) 1346.
D.J. Gross and F.A. Wilczek, Phys. Rev. Lett. 30 (1973) 1343.
[180] For a discussion of different definitions of $\Lambda$ and the experimental value in 1979, see:
J. Ellis, Proc. Int. Symp. on Lepton and Photon Interactions at High Energies, Fermilab 1979, eds. T.B.W. Kirk and H.D.I. Abarbanel (FNAL, Batavia, 1979) p. 412.
Since then the preferred value of $\Lambda$ has decreased; see for example:
D. Bollini et al., Phys. Lett. 104B (1981) 403.
J.J. Aubert et al., Phys. Lett. 105B (1981) 315, 322.
[181] D.A. Ross, Nucl. Phys. B140 (1978) 1.
[182] T. Goldman and D.A. Ross, Phys. Lett. 84B (1979) 208.
[183] W.J. Marciano, Phys. Rev. D20 (1980) 274.
[184] T. Goldman and D.A. Ross, Nucl. Phys. B171 (1980) 273.
[185] J. Ellis, M.K. Gaillard, D.V. Nanopoulos and S. Rudaz, Nucl. Phys. B176 (1980) 61.
[186] P. Binétruy and T. Schücker, Nucl. Phys. B178 (1981) 293, 307.
[187] L. Hall, Nucl. Phys. B178 (1981) 75.

[188] C.H. Llewellyn Smith, G.G. Ross and J. Wheater, Nucl. Phys. B177 (1981) 263.
[189] I. Antoniadis, C. Bouchiat and J. Iliopoulos, Phys. Lett. 97B (1980) 367.
[190] Strictly speaking the intermediate energy symmetry group is S(U(3)×U(2)) which has the same local structure as SU(3)×SU(2)×U(1) but a different global structure: see the discussion after eq. (6.13).
[191] However one could add to the reducible representation (6.12) an SU(5) singlet field which could be a $\nu_L^C$: see discussion in lecture 8.
[192] See also:
H. Georgi, Particles and Fields—1974, ed. C.E. Carlson (American Institute of Physics, New York, 1975).
H. Fritzsch and P. Minkowski, Ann. Phys. (NY) 93 (1975) 193.
H. Georgi and D.V. Nanopoulos, Phys. Lett. 82B (1979) 392; Nucl. Phys. B155 (1979) 52.
[193] See for example:
H. Georgi and D.V. Nanopoulos, Nucl. Phys. B159 (1979) 16.
T.A. Goldman and D.A. Ross, Nucl. Phys. B162 (1980) 102.
T. Rizzo and G. Senjanović, Phys. Rev. Lett. 46 (1981) 1315; Phys. Rev. D24 (1981) 704; D25 (1982) 235.
[194] F. Gürsey, P. Ramond and P. Sikivie, Phys. Lett. 60B (1975) 177.
For a review, see:
B. Stech, in: Unification of the Fundamental Particle Interactions, eds. S. Ferrara, J. Ellis and P. Van Nieuwenhuizen (Plenum, New York, 1980) p. 23.
[195] P.H. Frampton, Phys. Lett. 88B (1979) 299.
[196] P.H. Frampton and S. Nandi, Phys. Rev. Lett. 43 (1979) 1461.
[197] F. Wilczek and A. Zee, Phys. Rev. D25 (1982) 553.
M. Gell-Mann, P. Ramond and R. Slansky (1979) unpublished.
[198] For a careful and complete discussion, see references [181–189].
[199] J. Ellis and D.V. Nanopoulos, Nature 292 (1981) 436.
[200] W.J. Marciano and A. Sirlin, Phys. Rev. Lett. 46 (1981) 163.
For a review and update, see also:
W.J. Marciano and A. Sirlin, Proc. Second Workshop on Grand Unification, Ann Arbor, eds. J.P. Leveille, L.R. Sulak and D.G. Unger (Birkhäuser, Boston, 1981) p. 151.
[201] A.J. Buras, Proc. 1981 Int. Symp. on Lepton and Photon Interactions at High Energies, Bonn, ed. W. Pfeil (Univ. of Bonn, 1981) p. 636.
[202] H. Georgi and H.D. Politzer, Phys. Rev. D14 (1976) 1829.
[203] For a more complete description of the absorption of wave function renormalization factors, see ref. [68]. The effective mass parameter defined in this way is in general gauge-dependent, and one should refine this definition: see refs. [186,204,205] and references therein.
[204] D.V. Nanopoulos and D.A. Ross, Nucl. Phys. B157 (1979) 273.
[205] D.V. Nanopoulos and D.A. Ross, Phys. Lett. 108B (1982) 351. For previous estimates of $m_t$ using fixed point ideas, see:
M.E. Machacek and M.T. Vaughn, Phys. Lett. 103B (1981) 427, and especially:
C.T. Hill, Phys. Rev. D24 (1981) 691. For a recent complete two-loop calculation, see:
M. Fischler and J. Oliensis, Fermilab preprint PUB-82/63-THY (1982).
[206] S. Weinberg, I.I. Rabi Festschrift, Trans. N.Y. Acad. Sci. II, 38 (1977).
[207] For reviews and references, see:

S. Narison, Techniques of Dimensional Regularization, Phys. Rep. 84C (1982) 263; Z. Phys. C14 (1982) 263.
[208] H. Hamber and G. Parisi, Phys. Rev. Lett. 47 (1981) 1792.
[209] H. Georgi and C. Jarlskog, Phys. Lett. 86B (1979) 297.
[210] R. Barbieri and D.V. Nanopoulos, Phys. Lett. 95B (1980) 43.
S.M. Barr, Phys. Rev. D24 (1981) 1895.
M.J. Bowick and P. Ramond, Phys. Lett. 103B (1981) 338.
R. Barbieri, D.V. Nanopoulos and D. Wyler, Phys. Lett. 103B (1981) 433.
R. Barbieri, D.V. Nanopoulos and A. Masiero, Phys. Lett. 104B (1981) 194.
and for a review:
L.E. Ibáñez, Oxford Univ. preprint OUTP 81-83 (1981).
[211] J. Ellis and M.K. Gaillard, Phys. Lett. 88B (1979) 315.
[212] R. Barbieri and D.V. Nanopoulos, Phys. Lett. 91B (1980) 369.
S.L. Glashow, Phys. Rev. Lett. 45 (1980) 1914.
[213] Ya.B. Zeldovich, Phys. Lett. 59A (1976) 254.
[214] G. 't Hooft, Phys. Rev. Lett. 37 (1976) 8; Phys. Rev. D14 (1976) 3432.
[215] S. Weinberg, Phys. Rev. Lett. 43 (1979) 1566; Phys. Rev. D22 (1980) 1694.
[216] F.A. Wilczek and A. Zee, Phys. Rev. Lett. 43 (1979) 1571.
H.A. Weldon and A. Zee, Nucl. Phys. B173 (1980) 269.
[217] A. De Rújula, H. Georgi and S.L. Glashow, Phys. Rev. Lett. 45 (1980) 413.
[218] See for example:
J.C. Pati, in: Unification of the Fundamental Particle Interactions, eds. S. Ferrara, J. Ellis and P. Van Nieuwenhuizen (Plenum, New York, 1980) p. 267, and references therein.
[219] V.A. Kuzmin, Zh. Eksp. Teor. Fiz. 13 (1970) 335.
[220] For a recent review, see:
R.E. Marshak, Virginia Polyt. Inst. preprint VPI-HEP-81/3 (1981).
[221] S. Weinberg, Phys. Lett. 91B (1980) 51, contains a general discussion of the treatment of thresholds in the minimal subtraction scheme.
[222] For recent reviews, see:
P. Hasenfratz, CERN preprint TH-3157 and Proc. 1981 Int. Symp. on Lepton and Photon Interactions at High Energies, Bonn, ed. W. Pfeil (Univ. of Bonn, 1981) p. 866.
[223] J. Ellis, M.K. Gaillard and D.V. Nanopoulos, Phys. Lett. 88B (1980) 320.
[224] J. Ellis, M.K. Gaillard and D.V. Nanopoulos, Phys. Lett. 80B (1979) 360 (Erratum 82B (1979) 464).
[225] C. Jarlskog, Phys. Lett. 82B (1979) 40.
[226] V.S. Berezinsky and A.Yu. Smirnov, Phys. Lett. 97B (1980) 371.
[227] M. Daniel and J.A. Peñarrocha, Southampton Univ. preprints SHEP 80/81-5,6 (1981).
[228] C.T. Sachrajda (1981) private communication; Course 2, this volume.
[229] C. Jarlskog and F.J. Yndurain, Nucl. Phys. B149 (1979) 29.
[230] M. Machacek, Nucl. Phys. B159 (1979) 37.
[231] M.B. Gavela, A.Le Yaouanc, L. Oliver, O. Pène and J.C. Raynal, Phys. Lett. 98B (1981) 51; Phys. Rev. D23 (1981) 1580.
[232] A. Din, G. Girardi and P. Sorba, Phys. Lett. 91B (1980) 77.
[233] J.F. Donoghue, Phys. Lett. 92B (1980) 99.
E. Golowich, Phys. Rev. D22 (1980) 1148.
[234] For other recent calculations of nucleon decays, see:

Y. Tomozawa, Phys. Rev. Lett. 46 (1981) 463.
V.S. Berezinsky, B.L. Ioffe and Ya.I. Kogan, Phys. Lett. 105B (1981) 33.
[235] J. Learned, F. Reines and A. Soni, Phys. Rev. Lett. 43 (1979) 907.
E.N. Alekseev et al. (Baksan Laboratory) give a limit $\tau_N > 1.2 \times 10^{30}$ yr.
M.L. Cherry et al., Phys. Rev. Lett. 47 (1981) 1507, give a similar limit.
M.R. Krishnaswamy et al., Phys. Lett. 106B (1981) 339, report a possible signal corresponding to $\tau_N \approx 8 \times 10^{30}$ yr.
[236] For a review of new experiments, see:
L. Sulak, Seminar 4, this volume.
[237] G. Kane and G. Karl, Phys. Rev. D22 (1980) 1808.
[238] M.B. Wise, R. Blackenbecler and L.F. Abbott, Phys. Rev. D23 (1981) 1591.
[239] M. Baldo-Ceolin et al., ILL-research proposal 03-05-027 (1980).
G.R. Young et al., Oak Ridge proposal ORNL/PHYS-82/1 (1982).
NADIR proposal, Pavia (1981).
R. Ellis et al., LAMPF proposal 647 (1981).
[240] R.E. Marshak, R.N. Mohapatra and Riazuddin, Proc. TRIUMF workshop, eds. J.A. MacDonald, J.N. Ng and A. Strathdee (TRIUMF, Vancouver, 1981) p. 10.
[241] R.N. Mohapatra and R.E. Marshak, Phys. Lett. 94B (1980) 183.
K.G. Chetyrkin, M.V. Kazarnovsky, V.A. Kuzmin and M.E. Shaposhnikov, Phys. Lett. 99B (1981) 358.
Riazuddin, Phys. Rev. D25 (1982) 885.
[242] O. Sawada, M. Fukugita and J. Arafune, Astrophys. J. 248 (1981) 1162.
[243] A. Buffington, S.M. Schindler and C.R. Pennypacker, Astrophys. J. 248 (1981) 1179.
[244] P. Kiraly, J. Szabelski, J. Wdowczyk and A. Wolfendale, Nature 293 (1981) 120.
[245] Furthermore, the experiment of M. Baldo-Ceolin et al., also has a more direct limit $t > 1.2 \times 10^5$ s from a preliminary analysis of data:
G. Fidecaro, Proc. Neutrino 81, Hawaii, eds. R.J. Cence, E. Ma and A. Roberts (HEP group, Hawaii, 1981) vol. 1, p. 264.
[246] K.E. Bergkvist, Nucl. Phys. B39 (1972) 317.
[247] M. Daum et al., Phys. Lett. 74B (1978) 126.
[248] J. Kirkby, Proc. Int. Symp. on Lepton and Photon Interactions at High Energies, Fermilab 1979, eds. T.B.W. Kirk and H.D.I. Abarbanel (FNAL, Batavia, 1980) p. 107.
[249] V.A. Lyubimov et al., Yad. Fiz. 32 (1980) 30; Phys. Lett. 94B (1980) 266.
[250] R. Barbieri, J. Ellis and M.K. Gaillard, Phys. Lett. 90B (1980) 249.
[251] R. Barbieri, D.V. Nanopoulos, G. Morchio and F. Strocchi, Phys. Lett. 90B (1980) 91.
[252] M. Magg and C. Wetterich, Phys. Lett. 94B (1980) 61.
[253] M. Gell-Mann, P. Ramond and R. Slansky (1979), unpublished.
R. Slansky, Talk at the Sanibel Symposium, Caltech preprint CALT-68-709 (1979).
[254] T. Yanagida, Proc. Workshop on the Unified Theory and the Baryon Number in the Universe (KEK, Japan, 1979).
[255] G. Lazarides and Q. Shafi, Phys. Lett. 99B (1981) 113.
[256] E. Witten, Phys. Lett. 91B (1980) 81.
[257] B. Pontecorvo, Zh. Eksp. Teor. Fiz. 33 (1957) 549; 34 (1958) 247; 53 (1967) 1717.
Z. Maki, M. Nakagawa and S. Sakata, Prog. Theor. Phys. 28 (1962) 870.
[258] S. Bikenky and B. Pontecorvo, Phys. Rep. 41C (1978) 225.
[259] B. Kayser, Phys. Rev. D24 (1981) 110.
[260] L. Wolfenstein, Phys. Rev. D17 (1978) 2369.
[261] V. Barger, K. Whisnant, S. Pakvasa and R.J.N. Phillips, Phys. Rev. D22 (1980) 2718.

[262] A. De Rújula, Nucl. Phys. B188 (1981) 414.
[263] M.K. Moe and D.D. Lowenthal, Phys. Rev. C22 (1980) 2186.
[264] E.W. Hennecke, O.L. Manuel and D.D. Sabu, Phys. Rev. C11 (1975) 1378.
[265] H. Primakoff and S.P. Rosen, Phys. Rev. 184 (1969) 1925.
[266] M. Doi et al., Phys. Lett. 103B (1981) 219; Prog. Theor. Phys. 66 (1981) 1739, 1761.
[267] W.C. Haxton, G.J. Stephenson and D. Strottman, Phys. Rev. Lett. 47 (1981) 153.
J. Vergados, Phys. Lett. 109 (1982) 96.
L. Wolfenstein, Phys. Lett. 107B (1981) 77.
[268] For recent phenomenological reviews, see:
V. Barger, Proc. Neutrino 81, Hawaii, eds. R.J. Cence, E. Ma and A. Roberts (HEP group, Hawaii, 1981) vol. 2, p. 1.
D. Silverman and A. Soni, Invited Talk at 1981 APS meeting, Santa Cruz; UCLA preprint UCLA/81/TEP/25 (1981).
[269] J.K. Rowley et al., Brookhaven preprint BNL 27190 (1980).
For reviews, see:
J.N. Bahcall, Rev. Mod. Phys. 50 (1978) 88.
J.N. Bahcall and R. Davis, Contribution to the Festschrift for Willy Fowler (1980).
[270] H.P. Trautvetter, Comm. Nucl. Part. Phys. 10 (1981) 123 and references therein.
B.W. Filippone and D.N. Schramm, Astrophys. J. 253 (1982) 393 and references therein.
[271] D. Gough, Nature 293 (1981) 703.
[272] M.F. Crouch et al., Phys. Rev. D18 (1978) 2239.
[273] M.R. Krishnaswamy et al., Proc. Roy. Soc. A323 (1971) 489.
[274] A. Yu. Smirnov, Talk presented at the XVI Rencontre de Moriond, Les Arcs, France (1981), reporting on work at the Baksan Laboratory (unpublished).
[275] For recent reviews, see:
L.W. Jones, Proc. XVIth Rencontre de Moriond, Les Arcs, France, ed. J. Tran Thanh Van (Editions Frontières, Gif-sur-Yvette, 1981) p. 203.
J. Wotschack, Invited Talk at the IVth Warsaw Symposium on Elementary Particle Physics, Kazimierz, Poland; CERN preprint EP/81-81 (1981).
[276] E. Pasierb, H. Sobel and F. Reines, Phys. Rev. Lett. 45 (1980) 1307.
[277] F. Boehm et al., Phys. Lett. 97B (1980) 310 and Phys. Rev. D24 (1981) 1097.
[278] D. Silverman and A. Soni, Phys. Rev. Lett. 46 (1981) 467; Proc. 1981 Orbis Scientiae, eds. B. Kursunoglu and A. Perlmutter (Plenum, New York, 1981) p. 249, and ref. [268].
[279] For a recent review and references, see:
M.S. Turner, Talk presented at the Virginia Polytechn. Inst. Symp. on Weak Interactions as Probes of Unification, Blacksburg, VA, Enrico Fermi Inst. preprint 81-09 (1981).
[280] A.D. Sakharov, Zh. Eksp. Teor. Fiz. Pis'ma 5 (1967) 32.
[281] M. Yoshimura, Phys. Rev. Lett. 41 (1978) 381 (Erratum 42 (1979) 746); Phys. Lett. 88B (1979) 294.
A. Yu Ignatiev, N.V. Krasnikov, V.A. Kuzmin and A.N. Tavkhelidze, Phys. Lett. 76B (1978) 436.
S. Dimopoulos and L. Susskind, Phys. Rev. D18 (1978) 4500; Phys. Lett. 81B (1979) 416.
D. Toussaint, S.B. Treiman, F. Wilczek and A. Zee, Phys. Rev. D19 (1979) 1036.
S. Weinberg, Phys. Rev. Lett. 42 (1979) 850.
D.V. Nanopoulos and S. Weinberg, Phys. Rev. D20 (1979) 2484.

[282] G. Segrè and M.S. Turner, Phys. Lett. 99B (1981) 399.
[283] Unless one resorts to a very peculiar temperature dependence of the SU(2)-breaking vacuum expectation value of the Higgs field:
V.A. Kuzmin, M.E. Shaposhnikov and I.I. Tkachev, Phys. Lett. 105B (1981) 159.
[284] J. Ellis, M.K. Gaillard, D.V. Nanopoulos and S. Rudaz, Phys. Lett. 99B (1981) 101.
[285] J. Ellis, M.K. Gaillard and D.V. Nanopoulos, in: Unification of the Fundamental Particle Interactions, eds. S. Ferrara, J. Ellis and P. Van Nieuwenhuizen (Plenum, New York, 1980), p. 461.
[286] J. Yang, M.S. Turner, D.N. Schramm, G. Steigman and K.A. Olive, Enrico Fermi Inst. preprint, in preparation; results reported by
G. Steigman, ref. [87].
[287] J. Ellis, M.K. Gaillard, D.V. Nanopoulos and S. Rudaz, Nature 293 (1981) 41.
[288] For words of caution, see:
A. Masiero and R.D. Peccei, Phys. Lett. 108B (1982) 111.
[289] R. Barbieri, R.N. Mohapatra, D.V. Nanopoulos and D. Wyler, Phys. Lett. 107B (1981) 80.
[290] J. Ellis, M.K. Gaillard, D.V. Nanopoulos and S. Rudaz, Phys. Lett. 106B (1981) 298.
[291] J. Ellis, M.K. Gaillard, A. Peterman and C.T. Sachrajda, Nucl. Phys. B164 (1980) 253.
[292] S. Dimopoulos and H. Georgi, Nucl. Phys. B193 (1981) 150.
[293] N. Sakai, Z. Phys. C11 (1981) 153.
[294] R.K. Kaul, Phys. Lett. 109B (1982) 19.
[295] S. Weinberg, Phys. Rev. D26 (1982) 287.
N. Sakai and T. Yanagida, Nucl. Phys. B197 (1982) 533.
[296] For a general discussion of renormalization theory, see:
W. Zimmermann, Lectures on Elementary Particles and Quantum Field Theory (Brandeis Summer Institute, 1970) eds. S. Deser, M. Grisaru and H. Pendleton (M.I.T. Press, Cambridge, MA, 1970).
[297] M. Veltman, Acta Phys. Pol. B12 (1981) 437.
[298] P. Fayet and J. Iliopoulos, Phys. Lett. 31B (1974) 461.
[299] W. Fischler, H.P. Nilles, J. Polchninski, S. Raby and L. Susskind, Phys. Rev. Lett. 47 (1981) 757.
[300] K. Harada and N. Sakai, Prog. Theor. Phys. 67 (1982) 1877.
R.K. Kaul and P. Majumdar, Nucl. Phys. B199 (1982) 36.
[301] L. Girardello and M.T. Grisaru, Nucl. Phys. B194 (1982) 65.
[302] S. Dimopoulos, S. Raby and F.A. Wilczek, Phys. Rev. D24 (1981) 1681.
[303] L. Ibáñez and G.G. Ross, Phys. Lett. 105B (1981) 439.
[304] M.B. Einhorn and D.R.T. Jones, Nucl. Phys. B196 (1982) 475.
[305] S. Ferrara, L. Girardello and F. Palumbo, Phys. Rev. D20 (1979) 403.
[306] This may cause cosmological problems according to I. Affleck and P. Ginsparg (private communication from M. Sher).
[307] J. Wess and B. Zumino, Phys. Lett. 49B (1974) 52.
J. Iliopoulos and B. Zumino, Nucl. Phys. B76 (1974) 310.
S. Ferrara, J. Iliopoulos and B. Zumino, Nucl. Phys. B77 (1974) 413.
M.T. Grisaru, W. Siegel and M. Roček, Nucl. Phys. B159 (1979) 420.
[308] S. Dimopoulos and F.A. Wilczek, I.T.P. Santa Barbara preprint (1981), unpublished.
R. Barbieri, S. Ferrara and D.V. Nanopoulos (1981) private communication.
[309] S. Weinberg, and N. Sakai and T. Yanagida, ref. [295] and R. Barbieri et al., ref. [289].
[310] J. Ellis, D.V. Nanopoulos and S. Rudaz, Nucl. Phys. B202 (1982) 43.
See also S. Dimopoulos, S. Raby and F.A. Wilczek, Phys. Lett. 112B (1982) 133.

[311] A. Salam and J. Strathdee, Nucl. Phys. B80 (1974) 499.
P.H. Dondi and M. Sohnius, Nucl. Phys. B81 (1974) 317.
[312] D.Z. Freedman, P. Van Nieuwenhuizen and S. Ferrara, Phys. Rev. D13 (1976) 3214.
S. Deser and B. Zumino, Phys. Lett. 62B (1976) 335.
[313] For a review, see:
P. Van Nieuwenhuizen, Phys. Rep. 68C (1981) 189.
[314] B. De Wit and H. Nicolai, Nucl. Phys. B188 (1981) 98.
[315] B. De Wit and H. Nicolai, Phys. Lett. 108B (1982) 285.
[316] S.W. Hawking, Talk presented at the 1981 Heisenberg Symposium, Munich (1981).
[317] M. Gell-Mann, Talk at the 1977 Washington Meeting of the American Physical Society, unpublished.
[318] For an early attempt to use composite fields in $N = 8$ extended supergravity, see:
T. Curtright and P.G.O. Freund, in: Supergravity, eds. P. Van Nieuwenhuizen and D.Z. Freedman (North-Holland, Amsterdam, 1979) p. 167.
[319] And in fact many of them *do* acquire masses when one constructs a supergravity theory in four dimensions by using dimensional regularization: see
S. Ferrara and B. Zumino, Phys. Lett. 86B (1979) 279 and references therein.
[320] E. Cremmer and B. Julia, Phys. Lett. 80B (1978) 48; Nucl. Phys. B159 (1979) 141.
[321] A. D'Adda, P. Di Vecchia and M. Lüscher, Nucl. Phys. B146 (1978) 63; B152 (1979) 125.
E. Witten, Nucl. Phys. B149 (1979) 285.
[322] E.R. Nissimov and S.J. Pacheva, Compt. Rend. Acad. Bulg. Sc. 32 (1979) 1475.
[323] E.R. Nissimov and S.J. Pacheva, Lett. Math. Phys. 5 (1980) 67, 333.
[324] J. Ellis, M.K. Gaillard, L. Maiani and B. Zumino, in: Unification of the Fundamental Particle Interactions, eds. S. Ferrara, J. Ellis and P. Van Nieuwenhuizen (Plenum, New York, 1980) p. 69.
[325] J. Ellis, M.K. Gaillard and B. Zumino, Phys. Lett. 94B (1980) 343.
[326] J.P. Derendinger, S. Ferrara and C.A. Savoy, Nucl. Phys. B188 (1981) 77.
[327] J.E. Kim and H.S. Song, Phys. Rev. D25 (1982) 2996.
[328] H. Römer, Phys. Lett. 83B (1979) 172.
[329] S. Coleman and B. Grossman, Nucl. Phys. B203 (1982) 205.
[330] B. Zumino, Proc. 1980 Int. Conf. on High Energy Physics, Madison (1981), eds. L. Durand and L.G. Pondrom (American Institute of Physics, New York, 1981) p. 964;
B. Zumino, in: Superspace and Supergravity, eds. S.W. Hawking and M. Roček (Cambridge Univ. Press, Cambridge, 1981) p. 423.
M. Gell-Mann (1980) private communication.
[331] M. Glück and E. Reya, Phys. Lett. 105B (1981) 30.
[332] S. Weinberg and E. Witten, Phys. Lett. 96B (1980) 59.
[333] M.K. Gaillard and B. Zumino, Nucl. Phys. B193 (1981) 221.
[334] J.P. Ader, M. Capdeville and H. Navelet, Nuovo Cim. 56A (1968) 315.
[335] M.T. Grisaru, H.N. Pendleton and P. Van Nieuwenhuizen, Phys. Rev. D15 (1977) 996.
[336] D.Z. Freedman and A. Das, Nucl. Phys. B120 (1977) 221.
S. Deser and B. Zumino, Phys. Rev. Lett. 38 (1977) 1433.
[337] C.H. Llewellyn Smith, Phys. Lett. 46B (1973) 233.
J.S. Bell, Nucl. Phys. B60 (1973) 427.
J.M. Cornwall, D.N. Levin and G. Tiktopoulos, Phys. Rev. Lett. 30 (1973) 1268; Phys. Rev. D10 (1974) 1145.
[338] S. Weinberg, in: Lectures on Elementary Particles and Quantum Field Theory (Brandeis Summer Institute, 1970) eds. S. Deser, M. Grisaru and H. Pendleton

(M.I.T. Press, Cambridge, MA, 1970).
[339] G. Steigman, K.A. Olive and D.N. Schramm, Phys. Rev. Lett. 43 (1979) 239.
K.A. Olive, D.N. Schramm and G. Steigman, Nucl. Phys. B180 (FS2) (1981) 497.
[340] H. Haber, I. Hinchliffe and E. Rabinovici, Nucl. Phys. B172 (1980) 458.
[341] E. Rabinovici (1980) private communication.
[342] R. Barbieri, S. Ferrara and D.V. Nanopoulos, Phys. Lett. 107B (1981) 275.
[343] Generalized Graded Kac–Moody algebras may play a crucial rôle:
V. Kac, Math. U.S.S.R. Izv. Ser. Math. 32 (1968) 1271.
R. Moody, Bull. Ann. Math. Soc. 73 (1967) 217; J. of Algebra 10 (1968) 211.
[344] S.W. Hawking, Lucasian Inaugural Lecture, "Is the End in Sight for Theoretical Physics?" (Cambridge Univ. Press, 1980).
[345] J.G. Branson, MIT Technical Report 133 (1983).
[346] L.L. Chau, Brookhaven preprint (1982) to appear in Phys. Rep. C.
L.L. Chau, W.-Y. Keung and M.D. Tran, Brookhaven preprint BNL 31725 (1982).
[347] R.I. Dzhelyadin et al., Phys. Lett. 97B (1980) 471.
[348] J.F. Donoghue, E. Golowich and B.R. Holstein, Phys. Lett. 119B (1982) 412.
[349] F.J. Gilman and J.S. Hagelin, SLAC-PUB-3087 (1983).
[350] G. Arnison et al., Phys. Lett. 122B (1983) 103.
M. Banner et al., Phys. Lett. 122B (1983) 476.
[351] J. Ellis and J.S. Hagelin, CERN preprint TH-3390 (1982).
[352] S. Dimopoulos, H. Georgi and S. Raby, Harvard Univ. preprint HUTP 83/A002 (1983).
[353] TASSO Collaboration, M. Althoff et al., Phys. Lett. 122B (1983) 95.
[354] V.A. Rubakov, Zh. Eksp. Teor. Fiz. Pis'ma Red. 33 (1981) 658 [JETP Lett. 33 (1982) 644]; Nucl. Phys. B203 (1982) 311.
[355] J. Ellis, SLAC-PUB-3003 (1982) and references therein.
[356] Irvine–Michigan–Brookhaven collaboration, J. van der Velde, talk presented at 1983 Rencontre de Moriond, to be published (1983).
J. Bartelt et al., Phys. Rev. Lett. 50 (1983) 651.
[357] G. Battistoni et al., Phys. Lett. 118B (1982) 461.
Irvine–Michigan–Brookhaven collaboration, ref. [356].
[358] P. Hasenfratz and I. Montvay, Phys. Rev. Lett. 50 (1983) 309.
[359] S.J. Brodsky, G.P. Lepage and P.B. Mackenzie, SLAC-PUB-3011 (1982).
[360] C. Baltay, Proc. 1982 Rencontre de Moriond, to be published.
[361] J. Ellis and D.V. Nanopoulos, Phys. Lett. 110B (1982) 44.
[362] See for example:
J. Ellis, J.S. Hagelin, D.V. Nanopoulos and K. Tamvakis, SLAC-PUB-3042 (1983) and references therein.
[363] E. Cremmer et al., Nucl. Phys. B147 (1979) 105.
[364] A.C. Davis, A.J. Macfarlane and J.W. van Holten, Composite Gauge Bosons in Sigma Models and Supergravity, Princeton preprint (1983).
E. Rabinovici and Y. Cohen, CERN preprint TH-3502 (1983).
[365] J. Ellis, M.K. Gaillard, M. Günaydin and B. Zumino, SLAC-PUB-3065 (1983).

SEMINAR 1

# SOME COSMOLOGICAL CONSTRAINTS ON GAUGE THEORIES

David N. SCHRAMM

*Astronomy and Astrophysics Center*
*University of Chicago*
*Chicago, IL 60637, USA*

## Contents



*M.K. Gaillard and R. Stora, eds.*
*Les Houches, Session XXXVII, 1981*
*Théories de jauge en physique des hautes énergies/Gauge theories in high energy physics*

## 1. Introduction

In these lectures, a review will be made of various constraints cosmology may place on gauge theories. Particular emphasis will be placed on those constraints obtainable from Big Bang *Nucleo*synthesis, with only brief mention made of Big Bang *Baryo*synthesis, since that will be treated in detail by other lecturers (e.g. M. Turner). There will also be considerable discussion of astrophysical constraints on masses and lifetimes of neutrinos with specific mention of the "missing mass (light)" problem of galactic dynamics.

The recent high level of activity at this boundary of particle physics and cosmology comes from the combination of a general acceptance of the Big Bang model of the universe and the fact that recent developments in particle theory point toward effects which will be important only at energies far beyond those accessible with accelerators.

The general acceptance of a hot–dense early universe (the Big Bang) began to happen with the discovery of the 3 K background radiation by Penzias and Wilson [1]. However complete acceptance did not occur until the experiments of Richards and co-workers [2] showed that the background radiation had the appropriate thermal turnover at wavelengths of $\sim 1$ mm. The existence of this radiation tells us that the temperature of the universe was at one time at least $T \gtrsim 10^4$ K. At $T \approx 10^4$ K hydrogen would be ionized and the free electrons would easily scatter the photons. Thus, the present observed radiation is merely the last scattered thermal radiation from $T \approx 10^4$ K.

We actually have confidence that the universe was a good deal hotter than this. Gamow and co-workers predicted that there would be a thermal background on the basis of assuming that nuclear reactions occurred in the Big Bang. To have nuclear reactions requires that the temperature is higher than $\sim 10^9$ K. The verification of a temperature at least as high as $10^{10}$ K comes from the fact that the $^4$He abundance is about 25% by mass. This helium abundance comes as a natural consequence of the standard Big Bang if the temperature was greater than $\sim 10^{10}$ K (Schramm and Wagoner [3] and refs. therein).

There is at present no direct observational tie to earlier times in the universe when the temperature was even higher (unless one accepts the existence of matter as an indication of grand unification decoupling at $\sim 10^{15}$ GeV). We do have some feeling that it is not totally absurd to discuss temperatures in the early universe as high as $10^{15}$ GeV or maybe even $10^{19}$ GeV. This lack of complete fear comes from the singularity theorems of Hawking, Ellis and Penrose (Hawking and Ellis [4] and references, therein). The theorems state that if the universe has a net free energy then the world lines must have come out of a singularity. Since we know the universe is roughly homogeneous and isotropic then the singularity becomes the global one which we call the Big Bang. The crux of these theorems is the present epoch of the universe. We have the 3 K radiation so we know there exists positive free energy and we can extrapolate our world lines back to higher densities and temperatures.

The place where the positive energy conjecture and our extrapolation run into trouble is at the Planck time, $10^{-43}$ s after the Big Bang, at a temperature of $10^{19}$ GeV. At this temperature, gravity becomes quantized and all bets are off. It may be that at the Planck time all of space–time is

Table 1
The very early universe

| Time | Temperature | Event | Observable |
|---|---|---|---|
| ? | ? | Domain of quantum gravity | ? |
| $10^{-43}$ s | $10^{19}$ GeV | Planck time, decoupling of gravitons | Homogeneity–isotropy (?), $H_0$, $\Omega$(?) (gravitons) |
| $10^{-35}$ s | $10^{15}$ GeV | Grand Unification, decoupling of $X$, $Y$ | Matter-to-radiation ratio (perturbations ?) |
| $10^{-6}$ s | 200 MeV | Quark-hadron phase transition, quark-gluon confinement | Hadrons (n, p, $\pi$, K, ...) (perturbations ?) |
| 1 s | 1 MeV | Weak interaction freeze-out, decoupling of neutrinos | n/p (neutrino background) |
| 3 min | $10^9$ K | Big Bang Nucleosynthesis | H, D, $^4$He ($^7$Li), ($^3$He) |
| $10^5$ yr | $10^5$ K | Recombination decoupling of photons, transition from radiation-to-matter-domination | 3/K background |
| $10^7$–$10^8$ yr | | Star and galaxy formation, heavy element synthesis | |

a foam of mini black holes which are forming and exploding via the Hawking process on a Planck timescale. Such instantaneous black hole formation throughout space–time would violate the positive energy condition and our extrapolation to higher densities and temperatures must break down. However, we have already reached $10^{19}$ GeV, so let us proceed from the Big Bang in chronologic order from such temperatures (see table 1). (Much of what follows is based on a lecture given at the 10th Texas symposium in Relativistic Astrophysics [5].)

## 2. The Planck time

Because we know that gravity must become quantized for temperatures higher than $\sim 10^{19}$ GeV which correspond to Big Bang model times of $\sim 10^{-43}$ s after the classical Big Bang, and because there is no consistent quantum theory of gravity, it is obvious that we cannot talk in a scientific manner about times "earlier" than $10^{-43}$ s. In fact the whole concept of space and time becomes jumbled at this epoch. Hawking refers to space–time being a foam of mini black holes constantly bursting and reforming. Hartle and Hu [6], using a semi classical approach, have suggested that the universe may be horizonless at the Planck time. If this were the case then many, if not all of the causality problems associated with the Big Bang may be solvable. In particular, the standard model does not enable us to understand how the universe can be homogeneous and isotropic on scales that were never in causal connection. But if the Planck time results in a vanishing of the horizon, then all of the universe might have been in causal connection and become homogeneous and isotropic prior to the formation of the horizon.

One should remember that $10^{-43}$ s is a model time. Earlier times really do not exist in the normal sense or, for that matter, there may be an infinite amount of time "prior" to $10^{-43}$ s. Our time extrapolation breaks down as we try to push beyond the Planck barrier.

We will see later that everytime an interaction decouples from the rest of the universe during its expansion, we end up with an observable. The 3 K radiation, the n/p ratio ($^4$He abundance) and perhaps even the baryon/photon ratio are examples of this. One hope is that our expansion rate $H_0$ and the deceleration rate $q_0$ are consequences of the "freeze-out" of quantum gravity. However at the present time such hopes are more religious than scientific.

Before leaving the quantum gravity era, it is interesting to note a proposal by Cocconi [7] that at temperatures greater than $10^{19}$ GeV the gravitational constant may vary like other gauge coupling constants do when one goes above the appropriate critical energy. In the case of gravity, such variation may be what causes the Big Bang in the first place and may be what prevents singularities from actually occurring. Thus while $G$ may be constant for all normal interactions, its constancy above $10^{19}$ GeV is not obvious.

Let us move from the realm of total speculation in the direction of more developed ideas.

## 3. Grand unification and the origin of matter

In this section we will briefly mention how recent developments in Grand Unification theories may resolve the cosmological puzzle of $n_B/n_\gamma \approx 10^{-10}$–$10^{-9}$ (where $n_B$ is the baryon density and $n_\gamma$ is the photon density). Note that the fact that this number is far larger than the $10^{-18}$ which would be obtained in a homogeneous isotropic universe with equal amounts of matter and antimatter, argues strongly that our universe is asymmetric (cf. Steigman [8] and refs. therein). A little over a decade ago, another puzzle, the large observed mass fraction of $^4$He, was resolved by considering the role of nuclear physics in the early universe. Today, application of Grand Unified Theories of particle interactions may explain the origin of baryons in the universe.

The striking success of gauge theories to describe particle interactions has motivated Grand Unified Theories (GUTs), gauge theories which unify the weak, electromagnetic and strong interactions. A common feature of all these GUTs is that baryon number is no longer absolutely conserved.

In the simplest of the GUTs, SU(5), there are the six usual quark flavors (u, d, s, c, t, b), the six usual leptons ($e^-, \nu_e, \mu^-, \nu_\mu, \tau^-, \nu_\tau$) and 24 gauge particles: the photon, $W^\pm, Z^0$, 8 gluons, and 12 new superheavy gauge particles which mediate baryon nonconservation. In addition to the usual light ($\sim 200$ GeV) Higgs particles which generate masses for the quarks, leptons, $W^\pm$, and $Z^0$ there are superheavy Higgs particles which generate masses for the superheavy gauge particles. These superheavy Higgs particles can also mediate baryon nonconservation and for standard SU(5) models actually dominate the final results [52, 11, 12].

The mass of a superheavy Higgs or gauge boson, $m_x$, is of the order of $10^{15}$ GeV$/c^2$. At present energies (hundreds of GeV) baryon nonconserving processes are almost negligible (*almost*, but not quite, since GUTs *do* predict proton decay with lifetimes at $\sim 10^{32}$ yr; appendix 1 shows an alternative calculation of this lifetime inspired by a conversation with Hubert Reeves). However, at energies of the order of the unification mass, $m_x$, these processes will be roughly as strong as all the other interactions. Energies of the order of $10^{15}$ GeV occurred in the very early universe when the temperature was of the order of $10^{28}$ K. In the standard model this corresponds to a time of $10^{-35}$ s after the singularity. Even though the baryon non-conserving processes are strong at these early times, two additional ingredients are necessary to produce a net baryon number.

The first is particle–antiparticle asymmetry or $C$ and $CP$ violations. An arrow is needed to specify the direction of the violation. To see this, consider the following two reactions which violate baryon number,

$$A + B \rightarrow C + D, \quad \Delta B_1 \neq 0, \quad \text{rate } r_1,$$

$$\bar{A} + \bar{B} \rightarrow \bar{C} + \bar{D}, \quad \Delta B_2 \neq 0, \quad \text{rate } r_2.$$

In the second reaction, particles have been replaced by their antiparticles. Both processes violate $B$ (baryon number), and $\Delta B_1 = -\Delta B_2$. If $C$ and $CP$ are good symmetries, then the rates, $r_1$ and $r_2$, will be equal and the effect of the two reactions is no net baryon generation. If $C$ and $CP$ are violated then $r_1$ does not have to be equal to $r_2$ and there can be a net generation of baryon number. $CP$ and $C$ violation occur in nature (the $K^0$–$\bar{K}^0$ system) and also occur naturally in GUTs.

The final ingredient is departure from thermal equilibrium at some time in the early universe. Such departures occur naturally during the evolution of the universe. This ingredient too is essential, as it has been shown by several authors that if $CPT$ is a good symmetry and if a system is in equilibrium, then regardless of $B$, $C$, and $CP$ violations the net baryon number will remain equal to zero (Weinberg [9], Toussaint et al. [10], and Dimopoulos and Susskind [47] and references therein).

The exciting and amazing result is that when GUTs are used to describe the interactions in the very early universe ($kT \gtrsim m_x c^2$), an initially baryon-symmetrical universe can for some models evolve a net baryon excess of the observed magnitude. This point has now been worked out in detailed calculations by Fry et al. [11] and by Kolb and

Wolfram [12]. Much later, when the baryons and antibaryons annihilate ($T \approx 10^{12}$ K) this excess leaves the one baryon per $\sim 10^{10}$ photons we see today.

If one views the obtainment of $n_b/n_\gamma$ as a GUTs requirement, then only certain GUTs are allowed. The minimal SU(5) model with three generations (a **24** and a **5**) of Higgs will only get $n_b/n_\gamma \gtrsim 10^{-17}$. Thus if GUT does solve the problem something more than minimal SU(5) is needed. It has been shown that SO(10) works as does SU(5) with a slightly more complex Higgs sector, as may be required independently on other grounds. Another way of enabling SU(5) to work perhaps is to have four generations but keep the minimal **5** and **24** Higgs sector (Segre and Turner [13]).

Cosmological generation of baryons has some important astrophysical implications. In the scenario described above (and in all but one of the scenarios suggested) the baryon/entropy ratio is determined only by the parameters of the GUT. Therefore, independent of any primordial temperature fluctuations, the baryon/entropy ratio will be constant throughout the universe (unless the fluctuations were so large that some parts of the universe were never hot enough to have had X, $\overline{\mathrm{X}}$-bosons). This bears directly upon the question of galaxy formation where initial density fluctuations are needed. These fluctuations are of two basic types: adiabatic and isothermal. In an adiabatic fluctuation both radiation and matter participate, with the baryon/entropy ratio remaining constant. In an isothermal fluctuation radiation does not participate, i.e., the temperature remains constant. One can see that if the baryon were cosmologically produced as outlined above, only adiabatic initial fluctuations would be permitted. There are several proposed ways around this restriction, for example primordial black holes and the Hawking process (Turner and Schramm [14, 15] and Barrow and Turner [16]). However all appear somewhat ad hoc although not impossible.

The baryon/photon ratio can be turned upside down and viewed as a photon or entropy/baryon ratio of $10^{10\pm 1}$ indicating an apparently large entropy per baryon. Some years ago, Misner suggested an alternate explanation based on this idea. He proposed that the universe may have started with cold baryons and a very chaotic geometry (rather than the isotropic and homogeneous geometry it has today) and through dissipation, the large entropy per baryon of $\sim 10^{10}$ and the isotropy and homogeneity were produced. However, Penrose and others have argued that the amount of entropy per baryon that could have been produced by a chaotic geometry being smoothed by dissipation is more like $10^{40}$; and so, they conclude that our *apparently* large entropy per baryon is in fact very small, indicating that the initial geometry was very close to being isotropic and homogeneous. However, if the Grand Unification ideas are correct, this potential probe of the initial geometry, the baryon/entropy ratio, does not remain constant and could easily have been raised from 0 ($\approx 10^{-40}$) to the present $\sim 10^{-10}$ by cosmological baryon generation. Therefore, the baryon/entropy ratio cannot be used as a "footprint" of the initial geometry and so our relatively small (compared to $10^{40}$) entropy/baryon ratio cannot be used to infer that the universe has been isotropic and homogeneous *ab initio*. In fact, the initial geometry could have been quite chaotic (Turner [17]).

One cosmological problem associated with GUTs has been the generation of monopoles (cf. Presskill [18], and Fry and Schramm [19] and references therein). Since monopoles are not observed and yet they may be generated by the GUT in the early universe one would like some method of suppressing monopole production. A variety of schemes have been proposed and these are summarized by Presskill, and Fry and Schramm. If the phase transition from GUTs to SU(3)×SU(2)×U(1) is delayed in some way maybe the monopoles could be suppressed. Guth and others have suggested that it also may be that, due to an early rapid horizon growth prior to it, this phase transition could generate perturbations on scales much larger than the standard Friedman horizon at this epoch. Let us now skip from GUTs temperatures of $10^{15}$ GeV down to the MeV temperatures at the weak interaction. During this cooling expansion quarks will condense to hadrons (Olive [21] and Schramm et al. [22] and references therein).

## 4. The decoupling of the weak interaction and relic neutrinos

During the early evolution of the universe, all particles, including neutrinos, were produced copiously. Neutrinos with full strength, neutral current, weak interactions were produced by reactions of the type

$$e^+e^- \leftrightarrow \nu_i + \bar{\nu}_i \quad (i = e, \mu, \tau).$$

At high temperatures ($kT > m_\nu c^2$), these neutrinos were approximately as abundant as photons:

$$n_{\nu_i}/n_\gamma = \tfrac{3}{8} g_{\nu_i},$$

where $g_\nu$ is the number of neutrino helicity states. $g_\nu = 2$ for massless spin

1/2 neutrinos with $\bar{\nu}_i \neq \nu_i$. If the neutrinos have a mass, each spin 1/2 particle has two helicity states; thus for $\bar{\nu}_i \neq \nu_i$, $g_\nu = 4$. However, because known neutrinos appear to be left handed only, then, if massive, they may be of the Majorana type ($\nu_i \approx \bar{\nu}_i$), in which case $g_\nu$ is still 2. A numerical factor of 3/4 comes from the difference between Fermi–Dirac statistics (neutrinos) and Bose–Einstein statistics (photons); the remaining factor of 1/2 is from the number of photon spin states ($g_\gamma = 2$).

For light neutrinos ($m_\nu \ll 1$ MeV), equilibrium was maintained until $T \approx 1$ MeV. (Massive neutrinos with $m_\nu \gtrsim 1$ MeV will annihilate for 1 MeV $\lesssim T < m_\nu$ and thus will not be as abundant, see Gunn et al. [23] and references therein). At lower temperatures the weak interaction rate is too slow to keep pace with the universal expansion rate so that few neutrinos are produced and, equally important, few annihilate. Thus, for $T \approx 1$ MeV, the neutrinos decouple; at this stage their relative abundance is given above. When the temperature drops below the electron mass, electron–positron pairs annihilate heating the photons but not the decoupled neutrinos. The present ratio of neutrinos to photons must account for the extra photons produced when the $e^\pm$ pairs disappeared (cf. Steigman [8]). Because of this, the present neutrino temperature is ~ 2 K rather than the photon temperature of 3 K. From the present density of photons and the above, we obtain the present number density of neutrinos. If the neutrinos have mass we can obtain the mass density, $\rho_\nu$, by

$$\rho_\nu = \sum_i \left( g_{\nu_i} m_{\nu_i} / 200\, H_0^2 \right) (T_0/2.7)^3 .$$

In this equation and subsequently, $m_\nu$ is in eV, the sum is over all neutrino species with $m \ll 1$ MeV, and $H_0$ is the Hubble constant in units of 100 km s$^{-1}$ Mpc$^{-1}$. It has been implicitly assumed that the neutrinos still exist today and thus have a lifetime greater than the age of the universe for decay into anything other than neutrinos.

It is interesting to compare this density to the critical density of the universe, $\rho_c = 3H_0^2/8\pi G$. The density ratio is defined as $\Omega = \rho/\rho_c$. For $T_0 \lesssim 3$ K, $H_0 < 1$ and assuming Majorana mass neutrinos with $g_\nu = 2$ we find that

$$\Omega_\nu \gtrsim 0.014 \sum_i m_{\nu_i} .$$

Later, we will show from Big Bang nucleosynthesis, that $\Omega_b \lesssim 0.14$. We obtain (with $g_{\nu_i} = 2$):

$$\Omega_\nu / \Omega_b \gtrsim 0.1\, m_{\nu_i} ,$$

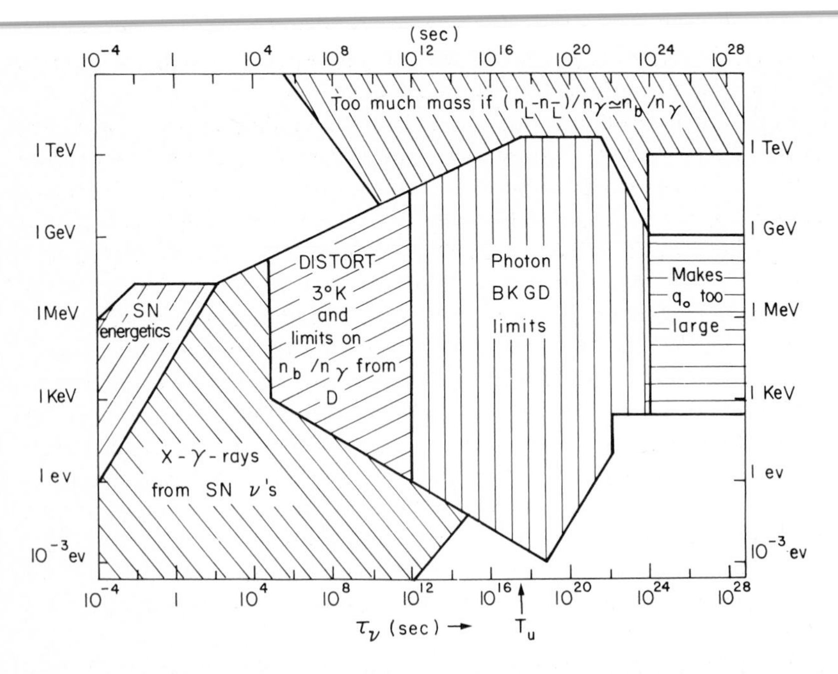


Fig. 1. Astrophysical constraints on combinations of mass and lifetime for neutrinos. From Turner [53] and arguments by refs. [23,25–32,54].

which is independent of $H_0$ and $T_0$. Therefore *if neutrinos have masses of the order of* 10 *eV or greater then neutrinos are the dominant mass component of the universe today.*

From the above arguments on numbers of neutrinos produced during the Big Bang one can put constraints on allowed combinations of neutrino mass and lifetime with non-neutrino decay modes, due to limits on intensity of photon backgrounds in various energy regimes and to limits on the mass density of the universe (cf. Turner [53] and refs. therein).

Figure 1 shows a summary of astrophysical constraints on neutrino lifetimes and masses. For very long lifetimes the constraints come from the limits on the cosmological mass density (Cowsik and McClelland [25] and Szalay and Marx [26]). We know the universe is not *too* closed. In fact, from the nucleochronologic constraint that the age of the universe must be $\gtrsim 8.5 \times 10^9$ yr we know that $\Omega H_0^2 \lesssim 1$. Since $\Omega_\nu \gtrsim 0.014 \Sigma_i m_{\nu_i}$ we obtain the limit that the sum of $m_{\nu_i}$'s must be less than $\sim 100$ eV for Majorana-mass neutrinos. For higher masses limits follow the arguments of Hut and Olive [27], and Lee and Weinberg [54].

For $\nu$'s with finite lifetimes greater than the radiation decoupling time limits can be obtained from photon backgrounds. The most interesting constraints here come from the UV. Recent discussion of this point comes from DeRujula and Glashow [28] and Stecker [29] with the most comprehensive discussion being that of Kimbal et al. [30] where it is shown that low mass neutrinos must have lifetimes longer than $\sim 10^{23}$ s. For lifetimes in the regime between Big Bang nucleo-synthesis ($t \approx 10^3$ s) and decoupling ($t \approx 10^{12}$ s) constraints come from the distortion of the 3 K background (Gunn et al. [23]) and the constraints from deuterium synthesis on $n_b/n_\gamma$ (Dicus et al. [31]). For lifetimes between $\sim 10^{-3}$ s and $\sim 10^3$ s Falk and Schramm [32] have ruled out all neutrinos with masses less than $\sim 10$ MeV by supernova dynamics arguments. Lifetimes less than $10^{-3}$ s are probably eliminated by reactor and accelerator data. It should be noted that with small modification these arguments also apply to "axion-like" particles. One should also be aware that most of the arguments do not apply if the decaying particle does not produce a photon sometimes.

From fig. 1 it appears that with the experimental limits $m_{\nu_\mu} < 700$ keV and $m_{\nu_e} < 50$ eV these both must be very long lived ($> 10^{23}$ s) and have $m_{\nu_\mu} + m_{\nu_e} < 100$ eV. For $m_{\nu_\tau}$ there is also this low mass possibility, however since the experimental upper limit on $m_{\nu_\tau}$ is only $\sim 250$ MeV, it cannot be ruled out that $m$ is between 10 and 250 MeV with a lifetime on the order of seconds.

Since it has been shown that finite but small neutrino masses cannot be ruled out even for $\nu_\mu$ and $\nu_\tau$ and there may even be some theoretical justification for such masses from Grand Unification, let us look at possible cosmological consequences. In fact, as we will see, these cosmological arguments were used by Schramm and Steigman [33, 34] to argue in favor of finite mass neutrinos.

A long standing problem in Astronomy has been the so-called "Missing Mass" problem (or as Schramm and Steigman [33, 34] say the "Missing Light" problem). This is the fact that when one measures the mass of a galaxy-size object by looking at the dynamics of the object, one tends to assign a progressively larger mass to the galaxy when one looks at the dynamics on larger scales. However, the amount of light the galaxy emits stays fixed. Table 2, based on the reviews of Faber and Gallagher [35] and Peebles [36], shows this trend. Note that while the mass-to-light

Table 2
Mass-to-light ratios on various scales and the *implied* density parameter

| Scale | $M/L$ | $\Omega$[a] |
|---|---|---|
| Stars | 1–2 | $\sim 0.001/h_0$ |
| Inner parts of spiral galaxies and ellipticals | $10h_0$ | $\sim 0.01$ |
| Binaries and small groups | $40\text{–}100h_0$ | 0.04–0.1 |
| Large clusters of galaxies | $100\text{–}800h_0$ | 0.1–0.8 |

[a] Based on multiplying $M/L$ by the Kirshner–Oemler–Schecter [49] luminosity density.

$(M/L)$ ratio associated with the actual light-emitting objects (stars) is only 1 to 2 solar units, the $M/L$ ratio associated with the central visible region of a galaxy is $\sim 10$ times this. When that same galaxy is observed in a binary pair or in a small group and the total mass of the system is calculated and distributed among the members, it is found that the $M/L$ ratio associated with each galaxy is 4 to 10 times that estimated from the galaxy's own internal motion. Similarly when a galaxy is in a large cluster and the virial theorem is applied to the cluster to estimate the mass of the galaxy it is found that the $M/L$ ratio is now $\sim 400$. Thus there seems to be more and more mass on larger scales with no more light emission. This additional invisible mass is what is called the "Missing Mass" problem. However, the mass seems to really be there, so what is missing is really the light from that mass, hence Schramm and Steigman's [33, 34] term "Missing Light".

Over the years, many things have been proposed for this missing mass. However, as Schramm and Steigman [34] showed, many of these can be eliminated *if* the missing mass truly must give an $\Omega > 0.1$ (implied by the $M/L$ for large clusters, with a current best estimable value of 0.4) and *if* these large clusters truly give the best estimate of the mass associated with the cosmological luminosity density.

As pointed out by Gott et al. [37], Schramm and Wagoner [3], and Yang et al. [38] and references therein, Big Bang nucleosynthesis cannot give consistent abundances for $^4$He nor D if the density of baryons is greater than about 10% of the critical value. Thus if $\Omega$ for the universe is $\gtrsim 0.1$ as implied by large clusters then the missing mass (light) for these clusters must *not be in the form of baryons*. This eliminates most possibilities except low and high mass neutral weakly interactive particles (neutrinos, axions, majorons,...), monopoles and black holes with masses less than $\sim 1\ M_{\odot}$. (Solar mass and larger black holes were in the form of baryons at the time of Big Bang nucleosynthesis.) Monopoles run into a variety of other problems such as their destruction of galactic magnetic fields. Small black holes ($M \lesssim 10^{15}$ g) have the problem of blowing up via the Hawking process and thus producing too many unobserved $\gamma$-rays. Thus all that remains are the weakly interacting neutrals and black holes with masses between $\sim 10^{15}$ g and $\sim 10^{33}$ g. Although those black holes may be generated at the quark–hadron transition [47], it appears that the most likely prospect would be weakly interacting neutrinos with non-zero rest mass. While very massive, as-yet-undiscovered neutrinos may do the trick (GeV-mass neutral leptons or axion–Majorana-type particles), without any experimental verification they appear extremely ad hoc. This, as pointed out by Schramm and Steigman [33, 34], only leaves low mass neutrinos.

Massive neutrinos gravitate and they will have participated in gravitational clustering (Gunn et al. [23]). However, since neutrinos are non-interacting, their phase space density is conserved and they will cluster only in the deepest potential wells; the slowest moving (i.e. the heaviest) will cluster most easily (Tremaine and Gunn [39]).

Tremaine and Gunn [39] point out that neutrinos lighter than $\sim 3$ eV will not cluster at all on even the larger scales, whereas neutrinos with $m_{\nu} > 3$ eV can be trapped in large clusters, those with $m_{\nu} \gtrsim 10$ eV can be trapped in binaries and small groups and those with $m_{\nu} \gtrsim 20$ eV can be trapped in single galaxies. As we will see, being confinable if trapped is not sufficient.

We have already seen that $m_{\nu}$ must be less than 100 eV. In fact if neutrinos cluster efficiently on the scales of single galaxies then $m_{\nu}$ must be $\lesssim 20$ eV or neutrinos will contribute too much mass on those scales. However, it may be that single galaxies, or even binaries and small groups are just inefficient at trapping neutrinos. But as we will see from nucleosynthesis arguments $\Omega_{\mathrm{b}} \gtrsim 0.02$ thus baryons probably account for most of the mass on scales as large as binaries and small groups. If isothermal perturbations produce neutrino trapping on all scales 3 eV $\lesssim m_{\nu} \lesssim 20$ eV, but if adiabatic perturbations, as implied by simple GUTs, dominate, 3 eV $\lesssim m_{\nu} \lesssim 100$ eV.

Since neutrinos would cluster like an isothermal gas they would have a density distribution that falls off with $1/r^2$. When compared with the fact that light from galaxies falls off like $1/r^3$ (Kron [51]) this yields $M/L$ proportional to $r$ as observed [48]. Thus, neutrinos may be ideal missing-mass candidates. Since they do not radiate they will not settle into the central disk regions of galaxies or into stars.

However, as pointed out by Bond et al. [40], it is extremely difficult to get neutrinos to cluster on scales smaller than the large clusters since their Jeans mass is $10^{15}\ M_{\odot}$. Thus while clustering on large scales is natural and is the only place where non-baryonic mass is needed [46], clustering on smaller scales is difficult [41, 42].

It is also interesting to note that the quadrupole anisotropy measurements of the 3 K background seem to eliminate adiabatic perturbations unless neutrinos have mass [41].

A possible way of verifying that the missing mass for galaxies is in neutrinos is to look at distant quasars through galactic halos and see if there are any gravitational lens effects due to low mass stars (Gott [43]). The possibility that one might also directly observe annihilation into photons or even $\nu_i\bar{\nu}_i \rightarrow \nu_j\bar{\nu}_j + \gamma$ in the density enhancements in clusters seems unlikely (Hill [48]).

Obviously, if $\Omega_\nu$ were $>1$ the Friedman universe would be closed by neutrinos. Current estimates put $\Omega_{\text{cluster}} < 1$ and thus $\Omega_\nu$ would probably also be constrained to be $<1$; however, the uncertainties are sufficiently large that closure by neutrinos cannot be completely excluded. Note that with $g_\nu = 2$, if

$$\sum_i m_{\nu_i} \gtrsim 10\ H_0^2 (T_0/3)^3,$$

then the Friedman universe with $\Lambda = 0$ is closed. With $H_0 \gtrsim 0.5$ and $T_0 \leqslant 3$ we see that for a $\Sigma_i m_{\nu_i}$ as small as 18 eV, closure is in principle conceivable. The total number of such species is constrained by arguments of the types to be discussed in the next section.

## 5. Big Bang nucleosynthesis

Big Bang nucleosynthesis has been described in detail in many places (cf. Schramm and Wagoner [3] and references therein). Thus, it will not be reviewed here otherwise than to show the possibility that neutrino mass effects the limits on the number of neutrino species, and to describe how the use of $^3$He and D can provide a lower limit to $\Omega_b$ to complement the upper limit [3] provided by D and $^4$He. The basic argument on the number of leptons was described by Steigman et al. [24] and the current details can be found in Olive et al. [44] and Yang et al. [46]. The point is that the greater the number of low mass ($m_\nu < 1$ MeV) neutrinos, the more $^4$He that is produced in the Big Bang.

In order to make use of these results, the mass fraction $Y$ must be known. This is a bit of a problem since $^4$He is made in stars as well as in the big bang. Stars forming now may be contaminated by as much as $\Delta Y \approx 0.06$ "new helium". The best estimates for $Y$ give a value of between 0.20 with 0.25 and 0.25 as an upper limit (Yang et al. [38]). The most recent value (Kundt and Sargent [57]) is $0.24 \pm 0.01$. The $^4$He abundance also has a dependence on the *baryon* density of the universe. If binaries and small groups of galaxies are made of baryons then a lower limit on the present density is about $2 \times 10^{-31}$ g cm$^{-3}$ ($\approx 0.04$ of closure density for $H_0 = 50$ km s$^{-1}$ Mpc$^{-1}$). It will be shown shortly that this is approximately the same lower limit on baryon density as obtained using $^3$He and D. A lower limit on the baryon density of this value and $Y \lesssim 0.25$ constrain the number of additional neutrino types (beyond $\nu_e$, $\nu_\mu$ and $\nu_\tau$) to $\lesssim 1$ (fig. 2). (These $^4$He constraints can also be generalized to massless particles that couple more weakly than the usual neutrinos (cf. Olive et al. [50]). In general the more weakly a particle couples, the more types are allowed. However, even for gravitinos the number allowed is only about 20.)

Although normally one only concentrates on the upper limits to baryon density from Big Bang nucleosynthesis, there are some lower limits which are of interest. In particular, $^3$He and D production increases dramatically with lower density. Since $^3$He is destroyed in normal stellar

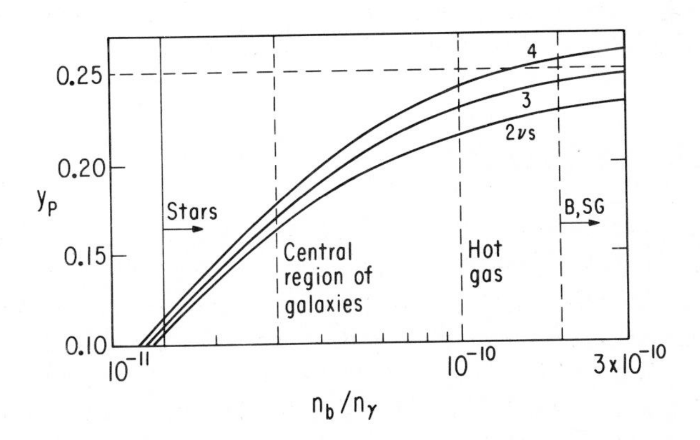


Fig. 2. Primordial $^4$He abundance, $Y_p$, versus baryon to photon ratio for different numbers of low mass ($m_\nu \ll 1$ MeV) neutrino species.

populations one can argue that observations (cf. Rood et al. [45]) on $^3$He place a lower limit to the baryon density. This argument is substantially strengthened when it is remembered that the probable way D is reduced from its Big Bang value to its present observed abundance is via $D+p \rightarrow {}^3He+\gamma$. Thus excess D adds to primordial $^3$He to give the non-stellar produced $^3$He. From present limits this implies $n_b/n_\gamma \gtrsim 2 \times 10^{-10}$ or $\Omega_b > 0.02$ which is not too different from the limit obtained using $\Omega_{B,SG}$. The one loop hole is the possibility that some pregalactic stellar generation burns up $^3$He without making other excesses. (These arguments are discussed in detail by Yang et al. [46].) It is obvious that the primordial helium-3 and -4 abundance are extraordinarily important to the understanding of this problem. It is important to note that this lower limit on $n_b/n_\gamma$ does not allow $Y$ to be less than 0.23 for 3 neutrino species (see fig. 2). If $Y$ were found to be lower then we would be forced to say that $^3$He was destroyed by some unknown early process and we would have to find some other lower limit on $n_b/n_\gamma$. A clear lower limit is from stellar matter but that only gives $\rho_b \gtrsim 10^{-32}$ g cm$^{-3}$ ($\Omega \gtrsim 10^{-3}$), which gives no limit to the number of neutrinos (Olive et al. [44]). If we note that the centers of galaxies are probably baryons then we can use internal galactic dynamics to argue that $\rho_b \gtrsim 5 \times 10^{-32}$ g cm$^{-3}$. This limit, coupled with $Y \leqslant 0.25$ allows a total of nine two-component neutrinos (e, $\mu$, $\tau$ and six more). If the neutrinos have Dirac masses and thus four components, the limit becomes four, and e, $\mu$, $\tau$ and one more are again all that are allowed. However, this limit of $5 \times 10^{-32}$ from internal galactic dynamics has some observational uncertainties. If it goes much lower then no limits are obtainable for $Y < 0.25$. Hot X-ray emitting gas from clusters argues that $n_b/n_\gamma \gtrsim 10^{-10}$ but there are possible loop holes depending on the universality of the gas.

Before leaving the subject, it is important to note that the ideas are testable by experiment. In particular, if the width of the neutral intermediate vector boson, $Z^0$, is measured, then it will tell us the number of neutrino flavors. It is fascinating that one of the most important tests of our cosmological ideas will come from accelerators rather than telescopes. Table 3 summarizes some tests and key observations regarding our model or the very early universe.

## 6. Conclusion

To date, the interdisciplinary effort involving cosmologists and nuclear physicists has produced some exciting results. The $^4$He abundance fixes

Table 3
Experimental and observational tests

| |
|---|
| *Particle physics* |
| —Lifetime of proton and branching ratio of decay, electric dipole moment of the neutron |
| -*CP* violation of GUTs |
| -Width of $Z^0$ |
| -Mass of neutrinos |
| *Astronomy* |
| -Primordial $^4$He and $^3$He abundances |
| -Is $\Omega \sim \Omega_{B,SG}$ or is $\Omega \sim \Omega_{cluster}$ or is $\Omega \sim \Omega_{B,SG} \sim \Omega_{cluster}$? |
| -Are there gravitational lens effects through galactic halos? |
| -Are galaxy formation perturbations and clustering characteristics adiabatic or isothermal? What are the anisotropies in the 3 K radiation? |

an upper limit on neutrino and quark flavors. The Grand Unification ideas may resolve the puzzle of one baryon for every $10^{10}$ photons. If neutrinos have mass then the bulk of the mass of the universe may be in the form of leptons. In fact one might say that we have taken the Copernican principle to the extreme. Copernicus showed that the earth is not the center of the solar system, Shapley showed that the sun is not the center of the galaxy and Hubble showed that our galaxy is not at the center of the universe—there is no center. Now, if neutrinos have mass, then our kind of matter may not even be the dominant matter of the universe. As our knowledge of the fundamental particles and their interactions increases, and as our determination of cosmological observables improves (or new observables are discovered) the close relationship of these two disciplines promises to continue to be an exciting one.

## Acknowledgements

I would like to thank my collaborators Mike Turner, Gary Steigman, Jim Fry, Keith Olive, Matt Crawford, Eugene Symbalisty, Jongmann Yang, and Bob Wagoner for enabling me to draw on much jointly generated material, and Hubert Reeves for the discussion leading to appendix 1. This work was supported by NSF grant AST 78-20402, by NASA grant NSG 7212 and by DOE grant DE AC02-80ER10773, all at the University of Chicago.

## Appendix 1. Alternative estimate of the proton lifetime.*

It is said that the Bhudda was once asked how long the universe would last. He replied that it would last as long as it would take an old man to rub away a mountain taller than the highest on earth by making one stroke at the mountain with a silk handkerchief every 100 years. To quantify that into modern units let us assume an old man's stroke contains $\sim 10^3$ erg and that the fraction of energy going into rubbing as opposed to kinetic energy is $\sim 10^{-3\pm1}$. (The uncertainty comes from our lack of quantitative knowledge of the friction of silk against rock.) The typical binding energy of atoms in minerals is about 0.5 eV which leads to $\sim 10^{12\pm1}$ atoms being rubbed off per stroke. If we take the size of the large mountain to be $\sim 10$ km high with a base of $\sim 10^3$ km$^2$ and assume it has an average density of $\sim 3$ and an average atomic weight of 20, then it contains $\sim 10^{42}$ atoms. Thus the age is $\sim (10^{42}/10^{12\pm1})\times 100$ yr $= 10^{32\pm1}$ yr.

## References

[1] A.A. Penzias and R.W. Wilson, A measurement of excess antenna temperature at 4080 Mc/s, Astrophys. J. 142 (1965) 419–421.

[2] D.P. Woody, J.C. Mather, N.S. Nishioka and P.L. Richards, Measurement of the spectrum of the submillimeter cosmic background, Phys. Rev. Lett. 34 (1975) 1036–1039.

[3] D.N. Schramm and R.V. Wagoner, Element production in the early Universe, Ann. Rev. Nucl. Sci. 27 (1977) 37–74.

[4] S.W. Hawking and G.F.R. Ellis, The large scale structure of space–time (Cambridge Univ. Press, 1973).

[5] D.N. Schramm, Particle physics in the very early Universe, 10th Texas Symposium on Relativistic Astrophysics, Ann. NY Acad. Sci. 375 (1981) Dec., 54–58.

[6] J.B. Hartle and B.L. Hu, Quantum effects in the early Universe. II. Effective action for scalar fields in homogeneous cosmologies with small anisotropy, Phys. Rev. D20 (1979) 1772–1782.

[7] G. Cocconi, Big and smaller bangs suggesting new physics, CERN preprint (1980).

[8] G. Steigman, Cosmology confronts particle physics, Ann Rev. Nucl. Part. Sci. 29 (1979) 313–337.

[9] S. Weinberg, Cosmological production of baryons, Phys. Rev. Lett. 42 (1979) 850–853.

[10] D. Toussaint, S.B. Tremaine, F. Wilczek and A. Zee, Matter–antimatter accounting, thermodynamics, and black hole radiation, Phys. Rev. D19 (1979) 1035–1045.

[11] J.N. Fry, K.A. Olive and M.S. Turner, Evolution of cosmological baryon asymmetries, Phys. Rev. D22 (1980) 2953–2988.

*From a discussion with H. Reeves.

[12] E.W. Kolb and S. Wolfram, Baryon number generation in the early Universe, Nucl. Phys. B172 (1980) 224.
[13] G. Segre and M.S. Turner, Baryon generation, the K–M mechanism and minimal SU(5), EFI preprint 80-43 (1981).
[14] M.S. Turner and D.N. Schramm, The origins of baryons in the Universe, Nature 279 (1979) 303–305.
[15] M.S. Turner and D.N. Schramm, Cosmology and elementary particle physics, Physics Today 32 (1979) 42–48.
[16] J. Barrow and M.S. Turner, Baryonsynthesis and the origin of galaxies, EFI preprint 81-02 (1981).
[17] M.S. Turner, On the isotropy and homogeneity of the Universe. Nature 281 (1979) 549–550.
[18] J.P. Presskill, Cosmological production of superheavy magnetic monopoles, Phys. Rev. Lett. 43 (1979) 1365–1368.
[19] J.N. Fry and D.N. Schramm, Unification, monopoles and cosmology, Phys. Rev. Lett. 44 (1980) 1361–1364.
[20] K.A. Olive, The thermodynamics of the quark–hadron phase transition in the early Universe, EFI preprint 80-52 (1981); Nuclear Physics B190 (1981) 483–503.
[21] K.A. Olive, A quark signature in the nuclear fireball model of heavy ion collisions, Phys. Lett. 89B (1980) 299–302.
[22] D.N. Schramm, M. Crawford and K.A. Olive, Astrophysics perspectives on high energy nucleus–nucleus collisions, in: Proceedings of the 1st Workshop on Ultra-relativistic Nuclear collisions, LBL 8957 (1979) 241–260.
[23] J.E. Gunn, B.W. Lee, I. Lerche, D.N. Schramm and G. Steigman, Some astrophysical consequences of the existence of a heavy stable neutral lepton, Astrophys. J. 223 (1978) 1015–1031.
[24] G. Steigman, D.N. Schramm and J.E. Gunn, Cosmological limits to the number of massive leptons, Phys. Lett. 66B (1977) 202–204.
[25] R. Cowsik and J. McClelland, An upper limit on the neutrino rest mass, Phys. Rev. Lett. 29 (1972) 669–670.
[26] A.S. Szalay and G. Marx, Limit on the rest masses from big bang cosmology, Acta. Phys. Acad. Sci. Hung. 35 (1979) 113–129.
[27] P. Hut and K.A. Olive, a cosmological upper limit on the mass of heavy neutrinos, Phys. Lett. 87B (1979) 144–146.
[28] A. DeRujula and S.L. Glashow, Galactic neutrinos and UV astronomy, Phys. Rev. Lett. 45 (1980) 942–944.
[29] F.W. Stecker, Have massive cosmological neutrinos already been detected? Phys. Rev. Lett. 45 (1980) 1460–1462.
[30] M. Kimball, S. Bowyer and B. Jacobson, Astrophysical constraints on the radiative lifetime of neutrinos with mass between 10 and 100 $eV/c^2$, Phys. Rev. Lett. 46 (1981) 80–83.
[31] D.A. Dicus, E.W. Kolb, V.L. Teplitz and R.V. Wagoner, Limits from primordial nucleosynthesis on the properties of massive neutral leptons, Phys. Rev. D17 (1978) 1529–1538.
[32] S.W. Falk and D.N. Schramm, Limits from supernovae on neutrino radiative lifetimes, Phys. Lett. 79B (1978) 511–513.
[33] D.N. Schramm and G. Steigman, Relic neutrinos and the density of the Universe, Astrophys. J. 243 (1981) 1–7.

[34] D.N. Schramm and G. Steigman, A neutrino dominated Universe, GRG 13 (1981) no. 2.
[35] S.M. Faber and J.S. Gallagher, Masses and mass-to-light ratios of galaxies, Ann. Rev. Astron. Astrophys. 17 (1979) 135–187.
[36] P.J. Peebles, Astrophys J. 84 (1979) 730.
[37] J.R. Gott, J.E. Gunn, D.N. Schramm and B.M. Tinsley, An unbound Universe? Astrophys. J. 194 (1974) 543–553.
[38] J. Yang, D.N. Schramm, G. Steigman and R.T. Rood, Constraints on cosmology and neutrino physics from Big Bang nucleo-synthesis, Astrophys. J. 227 (1979) 697–704.
[39] S. Tremaine and J.E. Gunn, Dynamical role of light neutral leptons in cosmology, Phys. Rev. Lett. 42 (1979) 407–410.
[40] R. Bond, R. Efstatiov and J. Silk, Massive neutrinos and the large-scale structure of the Universe, Phys. Rev. Lett. 45 (1980) 1980–1984.
[41] M. Doroshkevich, Yu. Khlopov, R.A. Sunyaev, A.S. Szalay and Ya. B. Zel'dovich, Ann. NY Acad. Sci. 375 (1981) 32–42.
[42] H. Sato, 10th Texas Symposium on Relativistic Astrophysics, Ann NY Acad. Sci. 375 (1981) 43–53.
[43] J.R. Gott, Are heavy halos made of low mass stars? Princeton preprint (1981).
[44] K.A. Olive, D.N. Schramm, G. Steigman, M.S. Turner and J. Yang, Big bang nucleosynthesis as a probe of cosmology and particle physics, Astrophys. J. 246 (1981) no. 3, part 1.
[45] R.T. Rood, T.L. Wilson and G. Steigman, The probable detection of interstellar $^3He^+$ and its significance, Astrophys. J. 227 (1979) L97–110.
[46] J. Yang, M.S. Turner, G. Steigman, D.N. Schramm and K.A. Olive (1983) in preparation.
[47] S. Dimopoulos and L. Susskind, Baryon number of the universe, Phys. Rev. D18 (1978) 4500–4509.
[48] C. Hill, private communication.
[49] R.P. Kirshner, A. Oemler, Jr. and P.L. Schecter, Astrophys. J. 84 (1979) 951.
[50] K.A. Olive, D.N. Schramm and G. Steigman, Limits on new superweakly interacting particles from primordial nucleosynthesis, Nucl. Physics B180 (1981) 497–515.
[51] R. Kron, private communication.
[52] J. Ellis, M.K. Gaillard and D. Nanopoulos, Phys. Lett. 80B (1979) 360.
[53] M. Turner, Proc. Neutrino 81, Hawaii, eds. R.J. Cence, E. Ma and A. Roberts (HEP group, Hawaii, 1981) p. 95–113.
[54] B.W. Lee and S. Weinberg, Phys. Rev. Lett. 39 (1977) 165.
[55] D.N. Schramm and M. Crawford, Fermi Inst. preprint (1981).
[56] M. Crawford and A. Jan Kevicks, Proc. Neutrino 81, Hawaii, eds. R.J. Cence, E. Ma and A. Roberts (HEP group, Hawaii, 1981) p. 82–94.
[57] D. Kunth and W.L.W. Sargent, Astron. Astrophys. Suppl. 36 (1979) 259.

SEMINAR 2

# BIG BANG BARYOSYNTHESIS

Michael S. TURNER

*Institute for Theoretical Physics, University of California*
*Santa Barbara, CA 93106, USA*
*and*
*Astronomy and Astrophysics Center, University of Chicago**
*Chicago, IL 60637, USA*

## Contents



*Permanent address.

*M.K. Gaillard and R. Stora, eds.*
*Les Houches, Session XXXVII, 1981*
*Théories de jauge en physique des hautes énergies/Gauge theories in high energy physics*

## 0. Introduction

The Universe possesses an overt matter–antimatter asymmetry, quantified as the present baryon-density to photon-density ratio, $n_b/n_\gamma \approx (2\text{–}8)\times 10^{-10}$ [1]. Grand Unified Theories (GUTs) make the startling prediction that baryon number is not conserved; in most GUTs this implies that the proton is unstable (an example of a GUT in which the proton is stable is given in ref. [2]), with a lifetime of $O(10^{30})$ yr. If the Universe is open, then GUTs predict the ultimate demise of matter. Perhaps the most impressive achievement of GUTs is the possible explanation of the Baryon Asymmetry of the Universe (BAU), as being due to *B*-, *C*-, and *CP*-nonconserving processes which occurred $\sim 10^{-35}$ s after the "Bang". In fact, until proton decay is detected, the BAU is the only "experimental evidence" for baryon nonconservation.

In these two lectures I shall briefly review Big Bang baryosynthesis. In the first lecture I shall discuss the evidence which exists for the BAU, the failure of non-GUT symmetrical cosmologies, the qualitative picture of baryosynthesis, and numerical results of detailed baryosynthesis calculations. In the second lecture I shall discuss the requisite *CP* violation in some detail, further the statistical mechanics of baryosynthesis, possible complications to the simplest scenario, and one cosmological implication of Big Bang baryosynthesis. Although the standard hot Big Bang model is an underlying assumption for everything that I will discuss, I will not review the model or the impressive body of evidence which supports it. Instead I refer the reader to the excellent reviews in refs. [3,4].

## 1. Big Bang baryosynthesis I

### 1.1. An asymmetrical Universe

Although the laws of physics are, to a high degree of precision, matter–antimatter symmetrical at the microscopic level (the only observed *CP* violation being in the $K^0\text{–}\bar{K}^0$ system [5]), the Universe possesses an overt matter–antimatter asymmetry [6]. Galactic cosmic rays provide us with samples of material from throughout the galaxy; the ratio of protons to antiprotons is $\sim 3\times 10^{-4}$, and the ratio of $^4\overline{\text{He}}$ to $^4$He is $<10^{-5}$ [7]. Antiprotons are expected to be produced as cosmic ray secondaries (e.g., $\text{p}+\text{p}\rightarrow \text{p}+\text{p}+\text{p}+\bar{\text{p}}$, etc.) at about the $10^{-4}$ level. However, the observed antiprotons ($\sim 40$ in total) have a different spectrum than is expected if they are produced as secondaries—there are far too many low-energy $\bar{\text{p}}$'s. At the moment the observation is very puzzling, although a variety of potential explanations exist: the antiprotons have been decelerated since their production; the 'standard model' of cosmic ray propagation through the galaxy is wrong; or perhaps the experiment is wrong. It has also been suggested that the detected antiprotons are cosmic rays from antimatter galaxies [8]; this idea runs into a bit of trouble, because in that case one expects $^4\overline{\text{He}}/^4\text{He}\simeq \bar{\text{p}}/\text{p}$. Although the question of the origin of the cosmic ray antiprotons is far from being settled, it is clear that there is no evidence for any appreciable quantity of antimatter in our galaxy. Steigman has argued that the nearby ($d\approx 20$ Mpc, 1 Mpc $\approx 3\times 10^6$ $\ell$yr $\approx 10^{24}$ cm) Virgo cluster, which contains several hundred galaxies must not have any antimatter galaxies, otherwise we would observe a strong $\gamma$-ray flux from nucleon–antinucleon annihilations [6]. As we shall see shortly, this observation, that matter and antimatter (if any exists) are separated at least on scales of galaxies or clusters of galaxies, will be devastating for any baryon-symmetrical cosmology.

Baryons account for only a tiny fraction of the observed particles in the Universe; their ratio to the number of 3 K microwave photons (the most abundant particles in the Universe) is $n_\text{b}/n_\gamma \approx (2\text{–}8)\times 10^{-10}$, as determined from Big Bang nucleosynthesis [1]. Since apparently $n_{\bar{\text{b}}} \ll n_\text{b}$, this is also the baryon number-density to photon-density ratio. The number of photons in the Universe increased as various particle species present at earlier epochs (e.g., $\text{e}^\pm$ pairs, $\pi^\pm$ pairs, etc.) annihilated. However, if the expansion of the Universe has been isentropic (i.e. no bulk entropy production), then the total entropy has remained constant. The contribution of a relativistic species to the specific entropy density is $(4/3)\rho/kT\sim n$, where $\rho$ and $n$ are the energy and number density of the species. Today the 'known' entropy of the Universe is about evenly divided between the 3 K photons and the three cosmic neutrino backgrounds (e, $\mu$, $\tau$). The baryon-number/entropy ratio, and baryon/photon ratio are related by:

$$n_\text{b}/s \approx (1/7)n_\text{b}/n_\gamma \approx (3\text{–}10)\times 10^{-11}. \tag{1}$$

Throughout, I shall employ units in which $\hbar = c = k = 1$.

When baryon number is effectively conserved and the expansion is isentropic, this ratio ($n_B/s$) remains constant. Although the matter–antimatter asymmetry in the Universe is maximal today, $n_B/s \approx (3–10)\times 10^{-11}$ implies that earlier than $\sim 10^{-6}$ s after the 'Bang' the asymmetry was tiny. In the standard hot Big Bang [Friedmann–Robertson–Walker (FRW)] cosmological model, the age and temperature of the Universe during the radiation-dominated epoch ($t \lesssim 1\times 10^{10}\Omega^{-2}h^{-4}$ s) are related by:

$$H^2 = (\dot{R}/R)^2 = (\dot{T}/T)^2 = (2t)^{-2} = \tfrac{4}{45}\pi^3 g_* T^4/m_P^2, \qquad (2a)$$

$$t = 2.4\times 10^{-6} g_*^{-1/2}(T/1\ \mathrm{GeV})^{-2}\ \mathrm{s}. \qquad (2b)$$

$\Omega$ is the ratio of the present energy density to the critical energy density, the Hubble parameter $H_0 = 100h$ km s$^{-1}$ Mpc$^{-1}$ ($1/2 \leqslant h \leqslant 1$), $m_P = G^{-1/2} \approx 1.2\times 10^{19}$ GeV is the Planck mass, and $g_*(T)$ counts the total number of effective degrees of freedom of all the relativistic species at temperature $T$ (i.e. those species with $m \ll T$):

$$g_* = \sum_{\text{bose}} g + \tfrac{7}{8}\sum_{\text{fermi}} g. \qquad (3)$$

For $t \lesssim 10^{-6}$ s, $T \gtrsim 1$ GeV and baryons and antibaryons were about as abundant as photons, $n_b \approx n_{\bar{b}} \approx n_\gamma$. The number of relativistic species $g_*$ was O(10), so that $s \sim g_* n_\gamma \sim 10 n_\gamma$. The present baryon/entropy ratio implies that for $t \lesssim 10^{-6}$ s, $(n_{\bar{b}} - n_b)/n_b \approx 10^{-9}$; at the earliest epochs the Universe was very nearly (but not quite) symmetrical!

### 1.2. *Failure of the non-GUT symmetrical cosmologies*

The observation that $n_b \gg n_{\bar{b}}$ appears to deal a deathblow to symmetrical cosmologies. However, one might argue that while locally $n_b \gg n_{\bar{b}}$, globally $n_b \equiv n_{\bar{b}}$ due to domains where $n_{\bar{b}} \gg n_b$. As I shall describe, the locally measured BAU makes even this possibility nearly untenable.

#### 1.2.1. *Locally symmetrical cosmology*

Suppose that initially the Universe was locally baryon symmetrical ($n_b \equiv n_{\bar{b}}$). For $t \lesssim 10^{-6}$ s, T $\gtrsim 1$ GeV $\gtrsim m_b$, and so baryons, antibaryons and photons were about equally abundant. When the temperature fell below $\sim m_b$ ($t \gtrsim 10^{-6}$ s) the equilibrium abundance of baryons and antibaryons was a factor of $\sim \exp(-m_b/T)$ smaller than that of the photons. Baryons and antibaryons annihilated faster than they were created in pairs, and naively one might expect this to have continued until $n_b/n_\gamma = n_{\bar{b}}/n_\gamma = 0$. However, for $T \lesssim 20$ MeV baryons and antibaryons were so rare that they could no longer find each other to annihilate, and their annihilations effectively ceased. In this model, due to the incompleteness of annihilations, one expects that today $n_b/n_\gamma = n_{\bar{b}}/n_\gamma \approx 10^{-18}$, which, even if one could subsequently separate matter from antimatter is a factor of $\sim 10^8$ too small—a discrepancy even in astrophysics [6]. To avoid the 'annihilation catastrophe' one must violate the condition of local baryon symmetry for $T \gtrsim 20$ MeV.

### *1.2.2. Statistical fluctuations*

One possible mechanism for doing this is statistical fluctuations (Poisson). Consider the comoving volume which contains our galaxy. Today, it contains $\sim 10^{12}\ M_\odot$ of baryons ($\approx 10^{69}$ baryons) and $\sim 10^{79}$ photons (1 $M_\odot \approx 2\times 10^{33}$ g $\approx 10^{57}$ baryons). For $t \lesssim 10^{-6}$ s ($T \gtrsim 1$ GeV), this comoving volume contained $\sim 10^{79}$ photons, $\sim 10^{79}$ baryons and $\sim 10^{79}$ antibaryons; the net baryon number of this volume must have been $n_b - n_{\bar{b}} \approx 10^{69}$ to avoid the 'annihilation catastrophe'. One would expect a net baryon number due to Poisson fluctuations of order $n_b^{1/2} \approx 3\times 10^{39}$—a mere $29\frac{1}{2}$ orders of magnitude too small!

### *1.2.3. Causality*

Clearly statistical fluctuations do not help, so consider a hypothetical interaction which separates matter and antimatter. In the FRW model the distance over which light signals (and hence causal effects) could have propagated since the 'Bang' is finite, and $\approx ct$. When $T \approx 20$ MeV ($t \approx 10^{-3}$ s), causally coherent regions encompassed only $\sim 10^{-5}\ M_\odot$. Therefore, causal processes could only have separated out $\lesssim 10^{-5}\ M_\odot$ chunks of matter and antimatter.

It should now be clear that the two observations, $n_b \gg n_{\bar{b}}$ and $n_b/n_\gamma \approx (2\text{–}8)\times 10^{-10}$, effectively render all conventional symmetrical cosmologies untenable. Until a few years ago it was necessary to specify $n_B/s \approx (3\text{–}10)\times 10^{-11}$ as an initial condition in the Standard Model—a very peculiar one at that! As most of you are aware, there now exist very attractive dynamical explanations for the BAU which involve GUTs. Baryosynthesis is probably the most significant development in cosmology since Big Bang nucleosynthesis.

## *1.3. Baryosynthesis—Qualitative picture*

As early as 1967 Sakharov and others [9] suggested that an initially baryon symmetrical Universe might dynamically evolve a baryon excess of $O(10^{-10})$, which, after essentially all the antibaryons and most of the baryons annihilated ($T \approx 1$ GeV–20 MeV), would leave the ~ one baryon per $10^{10}$ photons that we observe today.

As Sakharov pointed out in 1967 three ingredients are necessary for baryosynthesis: (i) $B$-nonconserving interactions—which of course are predicted by GUTs (for an excellent review, see ref. [10]); (ii) $C$- and $CP$-violations, an arrow to specify that a matter excess be produced—such violations exist in the $K^0$–$\bar{K}^0$ system, and can easily be accommodated by GUTs; (iii) a departure from thermal equilibrium. A heuristic argument for the necessity of (iii) goes as follows: equilibrium distributions are of the form $n(\boldsymbol{p}) = [\exp(\mu/T + E/T) \pm 1]^{-1}$; this follows from $CPT$ and unitarity alone. In equilibrium, processes like $\gamma + \gamma \rightleftarrows b + \bar{b}$ guarantee $\mu_b = -\mu_{\bar{b}}$, while processes like (but not literally) $b + b \rightleftarrows b + b + b$ require $\mu_b = 0$, and so $\mu_b = \mu_{\bar{b}} = 0$. Since $E^2 = p^2 + m^2$ and $m_b = m_{\bar{b}}$ ($CPT$), it follows that in equilibrium $n_b \equiv n_{\bar{b}}$.

### 1.3.1. *The out-of-equilibrium decay scenario*

The basic idea of baryosynthesis has been discussed by many authors [11]; the model which incorporates all three ingredients and has become the 'standard scenario' is the so called out-of-equilibrium decay scenario of Weinberg [12] and Wilczek [13], which I shall now describe in some detail.

For $t \gtrsim t_{\text{Planck}} \approx 10^{-43}$ s, quantum corrections to general relativity should be small, and the dynamics of the expansion should be accurately described by eq. (2). Although particle number densities should have been $\sim 10^{99}$ cm$^{-3}$ at $t \approx t_{\text{Planck}}$, the asymptotic freedom exhibited by gauge theories allows one to treat the early Universe as an ideal gas of weakly interacting, point-like particles. The expansion rate of the Universe [eq. (2)] sets the timescale for thermodynamics since $|\dot{T}/T| \sim H \sim t^{-1}$, and is given by

$$H = 1.66 g_*^{1/2} T^2 / m_P. \tag{4}$$

I shall denote by 'S', the superheavy ($\gtrsim 10^{14}$ GeV) boson whose interactions violate $B$ conservation. S might either be a gauge boson or a Higgs boson. Let its coupling strength to fermions and its mass be $\alpha$ and $M$, respectively. From dimensional considerations, its decay rate $\Gamma_D \sim \tau_S^{-1}$ should be given by

$$\Gamma_D \sim \alpha M. \tag{5}$$

At the Planck time the Universe is assumed to be baryon symmetrical ($n_{\mathrm{b}} = n_{\bar{\mathrm{b}}}$, $n_{\mathrm{B}}/s \equiv 0$), with all the fundamental particle species (fermions, gauge and Higgs bosons) present in equilibrium distributions. (The validity of the assumption of thermal equilibrium will be discussed in more detail in my second lecture.) The temperature at $t_{\mathrm{Planck}}$ is $T \approx g_*^{-1/4} 10^{19}$ GeV $\approx 3 \times 10^{18}$ GeV since $g_* \approx O(100)$. [In minimal SU(5), $g_* = 160$.] The temperature $T \gg M$, so S, $\bar{\mathrm{S}}$ bosons are ultrarelativistic and essentially as abundant (up to statistical factors) as photons. Nothing of interest occurs until $T$ falls to $T \lesssim M$.

For $T \lesssim M$, the equilibrium abundance of S, $\bar{\mathrm{S}}$ bosons relative to photons is:

$$N_{\mathrm{S}} \equiv n_{\mathrm{S}}/n_{\gamma} \approx (M/T)^{3/2} \exp(-M/T). \tag{6}$$

($N_{\mathrm{S}}$ is also the number of S, $\bar{\mathrm{S}}$ bosons per comoving volume.) In order for S, $\bar{\mathrm{S}}$ bosons to maintain an equilibrium abundance as $T$ falls below $M$, they must be able to diminish in number rapidly compared to $H \sim |\dot{T}/T|$. The most important process in this regard is decay; other processes (e.g., annihilation) are higher order in $\alpha$. If $\Gamma_{\mathrm{D}} \gg H$ for $T \approx M$, then S, $\bar{\mathrm{S}}$ bosons can adjust their abundance (by decay) rapidly enough so that $N_{\mathrm{S}}$ 'tracks' the equilibrium value. In this case thermal equilibrium is maintained; condition (iii) is not satisfied, and no asymmetry evolves.

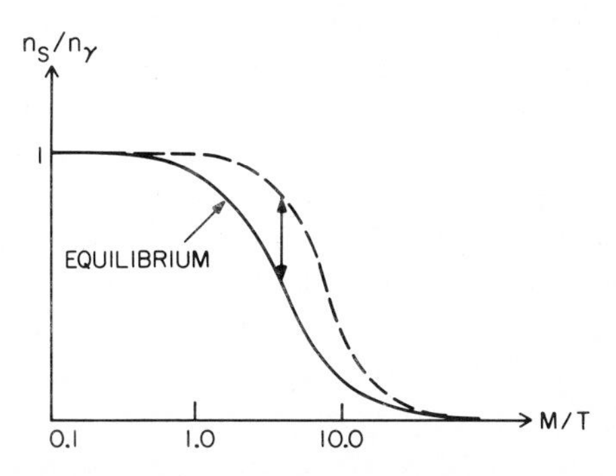


Fig. 1. The abundance of S, $\bar{\mathrm{S}}$ bosons relative to photons as a function of $M/T$. The solid curve shows the equilibrium value of $N_{\mathrm{S}} = n_{\mathrm{S}}/n_{\gamma}$, which, for $T \lesssim M$, is $\sim (M/T)^{3/2} \times \exp(-M/T)$. The dashed curve shows the actual abundance. If $\Gamma_{\mathrm{D}} < H$ for $T = M$, then S, $\bar{\mathrm{S}}$ bosons do not diminish rapidly enough to maintain an equilibrium abundance and are overabundant (indicated by arrow). This is the departure from equilibrium which drives baryosynthesis in the out-of-equilibrium decay scenario.

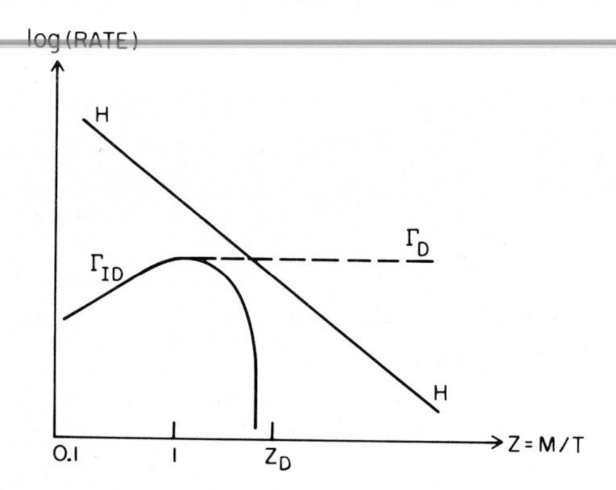


Fig. 2. The key rates for baryosynthesis as a function of $z = M/T$. The expansion rate $H \propto T^2$; for $T > M$, decays and inverse decays are suppressed by time dilation factors, and $\Gamma_D \approx \Gamma_{ID} \propto T^{-1}$. For $T \lesssim M$, $\Gamma_D \sim \tau_S^{-1}$, and $\Gamma_{ID} \sim \Gamma_D \exp(-M/T)$ since typical fermion pairs are not energetic enough to produce an S or $\bar{\mathrm{S}}$. If $\Gamma_D < H$ when $T = M$, then S,$\bar{\mathrm{S}}$ bosons will become overabundant. Eventually, when $\Gamma_D \approx H$ ($z = z_D$) they freely decay, inverse reactions being blocked by the $\exp(-M/T)$ factor.

More interesting is the case where $\Gamma_D < H$ for $T \approx M$, or equivalently $M > g_*^{-1/2} \alpha \times 10^{19}$ GeV. In this case S, $\bar{\mathrm{S}}$ bosons are *not* decaying on the expansion timescale ($\tau_S > t$), and so remain as abundant as photons ($N_S \approx 1$) for $T \approx M$, and hence are overabundant relative to their equilibrium number. This overabundance is the requisite departure from thermal equilibrium. Much later when $T \ll M$, $\Gamma_D \gtrsim H$ (i.e., $t \gtrsim \tau_S$), and S, $\bar{\mathrm{S}}$ bosons begin to disappear by decay. To a good approximation they decay freely, since the fraction of fermion pairs with CM energy $\gtrsim M$ is $\sim \exp(-M/T)$, thus greatly suppressing inverse decay, $\Gamma_{ID} \approx \exp(-M/T)\Gamma_D \ll H$. The time evolution of $N_S$ is summarized in fig. 1; figure 2 shows the relationship of the various rates ($\Gamma_D$, $\Gamma_{ID}$, and $H$) as a function of $z = (M/T) \sim t^{1/2}$.

Now consider the decay of S and $\bar{\mathrm{S}}$ bosons: suppose S decays to channels 1 and 2 with baryon numbers $B_1$ and $B_2$, and branching ratios $r$ and $(1-r)$. Denote the corresponding quantities for $\bar{\mathrm{S}}$ by $-B_1$, $-B_2$, $\bar{r}$, and $(1-\bar{r})$. [E.g., 1 = ($\overline{\mathrm{qq}}$), 2 = (q$\ell$), $B_1 = -2/3$, and $B_2 = 1/3$.]

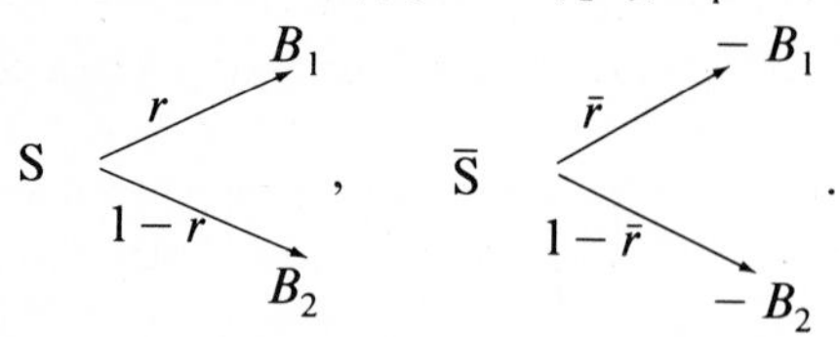

Then the mean net baryon numbers of the decay products of the S and $\bar{S}$ are respectively:

$$B_S = rB_1 + (1-r)B_2, \qquad B_{\bar{S}} = -\bar{r}B_1 - (1-\bar{r})B_2. \tag{7}$$

Hence the decay of an S, $\bar{S}$ pair on average produces a baryon number $\varepsilon$,

$$\varepsilon \equiv B_S + B_{\bar{S}} = (r-\bar{r})(B_1 - B_2). \tag{8}$$

Unless both $C$ and $CP$ are violated, $r = \bar{r}$. The quantity $\varepsilon$ measures the magnitude of the $C$-, $CP$-violation.

When the S, $\bar{S}$ bosons decay ($T \ll M, t \approx \tau_S$), $n_S = n_{\bar{S}}$, and they are still about as abundant as photons. Therefore, the total baryon excess produced is $n_B \approx \varepsilon n_\gamma$. The entropy is proportional to the number of relativistic particles present, and so $s \sim g_* n_\gamma$. The baryon asymmetry produced is:

$$n_B/s \approx \varepsilon/g_* \approx 10^{-2}\varepsilon, \tag{9}$$

since $g_* \sim O(10^2)$.

Recall that the condition for a departure from equilibrium to occur is: $\Gamma_D < H$ $(T \approx M)$ or $M > g_*^{-1/2}\alpha \times 10^{19}$ GeV $\approx \alpha \times 10^{18}$ GeV. If S is a gauge boson, then $\alpha \approx 1/45$, so that $M$ must be $M \gtrsim 10^{16}$ GeV. If S is a Higgs boson, then $\alpha \approx \alpha_{gauge}(m_{fermion}/M_W)^2 \approx 10^{-4}-10^{-6}$, so that $M$ must be $M \gtrsim 10^{12}-10^{14}$ GeV. If $M \gtrsim$ "the critical mass", then only a modest $CP$ violation ($\varepsilon \approx 10^{-8}$) is required to explain $n_B/s \approx (3\text{–}10)\times 10^{-11}$. As I shall discuss later, even when $M \lesssim$ "the critical mass", an asymmetry of non-negligible magnitude evolves.

### *1.4. Numerical results*

Two groups have developed numerical codes to follow the evolution of the BAU in detail [14–18]. The codes integrate the relevant Boltzmann equations with the following assumptions being made:

–The dynamics of the Universe are described by the radiation-dominated hot Big Bang model (FRW cosmology).

–$B$-conserving interactions are assumed to be happening rapidly enough ($\Gamma \gg H$) so that the distribution of all particle species in momentum space is thermal (*kinetic* equilibrium), although some species might be over- or underabundant (*chemical* equilibrium is not necessarily maintained).

–Interactions included in the network are: decays, inverse decays, and annihilations of superheavy bosons; $B$-nonconserving two-body collisions mediated by superheavy bosons.

–GUTs are used to compute the required matrix elements.

Thus far SU(5) [15–18] and SO(10) [15] have been studied. The two groups reach essentially the same conclusions. The results for a specific model can be fairly complicated and not particularly illustrative. Therefore, I will restrict my discussion to the effect of a single gauge or Higgs species on the evolution of the BAU.

It is extremely useful to define the quantity $K$ which describes the "effectiveness" of superheavy boson decays (and as I shall discuss in the second lecture, all the interactions involving the superheavy boson) [16–18]:

$$K = (\Gamma_{\mathrm{D}}/H)_{T=M} \approx 3\times 10^{17}\alpha\ \mathrm{GeV}/M, \tag{10}$$

where the numerical value applies specifically to minimal SU(5). For $K<1$, decays are *not* occurring rapidly on the expansion timescale at the critical epoch ($T\approx M$), and S, $\bar{\mathrm{S}}$ bosons become overabundant—supplying the requisite departure from equilibrium. On the other hand, for $K>1$, decays are occurring rapidly, and one expects equilibrium to be nearly maintained.

In fig. 3 the final value of $n_{\mathrm{B}}/s$ produced by a single gauge (3a) or Higgs (3b) species is shown as a function of $K$. There are several interesting features to take note of.

–For $K\lesssim 1$, the BAU which evolves is roughly independent of $K$ and equal to the "saturation value", $n_{\mathrm{B}}/s\approx 10^{-2}\varepsilon$—in agreement with the qualitative discussion in subsection 1.3.1.

–For $K>1$, one might have expected equilibrium to be very precisely maintained, and no baryon asymmetry to evolve—this is certainly not the case (cf. fig. 3). Baryosynthesis only falls off as $\sim K^{-1.3}$ (approximately). The reason for this is that equilibrium is difficult to maintain because the temperature of the Universe is always decreasing (this will be discussed in more detail in lecture 2).

–A gauge boson of mass $\sim 5\times 10^{14}$ GeV can produce the observed BAU if $\varepsilon\approx 10^{-4}$. A Higgs boson of mass $\gtrsim 3\times 10^{13}$ GeV and coupling $\alpha_{\mathrm{H}}\approx 10^{-4}$ can produce the observed BAU if $\varepsilon\approx 10^{-8}$.

Both gauge and Higgs bosons can damp pre-existing baryon asymmetries (i.e. those present prior to the GUT epoch). The damping is primarily by the two-step process: ID, D (e.g., $\mathrm{qq}\rightarrow\bar{\mathrm{S}}, \bar{\mathrm{S}}\rightarrow\overline{\mathrm{q}\ell}, \Delta B=-1$). The damping factor, $(n_{\mathrm{B}}/s)_{\mathrm{final}}/(n_{\mathrm{B}}/s)_{\mathrm{initial}}\sim \exp(-AK)$, where $A\sim$ O(1). This is discussed in more detail in refs. [16–18].

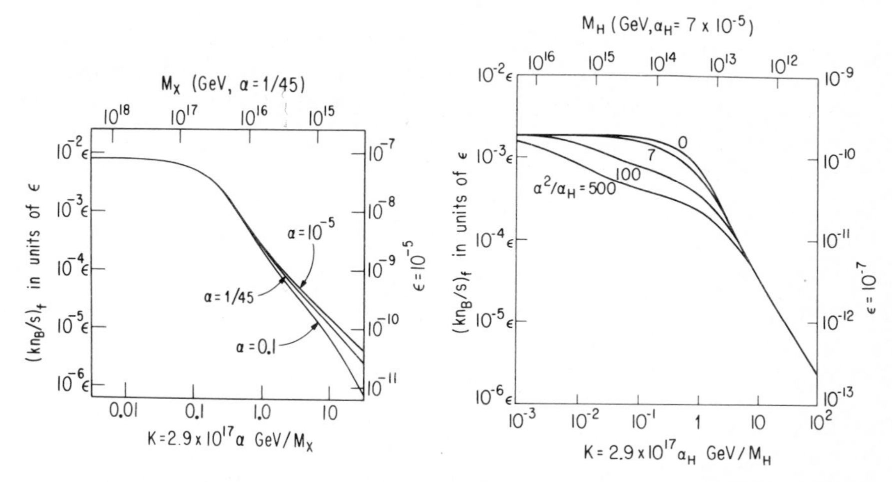


Fig. 3. The final value of $n_B/s$ produced by a superheavy gauge boson (a) or Higgs boson (b) in units of $\varepsilon$ as a function of $K$. (a) For fixed $K$ there is little dependence upon $\alpha$; the slight dependence upon $\alpha$ is due to $\not{B}$ scatterings [16]. (b) For Higgs bosons the coupling to fermions is $\alpha_H \sim \alpha_G(m_f/M_W)^2$, while for Higgs annihilations the coupling is $\alpha_G$. The dependence upon $\alpha_G^2/\alpha_H$ shows the small effect of Higgs annihilations on baryon production—the annihilations help to maintain an equilibrium abundance of Higgs bosons [17].

## 2. Big Bang baryosynthesis II

### 2.1. *The CP violation $\varepsilon$*

#### 2.1.1. *General remarks*

Since the baryon asymmetry which evolves scales with $\varepsilon$, a number of questions naturally arise, such as: Can $\varepsilon$ be calculated? What is the sign of $\varepsilon$? Can $\varepsilon$ be related to the $K^0$–$\bar{K}^0$ system (or some other parameter which can be measured at 'low energies'), etc. Recall from the first lecture that $\varepsilon = (r - \bar{r})(B_1 - B_2)$, where $r$ is the branching ratio of S → channel 1 and $\bar{r}$ is the branching ratio of $\bar{S}$ → channel $\bar{1}$ [cf. eq. (8)]. It is easy to show that $\varepsilon$ vanishes if the theory is $C$- or $CP$-symmetric. Let $B_{up}$ be the mean net $B$ number of the upward moving decay products of S, $B_{down}$ be the mean net $B$ number of the downward moving decay products of S, etc.:

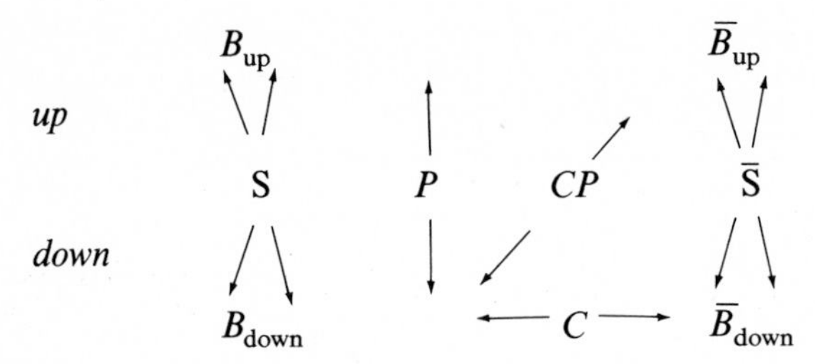


Suppose that the theory is $CP$ symmetric, but not C symmetric, then $B_{up} = -\bar{B}_{down}$ and $B_{down} = -\bar{B}_{up}$. The net baryon number produced in the up direction is $\varepsilon_{up} = B_{up} + \bar{B}_{up}$ and in the down direction $\varepsilon_{down} = B_{down} + \bar{B}_{down} = -\varepsilon_{up}$, so that averaged over all directions $\varepsilon = 0$. If the theory is $C$ symmetric, but not $CP$ symmetric, then $B_{up} = -\bar{B}_{up}$ and $B_{down} = -\bar{B}_{down}$. In this case $\varepsilon_{up} = \varepsilon_{down} = 0$, so that again $\varepsilon = 0$.

Under $C$ (charge conjugation) the fermion field, $\psi_{L\,(R)} \to \psi^c_{L\,(R)}$, and under $CP$ (charge conjugation composed with parity) $\psi_{L\,(R)} \to \gamma_0 \psi^c_{R\,(L)}$. Here $\psi_{L(R)}$ is the field which annihilates a left (right) handed particle, or creates a right (left) handed antiparticle, and $\psi^c_{L\,(R)}$ is the field which annihilates a left (right) handed antiparticle, or creates a right (left) handed particle. Under $C$ a lefthanded current $(\bar{\psi}_{1L}\gamma_\mu\psi_{2L}) \to$ righthanded current $(\bar{\psi}_{2R}\gamma_\mu\psi_{1R})$, so that a $C$-symmetric theory must be LR symmetric. SO(10) is a $C$-symmetric theory. This is easy to see. In SO(10) the fermions are placed in the 16-dimensional spinor representation, e.g., the first generation:

$$\mathbf{16}_L = (u^1_L u^2_L u^3_L d^1_L d^2_L d^3_L e^-_L \nu_{eL} \bar{u}^1_L \bar{u}^2_L \bar{u}^3_L \bar{d}^1_L \bar{d}^2_L \bar{d}^3_L e^+_L \bar{N}_{eL}), \tag{11}$$

where $N_{eR}$ is the righthanded partner of $\nu_{eL}$. Under $C$ $u^i_L \leftrightarrow \bar{u}^i_L$, $d^i_L \leftrightarrow \bar{d}^i_L$, $e^-_L \leftrightarrow e^+_L$, and $\nu_{eL} \leftrightarrow \bar{N}_{eL}$, and so $C$ just corresponds to a gauge transformation. Hence, an unbroken SO(10) theory is $C$ symmetric. Since SU(5) is not LR symmetric, it is not $C$ symmetric [19,20]. SU(5) and SO(10) are not in general $CP$ symmetric.

In general, $\varepsilon$ must be $\lesssim O(\alpha_G) \approx 10^{-2}$ [21–24]. This is because at the tree graph level, superheavy boson decays are $CP$ conserving. Put another way, in the Born approximation the Lagrangian is hermetian. Any $C$- and $CP$-violating effects arise from the imaginary part of the interference between the tree graph and higher order loop diagrams (see fig. 4, for example). If $\ell$ is the number of loops where at lowest order the interference term is nonzero, then $\varepsilon \approx O(\alpha^\ell)$, where $\alpha$ is the coupling strength

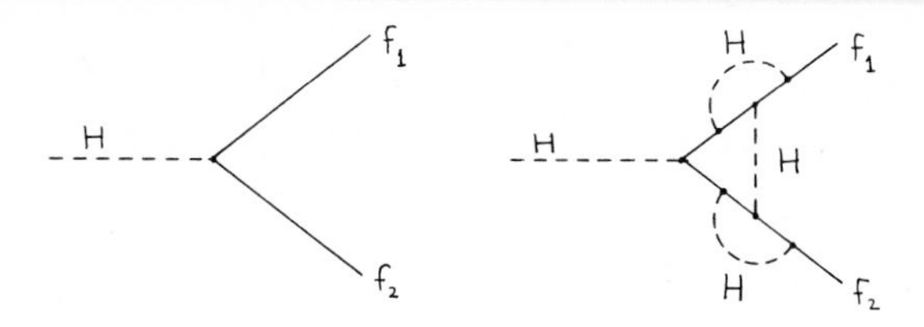


Fig. 4. The tree graph for superheavy Higgs decay and a three-loop diagram. In minimal SU(5) their interference is responsible for the lowest order *C*- and *CP*-violating effects in Higgs decay.

of the boson in the loop. In fact $\varepsilon \sim O(\alpha_H^\ell)$, where $\alpha_H^\ell$ is the appropriate combination of Higgs couplings; $\alpha_H \approx \alpha_G (m_f/M_W)^2$. This is easy to see: in the interaction (rather than mass eigenstate) basis the gauge–fermion couplings are real, and hence the requisite complexity must come from fermion–Higgs couplings. Thus the baryon asymmetry generated can be as large as $n_B/s \approx 10^{-2}\varepsilon \approx 10^{-2}\alpha_H \lesssim 10^{-4}$, which leaves plenty of leeway to produce the observed BAU, $n_B/s \approx (3–10)\times 10^{-11}$!

*2.1.2. SU(5)*

Consider the minimal SU(5) model (one **5** and one **24** of Higgs) with the usual three generations of fermions. Both the XY gauge bosons (color triplet, isodoublet), and the H which is the color triplet, isosinglet member of the **5** of Higgs are superheavy and mediate *B* nonconversation. In the H system the first nonzero contribution to $\varepsilon_H$ is from the interference of the tree graph and a three loop diagram (fig. 4), and

$$\varepsilon_H \approx 2^{-9}\alpha^3 (m_f/M_W)^6 f(\theta_i)\sin\gamma, \tag{12}$$

where $2^{-9}$ is a "factor of O(1)" which results from various loop integrals, etc., $m_f^6$ is a specific combination of fermion masses, $f(\theta_i)$ is a product of $\sin\theta_i$, $\cos\theta_i$ factors where $\theta_i$ are mixing angles, and $\gamma$ is a phase [22,24,25]. The phase $\gamma$ can arise in two ways: (i) from a mixing matrix which only involves the righthanded fermion fields and only affects the XY–fermion couplings, and so $\gamma$ is only observable at GUT energies; (ii) from the usual phase in the K–M matrix. In case (i), $f(\theta_i) = \sin^2 2\theta$ where $\theta$ is the mixing between the second and third generations, and $\sin^2 2\theta \sin\gamma$ can be of O(1); $m_f^6 = m_b^4 m_t m_c$ [24,25]. We then have $\varepsilon_H \approx 4\times 10^{-15}(m_t/M_W)$, which is clearly unacceptably small. In case (ii), $f(\theta_i)\sin\gamma \approx \sin\theta_1 \sin 2\theta_1 \sin 2\theta_2 \sin 2\theta_3 \sin\gamma$, which is certainly $\lesssim 10^{-3}$; $m_f^6 = m_t m_b^2 m_s^2 m_u$ [25]. We then have $\varepsilon_H \lesssim 4\times 10^{-22}$ $(m_t/M_W)$, which is also clearly unacceptably small.

One suggestion is to introduce a fourth (and possibly fifth) generation of heavy quarks ($m_f \approx 30$–$200$ GeV) [25]. Denote these new generations by (t′,b′) and (t″,b″) respectively. In case (i) with four generations, $\sin^2 2\theta \sin\gamma$ can again be of O(1), while $m_f^6 = m_{b'}^4 m_{t'} m_t$, so that $\varepsilon_H \approx 2\times 10^{-8}(m_{b'}^4 m_{t'} m_t)/M_W^6$ which can be just large enough to give $n_B/s \approx (3–10)\times 10^{-11}$ if $m_{b'}, m_{t'}, m_t \gtrsim M_W$. (One has to be a bit careful, since the masses which go into (12) are not the usual quark masses, but are probably smaller by a factor of $\sim 3$ due to the renormalization of fermion masses from $Q^2 = M_{GUT}^2$ to $Q^2 \approx 0$ [26].)

In case (ii), with four generations $\varepsilon_H$ is still too small, however with five generations $\varepsilon_H$ can be just large enough. Since the mixing angles and K–M phase involve only the three heaviest generations, $f(\theta_i)\sin\gamma$ can be of O(1); $m_f^6 = m_{b''}^2 m_{b'}^2 m_{t'} m_t$, so that $\varepsilon_H \approx 2\times 10^{-8}(m_{b''}^2 m_{b'}^2 m_{t'} m_t)/M_W^6$, which can be large enough to explain the BAU if $m_{b''}, m_{b'}, m_{t'}, m_t \gtrsim M_W$. I should also point out that in the five generation model, all the masses, mixing angles, and phases are in principle measurable in low energy experiments ($E \approx$ few TeV).

For the XY gauge bosons, the lowest order contribution to $\varepsilon_{XY}$ results from a four loop diagram involving Higgs bosons in the loops, so that $\varepsilon_{XY} \approx \alpha_H \varepsilon_H \lesssim 10^{-2}\varepsilon_H$. In addition, because of their stronger coupling, for a given superheavy boson mass, XY bosons are less efficient at producing a baryon asymmetry than Higgs bosons (cf. fig. 3a, 3b).

The message to be gleaned from the discussion above is that by adding additional heavy generations one can just squeak by with the minimal SU(5) model. One can also salvage SU(5) by enlarging the Higgs sector, with either an additional **5** or an additional **45** of Higgs [21,23]. In this case the lowest order contribution to $\varepsilon$ is at the one loop level, so that one expects $\varepsilon \lesssim O(\alpha_H) \approx 10^{-4}$–$10^{-6}$—which leaves plenty of room for *CP*-violating phase angles, etc. Harvey et al. [15] have investigated SU(5) models with an additional **5** and **45** in detail and find the interesting result that the sign of the BAU is not uniquely specified by *CP*-violating phase angles, but also depends upon the ratio of the masses of the two superheavy Higgs bosons. [It should be noted that additional Higgs multiplets are desireable for other reasons; with the minimal Higgs structure SU(5) makes the embarrassing prediction: $m_d/m_s = m_e/m_\mu \approx 1/207$.]

### 2.1.3. *SO(10)*

Generating a baryon asymmetry in an SO(10) model is tricky, because SO(10) is a *C*-symmetric theory. In order to do so, one must first break the $C$ symmetry. When the $C$ symmetry is broken, the LR symmetry is also broken, and $N_R$ (the righthanded partner of $\nu_L$), can acquire a large Majorana mass $m_N \approx O(V_R)$, while $\nu_L$ effectively acquires a small Majorana mass $m_\nu \approx m_\ell^2/m_N$ through the Dirac mass term coupling it to $N_R$ [27]. Here $V_R$ is the energy scale associated with the breaking of the $SU(2)_R \times SU(2)_L$ symmetry and $m_\ell$ is the mass of the lepton associated with $\nu_L$. If $M$ is the mass of the superheavy boson which generates the BAU, then $\varepsilon \approx O(\alpha_H^l)(V_R/M)$—which is exactly zero in the limit $V_R \to 0$ [19]. This general consideration argues strongly for $V_R \approx O(M) \approx O(10^{14}$ GeV), and therefore a large mass for $N_R$. ($N_R$ may not necessarily get its mass at the tree graph level, but instead may get it through radiative corrections, in which case $m_N \approx 10^{-2}$–$10^{-6}$ $V_R$, see e.g. ref [28].) Of course this also has implications for the masses of the lefthanded neutrinos since $m_\nu \approx m_\ell^2/m_N$ [29].

The simplest way to break the $C$ symmetry of SO(10) is by $SO(10) \to SU(5) \times U(1)$ or $SO(10) \to SU(3)_C \times SU(2)_L \times U(1)_Y$. Since $\mathbf{16} \otimes \mathbf{16} = \mathbf{10} + \mathbf{126} + \mathbf{120}$, the Higgs which couple to the fermions are the **10**, **120**, and **126**. Under SU(5) the $\mathbf{10} = \mathbf{5} + \bar{\mathbf{5}}$. With two **10**'s of Higgs, baryon generation in SO(10) proceeds in a similar manner as in an SU(5) model with two **5**'s, except that the final asymmetry is proportional to $(m_N/M)^2$—again a strong argument for $m_N$ to be not too different from $M$. Masiero and Mohapatra [30] have discussed a scheme in which the BAU is produced by the out-of-equilibrium decays of $N_R$ with $m_N \approx 10$ TeV. The main conclusion to be drawn about SO(10) is that producing the BAU is non-trivial because of the $C$ symmetry, and seems to require that $m_N$ not be too different from $M$. Harvey and coworkers have investigated baryosynthesis in SO(10) in detail [15].

*2.1.4. Is ε related to quantities which can be measured at low energies?*

In general $\varepsilon$ involves the interference between loop diagrams and the tree diagram, where the particles in the loops are superheavy Higgs bosons. Since $\alpha_H \approx \alpha_G(m_f/M_W)^2$ and the $K^0$–$\bar{K}^0$ system involves the lightest quarks it seems very unlikely that $\varepsilon$ is related to parameters which characterize the neutral kaon system. The quantitative discussion in subsection 2.1.2 bears this point out. However, it is not impossible that the same underlying mechanism might explain both instances of $CP$ violation (e.g., the five generation model discussed in subsection 2.1.2.). The result of Harvey et al. [15] is also of interest—in an SU(5) model with two **5**'s or a **5** and a **45** of Higgs, the sign of $\varepsilon$ also depends upon the mass ratio of the two superheavy bosons.

Recently, Ellis et al. [31] have suggested the intriguing possibility that the BAU and the electric dipole moment of the neutron are related. In their analysis they assume that the QCD $\theta$ parameter, in the absence of some automatic scheme (e.g. an axion, or massless quark) to allow it to be rotated away, is zero (for as of yet unknown reasons!) at some superunification scale $\gg M_{GUT}$. The $\theta$ parameter at 1 GeV receives renormalization contributions from the weak interactions, and from the GUT. They show that the contribution from the renormalization of $\theta$ at the GUT scale dominates, and that it receives contributions from the same class of diagrams which contribute to $\varepsilon$. Using this fact, they obtain the lower bound

$$d_n \gtrsim 2.5 \times 10^{-18}(n_B/n_\gamma)e \text{ cm}, \tag{13}$$

or $d_n \gtrsim 5 \times 10^{-28} e$ cm since $n_B/n_\gamma \approx (2\text{–}8)\times 10^{-10}$. The present experimental upper limit to $d_n$ is $1.6 \times 10^{-24} e$ cm [32]. Although this scheme is very attractive, it is not obligatory; e.g., (i) the $U(1)_{PQ}$ with its associated axion, or a massless quark allow $\theta_{QCD}$ to be rotated away; (ii) their analysis is valid only if the BAU is produced by superheavy Higgs, but perhaps (although it is less likely) it is produced by gauge bosons; (iii) there may be unforeseen cancellations. Finally, Masiero et al. [33] have pointed out that even in the scenario discussed above, it is probably not possible to relate $\varepsilon$ to $d_n$ directly, due to additional phases which arise from the SU(2)×U(1) phase transition.

### 2.1.5. Summary

The most important parameter for baryosynthesis, $\varepsilon$, is not well constrained at present. In general $\varepsilon \approx O(\alpha_H^l) \lesssim 10^{-2}$, and $\varepsilon$ vanishes in a *C*- or *CP*-symmetric theory. It seems highly unlikely that $\varepsilon$ can be related directly to the $K^0$–$\bar{K}^0$ system. The BAU might be produced in a minimal SU(5) model, but only if there exist very massive (30–200 GeV) fourth (or perhaps fifth) generations of fermions. With the addition of a second **5** or a **45** of Higgs to the minimal SU(5) model the BAU can be easily produced. Baryosynthesis in SO(10) is complicated by its *C* symmetry, and in general requires $m_N$ not too different from $M_{GUT}$, and hence $m_\nu \approx m_\ell^2/m_N$, so $m_\nu$ is very small.

## 2.2. Baryosynthesis in detail

For $K < 1$ the numerical results (discussed in section 1.4.) agree well with the qualitative discussion of section 1.3. When $T \lesssim M$, the S, $\bar{\text{S}}$

bosons remain as abundant as photons—a departure from equilibrium, eventually they decay freely ($T \ll M$) producing an asymmetry, $n_B/s \approx 10^{-2}\varepsilon$. For $K > 1$ one naively expects equilibrium to be maintained and no baryosynthesis; however the numerical results reveal that the efficiency of baryosynthesis falls off only as $K^{-1.3}$ (roughly). In this section I will explain this result by discussing a simplified set of Boltzmann equations; simplified in the sense that I have taken the exact equations and retained only the terms which are responsible for the basic results (for a more detailed discussion, see ref. [34]).

I shall consider only the case where $K \gg 1$, and discuss the evolution of the BAU for $z \equiv M/T \gtrsim 1$. For $z \lesssim 1$, the $S, \bar{S}$ bosons are relativistic and are present in equilibrium numbers—assuming they were initially. All the action occurs for $z \gtrsim 1$. Recall that $K \equiv (\Gamma_D/H)_{T=M}$ measures the 'effectiveness' of $B$-nonconserving ($\not{B}$) decays of S bosons. Suppose that the rate of $\not{B}$ decays $\Gamma_D = c_D \alpha_{\not{B}} M$, with $c_D$ a theory dependent constant of O(1), then:

$$K = \left. \frac{c_D \alpha_{\not{B}} M}{1.66[g_*(T)]^{1/2} T^2/m_P} \right|_{T=M} \approx \frac{c_D \alpha_{\not{B}} \times 10^{19}\ \mathrm{GeV}}{g_*^{1/2} M}. \tag{14}$$

There are two key equations for baryosynthesis: (i) for the abundance of S bosons, (ii) for the baryon to photon ratio. Define

$$\Delta \equiv n_S/n_\gamma - (n_S/n_\gamma)_{\mathrm{EQuilibr.}} = N_S - N_{EQ}, \tag{15}$$

which measures the departure from equilibrium in the abundance of $S, \bar{S}$ bosons. The equation for $\Delta$ is,

$$\Delta' = -zK(\gamma_D + \gamma_A z^{-3} N_{EQ})\Delta - N'_{EQ}, \tag{16}$$

where $'$ denotes $\mathrm{d}/\mathrm{d}z$ and $N_{EQ}$ is the equilibrium value of $n_S/n_\gamma$, which for $z \gtrsim 1$ is $N_{EQ} \approx (\pi/2)^{1/2} z^{3/2} \exp(-z)$. The quantities $\gamma_D$ and $\gamma_A$ are dimensionless and represent decays ($B$ conserving and $\not{B}$) and annihilations, normalized relative to $\Gamma_D = c_D \alpha_{\not{B}} M$. The total decay rate $\Gamma_T \approx (c_D \alpha_{\not{B}} + c'_D \alpha_B)$M, and $\gamma_D = (1 + c'_D \alpha_B / c_D \alpha_{\not{B}})$. Of course, if $S, \bar{S}$ have no $B$-conserving decay modes, $\gamma_D = 1$. The annihilation rate $\Gamma_A = n_S (\sigma v)_{\mathrm{ann}} = c_A N_S T^3 \alpha_A^2 M^{-2} \approx c_A N_{EQ} z^{-3} \alpha_A^2 M$ when $N_S$ is not too far from its equilibrium value $N_{EQ}$—which is a good approximation when $K > 1$. Therefore $\gamma_A = c_A \alpha_A^2 / \alpha_{\not{B}}$ where $\alpha_A$ is the relevant coupling for annihilations (e.g., if the annihilation channel is gauge-mediated, $\alpha_A \approx \alpha_{\mathrm{gauge}}$).

In the absence of the $N'_{\mathrm{EQ}}$ term in eq. (16), any departure from equilibrium would relax exponentially. However, because the Universe is cooling, $T \propto z^{-1}$, $N_{\mathrm{EQ}}(z)$ is changing, and $\mathrm{S}, \bar{\mathrm{S}}$ bosons are trying to 'track' an equilibrium abundance which is constantly decreasing (see fig. 1)—the $N'_{\mathrm{EQ}}$ term represents this effect. For $K \gg 1$, eq. (16) quickly relaxes to $\Delta' \approx 0$, so that to a good approximation, the solution is obtained by setting $\Delta' \approx 0$,

$$\Delta \approx \frac{-N'_{\mathrm{EQ}}}{zK\left(\gamma_{\mathrm{D}} + \gamma_{\mathrm{A}} z^{-3} N_{\mathrm{EQ}}\right)} \approx \frac{N_{\mathrm{EQ}}}{zK\left(\gamma_{\mathrm{D}} + \gamma_{\mathrm{A}} z^{-3} N_{\mathrm{EQ}}\right)}. \tag{17}$$

The equation governing the baryon-density to photon-density ratio, $B = (n_{\mathrm{b}} - n_{\bar{\mathrm{b}}})/n_{\gamma}$, is:

$$B' = -zK\left(c_{\mathrm{S}}\alpha_{\not{B}} z^{-5} + N_{\mathrm{EQ}}\right)B + zK(\varepsilon\Delta), \tag{18}$$

where again all the rates have been normalized to $\Gamma_{\mathrm{D}}$ [$c_{\mathrm{S}}$ is a theory dependent constant of O(1)]. The $-zKN_{\mathrm{EQ}}B$ term represents $B$ number destruction by inverse decay followed by decay (e.g., $\mathrm{q}+\mathrm{q} \to \bar{\mathrm{S}}, \bar{\mathrm{S}} \to \bar{\mathrm{q}} + \bar{\ell}$), and so its rate is controlled by the rate of ID's, $\Gamma_{\mathrm{ID}} \approx N_{\mathrm{EQ}}\Gamma_{\mathrm{D}}$ (only fermion pairs with CM energy $\gtrsim M$ can produce an S or $\bar{\mathrm{S}}$). The $-zKc_{\mathrm{S}}\alpha_{\not{B}} z^{-5}$ term represents $B$ number destruction by S-mediated $\not{B}$ scatterings (e.g., $\mathrm{q}+\mathrm{q} \to \bar{\mathrm{q}} + \bar{\ell}$). The rate for this process $\Gamma_{\mathrm{S}} \approx n_{\mathrm{B}}\sigma v \approx (BT^3)(c_{\mathrm{S}}\alpha_{\not{B}}^2 T^2/M^4)$, and $\Gamma_{\mathrm{S}}/\Gamma_{\mathrm{D}} = c_{\mathrm{S}}\alpha_{\not{B}} z^{-5}$. Finally, the $zK\varepsilon\Delta$ term represents $B$ number production by $B$-, $C$-, and $CP$-nonconserving decays of $\mathrm{S}, \bar{\mathrm{S}}$ bosons. Note that if $B$, $C$, or $CP$ is a good symmetry ($\varepsilon = 0$), or if equilibrium is maintained ($\Delta = 0$), then the $B$ production term vanishes. This is in accordance with the discussion in lecture one of the three conditions outlined by Sakharov. For $K \gg 1$, both the destruction and production terms in eq. (18) are large, so that $B'$ remains $\approx 0$, and therefore

$$B(z) \approx \frac{\varepsilon\Delta}{\left(c_{\mathrm{S}}\alpha_{\not{B}} z^{-5} + N_{\mathrm{EQ}}\right)}$$

$$\approx \frac{\varepsilon N_{\mathrm{EQ}}}{zK\left(\gamma_{\mathrm{D}} + \gamma_{\mathrm{A}} z^{-3} N_{\mathrm{EQ}}\right)\left(c_{\mathrm{S}}\alpha_{\not{B}} z^{-5} + N_{\mathrm{EQ}}\right)}, \tag{19a}$$

$$B(z) \approx \frac{(\varepsilon/\gamma_{\mathrm{D}})N_{\mathrm{EQ}}}{zK\left(c_{\mathrm{S}}\alpha_{\not{B}} z^{-5} + N_{\mathrm{EQ}}\right)}. \tag{19b}$$

In going from eq. (19a) to (19b) the annihilation term has been ignored relative to the decay term, which is a good approximation for $z > 1$ since, as $S, \bar{S}$ bosons become less abundant, pairs have difficulty 'finding' each to annihilate, and decay is the dominant process for reducing the number of $S, \bar{S}$ bosons ($\Gamma_A/\Gamma_T \approx \gamma_A/\gamma_D z^{-3/2} e^{-z}$).

Note that as long as the rates in eq. (18) are large ($\Gamma_{ID}, \Gamma_S > H$), $B(z)$ decreases at least as rapidly as $z^{-1}$. Eventually the $B$-destroying processes become ineffective ($\Gamma_{ID}, \Gamma_S < H$), and $B(z)$ ceases to decrease and 'freezes out'. The condition for freeze out is: $\Gamma = H$ (or equivalently $|B'z| = B$), and

$$z_{ID}^{7/2} \exp(-z_{ID}) K = 1, \tag{20a}$$

$$z_S = (c_S \alpha_{\not{B}} K)^{1/3}, \tag{20b}$$

define the 'freeze-out' values of $z$ for ID and $\not{B}$ scatterings respectively. The final value of $B$ is determined by $z_f = \max\{z_S, z_{ID}\}$: $B(\infty) = B(z_f)$. Unfortunately eq. (20a) is not easily solved (asymptotically $z_{ID} \to \ln K$). However, numerically I find that for $K \gtrsim K_{crit}$: $z_S \gtrsim z_{ID}$, and for $K \lesssim K_{crit}$: $z_{ID} \gtrsim z_S$, where

$$K_{crit} \approx 7000(c_S \alpha_{\not{B}})^{-5/4}. \tag{21}$$

$K_{crit}$ separates baryosynthesis into two qualitatively different regimes: ID dominated and scattering dominated.

### 2.2.1. *ID-dominated baryosynthesis:* $1 \lesssim K \lesssim K_{crit}$

In this regime ID's freeze out after $\not{B}$ scatterings do, so that the final value of $B$ is set by $B(z_{ID})$ (see fig. 5a):

$$B(z \gtrsim z_{ID}) \approx (\varepsilon/\gamma_D)/(z_{ID} K). \tag{22}$$

A reasonable numerical fit to eq. (22) is:

$$B(z \gtrsim z_{ID}) \approx (\varepsilon/\gamma_D) K^{-1.3}, \qquad n_B/s \approx (\varepsilon/\gamma_D g_*) K^{-1.3}, \tag{23}$$

since $s \sim g_* n_\gamma$. This is the slow fall off in baryon production discussed previously. Note that for the SU(5) XY gauge bosons $c_S \approx 1500$ [16] and $\alpha_{\not{B}} \approx 1/45$, so that $K_{crit} \approx 100$. For the superheavy isoscalar, color triplet Higgs boson of SU(5): $c_S \approx 10^3$, $\alpha_{\not{B}} \approx 10^{-4}$, and $K_{crit} \approx 10^5$.

### 2.2.2. *Scattering-dominated baryosynthesis:* $K \gtrsim K_{crit}$

In this regime $\not{B}$ scatterings freeze out after ID's, and thus the final value

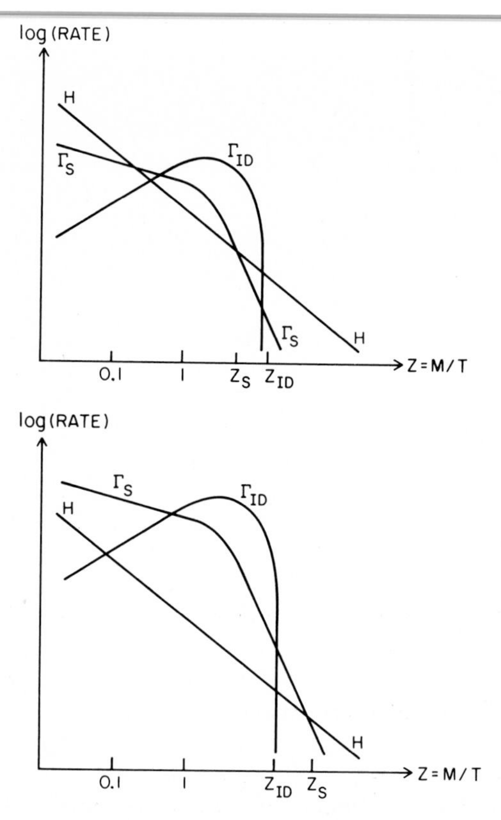


Fig. 5. Important rates for baryosynthesis as a function of $z = M/T$: $H \propto z^{-2}$; $\Gamma_{ID} \propto z^{3/2}Ke^{-z}$ ($z \gtrsim 1$)—damping by inverse decay; and $\Gamma_S \propto \alpha_{\not B} K z^{-5}$—damping by $\not B$ scatterings. Freeze out occurs when $\Gamma = H$. For $1 \lesssim K \lesssim K_{crit}$, $\not B$ scatterings freeze out first (a). For $K \gtrsim K_{crit}$ ID's freeze out first (b). The final value of $n_B/s$ is determined by the *last* $B$-destroying process to freeze out.

of $B$ is set by $B(z_S)$ (see fig. 5b):

$$B(z \gtrsim z_S) \approx (\varepsilon/\gamma_D)(c_S\alpha_{\not B}K)^{5/6}\exp\left[-(c_S\alpha_{\not B}K)^{1/3}\right],$$

$$n_B/s \approx (\varepsilon/g_*\gamma_D)(c_S\alpha_{\not B}K)^{5/6}\exp\left[-(c_S\alpha_{\not B}K)^{1/3}\right]. \tag{24}$$

When baryosynthesis is scattering dominated, production falls off *exponentially* with $K$. (A more careful analysis of freeze-out reveals that $z_{ID}^{5/2}\exp(-z_{ID})K = 1$, $z_{ID} \approx 4.2(\ln K)^{0.6}$, $z_S = (c_S\alpha_{\not B})^{1/4}$, $B(\infty) \approx z_f^2\exp(-4z_f/3)$, and $K_{crit}(\ln K_{crit})^{-2.4} \approx 300/c_S\alpha_{\not B}$, see ref. [34].)

Baryosynthesis as a function of $K$ is summarized in fig. 6.

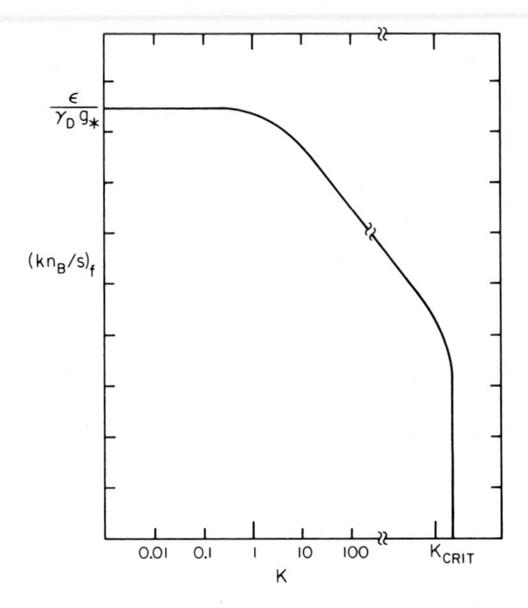


Fig. 6. Schematic plot of the final value of $n_B/s$ versus $K$.

## 2.3. *Variants of the "standard scenario"*

### 2.3.1. *Non-equilibrium initial distributions*

In the standard out-of-equilibrium decay scenario it is assumed that particle distributions are initially equilibrium distributions. In a statistical sense these distributions are certainly the 'most probable', and if such distributions were somehow initially established, a particle species would continue to maintain an equilibrium distribution as long as it was "effectively massless" ($m \ll T$), even in the absence of thermalizing interactions, with a redshifted temperature, $T \propto R(t)^{-1}$. [This is simply a result of the usual momentum redshift, $p \propto R(t)^{-1}$, and number density dilution, $n \propto R(t)^{-3}$, caused by the expansion.] However, this need not necessarily be the case. Ellis and Steigman [35] have discussed the question of whether or not particle interactions occur rapidly enough in the early Universe to establish thermal distributions. The answer is determined by whether or not such interactions are happening rapidly on the expansion timescale, i.e., $\Gamma > H$? For two-body reactions $\Gamma \propto \sigma n v$; $n \sim T^3$, $v \sim c$, and for renormalizable interactions $\sigma \sim \alpha^2/T^2$, so that $\Gamma \approx \alpha^2 T$. The expansion rate $H \approx T^2/m_P$. The condition $\Gamma > H$ requires

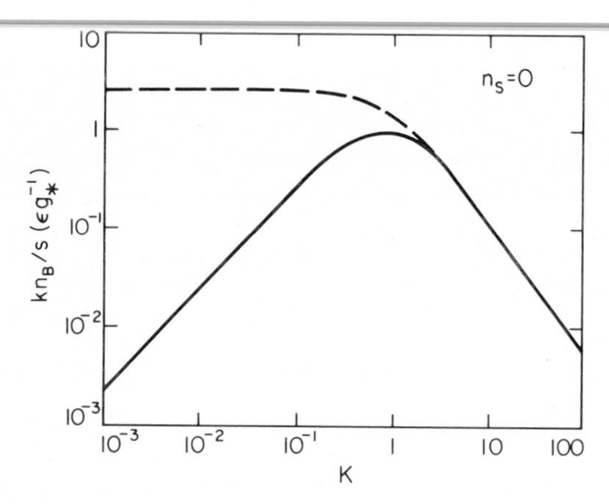


Fig. 7. Final value of $n_B/s$ as a function of $K$ for: $n_S/n_\gamma = 1$ $(T \gg M)$, "standard scenario": dashed curve; $n_S/n_\gamma = 0$ $(T \gg M)$, S and $\bar{\mathrm{S}}$ bosons initially absent: solid curve. If S and $\bar{\mathrm{S}}$ bosons are initially absent: for $K < 1$, $n_B/s \propto K^2(\varepsilon/g_*)$; for $K \gtrsim 1$, production is the same as in the standard scenario (see ref. [35]).

$T < \alpha^2 m_P \approx 10^{15}$ GeV. (By a more careful analysis they arrive at $T < 3 \times 10^{15}$ GeV). If particle distributions were not initially thermal, then thermal distributions will not be established until $T \approx 3 \times 10^{15}$ GeV. If the unification scale is $\lesssim 3 \times 10^{15}$ GeV [as it appears to be for SU(5)], this question is irrelevant for baryosynthesis.

However if the unification scale is $\gtrsim 3 \times 10^{15}$ GeV, then the possibility of non-equilibrium distributions, either due to initial conditions or due to a strongly first-order phase transition whose large entropy release resulted in nonthermal distributions, could be significant enough for baryosynthesis. The new effect is that other departures from equilibrium could drive baryosynthesis. The possibilities include: (i) a massless fermion species could be over- or underabundant, or could have a skewed distribution in momentum space [24, 36]; (ii) S and $\bar{\mathrm{S}}$ bosons might initially be absent (i.e. underabundant) [37]. Figure 7 shows the final value of $n_B/s$ as a function of $K$ for case (ii). Unlike the "standard scenario", baryosynthesis in these variants depends upon initial conditions which are likely to remain a mystery.

*2.3.2. Hierarchy of baryosynthesis*

Thus far I have only discussed the effect of a single superheavy species on the BAU of the universe. In general a given superheavy boson can produce a baryon excess, or damp a pre-existing baryon asymmetry; the magnitude of both effects can be well described in terms of one parameter: $K \approx 3 \times 10^{17}\alpha$ GeV$/M$. Baryon production by a single species is summarized in fig. 3. As mentioned earlier, damping is proportional to $\exp(-AK)$, where $A$ is of O(1). Depending on the species, there are certain modes of baryon asymmetry which are not damped. For example, the superheavy bosons of SU(5) all conserve $B-L$, and hence cannot damp any baryon asymmetry associated with a nonzero $B-L$. The gauge bosons of SU(5) conserve '5-ness' ['5-ness' is the property associated with being in the 5 dimensional representation of SU(5)], and cannot damp an asymmetry with net '5-ness'. The Higgs bosons in SU(5) do not conserve '5-ness' [38].

In order to compute the BAU today one must take into account the effects of *all* superheavy species. Even in minimal SU(5) there are two superheavy species which can change the BAU (the XY gauge bosons, and the color triplet, isosinglet member of the **5** of Higgs). This topic is discussed in some detail in refs. [15, 18].

*2.3.3. Out-of-equilibrium fermion decays*

The decays of superheavy gauge or Higgs bosons provide all the necessary ingredients for baryosynthesis: $B$-, $C$-, and $CP$-violations, and departure from equilibrium if $\Gamma_D < H$ when $T = M$. It has also been suggested that a superheavy (or not so superheavy) fermion could do the same. The most popular candidate is $N_R$, the righthanded partner of $\nu_L$ in LR-symmetric theories. The decay modes of $N_R$ include: $\rightarrow$ $qqq, \overline{qqq}, \ell_L^- \phi^+$, and (if $m_N \geq m_X$) $\rightarrow X\bar{q}, \bar{X}q$ where X is a superheavy Higgs, and $\phi$ is a W–S light Higgs doublet. Clearly the decays of $N_R$ violate $B$, and it is also possible for them to violate $C$ and $CP$. The difficulty is that the two-body decay $N_R \rightarrow \ell\phi$ tends to dominate, so that the decay of $N_R$ effectively conserves $B$ (see, e.g., ref. [29]). Two scenarios have been suggested in which $N_R$ is light ($\sim 10^4$–$10^{10}$ GeV), and the mode $N_R \rightarrow \ell\phi$ is suppressed so that it proceeds at the same rate as the three-body modes [30, 39]. The decay rate for the three-body mode, $\Gamma_D \sim \alpha^2 m_N^5/M^4$ ($M$ = mass of the boson which mediates the decay) and the condition for a departure from equilibrium ($\Gamma_D < H$ for $T = M$) implies

$$m_N \lesssim \alpha^{-2/3} g_*^{1/6} M^{4/3} m_P^{-1/3} \approx 10^{-5} M_{\rm GeV}^{4/3}. \tag{25}$$

*2.3.4. CP violation: intrinsic or spontaneous?*

Thus far, I have tacitly assumed that the $CP$ violation required for baryosynthesis is intrinsic to the theory, so that $\varepsilon$ is fixed by explicit

$CP$-violating terms which exist in the Lagrangian, and hence the BAU is the same throughout the Universe (as the observations suggest). If the requisite $CP$ violation arises spontaneously, then the sign of $\varepsilon$ can differ from place to place depending upon the direction chosen by the Higgs field. One then expects regions of the Universe which were causally disconnected at the epoch of baryosynthesis, to have baryon asymmetries of random sign, leading to a globally symmetrical universe [40]. It is difficult to see how the 'annihilation catastrophe' discussed earlier could be avoided when the different domains came into causal contact. In addition, the large density perturbations associated with the domain walls should lead to a proliferation of black holes or observable anisotropies [41]. Although one might argue that it is 'unnatural' to have $CP$-violating phases intrinsic to the Lagrangian, it is no more arbitrary to have complex Higgs–fermion couplings in the Lagrangian than to have the usual hierarchy of Higgs couplings $\alpha_{\mathrm{H}} \sim \alpha(m_{\mathrm{f}}/M_{\mathrm{W}})^2$ which are already required to explain fermion masses. Perhaps $CP$ violation *does* arise spontaneously, but at the Planck mass, where quantum gravity effects might remove the horizon and allow the Universe to exist as a single domain [42].

### 2.4. *Cosmological implications*

Clearly, if the GUT scenario for the origin of the BAU (and hence the origin of matter) proves to be correct, there are many cosmological implications. I will briefly discuss one example here which has to do with galaxy formation—the origin of density fluctuations. It is generally believed that the structure we observe today (planets, stars, galaxies, clusters, and superclusters) resulted from small initial inhomogeneities ($\delta\rho/\rho \lesssim 10^{-2}$) which grew, after matter decoupled from radiation ($T \sim 1$ eV) and was freed of its pressure support, due to the Jeans (gravitational) instability (for a review, see ref. [43]). During the radiation dominated epoch ($T \sim 1$ eV) primordial density fluctuations on an FRW background can be divided into two types: (i) adiabatic, characterized by $\delta T/T \neq 0$ but with $\delta(n_{\mathrm{B}}/s) = 0$; (ii) isothermal, characterized by $\delta T = 0$, but with $\delta(n_{\mathrm{B}}/s) \neq 0$. Which type of fluctuation is required to explain the observed structure today, and the origin of the fluctuations, are unresolved questions of keen interest.

The origin of isothermal fluctuations and baryosynthesis are obviously intimately connected. Baryosynthesis is described by $K$ and $\varepsilon$,

$$n_{\mathrm{B}}/s = f(\varepsilon, K), \tag{26}$$

where $K=(\Gamma_D/H)_{T=M}$. In a nearly FRW universe $H=g_*^{1/2}T^2/m_P$, and so $K=(c_D\alpha m_P)/(g_*^{1/2}M)$ [cf. eq. (14)], and hence, only depends upon physical constants and particle physics parameters. Therefore, $n_B/s$ has no *spatial* dependence, even if the temperature of the universe at a given epoch (i.e. time slice) had spatial dependence. Baryosynthesis occurs at different times in different regions, but always results in the same $n_B/s$. In a nearly FRW model baryosynthesis and isothermal perturbations are incompatible—thus favoring the adiabatic picture [44].

If the Universe was initially very non-Friedmannian, e.g., dominated by large-scale, inhomogeneous shear, the story is quite different [45,46]. In this case the expansion rate of the Universe is governed by

$$H^2 = g_*T^4/m_P^2 + \Sigma(\boldsymbol{x})^2/R^6, \tag{27}$$

where $\Sigma(\boldsymbol{x})^2$ parametrizes the shear energy density. Now $K = (\Gamma_D/H)_{T=M} = K(\boldsymbol{x})$, and it is possible to produce simultaneously a baryon excess and isothermal fluctuations (i.e. spatial variations in $n_B/s$). The spectrum of isothermal fluctuations just mirrors the initial spectrum of shear. Since the Universe is not today dominated by large-scale shear, one must insist that the initial shear be a dying-mode perturbation (i.e. one whose importance *decreases* with time). The model described by eq. (27) satisfies this condition. In this model the shear energy density decreases relative to the radiation energy density $\propto T^2$, and thus rapidly decays away adiabatically.

## Acknowledgements

I am grateful to Rocky Kolb for several helpful conversations. This work was supported in part by NSF Grant PHY77-27084 at Santa Barbara, and by DOE grant DE-AC02-80ER10773 at Chicago.

## References

[1] K.A. Olive, D.N. Schramm, G. Steigman, M.S. Turner and J. Yang, Astrophys. J. 246 (1981) 557.
J. Yang, M.S. Turner, G. Steigman, D.N. Schramm and K.A. Olive, Univ. of Chicago preprint, submitted to Astrophys. J. (1983).

[2] G. Segrè and H.A. Weldon, Phys. Rev. Lett. 44 (1980) 1737.

[3] S. Weinberg, Gravitation and Cosmology (Wiley, New York, 1972) ch. 15.

[4] R. Balian, J. Audouze and D.N. Schramm, eds., Physical Cosmology, Les Houches (1979), session 32 (North-Holland, Amsterdam 1980).

[5] J.H. Christenson, J.W. Cronin, V.L. Fitch and R. Turlay, Phys. Rev. Lett. 13 (1965) 138; Phys. Rev. B140 (1965) 74.

[6] G. Steigman, Ann. Rev. Astron. Astrophys. 14 (1976) 339.
[7] A. Buffington and S.M. Schindler, Astrophys. J. 247 (1981) L105.
R.L. Golden et al., Phys. Rev. Lett. 43 (1979) 16.
E.A. Bogomolov et al., Proc. 16th Int. Conf. on Cosmic Rays (1979).
[8] F.W. Stecker, R.J. Protheroe and D. Kazanas, Proc. 17th Int. Conf. on Cosmic Rays (1982).
[9] A. Sakharov, JETP Lett. 5 (1967) 24.
S. Weinberg, in: Lectures on Particles and Field Theory, eds. S. Deser and K. Ford (Prentice-Hall, Englewood Cliffs, 1964).
L. Parker, in: Asymptotic Structure of Space–Time, eds. F.P. Esposito and L. Witten (Plenum, New York, 1977).
S. Hawking (1975), unpublished.
[10] P. Langacker, Phys. Rep. 72 (1981) 185.
[11] M. Yoshimura, Phys. Rev. Lett. 41 (1978) 281.
A. Ignatiev, N. Krasnikov, V. Kuzmin and A. Tavkhelidze, Phys. Lett. 76B (1978) 486.
S. Dimopoulos and L. Susskind, Phys. Rev. D18 (1978) 4500.
J. Ellis, M.K. Gaillard and D.V. Nanopoulos, Phys. Lett. 80B (1979) 360.
N.J. Papastamatiou and L. Parker, Phys. Rev. D19 (1979) 2283.
[12] S. Weinberg, Phys. Rev. Lett. 42 (1979) 850.
[13] D. Toussaint, S.B. Treiman, F. Wilczek and A. Zee, Phys. Rev. D19 (1979) 1036.
[14] E.W. Kolb and S. Wolfram, Phys. Lett. 91B (1980) 217; Nucl. Phys. B172 (1980) 224.
[15] J.A. Harvey, E.W. Kolb, D.B. Reiss and S. Wolfram, Nucl. Phys. B201 (1982) 16.
[16] J.N. Fry, K.A. Olive and M.S. Turner, Phys. Rev. D22 (1980) 2953.
[17] J.N. Fry, K.A. Olive and M.S. Turner, Phys. Rev. D22 (1980) 2977.
[18] J.N. Fry, K.A. Olive and M.S. Turner, Phys. Rev. Lett. 45 (1980) 2074.
[19] V.A. Kuzmin and M.E. Shaposhnikov, Phys. Lett. 92B (1980) 115.
[20] T. Yanagida and M. Yoshimura, Phys. Rev. D23 (1981) 2048.
[21] D.V. Nanopoulos and S. Weinberg, Phys. Rev. D20 (1979) 2484.
[22] S. Barr, G. Segrè and H.A. Weldon, Phys. Rev. D20 (1979) 2484.
[23] A. Yildiz and P. Cox, Phys. Rev. D21 (1980) 306.
[24] J. Ellis, M.K. Gaillard and D.V. Nanopoulos, Phys. Lett. 80B (1979) 360 (Erratum 82B (1979) 464).
[25] G. Segrè and M.S. Turner, Phys. Lett. 99B (1981) 399.
[26] A.J. Buras, J. Ellis, M.K. Gaillard and D.V. Nanopoulos, Nucl. Phys. B135 (1978) 66.
D.V. Nanopoulos and D.A. Ross. Nucl. Phys. B157 (1979) 273.
[27] M. Gell-Mann, P. Ramond and R. Slansky, unpublished.
P. Ramond, Caltech preprint CALT 68-709 (1981).
[28] E. Witten, Phys. Lett. 91B (1980) 81.
[29] M. Fukugita, T. Yanagida and M. Yoshimura, Phys. Lett. 106B (1981) 183.
[30] A. Masiero and R.N. Mohapatra, Phys. Lett. 103B (1981) 343.
[31] J. Ellis, M.K. Gaillard, D.V. Nanopoulos and S. Rudaz, Phys. Lett. 99B (1981) 101.
[32] I.S. Altarev et al., Zh. Eksp. Teor. Fiz. Pis'ma 29 (1979) 794.
W.B. Dress et al., Phys. Rev. D15 (1977) 9.
N.F. Ramsey, Phys. Rep. 43 (1978) 409.
[33] A. Masiero, R.N. Mohapatra and R.D. Peccei, Phys. Lett. 108 (1982) 111.
[34] M.S. Turner and J.N. Fry, Phys. Lett. B (1983) in press.
[35] J. Ellis and G. Steigman, Phys. Lett. 89B (1980) 186.
[36] M.S. Turner, Phys. Rev. D25 (1982) 289.

[37] M.S. Turner and J.N. Fry, Phys. Rev. D24 (1981) 3341.
[38] S.B. Treiman and F. Wilczek, Phys. Lett. 95B (1980) 222.
[39] H. Harari and M.S. Turner (1981) unpublished.
[40] F.W. Stecker, Nature 273 (1978) 493.
[41] Ya. B. Zel'dovich, L.B. Okun and I. Yu. Kobzarev, Sov. Phys. JETP 40 (1974) 1.
[42] L. Parker and S.A. Fulling, Phys. Rev. D7 (1973) 2357.
M.V. Fischetti, J.B. Hartle and B.-L. Hu, Phys. Rev. D20 (1979) 1757.
A.A. Starobinsky, Phys. Lett. 91B (1980) 99.
A. Zee, Phys. Rev. Lett. 44 (1980) 703.
[43] P.J.E. Peebles, The Large-Scale Structure of the Universe (Princeton Univ. Press, 1980).
[44] M.S. Turner and D.N. Schramm, Nature 279 (1979) 303.
[45] J.D. Barrow and M.S. Turner, Nature 291 (1981) 469.
[46] J.R. Bond, E.W. Kolb, and J. Silk, Astrophys. J. 255 (1982) 341.

SEMINAR 3

# GENERALIZED HIGGS PHYSICS AND TECHNICOLOR

G.L. KANE

*Randall Laboratory of Physics, University of Michigan*
*Ann Arbor, MI 48109, USA*

## Contents



*M.K. Gaillard and R. Stora, eds.*
*Les Houches, Session XXXVII, 1981*
*Théories de jauge en physique des hautes énergies/Gauge theories in high energy physics*

## 1. Introduction

To have a renormalizable gauge theory with massive gauge bosons and massive fermions, it is apparently necessary to have some additional physics in the theory. In the standard electroweak theory, scalar bosons are introduced and a single, neutral, scalar boson, with non-gauge interactions, should occur in the particle spectrum. Since gauge bosons and fermions are massive, we already have the first experimental evidence for some new physics, which we will call "Higgs physics" (since in its simplest form it directly gives elementary Higgs bosons). Perhaps surprisingly, no further positive evidence for Higgs physics has emerged in a decade. The basic physics of the origin of gauge boson and/or fermion masses is not understood at all. In these lectures I will give a partial review of present approaches to the problem, present constraints from data, and hopes for new evidence from experiment in the next few years.

In the standard theory, Higgs scalars are introduced in an ad hoc way, with interactions of non-gauge origin. There is great difficulty in understanding why the physical scalars might have masses on the scale needed to give the correct $m_{\mathrm{W}}, m_{\mathrm{Z}}$. So far, two approaches have showed promise for making sense of the presence of scalars and their interactions: supersymmetric theories and so called "dynamical symmetry breaking" theories. In the former, scalars occur naturally in the theory, with interactions determined by the gauge structure, and scalar masses can in principle be determined by the structure of the theory and by the way the supersymmetry and $\mathrm{SU}(2)\times\mathrm{U}(1)$ are broken. The main testable predictions will be the existence and properties of a large number of spin 0 and spin 1/2 particles. So far, no model has been able to produce numerical predictions, but soon that should be remedied.

We will spend more time below on the dynamical theories, emphasizing the technicolor (or hypercolor) approach because it has led to several fairly clear predictions. Since the theory is covered in John Ellis' lectures (course 3, this volume) and well described in several recent reviews (by Lane and Peskin [1], Eichten [2], Fahri and Susskind [3], and Sikivie [4]), I will emphasize the phenomenological questions.

In addition, because there has been so little experimental input in this field, and because it has proved so difficult theoretically, I will cover briefly various ways Higgs physics might show up, such as scalar currents, flavor-changing currents, or high energy interactions.

This is not a field where one can proceed with a neat elegant development. As we go along, and some things get complicated, the reader should keep in mind: (i) In Higgs physics there are almost no relevant theorems... every statement has exceptions. (ii) However ugly the theory might get, some new physics *must* occur, and it may well be like one of our alternatives. (iii) So far Higgs physics has hardly been taken seriously; e.g. there are essentially no dedicated experimental proposals, and no limits on masses or couplings published by experimental groups.

## 2. The Standard Model

Let us briefly review [5] the standard model for Higgs from a slightly different viewpoint to help understand some of what comes later.

We have gauge bosons $W^{\pm}, Z^0$ to which mass must be given. We use one scalar doublet:

$$\phi = \begin{pmatrix} \phi^+ \\ \phi^0 \end{pmatrix}. \tag{1}$$

These are complex scalar fields: $\phi^+ = \phi_1 + i\phi_2, \phi^0 = \phi_3 + i\phi_4$, with $\phi_i$ real. We can write

$$\phi = \begin{pmatrix} \phi_1 \\ \phi_2 \\ \phi_3 \\ \phi_4 \end{pmatrix}, \tag{2}$$

and if nothing distinguishes among the components, so they could be rotated into one another, there is an O(4) symmetry, with six generators. Then $\phi_3$, the real part of $\phi^0$, gets a vacuum expectation value (vev):

$$\langle\phi\rangle = \begin{pmatrix} 0 \\ 0 \\ v/\sqrt{2} \\ 0 \end{pmatrix}, \tag{3}$$

leaving an unbroken O(3) symmetry (with three generators) since $\phi_1, \phi_2, \phi_4$ can still be mixed. There are three broken generators, and thus three Goldstone bosons, one each for $W^+$, $W^-$ and $Z^0$. A physical neutral Higgs boson is left.

One can write fermion Yukawa couplings which give fermions mass,

$$g_f \bar{f}_L \phi f_R \rightarrow g_f \bar{f}_L f_R (\phi + v/\sqrt{2}). \tag{4}$$

So $m_f = g_f v/\sqrt{2}$. Note that $g_f \sim m_f$ so heavier fermions couple more strongly to the Higgs bosons than do lighter fermions. In the Standard Model one can argue [5] that $m_H >$ (few GeV) to have a stable minimum.

How oversimplified is all this for the real world? How much is really known about Higgs bosons or the properties they would have if they existed? What Higgs particles might occur—charged ones also? What can be said, theoretically or phenomenologically, about their masses, couplings, decays, production mechanisms? We will discuss these questions below. The *only* general statement that can be made is that Higgs bosons have spin 0 (but all spin zero states need not be Higgs bosons) with production and decay properties constrained by angular momentum arguments.

If fermion couplings are proportional to masses, then decays will be into the heaviest allowed fermions, and universality will be violated ($e \neq \mu \neq \tau, d \neq s \neq b, u \neq c \neq t$). Unfortunately, production of Higgses will be small since we have to start from $e^+e^-$ or light quarks if we want intensive beams.

If no Higgs particles occur, the theory will have unitarity violations by $s^{1/2} \leqslant 1\,\text{TeV}$, so we must in principle see some sign of Higgs physics by that scale. But the effects are small, being screened [6] by powers of $g^2$, and made difficult to observe experimentally even in high energy processes by powers of $m_f/m_W$ where f is the fermion used for the beam in the experiment.

Higgs masses could be on a TeV scale, which would occur naturally in dynamical theories; or of order $m_W$, which could occur naturally in supersymmetric theories. Sometimes Higgs bosons or their dynamical equivalent might be very light because of symmetries or because they are pseudo-Goldstone bosons [8]—they could be massless Goldstone bosons for part of the theory, but they get some mass from interactions due to another part of the theory; we will give an example below in a simple model, and argue that this happens for many states in the Technicolor approach.

## 3. Two doublets

Whatever the origin of Higgs physics, one can argue (though not prove) that very likely there are at least two Higgs doublets in the low energy region. One arrives at that conclusion when one takes a more complete view of all the physics that needs to be done with the Higgs particles, and includes them in larger, more unified theories. There are several arguments; none are compelling but several are convincing. There will probably be at least two doublets at low energies if any of the following (not in any particular order) happens:

–It is possible to calculate [9] Cabibbo–Kobayashi–Maskawa quark mixing angles from quark mass ratios.

–Dynamical approaches such as technicolor are basically correct.

–Higgses are supersymmetric scalars.

–Left–right symmetric theories are basically correct, with right-handed bosons somewhat heavier than left-handed ones.

–*CP* violation is due to the Higgs sector [10].

–The strong *CP* problem is solved by the Peccei–Quinn mechanism with two Higgs doublets.

–Fermion masses are simply described in Grand Unified Theories.

–Gauge boson and fermion masses, with their very different scales, arise from different sources [11].

–Several fermion families might imply several Higgs families. If there are two doublets [11], there are eight real fields (with three for $W^{\pm}, Z^0$), leaving five physical bosons, a charged pair $H^{\pm}$, and three neutrals. There are two vacuum expectation values $v$, $v'$. Couplings can be enhanced or suppressed by $v/v'$. There are no restrictions on masses.

## 4. Higgs triplet example—pseudo-Goldstone bosons

To see several properties of Higgs systems in a nice example, consider a model [12] with a Higgs doublet $\phi$, and a Higgs triplet $\eta$ with hypercharge zero,

$$\eta = \begin{pmatrix} \eta^+ \\ \eta^0 \\ \eta^- \end{pmatrix}. \tag{5}$$

(Assume also that the theory is invariant under $\eta \to -\eta$ so there is no cubic term in the Higgs potential.) As above, $\phi$ has an O(4) invariance, and $\eta$ has an O(3) invariance, so initially the symmetry is O(4)×O(3) with 6+3 generators. After symmetry breaking, with $\phi^0$ and $\eta^0$ getting real vev's, the remaining symmetry is O(3)×O(2) with 3+1 generators, so there are five Goldstone bosons. Three become the longitudinal partners for $W^{\pm}, Z^0$, leaving a massless, charged pair of Higgs bosons!

Radiative corrections due to gauge boson loops break the symmetry and generate [8, 13] a mass for $H^{\pm}$; since it is an induced mass term it is calculable, giving

$$m_{H^{\pm}}^2 = \frac{3\alpha}{4\pi} m_Z^4 \cos^2\theta_W / (m_Z^2 - m_W^2) \ln m_Z^2/m_W^2 \approx \alpha m_W^2/\sin^2\theta_W . \tag{6}$$

Numerically: $m_{H^{\pm}} \approx 3.5\,\text{GeV}$. These are true pseudo-Goldstone bosons. Note that we did not have to set a bare mass term to zero as in the similar calculation [13] for doublets.

This illustrates how some of the scalar bosons can be light compared to the symmetry scale of the theory. They would have zero mass, but because they have electroweak quantum numbers $I_3 \neq 0$, they get some mass (of order $\alpha^2$ in fact, since $m_W^2 \sim \alpha$). A similar thing will happen below for technicolor, where there are initially many massless bosons, but some get mass from color, some from electroweak interactions, and two presumably get mass from presently unknown and subtle sources.

From the triplet example one can quickly see one further result. The covariant derivative on the triplet gives

$$\left(\partial_\mu \eta^a + g\varepsilon^{abc}\eta^b W_\mu^c\right)^2 = g^2\eta^{02}\left(W_1^2 + W_2^2\right), \tag{7}$$

so when $\langle \eta^0 \rangle = W \neq 0$ this gives additional mass to gauge bosons, but only to $W^{\pm}$. Then the quantity $\rho = m_W^2/m_Z^2\cos^2\theta_W$ is

$$\rho = 1 + 4W^2/v^2 \tag{8}$$

with a positive deviation from unity. In general [14], any number of Higgs doublets gives $\rho = 1$. Experimentally, $\rho$ is within 2% of unity [15], strongly suggesting that only doublet Higgses are responsible for symmetry breaking; such a result should be required to emerge from any dynamical theory. Explicitly here it gives $W/v \leqslant 0.07$.

## 5. How can Higgs physics be observed?

Before we turn to technicolor, let us discuss in a general way how Higgs physics might be observed [11, 16, 17]. One can either (A) observe an effect of scalar or pseudoscalar currents, or (B) observe direct production of a scalar or pseudoscalar boson. There are several possibilities for (A) and (B):

(A1) Scalars would show up in the muon g−2 [18]. Because scalars and pseudoscalars necessarily have cancellations, this is not always a good constraint. With normal couplings proportional to the fermion mass, and no concellations, one finds

$$\begin{array}{llll} \text{for } m_H = & 25\ \text{MeV} & 1\ \text{GeV} & 10\ \text{GeV} \\ C & <1.2 & <6 & <35 \end{array} \qquad (9)$$

where $L = 9(m_\mu/m_W)C\bar{\mu}\mu H^0$, so $C$ would be unity for standard couplings. Thus no real limit occurs. However, in any model with enhanced couplings (e.g. if all fermions couple proportionally to the heaviest mass in a Grand Unified or horizontal family, as could be arranged, so $m_\mu$ is replaced by $m_c$ or $m_\tau$) a useful lower limit on $m_H$ is obtained. Conversely, if $m_H$ were given, a useful limit on $C$ is obtained. For the pseudoscalar $H^0$ the limit is somewhat less good, and for mixed scalar and pseudoscalar each case must be analyzed separately. The limits come essentially from neutral Higgses; the charged contributions are always smaller.

(A2) In $\mu$ decay and $\beta$ decay there are limits on the amount of scalar or pseudoscalar current, but they are not very restrictive. If Higgses couple proportionally to the masses it will be very hard to see them here [11] as the quarks and leptons are so light; nevertheless, it would be worthwhile to have better limits.

(A3) Direct decays such as $\pi^\pm \to \ell^\pm \nu, \pi^0 \to e^+e^-, \eta \to \mu^+\mu^-, K \to \pi\ell^+\ell^-$ again are often not places where dramatic effects are expected, but surprises might occur. The familiar helicity-suppressed decays $\pi^\pm \to \ell^\pm \nu, \ell = \mu$ or e, are amusing [11] because they give a width proportional to $m_\ell^2$ whether mediated by pseudoscalar or axial vector currents, once from the coupling and once from the helicity flip. If $f_\pi^2$ were independently well measured the ratio $\Gamma(\pi^\pm \to \mu^\pm \nu)/\Gamma(\pi^\pm \to e^\pm \nu)$ would give a strong restriction on scalar currents with coupling not proportional to $m_\ell^2$.

(A4) Any violation of universality in fermion couplings could indicate a Higgs effect and should be constantly looked for. Possible examples are

$$\Gamma(\tau \to \mu\nu\bar{\nu})/\Gamma(\tau \to e\nu\bar{\nu}) \neq 1, \qquad \Gamma(b \to \mu X)/\Gamma(b \to eX) \neq 1,$$
$$\sigma(eN \to eX) \neq \sigma(\mu N \to \mu X).$$

(B1) Direct production of charged Higgses is most simply studied in $e^+e^-$ reactions. The cross section is standard for scalars, with $\beta^3/4$ units of $R$ [$\beta = (1-4m_H^2/s)^{1/2}$], and a $\sin^2\theta$ production angular distribution. As we will see below, the technicolor approach predicts a charged pseudo-Goldstone boson with mass around 8 GeV, which should be detectable at PETRA, PEP. No experimental results are published so far on this subject (which is surprising), but just recently (see below) two groups made preliminary analyses which fail to find the hoped-for signal. Whether a special pattern of decays can explain this is not yet clear.

(B2) Charged Higgses would occur semiweakly in decays. From the absence of the decay $\tau^\pm \to H^\pm \nu$, it is clear that $m_{H^\pm} > m_\tau$. Similarly, it appears [19] that the b-quark decays normally rather than through a two-body Higgs mode $b \to H^- u$, so $m_{H^-} \gtrsim m_b$. As soon as the t-quark is found we will immediately either see a charged Higgs because $t \to H^+ b$, or learn that $m_{H^+} > m_t - m_b$. There should be no difficulty in distinguishing this mode from the usual weak decay $t \to b\bar{f}f'$.

(B3) Amusingly, if charged Higgses exist, a number of states such as heavier neutral Higgses and heavy leptons could be harder to see, as semiweak decays like $H^0 \to H^+H^-$, $L^0 \to L^\pm H^\pm$. $L^\pm \to H^\pm \nu$, could dominate and mask better signals. If this situation holds, existing limits on these states do not apply.

(B4) Neutral Higgses are very difficult to produce or detect. The best method is the Wilczek mechanism [20], where a quarkonium resonance decays: $V \to H^0\gamma$, giving a monoenergetic photon. The branching ratio is calculable and is about $5\times10^{-5}$ for $\Psi$, $2\times10^{-4}$ for $\gamma$, assuming no significant phase space correction. The best measurement so far is from the SLAC Crystal Ball group, giving a limit [21] of about $10^{-3}$ for $\Psi$.

(B5) Drell–Yan production of an $H^0$ via heavier quarks such as $s\bar{s}$ is larger [11]. If an experiment could get mass resolution much better than 200 MeV a signal might be detectable. In any case, Drell–Yan experiments can put helpful limits on scalar couplings, but unfortunately no data has ever been *analyzed* from that point of view.

(B6) Finally, a clever or lucky experimenter might find Higgs in many interesting exclusive modes, e.g.

$$\begin{aligned} H^0 &\to \Psi\Psi, \phi\phi, \tau^+\tau^-, D\bar{D}, B\bar{B}, \gamma\gamma, \Psi D\bar{D}, \phi K\bar{K}, \bar{D}D^*, \\ &\quad \phi\pi, \Psi\Psi\pi^+\pi^-, \ldots, \\ H^\pm &\to \Psi F, \Psi F^*, \Psi K\bar{K}, \Psi\eta\pi, \Psi DK, DK, DK^*, F\eta, \\ &\quad F\phi, D\bar{B}, \Psi Be, \phi\pi, \ldots . \end{aligned} \tag{10}$$

We conclude this section by emphasizing that in the above we are using "Higgs bosons" and "Higgs physics" to denote the particles and interactions associated with spontaneously broken gauge theories, whether the bosons are dynamical or fundamental. Experimentally the general situation is identical. Now we turn to the TechniColor (TC) approach in detail, and study of the particular experimental situations it naturally suggests.

## 6. Technicolor

As discussed briefly above, and extensively in the reviews mentioned and in John Ellis' lectures, the motivation for technicolor (TC) is to make sense of scalar bosons, to give them interactions that originate in a gauge structure, and to understand their masses. For completeness we will very briefly review the main ideas.

Instead of new fundamental *bosons*, one postulates new, massless, fundamental *fermions* [techniQuarks (Q) and techniLeptons (L)], and a new fundamental non-Abelian force (TC). Becuase of the strong force, many pairs with vacuum quantum numbers (in particular, zero angular momentum) are produced, and lower the vacuum energy, giving mass to gauge bosons (by hypothesis). Since the fermions Q, L are assumed massless, the left-handed ones $Q_L, L_L$ can be transformed among themselves independently of the right-handed ones, and vice versa, giving rise to a large amount of chiral symmetry. But since the pairs that lower the vacuum energy are scalars (and not statistical mixtures of spin 0, 1) the left- and right-handed fermions become correlated and the symmetry is reduced. This gives rise to some massless Goldstone bosons. Now (as for the triplet example above) when the theory is extended to include color and electroweak interactions, the Goldstone bosons get some mass.

The basic mass scale of the TC interaction must be of order 1 TeV to get $M_W$ and $M_Z$ right (see refs. [1–4] and the simple argument below). Ordinarily all predictions of new physics would be on the TeV scale and we would not expect to see anything at lower energies. But the existence of the Goldstone modes saves us—the theory gives rise to *calculable* "low energy" predictions.

There is an important subtility we skipped over above, namely the fact that in this approach gauge bosons and fermions get masses in different ways. It is simplest to use a model (the usual one) to illustrate this and other points.

Consider two massless techniquarks U, D with the usual electroweak structure

$$Q_L = \begin{pmatrix} U \\ D \end{pmatrix}_L, U_R, D_R. \tag{11}$$

Then there are left- and right-handed currents, with weak isospin $a$,

$$J^{\mu a}_{L,R} \sim \bar{Q}\gamma^{\mu}(1 \pm \gamma_5)\tau^a Q, \tag{12}$$

and these are separately conserved. After the pairs form, so that $\overline{U}U, \overline{D}D$ get an expectation value, the quarks effectively get a constituent mass, and $J^{\mu a} = J^{\mu a}_L + J^{\mu a}_R$ is still conserved, while $J^{\mu a}_5 = J^{\mu a}_L - J^{\mu a}_R$ is not. The broken symmetry gives Goldstone bosons P which transform like $\bar{Q}\gamma_5\tau^a Q$ $(a = 1, 2, 3)$, and the current $J^{\mu a}_5$ can create them from the vacuum as is standard for the Goldstone modes (and familiar from the pion),

$$\langle 0 | J^{\mu a}_5 | P^b \rangle = i\delta^{ab} q^{\mu} F_T. \tag{13}$$

$F_T$ is the analogue of $F_\pi$, the pion decay constant. Because the current couples to P, the gauge boson propagator,

$$\Delta_{\mu\nu} = \frac{g_{\mu\nu} - g_\mu g_\nu / q^2}{q^2 [1 + \pi(q^2)]}, \tag{14}$$

has contributions

with a factor $gF_T q_\mu$ at each vertex, and $1/q^2$ propagators. These sum to give a factor:

$$1/(1 - g^2 F_T^2 / q^2) \tag{15}$$

for the $g_\mu g_\nu$ term (which must be in the transverse $g_{\mu\nu}$ term by gauge invariance). Finally then, the propagator has a denominator

$$q^2 - g^2 F_T^2, \tag{16}$$

so a mass $gF_T$ has been generated for the gauge boson. Since $g$ is the usual gauge coupling, $g \approx \frac{1}{2}$, and $gF_T \equiv m_W$, we must have $F_T \approx 200$ GeV. Then $F_T / f_\pi \approx 2000$, so the TC interaction is strong at a mass scale of order 2000 times that of QCD, or about 1 TeV.

It is important to emphasize here that: (a) the gauge bosons got mass, with no left over low mass states; (b) the fermions are still massless. The dynamical approach separates the mechanisms of mass generation for gauge bosons and fermions!

## 7. Fermion masses

Now it is necessary to give mass to the fermions. Unfortunately, that remains a challenge to theorists; in spite of a lot of work it has not been accomplished in a way which includes fully the effects of quark mixing angles and the absence of flavor changing neutral currents. Whether the fault lies in a weakness of the basic approach or in the theoretical efforts carried out so far is unclear. I will only observe that so far, the theoretical efforts have relied on inspiration rather than systematic study, and perhaps there is room yet for optimism.

An approach that goes some distance toward what is desired was introduced independently by Dimopoulos and Susskind ("extended TechniColor," [22]), and Eichten and Lane ("Sideway interactions," [23]). They postulate another gauge group, with Extended Technicolor (ETC) gauge bosons that couple ordinary fermions to technifermions, as shown.

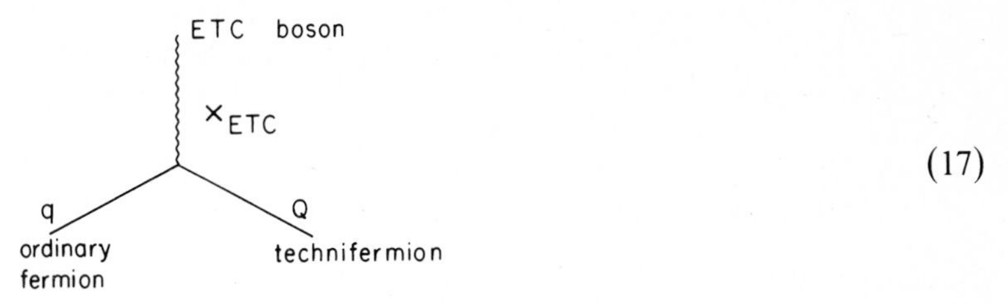


(17)

Then one can have contributions

×ETC
$q_L$ $Q_L$ $Q_R$ $q_R$

(18)

where a left-handed ordinary fermion turns into a right-handed one, which means that this fermion is massive. The $Q_L - Q_R$ transition inside the diagram is the $\langle \bar{Q}Q \rangle$ "condensate" postulated before.

## 8. Pseudo-Goldstone bosons

To see where this leads us, we have to make a somewhat realistic model, and we follow one first studied by Dimopoulos [24]; more detailed calculations were done in ref. [25] and are used extensively in the following. We assume we have a set of massless colored techniquarks $U_c$,

$D_c$ and technileptons N, E, all with standard electroweak structure. *Assuming* the technicolor force does not distinguish among these, we can again rotate the eight left-handed states into one another, and separately the eight right-handed states. As before the condensates $\langle \bar{U}_c U_c \rangle = \langle \bar{D}_c D_c \rangle = \langle \bar{N} N \rangle = \langle \bar{E} E \rangle$ break the symmetry, giving $63 + 63 - 63$ Goldstone bosons. Three of these are for the $W^{\pm} Z^0$, sixty massless pseudoscalars are left to find experimentally. As above, those with nonzero color or electroweak quantum numbers will get some *calculable* mass.

To see the quantum numbers of some of these states, let $\lambda^a$ be the $SU(3)_c$ matrices, $\tau^i$ the $SU(2)_L$ matrices. All are pseudoscalars. Then we have:

a color octet and SU(2) singlet:

$$\eta_t^c = \bar{Q}\gamma_5 \lambda^c Q, \tag{19}$$

some charged, colored states:

$$\Pi_{Ti}^c = \bar{Q}\gamma_5 \lambda^c \tau^i Q, \tag{20}$$

some leptoquark triplets:

$$\bar{L}\gamma_5 \tau^i Q^c, \tag{21}$$

and an electroweak triplet and singlet of color singlet states:

$$\bar{Q}\gamma_5 \tau^i Q - 3\bar{L}\gamma_5 \tau^i L, \qquad \bar{Q}\gamma_5 Q - 3\bar{L}\gamma_5 L. \tag{22}$$

These are written with a 3 (because of color) so they are orthogonal to the states $\bar{Q}\gamma_5 \tau^i Q + \bar{L}\gamma_5 \tau^i L$, which become the longitudinal $W^{\pm}, Z^0$. However, these choices are not rigidly determined by the physics introduced so far. They affect the decay branching ratios of the particles we will talk about, and it is an important weakness of the approach that no one knows how to determine these wave functions.

These states [which we call collectively pseudo-Goldstone bosons (PGB's)] get mass on three scales.

(A) The colored PGB's get mass from a gluon interaction. It must be

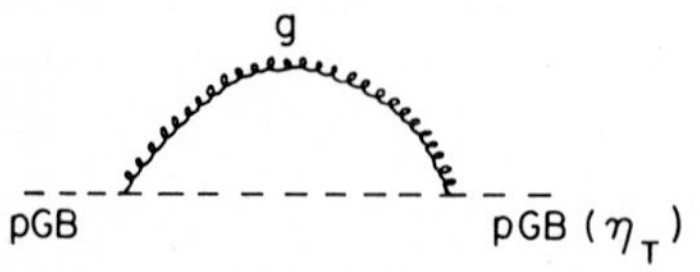

g

pGB pGB ($\eta_T$) (23)

$m^2 \sim \alpha_s \Lambda_{TC}^2$, where $\Lambda_{TC}$ is the TC mass scale. We guess $\Lambda_{TC} \approx 2000$

$\Lambda_{\text{QCD}} \approx 600$ GeV, and with $\alpha_s = 0.15$, we find:

$$m \approx 250 \text{ GeV}. \tag{24}$$

Calculations [25] indeed give $m(\eta_T^c) \approx 250$ GeV, with similar results for the charged color octets, and $m \approx 160$ GeV for the triplet leptoquark states with small Clebsch–Gordan factors. This state is one of the most important predictions of the theory, and can be looked for at FNAL and ISABELLE (see below). It will occur in any dynamical theory where the new fermions carry ordinary color as well as a new force—and it is not easy to imagine giving fermions mass without colored technifermions. So the existence of such a state, while not a theorem, is a rather general prediction of any dynamical approach.

(B) Some PGB's are color singlet but charged, so they get mass from electroweak interactions. This is similar to the pion electromagnetic mass calculations, with a $\rho_{\text{TC}}$ intermediate state, since $\pi \leftrightarrow \rho\gamma$,

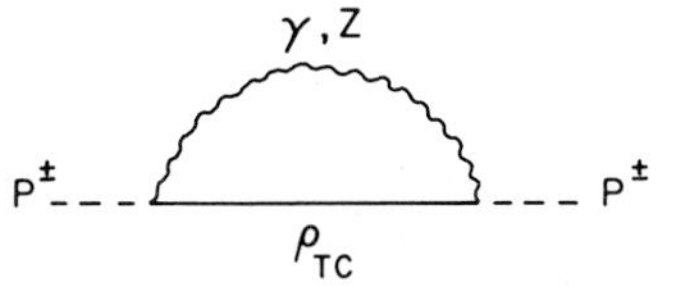

γ, Z
P± P±
ρTC
(25)

and gives:

$$m_\pm^2 \approx \frac{3\alpha}{4\pi} m_Z^2 \ln\left(m^2\rho_T/m_Z^2\right) \approx (7 \text{ GeV})^2. \tag{26}$$

It is important to notice that there would be a cancellation of $\gamma$,Z if $m_Z = m_\gamma = 0$, so this is of order $\alpha^2$ (directly, $m_Z \sim \alpha$). Numerically it gives [23, 25, 26] $m_\pm \approx 7$ GeV (the uncertainty is of order $\pm 1$ GeV); Peskin and Chadha [27] say that in some models one can find $m_\pm$ in the range 10–14 GeV. On our mass scale these states are indistinguishable from any charged Higgses. We will discuss below how to detect them.

(C) Have all the PGB's gotten mass at this stage? Surprisingly not [23]. How do we tell? We have to stare at the states $\bar{Q}\gamma_5 TQ, Q\gamma_5 TL, \ldots$ where $T$ is $\lambda^c$ or $\tau^i$, and see if any combinations that we can form still commute with $SU(3)_c \times SU(2)_L \times U(1)$.

They will commute with $SU(3)_c$ if they are color singlets. If we make them right handed they will commute with $SU(2)_L$. If they are diagonal they will commute with U(1), and there are two diagonal color singlet generators, $\tau^3$ and 1. Consequently the two currents

$$J_\mu^{3,0} = \bar{Q}\gamma_\mu(1+\gamma_5)\begin{Bmatrix}\tau^3\\1\end{Bmatrix}Q - 3\bar{L}\gamma_\mu(1+\gamma_5)\begin{Bmatrix}\tau^3\\1\end{Bmatrix}L, \tag{27}$$

are both still conserved, and their fourth components have a pseudoscalar piece, so they couple to the neutral PGB's $P^3, P^0$, with

$$\langle 0|J_\mu^{3,0}|P^{3,0}\rangle = iF_T q_\mu. \tag{28}$$

If we do a standard thing and take the divergence of this, we have $q_\mu J_\mu^{3,0} = 0$. But $q^2 = m_{3,0}^2$, so $m_{3,0}^2 = 0$ as expected. This is not allowed—$P^{3,0}$ couple to ordinary q, $\ell$ and are like ordinary axions. They would be produced in hadron reactions and seen by secondary interactions in beam dump experiments, but they are not. They are probably also excluded by energy loss mechanisms in stars.

Now the trouble really starts. We must give the $P^{3,0}$ some mass, so that there is no conflict with experiment. But no one knows a good way to do it, and the few ways tried so far have led to a contradiction. Because of that, I will mention one, and the implications, but not describe it in detail.

We can introduce an additional set of gauge bosons, which couple TQ and TL, perhaps along the lines suggested by Pati and Salam. In general this gives a mass to $P^{3,0}$. If one estimates that mass, assuming the associated gauge boson is somewhat heavier than the ETC bosons, one finds $m_{3,0} \sim 50$ GeV, to within perhaps a factor of 2 or so. But the same physics will also lead to flavor changing neutral currents (FCNC) that do not occur, e.g.

(29)

will give $K_L \to \mu e$. The observed limit on that process requires:

$$m_{0,3} \leqslant 3 \text{ GeV}. \tag{30}$$

The same mechanism will give contributions to the charged $P^{\pm}$. If the contribution is $\leqslant 3$ GeV, they add quadratically, and there is little effect on $m^{\pm}$. But if the way to avoid FCNC is through a symmetry and the contribution is more like 50 GeV, the charged mass goes up accordingly. Any dynamical theory where electrically charged, color singlet, Goldstone bosons occur will give $P^{\pm}$. The electroweak part of their mass is about 7 GeV. Whether other, weaker interactions give them more mass is presently unknown.

Finally, the best numbers we can presently estimate for the predicted particle states are

$$m_{\pm} \approx 8\ \text{GeV}, \qquad m_{0,3} \approx 3\ \text{GeV},$$
$$m(\eta_{\mathrm{T}}^{\mathrm{c}}) \approx 250\ \text{GeV}, \qquad m(Q\gamma_5 L) \approx 160\ \text{GeV}, \tag{31}$$

where we mentioned only the states that we will discuss below. How do we find these?

## 9. Detecting pseudo-Goldstone bosons

### 9.1. *Light color singlet neutrals $P^0, P^3$*

The situation here is essentially identical to that for any neutral Higgs, and the remarks in section 5 directly apply [32].

### 9.2. *Heavier color octet neutrals; $\eta_T^c$*

The most important state is probably the colored octet of electroweak singlets, $\eta_{\mathrm{T}}^{\mathrm{c}}$. It was first discussed by Dimopoulos [24], and then studied in more detail in refs. [25, 26, 28, 29]. It is significant because: (i) It will be present in any dynamical theory with colored new fermions. (ii) Its mass is not so model dependent. For instance, since masses add in the square, a 50 GeV effect that was extreme for $P^{\pm}$ only gives a 5 GeV shift here. (iii) If the whole approach makes any sense at all, we can use the chiral symmetry techniques of soft pion physics and related theorems.

In particular, the coupling of $\eta_{\mathrm{T}}^{\mathrm{c}}$ to two gluons is analogous to the coupling $\pi \to \gamma\gamma$, i.e. it is through the anomaly diagram to the same approximation, as shown:

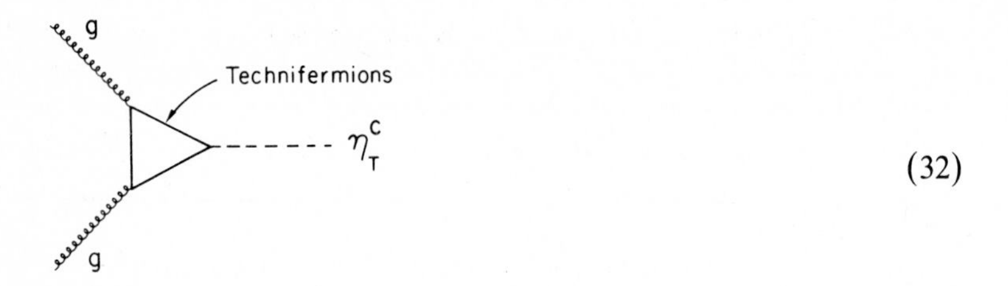


(32)

This determines [25] the $\eta_T^c$ partial width to two gluons, and more important, the production cross section in hadron reactions, by attaching hadron quark lines to gluons. This gives [25]:

$$\left.\frac{d\sigma}{dy}\right|_{y=0} = \pi^2 \frac{\Gamma(\eta_T^c \to gg)}{m^3(\eta_T^c)} G(x_1)G(x_2) \tag{33}$$

where $y$ is the rapidity, and $G(x)$ the gluon distribution function in the hadron. From the appropriate generalization of $\pi^0 \to \gamma\gamma$,

$$\Gamma(\eta_T^c \to gg) = \frac{5}{384}\frac{N^2}{\pi^2}\frac{m^3(\eta_T^c)}{F_T^2}\alpha_s^2 \tag{34}$$

where there are three colors, two flavors, and $N$ technicolors summed over in the fermion loop, and $\alpha_s$ is the strong coupling, evaluated at $m(\eta_T^c)$. Numerically, $\Gamma(gg) \approx 60$ MeV. One can take gluon distribution functions $G(x) \approx 3(1-x)^5$ if scaling violations can be ignored, which is true for $x \leqslant 0.15$. Near $y = 0$, $x \approx m(\eta_T^c)/s$. If $x \ll 0.15$, then scaling violations increase the cross section, and if $x \gg 0.15$, scaling violations decrease the cross section. Because $\eta_T^c$ is colored, any member of the octet can be produced and this cross section is eight times as large as that of an uncolored Higgs of the same mass. Since there are $N$ technicolors in the loop in addition to flavors and colors, this cross section is an additional factor $N^2$ as large as it would be for a fundamental, standard model Higgs ($N \geqslant 4$ so the TC interaction gets strong at a higher mass scale). Thus $\sigma(\eta_T^c)$ is at least 128 times that for a standard model Higgs of the same mass.

Note that $\Gamma(gg) \sim m^3$, and $\sigma \sim \Gamma/m^3$, so apart from the reasonably well known gluon distribution functions, $\sigma$ is independent of $m(\eta_T^c)$. This makes comparison of different machines easy. Figure 1 shows a plot of $\sigma \mathcal{L} T =$ number of events, for interesting luminosities $\mathcal{L}$ at several machines, and $T = 10^7$ s, about what could be hoped for in a year.

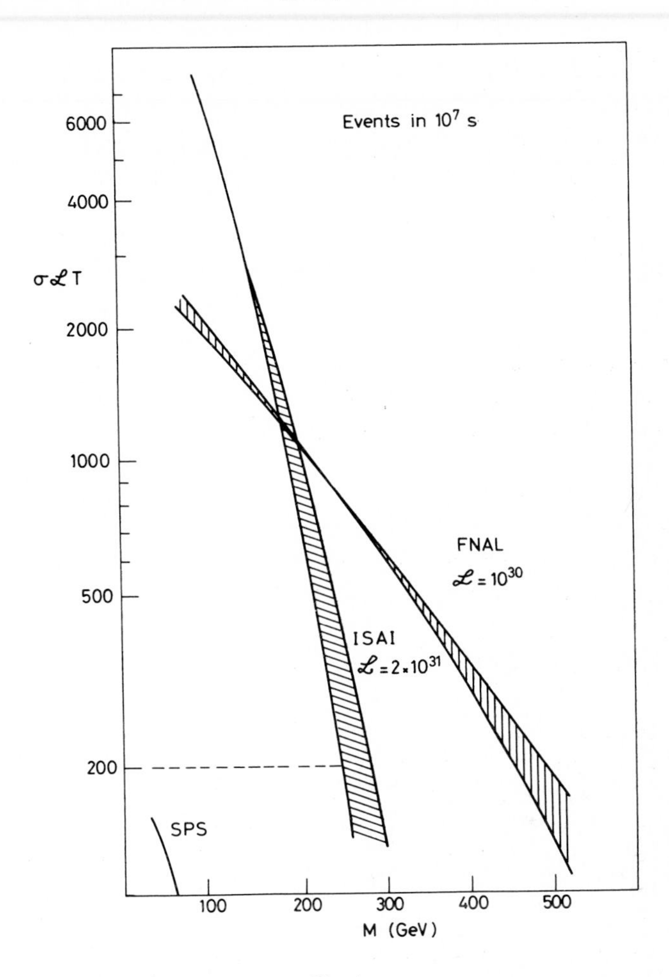


Fig. 1.

Depending on detectors, perhaps 200 events would be enough to guarantee seeing $\eta^c_T$. The shaded regions represent an estimate of the effect of scaling violations and probably one should use the lower side. The reader can correct to other $\mathcal{L}$ or $T$ easily. It is clear that FNAL and, depending on the mass, ISABELLE, can find $\eta^c_T$ (the ISABELLE curve is drawn for $E_{TOT} = 710$ GeV, FNAL for 2000 GeV).

How will $\eta^c_T$ be detected? Its dominant decay is expected to be to the heaviest fermion, as for all Higgs-like objects. Assuming a Yukawa coupling at the $\eta_T t\bar{t}$ vertex with a factor $m_t/F_T$, gives a partial width $\Gamma(\eta_T^c \to t\bar{t}) \approx 1$ GeV if $m_t = 25$ GeV. This is essentially zero on the scale of most detectors, though perhaps eventually such a width could be detected. As with all fermion couplings, the result is model dependent. Similarly, $\Gamma(\eta_T^c \to b\bar{b}) \approx 60$ MeV $\approx \Gamma(\eta_T^c \to gg)$. The decays to $b\bar{b}$ and gg will have lots of background and will be difficult to separate out, though it is possible.

The $t\bar{t}$ mode will be easier. If there is a charged Higgs for any reason, e.g. if $m(P^{\pm}) \approx 8$ GeV, then $t \to P^+ b$ is the dominant mode, and:

$$\begin{array}{l} \eta_T^c \longrightarrow t\bar{t} \searrow \\ \qquad\quad \swarrow \qquad \bar{b}P^- . \\ \quad bP^+ \end{array} \tag{35}$$

Then one has either:

(a) if $P^+$ decays hadronically there are four jets. They reconstruct in pairs to a t quark mass, and then, holding $m_t$ fixed, they give a resonance at $m\,(\eta_T^c)$, or

(b) if $P^+$ decays via $P^+ \to \tau^+ \nu$ then there is (about 2/3 of the time) a single energetic charged track from the $\tau$ and significant missing energy in the neutrinos. About half the time the single charged track is a lepton ($\mu$ or e).

With detectors that have been planned to look for such physics and have appropriate capabilities built in, it should be fairly easy to detect such events. With $BR(\eta_T^c \to t\bar{t}) \sim 90\%$, perhaps the effective detection efficiency for $\eta_T^c$ could be as high as $\frac{2}{3}$.

The only things that affect such estimates are $m_t$ and the interesting possibility that an even heavier fermion exists and could dominate the decays—production via $\eta_T^c$ might be the dominant process for a fermion, and once $\eta_T^c$ is found we will immediately know whether additional fermions exist up to half $m(\eta_T^c)$.

There are several other interesting decay modes of $\eta_T^c$, such as $GG, GZ^0, G\gamma, GW^+, W^-$, etc, but none are larger than a few percent so they will be for later study. Once a signal is found it will not be difficult to decide what it is. The size of the cross section is characteristic. It can easily be distinguished from $Z^0$ or a vector quarkonium state by the decay pattern. $Z^0$'s have equal and large branching ratios for $\mu^+\mu^-, e^-e^-, \nu\bar{\nu}$ while $\eta_T^c$ does not have these modes. It will violate $t\bar{t}, b\bar{b}, c\bar{c}$ universality. It has a gg mode while a vector quarkonium state has a ggg mode, and so on.

There have been useful calculations of various distributions by Hayot and Napoly [28], and by Girardi et al. [29], particularly the latter paper has a number of helpful analyses for studying the details of detecting $\eta_T^c$.

### 9.3. *Light, charged pseudo-Goldstone bosons*

The 8 GeV $P^{\pm}$ (see comments about its mass above) will be produced [23, 25, 31] in $e^+e^-$ reactions:

(36)

The cross section is standard for spin zero particles, giving $\frac{1}{4}$ unit of $R$ ($\frac{1}{4}$ as much as a $\mu^+\mu^-$ pair), rising from threshold as $\beta^3 = (1-4m_{\pm}^2/s)^{3/2}$, and with a $\sin^2\theta$ angular production distribution. The factor $\beta^3$ does not quickly rise to unity, giving almost a factor two suppression at $E_{TOT} = 30$ GeV. $P^{\pm}$ being scalars will decay with flat distributions in their rest frames.

$P^{\pm}$ can decay into various fermion pairs, and estimates of branching ratios are model dependent. The pair $c\bar{b}$ is heaviest, but might occur with a mixing angle factor (say $\sin\varepsilon$) that suppresses it. The decays proceed via t-channel ETC bosons,

(37)

so there are no color factors if the ETC bosons are color singlets. A reasonable guess for the decay pattern is as in table 1. The branching ratio $x$ is unknown, but I like the model wave functions given above, which give $x \lesssim 0.1$.

While searching for charged Higgses and $P^{\pm}$ has been given a very low priority at PETRA, recently two groups have presented informal negative results. H. Meyer [33], from the Pluto detector, has estimated that in fig. 3 of ref. [34], the region with thrust below 0.8 and $\Sigma|\boldsymbol{P}_T|/\Sigma|\boldsymbol{P}| \gtrsim 0.6$

Table 1
Width and branching ratio of various $P^+$ decays.

| Mode | $\Gamma \sim$ | *BR* (guess) |
|---|---|---|
| $P^+ \to c\bar{b}$ | $(m_b + m_c)^2 \sin^2 \varepsilon$ | $x$ |
| $P^+ \to c\bar{s}$ | $(m_c + m_s)^2$ | $x' \approx x$ |
| $P^+ \to \tau\nu$ | $m_\tau^2$ | $1-2x$ |
| $P^+ \to \mu\nu$ | $m_\mu^2$ | $\sim 0.1\%$ |
| $P^+ \to e\nu$ | $m_e^2$ | $10^{-5}$ |
| $P^+ \to \gamma\ell^+\nu$ | | ? |

should have had 8–10 events from $P^\pm$, while none are observed. An analysis of data by the JADE group has been presented [35], which claims to exclude a mass range 5–15 GeV and $\tau\nu$ branching ratios of 5–95%, at the 90% confidence level. It is important that these results be explained in detail and published. Do they (or can they soon) exclude a 90%–100% $\tau\nu$ branching ratio?

How disastrous is this for TC? As discussed above, the electroweak contribution to the $P^\pm$ mass is about 7 GeV (or perhaps up to twice that in some models). That does not seem to be in question. The problem is that to avoid massless neutral PGB's there must be another interaction, and it is not clear how much mass it gives. If its strength is directly limited [25] to avoid FCNC, it gives very little mass. If somehow FCNC are avoided by clever symmetries, which would not be unreasonable, the new interaction could give more mass. Since it is a weak and unknown interaction, it presumably does not give much mass.

So certainly I expect $m_\pm$ is not too large, and its seems reasonable that it should be below 15 GeV. But since no one understands the detailed structure of the theory well, it is very important to search seriously up to masses $\lesssim 50$ GeV. Remember, so far TC is one of the two approaches that may help to understand the origin of mass, and any possible experimental input is very important.

## 10. t Decay and $P^\pm$

As remarked briefly in the discussion of $\eta_T^c \to t\bar{t}$, if there exists any charged Higgs object, whatever its origin, it will couple t and b semiweakly,

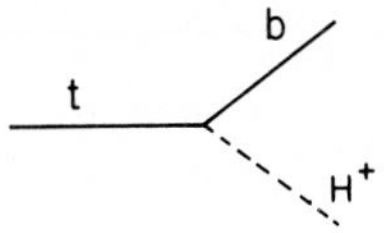

(38)

and completely dominate t decay. Then the usual mode,

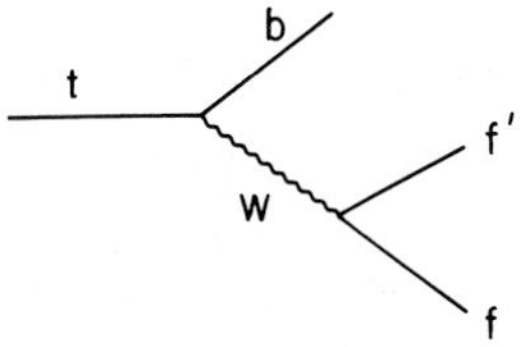

(39)

will not be observable, and all searches for t decay must look for the former mode. Because of this connection there is a good chance that the t quark and the charged Higgs(es) will be discovered simultaneously by the first machine to produce t's.

At lower energy $e^+e^-$ machines, the t searches would proceed a little differently if $t \to bH^+$ dominated, especially if $H^+$ decays like $H^+ \to \tau\nu$ most of the time so that on average $\frac{1}{3}$ of the energy is missing into neutrinos; but because of the high multiplicity b decays, no basic change in results would be expected.

## 11. TC and the parity of pseudo-Goldstone bosons

One way to establish that one is observing a dynamical theory like TC, is to study the parity properties of neutral PGB's after they are discovered. However, the analysis is very subtle and most early remarks in the literature were oversimplified.

A PGB decays into ordinary fermions as:

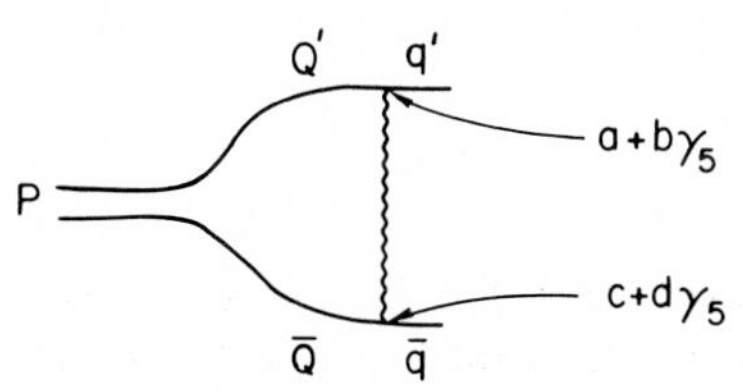

(40)

At each vertex there is a coupling, related to the fermion masses, of the general form $a + b\gamma_5$, perhaps different at each vertex. Then the effective Hamiltonian for the decay will be $\bar{q}(A + B\gamma_5)qP$ where $A$ and $B$ can have any relative value. In particular, though $P$ is a pseudoscalar in the space of $Q\bar{Q}$, it can have any apparent parity for $\bar{q}q'$, including scalar. So the best one could hope for to detect the presence of a dynamical boson from its fermion decays, is to *compare* decays into several fermion channels and find that the parities are *different*. No single measurement can clarify the situation. For example, one could observe both $K\bar{K}$ and $K^*\bar{K}$ (or $\rho\pi$) from a given state, showing it has a mixed parity, or compare $K\bar{K}$ and $\mu^+\mu^-$ modes. This applies to $\eta_T^c$ as well as $P^0, P^3$.

The decays to a pair of gauge bosons ($\gamma\gamma$ or gg or $W^+W^-$ or $Z^0Z^0$ or $gZ^0$) are via the coupling to the actual technifermions, so they do directly reflect the parity:

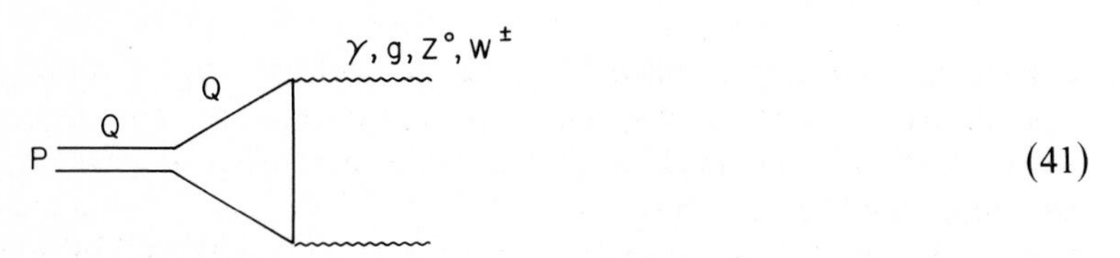


(41)

Thus a measurement of the parity is in principle possible here, by comparing [36] the polarization planes of the two final gauge bosons, just as for $\pi^0 \to \gamma\gamma, \gamma \to e^+e^-$, one can measure the $\pi^0$ parity. Alternatively, one could produce P (or $\eta_T^c$) with polarized $\gamma$'s or polarized gluons. While some valid statements can be made, it is doubtful that a practical distinction between theories can be made from any method.

## 12. Flavor-changing neutral currents

This subject will be covered in other lectures, so I only mention it here for completeness. As it stands, there is no guarantee that TC interactions —exchange of leptoquark PGB's $\bar{Q}\gamma_5 L$, or exchange of Pati–Salam bosons, or exchange of light neutral $P^0, P^3$—will not cause reactions [25, 39] which violate existing limits on $\Delta m(K_L - K_s)$, $\Gamma(K_L \to \mu e)$, etc. Indeed, naive estimates suggest such violations will occur. The main lessons are: (i) In any given model, look carefully for FCNC interactions. (ii) Without interesting cancellations, probably there will be contradictions with some existing experimental data. Some clever theory leading to XYZ... mechanisms is required.

Speaking more generally, as mentioned earlier it is expected that whatever their origin there will be two or more Higgs doublets. This can lead to rare flavor changing decays, and any search for them should be strongly encouraged.

## 13. What if $m_H \sim$ TeV—strong interactions at 1 TeV?

To conclude we briefly discuss the situation if there are no low energy phenomena associated with Higgs physics. How can we learn about it?

One might think that, because the Higgs role is to make the theory renormalizable, with no Higgs or with very heavy Higgs one gets large unitarity violations and perhaps "strong" weak interactions at TeV energies. Unfortunately, the latter is probably not the case.

To see why, recall [5] that in a reaction such as $f\bar{f} \to W^+W^-$ the unitarity violations show up in the production of longitudinally polarized $W_L^{\pm}$ states. With just the diagrams of fig. 2a the amplitude $M$ grows like $s$. When the $Z^0$ contribution (fig. 2b) is added with the gauge theory couplings, the leading power is cancelled, but there is still a term growing like $\sqrt{s}$, $M \sim m_f\sqrt{s}$, and one can show that the $\sqrt{s}$ is replaced by a factor $m_f$. It is this $\sqrt{s}$ growth that is cancelled by the Higgs contribution of fig. 2c, which is related to the factor of $m_f$ in the Higgs–fermion coupling and to the fermion getting its mass from the Higgs interaction. Unfortunately, it means that numerically the strong effects of a large Higgs interaction are masked [7] by factors of $m_f/m_W$ to a power, and this is very serious, since all the beams we have to initiate experiments are composed of very light fermions, with $m_f/m_W \lesssim 10^{-3}$. [This screening is different from that of Veltman [6] or of Appelquist [37] and collaborators, where powers of $g^2$ also come in to reduce the effect of Higgs interactions on low energy parameters.]

One might think that the effects could be propagated through a box diagram such as fig. 2d, but one can still show [7] that a factor of $(m_f/m_W)^2$ enters when the full calculation is done. So far the only way to get a bigger effect that has been studied [7, 11], is to introduce models with at least two Higgs doublets, where some couplings are enhanced because the ratio of vacuum expectation is not unity, and some effects correspondingly larger, but there is no nice model of this nature; in supersymmetric theories with two doublets the ratio of vacuum expectation values is constrained [38] by the mass spectrum to be unity or very nearly so, in some models but not in others.

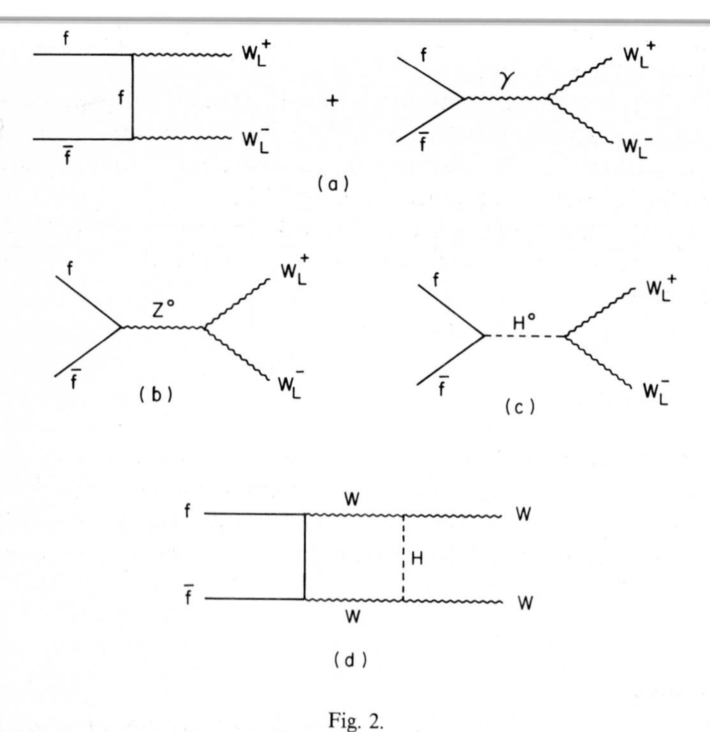


Fig. 2.

With machines that are coming the situation is not hopeless for finding such effects if they could be seen. The constituent subenergy is $\hat{s} = x_1 x_2 s$, so the effective center-of-mass energy is about $1/5$ of the actual $\sqrt{s}$, which might be large enough for the tail of an effect. But more thinking is needed to understand whether any way can be found to expect such an effect, and also on how it might show up. Will we see unusual violations of *CP*, or multiple productions of $W^{\pm}, Z^0$? Strong isospin violations? In trying to understand fermion masses one is led to interesting notions.

## 14. Conclusions

To understand mass some new physics is needed. Technicolor is a serious possibility of some attractiveness, although as the theory develops it gets more complicated than one would like.

Any dynamical theory with color singlet pseudo-Goldstone bosons will have a charged pair $P^{\pm}$ whose mass is likely to be of order 8 GeV.

PETRA and PEP can definitively find or exclude these states up to their top mass in the near future.

Any dynamical theory with new colored fermions will have a colored $\eta$-like state with mass of order 250 GeV—its properties are rather reliably predicted by the theory, and it can surely be found or excluded by the FNAL Tevatron collider or ISABELLE.

So far, the development of the technicolor theory has been good, but not subtle or powerful. There is considerable room for theorists to build a better theory.

## Acknowledgements

I would like to thank the organizers of the "Les Houches" school for the opportunity to give these lectures, and particularly M. K. Gaillard for generous hospitality at "Les Houches." I am grateful to Stuart Raby, Savas Dimopoulos, and J. D. Bjorken for many discussions on technicolor physics.

## References

[1] K. Lane and M. Peskin, Proc. XVth Rencontre de Moriond (1980), ed. J. Tran Thanh Van.
[2] E. Eichten, Proc. 1980 Vanderbilt $e^+e^-$ conf.
[3] E. Fahri and L. Susskind, CERN preprint TH 2975 (1980).
[4] P. Sikivie, Varenna lectures (1980).
[5] See the lectures of J. Ellis (course 3, this volume) for a complete treatment.
[6] M. Veltman, Acta. Phys. Pol. B8 (1977) 475.
[7] H.E. Haber and G.L. Kane, Nucl. Phys. B144 (1978) 525.
[8] S. Weinberg, Phys. Rev. D7 (1973) 2887.
[9] See R. Gatto and collaborators, and G. Segrè and collaborators.
[10] Recent papers by A. Sanda (Rockefeller preprint) and N. Deshpande (Oregon preprint) suggest this is unlikely.
[11] H.E. Haber, G.L. Kane and T. Sterling, Nucl. Phys. B161 (1979) 493.
[12] D.R.T. Jones, G.L. Kane and J.P. Leveille, Univ. of Michigan preprint UM HE 81-45 (1981).
[13] S. Coleman and E. Weinberg, Phys. Rev. D7 (1973) 1888.
[14] D.A. Ross and M. Veltman, Nucl. Phys. B95 (1975) 135.
[15] J.E. Kim et al., Rev. Mod. Phys. 53 (1980) 211.
[16] J. Ellis, M.K. Gaillard and D.V. Nanopoulos, Nucl. Phys. B106 (1976) 292.
[17] J.F. Donoghue and L.-F. Li, Phys. Rev. D19 (1979) 945.
[18] J. Leveille, Nucl. Phys. B137 (1978) 63.
[19] See the analysis of ref. [12].

[20] F. Wilczek, Phys. Rev. Lett. 39 (1977) 1304. The results hold for scalar or pseudoscalar bosons.
[21] E. Bloom, private communication.
[22] S. Dimopoulos and L. Susskind, Nucl. Phys. B155 (1979) 237.
[23] E. Eichten and K. Lane, Phys. Lett. 90B (1980) 125.
[24] S. Dimopoulos, Nucl. Phys. B168 (1980) 69.
[25] S. Dimopoulos, S. Raby and G.L. Kane, Nucl. Phys. B182 (1981) 77.
[26] J.D. Bjorken, unpublished.
[27] S. Chadha and M. Peskin, CERN preprints TH-3023, TH-3038 (1980).
[28] F. Hayot and O. Napoly, Saclay preprint.
[29] G. Girardi, P. Mery, and P. Sorba, CERN preprint.
[30] M.A.B. Bèg, in: Proc. VPI Weak Interaction Workshop, Blacksburg, VA (1980).
[31] M.A.B. Bèg, H.D. Politzer and P. Ramond, Phys. Rev. Lett. 43 (1979) 1701.
[32] See also J. Ellis, D.V. Nanopoulos, and P. Sikivie, Phys. Lett. 101B (1981) 387.
J. Ellis, M.K. Gaillard, D.V. Nanopoulos and P. Sikivie, Nucl. Phys. B182 (1981) 529.
[33] H. Meyer, private communication.
[34] Ch. Berger et al., Phys. Lett. 99B (1980) 489.
[35] P. Duinker, talk at ISABELLE study (July 1981).
[36] See, for example, G.J. Gounaris and A. Nicolaidis, Phys. Lett. 102B (1981) 144. The expected rates are very small.
[37] T. Appelquist and C. Bernard, Phys. Rev. D22 (1980) 200.
A.C. Longhitano, Phys. Rev. D22 (1980) 1166.
[38] I thank S. Raby for emphasizing this point.
[39] S. Dimopoulos and J. Ellis, Nucl. Phys. B182 (1980) 505.

SEMINAR 4

# WAITING FOR THE PROTON TO DECAY

## *A Comparison of the New Experiments*

Lawrence R. SULAK

*Randall Laboratory, University of Michigan*
*Ann Arbor, MI 48109, USA*

## Contents



*M.K. Gaillard and R. Stora, eds.*
*Les Houches, Session XXXVII, 1981*
*Théories de jauge en physique des hautes énergies/Gauge theories in high energy physics*

## 1. Introduction

In the past three years, at least nine dedicated attempts to measure the proton lifetime have been initiated. This burgeoning effort has been stimulated by recent compelling physics arguments that challenge the permanence of the proton.

Speculation on the possible demise of the nucleon was absent until Sakharov's precocious work [1] in 1966, where he argued that the baryon excess in the universe implied that the proton is unstable. An independent line of reasoning, based on a desire to unify the theories of the strong force and the electroweak force, has inspired several authors (Pati and Salam; Georgi and Glashow; Georgi et al., [2]) to include both quarks and leptons in the same multiplets. The virtual transitions of quarks into leptons and antiquarks within these multiplets, albeit slow, naturally give rise to nucleon instability.

The lifetimes which are predicted by Grand Unified Theories ($10^{31\pm1}$ yr) are tantalizingly close to the old experimental limits. Moreover, they are just within reach of the new detector technologies developed for large neutrino experiments at accelerators. Therefore, many groups are now in the process of preparing or proposing experiments. This paper will present a comparative study of the new experiments.

## 2. General features of the new proton decay detectors

The new detectors fall into two classes: totally-active water Čerenkov detectors with light collected by phototubes, and sampling calorimeters with particle ionization tracked by gas tube arrays. I will discuss one of each of these detectors in detail, point out the features of other detectors in the same class, and compare them all in tables 1 and 2.

But first let us say a word about the depth of the detectors. The greater the shielding above an experiment, the lower the cosmic ray muon rate, and the slower the electronics and data acquisition system must work. On the other hand, the detectors at the greatest depths are generally forced (by the limited size of the cavities found there) to have a source material with a high density to achieve a given minimum detector mass. Another depth dependent consideration is the necessity of distinguishing cosmic ray muon-induced background from potential proton decay events. At the relatively shallow depth of three of the mine experiments in the USA, typically $10^8$ muons traverse a detector per year. This places a burden on pattern recognition if one is eventually to see a few proton decay events per year. In the deep Alpine tunnels of Europe, and the still deeper Kolar gold fields, muon-induced background is no problem. On the other hand, monitoring the minute by minute integrity of the detector is aided by frequent muon traversals.

## 3. Čerenkov devices

### *3.1. The Irvine–Michigan–Brookhaven detector in the Morton Salt Mine*

This review of dedicated counter experiments begins with a discussion of the one with which I am associated, the Irvine–Michigan–Brookhaven (IMB) experiment [3]. The detector is a 10 kiloton cube of water that will be viewed by 2048 PM tubes on the faces of the cube.

The size of this detector is natural for two reasons. First, to test definitively the predictions for the proton lifetime based on SU(5) and related higher groups, one needs $\geqslant 10^{33}$ nucleons in the fiducial volume to reach a lifetime limit of $10^{33}$ yr in a finite experimental period. Experience with total absorption neutrino detectors shows that about one half of the events are unambiguously contained within the total volume. Applying this rule of thumb to a proton decay calorimeter, a 10 kt detector includes 5 kt (i.e. $4\times10^{33}$ nucleons) of fiducial mass. Secondly, at this size of detector, the inherent neutrino-induced background mimics proton decay at the rate of one event per year. Any larger detector would require a statistical subtraction of the background, and therefore, its sensitivity would only grow as the square root of its mass.

Water was a natural choice for the detection medium. We needed a very large mass of material at a reasonable cost and simultaneously desired a totally-active device with information on particle directionality. If a proton were to decay in water, energetic, charged secondaries would produce Čerenkov radiation. This light would travel to the walls of the detector where it could be intercepted by PM tubes. By detecting the time of arrival and the intensity of the radiation, one could reconstruct the decay event.

Having chosen a large mass of low density material, we began searching for a large, deep cavity. After an extensive study of possible sites, we

Table 1
Status of nucleon lifetime experiments—Čerenkov technique (October 1981)

| Aspect | Collaborative Institutions | | | |
|---|---|---|---|---|
| | Univ. of California, Irvine<br>Univ. of Michigan<br>Brookhaven National Lab. | Univ. of Pennsylvania<br>Brookhaven National Lab. | Harvard Univ.<br>Purdue Univ.<br>Univ. of Wisconsin | KEK<br>Univ. of Tokyo<br>Univ. of Tsukuba |
| Location | Morton Salt Mine<br>Painesville, OH | Homestake Gold Mine<br>Lead, SD | Silver King Mine<br>Park City, UT | Kamioka |
| Depth (kmwe)[a] | 1.7 | 4.4 | 1.7 | 2.7 |
| Detector mass (kiloton) | 10 kt total<br>~ 5 kt fiducial | 0.3 kt total<br>0.15 kt fiducial | 0.8 kt water, surrounded by mirrors and lead/gas tube shower counter | 3.4 kt total<br>1.0 kt fiducial |
| Cosmic muon flux | $3\mu/s$,<br>$10^8\mu/yr$ | $\sim 3\times 10^5\mu/yr$ | $\sim 1\mu/s$,<br>$\sim 3\times 10^7\mu/yr$ | $6\text{–}7\times 10^6\mu/yr$ |

| | | | | |
|---|---|---|---|---|
| Nucleon source | cube of water, high transparency | water | cylinder of water | cylinder of pure water |
| Detection method | 2048 5″ PM's on 1.2 m surface grid | 144 5″ PM's, 4 each, in 4 $m^3$ optically segmented units; waveshifted; separate cosmic veto on top face | 800 5″ PM's on 1 m ~ cubic lattice; waveshifter (both with and without) | 1044 20″ PM's on 1 m surface grid |
| Ultimate lifetime | $\sim 10^{33}$ yr | $\sim 10^{31}$ yr | $\sim 10^{32}$ yr | $\sim 10^{32}$ yr |
| Present status | Cubical reservoir completed; operational ~ winter 1981 | Current result: $\tau > 2.4 \times 10^{31}$ $B\mu$ yr; For $B\mu = 5.1\%$, $\tau > 1.2 \times 10^{30}$ yr; Proposal for 1 kt detector submitted | Installation of PM's into cylindrical tub underway; operational ~ winter 1981 | Under construction; operational ~ summer 1982 |
| Ref. | [3] | [5] | [6] | [7] |

[a] km water equivalent.

Table 2

Status of nucleon lifetime experiments—sampling calorimeter technique (October 1981)

| Aspect | Collaborative Institutions | | | | |
|---|---|---|---|---|---|
| | Tata Inst. Bombay<br>Oska City Univ.<br>Univ. of Tokyo | CERN<br>INFN Frascati<br>Univ. of Milan<br>Univ. of Turin | Orsay<br>Ecole Polytechnique<br>Saclay<br>Wuppertal | Univ. of Minnesota<br>Argonne National Lab<br>Oxford Univ. | INFN Frascati<br>Univ. of Milan<br>Univ. of Rome<br>Univ. of Turin |
| Location | Kolar Gold Fields,<br>South India | Mont Blanc Auto Tunnel<br>(Garage 17)<br>Franco–Italian Border | Frejus Auto Tunnel<br>Franco–Italian Border | Soudan Iron Mine<br>Vermillion, MI | Grand Sasso<br>Auto Tunnel<br>150 km from Rome |
| Depth (kmwe) | 7.6 | 5.0 | 4.5 | 1.8 | 4.5 |
| Detector mass (kt) | 0.14 kt total<br>~ 0.10 kt fiducial | 0.15 kt total | 1.5 kt total<br>1.0 kt fiducial | 30 t total<br>(prototype) | 1 kt/module<br>10 kt total mass |
| Cosmic muon flux | $2\mu$/day<br>$700\mu$/yr | $\sim 6\mu$/day<br>$2\times10^3\mu$/yr | $1\mu$/min.<br>$5\times10^5\mu$/yr | $\sim 3\times10^6\mu$/yr | $\sim 1\mu$/min/module |
| Nucleon source | 34 horizontal Fe plates,<br>1.2 cm × 4 m × 6 m | 134 horizontal Fe plates,<br>1 cm × 3.5 m × 3.5 m | 1500 vertical 3 mm<br>Fe plates | 48 horizontal slabs of<br>taconite (iron ore)<br>concrete; average<br>density = 1.85 | 650 vertical 3mm Fe<br>plates (8 m square) per<br>1 kt module |

| Detection method | 1600 proportional gas tubes, $(10\ cm)^2 \times \sim 5$ m | 47000 limited streamer tubes, $(1\ cm)^2 \times 3.5$ m | 1500 plastic flash tube planes $(0.5\ cm)^2$ 200 Geiger tube (1.5 cm) planes | 3456 proportional gas tubes, 1″ diam. $\times$ 3 m | $10^6$ flash tube cells $(0.4\ cm)^2$ in 650 $x-y$ planes; 70 limited streamer trigger planes per 1 kt module |
|---|---|---|---|---|---|
| Ultimate lifetime | $\sim 3 \times 10^{31}$ yr | $\sim 4 \times 10^{31}$ yr | $\sim 10^{32}$ yr | $\sim 2 \times 10^{30}$ yr | $\sim 10^{33}$ yr |
| Present status | Operational for 205 days; current results: $\tau = 8.4^{+11}_{-5.5} \times 10^{30}$ yr | Prototype calibrated in CERN beams; detector operational ~ winter 1981 | Prototype calibrated in CERN beams; staging area excavated in tunnel | Operational since April 1981; Proposal for 1 kt detector submitted | Excavation of cavity underway; detector details under discussion. |
| Ref. | [8] | [9] | [10] | [11] | [12] |

chose the Morton Salt Mine near Cleveland, OH. It provided the deepest location in which one could economically excavate a (21 m)$^3$ reservoir. Primary in our considerations was the fact that the excavation of salt costs an order of magnitude less than mining granite. The Morton mine had previously been used by physicists (including Reines and Madame Wu) for a long time, and proved by far to be the most suitable site for a 10 kt device.

Relying on the long distance transmission of light in water presents a unique technological challenge. Although the Čerenkov spectral intensity increases as the frequency squared, ultraviolet light is very rapidly absorbed in water. After the transmission of Čerenkov light through 10 m of water (the mean "radius" of the cubically shaped IMB detector), the light which is left is in the deep blue part of the spectrum (350 nm and 550 nm half-power points), where specially purified water is very transparent. This surviving photon spectrum closely matches the wavelength sensitivity of phototubes with a bialkali photocathode.

To verify these expectations, we constructed a water tank with a 10 m baseline at the University of Michigan. Čerenkov light was produced by cosmic ray muons traversing the water. The muon trajectory was defined by a coincidence between two scintillation counters. Phototubes immersed in the water recorded the light from the Čerenkov cone of the muons. We predicted a signal of 2.5 photoelectrons at 10 m. Three photoelectrons were seen. The water had been purified by the same scheme planned for the mine, while the amount of water circulating per unit time and the surface area of materials in contact with the water were both scaled to match those anticipated for the final detector. In other tests we have established that the mean free path at the wavelength of maximum transmission ($\sim 440$ nm) is $> 40$ m in water purified by reverse osmosis technology.

The eventual installation of the PM's in the detector has been mimicked in a vertical test tank, 21 m deep. The data acquisition system has also been put through its paces at Michigan on a scaled down version of the detector, consisting of 128 PM's in a black room. The tubes have been stimulated by light from computer-controlled LED's (550 nm) and a nitrogen laser (330 nm). Alternatively, mock proton decay events have been generated by cosmic ray muons passing through small pieces of Lucite and scintillator in the room. Reconstruction of these mock events confirm our calculation of detector resolutions.

To determine the optimal number of phototubes ($\sim 2000$) necessary for the 10 kt detector, we have compared the calculated performance of different configurations of PM's. A comparison of the initially generated variables with those obtained after reconstructing the events shows that the angular error for idealized tracks (2°) is much less than that obtained from tracks with the inclusion of multiple scattering (7°), showering (12°), or the Fermi motion of oxygen (20°). The error (70 cm) in the inferred starting and ending positions of a track, and in the total range (energy) is a strong function of the tube spacing. Using Čerenkov light yields consistent with the long baseline tests mentioned above and the measured time response of the phototubes ($\sigma = 5$ ns), an optimization shows that a tube spacing of $\sim 1$ m achieves an energy resolution for idealized tracks of $\sim 10\%$ with 140 photoelectrons per $e^+$ or $\pi^0$ track. A larger tube spacing results in significant deterioration of the resolution, because the grid size becomes comparable to the range of the tracks. (Also, the efficiency for detecting the electron of a $\pi \to \mu \to e$ decay sequence drops below 50%.) When energy smearing due to shower fluctuations is included, the energy resolution becomes 15% per track. Including Fermi motion increases the smearing to 20% per track. If the outputs of all the PM's are summed to yield a measure of the total decay energy, the Fermi motion smearing cancels and the anticipated monochromatic line at $\sim m_p$ has a width of $\sim 10\%$ after reconstruction of events from the $\pi^0 e^+$ decay mode.

When operational, the most common background is expected to come from entering cosmic ray muons ($10^8$/yr) or the remnants of their interactions in the rock nearby the detector. We believe that we can reject these backgrounds by insisting that the observed Čerenkov light emanate only from the inner part of the water cube (the fiducial volume). The outer 2 m of the detector provide a protective shell which will veto entering charged particles at the software level. However, neutrinos can penetrate this shell and initiate, inside the fiducial volume, events which might sometimes mimic proton decay. Here we have used the events [4] recorded in the Gargamelle neutrino experiment to simulate this ultimate background. The CERN neutrino beam had roughly the same energy spectrum as that expected from atmospheric neutrinos in our detector. Assuming the energy (20%) and angular (20°) resolution per track discussed above, we calculate that less than one neutrino event per year will mimic the $N \to e\pi$ decay mode. This background would limit the sensitivity of the detector to $10^{33}$ yr after several years of running.

The precise kinematic cuts and the surviving background will be decay-mode dependent. For example, $\nu_\mu$ charged-current background can be eliminated from searches for $e^\pm$ modes by the signal from the muon decay. This should be reliable enough to reject 70% to 80% of the $\mu^+$'s, and a somewhat smaller fraction of the $\mu^-$'s (since 18% of the $\mu^-$'s are captured in oxygen). In addition, $\pi^+$ backgrounds in $\pi^-$ final states may be eliminated by this technique. Even some multi-body modes are distinctive. For example,

$$p \to e^+\omega^0, \quad \omega^0 \to \pi^+\pi^-\pi^0$$

has a signature comprised of one monoenergetic $e^+$ shower and two other showers consistent with a $\pi^0$, as well as at least one delayed muon decay. Although this is a fairly restrictive signature, it can be mimicked by:

$$\nu_e N \to eN^*, \quad N^* \to \pi^0\pi^+ X \qquad \text{or} \qquad nN \to \pi^0\pi^0\pi^+ X.$$

Table 3 summarizes our best estimates of the sensitivity of the detector to decay modes satisfying several possible selection rules. The second column of the table shows that for each selection rule at least one nucleon decay mode exists that is unique to that selection rule and recognizable in a Čerenkov detector.

Table 3
Čerenkov detector sensitivity to various selection rules

| $\Delta(b-\ell)$ | Signature | Decay Mode proton | neutron | Lifetime limit (yr) |
|---|---|---|---|---|
| $-4$ | No line<br>1 shower<br>2 $\mu$ decays | $\nu\nu\nu\pi^+$<br>$\nu\nu e^-\pi^+\pi^-$ | $\nu\nu\nu\pi^0$<br>$\nu\nu e^-\pi^+$ | $\gtrsim 10^{30}$<br>$\gtrsim 10^{30}$ |
| $-2$ | line at 1 GeV<br>2 $\mu$ decays | $\ell^-\pi^+\pi^+$<br>$\nu\pi^+$<br>$\nu\nu e^+$ | $\ell^-\pi^+$<br>$\nu\pi^0$<br>$\nu\nu e^+\pi^-$ | $\sim 10^{33}$<br>$\sim 3\times10^{31}$<br>$\gtrsim 10^{30}$ |
| 0 | line at 1 GeV<br>2 body | $\ell^+\pi^0$<br>$\bar{\nu}\pi^+$<br>$\ell^+K^0$ | $\ell^+\pi^-$<br>$\bar{\nu}\pi^0$<br>$\ell^+K^-$ | $\sim 10^{33}$<br>$\sim 3\times10^{31}$<br>$\sim 10^{32}$ |
| $+2$ | event deep<br>in detector | $\overline{\nu\nu\nu}\pi^+$<br>$\overline{\nu\nu}e^+$ | $\overline{\nu\nu\nu}\pi^0$<br>$\overline{\nu\nu}e^+\pi^-$ | $\gtrsim 10^{30}$<br>$\gtrsim 10^{30}$ |

The IMB detector should be operational in the winter of 1981. The double-lined cavity is fully outfitted and at the writing of this article is being filled with water. The PM detectors are all constructed and are at the mine awaiting installation. The 2048 PM data acquisition system has been operational in the mine for several months.

### *3.2. The Penn–Brookhaven detector in the Homestake Gold Mine*

An alternative Čerenkov technique is used by a group [5] from the University of Pennsylvania and Brookhaven National Laboratory. They dissolve a waveshifting dye in the water, which shifts the ultraviolet part of the spectrum to the visible before it is absorbed appreciably. Although a factor of three or four in detected photons can be gained in this way, the price paid is twofold. First, the timing resolution is degraded by the waveshifter lifetime, and secondly, all directional information is lost since the reemission of the light is isotropic. The Penn–BNL group overcomes this disadvantage by optically segmenting their detector into 4 $m^3$ units.

The group has been collecting data for the last year. This is due to the happy circumstance that some time ago they built a supernova neutrino detector in the Homestake Mine where Davis' solar neutrino experiment is located. In fact their detector is the water shield which surrounds Davis' chlorine tank. In the meantime, the Penn–BNL supernova detector has been modified into a proton decay detector. They look for $\mu \rightarrow e$ decay, where a muon could either be created directly from nucleon decay or indirectly as a decay product of the charged pions that may be produced. The muon decay signature is recognized as one or two modules firing, with a subsequent delayed pulse in one of them.

Using a Monte Carlo study of the energy distribution expected from nucleon decays which lead to a $\mu \rightarrow e$ decay signature, they conclude that none of their observed candidates are due to proton decay. To interpret the data they use the theoretical branching ratio predictions of SU(5) into muons and report a limit of $\geqslant 1 \times 10^{30}$ yr for the nucleon lifetime. They now hope to push the limit to $\sim 10^{31}$ yr.

The advantage of this detector over the other US Čerenkov detectors is its depth. However, it suffers from the lack of a veto other than that on the top face of the detector ($10^5$ muons per year enter the sides of the detector). Recognizing the limitations of this technology, the Penn–BNL team has proposed a new 1 kt segmented liquid scintillator detector. It would be housed in an adjacent new cavity that would have to be excavated in the Homestake mine. This would provide a major detector in the USA at a depth comparable to that of the Alpine tunnels of Europe.

### *3.3. The Harvard – Purdue – Wisconsin detector in the Silver King Mine*

The Harvard–Purdue–Wisconsin Group [6] is building an 800 ton water detector in the old silver mine at Park City, UT which Keuffel and his group used for their cosmic ray research. It has about the same depth as the IMB detector and will also use the water Čerenkov technique, both with and without a wavelength shifter. Their phototubes are distributed throughout the counter volume so as to be as close as possible to the emission point of the light. They hope to obtain superior energy resolution by this technique. Their fiducial volume is limited by the small size of the cavity. In an attempt to use as much as possible of the 800 tons as fiducial volume, the water tanks will eventually be surrounded by an external shower counter of lead, concrete, and tube counters. The water portion of the detector will be operational in the winter of 1981.

### *3.4. The Tokyo – KEK – Tsukuba detector in the Kamioka Metal Mine*

A large new underground project is being developed in Japan by a Tokyo–KEK–Tsukuba group [7]. Containing 3400 total tons of water, the Čerenkov detector is surrounded by 1056 specially developed 20″ phototubes. It is expected to be in operation at 2.7 kmwe in the summer of 1982. The remarkable advantage of this detector will be its 20% photocathode coverage of the surfaces of the detector, a factor of ten better than the IMB detector. Moreover, its greater depth will decrease the muon flux by a factor of 15 relative to the IMB device.

## 4. Dense tracking devices

Let us turn to the dense detectors, which are complementary to the Čerenkov devices in that they respond to $\mathrm{d}E/\mathrm{d}x$ (rather than $\beta$), and produce event "pictures" of the tracks of the event rather than the less familiar rings of light produced by tracks in the Čerenkov detectors. The trade off is that the dense detectors do not determine the sense of direction of a track. Therefore it is harder to locate the vertex and to prove the back-to-back nature of a decay. One recognizes the vertex of a two-body decay originating inside a heavy nucleus by the angle between the two exiting particles due to the Fermi motion of the decaying nucleon. Most of the dense detectors also lack muon decay sensitivity. Further, their high cost per unit of mass has made, until recently, large fine-grain calorimeters prohibitively expensive. However, for a given detector mass, dense calorimeters can be located much deeper since they require smaller cavities. Also, their sensitivity to multibody decay modes can be superior to that of Čerenkov detectors by virtue of their fine granularity. The ability to define a vertex in a calorimeter (and to distinguish the tracks in multibody events) is determined by cell sizes as small as 0.5 cm, which can be contrasted with the ~ 50 cm resolution of Čerenkov devices. An advantage of dense detectors using iron or iron loaded concrete is the eventual possibility of implementing a magnetic field throughout the detector. This could facilitate charge determination and measurement of muon polarization.

### *4.1. The Indian–Japanese detector in the Kolar Gold Fields*

The Indian–Japanese collaboration has reported [8] the data from 205 days of operation of their 140 t detector in the Kolar gold fields. The device consists of horizontal slabs or iron, 1/2″ thick, separated by proportional tubes (10 cm × 10 cm in cross section) in an alternating $x-y$ grid. The 1600 tube detector has a horizontal area of 4 m × 6 m and a height of 4 m. The discriminators on the tubes are set to fire if a particle deposits an energy $\geqslant 1/2$ of that of a minimum ionizing particle. The noise rate of a tube near the edge of the detector is 100 Hz due to the radioactivity in the rock. Inside the detector, this rate drops to 2 Hz because of the self shielding of the iron. The radioactivity-induced rate is the basic measure of the health of the detector. (The cavity is so deep that only two muons per day traverse it!) Although the muon rate is insufficient for monitoring, the background that it induces is orders of magnitude lower than in any other detector. The pulse heights on each tube are recorded if a fivefold coincidence occurs between any five layers. The resolving time of the tubes is 1 $\mu$s. Therefore, no time-of-flight (directional) information (nor muon stop signature) is available.

The authors have recorded three neutrino interactions in the detector, while 4–6 were expected. On the other hand, they have reported two events which do not fit any expected hypothesis; they can be considered as proton decay candidates. However the limited detector volume renders the events hard to interpret. In particular, the tracks have segments at the edge of the detector, with at least one particle possibly entering or exiting. Further, the events are vertically oriented. This has led some skeptics to challenge the candidates as remnants of cosmic ray muons. On the other hand, the statistics are low. Also the vertical orientation may be a result of the greater ease with which the five-layer trigger may be satisfied in a horizontally oriented detector stack.

Both proton decay candidates have tracks that include an abnormally high number (relative to straight-through muon tracks) of proportional tubes that did not fire, particularly next to the kink that is presumably the vertex. These gaps may be characteristic of electromagnetic showers in the detector. Perhaps the candidates could have been induced by electron neutrinos near the edge of the detector. These are expected at the same rate as the observed candidates.

In any case, the candidates show that a 140 t detector, even when very well shielded by great depth, is limited in its ability to contain proton decay candidates (only an event in the inner 1/10 of the detector would be fully contained). If the two events are considered as proton decay candidates, the lifetime would be $8.4^{+11}_{-5.5} \times 10^{30}$ yr, where the authors assume that the detection efficiency × branching ratio is 50%.

### *4.2. The CERN–Frascati–Milan–Turin detector in the Mont Blanc Tunnel*

A dense detector of similar mass is being constructed in the Mont Blanc Tunnel by the CERN–Frascati–Milan–Turin collaboration [9]. Their advantage is a grid size of $(1\ \text{cm})^2$ which is substantially finer than that of the Indian–Japanese $(10\ \text{cm})^2$. Further, they have studied showers and tracks from CERN calibration beams of electrons, pions, and neutrinos in a prototype module. This should strengthen the confidence in their interpretation of any events. The depth of this detector is a factor of three deeper than that of the detectors in the USA (Alpine tunnels are an asset!), but a factor of two shallower than the depth of the Kolar gold fields detector. The device is scheduled to be on-line in the winter of 1981.

### *4.3. The Orsay–Ecole Polytechnique–Saclay–Wuppertal detector in the Fréjus Tunnel*

An order of magnitude improvement in mass over the Italian and Indian–Japanese detectors will be achieved in another Alpine tunnel by the collaboration of Orsay–Ecole Polytechnique–Saclay–Wuppertal [10].

They will further strengthen the dense detector technology by interspersing planes of Geiger tubes in the detector. The Geiger tubes trigger flash chambers that will be used for tracking. In addition, the time resolution of the Geiger tubes is sufficient to tag delayed muon decays. This detector will take a few years to get on the air.

### *4.4. The Minnesota–Argonne–Oxford detector in the Soudan Iron Mine*

The University of Minnesota–Argonne group has constructed a prototype 30 t detector in an iron mine in Minnesota [11]. They use an inexpensive ferro-concrete medium which can be built in small modules. Gas proportional counters made from thin steel tubes are embedded in the concrete. The detector is currently taking data, but suffers because of its small volume at a relatively shallow depth. With the addition of Oxford University, the group has proposed a 1 kt version of their prototype. However, the gas tubes do not scale from 30 t to 1000 t; they would be replaced with drift chambers that have a 1 m drift length and readout at both ends.

### *4.5. The Grand Unified Detector (GUD) in the Grand Sasso Tunnel*

The most ambitious proton decay project [12], intended as a laboratory open to the international physics community, has been officially approved by the Italian Government. To be built under the Grand Sasso mountain (150 km from Rome) between two almost completed auto tunnels, this facility would provide an enormous underground experimental area measuring $50 \times 50 \times 20$ meters at a depth (4.5 kmwe) typical of the Alpine tunnels. A modular iron detector with a mass of $\lesssim 10$ kt is under consideration. The detector would consist of 3 mm thick iron plates interspersed with planes of flash tubes $(4 \text{ mm})^2$ alternated in an $x-y$ grid. They are triggered by limited streamer chambers that appear every ninth layer. This detector offers fine granularity, vertical plates, large mass, and great depth. Since the systematics will be very different, it will be a valuable complement to the big Čerenkov detectors.

Within the next few years, we will clearly have a plethora of data concerning proton decay limits or measurements from a variety of detectors with very different systematic errors. In addition, the advent of highly instrumented and shielded $10^4$ t detectors—with sensitivities three orders of magnitude greater than previous efforts—may reap rewards never dreamed by their early proponents.

## Acknowledgements

I would like to extend special gratitude to one of my IMB collaborators, Maurice Goldhaber; this paper is an expanded version of the last half of a paper that we wrote together for "Comments in Nuclear and Particle Physics." I also acknowledge the long hours of learning together with each of my other collaborators on the IMB proton decay project: C. Bratton, C. Cory, W. Gajewski, W. Kropp, J. Learned, F. Reines, J. Schultz, D. Smith, H. Sobel, and C. Wuest, associated with the University of California at Irvine, and R. Bionta, B. Cortez, S. Errede, G. Foster, J. Greenberg, T.W. Jones, J. LoSecco, R. Murthy, H.-Y. Park, E. Shumard, D. Sinclair, J. Stone, and J. van der Velde, associated with the University of Michigan.

This paper was largely written at the 1981 Les Houches Summer School of Physics. My great thanks to Mary K. Gaillard and her schoolmates for stimulating days and dancing nights in the tranquility of the Haute Savoie mountains.

This work has been sponsored in part by the U.S. Department of Energy.

## References

[1] A.D. Sakharov, Zh. Eksp. Teor. Fiz. Pis'ma 5 (1967) 32. See also
L. Susskind's article in Andrei Sakharov's Collected Works (in print).

[2] J.C. Pati and A. Salam, Phys. Rev. Lett. 31 (1973) 661.
H. Georgi and S. Glashow, Phys. Rev. Lett. 89B (1974) 438.
H. Georgi, H.R. Quinn and S. Weinberg, Phys. Rev. Lett. 33 (1974) 451.

[3] For details, see
L.R. Sulak, in: Unification of the Fundamental Interactions, eds. S. Ferrara, J. Ellis and P. van Nieuwenhuizen (Plenum, 1980) p. 639, and T. Jones and D. Sinclair, in: Proc. 2nd Workshop on Grand Unification, Ann Arbor, eds. J. Leveille, L. Sulak and D. Unger (Birkhauser Press, 1981) p. 84. See also J. van der Velde, Proc. Summer Inst. on Particle Physics, SLAC 239 (198X) 457.

[4] G. Martin and M. Pohl, private communication.

[5] R. Steinberg, in: Proc. 2nd Workshop on Grand Unification, Ann Arbor, eds. J. Leveille, L. Sulak and D. Unger (Birkhauser Press, 1981) p. 22; in: Proc. 1st Workshop on Grand Unification, eds. P.H. Frampton, S.L. Glashow and A. Yildiz, (Math. Sci. Press, 1980) p. 313.

[6] C. Wilson, in: Proc. 2nd Workshop on Grand Unification, Ann Arbor, eds. J. Leveille, L. Sulak and D. Unger (Birkhauser Press, 1981) p. 80.
D. Winn, in: Proc. 1st Workshop on Grand Unification, eds. P.H. Frampton, S.L. Glashow and A. Yildiz (Math. Sci. Press, 1980) p. 189.
D. Cline, presentation at Interaction of Physics and Astrophysics, Santa Barbara (1981), unpublished.

[7] "KEK–(UT)$^2$ Experiment on Proton Decay," H. Ikeda et al., KEK, University of Tokyo, and Tsukuba preprint.
[8] S. Miyake and V.S. Narasimham, in: Proc. 2nd Workshop on Grand Unification, eds. J. Leveille, L. Sulak and D. Unger (Birkhauser Press, 1981) p. 11.
V.S. Narasimham, Presentation at Vth High Energy Symposium, Cochin, Dec. 1980, Tata Institute Preprint.
M.R. Krishnishswamy et al., XVII Int. Cosmic Ray Conf., Paris, July 1981. Data from this last report was kindly provided to me by R. Murthy and L.W. Jones.
[9] E. Fiorini, in: Proc. 2nd Workshop on Grand Unification, eds. J. Leveille, L. Sulak and D. Unger (Brikhauser Press, 1981) p. 55, see also
E. Bellotti, in: Unification of the Fundamental Particle Interactions (Plenum, 1980) p. 673.
[10] R. Barloutaud, in: Proc. 2nd Workshop on Grand Unification, eds. J. Leveille, L. Sulak and D. Unger (Birkhauser Press, 1981) p. 70.
Also, see the second paper cited in ref. 9.
[11] M.A. Shupe, in: Proc. 2nd Workshop on Grand Unification, eds. J. Leveille, L. Sulak and D. Unger (Birkhauser Press, 1981) p. 41. See also:
M. Marshak, in: Proc. 1st Workshop on Grand Unification, eds. P.H. Frampton, S.L. Glashow and A. Yildiz (Math. Sci. Press, 1980) p. 305.

**Other seminars related to Course 3**

3.5. Deeply bound composite fermions with Dirac magnetic moments, by Ting-Wai Chiu (Univ. of California, Irvine).

3.6. Weak production and decay of short-lived particles, by David Bailey (McGill) and Dale Pitman (Toronto).

3.7. New ways to measure the neutrino masses, by Alvaro de Rújulà (CERN).

3.8. A confining model of electroweak interactions, by Eddy Fahri (MIT).

3.9. Neutrino mass and global lepton number violation, by Graciela Gelmini (Munich).

3.10. Measuring the $\gamma - Z^0$ interference in muon scattering, by Michel Goossens (CERN).

3.11. The determination of the weak angle from experiments and comparison with GUTs, by John Wheater (Oxford).

3.12. Supersymmetric GUTs, by Bruno Zumino (CERN).

COURSE 4

# LE MONOPOLE MAGNETIQUE CINQUANTE ANS APRES*

Sidney COLEMAN**

*Lyman Laboratory of Physics, Harvard University*
*Cambridge, MA 02138, USA*

*Recherche partiellement subventionnée par la N.S.F. des Etats Unis d'Amérique, allocation No. PHY 77-22864
**Traduit par R. STORA

*M.K. Gaillard and R. Stora, eds.*
*Les Houches, Session XXXVII, 1981*
*Théories de jauge en physique des hautes énergies/Gauge theories in high energy physics*

## Table

*NOTE DU TRADUCTEUR*: La version en langue Française qu'on trouvera ici a tenté de respecter le style direct et imagé qui caractérise la version originale en langue Anglaise.

## 1. Introduction

Nous sommes dans une année d'anniversaire. En 1931, P. A. M. Dirac [1] fonda la théorie des monopôles magnétiques. Au cours des cinquante années qui ont suivi aucun monopôle n'a été observé; néanmoins, jamais l'intérêt dans le sujet n'a été aussi élevé que maintenant.

Il y a une bonne raison à cela. Pendant plus de quarante ans le monopôle magnétique a été un accessoire superflu. Dirac avait montré comment construire des théories contenant des monopôles, mais personne n'était tenu de les utiliser à moins d'en avoir envie. La situation a changé il y a sept ans; 't Hooft et Polyakov [2] ont montré que les monopôles magnétiques sont inévitables dans certaines théories de jauge. En particulier, toutes les Théories de Grande Unification contiennent nécessairement des monopôles. (Dans ce contexte, une Théorie Grande Unifiée est une théorie dans laquelle un groupe de symétrie interne est spontanément brisé jusqu'au groupe U(1) électromagnétique). Beaucoup d'entre nous sont convaincus que les Théories Grandes Unifiées décrivent la nature, au moins jusqu'à des distances de l'ordre de la longueur de Planck. Alors, où sont les monopôles?

Comme on le verra, les monopôles de Grande Unification sont très lourds; la masse typique est en gros cent fois plus grande que l'échelle de Grande Unification. Ainsi, il y a une faible probabilité qu'ils soient produits par les accélérateurs contemporains ou les supernovae. Par contre, l'énergie était plus abondante peu après le "Big Bang." Comme l'a remarqué Preskill [3], des estimations naïves nous conduiraient à penser que les monopôles ont été produits si abondamment au tout début de l'Univers, et annihilés avec si peu d'efficacité par la suite, qu'ils fourniraient à présent la contribution dominante de la masse de l'Univers.

Ainsi, l'absence de monopôles est chargée de sens. Elle nous renseigne soit sur le tout début de l'Univers, soit sur les confins de la microphysique, soit sur les deux. Comme nous manquons d'indice irréfutable sur l'un et l'autre de ces deux sujets, les monopôles sont importants.

C'est la dernière fois que je parlerai de cosmologie dans ces conférences, ainsi d'ailleurs que de quelque raison que ce soit qui justifie l'étude des monopôles; le développement des bases de théorie des monopôles suffira bien à nous occuper.

J'ai essayé dans ces conférences d'aller du plus simple au plus complexe, du monopôle tel qu'on le perçoit à grande distance, à la structure interne du monopôle, des aspects classiques aux aspects quantiques. Bien entendu, je n'ai pas réussi à m'en tenir rigoureusement à ce programme; par exemple, nous serons sans cesse en contact avec la mécanique quantique élémentaire depuis le tout début, bien que la complexité de la théorie quantique des champs ne soit pas envisagée dans sa totalité avant la toute dernière conférence.

Ces conférences sont organisées comme suit:

Dans la section 2, je discuterai la théorie du monopôle magnétique classique "vu de loin." Autrement dit, je ne m'intéresserai qu'aux questions auxquelles on peut répondre sans regarder l'intérieur du monopôle, sans se demander, par exemple, s'il y a une singularité au coeur du monopôle ou seulement quelque excitation compliquée des degrés de liberté qui ne se propagent pas à de grandes distances, comme par exemple des champs de jauge massifs.

Dans la section 3, j'étendrai l'analyse précédente au monopôle non Abélien classique. J'insiste sur le fait qu'il ne s'agit pas ici d'un objet qui contient dans son coeur des champs non Abéliens massifs, mais d'un objet qui est entouré par des champs de jauge Abéliens non massifs qui s'étendent à grande distance. Comme il n'existe pas de tels champs dans la nature, cela peut ressembler à un exercice stupide mais il y a deux bonnes raisons pour le faire: il y a d'abord une raison pédagogique. Dans l'étude des monopôles dans des théories de jauge non Abéliennes non brisées, nous rencontrerons des structures mathématiques qui seront utiles dans la suite au cours de l'étude du cas plus compliqué où il y a brisure spontanée de symétrie. Il y a en outre une raison physique. Certains des monopôles qui apparaissent dans les Théories Grandes Unifiées sont entourés par des champs de jauge colorés; ce sont des monopôles magnétiques colorés aussi bien que des monopôles électromagnétiques ordinaires. Il est vrai que ces champs colorés sont atténués par des effets de confinement à des distances supérieurs à $10^{-13}$ cm. Néanmoins, le coeur de ces monopoles est de l'ordre de l'échelle de Grande Unification, de l'ordre de $10^{-28}$ cm. On a donc environ quinze ordres de grandeur de taille où ils apparaissent comme des monopôles non Abéliens: pas mal de place pour de la physique intéressante.

Dans la section 4, on va scruter l'intérieur de la bête et voir dans quelles conditions on peut prolonger à courtes distances les structures qu'on a découvertes à grandes distances sans rencontrer de singularité.

C'est le sujet que j'ai discuté dans mes conférences d'Erice en 1975 [4] (avec un point de départ différent), et bien que je tenterai de réduire les répétitions au minimum, on trouvera ici un recyclage inévitable de ces conférences.

Finalement, dans la section 5, j'envisagerai les aspects quantiques et je discuterai par exemple les excitations dyoniques et les effets de confinement. Il y a beaucoup de choses qui ne se trouvent pas dans ces notes. Comme je l'ai dit, je ne m'occuperai pas de cosmologie, je n'aurai rien à dire sur les interactions des monopôles avec la matière ordinaire, la façon dont ils produisent des traces dans les émulsions et atténuent les champs magnétiques galactiques. Sur un plan plus théorique, je ne parlerai pas des solutions exactes, des théorèmes d'index, ou de la fractionalisation fermionique. Finalement, je n'ai pas essayé de composer une bibliographie complète et même équitable. Si je ne cite pas un article en référence, il faut en déduire que c'est le résultat de mon ignorance ou de ma paresse, et non d'un jugement critique informé [5].

La plupart de mes connaissances sur ce sujet proviennent de conversations avec Curtis Callan, Murray Gell-Mann, Jeffrey Goldstone, Roman Jackiw, Ken Johnson, David Olive, Gerard 't Hooft et Erick Weinberg. C'est un plaisir de reconnaître ma dette.

*Notations*: Comme d'habitude, les indices grecs vont de 0 à 3, les indices latins à partir du milieu de l'alphabet, de 1 à 3. Pour l'analyse vectorielle ordinaire (comme dans toute la section 2) la signature de la métrique tri-dimensionnelle est $(+++)$; à quatre dimensions, la métrique est $(+---)$. Je poserai en général $\hbar = c = 1$, bien qu'à l'occasion il m'arrive de réinsérer $\hbar$ explicitement quand je discuterai de l'approche vers la limite classique. J'utiliserai les unités rationalisées pour l'électromagnétisme; ainsi la densité Lagrangienne électromagnétique est $\frac{1}{2}(E^2 - B^2)$ et il n'y a pas de $\pi$ dans les équations de Maxwell.

## 2. Les monopôles Abéliens vus de loin

### 2.1. *Le tour du monopôle, et une première approche à la condition de quantification de Dirac*

Considérons une distribution de charges et de courants stationnaires restreints à une région bornée. En dehors de cette région il n'y a que des champs électromagnétiques stationnaires qui s'annulent à l'infini dans l'espace. Que peut-on dire des champs extérieurs sans connaître en détail la distribution de charges et de courants?

La réponse est connue de quiconque a suivi un cours d'électromagnétisme. Les champs extérieurs consistent en une somme de monopôle électrique, dipôle magnétique, quadrupôle électrique, etc. Les termes successifs de cette série décroissent de plus en plus vite à l'infini; le terme dominant à grande distance est le monopôle électrique:

$$\boldsymbol{E} = \frac{e\boldsymbol{r}}{4\pi r^3}, \qquad \boldsymbol{B} = 0, \tag{2.1}$$

où $e$ est un nombre réel appelé charge électrique (du système à l'intérieur de la région).

Cette analyse dépend de l'hypothèse qu'il n'y a à l'intérieur de la région que des charges et des courants, autrement dit que les équations de Maxwell dans le vide,

$$\nabla \cdot \boldsymbol{B} = 0 = \nabla \times \boldsymbol{E} + \partial \boldsymbol{B}/\partial t, \tag{2.2}$$

restent vraies à l'intérieur de la région. Si nous abandonnons cette hypothèse, nous devons ajouter à la série précédente une série duale consistant en monopôle magnétique, dipôle électrique, etc. Le terme dominant à grande distance est le monopôle magnétique:

$$\boldsymbol{E} = 0, \qquad \boldsymbol{B} = \frac{g\boldsymbol{r}}{r^3}, \tag{2.3}$$

où $g$ est un nombre réel appelé charge magnétique (du système dans la région) [6]. Les systèmes qui portent une charge magnétique non nulle sont aussi appelés de façon quelque peu confuse monopôles magnétiques.

Ainsi que je l'ai dit dans mes remarques d'introduction, personne n'a jamais observé de monopôle magnétique. Imaginons, cependant, que nous essayions de mystifier un expérimentateur naïf en lui faisant croire qu'il a découvert un monopôle. Pour ce faire, prenons un solénoïde très long et très fin; le mieux est qu'il ait plusieurs kilomètres de long et qu'il soit considérablement plus fin qu'un fermi. (Il s'agit d'une "Gedanken" supercherie.) Nous mettons un bout du solénoïde dans le laboratoire de l'expérimentateur à Paris* et l'autre ici aux Houches*. Nous branchons le courant. L'expérimentateur voit un flux magnétique de $4\pi g$ sortir de la table de laboratoire; il ne peut pas se rendre compte qu'il provient du solénoïde, il pense qu'il a un monopôle.

Y a-t-il un moyen quelconque par lequel il puisse dire qu'il a été roulé, que le monopôle est truqué? Certainement pas, si tout ce qu'il peut détecter se résume à des particules classiques chargées, car celles-ci ne voient que $\boldsymbol{E}$ et $\boldsymbol{B}$, qui sont les mêmes que ceux d'un réel monopôle. (Sauf à l'intérieur du solénoïde qui par hypothése est si fin qu'il est indétectable!)

*Mutatis Mutandis (note du traducteur).

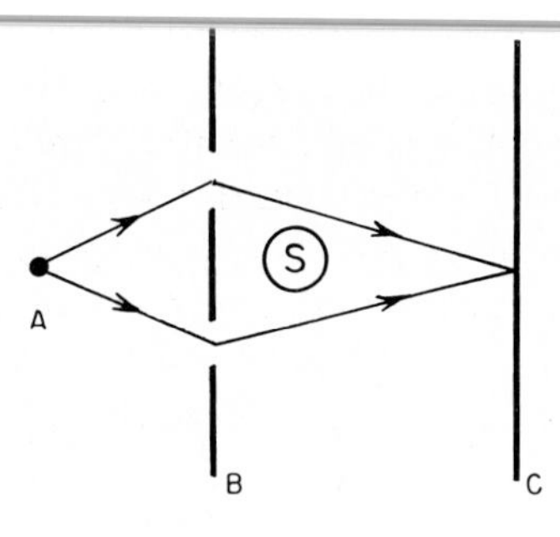


Fig. 1.

Cependant, la situation est très différente si notre expérimentateur a accès à des particules quantiques chargées. Avec elles, un expérimentateur astucieux peut chercher le solénoïde via l'effet de Bohm–Aharonov [7].

Ceci est une variante de la fameuse expérience de diffraction à deux fentes décrite dans la fig. 1. Les particules chargées émises par la source A passent à travers les deux fentes de l'écran B et sont détectées sur l'écran C. Ainsi qu'il est connu de quiconque a lu six pages d'un livre de mécanique quantique, les amplitudes de passage à travers chacune des fentes s'ajoutent de façon cohérente; la densité de probabilité en C est

$$|\psi_1 + \psi_2|^2, \tag{2.4}$$

où $\psi_1$ est l'amplitude de passage à travers la première fente et $\psi_2$ l'amplitude de passage à travers la seconde.

Si un solénoïde S, montré par le bout sur la figure, est placé entre les fentes, la densité de probabilité en C est changée; elle devient:

$$|\psi_1 + e^{ie\Phi}\psi_2|^2 \tag{2.5}$$

où $e$ est la charge de la particule et $\Phi$ est le flux à travers le solénoïde (dans notre cas $4\pi g$). Ainsi en déplaçant l'appareillage et en observant le changement du schéma d'interférences, notre expérimentateur peut détecter le solénoïde à moins que $eg$ ne soit un demi entier,

$$eg = 0, \pm\tfrac{1}{2}, \pm 1, \ldots. \tag{2.6}$$

Dans ce cas, les expressions (2.4) et (2.5) sont identiques. Le solénoïde est indétectable et notre tour a réussi. L'équation (2.6) est la fameuse condition de quantification de Dirac. Comme on le verra, lorsqu'elle est remplie, le solénoïde est indétectable, non seulement par l'effet de

Bohm–Aharonov, mais par n'importe quelle méthode concevable. Il s'élimine complètement et se réduit à une singularité mathématique sans conséquence physique, la corde de Dirac; tout ce qui reste est un authentique monopôle. La démonstration de ces affirmations demande une analyse plus fine de la physique d'une particule chargée dans un champ de monopôle à laquelle nous allons maintenant nous consacrer.

### 2.2. *L'invariance de jauge et une deuxième approche à la condition de quantification*

Considérons une particule non relativiste sans spin dans un champ magnétique statique (décrit comme d'habitude par un potentiel vecteur $\boldsymbol{A}$), et peut être aussi dans un potentiel scalaire $V$. L'équation de Schrödinger pour ce système est:

$$\mathrm{i}\frac{\partial \psi}{\partial t} = -\frac{1}{2m}(\nabla - \mathrm{i}e\boldsymbol{A})^2\psi + V\psi. \tag{2.7}$$

Ce système admet une invariance de renom, l'invariance de jauge. Pour ce qui nous occupe, il nous suffit de considérer les transformations de jauge indépendantes du temps:

$$\psi \to \mathrm{e}^{-\mathrm{i}e\chi}\psi, \tag{2.8a}$$

$$\boldsymbol{A} \to \boldsymbol{A} - \nabla\chi = \boldsymbol{A} - \frac{\mathrm{i}}{e}\mathrm{e}^{\mathrm{i}e\chi}\nabla\mathrm{e}^{-\mathrm{i}e\chi}. \tag{2.8b}$$

J'ai écrit la seconde de ces équations de façon quelque peu non conventionnelle pour insister sur le fait que la seule quantité significative est $\mathrm{e}^{-\mathrm{i}e\chi}$. Ainsi, par exemple, $\chi$ et $\chi + 2\pi/e$ sont des fonctions différentes, mais elles définissent la même transformation de jauge [8].

Après ces généralités venons en au cas qui nous intéresse, un monopôle ponctuel à l'origine des coordonnées. On peut considérer le champ comme celui d'un objet véritablement ponctuel ou comme celui d'un objet étendu vu à très grande distance; c'est sans importance. Bien sûr, il est impossible de trouver un potentiel vecteur dont le rotationnel soit le champ du monopôle, puisque le champ du monopôle a une divergence non nulle. Cependant, il est possible de trouver un potentiel qui fasse l'affaire en dehors d'une ligne qui s'étend du monopôle jusqu'à l'infini, "la corde de Dirac." Un tel potentiel est simplement celui de notre tour de prestidigitation, dans la limite où le solénoïde devient infiniment long et infiniment fin.

Par exemple, si j'arrange la corde le long du demi axe des $Z$ négatifs, $\boldsymbol{A}$ est donné par

$$\boldsymbol{A} \cdot \mathrm{d}\boldsymbol{x} = g(1-\cos\theta)\,\mathrm{d}\varphi. \tag{2.9}$$

Comme promis $\boldsymbol{A}$ n'est pas défini sur la corde, $\theta = \pi$. Vérifions que son rotationnel est correct en dehors de la corde. La façon la plus simple de mener le calcul consiste à utiliser les méthodes de l'analyse tensorielle. La seule composante covariante non nulle de $\boldsymbol{A}$ est:

$$A_\varphi = g(1-\cos\theta). \tag{2.10}$$

Ainsi, la seule composante non nulle du tenseur de champ est

$$F_{\theta\varphi} \equiv \partial_\theta A_\varphi - \partial_\varphi A_\theta = g\sin\theta. \tag{2.11}$$

Cela correspond bien à un champ magnétique radial. Pour vérifier qu'il a la bonne intensité, nous calculons le flux à travers un élément infinitésimal d'angle solide:

$$F_{\theta\varphi}\,\mathrm{d}\theta\,\mathrm{d}\varphi = \frac{g}{r^2} r^2 \sin\theta\,\mathrm{d}\theta\,\mathrm{d}\varphi. \tag{2.12}$$

C'est le résultat désiré, $g/r^2$ fois l'élément d'aire infinitésimal.

J'insiste sur le fait que je n'ai pas essayé de définir $\boldsymbol{A}$ ou de calculer $\boldsymbol{B}$ sur la corde. Comme on le verra il n'est pas besoin de se préoccuper de ces détails.

Bien sûr, le choix du demi-axe des $Z$ négatifs est totalement arbitraire. Par exemple, j'aurais pu aussi bien mettre la corde le long du demi-axe des $Z$ positifs:

$$\boldsymbol{A}' \cdot \mathrm{d}\boldsymbol{x} = -g(1+\cos\theta)\,\mathrm{d}\varphi. \tag{2.13}$$

Il sera utile dans la suite de noter que les deux potentiels que nous avons introduits sont simplement reliés l'un à l'autre:

$$\boldsymbol{A}'(\boldsymbol{x}) = \boldsymbol{A}(-\boldsymbol{x}). \tag{2.14}$$

Dans son article classique sur les monopôles, Dirac a montré que si la condition de quantification était remplie, la corde était inobservable. Il l'a fait en montrant qu'il y avait une transformation de jauge qui permettait de déplacer la corde du demi-axe des $Z$ négatifs dans n'importe quelle autre position, qui transformait le potentiel $\boldsymbol{A}$, par exemple, dans le potentiel $\boldsymbol{A}'$. Cependant cet argument présente des aspects déplaisants: la transformation de jauge est nécessairement singulière à la fois à l'ancienne et à la nouvelle position de la corde, et les transformations de jauge singulières laissent planer une impression désagréable.

Je vais maintenant donner un raffinement de l'argument de Dirac, dû à Wu et Yang [9], et qui évite ces difficultés. Dans la construction de Wu et Yang on n'a jamais affaire à des transformations de jauge singulières, ni même à des potentiels singuliers (excepté, bien entendu, à l'origine, où $\boldsymbol{B}$ explose et où il y a une vraie singularité). Le prix à payer est la nécessité d'utiliser différents potentiels dans différentes régions de l'espace.

Définissons la région supérieure de l'espace comme l'ouvert $\theta < 3\pi/4$, et la région inférieure comme l'ouvert $\theta > \pi/4$. L'union de ces deux régions est la totalité de l'espace (excepté l'origine des coordonnées qui est de toutes façons un point singulier). Définissons le champ monopolaire en utilisant le potentiel $\boldsymbol{A}$ dans la région supérieure et le potentiel $\boldsymbol{A}'$ dans la région inférieure. Ainsi, ni l'un ni l'autre des deux potentiels n'a de singularité dans sa région de définition; dans chaque cas la corde se trouve en dehors de la région.

Ceci est une adaptation d'un stratagème de fabriquant de cartes. Il n'y a pas de façon d'appliquer la surface sphérique de la terre sur une portion de plan sans introduire de singularité. Par exemple, sur le drapeau des Nations Unies qui est une projection nord-polaire, le pôle sud, qui est un seul point, est transformé dans la circonférence de la carte. Cependant nous pouvons aisément faire le travail avec deux cartes. Par exemple, nous pourrions utiliser une projection nord-polaire pour la région au nord du tropique du Capricorne et une projection sud-polaire pour la région au sud du tropique du Cancer. Les deux cartes prises ensemble formeraient une représentation du globe sans singularité. Cependant, nous devons nous assurer que les deux cartes se recollent de façon appropriée, que nous n'avons pas reçu par erreur une carte de la partie nord de la terre et de la partie sud de Mars. Autrement dit les deux cartes doivent décrire la même géographie dans la région équatoriale où elles se recouvrent.

Lorsqu'on transpose à la théorie des champs ce critère de la cartographie, on doit s'assurer que les deux potentiels vecteurs décrivent la même physique dans l'intersection des régions inférieure et supérieure, autrement dit que $A'$ est un transformé de jauge de $A$. Ainsi, dans l'intersection des deux régions on doit trouver $\chi$ tel que:

$$\boldsymbol{A} - \boldsymbol{A}' = \nabla\chi. \tag{2.15}$$

C'est chose facile:

$$(\boldsymbol{A} - \boldsymbol{A}')\cdot \mathrm{d}\boldsymbol{x} = 2g\,\mathrm{d}\varphi, \tag{2.16}$$

d'où

$$\chi = 2g\varphi. \tag{2.17}$$

Cette fonction n'est pas bien définie dans l'intersection; elle est multivaluée. Heureusement, comme je l'ai expliqué au début de cette section, l'objet intéressant n'est pas $\chi$ mais $\exp(-ie\chi)$ qui est univalué si

$$eg = 0, \pm\tfrac{1}{2}, \pm 1 \ldots . \tag{2.6}$$

C'est la condition de quantification de Dirac, ce qui conclut l'argument.

En principe, la description sans singularité de Wu–Yang pour le champ monopolaire est beaucoup plus propre que la description de Dirac avec sa corde singulière. Cependant, en pratique, il est malaisé de changer continuellement de jauge suivant la région de l'espace, de sorte que, dans le reste de ces notes, je me servirai la plupart du temps de la description de Dirac en portant à la corde toute l'attention nécessaire, de la même façon qu'on manie les singularités dûes au choix des coordonnées quand on travaille en coordonnées polaires.

### 2.3. *Remarques sur la condition de quantification*

1. Toutes les charges électriques observées sont des multiples entiers d'une unité commune; cela s'appelle la quantification de la charge. Il n'y a pas d'explication de ce phénomène frappant ni en électrodynamique classique, ni en électrodynamique quantique; par exemple rien de mal ne se produirait si la charge du proton était $\pi$ fois celle de l'électron. L'un des aspects les plus attrayants de la théorie des monopôles de Dirac, quand elle fut proposée pour la première fois, était qu'elle expliquait la quantification de la charge. S'il y avait des monopôles dans l'univers (en fait, même s'il n'y avait qu'un seul monopôle dans l'univers), toutes les charges électriques seraient forcées d'être des multiples entiers d'une unité commune, à savoir, l'inverse du double de la charge magnétique minimale.

Bien entendu, à l'heure actuelle, personne ne croit plus à l'électrodynamique pure. Nous pensons que le groupe U(1) de l'électrodynamique est une partie d'un groupe plus grand qui se brise spontanément. Cependant, ceci n'explique la quantification de la charge que si le groupe plus grand est semi-simple. (En fait, le groupe plus grand peut même être le produit d'un groupe semi-simple et d'un certain nombre de facteurs U(1) pourvu que le groupe électrodynamique appartienne complètement à un facteur semi-simple). Comme je l'ai dit dans l'introduction et comme je le démontrerai plus tard, la théorie dans ce cas contient inévitablement des monopôles.

Ainsi, la relation entre monopôles et quantification de la charge est plus solide que jamais, bien que les détails diffèrent profondément de ce qui avait été envisagé au départ. L'un et l'autre sont conséquence d'une cause commune, la brisure spontanée d'une symétrie semi-simple.

2. Même si je n'ai pas grand chose à en dire dans ces conférences, des efforts considérables ont été consacrés au cours des années au développement de l'électrodynamique quantique avec des monopôles, une théorie des champs relativistes avec des particules fondamentales à la fois électriquement et magnétiquement chargées. Un des aspects particuliers de cette théorie est qu'on ne peut pas l'étudier en utilisant la théorie diagrammatique des perturbations, méthode si féconde en électrodynamique quantique ordinaire. Ceci est dû à ce que n'importe quelle sorte de théorie perturbative conduit à un non-sens. Les deux constantes de couplage de la théorie ne peuvent pas être rendues arbitrairement petites simultanément, à cause de la condition de quantification; lorsque $e$ devient petit, $g$ devient grand.

Cela ne veut pas dire que les effets des monopôles virtuels sont nécessairement grands, même pour $e$ très petit. Les grands couplages amplifient les effets des particules virtuelles, mais les grandes masses les diminuent à la fois par le jeu des grands dénominateurs d'énergie et par celui de la décroissance des facteurs de forme pour les grands transferts de moment. Ainsi, même des particules très fortement couplées peuvent avoir des effets petits aux basses énergies si elles sont suffisamment massives. Comme on le verra, c'est ce qui arrive pour les monopôles dans les théories de jauge. La masse de ces particules est proportionnelle à $1/e^2$; quand $e$ tend vers zéro, leurs effets sont négligeables à n'importe quelle énergie fixée.

3. Les dyons sont définis comme des objets qui portent des charges électriques aussi bien que magnétiques. On peut imaginer pour construire des dyons qu'on lie ensemble des particules chargées électriquement et magnétiquement, ou bien on peut les imaginer comme des entités fondamentales qui ne sont pas composées d'objets plus primitifs. Dans les deux cas on peut rapidement obtenir les propriétés des dyons en exploitant l'invariance des équations de Maxwell par rotation de dualité. Une rotation de dualité est définie par:

$$\boldsymbol{E} \to \boldsymbol{E}\cos\alpha + \boldsymbol{B}\sin\alpha, \qquad \boldsymbol{B} \to -\boldsymbol{E}\sin\alpha + \boldsymbol{B}\cos\alpha, \tag{2.18}$$

où $\alpha$ est un nombre réel. Il est facile de vérifier que cette transformation laisse invariantes les équations de Maxwell dans le vide. Nous décrirons les éqs. (2.18) en disant que $(\boldsymbol{E}, \boldsymbol{B})$ est un vecteur de dualité. D'après les éqs. (2.1) et (2.3), $(e, 4\pi g)$ est aussi un vecteur de dualité.

Etant donné deux dyons de charges électrique et magnétique, $e_i$ et $g_i$ $(i=1,2)$, on peut former les deux invariants de dualité bilinéaires dans les charges:

$$\text{Le produit scalaire:} \quad e_1e_2 + 16\pi^2 g_1 g_2, \tag{2.19a}$$

$$\text{et le produit extérieur:} \quad 4\pi(e_1g_2 - g_1e_2). \tag{2.19b}$$

Toute propriété observable du système des deux dyons peut s'exprimer en fonction de ces invariants.

Par exemple la force exercée par un dyon fixé à l'origine sur un autre dyon mobile doit être une fonction linéaire de ces invariants. Ainsi, dans la limite non relativiste, par exemple,

$$\boldsymbol{F} = \left(\frac{e_1e_2}{4\pi} + 4\pi g_1 g_2\right)\boldsymbol{r}/r^3 + (e_1g_2 - g_1e_2)\boldsymbol{v} \times \boldsymbol{r}/r^3, \tag{2.20}$$

où "1" est le dyon mobile, et "2" le dyon fixé. La raison en est que c'est la seule expression qui coïncide avec les expressions connues à la fois pour $g_1 = g_2 = 0$ et pour $g_1 = e_2 = 0$.

Par un raisonnement similaire la condition de quantification pour les dyons est:

$$e_1g_2 - g_1e_2 = 0, \pm\tfrac{1}{2}, \pm 1 \ldots . \tag{2.21}$$

Cette équation a des solutions bizarres. Par exemple, elle est satisfaite par:

$$e_i = n_i e + m_i f,$$
$$g_i = m_i/2e, \tag{2.22}$$

où $n_i$ et $m_i$ sont des entiers, $e$ un nombre réel arbitraire *ainsi que* $f$. Autrement dit, il est parfaitement cohérent avec la condition de quantification que toutes les particules chargées magnétiquement aient des charges électriques fractionnaires (nous verrons à l'occasion que, non seulement c'est cohérent, mais qu'il y a des circonstances où c'est inévitable).

4. Au cours de la discussion du "Tour du monopôle", j'ai dit que le solénoïde était invisible pour les particules classiques chargées, mais détectable (à moins que la condition de quantification ne soit satisfaite) par les particules quantiques. Cependant, le solénoïde pourrait aussi être détecté par un autre genre d'entité physique, un champ classique chargé.

Nous avons déjà essentiellement fait cette observation. J'ai parlé de l'équation de Schrödinger comme si c'était l'équation pour une amplitude de probabilité quantique (ce qu'elle est), mais tous nos arguments resteraient tout aussi valables si $\Psi$ était juste un champ classique comme n'importe quel autre (comme Schrödinger l'avait pensé pendant quelque temps).

Ainsi la condition de quantification de Dirac peut être une conséquence de la physique classique si la physique classique contient des champs chargés. Je fais cette remarque maintenant parce que nous allons bientôt avoir affaire à des théories de champs classiques qui ne contiennent pas de champ chargé: les théories de champs de Yang–Mills. Là, on trouvera à nouveau la condition de quantification et je ne veux pas que vous soyiez désorientés en vous demandant ce qu'un effet quantique fait dans un contexte purement classique.

(Il y a une différence entre la version champ classique et la version particule quantique de la condition de quantification, qui a été obscurcie par mon utilisation d'unités où $\hbar$ égale un. La charge électrique d'un champ classique, quantité qui gouverne la force de son interaction électromagnétique, est très différente de la charge électrique d'une particule classique chargée; elles ont même des dimensions différentes. Elles sont liées uniquement par la dualité onde quantique/corpuscule; $e_{\text{champ}} = e_{\text{particule}}/\hbar$. Ainsi, pour les champs la condition dit que $eg$ est demi-entier; pour les particules, que $eg$ est un multiple demi-entier de $\hbar$).

### *2.4. Drôle de drame avec le moment angulaire*

L'invariance par rotation joue un rôle simplificateur important des problèmes dynamiques. Par exemple, pour une particule sans spin non relativiste dans un potentiel central,

$$H = -\frac{1}{2m}\nabla^2 + V(r). \tag{2.23}$$

Cet opérateur commute avec le moment angulaire. Dans un sous-espace d'états de moment angulaire total donné, $H$ se simplifie:

$$H_l = -\frac{1}{2m}\left(\frac{\partial^2}{\partial r^2} + \frac{2}{r}\frac{\partial}{\partial r}\right) + \frac{l(l+1)}{2mr^2} + V, \tag{2.24}$$

où $l = 0, 1, 2, \ldots$ .

Dans cette sous-section, je vais étendre ce résultat à une particule dans le champ d'un monopôle [10].

$$H = -\frac{1}{2m}(\nabla - \mathrm{i}e\boldsymbol{A})^2 + V(r), \tag{2.25}$$

où $\boldsymbol{A}$ est le potentiel vecteur du monopôle, éq. (2.9). Le premier obstacle à l'analyse est que le potentiel monopolaire détruit l'invariance manifeste par rotation. On évitera ce problème en introduisant

$$\boldsymbol{D} \equiv \nabla - \mathrm{i}e\boldsymbol{A}. \tag{2.26}$$

C'est un opérateur invariant de jauge. De plus $\boldsymbol{H}$ s'exprime en fonction de $\boldsymbol{D}$ et de l'opérateur position, $\boldsymbol{r}$,

$$H = -\frac{1}{2m}\boldsymbol{D}^2 + V(r), \tag{2.27}$$

et ces opérateurs satisfont à l'ensemble de relations de commutation invariant par rotation

$$[r_i, r_j] = 0, \tag{2.28}$$

$$[D_i, r_j] = \delta_{ij}, \tag{2.29}$$

$$[D_i, D_j] = -\mathrm{i}eg\varepsilon_{ijk}r_k/r^3. \tag{2.30}$$

Notre méthode consistera à travailler autant que possible avec les éq. (2.27) à (2.30) et à éviter d'utiliser la forme explicite de $\boldsymbol{A}$.

La première étape consiste à construire un opérateur moment angulaire, $\boldsymbol{L}$, une fonction vectorielle de $\boldsymbol{D}$, $\boldsymbol{r}$ satisfaisant:

$$[L_i, D_j] = \mathrm{i}\varepsilon_{ijk}D_k, \tag{2.31}$$

$$[L_i, r_j] = \mathrm{i}\varepsilon_{ijk}r_k. \tag{2.32}$$

Par conséquent, $\boldsymbol{L}$ satisfera automatiquement:

$$[L_i, L_j] = \mathrm{i}\varepsilon_{ijk}L_k, \tag{2.33}$$

$$[L_i, H] = 0. \tag{2.34}$$

Un candidat naturel est:

$$\boldsymbol{L} \overset{?}{=} -\mathrm{i}\boldsymbol{r} \times \boldsymbol{D}. \tag{2.35}$$

Mais ça ne marche pas; les commutateurs avec $\boldsymbol{r}$ sont corrects, mais pas ceux avec $\boldsymbol{D}$. La réponse correcte se trouve être:

$$\boldsymbol{L} = -\mathrm{i}\boldsymbol{r} \times \boldsymbol{D} - eg\boldsymbol{r}/r. \tag{2.36}$$

Le second terme paraît bien étrange; suivant les paroles immortelles de Rabi au sujet de tout autre chose, "Qui a passé cette commande?".

Je connais trois façons de répondre à cette question. J'esquisserai deux d'entre elles (laissant les détails en exercice), et donnerai la troisième in extenso.

1. J'ai déjà donné la première réponse: L'équation (2.36) donne les règles de commutation correctes. Je laisse la vérification de cette affirmation en exercice.

2. Des équations de mouvement de Heisenberg,

$$\dot{\boldsymbol{r}} = -\mathrm{i}\boldsymbol{D}/m, \tag{2.37}$$

il s'ensuit que:

$$\boldsymbol{L} = m\boldsymbol{r} \times \dot{\boldsymbol{r}} - eg\boldsymbol{r}/r. \tag{2.38}$$

Ainsi le système possède du moment angulaire même si la particule est au repos. La seule source possible de ce moment angulaire est le moment angulaire du champ électromagnétique,

$$\boldsymbol{L}_{\mathrm{em}} = \int \mathrm{d}^3x\, \boldsymbol{x} \times (\boldsymbol{E} \times \boldsymbol{B}). \tag{2.39}$$

Il suffit d'un tout petit peu de travail pour aller très loin dans l'évaluation de cette intégrale. D'abord elle doit être proportionnelle à $eg$. Deuxièment, par analyse dimensionnelle elle doit être homogène d'ordre zéro en $r$. Troisièmement, en vertu de l'invariance par rotation, elle doit être proportionnelle à $\boldsymbol{r}$, le seul vecteur dans le problème. Ainsi,

$$\boldsymbol{L}_{\mathrm{em}} = \beta eg\boldsymbol{r}/r, \tag{2.40}$$

où $\beta$ est un coefficient numérique. L'évaluation de $\beta$ est laissée en exercice.

3. Dans mon premier cours de physique, je tenais une roue de bicyclette en rotation par son axe et j'essayais de faire tourner l'axe. A ma grande surprise, je sentais une force orthogonale à la force appliquée; plus tard, j'appris que c'était parce que le système était doué de moment angulaire.

C'est exactement ce qui se passe si on essaye de faire bouger une particule chargée au repos dans un champ monopolaire. La force de Lorentz tire la particule dans une direction orthogonale à l'impulsion initiale. Ceci suggère que ce système a, lui aussi, du moment angulaire.

Essayons de calculer ce moment angulaire en identifiant son taux de variation temporelle avec le moment extérieur appliqué au système. Par analogie avec l'argument précédent, faisons l'Ansatz:

$$\boldsymbol{L} = m\boldsymbol{r} \times \dot{\boldsymbol{r}} + \beta e g \boldsymbol{r}/r, \tag{2.41}$$

où $\beta$ est une constante qui sera fixée au cours du calcul.

L'équation du mouvement est:

$$m\ddot{\boldsymbol{r}} = \boldsymbol{F}^{\text{ext}} + eg\dot{\boldsymbol{r}} \times \boldsymbol{r}/r^3, \tag{2.42}$$

où $\boldsymbol{F}^{\text{ext}}$ est la force extérieure. Utilisant en outre l'identité

$$\frac{\mathrm{d}}{\mathrm{d}t}(\boldsymbol{r}/r) = \dot{\boldsymbol{r}}/r - (\boldsymbol{r}\cdot\dot{\boldsymbol{r}})\boldsymbol{r}/r^3, \tag{2.43}$$

on trouve

$$\frac{\mathrm{d}\boldsymbol{L}}{\mathrm{d}t} = \boldsymbol{r} \times \boldsymbol{F}^{\text{ext}} + \boldsymbol{r} \times (eg\dot{\boldsymbol{r}} \times \boldsymbol{r})/r^3 + eg\beta\left[r^2\dot{\boldsymbol{r}} - (\boldsymbol{r}\cdot\dot{\boldsymbol{r}})\boldsymbol{r}\right]/r^3. \tag{2.44}$$

Ainsi, si $\beta = -1$:

$$\mathrm{d}\boldsymbol{L}/\mathrm{d}t = \boldsymbol{r} \times \boldsymbol{F}^{\text{ext}}. \tag{2.45}$$

Ceci complète la discussion du terme supplémentaire dans l'expression du moment angulaire. Retournons maintenant à l'analyse de l'Hamiltonien (2.27).

Nous commençons par écrire l'identité

$$\boldsymbol{D}\cdot\boldsymbol{D} = \boldsymbol{D}\cdot\boldsymbol{r}\frac{1}{r^2}\boldsymbol{r}\cdot\boldsymbol{D} - \boldsymbol{D}\times\boldsymbol{r}\cdot\frac{1}{r^2}\boldsymbol{r}\times\boldsymbol{D}. \tag{2.46}$$

Ici, j'ai ordonné les termes de façon que l'identité soit vraie indépendamment des commutateurs entre $\boldsymbol{D}$ et $\boldsymbol{r}$ [11]. Nous allons analyser les deux termes de cette expression séparément.

Comme $A_r = 0$,

$$\boldsymbol{r}\cdot\boldsymbol{D} = \boldsymbol{r}\cdot\nabla = r\frac{\partial}{\partial r}, \tag{2.47}$$

$$\boldsymbol{D}\cdot\boldsymbol{r} = \boldsymbol{r}\cdot\boldsymbol{D} + 3. \tag{2.48}$$

Ainsi:

$$\boldsymbol{D}\cdot\boldsymbol{r}\frac{1}{r^2}\boldsymbol{r}\cdot\boldsymbol{D} = \left(r\frac{\partial}{\partial r} + 3\right)\frac{1}{r}\frac{\partial}{\partial r} = \frac{\partial^2}{\partial r^2} + \frac{2}{r}\frac{\partial}{\partial r}. \tag{2.49}$$

A partir des commutateurs de $\boldsymbol{r}$ et $\boldsymbol{D}$, $\boldsymbol{r}\times\boldsymbol{D} = -\boldsymbol{D}\times\boldsymbol{r}$ commute avec $r^2$.

Ainsi:

$$-\boldsymbol{D}\times\boldsymbol{r}\frac{1}{r^2}\boldsymbol{r}\times\boldsymbol{D} = \frac{1}{r^2}(\boldsymbol{r}\times\boldsymbol{D})^2. \tag{2.50}$$

Si on élève au carré l'expression du moment angulaire, eq. (2.36), on trouve

$$\boldsymbol{L}\cdot\boldsymbol{L} = -(\boldsymbol{r}\times\boldsymbol{D})^2 + e^2g^2. \tag{2.51}$$

Mettant ensemble ces résultats, on trouve que sur un sous-espace d'états de moment angulaire total donné,

$$H_l = -\frac{1}{2m}\left(\frac{\partial^2}{\partial r^2}+\frac{2}{r}\frac{\partial}{\partial r}\right)+\frac{l(l+1)-e^2g^2}{2mr^2}+V, \tag{2.52}$$

où, comme d'habitude $l(l+1)$ est la valeur propre de $\boldsymbol{L}\cdot\boldsymbol{L}$. Comme $\boldsymbol{L}$ satisfait à l'algèbre du moment angulaire, nous savons que $l$ doit être entier ou demi-entier, mais on a besoin d'une analyse supplémentaire pour déterminer quelles valeurs de $l$ apparaissent réellement.

Nous connaissons tous la solution de ce problème lorsque $eg = 0$. Les représentations apparaissent avec $l = 0, 1, 2, \ldots$ et, pour $r$ fixé, chacune apparaît une seule fois. Dans les ouvrages élémentaires, ce résultat est établi en étudiant les solutions de la partie angulaire de l'équation d'onde. Cette méthode peut certainement être étendue au cas $eg \neq 0$, mais elle devient un peu délicate; on doit se préoccuper des singularités sur la corde, ou, alternativement, du recollement des deux solutions dans l'intersection des deux régions. En conséquence, je résoudrai ici le problème par une méthode légèrement plus abstraite qui évite ces difficultés.

Le problème général est le suivant: étant donné un espace d'états qui se transforment par rotation de façon spécifiée, trouver un ensemble de vecteurs de base qui se transforment suivant les représentations irréductibles du groupe des rotations. Si on spécifie une rotation générale de la façon usuelle, par les trois angles d'Euler $\alpha, \beta, \gamma$, les états que nous cherchons satisfont:

$$\mathrm{e}^{-\mathrm{i}L_z\alpha}\mathrm{e}^{-\mathrm{i}L_y\beta}\mathrm{e}^{-\mathrm{i}L_z\gamma}|l, m\rangle = \sum_{m'} D^{(l)}_{m'm}(\alpha,\beta,\gamma)|l, m'\rangle, \tag{2.53}$$

où

$$D^{(l)}_{m'm}(\alpha,\beta,\gamma) = \mathrm{e}^{\mathrm{i}m'\alpha}d^{(l)}_{m'm}(\beta)\mathrm{e}^{-\mathrm{i}m\gamma}, \tag{2.54}$$

et $d^{(l)}$ est une matrice qu'on peut trouver dans n'importe quel ouvrage de mécanique quantique. (Nous n'aurons pas besoin de sa forme explicite ici.)

Pour nous mettre entrain, regardons le cas où nous connaissons déjà la réponse, $eg = 0$. Nous travaillons avec $r$ fixé, de sorte qu'un ensemble complet de vecteurs de base est constitué des vecteurs propres de la position angulaire de la particule, que nous décrivons comme d'habitude par les deux angles $\theta$ et $\varphi$. N'importe lequel de ces états peut être obtenu en appliquant la rotation appropriée à l'état où la particule est au pôle nord:

$$|\theta, \varphi\rangle = e^{-iL_z\varphi} e^{-iL_y\theta} |\theta = 0\rangle. \tag{2.55}$$

(Pour $\theta = 0$ nous n'avons pas besoin de spécifier $\varphi$.)

Nous connaissons les états que nous cherchons si nous connaissons leurs fonctions d'ondes dans l'espace de configuration, $\langle \theta, \varphi \mid l, m\rangle$. D'après les équations précédentes:

$$\begin{aligned} \langle \theta, \varphi \mid l, m\rangle &= \langle \theta = 0 | e^{iL_y\theta} e^{iL_z\varphi} | l, m\rangle \\ &= \sum_{m'} e^{-im\varphi} d^{(l)}_{m'm}(\theta) \langle \theta = 0 \mid l, m'\rangle. \end{aligned} \tag{2.56}$$

Ainsi, nous connaissons tout pourvu que nous connaissions $\langle \theta = 0 \mid l, m'\rangle$. Ces coefficients satisfont à une importante condition de cohérence qui est une conséquence de

$$e^{-iL_z\alpha} |\theta = 0\rangle = |\theta = 0\rangle, \tag{2.57}$$

ceci implique

$$\langle \theta = 0 \mid l, m'\rangle = 0, \qquad m' \neq 0. \tag{2.58}$$

Ainsi, nous pouvons nous borner à construire $\langle \theta, \varphi \mid l, m\rangle$ pour $l = 0, 1, 2 \ldots$, et pour chacune de ces valeurs de $l$, la solution est unique, mise à part une normalisation sans importance. Une fois construites ces fonctions, il est facile de s'assurer, en utilisant les règles de multiplication pour les matrices $D$, qu'elles se transforment bien de la façon désirée.

Etendons maintenant cette analyse au cas où $eg \neq 0$. Les commutateurs de $\boldsymbol{L}$ et $\boldsymbol{r}$ sont les mêmes qu'auparavant, de sorte que,

$$|\theta, \varphi\rangle = e^{-iL_z\varphi} e^{-iL_y\theta} |\theta = 0\rangle \times (\text{facteur de phase}). \tag{2.59}$$

Le facteur de phase dépend de la jauge dans laquelle on opère. Heureusement, il n'est pas nécessaire de connaître sa forme explicite; quelle qu'elle soit, la principale conclusion de l'analyse précédente reste inchangée: nous connaissons tout pourvu que nous connaissions $\langle \theta = 0 \mid l, m'\rangle$. Cependant la condition de cohérence, éq. (2.58) est changée:

$$L_z |\theta = 0\rangle = (-i\boldsymbol{r} \times \boldsymbol{D} - eg\boldsymbol{r}/r)_z |\theta = 0\rangle = -eg|\theta = 0\rangle. \tag{2.60}$$

Ainsi:

$$\langle \theta = 0 \mid l, m' \rangle = 0, \qquad m' \neq -eg, \tag{2.61}$$

et les valeurs autorisées pour le moment angulaire total sont:

$$l = |eg|, |eg| + 1, |eg| + 2 \ldots. \tag{2.62}$$

Comme auparavant elles apparaissent chacune une seule fois. Ceci termine notre analyse.

Remarques: (i) L'effet du monopôle est d'une simplicité surprenante. Il ne fait que changer légèrement le potentiel centrifuge dans l'équation de Schrödinger radiale. Si on peut résoudre l'équation de Schrödinger pour un potentiel radial sans monopôle, on peut la résoudre avec un monopôle. (ii) Comme $l$ est toujours supérieur ou égal à $|eg|$, le potentiel centrifuge est toujours positif, et le monopôle par lui-même ne lie pas les particules chargées sans spin. Comme on s'y attendrait et comme on le verra dans un cas particulier, la situation est très différente quand la particule est douée de spin. (iii) La condition de quantification permet à $|eg|$ d'être demi-entier. De la sorte deux particules sans spin dont l'une porte une charge électrique et l'autre une charge magnétique peuvent très bien se lier pour constituer un dyon doué d'un moment angulaire demi-entier. Ce phénomène est embarassant du point de vue du théorème spin-statistique. Je vais maintenant expliquer la solution de ce paradoxe.

### *2.5. La solution du paradoxe spin-statistique*

On pourrait penser qu'il n'y a rien à dire sur la connection spin-statistique dans le cadre de la mécanique quantique non relativiste. Il n'en est rien; bien que la théorie relativiste des champs soit en effet nécessaire pour montrer que des particules sans spin sont des bosons, la théorie non relativiste est alors suffisante pour déduire la statistique d'états composés de ces bosons. Je vais maintenant montrer qu'un dyon composé d'une particule sans spin chargée électriquement ("électron") et d'une particule sans spin chargée magnétiquement ("monopôle") obéit à la statistique de Bose–Einstein si $eg$ est entier et de Fermi–Dirac si $eg$ est demi-entier [12].

Nous connaissons déjà l'Hamiltonien d'un électron dans le champ d'un monopôle,

$$H = \frac{[\boldsymbol{p}_e - eg\boldsymbol{A}_D(\boldsymbol{r}_e - \boldsymbol{r}_m)]^2}{2m_e} + \cdots. \tag{2.63}$$

Ici, les points de suspension représentent les éventuelles interactions non électromagnétiques et $\boldsymbol{A}_{\mathrm{D}}$ est le potentiel de corde de Dirac standard:

$$\boldsymbol{A}_{\mathrm{D}}(\boldsymbol{x})\cdot \mathrm{d}\boldsymbol{x} = (1-\cos\theta)\,\mathrm{d}\phi. \tag{2.64}$$

Par une rotation de dualité nous connaissons donc l'Hamiltonien pour un monopôle dans le champ d'un électron,

$$H = \frac{[\boldsymbol{p}_{\mathrm{m}} + eg\boldsymbol{A}'_{\mathrm{D}}(\boldsymbol{r}_{\mathrm{m}} - \boldsymbol{r}_{\mathrm{e}})]^2}{2m_{\mathrm{m}}}. \tag{2.65}$$

Ici le signe a changé car la rotation de dualité qui transforme $e$ en $g$ transforme $g$ en $(-e)$, et $\boldsymbol{A}'_{\mathrm{D}}$ est un potentiel équivalent de jauge à $\boldsymbol{A}_{\mathrm{D}}$.

Pour déterminer $\boldsymbol{A}'_{\mathrm{D}}$, nous considérons un système composé d'un monopôle et d'un électron. Pour tout choix de $\boldsymbol{A}'_{\mathrm{D}}$, $\boldsymbol{p}_{\mathrm{e}} + \boldsymbol{p}_{\mathrm{m}}$ est une constante du mouvement car $H$ est invariant par translation. Cependant comme:

$$m_{\mathrm{e}}\boldsymbol{v}_{\mathrm{e}} = \boldsymbol{p}_{\mathrm{e}} + eg\boldsymbol{A}_{\mathrm{D}}(\boldsymbol{r}_{\mathrm{e}} - \boldsymbol{r}_{\mathrm{m}}), \tag{2.66a}$$

et

$$m_{\mathrm{m}}\boldsymbol{v}_{\mathrm{m}} = \boldsymbol{p}_{\mathrm{m}} - eg\boldsymbol{A}'_{\mathrm{D}}(\boldsymbol{r}_{\mathrm{m}} - \boldsymbol{r}_{\mathrm{e}}), \tag{2.66b}$$

$m_{\mathrm{e}}\boldsymbol{v}_{\mathrm{e}} + m_{\mathrm{m}}\boldsymbol{v}_{\mathrm{m}}$ n'est une constante du mouvement (comme l'équation classique du mouvement indique que cela doit être) que si

$$\boldsymbol{A}'_{\mathrm{D}}(\boldsymbol{x}) = \boldsymbol{A}_{\mathrm{D}}(-\boldsymbol{x}). \tag{2.67}$$

$\boldsymbol{A}'_{\mathrm{D}}$ est bien équivalent de jauge à $\boldsymbol{A}_{\mathrm{D}}$. (voir section 2.2.). En résumé, l'Hamiltonien correct pour un monopôle dans le champ d'un électron est:

$$H = \frac{[\boldsymbol{p}_{\mathrm{m}} + eg\boldsymbol{A}_{\mathrm{D}}(\boldsymbol{r}_{\mathrm{e}} - \boldsymbol{r}_{\mathrm{m}})]^2}{2m_{\mathrm{m}}} + \cdots. \tag{2.68}$$

Si on interchange les monopôles et les électrons on change le signe devant le potentiel vecteur mais pas l'ordre des termes dans l'argument du potentiel vecteur.

On peut maintenant passer au système qui nous occupe, à savoir deux dyons, chacun composé d'un monopôle sans spin et d'un électron sans spin. Exactement de la même façon que lorsqu'il s'agit de deux atomes, on décrit les états du système par une fonction d'onde de Schrödinger

$$\psi_{A_1A_2}(\boldsymbol{r}_1, \boldsymbol{r}_2).$$

Ici les $\boldsymbol{r}$'s sont les positions des dyons, et les $A_i$ sont des variables discrètes qui décrivent l'état interne des dyons, spin, énergie d'excitation, etc. Comme les électrons et les monopôles sont des bosons, les arguments standards conduisent à:

$$\psi_{A_1A_2}(\boldsymbol{r}_1, \boldsymbol{r}_2) = \psi_{A_2A_1}(\boldsymbol{r}_2, \boldsymbol{r}_1). \tag{2.69}$$

Normalement on dirait que ça implique que les dyons sont des bosons. Cependant, considérons plus en détail l'Hamiltonien pour le système de deux dyons

$$\begin{aligned} H = & \left[\boldsymbol{p}_1 - eg\boldsymbol{A}_\mathrm{D}(\boldsymbol{r}_1 - \boldsymbol{r}_2) + eg\boldsymbol{A}_\mathrm{D}(\boldsymbol{r}_2 - \boldsymbol{r}_1)\right]^2/2m \\ & + \left[\boldsymbol{p}_2 - eg\boldsymbol{A}_\mathrm{D}(\boldsymbol{r}_2 - \boldsymbol{r}_1) + eg\boldsymbol{A}_\mathrm{D}(\boldsymbol{r}_1 - \boldsymbol{r}_2)\right]^2/2m + \cdots, \end{aligned} \tag{2.70}$$

où les points de suspension représentent les interactions de Coulomb aussi bien que les interactions non électromagnétiques éventuelles. J'espère que l'origine des termes de cette équation est claire; l'électron du premier dyon voit le monopôle du second dyon, le monopôle du premier dyon voit l'électron du second dyon etc...

L'équation (2.70) semble décrire les forces les plus horriblement dépendantes des vitesses, mais il ne peut y avoir de telles forces entre deux dyons identiques car il existe une rotation de dualité qui annule simultanément leurs charges magnétiques. En fait, de l'eq. (2.16) on déduit,

$$A_\mathrm{D}(\boldsymbol{x}) - A_\mathrm{D}(-\boldsymbol{x}) = 2\nabla\varphi. \tag{2.71}$$

Ainsi, si on effectue la transformation de jauge

$$\psi \to \psi' = \psi \exp(2\mathrm{i}eg\varphi_{12}), \tag{2.72}$$

l'horrible interaction disparaît:

$$H \to H' = \boldsymbol{p}_1^2/2m + \boldsymbol{p}_2^2/2m + \cdots. \tag{2.73}$$

Mais cette transformation peut changer la symétrie de la fonction d'onde, car lorsque $\boldsymbol{r}_1$ et $\boldsymbol{r}_2$ sont échangés, $\varphi_{12}$ se transforme en $\varphi_{12} + \pi$. Par suite,

$$\psi'_{A_1A_2}(\boldsymbol{r}_1, \boldsymbol{r}_2) = \psi'_{A_2A_1}(\boldsymbol{r}_2, \boldsymbol{r}_1)\exp(2\pi \mathrm{i}eg).$$

C'est à dire que $\psi'$ est symétrique dans l'échange des dyons si $eg$ est entier et antisymétrique si $eg$ est demi-entier.

Nous avons ainsi deux descriptions du même système. L'une, donnée par $\Psi$ et $H$, dit que nos dyons sont des bosons quel que soit leur spin mais qu'ils échangent des forces d'extrêmement longue portée. L'autre, donnée par $\Psi'$, $H'$, dit qu'il n'y a pas de force extraordinaire mais que les dyons obéissent à la statistique appropriée à leur spin. Ces deux descriptions sont reliées par une transformation de jauge et font par conséquent les mêmes prédictions pour les observables. Néanmoins, il n'y a pas d'ambiguité; ce serait fou de choisir la première description alors que nous avons la seconde à notre disposition. (Après tout, nous pourrions utiliser la même transformation de jauge pour faire apparaître n'importe quelle paire de fermions comme des bosons, quelle que soit leur charge électrique ou magnétique). Les dyons satisfont de façon non-ambigue au théorème spin-statistique.

## 3. Monopôles non Abéliens à grande distance

### *3.1. Theorie de champs de jauge — une revue éclair*

Cette sous section est une collection de définitions et de formules appartenant à la théorie classique des champs de jauge, avec des commentaires occasionnels. Je l'ai introduite ici pour établir les notations et nous rappeler certains des aspects du sujet qui seront importants dans la suite. Elle est beaucoup trop abrégée pour tenir lieu d'exposé pédagogique; si vous ne connaissez pas déjà les éléments des théories de jauge, ce n'est pas ici que vous les apprendrez.

#### *3.1.1. Transformations de jauge*

Dans les théories de jauge, les variables dynamiques se scindent en deux classes, les champs de jauge et les champs de matière.

Nous commençons par les champs de matière que nous mettons ensemble dans un grand vecteur $\mathbf{\Phi}$. Une transformation de jauge est spécifiée par une fonction $g(x)$ de l'espace–temps dans un groupe de Lie connexe, compact $G$. Par une de ces tranformations, les champs de matière se tranforment suivant une représentation fidèle, unitaire de $G$, soit $D(g)$:

$$g(x) \in G\colon \mathbf{\Phi}(x) \to D(g(x))\mathbf{\Phi}(x). \tag{3.1}$$

Il nous sera commode d'identifier l'élément de groupe $g$ avec la matrice $D(g)$ et d'écrire cette équation sous la forme:

$$g(x) \in G\colon \mathbf{\Phi}(x) \to g(x)\mathbf{\Phi}(x). \tag{3.2}$$

Dans le voisinage de l'identité de $G$, un élément du groupe peut être développé en série:

$$g = 1 + \sum_{a=1}^{\dim G} \varepsilon^a T^a + \mathrm{O}(\varepsilon^2). \tag{3.3}$$

Ici les $\varepsilon$ sont des coordonnées pour le groupe, et les $T$ sont des matrices appelées les générateurs infinitésimaux du groupe. Les générateurs sous-tendent un espace linéaire appelé l'algèbre de Lie de $G$. Comme $g$ est unitaire les $T$'s sont antihermitiques.

$$T^a = -T^{a\dagger}. \tag{3.4}$$

(Nous suivons ici les conventions des mathématiciens; les physiciens insèrent fréquemment un facteur $i$ dans la définition des générateurs pour les rendre hermitiens.)

Pour un élément général du groupe, $g$,

$$gT^a g^\dagger = {}^{(\mathrm{adj})}D_b{}^a(g) T^b, \tag{3.5}$$

où ${}^{(\mathrm{adj})}D$ est une représentation du groupe appelée la représentation adjointe, et où on a utilisé la convention de sommation sur les indices répétés.

Le commutateur de deux quelconque générateurs est une combinaison linéaire des générateurs

$$[T^a, T^b] = c^{abc} T^c, \tag{3.6}$$

où les $c$'s sont des coefficients réels appelés constantes de structure. Ils dépendent uniquement du groupe abstrait $G$ et pas de la représentation particulière utilisée pour le réaliser. Nous choisirons toujours les $T$'s de sorte que:

$$\mathrm{Tr}\, T^a T^b = -N\delta^{ab}, \tag{3.7}$$

où $N$ est une constante de normalisation. Ainsi,

$$c^{abc} = -N^{-1} \mathrm{Tr}[T^a, T^b] T^c. \tag{3.8}$$

Il s'en suit que $c^{abc}$ est invariant par les permutations paires des indices et change de signe par les permutations impaires.

Par exemple, si $G$ est SU(2), le groupe des matrices $2\times 2$ unitaires unimodulaires, le choix standard est $T^a = \mathrm{i}\sigma^a/2$, où les $\sigma$'s sont les matrices de spin de Pauli. La représentation adjointe est la représentation vectorielle, $N$ vaut $1/2$ et $c^{abc}$ est $\varepsilon^{abc}$.

### *3.1.2. Champs de jauge*

Les champs de jauge sont un ensemble de champs vectoriels $A^a_\mu$, $a = 1 \ldots \dim G$. Il sera convenable d'assembler ces vecteurs en un seul champ vectoriel matriciel,

$$A_\mu = A_\mu^a T^a. \tag{3.9}$$

Les propriétés de transformation de ces champs par transformation de jauge sont définies par:

$$g(x): A_\mu \to g A_\mu g^{-1} + g \partial_\mu g^{-1}. \tag{3.10}$$

Si on définit la dérivée covariante du champ de matière par

$$D_\mu \Phi \equiv (\partial_\mu + A_\mu)\Phi, \tag{3.11}$$

alors, dans un transformation de jauge

$$g(x): D_\mu \Phi \to g D_\mu \Phi. \tag{3.12}$$

Le tenseur d'intensité de champ est un champ à valeur matricielle défini par:

$$[D_\mu, D_\nu]\Phi = (\partial_\mu A_\nu - \partial_\nu A_\mu + [A_\mu, A_\nu])\Phi \equiv F_{\mu\nu}\Phi. \tag{3.13}$$

Par une transformation de jauge, l'intensité de champ se transforme suivant la représentation adjointe de $G$,

$$g(x): F_{\mu\nu} \to g F_{\mu\nu} g^{-1}. \tag{3.14}$$

La dérivée covariante de l'intensité de champ est définie par

$$D_\lambda F_{\mu\nu} = \partial_\lambda F_{\mu\nu} + [A_\lambda, F_{\mu\nu}]. \tag{3.15}$$

Par transformation de jauge, elle se transforme de la même façon que $F$ elle-même.

Parallèlement à l'éq. (3.9),

$$F_{\mu\nu} = F_{\mu\nu}^a T^a, \tag{3.16}$$

où

$$F_{\mu\nu}^a = \partial_\mu A_\nu^a - \partial_\nu A_\mu^a + c^{abc} A_\mu^b A_\nu^c. \tag{3.17}$$

*3.1.3. Dynamique*

La densité Lagrangienne est:

$$\mathcal{L} = \frac{1}{4Nf^2} \operatorname{Tr} F_{\mu\nu} F^{\mu\nu} + \mathcal{L}_{\mathrm{m}}(\Phi, D_\mu \Phi), \tag{3.18}$$

où $\mathcal{L}_{\mathrm{m}}$ est une fonction invariante

$$\mathcal{L}_{\mathrm{m}}(g\Phi, g D_\mu \Phi) = \mathcal{L}_{\mathrm{m}}(\Phi, D_\mu \Phi), \tag{3.19}$$

et $f$ est un nombre réel appelé la constante de couplage. La partie de $\mathcal{L}$ spécifique au champ de jauge peut aussi s'écrire

$$\frac{1}{4Nf^2}\operatorname{Tr}F_{\mu\nu}F^{\mu\nu} = -\frac{1}{4f^2}F^a_{\mu\nu}F^{\mu\nu a}. \tag{3.20}$$

En l'absence de champs de matière, les champs de jauge satisfont à

$$D_\mu F^{\mu\nu} = 0. \tag{3.21}$$

Ces équations sont appelée équations de Yang–Mills sans source.

Si on définit de nouveaux champs, notés par un prime, par

$$A'^a_\mu = f^{-1}A^a_\mu, \tag{3.22}$$

et

$$F'^a_{\mu\nu} = f^{-1}F^a_{\mu\nu} = \partial_\mu A'^a_\nu - \partial_\nu A'^a_\mu + fc^{abc}A^{b'}_\mu A^{c'}_\nu, \tag{3.23}$$

alors

$$-\frac{1}{4f^2}F_{\mu\nu a}F^{\mu\nu a} = -\frac{1}{4}F'^a_{\mu\nu}F^{\mu\nu' a} \tag{3.24}$$

et

$$D_\mu\Phi = \partial_\mu\Phi + fA'^a_\mu T^a\Phi. \tag{3.25}$$

de là on voit que $f$ est bien une constante de couplage, absente des termes quadratiques du Lagrangien mais présente dans les termes d'ordre plus élevé.

Si le groupe de jauge est un produit de facteurs, groupes simples et groupes U(1), on peut avoir une constante de couplage indépendante pour chaque facteur. Le terme de jauge du Lagrangien devient:

$$-\frac{1}{4}\sum_a f_a^{-2}F^a_{\mu\nu}F^{\mu\nu a}, \tag{3.26}$$

où $f_a$ est le même pour les champs associés au même groupe facteur. De même, l'éq. (3.25) devient:

$$D_\mu\Phi = \partial_\mu\Phi + \sum_a f_a A^{a'}_\mu T^a\Phi. \tag{3.27}$$

### *3.1.4. La jauge temporelle*

La jauge temporelle est définie par $A_0 = 0$. Pour transformer une configuration du champ de jauge dans la jauge temporelle, on a besoin d'une fonction de jauge $g(x)$ telle que:

$$gA_0g^{-1} + g\partial_0 g^{-1} = 0. \tag{3.28}$$

Une solution de cette équation est:

$$g^{-1}(x,t) = T\exp\left[-\int_0^t \mathrm{d}t' A_0(\boldsymbol{x},t')\right], \tag{3.29}$$

où $T$ indique que l'intégrale est ordonnée dans le temps.

La jauge temporelle est utile car le terme de jauge de la densité Lagrangienne est:

$$-\partial_0 A_i^a \partial_0 A^{ia}/2f^2,$$

plus des termes sans dérivée par rapport au temps. Ainsi, la structure du problème aux valeurs initiales ressemble de près à celui d'une théorie de champ scalaire sans couplage dérivatif, de même que la forme des équations d'Hamilton.

L'adoption de la jauge temporelle ne détruit pas toute liberté de jauge; on peut encore faire des transformations de jauge indépendantes du temps.

### *3.1.5. Elements du groupe associés a des chemins*

Un chemin dans l'espace–temps est décrit par une fonction $X(s)$ ou $s$ varie de 0 à 1. A tout chemin, on associe l'élément $g$ du groupe défini par:

$$g = P\exp\left(-\int_0^1 \mathrm{d}s\, A_\mu \frac{\mathrm{d}X^\mu}{\mathrm{d}s}\right), \tag{3.30}$$

où $P$ implique que l'intégrale est ordonnée le long du chemin. $g$ peut aussi être identifié avec $g(1)$, où $g(s)$ est la solution de

$$\frac{Dg}{Ds} \equiv \frac{\mathrm{d}g}{\mathrm{d}s} + A_\mu \frac{\mathrm{d}X^\mu}{\mathrm{d}s} g = 0, \tag{3.31}$$

avec la condition aux limites $g(0) = 1$. [Notez la ressemblance avec les éq. (3.28) et (3.29).]

Par la transformation de jauge $h(X)$

$$h(X) \in G: g \to h(X(1))gh(X(0))^{-1}. \tag{3.32}$$

La situation est encore plus simple si le chemin est fermé, $X(0) = X(1)$

$$h(X) \in G: g \to h(X(0))gh(X(0))^{-1}. \tag{3.33}$$

Dans ce cas, en dépit du fait que c'est une intégrale non locale, $g$ se transforme exactement comme $F_{\mu\nu}$ qui est un champ local.

Pour une courbe fermée, $\mathrm{Tr}\, g$ est le fameux facteur de la "boucle de Wilson". C'est un invariant de jauge, indépendant du point de la boucle où on choisit l'origine et l'extrêmemité de la courbe. Il dépend par contre de la représentation matricielle utilisée; dans le langage de la théorie des groupes, des représentations différentes ont des caractères différents.

Pour l'électromagnétisme usuel, l'élément du groupe associé à une courbe fermée est:

$$g = \exp(\mathrm{i}e\Phi), \qquad (3.34)$$

où $\Phi$ est le flux *magnétique* qui passe à travers la surface limitée par la courbe. Ainsi, le concept d'élément du groupe associé à une courbe est une généralisation non Abélienne possible du concept de flux *magnétique*. (L'alternative évidente qui consiste à intégrer le champ *magnétique* ne marche pas; cela ne mène à rien d'additioner des quantités qui se transforment différemment par transformation de jauge, comme c'est le cas pour les champs *magnétiques* non Abéliens pris à des points différents).

### *3.2. La nature de la limite classique*

Nous allons passer quelque temps et dépenser quelque énergie à l'étude des propriétés des monopôles dans les théories non Abéliennes classiques. Il est, par conséquent, raisonnable de commencer par se demander dans quelles conditions on peut s'attendre à ce que la physique classique décrive la réalité quantique.

Soit $S(\mathbf{\Phi})$ la fonctionnelle d'action pour une théorie classique de champs quelconques, où $\mathbf{\Phi}$ représente l'ensemble des champs de la théorie, champs de jauge, ou autres. La forme Euclidienne de l'intégrale de chemin de Feynman nous dit que la version quantique de la théorie est définie par l'intégrale fonctionnelle

$$\int (\mathrm{d}\mathbf{\Phi}) \exp[-S(\mathbf{\Phi})/\hbar]. \qquad (3.35)$$

Si on définit de nouveaux champs $\mathbf{\Phi}' = \mathbf{\Phi}/\sqrt{\hbar}$, alors, à part une constante de normalisation sans importance, on peut récrire:

$$\int (\mathrm{d}\mathbf{\Phi}') \exp[-S(\mathbf{\Phi}'\sqrt{\hbar})/\hbar]. \qquad (3.36)$$

Cette transformation triviale met en évidence le rôle de $\sqrt{\hbar}$ comme constante de couplage, absent des termes quadratiques dans l'argument de l'exponentielle mais présent dans les termes d'ordre plus élevé. La limite classique est une limite de couplage faible. Faire tendre $\hbar$ vers 0, en gardant constantes toutes les constantes de couplage revient au même que de faire tendre vers 0 toutes les constantes de couplage (à une vitesse appropriée), à $\hbar$ fixé.

Ainsi, la théorie des champs de jauge classique (sans brisure spontanée), qui a été le sujet de la section 3.1, devrait être un bon guide à la théorie de jauge quantique faiblement couplée (également, sans brisure spontanée). Je connais deux régimes où de telles théories sont applicables. L'un a été mentionné dans l'introduction; c'est la chromodynamique quantique à des distances plus petites que la longueur de confinement. Comme la théorie est asymptotiquement libre il s'agit d'un régime de couplage faible. L'autre a trait au début de l'Univers. Les champs de jauge faiblement couplés qui sont maintenant massifs pourraient bien avoir été dépourvus de masses auparavant, quand la structure de la brisure de symétrie était différente.

Voici donc les domaines où on pourrait trouver les monopôles non Abéliens: bien à l'intérieur des hadrons ou, il y a bien longtemps. Maintenant, voyons à quoi ils ressembleraient si on les trouvait.

### *3.3. Classification dynamique (GNO) des monopôles*

Au début de la section 2, j'ai décrit la collection standard de champs électromagnétiques qui pourraient exister à l'extérieur d'une boîte noire de contenu inconnu; les termes dominants aux grandes distances étaient les champs électriques et magnétiques monopolaires. La classification dynamique des monopôles, inventée par Goddard, Nuyts, et Olive (GNO) [13], peut être considérée comme l'extension de cette analyse aux théories de jauge non Abéliennes. Je vais maintenant donner une description déductive de la classification GNO.

Exactement comme dans le cas de l'électromagnétisme, l'analyse utilisera exclusivement les équations en dehors de la boîte noire, à savoir, les équations de champs sans source. Comme ces équations sont non linéaires et compliquées, je ne m'attacherai pas à la construction d'une famille complète de solutions, j'essaierai au contraire de trouver la généralisation non Abélienne du champ magnétique monopolaire.

J'exclurai les champs électriques dès le début en cherchant des solutions qui ne sont pas seulement indépendantes du temps mais aussi invariantes par renversement du temps (dans une jauge appropriée). Le renversement du temps est l'opération:

$$T\colon A_0(\boldsymbol{x},t) \to -A_0(\boldsymbol{x},-t), \qquad \boldsymbol{A}(\boldsymbol{x},t) \to A(\boldsymbol{x},-t). \tag{3.37}$$

(La définition alternative autorisée dans le cas Abélien, c'est à dire cette opération composée avec "moins un", n'est pas une invariance de la théorie non Abélienne.) Ainsi,

$$A_0 = 0, \qquad \partial_0 A_i = 0, \tag{3.38}$$

et $F_{0i}$ est nul.

L'équation (3.38) nous laisse encore la liberté de transformations de jauge indépendantes du temps. Je profiterai de cette liberté pour imposer:

$$A_r = 0. \tag{3.39}$$

Exactement de la même façon qu'on intègre le long des lignes temporelles pour construire la transformation de jauge qui conduit à la jauge temporelle, éq. (3.29), on construit ici la transformation de jauge désirée en intégrant le long de lignes radiales en partant, par exemple, de la sphère unité. Evidemment, ce procédé pourrait créer des ennuis à l'origine, où les lignes radiales se coupent, mais nous n'avons pas à nous en préoccuper; l'origine est à l'intérieur de la boîte noire où de toutes façons nous ne voulons pas utiliser les équations de champ. Notez que nous avons encore la liberté de faire des transformations de jauge qui dépendent seulement de $\theta$ et de $\varphi$, liberté dont nous profiterons sous peu.

Nous supposons maintenant que pour $r$ grand, $\boldsymbol{A}$ peut être développé en puissances de $1/r$:

$$\boldsymbol{A} = \frac{\boldsymbol{a}(\theta,\varphi)}{r} + \mathrm{O}\left(\frac{1}{r^2}\right). \tag{3.40}$$

Comme les équations de Yang-Mills sont non linéaires, elles mettront en général en jeu des termes croisés entre les termes dominants de ce développement et les termes d'ordre plus élevé. Cependant, seuls les termes dominants entrent dans la partie des équations proportionnelle à $1/r^3$. Ainsi, si nous nous intéressons seulement aux termes dominants, comme c'est le cas, il est légitime d'ignorer les termes d' ordre plus élevé, ce que nous ferons.

Exactement comme dans la discussion du cas Abélien, il est convenable d'écrire $\boldsymbol{A}$ en fonction de ses composantes covariantes,

$$\boldsymbol{A}\cdot \mathrm{d}\boldsymbol{x} = A_\theta(\theta,\varphi)\mathrm{d}\theta + A_\varphi(\theta,\varphi)\mathrm{d}\varphi. \tag{3.41}$$

Pour que le champ soit non singulier au pôle nord et au pôle sud, $A_\varphi(0, \varphi)$ et $A_\varphi(\pi, \varphi)$ doivent s'annuler tous deux [14]. Je ferai maintenant une transformation de jauge de plus pour assurer

$$A_\theta = 0. \tag{3.42}$$

Cette transformation peut être construite en intégrant le long des lignes à $\varphi$ constant, les méridiens, en commençant par le pôle nord. Bien entendu, ce choix de jauge peut conduire à une singularité artificielle à l'endroit où les méridiens se recoupent, c'est à dire le pôle sud. En particulier il peut conduire à une valeur non nulle (et dépendant de $\varphi$) pour $A_\varphi(\pi, \varphi)$, c'est à dire à une singularité de corde de Dirac. Quand nous aurons fini de résoudre les équations, nous vérifierons que la singularité est bien un artifice de jauge, que la corde est indétectable.

Nous sommes maintenant en bonne position pour (finalement) utiliser les équations de champ. Le tenseur intensité de champ a une seule composante non nulle,

$$F_{\theta\varphi} = \partial_\theta A_\varphi. \tag{3.43}$$

En coordonnées curvilignes, les équations de Yang–Mills sans source prennent la forme

$$\partial_\mu \sqrt{g}\, F^{\mu\nu} + \left[A_\mu, \sqrt{g}\, F^{\mu\nu}\right] = 0. \tag{3.44}$$

Dans notre cas:

$$\sqrt{g}\, F^{\theta\varphi} = \left|\frac{1}{r^2 \sin\theta}\right| \partial_\theta A_\varphi. \tag{3.45}$$

Il y a donc deux équations de champ non triviales. L'une d'elles est

$$\partial_\theta \frac{1}{\sin\theta} \partial_\theta A_\varphi = 0, \tag{3.46}$$

dont la solution générale, sous la condition d'annulation de $A_\varphi(0, \varphi)$, est

$$A_\varphi = Q(1-\cos\theta), \tag{3.47}$$

où $Q$ est une fonction arbitraire (à valeurs matricielles) de $\varphi$. Cependant, l'autre équation de champ,

$$\partial_\varphi \sqrt{g}\, F^{\varphi\theta} + \left[A_\varphi \sqrt{g}\, F^{\varphi\theta}\right] = -\partial_\varphi Q = 0, \tag{3.48}$$

nous dit que $Q$ doit être une constante.

Ainsi, à une transformation de jauge près, le champ monopolaire non Abélien est construit en multipliant le champ monopolaire Abélien par une matrice constante. Il est trivial que ce procédé conduit à des monopôles non Abéliens; ce qui est non trivial est qu'il les donne tous.

Puisque le monopôle non Abélien est si simplement relié au monopôle Abélien, il est facile d'obtenir la condition de quantification. Nous n'avons qu'à suivre les raisonnements de la section 2.2, moyennant des modifications triviales. Le potentiel avec la corde pointant vers le nord, est défini par:

$$A'_\varphi = -Q(1+\cos\theta). \tag{3.49}$$

$A$ et $A'$ sont tranformés l'un de l'autre par la transformation de jauge

$$g = \exp 2Q\varphi, \tag{3.50}$$

qui est univaluée si

$$\exp 4\pi Q = 1. \tag{3.51}$$

Ceci est la condition de quantification. Comme petit test de cohérence, vérifions que l'on retrouve la condition correcte dans le cas particulier de la théorie Abélienne de la section 2. Il y a eu depuis un léger changement de notation; ce que nous appelons maintenant $\boldsymbol{A}$ était alors appelé $-\mathrm{i}e\boldsymbol{A}$ (voir l'expression pour la dérivée covariante, éq. (3.11). Ainsi $Q$ devient $-\mathrm{i}eg$, et l'éq. (3.51) devient la condition de quantification de Dirac, comme cela se doit.

Dans la théorie de haute énergie, il y a une tendance marquée à se concentrer sur l'algèbre de Lie d'un groupe et à ignorer sa structure globale; par exemple, on se refère indifféremment au groupe d'isospin comme SU(2) ou SO(3). C'est une habitude particulièrement néfaste dans la théorie des monopôles parce que la condition de quantification est sensible à la structure globale de $G$. Par exemple, supposons que $Q$ soit proportionnelle à $I_3$, le générateur des rotations autour du troisième axe dans l'espace d'isospin. Si $G$ est SO(3), $Q$ peut être n'importe quel multiple demi entier de $I_3$, puisqu'une rotation de $2\pi$ est l'identité. Si $G$ est SU(2), toutefois, seuls les multiples entiers sont autorisés, puisqu'une rotation de $4\pi$ est nécessaire pour retourner à l'identité.

Il n'y a ici rien de profond; tout cela peut être compris dans les termes élémentaires de la section 2.1. Dire que $G$ est SO(3) revient à dire que toutes les particules ont un isospin entier, et par conséquent une valeur entière de $I_3$. Dire que $G$ est SU(2) revient à autoriser les valeurs demi-entières. Avec un ensemble plus riche de particules d'essai on peut exécuter un ensemble plus riche d'expériences de Bohm–Aharanov, et détecter des solénoïdes auparavant indétectables.

Afin d'avoir des exemples à utiliser ultérieurement, je vais expliciter la forme de $Q$ pour quelques échantillons de groupes.

Si $G$ est SU($n$), $Q$ doit être une matrice $n \times n$ sans trace, anti-hermitienne. On peut toujours diagonaliser une telle matrice par une transformation unitaire, c'est à dire, par une transformation de jauge constante. Ainsi,

$$Q = -\tfrac{1}{2}\mathrm{i}\,\mathrm{diag}(q_1, q_2, \ldots, q_n). \tag{3.52}$$

La condition de trace nulle signifie que la somme des $q$'s est nulle, tandis que la condition de quantification implique que chaque $q$ est un entier. Comme on peut arbitrairement permuter les $q$'s par une transformation de jauge, seul l'ensemble des $q$'s est significatif, pas leur ordre.

La situation est particulièrement simple pour SU(2). Ici,

$$Q = -\tfrac{1}{2}\mathrm{i}\,\mathrm{diag}(q, -q) = -\tfrac{1}{2}\mathrm{i}q\sigma_3. \tag{3.53}$$

On peut utiliser la liberté de permuter les valeurs propres de $Q$ pour imposer à $q$ d'être toujours non négatif, $q = 0, 1, \ldots$.

Certaines théories de jauge construites sur l'algèbre de Lie de SU($n$) admettent comme groupe global, $G$, non SU($n$) mais SU($n$)/$Z_n$. [$Z_n$ est le groupe fini à $n$ éléments consistant en les racines $n$-ièmes de l'unité, c'est à dire, les puissances entières de $\exp(2\pi\mathrm{i}/n)$.] L'un des exemples est la théorie des champs de jauge seuls, sans champs de matière, des gluons sans quarks. Les champs de jauge se transforment suivant la représentation adjointe du groupe; deux matrices de SU($n$) qui diffèrent seulement par un facteur appartenant à $Z_n$ seront représentées par la même matrice dans la représentation adjointe.

Afin de traiter ce cas sans que les équations deviennent trop longues, introduisons quelques notations pour cela. Je dénoterai par $Q_{\mathrm{fund}}$ la matrice qui représente un générateur de groupe abstrait donné dans la représentation fondamentale de dimension $n$ du groupe, et par $Q_{\mathrm{adj}}$ la matrice qui représente le même générateur dans la représentation adjointe de dimension $(n^2-1)$. La condition de quantification dans le cas présent est $\exp(4\pi Q_{\mathrm{adj}}) = 1$. Ceci est vrai si et seulement si

$$\exp(4\pi Q_{\mathrm{fund}}) \in Z_n. \tag{3.54}$$

Ainsi, comme précédemment,

$$Q_{\mathrm{fund}} = -\tfrac{1}{2}\mathrm{i}\,\mathrm{diag}(q_1, \ldots, q_n), \tag{3.55}$$

mais maintenant:

$$q_m = r/n + \text{entier}, \tag{3.56}$$

où $r$ est un entier, indépendant de $m$. Comme auparavant, la somme des $q$'s doit être nulle, et seulement l'ensemble des $q$'s est significatif, pas leur ordre.

Ceci devient particulièrement simple pour $SU(2)/Z_2 = SO(3)$. Exactement comme dans l'éq. (3.53), $Q_{\text{fund}}$ est $-\frac{1}{2}iq\sigma_3$, avec $q \geqslant 0$. Cependant, $q$ peut maintenant être un demi-entier.

### 3.4. *Classification topologique (Lubkin) des monopôles*

La classification topologique des champs monopolaires a été développée il y a presque trente ans par Lubkin dans un article important et injustement ignoré [15]. Nous commençons par analyser une situation dans laquelle une boîte noire de contenu inconnu est entourée par des champs de jauge, situation identique à celle que nous avons étudiée dans la classification dynamique. Toutefois, nous n'allons pas maintenant supposer que les champs de jauge sont des solutions stationnaires des équations de Yang–Mills sans source, ou quelque type de solutions que ce soit d'équations dynamiques.

A la place, nous associerons à la configuration du champ de jauge une charge topologique, quantité qui reste inchangée par une déformation continue arbitraire de la configuration, et ainsi reste en particulier inchangée par l'évolution temporelle, en accord avec n'importe quelle équation de mouvement sensée.

Je vais d'abord décrire la construction de Lubkin et développer quelques unes de ses propriétés importantes. Ceci va se résumer à un (très) bref cours de théorie de l'homotopie (très peu). Si faible que soit votre tolérance aux mathématiques abstraites, je vous prie de faire très attention ici; autrement, vous n'allez rien comprendre à ce qui suit. Par la suite, j'étudierai la relation qui existe entre la classification topologique et la classification dynamique.

La figure 2a montre une famille de courbes fermées tracées sur une sphère qui entoure la boîte noire. Chaque courbe commence et finit au pôle nord; le chemin de retour est sur le côté caché de la sphère et n'apparaît pas sur la figure. Ces courbes sont indexées par un paramètre $\tau$ qui varie de 0 à 1; la première et la dernière courbe de la famille, $\tau = 0$, $\tau = 1$, sont les courbes triviales qui ne quittent pas le pôle. Au fur et à mesure que $\tau$ varie dans son intervalle les courbes balayent le pourtour

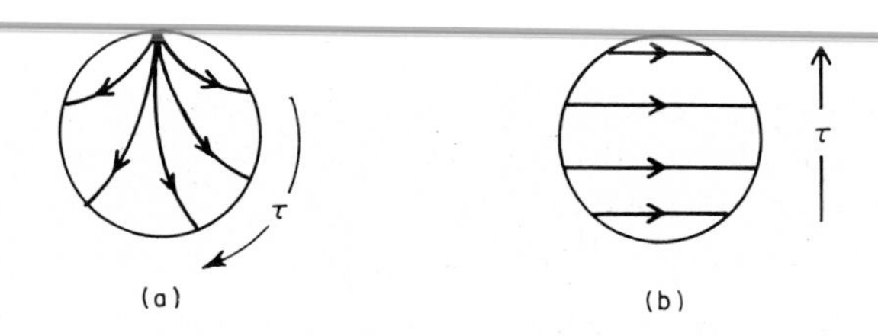


Fig. 2.

de la boîte noire comme le cerceau d'un magicien autour d'une dame en lévitation.

Pour le cas où vous auriez des difficultés à voir ce qui se passe sur la fig. 2a, j'ai dessiné la même structure d'une façon différente sur la fig. 2b. C'est une projection sud-polaire de la sphère; le pôle nord est représenté par la circonférence du disque. Les courbes sont représentées par les lignes horizontales.

Le long de chaque courbe, on peut intégrer le champ de jauge de façon à obtenir l'élément du groupe correspondant. De la sorte, nous obtenons à partir de notre famille de courbes, une courbe dans l'espace du groupe, $g(\tau)$, qui commence et se termine à l'identité. (Nous n'avons pas à nous faire de souci quant à la rencontre de singularités dependant de la jauge, telles que des cordes de Dirac, au cours du balayage de la sphère. On peut toujours faire des transformations de jauge pour se débarasser des singularités rencontrées; grâce à l'éq. (3.33), $g(\tau)$ n'en sera pas affecté, aussi longtemps qu'on prend soin de s'assurer que la transformation de jauge se réduit à l'identité du pôle nord.)

La courbe $g(\tau)$ dépend pratiquement de toutes les données du problème: la sphère choisie, le détail de la construction de la famille de courbes fermées sur la sphère, l'instant auquel nous faisons le calcul (si le champ de jauge dépend du temps), la jauge choisie. Cependant elle dépend continûment de toutes ces données et, en conséquence, l'élément correspondant du premier groupe d'homotopie de $G$, $\pi_1(G)$ reste inchangé.

L'explication de la phrase précédente requiert le bref cours de théorie de l'homotopie que nous avions annoncé [16].

Pour tout espace topologique $X$, une courbe, $X(t)$ est une fonction continue de l'intervalle $[0, 1]$ dans $X$. Soient $X(t)$ et $X'(t)$ deux courbes de mêmes extrémités, à savoir $X(0) = X'(0)$ et $X(1) = X'(1)$. Alors nous disons que les deux courbes sont homotopes, ou appartiennent à la même classe d'homotopie, si on peut les transformer continûment l'une dans l'autre en gardant les extrémités fixes. En termes analytiques, $X(t)$ et $X'(t)$ sont homotopes s'il existe une fonction continue de deux variables $F(s,t)$, $0 \leqslant s, t \leqslant 1$, telle que:

$$F(0,t) = X(t), \qquad F(1,t) = X'(t),$$
$$F(s,0) = X(0) = X'(0), \qquad F(s,1) = X(1) = X'(1). \tag{3.57}$$

Un exemple est illustré par la fig. 3. L'espace topologique $X$ est une portion de plan dont un disque (la région hachurée) à été enlevé. Quatre courbes sont tracées; elles ont toutes le même point de départ et d'arrivée $X_0$. La courbe $C$ est homotope à la courbe $D$ et aussi à la courbe triviale, $X(t) = X_0$ pour tout $t$. Autrement, il n'y a pas deux courbes tracées qui soient homotopes. (J'espère que ces affirmations vous paraissent évidentes, car elles sont plutôt difficiles à montrer analytiquement.)

Etant donné deux courbes dont l'une se termine là ou l'autre commence, le produit de ces courbes est défini en parcourant d'abord la première courbe, puis la seconde. Analytiquement:

$$X \cdot X'(t) = X(2t), \quad 0 \leqslant t \leqslant \tfrac{1}{2},$$
$$= X'(2t-1), \quad \tfrac{1}{2} \leqslant t \leqslant 1. \tag{3.58}$$

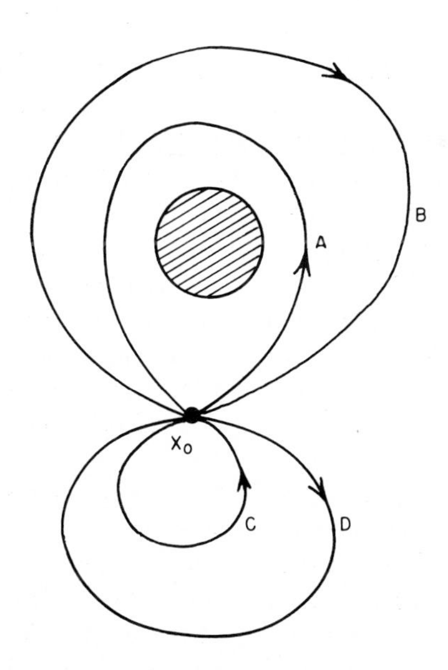


Fig. 3.

Les produits de courbes homotopes sont évidemment homotopes. L'inverse d'une courbe est définie en parcourant la courbe dans le sens opposé. Analytiquement,

$$X(t)^{-1} = X(1-t). \tag{3.59}$$

Le produit d'une courbe et de son inverse (dans n'importe quel ordre) est clairement homotope à une courbe triviale (i.e. constante). Ces concepts sont illustrés sur la fig. 3; A est homotope à l'inverse de B.

Si on considère toutes les classes d'homotopie de courbes fermées, de point initial–final $X_0$, ces opérations de multiplication et d'inversion définissent une structure de groupe. Le groupe ainsi obtenu s'appelle le premier groupe d'homotopie de $X$ et est noté $\pi_1(X)$. Nous n'avons pas besoin de spécifier $X_0$ parce que $\pi_1(X)$ est indépendant de $X_0$ si $X$ est connexe. Preuve: soit $y$ une courbe allant de $X_0$ à un autre point $X_1$. L'application $x \to y \cdot x \cdot y^{-1}$ transforme toute courbe fermée qui commence et se termine en $X_0$, de telle sorte que les opérations du groupe soient respectées.

Pour l'exemple de la fig. 3, les classes d'homotopie sont indexées par un entier, le nombre d'enlacement, à savoir le nombre entier de fois que la courbe s'enroule autour du disque hachuré. Par exemple, la courbe $A$ admet le nombre d'enlacement 1, la courbe $B$, $-1$, les courbes $C$ et $D$, 0. Le nombre d'enlacement d'un produit de courbes est la somme des nombres d'enlacement des facteurs. Ainsi $\pi_1$ est le groupe additif des entiers quelquefois dénoté $Z$.

Si $X$ est connexe et $\pi_1(X)$ trivial, on dira que $X$ est simplement connexe. Il est facile de voir que, dans ce cas deux courbes arbitraires qui relient les mêmes deux points sont homotopes.

Ainsi se termine la première partie du bref cours de théorie de l'homotopie. Il y aura une seconde partie à ce cours sur le calcul de $\pi_1$ pour les groupes de Lie compacts. Cependant, même à ce stade on peut déjà comprendre certaines choses sur les monopôles:

1. Vous devriez comprendre maintenant mon affirmation cryptique, quelques paragraphes plus haut, que la construction de la boucle de Lubkin associe aux champs de jauge en dehors de la boîte noire un élément de $\pi_1(G)$. Cet élément est la charge topologique à laquelle je me suis reféré au début de cette section.

2. On peut voir que la charge topologique est invariante de jauge. Si on fait une transformation de jauge égale à $h$ au pôle nord, alors $g(\tau)$ se transforme en $hg(\tau)h^{-1}$. Comme $G$ est connexe, $h$ peut être continûment déformé en 1, et la courbe transformée est homotope à la courbe initiale.

3. Considérons un monde qui contient deux boîtes noires. On peut calculer la charge topologique de chacune des boîtes en l'entourant d'une sphère qui ne contient pas l'autre, ou bien, on peut calculer la charge topologique de l'ensemble en entourant les deux boîtes par une grande sphère. Dans le dernier cas on peut continûment déformer la sphère en quelque chose comme un sablier, deux sphères, chacune entourant une boîte, reliées par un tube pincé en un point dans son milieu. Choisissons le "pôle nord" de la sphère déformée au point de pincement.

Maintenant, quand on balaye la sphère déformée par des boucles, elles passent d'abord autour d'une sphère, puis autour de l'autre; le chemin dans l'espace du groupe est le produit des chemins pour chacune des sphères prise individuellement. Ainsi, la charge topologique du système composé est le produit (au sens groupal) des charges topologiques des composantes. Comme sous produit, nous avons obtenu un argument très indirect indiquant que $\pi_1(G)$ est Abélien puisque l'ordre dans lequel on fait les choses est évidemment indifférent. Nous aurons bientôt une démonstration plus directe du même résultat.

4. On voit que la région interdite, la boîte noire est essentielle pour qu'on puisse obtenir une structure non triviale. En effet, si la boîte n'était pas là, on pourrait réduire la sphère à un point. Dans cette limite, l'élément correspondant de $\pi_1$, serait l'identité puisque tous les chemins sur la sphère seraient triviaux. Mais comme c'est un invariant topologique, s'il est l'identité à la limite, c'était déjà l'identité au début.

Pour qu'on obtienne une charge topologique non triviale, il faut qu'il y ait quelque chose d'autre dans la boîte que seulement des champs de jauge non singuliers. Nous verrons dans la section 4 de quoi il s'agit.

J'en reviens à ce cours. Comme promis je vais vous dire comment calculer $\pi_1$ pour n'importe quel groupe de Lie compact connexe. La meilleure attaque de ce problème est indirecte; je vais commencer par classer tous les groupes de Lie d'algèbre de Lie donnée. Ceci étant fait, le calcul de $\pi_1$ se trouvera être trivial.

Je suppose que vous savez que l'algèbre de Lie de tout groupe de Lie compact est la somme directe de copies de certaines algèbres de Lie fondamentales. Il s'agit des algèbres de Lie de U(1), des trois familles infinies des groupes classiques SO($n$), SU($n$), et Sp($n$) et des cinq groupes de Lie exceptionnels.

Pour chacune d'elle, il existe un groupe simplement connexe qui l'admet comme algèbre de Lie. Pour l'algèbre de U(1), c'est $R$, le groupe additif des réels. Pour les algèbres de SU($n$) et Sp($2n$) ce sont ces groupes eux-mêmes. Pour l'algèbre de SO($n$) c'est le recouvrement double de SO($n$) quelquefois appelé Spin($n$). Je ne me préoccuperai pas ici des cinq groupes exceptionnels; nous ne les utiliserons jamais dans ces conférences.

Ainsi en prenant des produits directs de ces groupes, on peut construire un groupe simplement connexe dont l'algèbre de Lie est isomorphe à celui d'un groupe de Lie compact connexe $G$. Je noterai ce groupe $\bar{G}$. $\bar{G}$ est appelé groupe de recouvrement de $G$; la raison en est qu'il recouvre $G$ de la même façon que SU(2) recouvre SO(3) = SU(2)/$Z_2$. Pour être précis on peut montrer que $G$ est isomorphe au groupe quotient $\bar{G}/K$ où $K$ est un sous groupe discret du centre de $\bar{G}$. (Je vous rappelle que le centre du groupe est le sous groupe formé de tous les éléments qui commutent avec tous les éléments du groupe.) C'est un théorème standard de la théorie des groupes de Lie, et je vous demande de me croire sur parole.

Ainsi, pour classer tous les groupes d'algèbre donnée, il faut trouver tous les sous-groupes discrets du centre de $\bar{G}$. Dans tous les cas que nous rencontrerons il sera trivial de trouver les sous groupes une fois connu le centre.

Le centre de $\bar{G}$ est le produit des centres de ses facteurs. Ils sont tous faciles à calculer. $R$ est Abélien, donc son propre centre. Le centre de SU($n$) est $Z_n$. Le centre de Sp($n$) consiste en 1 et $-1$. Le centre de Spin($n$) pour $n$ pair est formé des deux éléments qui se projettent sur 1 de SO($n$) et des deux éléments qui se projettent sur $-1$. Pour $n$ impair, $-1$ n'appartient pas à SO($n$), de sorte que le centre de Spin($n$) est formé seulement des deux éléments qui se projettent sur 1. A titre d'exemple, utilisons cet équipement pour expliciter tous les groupes dont l'algèbre de Lie est isomorphe à celui de $\bar{G}$ = SU(2)×SU(2). Pour mettre en relief les différences entre ces groupes, en même temps que je construirai les groupes je décrirai aussi leurs représentations irréductibles. Si je décris les éléments de $\bar{G}$ de la façon usuelle par $(g_1, g_2)$, le centre comprend $(1,1), (1,-1), (-1,1), (-1,-1)$. Il y a cinq sous groupes, de sorte que nous avons cinq groupes avec cette algèbre.

a) $K$ est $(1,1)$, $G$ est SU(2)×SU(2). Les représentations de $G$ sont de la forme $D^{(S_1)}(g_1) \otimes D^{(S_2)}(g_2)$, où les $D$'s sont les représentations de SU(2) et $S_1$ et $S_2$ sont entiers ou demi-entiers.

b) $K$ consiste en $(1,1)$ et $(1,-1)$. $G$ est SU(2)×SO(3). Les représentations sont de la même forme que précédemment; $S_1$ est entier ou demi-entier, $S_2$ est entier.

c) $K$ consiste en $(1,1)$ et $(-1,1)$. La situation est la même que pour b), avec les deux facteurs interchangés,

d) $K$ est formé des quatre éléments du centre. $G$ est SO(3)×SO(3). $S_1$ et $S_2$ sont tous deux entiers.

e) $K$ consiste en $(1,1)$ et $(-1,-1)$. Ce cas est de loin le plus intéressant. $G$ n'est pas un produit direct. $S_1$ et $S_2$ sont soit tous deux entiers, soit tous deux demi-entiers. Ceci rappelle le genre de structures que nous trouvons dans les théories réalistes où le groupe de jauge non brisé a l'algèbre de SU(3) (couleur)×U(1) (électromagnétique) mais n'est pas un produit direct, parce que seules les particules de trialité non nulle ont une charge fractionnaire.

Maintenant que nous avons classé les groupes de Lie, il est facile de calculer leur premier groupe d'homotopie. Nous raisonnons comme suit. En général l'application de $\bar{G}$ sur $G$ est multivaluée. Il n'y a pas un élément unique de $\bar{G}$ qui corresponde à un élément donné de $G$. Cependant, comme $K$ est discret, il y a pour chaque chemin continu dans $G$ commençant à l'identité un chemin *continu* unique dans $\bar{G}$ commençant à l'identité. Si le chemin dans $G$ est fermé, c'est-à-dire se termine à l'identité, le chemin dans $\bar{G}$ doit se terminer à un élément qui se projette sur l'identité, c'est-à-dire, un élément de $K$. Considérons deux chemins fermés dans $G$ tels que les chemins correspondants dans $\bar{G}$ se terminent au même élément de $K$. Comme $\bar{G}$ est simplement connexe les chemins de $\bar{G}$ peuvent être continûment déformés l'un dans l'autre; ainsi, il en est de même pour les chemins correspondants dans $G$. Par suite, les classes d'homotopie de chemins fermés dans $G$ sont en correspondance biunivoque avec les éléments de $K$. Il est facile de voir que le produit de deux chemins fermés correspond au produit des deux éléments du groupe; autrement dit $\pi_1(G)$ est isomorphe à $K$.

Ainsi se conclut le bref cours de théorie de l'homotopie. Retournons maintenant aux monopôles magnétiques.

La construction de Lubkin est complètement invariante de jauge, mais sa réalisation peut être ennuyeuse; nous avons à résoudre une équation différentielle (ou, de façon équivalente calculer une intégrale ordonnée le long d'une courbe), pour chaque valeur de $\tau$. Les choses se simplifient considérablement dans une jauge spéciale, la "jauge de corde." C'est tout à fait comparable à la jauge que nous avons utilisée dans la classification dynamique des monopôles. Sur chaque sphère à $r$ fixé, nous commençons au pôle nord et transformons de jauge le long des méridiens pour annuler $A_\theta(\theta,\varphi)$. Ceci peut induire un $A_\varphi(\pi,\varphi)$ non nul, une corde de Dirac le long de l'axe du pôle sud. Quand nous avons fait cette opération dans la section 2.3, nous avons supposé que le potentiel vecteur était proportionnel à $1/r$; il suffisait donc de faire une transformation de jauge sur une

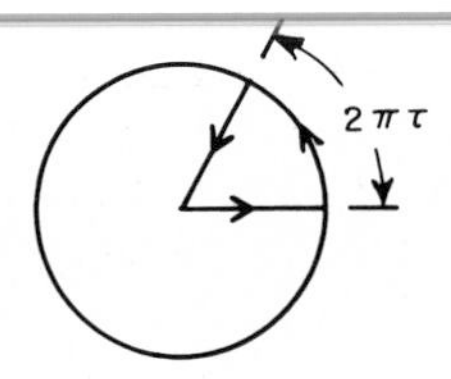


Fig. 4.

seule sphère. Ici nous ne faisons pas de telle hypothèse de sorte que nous devons faire une transformation de jauge indépendante sur chaque sphère. Ceci peut induire un $A_r$ non nul, mais ce n'est pas un problème; $A_r$ n'entre nulle part dans la construction de Lubkin.

Maintenant, nous choisissons notre famille de courbes fermées comme il est indiqué sur la fig. 4. C'est une projection polaire nord de la sphère; la circonférence du disque est le pôle sud $\theta = \pi$. Le chemin indexé par $\tau$ va du pôle nord au pôle sud le long du méridien $\varphi = 0$ et retourne le long de $\varphi = 2\pi\tau$.

Comme $A_\theta$ est nul,

$$g(\tau) = P\exp - \int_0^{2\pi\tau} A_\varphi(\pi, \varphi)\,\mathrm{d}\varphi. \tag{3.60}$$

Il est ainsi suffisant d'évaluer l'intégrale pour un seul chemin, un circuit infinitésimal entourant la corde de Dirac. Cette expression devient spécialement simple pour une des solutions de Goddard–Nuyts–Olive, éq. (3.47)

$$g(\tau) = \exp(-4\pi\tau Q). \tag{3.61}$$

Cette équation nous permet de calculer immédiatement la classe topologique à laquelle appartient un champ monopolaire de GNO. Commençons par l'électromagnétisme ordinaire, pour lequel $G$ est U(1) et $Q = -\mathrm{i}eg$. Pour $eg = n/2$:

$$g(\tau) = \exp(\mathrm{i}2\pi n\tau). \tag{3.62}$$

U(1) est recouvert $n$ fois quand $\tau$ varie de 0 à 1. Ainsi, pour chaque élément de $\pi_1$ il y a un et un seul champ monopolaire. La charge topologique est la charge magnétique. La situation est par contre tout à fait différente pour d'autres groupes. Par exemple, pour SU($n$) nous avons trouvé un nombre infini de champs monopolaires GNO, mais SU($n$) est simplement connexe: $\pi_1$ a un seul élément. De même pour

$SU(n)/Z_n$ il y a un nombre infini de champs monopolaires GNO mais seulement $n$ éléments dans $\pi_1 = Z_n$. (A titre d'exercice vous pouvez calculer l'élément de $Z_n$ associé à chacun de ces champs.)

Ainsi, en général, la classification topologique est plus grossière que la classification dynamique. C'est ce à quoi on pourrait s'attendre; la classification topologique est basée sur moins d'hypothèses que la classification dynamique et doit par conséquent contenir moins d'information.

Je vous ai induit en erreur. Non parce que j'ai dit un quelconque mensonge dans le paragraphe précédent, mais parce que j'ai gardé pour moi une information importante: la plupart des champs monopolaires GNO sont instables.

### 3.5. *Réduction de la classification dynamique*

L'analyse de la stabilité est triviale pour le monopôle Abélien parce que les équations de champ sont linéaires. Il n'en va pas de même pour un monopôle non Abélien. Même à grande distance les équations de champs ne se linéarisent pas; si le champ de jauge décroit comme $1/r$, les dérivés et commutateurs sont de grandeur comparable, tous deux $O(1/r^2)$.

Je vais maintenant étudier les petites vibrations autour d'un champ monopolaire de SO(3) [17]. (Bien entendu ceci résoudra aussi le problème pour les monopôles de SU(2) qui forment un sous ensemble de ceux de SO(3).) Une fois le calcul fait, je discuterai l'extension à un champ monopolaire GNO quelconque pour un groupe de jauge général. Le calcul est long et j'en présenterai seulement l'esquisse; vous ne devriez pas avoir de mal à compléter les détails si vous le désirez.

Nous travaillons dans la jauge temporelle, $A_0 = 0$ et écrivons le champ de jauge comme un monopôle SO(3) plus une petite perturbation,

$$\boldsymbol{A} = -\tfrac{1}{2}\mathrm{i}q\sigma_3\boldsymbol{A}_{\mathrm{D}} + \delta\boldsymbol{A}, \qquad (3.63)$$

où, comme on l'a expliqué dans la section 3.3, $q = 0, 1/2, \ldots$ et $\boldsymbol{A}_{\mathrm{D}}$ est le potentiel de corde de Dirac, éq. (2.64). $\pi_1(SO(3))$ est $Z_2$, de sorte qu'il n'y a que deux classes topologiques. Il est facile de vérifier que les monopôles avec $q$ entier sont tous dans une même classe, et ceux avec $q$ demi-entier tous dans l'autre.

Nous écrivons la perturbation comme une matrice $2\times 2$ explicite:

$$\delta\boldsymbol{A} = -\tfrac{1}{2}\mathrm{i}\begin{pmatrix} \boldsymbol{\phi} & \boldsymbol{\psi} \\ \boldsymbol{\psi}^* & -\boldsymbol{\phi} \end{pmatrix} \qquad (3.64)$$

où $\boldsymbol{\phi}$ est un vecteur réel et $\boldsymbol{\psi}$ un vecteur complexe. Si on linéarise les équations de champ par rapport à la perturbation un calcul facile montre que $\phi$ satisfait à une équation d'onde libre

$$-\partial_0^2\phi = \nabla \times (\nabla \times \phi), \tag{3.65}$$

tandis que $\psi$ satisfait à une équation plus compliquée:

$$-\partial_0^2\psi = \boldsymbol{D} \times (\boldsymbol{D} \times \psi) + \frac{\mathrm{i}q\boldsymbol{r}}{r^3} \times \psi \equiv H\psi, \tag{3.66}$$

où

$$\boldsymbol{D} = \nabla - \mathrm{i}q\boldsymbol{A}_{\mathrm{D}}. \tag{3.67}$$

(Notez que $q$ joue ici le rôle de $eg$ de la section 2.)

La stabilité du système par rapport aux petites perturbations est déterminée par le spectre des valeurs propres de l'opérateur différentiel $H$. Si $H$ a une valeur propre négative, le mode propre associé a une évolution temporelle exponentielle et le champ monopolaire est instable.

$H$ admet un grand ensemble de modes propres avec valeur propre nulle. C'est une conséquence de l'invariance des équations du mouvement dans la jauge temporelle par rapport aux transformations de jauge indépendantes du temps. Ecrivons une telle transformation comme:

$$g(\boldsymbol{x}) = 1 - \tfrac{1}{2}\mathrm{i}\begin{pmatrix} \lambda(\boldsymbol{x}) & \chi(\boldsymbol{x}) \\ \chi^*(\boldsymbol{x}) & -\lambda(\boldsymbol{x}) \end{pmatrix} + \cdots, \tag{3.68}$$

où les points de suspension indiquent les termes quadratiques et d'ordre plus élevé en $\lambda$ et $\chi$. Un calcul facile montre que, par cette transformation,

$$\psi \to \psi + \boldsymbol{D}\chi + \cdots. \tag{3.69}$$

Comme $\psi = 0$ est certainement une solution de l'éq. (3.66) il doit en être de même pour sa transformée de jauge. Ainsi

$$H\boldsymbol{D}\chi = 0, \tag{3.70}$$

quelle que soit $\chi$.

Nous appellerons modes de jauge ces modes propres triviaux. Toute la physique intéressante se trouve dans les modes physiques, c'est à dire les modes orthogonaux aux modes de jauge. Pour un mode physique,

$$\int \mathrm{d}^3\boldsymbol{x}\, \psi^* \cdot \boldsymbol{D}\chi = 0, \tag{3.71}$$

quelle que soit $\chi$. De façon équivalente

$$\boldsymbol{D} \cdot \psi = 0. \tag{3.72}$$

Comme toujours l'invariance par rotation est un élément de simplification important. $H$ commute avec

$$\boldsymbol{J} = \boldsymbol{L} + \boldsymbol{S}, \tag{3.73}$$

où, exactement comme dans la section 3.4,

$$\boldsymbol{L} = -\mathrm{i}\boldsymbol{r} \times \boldsymbol{D} - q(\boldsymbol{r}/r), \tag{3.74}$$

et $\boldsymbol{S}$ est l'opérateur de spin usuel pour les champs vectoriels, defini par

$$(\boldsymbol{a} \cdot \boldsymbol{S})\boldsymbol{b} = \mathrm{i}\boldsymbol{a} \times \boldsymbol{b}, \tag{3.75}$$

pour toute paire de vecteurs $\boldsymbol{a}$, $\boldsymbol{b}$. Comme on l'a expliqué dans la section 2.4, le moment angulaire orbital prend les valeurs

$$l = q, q+1, q+2 \ldots . \tag{3.76}$$

Ainsi, d'après les règles usuelles d'addition des moments angulaires, le moment angulaire total prend les valeurs

$$\begin{aligned} j &= q-1, q, q+1 \ldots, \quad q \geqslant 1, \\ &= q, q+1 \ldots, \quad q = 0 \text{ ou } 1/2. \end{aligned} \tag{3.77}$$

(Je ne me préoccuperai pas de la multiplicité de chaque valeur de $j$.)

On peut utiliser l'éq. (3.75) pour débarasser $H$ des termes croisés:

$$H = -(\boldsymbol{S} \cdot \boldsymbol{D})^2 + q(\boldsymbol{S} \cdot \boldsymbol{r}/r^2). \tag{3.78}$$

Il est alors élémentaire d'utiliser les méthodes de la section (2.4) pour montrer que, sur les fonctions de $j$ défini,

$$H\psi = \left( -\frac{\partial^2}{\partial r^2} - \frac{2}{r}\frac{\partial}{\partial r} + \frac{j(j+1) - q^2}{r^2} \right)\psi + \boldsymbol{X}(\boldsymbol{D} \cdot \psi) + \boldsymbol{D}(\boldsymbol{Y} \cdot \psi). \tag{3.79}$$

Ici $\boldsymbol{X}$ et $\boldsymbol{Y}$ sont des objets affreux [24] dont la forme explicite n'a aucun intérêt car les termes dans lesquels ils figurent ne contribuent pas aux éléments de matrice de $H$ entre des modes physiques. (Le terme en $\boldsymbol{X}$ annihile le mode à droite et $\boldsymbol{Y}$ annihile le mode à gauche.)

Nous avons maintenant terminé. Pour $q \geqslant 1$, $j$ peut prendre la valeur $q-1$ et

$$j(j+1) - q^2 = -q. \tag{3.80}$$

Autrement dit le potentiel centrifuge est attractif. C'est une mauvaise nouvelle. Le degré de la catastrophe peut se mesurer en calculant la valeur moyenne de $H$ pour la fonction radiale

$$\begin{aligned}\psi &= 0, \quad r < R,\\ &= \frac{1}{r}(\sqrt{r}-\sqrt{R})\mathrm{e}^{-r/a}, \quad r \geqslant R,\end{aligned} \tag{3.81}$$

où $R$ et $a$ sont des nombres positifs. La valeur moyenne est donnée par:

$$\begin{aligned}\langle H\rangle &= \int_0^\infty r^2 \mathrm{d}r\,\psi^*\left(-\frac{\mathrm{d}^2}{\mathrm{d}r^2}-\frac{2}{r}\frac{\mathrm{d}}{\mathrm{d}r}-\frac{q}{r^2}\right)\psi\\ &= \int_0^\infty \mathrm{d}r\left[r^2(\mathrm{d}\psi/\mathrm{d}r)^2-q\psi^2\right]\\ &= \left(\tfrac{1}{4}-q\right)\ln a+\cdots,\end{aligned} \tag{3.82}$$

où les points de suspension dénotent des termes qui ont une limite finie quand $a$ tend vers l'infini. Quel que soit $R$ fixé, cette expression devient négative pour $a$ suffisamment grand.

Ainsi, non seulement $H$ a des valeurs propres négatives, mais l'existence de ces valeurs propres est totalement insensible à la forme du champ de jauge à courte distance (et, en fait, à n'importe quelle distance finie). C'est bon à savoir parce que nous n'avons pas confiance dans nos expressions à courte distance, dans la boite noire. Tous les champs monopolaires GNO avec $q \geqslant 1$ sont instables par rapport à des perturbations arbitrairement petites à des distances arbitrairement grandes; autrement dit, ils se désintègrent par émission de rayonnement non Abélien. (Pour le cas où vous vous seriez fait du souci à ce sujet, tous les modes que nous avons étudiés sont bien des modes physiques. Le calcul de la divergence covariante est invariant par rotation. Ainsi si $\psi$ appartient à $j = q-1$ il en est de même pour $\boldsymbol{D}\cdot\psi$. Mais nous avons montré dans la section 2.4. que chaque fonction scalaire non nulle appartient à $j \geqslant q$. Donc $\boldsymbol{D}\cdot\psi = 0$.)

Pour SO(3) il n'y a que deux champs monopolaires GNO stables, $q = 0$ (pas de monopôle), et $q = 1/2$, un pour chacune des deux classes topologiques. Il est raisonnable qu'il y ait au moins un monopôle stable dans chaque classe topologique: on ne peut éliminer la charge topologique par radiation; le passage de la radiation à travers une sphère de Lubkin n'est rien d'autre qu'encore une déformation topologique. Ce qui est surprenant c'est qu'il n'y a qu'un monopôle stable pour chaque classe topologique. Pour SU(2), il est aussi vrai qu'il n'y a qu'un monopôle stable par classe topologique. SU(2) est simplement connexe, de sorte qu'il n'y a qu'une classe topologique, et seul $q=0$ est une solution GNO de SU(2).

Une fois l'analyse de la stabilité faite pour SO(3), il n'est pas besoin de recourir à nouveau à des équations différentielles pour des groupes plus généraux. Faisons par exemple l'analyse pour $\mathrm{SU}(n)/Z_n$.

Comme dans l'éq. (3.55), nous écrivons le champ de jauge en fonction des matrices $n \times n$ de la représentation fondamentale. En composantes:

$$\boldsymbol{A}_{ij} = -\tfrac{1}{2}\mathrm{i}\left[q_i \delta_{ij} \boldsymbol{A}_{\mathrm{D}}(\boldsymbol{x}) + \delta A_{ij}\right], \tag{3.83}$$

où il n'y a pas de sommation sur $i$. Dans cette expression

$$q_i = r/n + \text{entier}, \tag{3.56}$$

et la somme des $q$'s est nulle. Il est facile de vérifier que $r = 0, 1, \ldots, n-1$ indexe la charge topologique du monopôle. Il est aussi facile de vérifier que $\delta A_{ij}$ satisfait à une équation différentielle identique en forme à celle satisfaite par $\psi$, éq. (3.66), moyennant la substitution

$$q \to q_i - q_j. \tag{3.84}$$

Ainsi, la condition de stabilité infinitésimale est

$$q_i - q_j = 0, \pm 1, \quad \forall i, j. \tag{3.85}$$

La seule façon pour que cette condition soit vérifiée est que les $q$'s n'aient que deux valeurs. Comme la somme des $q$'s est nulle, une de ces deux valeurs doit être non négative. J'ordonnerai les $q$'s de telle sorte que les $s$ premiers aient cette valeur

$$q_i = q_1 \geqslant 0, \quad 1 \leqslant i \leqslant s. \tag{3.86a}$$

Par suite, par l'éq. (3.85):

$$q_i = q_1 - 1, \quad s+1 \leqslant i \leqslant n. \tag{3.86b}$$

La nullité de la somme implique

$$q_1 = (n-s)/n. \tag{3.87}$$

Ainsi $r$ est égal à $n-s$; encore une fois, il y a un et un seul monopôle stable dans chaque classe topologique.

Il est élémentaire d'étendre cette analyse à tous les groupes classiques, et, si vous vous souvenez de leur définition, aux cinq groupes de Lie exceptionnels. Comme alternative, si vous êtes entraînés à lever des poids et à extraire des racines, vous pouvez avoir raison du problème d'un coup en utilisant la théorie de structure des groupes de Lie [18]. Je n'utiliserai ici aucune des deux méthodes, mais vous dirai seulement que, quelle que soit la méthode utilisée, le résultat est le même: il y a un seul champ monopolaire GNO stable par classe topologique; la seule stabilité est la stabilité topologique.

Vous pouvez trouver utile, pour réfléchir à ce problème, de considérer un problème mécanique simple qui a la même propriété, une boucle élastique contrainte à rester sur la surface d'une sphère. Pour les besoins du problème, une boucle élastique est un système dont l'énergie potentielle est proportionnelle à sa longueur. Les solutions stationnaires du problème sont donc les géodésiques fermées sur la sphère. Il y a la géodésique triviale (la boucle ramassée en un point), une fois le parcours d'un grand cercle, deux fois le parcours d'un grand cercle, etc. Comme une sphère est simplement connexe, il y a une seule classe topologique. Il est facile de voir ici que la seule stabilité est la stabilité topologique. Si la boucle est enroulée autour de la sphère, il suffit de la bouger un tant soit peu et elle se détendra, en se recroquevillant sur un seul point.

### 3.6. *Une application*

La topologie est une puissance. Si on comprend un système dynamique en termes topologiques, on peut souvent déduire ses aspects qualitatifs sans s'embrouiller avec des calculs quantitatifs détaillés. Comme exemple, je discuterai ici la force entre des monopôles non Abéliens à grande distance l'un de l'autre. ("A grande distance", seulement parce que nous ne savons pas encore à quoi ressemblent les monopôles à petites distances.)

Pour des monopôles Abéliens séparés par une grande distance, la force peut être attractive ou répulsive, selon que les charges magnétiques sont de même signe ou de signes opposés. Comme nous le verrons, la situation est différente pour des monopôles non Abéliens; si $G$ est semi-simple, la force est toujours attractive. ("Semi-simple" veut dire dépourvu de facteur U(1); si $G$ contient des facteurs U(1), il peut y avoir une répulsion Abélienne qui surpasse l'attraction dûe aux autres facteurs.)

Je vais commencer par SO(3) et par la suite je généraliserai le résultat. Comme nous venons de le voir, il y a plusieurs façons équivalentes de jauge d'écrire le résultat; en particulier l'opposé de cette expression décrit exactement le même monopôle dans une autre jauge. Il s'ensuit qu'il y a deux façons d'écrire le champ de deux monopôles

$$\boldsymbol{A} = -\tfrac{1}{4}\mathrm{i}\sigma_3\left[\boldsymbol{A}_{\mathrm{D}}(\boldsymbol{x}-\boldsymbol{r}_1) \pm \boldsymbol{A}_{\mathrm{D}}(\boldsymbol{x}-\boldsymbol{r}_2)\right]. \tag{3.88}$$

Cette superposition de deux solutions est une solution puisque tout appartient à un même sous groupe Abélien; ainsi, le terme de commutateur non linéaire des équations de Yang–Mills n'apparaît jamais. Si, par exemple, on avait essayé d'ajouter un monopôle pointant dans la direction 3 et son transformé de jauge pointant dans la direction 2, les termes non linéaires auraient créé des ennuis. Il peut y avoir d'autres façons, moins triviales que celles-là, de mettre ensemble deux monopôles, mais je n'ai pas réussi à en trouver. Pour orienter la discussion, je supposerai que ce sont les deux seules.

Dans le cas Abélien, une expression telle que l'éq. (3.88) pourrait s'interpréter soit comme la superposition de deux monopôles (signe plus), soit comme la superposition d'un monopôle et d'un anti-monopôle (signe moins). Au risque d'être répétitif j'insiste sur le fait que ce n'est pas le cas ici. Le signe moins est simplement un transformé de jauge du signe plus; les deux signes correspondent à deux façons différentes de mettre ensemble les mêmes deux monopôles, exactement comme un spin un et un spin zéro sont deux façons différentes de mettre ensemble la même paire d'objets de spin 1/2.

En dépit de la différence d'interprétation, le calcul de l'énergie d'interaction emmagasinée dans le champ magnétique est le même que dans la théorie de jauge Abélienne:

$$E_{\text{int}} \propto \pm 1/|\boldsymbol{r}_1 - \boldsymbol{r}_2|, \tag{3.89}$$

répulsive dans le cas plus, attractive dans le cas moins. La différence avec le cas Abélien apparaît quand on étudie le champ à grande distance,

$$\boldsymbol{A} = -\tfrac{1}{4}\mathrm{i}\sigma_3 \boldsymbol{A}_{\mathrm{D}}(\boldsymbol{x})[1 \pm 1] + \mathrm{O}(1/r^2). \tag{3.90}$$

Tous deux sont des champs monopolaires GNO; il doit en être ainsi, car ce sont des solutions indépendantes des équations de Yang–Mills. Elles sont toutes deux dans la même classe topologique; il doit en être ainsi, parce que la charge topologique du système des deux monopôles est toujours le produit des charges topologiques des deux monopôles, quelle que soit la façon dont nous raccordons les champs. Mais, un seul des deux est stable puisqu'il n'y a qu'un champ GNO stable par classe topologique, et c'est le champ "moins."

Ainsi, même si nous étions capables de mettre ensemble les deux monopôles dans la configuration répulsive "plus," ils n'y resteraient pas un instant; ils émettraient du rayonnement non Abélien et se stabiliseraient dans la configuration attractive "moins." Cette situation ressemble beaucoup à celle de deux barreaux magnétiques, libres chacun de pivoter autour de leur centre d'inertie. L'orientation initiale des aimants est sans importance; ils se réaligneront jusqu'à ce qu'ils se trouvent dans la configuration la plus attractive (anti-alignés). Ici, le réalignement a lieu dans un espace interne plutôt que dans l'espace géométrique, mais la physique est essentiellement la même.

Passons maintenant à $SU(n)/Z_n$. Ici nous avons $n$ champs monopolaires stables, de sorte que les deux monopôles peuvent être inéquivalents de jauge. Néanmoins, comme nous le verrons, la force est encore attractive. On écrira le champ des deux monopôles sous la forme

$$\boldsymbol{A} = Q_1 \boldsymbol{A}_{\mathrm{D}}(\boldsymbol{x}-\boldsymbol{r}_1) + Q_2^P \boldsymbol{A}_{\mathrm{D}}(\boldsymbol{x}-\boldsymbol{r}_2). \tag{3.91}$$

Ma notation ici appelle quelque explication. $Q_1$ et $Q_2$ sont les deux matrices $n \times n$ qui apparaissent dans les champs monopolaires individuels. Comme avant, nous souhaitons éviter les termes non linéaires des équations de Yang–Mills, de sorte que nous choisissons $Q_1$ et $Q_2$ commutant; ceci implique qu'elles sont simultanément diagonalisées. Il nous reste la liberté de permuter les valeurs propres de $Q_2$, laissant celles de $Q_1$ fixes. Choisissons un ordre de valeurs propres comme ordre de référence et notons les autres $Q_2^P$, où $P$ est l'une des $n!$ permutations de $n$ objets. Les différentes façons de mettre ensemble les deux monopôles correspondent à des choix différents de la permutation $P$, mais, exactement comme auparavant, la charge topologique est indépendante de $P$. Comme il n'y a qu'un champ GNO stable par charge topologique, la plupart des choix de $P$ conduit à des champs instables à grande distance. Nous aimerions trouver la configuration stable unique.

Pour tout $P$, comme auparavant,

$$E_{\mathrm{int}} \propto -\mathrm{Tr}\, Q_1 Q_2^P / |\boldsymbol{r}_1 - \boldsymbol{r}_2|. \tag{3.92}$$

(Le signe moins vient de ce que les $Q$'s sont antihermitiens.) Si on somme sur les permutations

$$\sum_P Q_2^P = (n-1)!\, \mathrm{Tr}\, Q_2 = 0. \tag{3.93}$$

Ainsi la moyenne de l'énergie sur toutes les configurations est nulle, et, par conséquent, l'énergie de la configuration stable, unique, la configuration d'énergie minimale doit être négative. Une fois de plus la force est attractive. Si vous connaissez quelque chose à la théorie générale des groupes de Lie vous pouvez voir que ce raisonnement se généralise de façon évidente. La seule propriété des $Q$'s dont nous avons eu besoin a été la propriété de trace, propriété qui se généralise à un groupe semi-simple arbitraire. (Pour les experts de théorie des groupes, la propriété précise dont on a besoin est que seul l'élément nul d'une sous algèbre de Cartan est invariant par le groupe de Weyl.) Ainsi, même dans ce cas, n'importe quelle paire de monopôles s'attire.

## 4. A l'intérieur du monopôle

### *4.1. Brisure spontanée de symétrie — Une revue éclair*

Jusqu'ici, nous avons gardé nos distances avec le monopôle. Maintenant, nous allons ouvrir la boîte noire pour voir si les structures que nous avons trouvées à grande distance peuvent être prolongées jusqu'à $r = 0$.

Nous travaillerons dans le contexte des théories de champs de jauge avec brisure spontanée de la symétrie. Ces théories sont au menu de tous les jours en physique contemporaine des hautes énergies; néanmoins je vais en donner ici une revue éclair, en gros dans le style de la revue de la section 3.1, mais encore plus comprimé. Mon but est le même qu'auparavant, établir la notation commune et insister sur les points essentiels. Dans la section 3.1, nous avons divisé les champs de notre théorie en champs de jauge et champs de matière; il sera maintenant plus convenable de diviser les champs de matière en champs scalaires (tous supposés réel, pour la commodité), et les autres, typiquement les champs de spineurs. Nous changerons légèrement la notation et utiliserons $\mathbf{\Phi}$ pour désigner un grand vecteur composé uniquement des champs scalaires.

Nous nous limiterons aux théories pour lesquelles la densité Lagrangienne de la matière est de la forme

$$\mathcal{L}_{\mathrm{m}} = \tfrac{1}{2} D_\mu \mathbf{\Phi} \cdot D^\mu \mathbf{\Phi} - U(\mathbf{\Phi}) + \cdots, \tag{4.1}$$

où $U$ est une fonction invariante par $G$ et les points de suspension indiquent des termes mettant en jeu les champs non scalaires.

A une transformation de jauge près, les états fondamentaux de la théorie sont des états dans lesquels tous les champs non scalaires sont nuls, et les champs scalaires sont indépendants de l'espace–temps, et à un minimum de $U$. Nous utiliserons le langage quantique pour cette situation classique et appellerons ces états fondamentaux " vides", et la valeur correspondante de $\mathbf{\Phi}$ " valeur moyenne de $\mathbf{\Phi}$ dans le vide". Nous supposerons toujours qu'on a ajouté à $U$ une constante de sorte que l'énergie de l'état fondamental soit nulle.

Soit $\langle \Phi \rangle$ un minimum (arbitrairement choisi) de $U$; alors $g\langle \Phi \rangle$ est aussi un minimum quel que soit $g$ dans $G$. Nous supposerons que tous les minima de $U$ sont de cette forme. Alors tous les fondamentaux sont physiquement équivalents, et, sans perte de généralité, on peut se restreindre au cas où la valeur moyenne de $\Phi$ est $\langle \Phi \rangle$. (Cette hypothèse exclut des phénomènes intéressants tels qu'une dégénérescence accidentelle, dans laquelle les fondamentaux ne sont pas reliés par la symétrie, et l'existence de bosons de Goldstone liés à une symétrie qui n'est pas une symétrie de jauge. L'hypothèse faite ici permet de garder la simplicité de l'argumentation; il n'est pas difficile d'étendre l'analyse au cas général.)

Soit $H$ le sous groupe de $G$ qui laisse $\langle \Phi \rangle$ invariant:

$$h \in H \text{ si et seulement si } h\langle \Phi \rangle = \langle \Phi \rangle. \tag{4.2}$$

Nous disons que le groupe de symétrie $G$ est spontanément brisé sur $H$. Les champs de jauge associés à $H$ restent sans masse; les autres se combinent avec les champs scalaires pour former des champs vectoriels massifs. (C'est le fameux mécanisme de Higgs.) La matrice de masse vectorielle est donnée par

$$\mu_{ab}^2 = f_a T_a \langle \Phi \rangle \cdot f_b T_b \langle \Phi \rangle, \tag{4.3}$$

(pas de somme sur les indices répétés). Les champs scalaires absorbés dans les champs vectoriels massifs sont ceux qui correspondent aux perturbations du vide de la forme $\delta \Phi = T_a \langle \Phi \rangle$.

Si on avait fait un choix différent, arbitraire de $\langle \Phi \rangle$, $H$ aurait été un sous groupe différent dans $G$, bien qu'isomorphe à notre $H$ initial. A l'occasion, nous travaillerons dans une jauge telle que la valeur moyenne de $\Phi$ varie dans l'espace; il sera important de se rappeler alors que $H$ aussi varie.

Pour illustrer ces idées, prenons pour $G$, SO($n$), $\Phi$ un $n$-vecteur et

$$U = \frac{\lambda}{4}(\Phi \cdot \Phi - c^2)^2. \tag{4.4}$$

Les choix possibles pour le vecteur $\langle \Phi \rangle$ sont tous les vecteurs de longueur $c$. Si on choisit

$$\langle \Phi^a \rangle = c\delta^{an}, \quad a = 1, \ldots, n, \tag{4.5}$$

alors, $H$ est le sous groupe des rotations des $n-1$ premières coordonnées SO($n-1$). Des $n(n-1)/2$ champs de jauge initiaux, $(n-1)(n-2)/2$ restent sans masse; les $n-1$ autres se combinent avec $n-1$ champs scalaires pour acquérir une masse; un seul champ scalaire reste inchangé.

Dans ce qui suit on fera usage de cette théorie pour $n = 3$. Ce cas ressemble vaguement à la réalité en ce qu'il n'y a qu'un méson de jauge sans masse, qu'on peut identifier avec le photon; c'est pour cette raison qu'il a été incorporé par Georgi et Glashow dans une alternative, ingénieuse mais erronnée, au modèle de Weinberg–Salam, souvent appelée le modèle de Georgi–Glashow.

### *4.2. Construction de monopôles*

Je vais maintenant expliquer comment les monopôles magnétiques apparaissent spontanément dans les théories de jauge brisées. Ce phénomène a été découvert pour la première fois par 't Hooft et Polyakov [2], dans le modèle SO(3) que je viens de décrire. C'est l'un des effets les plus éblouissants de la théorie des champs; les monopôles magnétiques apparaissent miraculeusement dans des théories qui ne contiennent aucun champ fondamental magnétiquement chargé.

Il se trouve que les champs de matière non scalaires ne jouent aucun rôle dans la constitution des monopôles, de sorte que, par souci de simplicité, je supposerai que nous avons à faire à une théorie de champs de jauges et de champs scalaires, exclusivement. Dans cette théorie, nous étudierons les configurations de champs non singulières d'énergie finie à temps fixé. Par la suite, nous nous pencherons sur l'évolution temporelle de ces configurations.

La densité d'énergie est:

$$\Theta^{\infty} = \tfrac{1}{2}\boldsymbol{D\Phi}\cdot\boldsymbol{D\Phi} + U(\boldsymbol{\Phi}) + \text{autres termes positifs}. \tag{4.6}$$

Pour que l'intégrale d'énergie converge, il faut que chacun des deux termes explicités s'annulent pour $r$ grand. Pour ne pas compliquer le raisonnement, je supposerai qu'ils sont strictement nuls en dehors d'un rayon $R$.

$$U = \boldsymbol{D\Phi} = 0, \quad r \geqslant R. \tag{4.7}$$

J'insiste sur le fait que cette hypothèse est une pure lubie. Ce n'est pas une conséquence de la finitude de l'énergie et ce n'est même pas vrai pour n'importe laquelle des solutions connues d'énergie finie des équations de champs. Je la fais ici uniquement pour des raisons pédagogiques, afin de pouvoir construire un raisonnement dans lequel la structure sous-jacente n'est pas enfouie sous l'analyse de la rapidité et de l'uniformité de l'approche des limites. J'invite ceux d'entre vous qui sont habiles en analyse réelle à étendre le raisonnement à des hypothèses plus faibles et plus sensées.

L'équation (4.7) implique que $\mathbf{\Phi}$ doit être à un minimum de $U$ pour $r \geqslant R$, mais peut être à des minima différents suivant le point. Toutefois, on peut toujours faire une transformation de jauge telle que

$$\mathbf{\Phi} = \langle \mathbf{\Phi} \rangle, \quad r \geqslant R. \tag{4.8}$$

Deux étapes suffisent. On fait d'abord une transformation de jauge dépendant de $r$ seulement qui transforme $\mathbf{\Phi}$ en $\langle \mathbf{\Phi} \rangle$ le long de l'axe du pôle nord pour $r \geqslant R$. Ensuite, à $r$ fixé, on commence au pôle nord et on transforme le long des méridiens pour avoir $\mathbf{\Phi} = \langle \mathbf{\Phi} \rangle$ partout. Cette seconde étape peut introduire une singularité de corde de Dirac le long de l'axe du pôle sud, où les méridiens se recoupent.

Les équations (4.7) et (4.8) impliquent

$$\boldsymbol{D}\mathbf{\Phi} = \boldsymbol{A}\langle \mathbf{\Phi} \rangle = 0, \quad r \geqslant R. \tag{4.9}$$

Ainsi, après notre transformation de jauge, les seuls champs de jauge qui existent pour $r \geqslant R$ sont ceux qui sont associés avec le sous groupe $H$. C'est bien entendu exactement ce à quoi nous nous attendions; seuls les champs sans masse s'étendent aux grandes distances.

Ainsi, nous avons la même structure que dans la section 3; en dehors d'une boîte noire, la sphère de rayon $R$, il n'y a rien d'autre que des champs de jauge sans masse. Le seul changement est un léger changement de notation; le groupe que nous avons appelé $G$ dans la section 3 s'appelle maintenant $H$. Ainsi, en dehors de la sphère nous avons la classification habituelle des configurations de champs par leurs charges topologiques, les éléments de $\pi_1(H)$.

S'il n'y avait à l'intérieur de la sphère que des champs de jauge $H$, un champ non singulier aurait toujours une charge topologique triviale. J'ai donné l'argumentation correspondante dans la section 3.4 mais je vais la répéter ici: à chaque sphère nous associons un chemin dans l'espace du groupe. Si on prend une sphère autour d'un monopôle et qu'on la réduise à un point, le chemin associé se réduit au chemin constant. Il y a alors seulement deux possibilités: ou bien le chemin se déforme continûment, auquel cas il appartenait déjà au début à la classe d'homotopie triviale, ou bien il se déforme de façon discontinue auquel cas nous avons rencontré une singularité. (Notez que puisque la charge topologique est invariante de jauge, les singularités qui relèvent d'un simple artifice de jauge, comme les cordes de Dirac, ne conviennent pas; une authentique singularité invariante de jauge est requise.) On voit maintenant ce qui diffère dans le cas présent. A l'intérieur de la boite noire nous avons des champs de jauge $G$, pas seulement des champs de jauge $H$, et le chemin peut sortir de $H$ dans le groupe plus grand $G$. Il est tout à fait possible qu'un chemin qui est homotopiquement non trivial dans $H$ soit homotopiquement trivial dans $G$; un noeud topologique qui ne peut être dénoué dans un espace plus petit peut se défaire dans un espace plus grand. De cette façon nous pouvons avoir une charge topologique sans singularité.

On peut redire la même chose dans un langage plus abstrait. Comme $H$ est un sous groupe de $G$, chaque chemin dans $H$ est un chemin dans $G$. Ceci induit une application de $\pi_1(H)$ dans $\pi_1(G)$. Le noyau de cette application est défini, comme toujours, comme le sous groupe qui s'applique sur l'identité. Ainsi notre résultat peut se formuler comme suit:

| La condition d'existence d'un monopôle non singulier est que la charge topologique soit dans le noyau de l'application $\pi_1(H) \to \pi_1(G)$. |
|---|

Nous avons vu que cette condition est nécessaire. Je vais maintenant montrer qu'elle est suffisante, en construisant une configuration de champ non singulière d'énergie finie, pour toute charge topologique dans le noyau.

Pour toute charge topologique, il y a un champ à grande distance pourvu de cette charge, le champ GNO approprié,

$$\boldsymbol{A}_{\text{GNO}} \cdot \mathrm{d}\boldsymbol{x} = Q(1-\cos\theta)\,\mathrm{d}\varphi. \tag{4.10}$$

Ici, $Q$ est dans l'algèbre de Lie de $H$,

$$Q\langle\boldsymbol{\Phi}\rangle = 0, \tag{4.11}$$

et le chemin

$$g(\tau) = \exp(4\pi Q\tau), \quad 0 \leqslant \tau \leqslant 1, \tag{4.12}$$

appartient à la classe d'homotopie spécifiée. Dans $G$, $g(\tau)$ est homotope au chemin trivial. Par conséquent, il existe une fonction continue de deux variables $g(\theta,\varphi) \in G, \theta \in [0,\pi], \varphi \in [0,2\pi]$ telle que

$$g(0,\varphi) = g(\theta,0) = g(\theta,2\pi) = 1, \tag{4.13a}$$

$$g(\pi,\varphi) = \exp 2Q\varphi. \tag{4.13b}$$

Nous l'utiliserons pour définir nos champs vectoriel et scalaire pour $r \geqslant R$,

$$\boldsymbol{\Phi} = g\langle\boldsymbol{\Phi}\rangle, \tag{4.14a}$$

$$\boldsymbol{A} = g\boldsymbol{A}_{\text{GNO}}g^{-1} + g\nabla g^{-1}. \tag{4.14b}$$

C'est simplement un transformé de jauge du champ GNO, de sorte que c'est encore une configuration de champs d'énergie finie. Cependant, la corde de Dirac a complètement disparu; tous nos champs sont maintenant manifestement non singuliers à tous angles. Le prix que nous avons payé est que nous avons fait dépendre des angles la valeur moyenne dans le vide $\boldsymbol{\Phi}$.

Il est maintenant trivial de prolonger cette configuration pour $r \leqslant R$:

$$\boldsymbol{\Phi}(r,\theta,\varphi) = \frac{r^2}{R^2}\boldsymbol{\Phi}(R,\theta,\varphi), \tag{4.15a}$$

$$\boldsymbol{A}(r,\theta,\varphi) = \frac{r^2}{R^2}\boldsymbol{A}(R,\theta,\varphi). \tag{4.15b}$$

Cette configuration est manifestement d'énergie finie et sans singularité jusqu'à $r = 0$. Notez qu'on n'aurait pas pu prolonger jusqu'à l'intérieur de cette façon si on n'avait pas d'abord éliminé la corde. Si on avait simplement essayé de réduire à l'échelle le champ GNO, on aurait violé la condition de quantification.

Ainsi se conclut la preuve de la suffisance.

Le fait que la condition soit à la fois nécessaire et suffisante en fait le joyau de la théorie des monopôles qui lui vaut bien l'honneur d'être encadré. Elle nous permet de dire instantanément, sans résoudre une seule équation différentielle si une théorie donnée admet des monopôles non singuliers, et de quelle sorte. Je vais donner trois exemples:

1. $G$ est SO(3) et $H$ est SO(2). C'est la théorie décrite à la fin de la section 4.1, le modèle de Georgi–Glashow. Si nous identifions le sous groupe non brisé avec le groupe électromagnétique, alors l'analyse aux grandes distances permet à $eg$ d'être égal à 0, $\pm\frac{1}{2}$, $\pm 1$, etc. Topologiquement ces valeurs correspondent à des chemins qui tournent $2eg$ fois autour de SO(2). Seuls les chemins qui tournent un nombre pair de fois autour de SO(2) peuvent se déformer dans le chemin trivial de SO(3); ainsi, un terme sur deux de la série correspond à une valeur permise, $eg = 0$, $\pm 1$, etc.

2. $G$ est U(2) et $H$ est U(1), plongé comme le sous-groupe de U(2) qui laisse invariant le premier vecteur de base de l'espace de Hilbert à deux dimensions. C'est le modèle de Weinberg–Salam. $G$ est localement isomorphe à U(1)×SU(2) et tout chemin qui tourne autour de $H$ tourne aussi autour du facteur U(1) de $G$. Ainsi, seule la charge topologique triviale est permise, $eg = 0$.

3. $G$ est n'importe quel groupe semi-simple et $H$ est n'importe quel groupe avec un facteur U(1). Il s'agit des modèles Grand-Unifiés cités

dans la section 1. $\pi_1(G)$ est fini et $\pi_1(H)$ est infini, de sorte que le noyau de l'application est infini et la théorie doit contenir des monopôles comme je l'ai affirmé dans la section 1. Bien entendu, on ne peut pas dire avec précision quelles sont les charges magnétiques permises tant qu'on ne connait pas les détails de la théorie.

Nous nous sommes occupés ici des configurations de champs non singulières, d'énergie finie à temps fixé. Si le problème aux données initiales est bien posé, n'importe laquelle de ces configurations définira une solution des équations du mouvement, pas nécessairement statique. Ainsi, rien dans notre analyse ne démontre la possibilité de construire des monopôles statiques. Ce n'est pas important. Nous autres, théoriciens, aimons les solutions statiques, mais c'est parce qu'elles sont faciles à analyser et que nous sommes paresseux. Si un expérimentateur trouve une boite noire environnée par un champ monopolaire, c'est intéressant que la boite contienne des champs statiques, oscillants, tournants, ou fluctuant ergodiquement.

Il y a une chose que nous savons sur l'évolution temporelle de configurations avec une charge topologique non triviale. Quoiqu'il arrive, elles ne peuvent pas se dissiper complètement, simplement se répandre en dehors de la boîte sous la forme de champs de radiation ordinaire, avec ou sans masse. Ceci tient au fait que la radiation ne transporte pas de charge topologique, et la charge topologique est conservée. La topologie est puissance.

### 4.3. *L'objet de 't Hooft–Polyakov*

En dépit de ces mots aigre-doux sur les solutions statiques, je consacrerai quelque temps ici à la discussion du fameux monopôle statique trouvé dans le modèle de Georgi–Glashow, éq. (4.4) par 't Hooft et Polyakov [2]. La recherche de la solution statique la plus générale est difficile car on a affaire à plusieurs fonctions de trois variables; l'astuce consiste à simplifier la situation en cherchant des solutions symétriques par rapport à un sous-groupe intelligemment choisi du groupe de symétrie de la théorie. Il se trouve qu'un sous-groupe bien adapté à cet effet est le sous-groupe SO(3) formé de rotations spatiales et internes simultanées. Les champs scalaires invariants sont nécessairement de la forme

$$\boldsymbol{\Phi}^a = f(r^2)r^a/r, \tag{4.16}$$

où l'indice interne $a$ va de 1 à 3. Pour maintenir l'énergie finie à l'infini, $f(\infty)$ doit être égal à $c$. Pour assurer la régularité du champ à $r = 0$, $f(0)$

doit être nul. Cette forme de configuration est non seulement invariante par le groupe SO(3) mentionné, mais aussi par la parité si on définit $\mathbf{\Phi}$ comme un pseudoscalaire. Les seuls champs de jauge invariants par rapport à ces deux symétries sont de la forme

$$A^{ia} = h(r^2)\varepsilon^{iak}r_k. \tag{4.17}$$

La finitude de l'énergie impose des restrictions sur le comportement de $h$ pour $r$ grand, mais je ne me soucierai pas de les analyser ici.

Ainsi, la construction de solutions statiques devient un simple problème de calcul des variations; on doit minimiser l'énergie comme fonctionnelle de deux fonctions d'une seule variable. Ce problème n'est pas hors d'atteinte de l'analyse fonctionnelle ou numérique; j'espère que vous trouverez plausible que je vous dise qu'il est possible aussi bien de démontrer qu'il existe une solution que de la calculer avec une bonne précision sur un calculateur de poche.

Nous avons une solution, mais est-ce un monopôle? La façon la plus commode de répondre à cette question est de transformer la solution dans la jauge de corde. De l'éq. (4.16), il s'ensuit

$$\mathbf{\Phi}(r,\theta,\varphi) = g(\theta,\varphi)\mathbf{\Phi}(r,\theta=0), \tag{4.18}$$

où

$$g(\theta,\varphi) = \exp T_3\varphi \exp T_2\theta \exp - T_3\varphi, \tag{4.19}$$

et les $T$'s sont les générateurs de SO(3). L'équation (4.18) serait vraie même si j'omettais le dernier facteur de l'éq. (4.19); je l'ai introduit pour que $g$ soit bien défini à $\theta = 0$.

Faisons maintenant une transformation de jauge, utilisant $g^{-1}$:

$$\mathbf{\Phi}(r,\theta,\varphi) \rightarrow g^{-1}\mathbf{\Phi} = \mathbf{\Phi}(r,\theta=0). \tag{4.20}$$

La valeur moyenne dans le vide se trouve ainsi alignée dans tout l'espace; dans toutes les directions, le groupe non brisé $H$ est le sous-groupe SO(2) engendré par $T_3$. Par la même transformation de jauge,

$$\boldsymbol{A} \rightarrow g^{-1}\boldsymbol{A}g + g^{-1}\nabla g. \tag{4.21}$$

Seul le second terme de cette expression produit une singularité de corde de Dirac sur l'axe polaire sud. Ici

$$g(\pi,\varphi) = \exp T_2\pi \exp -2T_3\varphi, \tag{4.22}$$

$$A_\varphi(\pi,\varphi) \rightarrow -2T_3. \tag{4.23}$$

Par intégration, on voit qu'on fait deux tours du sous-groupe SO(2) pour un autour de la corde. C'est un monopôle avec $eg = 1$, la valeur minimale autorisée par les considérations topologiques de la section 4.2.

### 4.4. *Pourquoi les monopôles sont lourds*

J'ai dit dans l'introduction que les monopôles sont typiquement très lourds. Maintenant que nous comprenons leur structure topologique, il est possible de voir pourquoi il en est ainsi.

Commençons par considérer une théorie dans laquelle tous les champs de jauge lourds ont des masses comparables, de l'ordre d'une masse typique, $\mu$, et où tous les couplages de jauge sont de l'ordre d'un couplage typique, $e$. Dans n'importe quelle configuration non radiative, les champs de jauge lourds décroissent avec la distance comme $\exp(-\mu r)$. Par conséquent, nous nous attendons à ce que le coeur du monopôle, c'est à dire la région où les champs lourds sont significatifs, ait une taille de l'ordre de $1/\mu$. En dehors du coeur, seuls les champs sans masse devraient être significatifs, et le monopôle devrait ressembler à un champ Coulombien magnétique (dans le cas Abélien) ou à un champ GNO (dans le cas général).

Il est facile d'estimer l'énergie emmagasinée en dehors du coeur. Pour fixer les idées, faisons le calcul pour un monopôle Abélien de charge magnétique $e$,

$$E_{\text{magnetique}} = \tfrac{1}{2}\int_{r > \mathrm{O}(1/\mu)} \mathrm{d}^3\boldsymbol{x}\,|\boldsymbol{B}|^2 = 2\pi g^2 \int_{\mathrm{O}(1/\mu)}^{\infty} \frac{\mathrm{d}r}{r^2} = 2\pi g^2 \mathrm{O}(\mu). \tag{4.24}$$

Comme la densité d'énergie de la théorie est positive, l'énergie à l'intérieur du coeur ne peut que s'ajouter à celle-ci. Ainsi,

$$m \geqslant \mathrm{O}(\mu/e^2). \tag{4.25}$$

Sous cette forme d'ordre de grandeur, la borne est aussi clairement valable pour les monopôles non Abéliens. Les monopôles sont lourds parce que l'intégrale de l'énergie électromagnétique d'un champ de Coulomb diverge à courte distance. A condition de bien vouloir faire appel à une intuition raisonnable, on peut remplacer l'inégalité par une égalité. Si le coeur augmente de rayon, l'énergie magnétique décroît. Comme le monopôle est en équilibre, il doit y avoir compensation par un accroissement de l'énergie interne. (Si le monopôle n'est pas stationnaire c'est encore vrai, en moyenne dans le temps.) Il est donc plausible de supposer que l'énergie du coeur est du même ordre de grandeur que l'énergie magnétique.

Venons en maintenant à une théorie qui admet plusieurs échelles de masse. C'est ce qui se passe typiquement quand il y a une hiérarchie de brisures de symétrie,

$$H \subset G_1 \subset G_2 \subset G_3 \ldots, \tag{4.26}$$

à laquelle est associée une hiérarchie de masses

$$\mu_1 \ll \mu_2 \ll \mu_3 \ldots, \tag{4.27}$$

où $\mu_i$ est la masse typique d'un champ de jauge dans $G_i$ (mais pas dans $G_{i+1}$). Par exemple, dans le modèle Grand Unifié de Georgi et Glashow [20], SU(5) se brise en SU(3) (couleur)×U(2) (électrofaible); les champs de jauge associés acquièrent des masses de l'ordre de $10^{14}$ GeV. U(2) se brise à son tour jusqu'à U(1), comme dans le modèle de Weinberg–Salam; là, les champs de jauge acquièrent des masses de l'ordre de $10^2$ GeV.

Associée à la suite (4.26), il y a une suite d'applications

$$\pi_1(H) \to \pi_1(G_1) \to \pi_1(G_2) \ldots. \tag{4.28}$$

Nous pouvons construire un monopôle avec une charge topologique arbitraire qui s'applique par la suite sur l'identité. Supposons que ceci arrive pour la première fois pour le groupe $G_i$. Alors, le champ monopolaire reste coulombien jusqu'à des distances de l'ordre $1/\mu_i$ et la masse du monopôle satisfait à

$$m \geqslant \mathrm{O}(\mu_i/e^2). \tag{4.29}$$

Comme auparavant, pourvu que nous consentions à faire appel à une intuition raisonnable, cette inégalité peut être remplacée par une égalité.

A titre d'exemple, dans la Théorie Grande Unifiée de Georgi–Glashow, nous devons aller à SU(5), comme nous l'avons vu dans la section 4.3; la masse du monopôle est de l'ordre de $10^{16}$ GeV, ou $10^{-8}$ g.

Bien entendu, il peut arriver que différents éléments de $\pi_1(H)$ s'appliquent sur l'identité à des étapes différentes de la hiérarchie; ceci offre la possibilité d'une large variété d'échelles de masse pour les monopôles. Considérons par exemple une théorie dans laquelle SU(3) se brise sur son sous-groupe réel SO(3) à une échelle de masse élevée; à une échelle beaucoup plus petite SO(3) se brise sur SO(2) de la façon qui nous est familière. Il y a des monopôles pour des valeurs entières et demi-entières de $eg$, ces derniers étant beaucoup plus lourds que les premiers.

### 4.5. *La borne de Bogomol'nyi et la limite de Prasad–Sommerfield*

Nous avons déduit des estimations grossières des masses des monopôles. Pour certaines théories, il est possible de déduire une borne inférieure rigoureuse, la borne de Bogomol'nyi [21]. Je vais donner ici l'argumentation pour le modèle de Georgi–Glashow; on peut l'étendre directement à n'importe quelle théorie dans laquelle les champs scalaires se transforment selon la représentation adjointe du groupe.

Nous écrirons l'énergie de la théorie sous la forme

$$E = \tfrac{1}{2}\int \mathrm{d}^3\boldsymbol{x}\Big[\boldsymbol{E}^a\cdot\boldsymbol{E}^a + \boldsymbol{B}^a\cdot\boldsymbol{B}^a + (D_0\Phi^a)^2 + \boldsymbol{D}\Phi^a\cdot\boldsymbol{D}\Phi^a + \tfrac{1}{2}\lambda(\Phi^a\Phi^a - c^2)^2\Big]. \tag{4.30}$$

Ici les vecteurs indiquent les propriétés de transformation spatiale, tandis que les indices se réfèrent aux symétries internes. Aussi $E^a$ et $B^a$ sont définis comme dans l'électromagnétisme

$$E_i^a = F_{0i}^a, \qquad B_i^a = \tfrac{1}{2}\varepsilon_{ijk}F_{jk}^a, \tag{4.31}$$

et nous avons redéfini les champs comme dans l'éq. (3.23) de sorte que l'expression de l'énergie ne contient pas la constante de couplage.

Pour obtenir la borne, nous aurons besoin de deux propriétés du champ magnétique. L'une d'elle est

$$\boldsymbol{D}\cdot\boldsymbol{B}^a = 0. \tag{4.32}$$

C'est une conséquence de l'identité de Jacobi pour la différentiation covariante. L'autre est

$$\lim_{r\to\infty}\int \mathrm{d}^2S\,\boldsymbol{n}\cdot\boldsymbol{B}^a\Phi^a = 4\pi gc, \tag{4.33}$$

où $\mathrm{d}^2S$ est l'élément d'aire usuel et $\boldsymbol{n}$ la normale extérieure. La façon la plus facile de le voir est dans la jauge introduite dans la section 4.2 où, à grande distance

$$\Phi^a = \langle\Phi^a\rangle = \delta^{a3}c, \tag{4.34}$$

et la seule composante non nulle de $\boldsymbol{B}^a$ est $\boldsymbol{B}^3$, un champ magnétique monopolaire.

Nous sommes maintenant prêts pour le démarrage

$$E \geqslant \tfrac{1}{2}\int \mathrm{d}^3\boldsymbol{x}(\boldsymbol{B}^a\cdot\boldsymbol{B}^a + \boldsymbol{D}\phi^a\cdot\boldsymbol{D}\phi^a)$$

$$= \tfrac{1}{2}\int \mathrm{d}^3\boldsymbol{x}(\boldsymbol{B}^a \pm \boldsymbol{D}\phi^a)^2 \mp 4\pi gc, \tag{4.35}$$

par intégration par parties et en utilisant les éq. (4.32) et (4.33). Ainsi

$$E \geqslant |4\pi g c|, \tag{4.36}$$

c'est le résultat désiré.

Comme $g$ est $O(1/e)$ et $\mu$ est $O(ce)$ le membre de droite de cette équation est $O(\mu/e^2)$, en accord avec les arguments de la section 4.4. En fait, ceci n'est que de l'analyse dimensionnelle; une fois qu'on a éliminé la constante de couplage scalaire $\lambda$, la seule quantité qu'on puisse construire avec la dimension d'une énergie est $\mu/e^2$.

Il est en fait possible de saturer la borne dans une limite appropriée, étudiée pour la première fois par Prasad et Sommerfield [22]. En déduisant la borne, nous avons laissé de côté trois des cinq termes de l'éq. (4.30). Deux d'entre eux, le terme en $\boldsymbol{E}^2$ et le terme $(D_0\boldsymbol{\Phi})^2$, s'annulent automatiquement si on suppose l'invariance par renversement du temps. On peut annuler le troisième, le potentiel scalaire, en passant à la limite $\lambda \to 0^+$. En termes moins formels, nous éliminons tous les termes dépendant de $\lambda$ des équations du mouvement, mais gardons la condition aux limites $\boldsymbol{\Phi}^a\boldsymbol{\Phi}^a = c^2$ aux grandes distances.

Dans cette limite, l'éq. (4.35) est une égalité, non une inégalité, et on peut saturer la borne si on trouve des solutions de

$$\boldsymbol{B}^a \pm \boldsymbol{D}\boldsymbol{\Phi}^a = 0, \tag{4.37}$$

où le signe dans l'équation dépend du signe de $g$. En prime, toute solution de (4.36) est aussi une solution stationnaire des équations du mouvement; un minimum de la fonctionnelle énergie est **a fortiori** un point stationnaire.

L'équation (4.37) est du premier ordre, et considérablement plus facile à analyser que les équations du mouvement du second ordre. Je n'ai pas le temps d'en donner l'analyse ici ou même d'en décrire les résultats principaux dans quelque détail que ce soit. Il se trouve cependant que l'équation a des solutions et, à notre grande joie, pas seulement des solutions décrivant des monopôles simples mais aussi des monopôles multiples. Ce n'est pas aussi incroyable qu'il semble. Dans la limite de Prasad–Sommerfield, le champ scalaire devient sans masse et, par suite, plusieurs monopôles peuvent exister en équilibre statique, l'interaction scalaire équilibrant la répulsion magnétique.

## 5. La théorie quantique

### *5.1. Monopôles quantiques et excitations isorotationnelles**

Dans la section 3.2 j'ai fait le raisonnement qu'une théorie quantique de champs devrait ressembler le plus à la théorie classique correspondante

*Voir [19].

dans la limite du couplage faible. Dans cette section, je vais élaborer cette observation pour développer un schéma d'approximation quantitatif des théories dont la limite classique admet des solutions monopolaires statiques [23].

Pour être spécifique, on travaillera avec le modèle de Georgi–Glashow,

$$\mathcal{L} = -\frac{1}{4f^2}F^a_{\mu\nu}F^{\mu\nu a} + \tfrac{1}{2}D_\mu\mathbf{\Phi}^a \cdot D^\mu\mathbf{\Phi}^a - \tfrac{1}{4}\lambda(\mathbf{\Phi}^a\cdot\mathbf{\Phi}^a - c^2)^2. \tag{5.1}$$

On définit les nouvelles variables $\mathbf{\Phi}' = \mathbf{\Phi}f$, $c' = cf$, $\lambda' = \lambda/f^2$. En fonction de ces variables,

$$\mathcal{L} = \frac{1}{f^2}\left[-\tfrac{1}{4}F^a_{\mu\nu}F^{\mu\nu a} + \tfrac{1}{2}D_\mu\mathbf{\Phi}'^a D^\mu\mathbf{\Phi}'^a - \tfrac{1}{4}\lambda'(\mathbf{\Phi}'^a\cdot\mathbf{\Phi}'^a - c'^2)\right]. \tag{5.2}$$

Je propose d'étudier cette théorie dans la limite où $f$ est petit, à $c'$ et $\lambda'$ fixés. C'est la limite classique discutée dans la section 3.2; la quantité importante pour la théorie quantique est $\mathcal{L}/\hbar$, de sorte que la limite où $f$ est petit à $\hbar$ fixé est la même que celle où $\hbar$ est petit à $f$ fixé. De plus, quand les choses sont écrites de cette façon, la solution monopolaire classique est indépendante de $f$; un facteur multiplicatif devant le Lagrangien total disparaît des équations du mouvement. (Nous utiliserons les nouveaux paramètres dans tout le reste de la discussion de sorte qu'à partir de maintenant nous omettrons les primes.)

La théorie est spécialement facile à traiter dans la jauge temporelle, $A_0 = 0$,

$$\mathcal{L} = \frac{1}{f^2}\Big[\tfrac{1}{2}\partial_0\boldsymbol{A}^a\cdot\partial_0\boldsymbol{A}^a + \tfrac{1}{2}\partial_0\mathbf{\Phi}^a\partial_0\mathbf{\Phi}^a$$
$$+ \text{termes sans dérivées par rapport au temps}\Big] \tag{5.3}$$

Ainsi, le moment conjugué à $\mathbf{\Phi}^a$ est

$$\pi^a = f^{-2}\partial_0\mathbf{\Phi}^a, \tag{5.4}$$

celui conjugué à $A^a$ est

$$\boldsymbol{\pi}^a = f^{-2}\partial_0\boldsymbol{A}^a, \tag{5.5}$$

et l'Hamiltonien est donnée par

$$f^2H = \tfrac{1}{2}f^4\int[\boldsymbol{\pi}^a\cdot\boldsymbol{\pi}^a + \pi^a\pi^a]\,\mathrm{d}^3\boldsymbol{x} + V, \tag{5.6}$$

où $V$ est moins l'intégrale des termes sans dérivée par rapport au temps dans l'éq. (5.3). La solution monopolaire classique est un point stationnaire de la fonctionnelle $V$; en fait, si le monopôle est stable (ce que nous supposerons), c'est un minimum local de $V$. J'insiste sur le fait que toutes les puissances de $f$ sont explicitées dans l'éq. (5.6); $V$ est indépendant de $f$.

L'équation (5.6) définit un Hamiltonien qui offre des particularités du point de vue de la théorie des perturbations ordinaires. D'abord il y a un $f^2$ explicite dans le membre de gauche de l'équation. Bien sûr c'est une particularité triviale; si on peut trouver un développement pour les fonctions propres et valeurs propes de $f^2H$, on peut aussi en trouver pour ceux de $H$. Deuxièmement le petit paramètre multiplie l'énergie cinétique, le terme quadratique dans les moments canoniques, plutôt que l'énergie potentielle, le terme indépendant des moments canoniques. C'est très étrange; avons-nous jamais rencontré un tel système auparavant? Bien entendu. C'est en effet la situation pour une molécule diatomique:

$$H = \boldsymbol{P}^2/2M + V(r), \tag{5.7}$$

où $M$ est la masse nucléaire réduite. Le développement standard dans l'étude des spectres des molécules diatomiques utilise le petit paramètre $1/M$, coefficient de l'énergie cinétique. Bien sûr notre système n'est pas exactement une molécule polyatomique,

$$H = \sum_{i=1}^{N} \frac{\boldsymbol{P}_i^2}{2M_i} + V(\boldsymbol{r}_1, \dots, \boldsymbol{r}_N). \tag{5.8}$$

Pour être précis, c'est une molécule infiniment polyatomique où tous les noyaux ont la masse $1/f^4$. Ainsi, le problème de la construction de monopôles quantiques a été résolu complètement il y a une cinquantaine d'années.

Je vais maintenant expliquer la solution, d'abord en rappelant les résultats familiers pour une molécule diatomique, ensuite en donnant l'extension triviale à une molécule polyatomique, et finalement en donnant la transcription encore plus triviale de cette extension dans le langage de la théorie des champs.

Pour la molécule diatomique, nous supposons que le potentiel interatomique est comme indiqué sur la fig. 5. Le minimum de $V$ est à $r = r_0$ et $V(r_0) = E_0$. Les trois premières approximations aux états propres les plus bas et à leurs valeurs propres sont données par la table 1.

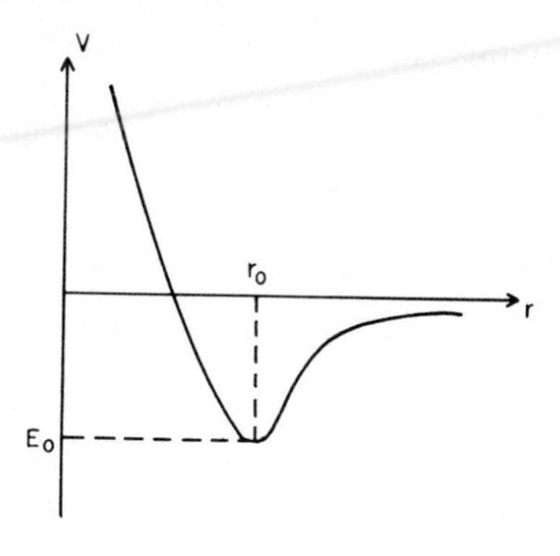


Fig. 5.

Comme on le voit à partir de la table, le paramètre correct de développement pour les valeurs propres de l'énergie est $1/\sqrt{M}$, qui deviendra $f^2$ dans le problème de théorie des champs. (La colonne de droite de la table est cumulative; autrement dit, l'énergie au premier ordre est la somme des deux premières lignes etc.) Je vais maintenant expliquer l'origine de la table.

A l'ordre 0, on néglige complètement l'énergie cinétique. La particule est placée au fond du puits de potentiel dans un état propre de l'opérateur position $\boldsymbol{r}$. Le module de la position est fixé à $r_0$, mais la position angulaire est arbitraire. Ceci ne ressemble pas beaucoup au spectre réel révélé par la spectroscopie moléculaire; en particulier il y a une dégénérescence angulaire infinie complètement parasite. Comme on le verra cette dégénérescence n'est levée qu'au second ordre. Au premier ordre, on commence à voir les effets de la vibration de la particule autour de sa position d'équilibre. Comme $M$ est très grand, la particule ne vibre pas très loin, et, au premier ordre, on peut remplacer le potentiel près de

Table 1
La molécule diatomique

| Ordre de l'approximation | Etat propre de l'énergie | Valeur propre de l'énergie |
|---|---|---|
| 0 | $\|r_0, \theta, \varphi\rangle$ | $E_0$ |
| 1 | $\|n, \theta, \varphi\rangle$ | $+(n+1/2)[V''(r_0)/M]^{1/2}$ |
| 2 | $\|n, l, m\rangle$ | $+l(l+1)/2Mr_0^2 + \cdots$ |

l'équilibre par un potentiel harmonique,

$$V(r) = E_0 + \tfrac{1}{2}V''(r_0)(r - r_0)^2. \tag{5.9}$$

Les fonctions propres sont maintenant des fonctions d'onde d'oscillateur harmonique en $r$, mais toujours des fonctions $\delta$ angulaires. Elles sont indexées par le nombre d'excitation usuel de l'oscillateur, $n$, et ont les énergies de l'oscillateur usuel. Ce sont les fameux niveaux vibrationnels de la spectroscopie moléculaire.

C'est seulement au second ordre qu'on commence à voir les effets de la rotation; c'est parce que le moment d'inertie à l'ordre 0 de la molécule est $Mr_0^2$. La dégénérescence angulaire est levée; les états propres angulaires sont remplacés par les états propres du moment angulaire

$$|n, l, m\rangle = \int \mathrm{d}\Omega\, Y_{lm}(\theta, \varphi)|n, \theta, \varphi\rangle, \tag{5.10}$$

et un terme de rotateur rigide s'ajoute à l'énergie. Ce sont les fameux états rotationnels de la spectroscopie moléculaire. Notez que la structure rotationnelle ne met en jeu aucune propriété de $V$ qui n'ait été utilisée dans les approximations précédentes. De plus, on commence à voir les effets de déviation de l'approximation harmonique éq. (5.9). J'ai indiqué ces termes (couplage des modes de vibration) dans la table par des points de suspension. Ils dépendent du détail de la forme de $V$ (en particulier de ses dérivées troisième et quatrième par rapport à $r_0$). A l'opposé du terme rotationnel, ils n'affectent pas les aspects qualitatifs du problème, pas plus que les termes plus élevés du développement.

L'extension de tout ceci à une molécule polyatomique est trivial. A moins que la configuration d'équilibre de la molécule ne voit tous les noyaux alignés, les configurations d'équilibre sont indexées, comme les positions d'un corps solide, par trois angles d'Euler plutôt que deux angles polaires. En conséquence, le spectre rotationnel, une fois qu'il apparaît au second ordre, sera celui d'un corps rigide plutôt que celui d'un rotateur rigide. Aussi, il y a de nombreuses façons de vibrer autour de l'équilibre, et l'entier $n$ est remplacé par une collection d'entiers $n_i$, un par mode normal.

On peut facilement transcrire tout ceci en théorie des champs; la seule question est de savoir ce qui remplace le spectre rotationnel. Les états moléculaires propres d'énergie à l'ordre zéro étaient dégénérés par rotation parce que l'état d'équilibre classique n'était pas invariant par rotation. Nous aurons des phénomènes analogues en théorie des champs chaque fois que l'état d'équilibre, le monopôle, n'est pas invariant par le groupe de symétrie non brisé de la théorie. La richesse d'un "spectre rotationnel" dépend du degré d'assymétrie du monopôle, de la multiplicité des solutions qu'on peut engendrer par application du groupe de symétrie.

Juste au minimum, la solution monopolaire n'est pas invariante par translation, de sorte que nous avons une famille de solutions indépendantes dépendant d'au moins trois paramètres, indexée par la position du centre du monopôle $\boldsymbol{r}$. J'ai construit la table 2, dans l'hypothèse où c'est la seule dégénérescence. C'est essentiellement une transcription de la table 1. Le petit paramètre $1/M$, le coefficient de l'énergie cinétique dans l'éq. (5.7) a été remplacé par le petit paramètre $f^4$, le coefficient de l'énergie cinétique dans l'éq. (5.6) et toutes les valeurs propres ont été divisées par $f^2$, puisque $f^2$ apparaît au membre de gauche de l'éq. (5.6) mais, autrement la colonne de droite est presque identique dans les deux tables.

Passons maintenant la table en revue dans le détail.

A l'ordre zéro, les états propres d'énergie sont des états propres des opérateurs de champ, avec, pour valeurs propres, les solutions classiques des équations de champ.

Analytiquement,

$$\phi^a_{\text{op}}(\boldsymbol{x})|\boldsymbol{r}\rangle = \phi^a_{\text{cl}}(\boldsymbol{x}-\boldsymbol{r})|\boldsymbol{r}\rangle, \tag{5.11}$$

où "op" indique un opérateur et "cl" la solution classique. Bien entendu, on a une équation similaire pour $\boldsymbol{A}^a$. $V_0$ est la valeur de la fonctionnelle $V$ à la solution monopolaire. On voit qu'à l'ordre dominant la masse du monopôle est proportionnelle à $1/f^2$, ce que nous savions déjà par d'autres arguments.

Table 2
Le monopôle quantique

| Ordre de l'approximation | Etat propre de l'énergie | Valeur propre de l'énergie |
|---|---|---|
| 0 | $\|\boldsymbol{r}\rangle$ | $V_0/f^2$ |
| 1 | $\|n_1, n_2 \ldots \boldsymbol{r}\rangle$ | $+\sum_i \omega_i(n_i+1/2)$ |
| 2 | $\|n_1, n_2 \ldots \boldsymbol{P}\rangle$ | $+f^2\boldsymbol{P}^2/2V_0+\cdots$ |

Au premier ordre, nous avons une somme sur les modes normaux, les modes propres des petites vibrations classiques autour de la solution monopolaire. Bien entendu, nous avons un système avec un nombre infini de degrés de liberté, de sorte que nous pouvons avoir des modes propres continus aussi bien que discrets; pour ceux-là, on doit remplacer la somme par une intégrale. Comme d'habitude quand on passe de la mécanique du point à la théorie des champs, nous réinterprètons les nombres d'excitation d'oscillateur harmonique comme les nombres d'occupation des modes normaux mésoniques. L'état où tous les $n$'s sont nuls est un monopôle isolé; les états avec des $n$'s non nuls correspondent à des états liés d'un ou plusieurs mésons dans le champ du monopôle (mode propre discret) ou à des états de diffusion dans le monopôle (état propre continu). A cet ordre, il n'y a pas de signe de l'interaction méson–méson dans l'énergie car les interactions méson–méson sont de l'ordre de $f^2$; leurs effets sont analogues au couplage des modes vibrationnels dans la molécule, et comme lui ils sont cachés derrière les points de suspension de l'énergie du second ordre.

Il y a une correction du premier ordre à la masse du monopôle, $\Sigma\frac{1}{2}\omega_i$. Cette somme est divergente, mais, au moins dans une théorie renormalisable, la différence entre cette somme et la somme correspondante pour le vide est finie et est une correction authentique à la masse du monopôle classique. Au second ordre, la dégénérescence est levée. Comme la dégénérescence est dûe à l'invariance par translation et non à l'invariance par rotation, les états propres de l'énergie ne sont pas des états propres du moment angulaire, comme dans l'éq. (5.10), mais de l'impulsion,

$$|n_1 \dots \boldsymbol{P}\rangle = \int \frac{\mathrm{d}^3 \boldsymbol{r}}{(2\pi)^{3/2}} \mathrm{e}^{\mathrm{i}\boldsymbol{P}\cdot\boldsymbol{r}} |n_1 \dots \boldsymbol{r}\rangle. \tag{5.12}$$

Par raccroc, nous connaissons la forme de l'énergie du second ordre sans avoir à faire de calcul; l'expression de la table vient simplement de

$$(\boldsymbol{P}^2 + M^2)^{1/2} = M + f^2\boldsymbol{P}^2/2V_0 + \mathrm{O}(f^4). \tag{5.13}$$

Abandonnons maintenant notre hypothèse que la seule dégénérescence est translationnelle. Il peut y avoir d'autres dégénérescences géométriques ou dégénérescence de symétrie interne.

La dégénérescence géométrique conduit à un spectre qui ressemble à celui de la physique moléculaire. Si la solution monopolaire n'est pas invariante par rotation mais admet un axe de rotation, les solutions classiques sont indexées par deux angles polaires, et, au second ordre se développe un spectre de rotateur rigide, comme dans les molécules diatomiques. Si le système n'admet aucun axe de symétrie, les trois angles d'Euler sont nécessaires et le spectre est un spectre de corps rigide comme celui d'une molécule polyatomique. D'intéressantes variations sur ce thème peuvent apparaître si la solution classique est invariante sous un sous-groupe discret de O(3). Par exemple, s'il y a à la fois un axe de symétrie de rotation et un plan orthogonal de symétrie de réflection, les moments angulaires impairs n'apparaissent pas dans le spectre.

Il apparaît une dégénérescence de symétrie interne si la solution classique n'est pas invariante par $H$, le sous-groupe non brisé de $G$. Comme toujours, les états propres de l'énergie au second ordre sont des combinaisons linéaires des états propres des opérateurs de champs. Cependant, les combinaisons linéaires se transforment maintenant selon des représentations irréductibles de $H$, plutôt que du groupe des translations ou des rotations. Le "spectre rotationnel" intéresse l'espace de symétrie interne et les "niveaux rotationnels" ont des nombres quantiques relatifs à $H$. J'appellerai ces états "niveaux isorotationnels." (Le spin est à l'isospin ce que les rotations sont aux isorotations.)

Le calcul de l'énergie des niveaux isorotationnels nécessite la connaissance de la forme de l'opérateur de l'énergie cinétique sur des fonctions d'onde restreintes à la surface des minima de $V$. C'est une généralisation de l'une des façons les plus directes pour résoudre le rotateur rigide, par l'analyse de la partie angulaire de l'opérateur de Laplace. Nous n'avons pas besoin du calcul pour voir que l'état $H$ singlet est *toujours* l'état le plus bas; cet état a une fonction d'onde constante et est toujours annihilé par n'importe quel opérateur de Laplace généralisé, quels que soient les détails de sa forme [25].

Considérons, par exemple, le monopôle de 't Hooft–Polyakov. Pour éviter la confusion dûe à la dépendance spatiale de $H$, nous nous placerons dans la jauge de corde définie dans les éqs. (4.19–21); $\mathbf{\Phi}$ est partout le long de la direction 3 et $H$ est partout le sous-groupe SO(2) des rotations autour de l'axe 3. Comme d'habitude, nous identifions ce sous-groupe avec le groupe de l'électromagnétisme. Les champs de la théorie comprennent un scalaire neutre, un vecteur neutre sans masse (le photon) et des vecteurs massifs chargés positivement et négativement. La solution monopolaire n'est invariante sous $H$ que si les champs chargés sont partout nuls. Mais c'est impossible; si tous les champs chargés étaient nuls, le champ électromagnétique satisferait aux équations de Maxwell dans le vide, et un champ monopolaire à grande distance impliquerait une singularité invariante de jauge à l'origine. Notez que cet argument est indépendant des détails du modèle; il est valable pour n'importe quel monopôle magnétique dans n'importe quelle théorie de champs spontanément brisée.

Ainsi les états propres à l'ordre zéro sont spécifiés non seulement par la position du centre, mais aussi par un nombre complexe de module un, $\exp(i\alpha)$, à savoir, la valeur du champ chargé en un point de référence.

Sous les rotations de SO(2),

$$\exp(-\mathrm{i}Q_{\mathrm{op}}\lambda)|\mathrm{e}^{\mathrm{i}\alpha}, \boldsymbol{r}\rangle = |\exp[\mathrm{i}(\alpha+e\lambda)], \boldsymbol{r}\rangle, \tag{5.14}$$

où j'ai normalisé l'opérateur de charge électrique, $Q_{\mathrm{op}}$, de telle sorte que le champ chargé ait la charge $e$. Les niveaux isorotationnels sont

$$|m, \boldsymbol{P}\rangle = \int_0^{2\pi} \frac{\mathrm{d}\alpha}{\sqrt{2\pi}} \mathrm{e}^{-\mathrm{i}m\alpha} \int \frac{\mathrm{d}^3\boldsymbol{r}}{(2\pi)^{3/2}} \mathrm{e}^{-\mathrm{i}\boldsymbol{P}\cdot\boldsymbol{r}} |\mathrm{e}^{\mathrm{i}\alpha}, \boldsymbol{r}\rangle, \tag{5.15}$$

où $m$ est un entier et où les nombres quantiques vibrationnels sont omis. Ce sont des états propres de la charge aussi bien que du moment

$$Q_{\mathrm{op}}|m, \boldsymbol{P}\rangle = me|m, \boldsymbol{P}\rangle. \tag{5.16}$$

Le monopôle a fabriqué des dyons [26].

J'insiste sur le fait que ce ne sont pas des dyons fondamentaux; ils sont inévitables et non "au choix," et leurs propriétés sont calculables, non ajustables. Ce ne sont pas non plus des états liés d'un monopôle et d'un méson chargé; leur énergie d'excitation est proportionnelle à $f^2$, tandis que la masse du méson est O(1). Pour la même raison, les dyons sont stables, au moins jusqu'à une masse très élevée $m$; ils ne peuvent pas se désexciter en émettant un méson chargé.

Seule la dernière de ces affirmations n'est pas vraie en général. Dans une théorie où il y a une hiérarchie de brisures spontanées et d'échelles de masse, $f^2$ fois une grande échelle de masse peut encore être beaucoup plus grande que la masse des particules légères de la théorie. Si ces particules légères ont les bons nombres quantiques, les dyons peuvent tous se désintégrer dans le monopôle fondamental.

### *5.2. L'effet Witten*

Il y a un célèbre terme de violation de $CP$ qui peut être ajouté au Lagrangien d'une théorie de jauge, le terme $\theta$,

$$\mathfrak{L}' = \frac{\theta}{32\pi^2} \varepsilon^{\mu\nu\lambda\sigma} F^a_{\mu\nu} F^a_{\lambda\sigma}, \tag{5.17}$$

où $\theta$ est un nombre réel, et les champs sont normalisés comme dans la section 3.1. Ce terme est une divergence, de sorte qu'il est sans effet sur les équations du mouvement; néanmoins il a des effets profonds sur la physique de la théorie. Son origine appartient à la théorie des instantons que je n'ai pas l'intention de passer en revue ici. Cependant, j'aurai besoin d'un résultat de cette théorie: $\theta$ est un angle, c'est à dire que tous les phénomènes physiques sont des fonctions périodiques de $\theta$, de période $2\pi$.

Witten a montré que pour les monopôles Abéliens le terme $\theta$ a une influence frappante sur le spectre des dyons [27]. Je vais donner ici une version de l'effet Witten qui est indépendante de la dynamique des champs lourds à l'intérieur du monopôle. Les champs lourds sont seulement importants en ce qu'ils assurent l'existence du monopôle; tout ce qui arrive après met uniquement en jeu les champs en dehors du monopôle, c'est à dire les champs électromagnétiques. Si j'en avais eu le courage, j'aurais pu discuter l'effet Witten dans la section 2. Ecrivons le terme $\theta$ en ignorant complètement les champs lourds. Autrement dit, on remplace $F^a_{\mu\nu}$ par le seul champ $F_{\mu\nu}$, où

$$F_{0i} = eE_i, \qquad F_{ij} = e\varepsilon_{ijk}B_k, \tag{5.18}$$

et $\boldsymbol{E}$ et $\boldsymbol{B}$ sont les champs électrique et magnétique conventionnellement normalisés.

En fonction de $\boldsymbol{E}$ et $\boldsymbol{B}$, on a

$$\mathfrak{L}' = \frac{\theta e^2}{4\pi^2}\boldsymbol{E}\cdot\boldsymbol{B}. \tag{5.19}$$

Ecrivons maintenant ces champs comme des champs électromagnétiques ordinaires (que nous supposerons statiques pour plus de simplicité), plus un fond monopolaire:

$$\boldsymbol{E} = \nabla\Phi, \qquad \boldsymbol{B} = \nabla\times\boldsymbol{A} + g\boldsymbol{r}/r^3. \tag{5.20}$$

où $\boldsymbol{A}$ et $\Phi$ sont les potentiels vecteur et scalaire conventionnels. (Nous généraliserons sous peu à un fond de dyon).

On trouve

$$L' = \int \mathrm{d}^3\boldsymbol{r}\,\mathfrak{L}' = \frac{e^2 g\theta}{\pi}\int \mathrm{d}^3\boldsymbol{r}\,\Phi(\boldsymbol{r})\delta^{(3)}(\boldsymbol{r}), \tag{5.21}$$

après une intégration par parties. C'est le couplage standard du potentiel scalaire avec une charge électrique $e^2 g\theta/\pi$ localisée au monopôle. En présence du terme $\theta$ la charge magnétique induit une charge électrique.

On peut grossièrement comprendre ce qui se passe. Si on avait ajouté un terme $\boldsymbol{E}\cdot\boldsymbol{E}$ à la densité Lagrangienne, cela aurait représenté une constante diélectrique, un terme qui permet à une charge électrique donnée d'induire une charge électrique additionnelle (typiquement, un effet d'écran). Un terme $\boldsymbol{B}\cdot\boldsymbol{B}$ aurait un effet similaire sur la charge magnétique. Il n'est pas surprenant qu'un terme $\boldsymbol{E}\cdot\boldsymbol{B}$ décrive l'induction d'une charge électrique par une charge magnétique.

Le problème avec ce raisonnement est que l'effet n'est pas symétrique; si on introduit un fond électrique monopolaire, sa contribution à $L'$ s'annule par intégration par parties. Cependant cela permet facilement l'extension aux dyons. En présence du terme $\theta$ la charge des dyons est donnée par

$$Q = e(m + eg\theta/\pi), \tag{5.22}$$

avec $m$ entier. Ceci viole $CP$, mais pas la condition de quantification [voir la discussion après l'éq. (2.22)].

Comme $eg$ est toujours entier ou demi-entier la série initiale de charges se reproduit quand $\theta$ augmente de $2\pi$, chaque terme augmentant de $2eg$. Avec les méthodes utilisées ici, on ne peut rien dire sur la dépendance en $\theta$ de l'énergie des dyons, mais Witten a étudié l'intérieur des monopôles et il nous dit que lorsqu'un dyon en a remplacé un autre, son énergie devient celle appropriée à la nouvelle charge. En un mot, $\theta$ est un angle; la physique est périodique en $\theta$ de période $2\pi$. Ce résultat est charmant, d'autant que les instantons n'ont jamais été utilisés dans l'argumentation.

### *5.3. Davantage sur les monopôles de SU(5)*

Plus haut dans ces conférences, j'ai fait quelques commentaires sur les monopôles de la théorie SU(5) de Georgi–Glashow. Nous allons maintenant regarder ces objets [28], d'un peu plus près. Il y a à cela deux raisons. D'abord la théorie SU(5) est intéressante en elle-même. C'est la plus simple d'une famille de Théories Grand-Unifiées dont l'une pourrait bien décrire la réalité, au moins aux énergies inférieures à la masse de Planck. Deuxièmement, la théorie est suffisamment riche en structure pour servir de bon exemple au jeu des idées fondamentales de la théorie des monopôles.

Je commencerai par résumer les aspects significatifs de la théorie. La théorie est une théorie de champs de jauge avec groupe de jauge SU(5). C'est un groupe simple, de sorte qu'il n'y a qu'une constante de couplage $f$; $f$ est de l'ordre de $e$, le couplage électromagnétique usuel. (Ce n'est pas exactement $e$ à cause de coefficients de Clebsch–Gordan et d'effets de renormalisation quand on passe des grandes distances où $e$ est défini à $10^{-28}$ cm, où vivent les monopôles.) Tout ce qu'on a besoin de savoir des champs scalaires de la théorie est que leurs interactions mettent en jeu deux échelles de masse, ou, de façon équivalente, deux échelles de valeur moyenne scalaire dans le vide.

La grande valeur moyenne brise SU(5) en $[\mathrm{SU}(3)\times\mathrm{SU}(2)\times\mathrm{U}(1)]/Z_6$. Si on réalise SU(5) par des matrices $5\times5$, SU(3) consiste dans les transformations des trois premières coordonnées, SU(2), des deux dernières, U(1), des matrices diagonales de la forme

$$\mathrm{diag}(\mathrm{e}^{2\mathrm{i}\theta},\mathrm{e}^{2\mathrm{i}\theta},\mathrm{e}^{2\mathrm{i}\theta},\mathrm{e}^{-3\mathrm{i}\theta},\mathrm{e}^{-3\mathrm{i}\theta}). \tag{5.23}$$

Si $\mathrm{e}^{6\mathrm{i}\theta}=1$, c'est un élément de $\mathrm{SU}(3)\times\mathrm{SU}(2)$; c'est la raison pour laquelle on doit prendre le quotient du produit direct par $Z_6$. Le sous-groupe SU(3) est identifié au groupe de couleur; $\mathrm{SU}(2)\times\mathrm{U}(1)$ est identifié au groupe du modèle de Weinberg–Salam. Il résulte de cette brisure de symétrie que douze des vingt-quatre champs de jauge initiaux acquièrent des masses de l'ordre de $10^{14}$ GeV. (C'est seulement à cette échelle de masse élevée que le couplage de jauge de couleur est comparable en intensité aux couplages électromagnétiques.) Ces mésons superlourds se transforment selon la représentation $(3,\bar{2})\oplus(\bar{3},2)$ de $\mathrm{SU}(3)\times\mathrm{SU}(2)$.

La valeur moyenne la plus petite brise le groupe encore une fois, jusqu'à SU(3) (couleur)$\times$U(1) (électromagnétisme); il résulte de cette brisure de symétrie que trois champs de jauge acquièrent des masses de l'ordre de $10^2$ GeV. Le groupe électromagnétique est engendré par

$$Q_{\mathrm{em}} = \mathrm{i}\,\mathrm{diag}(\tfrac{1}{3},\tfrac{1}{3},\tfrac{1}{3},-1,0), \tag{5.24}$$

où le générateur a été normalisé de sorte que la charge du proton soit l'unité. Les facteurs 1/3 signalent que la théorie contient des particules de charge fractionnaire: les quarks, bien entendu, mais aussi les mésons vectoriels superlourds. A cause de ces facteurs,

$$\exp(2\pi Q_{\mathrm{em}}) \neq 1. \tag{5.25}$$

Nous devons continuer trois fois plus loin,

$$\exp(6\pi Q_{\mathrm{em}}) = 1. \tag{5.26}$$

Cependant $\exp(2\pi Q_{\mathrm{em}})$ est une matrice de SU(3), de sorte que le groupe $H$ n'est pas $\mathrm{SU}(3)\times\mathrm{U}(1)$ mais $\mathrm{SU}(3)\times\mathrm{U}(1)/Z_3$. Ceci est simplement une façon sophistiquée de dire que seules les particules de trialité non nulle ont une charge fractionnaire (cf. l'exemple (e) à la fin de la section 3.4).

En plus de ces symétries de jauge, la théorie possède une symétrie interne globale qui, une fois la poussière de la brisure de symétrie retombée, conduit à la conservation de la différence des nombres de baryons et de leptons, $B - L$. ($B$ et $L$ ne sont pas séparément conservés; la théorie est bien connue pour sa prédiction de la désintégration du proton).

Nous savons déjà des choses sur cette théorie. Nous savons qu'il y a des monopôles (section 4.2), que leur masse est de l'ordre de $10^{16}$ GeV, que leur coeur a une taille de l'ordre de $10^{-28}$ cm (section 4.4), et qu'il y a un spectre d'excitations isorotationnelles, des dyons, avec une séparation de niveaux de l'ordre de $10^{12}$ GeV (section 5.1).

Regardons maintenant les monopôles avec plus de détail. Aux grandes distances le champ monopolaire doit avoir la forme standard,

$$\boldsymbol{A} = Q\boldsymbol{A}_{\mathrm{D}}, \tag{5.27}$$

où $Q$ est un générateur de SU(3)×U(1). La première pensée serait que, puisque la théorie contient des particules de charge fractionnaire, $g = \frac{1}{2}e$ n'est pas autorisé et qu'on doit avoir $g = \frac{3}{2}e$. En fait, $Q = Q_{\mathrm{em}}/2$ ne satisfait pas à la condition de quantification:

$$Q = 3Q_{\mathrm{em}}/2 \tag{5.28a}$$

est nécessaire comme on le voit à partir des éq. (5.25) et (5.26).

(Laissez-moi exhorter et apprivoiser un spectre. Toutes les particules de charge fractionnaire sont confinées; par exemple, on ne peut pas séparer les composantes de paires quark–antiquark de plus qu'environ $10^{-13}$ cm. Ne faisons-nous pas une erreur en utilisant des particules confinées pour établir la condition de quantification? Il n'en est rien. La corde de Dirac est infiniment fine et infiniment longue. On peut imaginer approcher la paire à $10^{-14}$ cm de la corde, à un millier d'années-lumières du monopôle, et diffracter le quark autour de la corde en tenant l'antiquark fixe. Le confinement ne joue aucun rôle ici.)

Il y a cependant une seconde possibilité [29]

$$Q = (Q_{\mathrm{em}} + Q_Y)/2 \tag{5.28b}$$

où $Q_y$ est le générateur de l'hypercharge de couleur

$$Q_Y = -\mathrm{i}\,\mathrm{diag}(\tfrac{2}{3}, -\tfrac{1}{3}, -\tfrac{1}{3}, 0, 0). \tag{5.29}$$

Ainsi

$$Q = -\mathrm{i}\,\mathrm{diag}(1, 0, 0, -1, 0)/2, \tag{5.30}$$

et la condition de quantification est satisfaite.

C'est une combinaison d'un monopôle électromagnétique et d'un monopôle chromomagnétique. Les particules sans couleur, telles que les leptons et les hadrons, voient seulement le champ électromagnétique et trouvent $g = \frac{1}{2}e$. Cependant, comme toutes les particules de charge fractionnaire sont colorées, une particule de charge fractionnaire voit à la fois le champ électromagnétique et le champ chromomagnétique, et l'effet combiné de ces deux champs rend la corde indétectable.

Bien entendu, (5.28a) et (5.28b) ne sont pas les seuls champs monopolaires possibles. A une transformation de jauge près, la solution de la condition de quantification et de la condition de stabilité non Abélienne est:

$$Q = -\mathrm{i}\,\mathrm{diag}(r, s, s, -r-2s, 0)/2, \tag{5.31}$$

avec $r$ et $s$ entiers tels que

$$r - s = 0, \pm 1. \tag{5.32}$$

Comme SU(3) est simplement connexe, la seule charge topologique est la charge magnétique Abélienne $r + 2s$.

On peut estimer l'énergie de ces configurations par la méthode de la section 4.4. L'énergie emmagasinée dans le champ en dehors du coeur est proportionnelle à

$$-\mathrm{Tr}\, Q^2 = s^2 + \tfrac{1}{2}(r+s)^2. \tag{5.33}$$

Le monopôle non trivial d'énergie la plus basse est $r = 1, s = 0$; c'est le monopôle combiné, éq. (5.28b). Si on prend l'éq. (5.33) au sérieux pour estimer l'énergie, il est facile de voir que n'importe quel autre monopôle non trivial a une énergie largement suffisante pour se désintégrer en un nombre approprié de monopôles combinés. Par exemple, le monopôle électromagnétique pur (éq. 5.28a) avec $r = s = 1$ a une énergie six fois plus grande que celle du monopôle combiné et est donc instable vis-à-vis de la désintégration en trois tels objets, le nombre minimum compatible avec la conservation de la charge topologique.

(Un autre spectre: dans notre estimation, nous avons négligé les constantes de couplage. Ne faisons-nous pas une erreur en traitant la constante de couplage de couleur comme si c'était la constante de couplage électromagnétique? Il n'en est rien. Les deux couplages sont en fait très différents à grande distance, mais identiques à $10^{-28}$ cm, près du coeur du monopôle, région qui domine l'intégrale d'énergie.)

Dokos et Tomaras [28] ont construit un monopôle combiné qui est une solution statique des équations de champs; leur méthode est une transposition de la construction de 't Hooft–Polyakov à un sous-groupe approprié SU(2) de SU(5).

Tout ce que nous avons besoin de savoir sur cette solution est sa dégéneréscence, car cela suffit pour déterminer le spectre des niveaux isorotationnels. Il se trouve que l'un des champs vectoriels supermassifs, $X^a$ $(a=1,2,3)$ est non nul dans la construction de Dokos–Tomara. Ce champ a une couleur 3, une charge électrique $-4/3$, et $B-L=-2/3$. De plus, une fois sa valeur connue au point de référence, la solution est déterminée de façon unique.

Comme tout trois-vecteur complexe peut se transformer en n'importe quel autre de même longueur par une matrice de SU(3), la multiplicité des solutions est en correspondance biunivoque avec l'ensemble de tous les trois-vecteurs unitaires,

$$X^a\overline{X}_a = 1, \tag{5.34}$$

et le problème de la construction des niveaux isorotationnels est le même que celui de la construction des fonctions de ces vecteurs qui se transforment suivant les représentations irréductibles du groupe de symétrie.

C'est facile à faire. Un ensemble complet de fonctions sur la variété est formé de tous les monômes en $X$ et $\overline{X}$. On peut les écrire comme des tenseurs

$$X^{a_1\dots a_n}_{b_1\dots b_m} = X^{a_1}\dots X^{a_n}\overline{X}_{b_1}\dots\overline{X}_{b_m}. \tag{5.35}$$

Ces tenseurs sont presque les objets qui forment l'espace de base de la représentation irréductible de SU(3) appelée $(n,m)$. La seule différence est que les tenseurs irréductibles sont sans traces. Cependant, il est immédiat de soustraire la trace de ces expressions. En vertu de l'éq. (5.34), les tenseurs de faible rang obtenus en prenant des traces ne sont pas de nouvelles fonctions, mais seulement des objets déjà construits comme monômes de degrés plus bas; le calcul de la charge électrique et de $B-L$ pour ces tenseurs est trivial: on obtient juste $n-m$ fois la contribution d'un $X$.

Ainsi: les niveaux isorotationnels se transforment comme la représentation $(n,m)$ de SU(3) couleur, chaque représentation apparaissant une et une seule fois. Ces niveaux ont une charge électrique $4(n-m)/3$ et un $B-L$ de $2(m-n)/3$. Notez le couplage entre l'excitation de charge et l'excitation de couleur. Ce sont les chromodyons.

Regardons de plus près la physique des chromodyons. Pour commencer, j'ignorerai l'existence de fermions.

Etant donné deux niveaux avec la même valeur de $m-n$, le plus élevé peut toujours se désintégrer dans le plus bas par émission de gluons colorés sans masse. (Le fait que les gluons sont confinés est sans importance. Ils sont confinés à $10^{-13}$ cm et la désintégration a lieu à $10^{-28}$ cm. Se soucier des effets du confinement des gluons sur la désintégration d'un chromodyon revient à peu près au même que de se soucier des effets des murs du laboratoire sur la désintégration d'un noyau radioactif). Ainsi, nous nous attendons à un seul niveau stable pour chaque valeur de $m - n$. Ces niveaux ne peuvent pas se désintégrer par émission de mésons vectoriels superlourds, parce que leur énergie d'excitation est trop faible; ils ne peuvent se désintégrer par l'émission de gluons colorés parce qu'ils n'ont pas de charge; ils ne peuvent se désintégrer par émission de mésons de jauge de Weinberg–Salam parce qu'ils n'ont pas de couleur; ils sont stables.

Bien qu'ils soient stables, ils sont colorés (excepté pour le monopôle fondamental). Ainsi, un $(n, m)$ chromodyon et un $(m, n)$ chromodyon forment une paire sans couleur liée par les forces confinantes. Le sort de cette paire dépend de l'interaction entre ses composantes à courte distance. (Néanmoins, pas assez courte pour que les coeurs s'interpénètrent).

L'interaction magnétique est dominante. Si les composantes ont des charges magnétiques Abéliennes opposées, c'est à dire, si ce sont des excitations de monopôle et d'anti-monopôle, la force magnétique est attractive et les composantes s'annihilent l'une l'autre.

On a besoin d'un calcul rapide si les composantes ont la même charge Abélienne; il faut se soucier de la compétition entre la répulsion Abélienne et l'attraction non Abélienne, ainsi que nous l'avons expliqué dans la section 3.6. Si on représente le champ d'un monopôle par la matrice $Q_1$ et l'autre par la matrice $Q_2$, l'énergie d'intéraction magnétique est proportionnelle à $-\operatorname{Tr} Q_1 Q_2 / r_{12}$. On peut toujours choisir la jauge de sorte que

$$Q_1 = -\mathrm{i}\,\mathrm{diag}(1, 0, 0, -1, 0)/2. \tag{5.36}$$

$Q_2$ peut être n'importe quelle matrice déduite de la précédente par une permutation des trois premiers éléments (éléments de couleur). On minimise l'énergie d'interaction en choisissant

$$Q_2 = -\mathrm{i}\,\mathrm{diag}(0, 1, 0, -1, 0)/2, \tag{5.37}$$

mais, même au minimum, l'énergie est encore positive. La répulsion Abélienne l'emporte sur l'attraction non Abélienne.

Ainsi, nous avons un système composé de deux particules en interaction attractive à grande distance, répulsive à courte distance. C'est à nouveau une molécule diatomique, pas par analogie, cette fois-ci, mais bien comme un objet réel avec toute la panoplie des états vibrationnels et rotationnels, bien qu'à une échelle d'énergie beaucoup plus grande que d'habitude en physique moléculaire.

Mais tout ceci est une simple fantaisie. Dans le monde réel, il y a des fermions légers, quarks et leptons, et tous les chromodyons peuvent se désintégrer vers le fondamental en émettant des paires de quarks et leptons. Quel dommage.

### *5.4. Renormalisation de la charge magnétique Abélienne*

Il y a quelques années, dans une série d'articles fameux, Schwinger généralisa l'électrodynamique quantique pour inclure des monopôles magnétiques fondamentaux [30]. Au cours de ce travail, il étudia la renormalisation de la charge magnétique. Bien que les monopôles fondamentaux ne soient pas l'objet de notre premier intérêt, je vais passer en revue les résultats de Schwinger ici en vue d'une comparaison ultérieure avec les nôtres.

Schwinger trouva une condition nécessaire pour la cohérence de la quantification de la théorie:

$$e_0 g_0 = n_0/2, \tag{5.38}$$

où $e_0$ et $g_0$ sont les constantes de couplage nues, les paramètres apparaissant dans le Lagrangien de la théorie, et $n_0$ est un entier. (En fait, Schwinger avait d'abord proposé que $n_0$ soit pair, mais ce point est sans importance pour le débat et je l'ignorerai ici.)

Les raisonnements standards conduisant à la condition de quantification de Dirac n'ont rien à faire avec le caractère fondamental ou composite des monopôles; ils nous disent que

$$eg = n/2, \tag{5.39}$$

où $e$ et $g$ sont les constantes de couplage physiques et $n$ un entier.

L'analyse de la renormalisation de charge électrique se développe de façon très analogue à l'électrodynamique. Comme il y a des courants magnétiques aussi bien que des courants électriques, le propagateur du champ électromagnétique renormalisé s'exprime comme une intégrale mettant en jeu trois fonctions spectrales au lieu d'une; pourtant, il n'y a toujours qu'un photon et qu'un terme de pôle à un photon. Il n'y a par conséquent pas de problème à suivre le développement habituel et à définir $Z_3$ comme le coefficient de ce terme et à montrer que

$$e = Z_3^{1/2} e_0. \tag{5.40}$$

Comme d'habitude, la représentation spectrale implique que si la théorie est non triviale, $Z_3$ est strictement plus petit que 1.

L'analyse de la renormalisation de charge magnétique s'obtient trivialement en réalisant la dualité de rotation qui échange $\boldsymbol{E}$ et $\boldsymbol{B}$. Les fonctions de poids spectrales s'interchangent mais le photon reste le photon. Par conséquent $Z_3$ reste inchangé et

$$g = Z_3^{1/2} g_0. \tag{5.41}$$

d'où

$$eg = Z_3 e_0 g_0. \tag{5.42}$$

C'est l'origine de l'affirmation de Schwinger que $Z_3$ est un nombre rationnel.

Ceci est fondamental pour les monopôles. Cependant, nous nous intéressons aux monopôles composés, comme l'objet de 't Hooft–Polyakov. Est-ce que l'éq. (5.42) reste vraie dans ce cas? Evidemment non, mais ce n'est pas non plus faux; ça n'a pas de sens. Dans une théorie avec des monopôles composés, $e_0$ apparaît encore comme paramètre dans le Lagrangien, mais on ne trouve plus $g_0$ nulle part.

Néanmoins tout n'est pas perdu. Si on définit $e_\lambda$ comme la quantité qui décrit l'intéraction d'un électron avec une petite charge d'essai à la distance $\lambda$, alors, $e_0$ est la limite de $e_\lambda$ quand $\lambda$ tend vers zéro. Ceci suggère que nous définissions $g_\lambda$ de façon analogue, comme le paramètre qui décrit l'intéraction entre un monopôle et une petite boucle de courant à la distance $\lambda$. Dans ce cas, cependant, on ne peut faire tendre $\lambda$ vers zéro; dès que $\lambda$ est plus petit que la taille du coeur du monopôle, il n'y a pas de façon non-ambiguë de séparer l'électromagnétisme de la plus grande structure non Abélienne dans laquelle il est plongé, pas de façon non-ambiguë de définir une boucle de courant. Cependant, jusqu'à ce qu'on atteigne le coeur, il n'y a pas de problème. Aussi, on peut se demander comment $e_\lambda$ et $g_\lambda$ sont reliés pour $\lambda$ plus grand que le rayon du coeur à $e$, et $g$, leurs limites à grande distance. Cette question n'est pas la même que celle posée par Schwinger, mais elle est intéressante; il y a plein de place pour de gros effets de renormalisation entre la distance infinie et le coeur du monopôle.

Cependant, il n'y a pas de place pour les effets de renormalisation causés par les monopôles virtuels. Pour de faibles couplages, la longueur d'onde de Compton est beaucoup plus petite que la taille géométrique des monopôles. (Autrement dit, les monopôles sont beaucoup plus lourds que les mésons vecteurs massifs.) Aussi, aux échelles de distance auxquelles nous travaillons, entre $\lambda$ et l'infini, les effets des monopôles virtuels sont négligeables. (J'insiste sur le fait que "couplage faible" ne signifie pas ici "ordre le plus bas en $e$". Si on raisonne en fonction d'un développement en puissances de $\hbar$, les effets des monopôles virtuels sont des effets de pénétration de barrière qui sont exponentiellement supprimés.) Ainsi, pour notre propos, il est tout à fait légitime de remplacer la théorie complète par une théorie simplifiée dans laquelle les seules variables dynamiques sont le champ électromagnétique et les champs des particules chargées ordinaires.

Pour simplifier les notations, nous nous restreindrons au cas où le seul champ chargé est celui d'un électron de Dirac; la généralisation est triviale. La théorie que nous allons étudier est définie par

$$\begin{aligned}\mathcal{L} = & -\tfrac{1}{4}(\partial_\mu A_\nu - \partial_\nu A_\mu)^2 + \frac{1}{2\alpha}(\partial_\lambda A^\lambda)^2 \\ & + \bar{\psi}(i\not{\partial} - e_\lambda \not{A} - e_\lambda g_\lambda \not{A}_{\mathrm{D}} - m_0)\psi + e_\lambda J_\mu(A^\mu + g_\lambda A^\mu_{\mathrm{D}}). \quad (5.43)\end{aligned}$$

Cette expression requiert quelques explications. D'abord pour ce qui est des notations: tous les champs sont non renormalisés, $\alpha$ est le paramètre usuel qui fixe la jauge, $A^\mu_{\mathrm{D}} = (0, \boldsymbol{A}_{\mathrm{D}})$ est le potentiel du monopôle de Dirac et $J^\mu$ est un courant conservé classique,

$$\partial_\mu J^\mu = 0, \quad (5.44)$$

que nous utiliserons pour construire des charges et des boucles de courant d'essai. Deuxièmement, pour ce qui est de la physique: ceci n'est supposé contenir que les degrés de liberté qui sont significatifs à des distances plus grandes que $\lambda$. Ainsi, bien que ce ne soit pas indiqué explicitement, la théorie est supposée régularisée—par exemple par des champs régulateurs—de façon invariante de jauge, à une distance de l'ordre de $\lambda$. C'est pour cette raison que $e_\lambda$ et $g_\lambda$ occupent la place des constantes de couplage nues et que le monopôle est remplacé par un potentiel de Dirac extérieur.

J'insiste sur le fait que l'éq. (5.43) n'est pas destinée à donner une description exacte de la théorie à des distances plus grandes que $\lambda$. Il est certainement possible de construire un Lagrangien effectif qui permettrait une telle description, mais ce serait beaucoup plus compliqué, rempli d'interactions non locales et non polynomiales. L'équation (5.43) vise simplement à donner un modèle simple de la physique qui nous concerne.

Des arguments qui devraient maintenant être excessivement familiers montrent que la singularité de corde du potentiel monopolaire est indétectable à condition que $e_\lambda g_\lambda$ soit demi-entier. Je supposerai donc qu'il en est ainsi. Il n'est pas clair du tout à ce stade qu'il s'ensuive que $eg$ soit demi-entier, mais nous avons un Lagrangien bien défini qui nous permet d'avancer et de calculer.

Pour nous mettre entrain, assurons-nous que notre équipement d'essai est adéquatement calibré, en calculant les effets du courant extérieur loin du monopôle, ou, de façon équivalente, pour $g_\lambda$ égal à zéro. Dans ce cas, l'élément de matrice de transition vide–vide est

$$\langle 0|S|0\rangle = 1 - \tfrac{1}{2}e_\lambda^2 \int \mathrm{d}^4x\,\mathrm{d}^4y\,J^\mu(x)J^\nu(y)T\langle 0|A_\mu(x)A_\nu(y)|0\rangle + \mathrm{O}(J^4). \tag{5.45}$$

J'insiste sur le fait qu'il ne s'agit pas d'un développement (illégitime) en puissance de $e_\lambda$; c'est un développement en puissance de $J^\mu$. Il n'y a rien de mal à cela; les intéractions avec $J^\mu$ sont invariantes de jauge et par conséquent incapables de détecter la corde de Dirac, quelle que soit la grandeur de $J_\mu$, pour autant qu'il soit conservé. (Ceci n'est qu'une autre façon de dire qu'il n'y a pas de condition de quantification pour des particules chargées classiques.)

L'objet important dans l'éq. (5.45) est le propagateur du champ non renormalisé,

$$T\langle 0|A_\mu(x)A_\nu(y)|0\rangle = Z_3^\lambda D_{\mu\nu}^0(x-y) + \cdots, \tag{5.46}$$

où $D_{\mu\nu}^0$ est le propagateur libre, les points de suspension indiquent les termes qui décroissent à grande distance plus vite que le terme explicité, et l'indice de $Z_3$ rappelle sa dépendance dans le paramètre de coupure.

Ainsi, deux courants interagissent à grande distance comme ils le feraient dans l'électromagnétisme pur avec une force d'interaction donnée par

$$e_\lambda^2 Z_3^\lambda = e^2. \tag{5.47}$$

C'est le résultat correct fameux.

Retournons maintenant à $g_\lambda$ non nul. Il y a maintenant un terme linéaire dans l'élément de matrice de transition,

$$\langle 0|S|0\rangle = 1 - \mathrm{i}e_\lambda \int \mathrm{d}^4x\,J^\mu(x)\left[g_\lambda A_\mu^{\mathrm{D}}(x) + \langle 0|A_\mu(x)|0\rangle\right] + \mathrm{O}(J^2). \tag{5.48}$$

J'insiste sur le fait que $|0\rangle$ est maintenant l'état fondamental de la théorie calculé à l'ordre zéro en $J^\mu$ mais à tous ordres dans le champ monopolaire externe. Nous voulons calculer $eg$ de la même façon que nous avons calculé $e^2$ dans le paragraphe précédent. Ceci est faisable pourvu que la totalité de l'intégrale dans l'éq. (5.48) ait la même forme que son premier terme, lorsque le support du courant est éloigné du monopôle; nous pouvons alors identifier $eg$ avec le coefficient de toute l'expression. Analytiquement,

$$\langle 0|S|0\rangle - 1 \underset{\text{grande distance}}{\overset{?}{\rightarrow}} -\mathrm{i}eg\int \mathrm{d}^4x\, J^\mu(x) A^{\mathrm{D}}_\mu(x) + \mathrm{O}(J^2). \quad (5.49)$$

J'ai mis un point d'interrogation car nous ne savons pas encore si nous obtiendrons la forme correcte dans cette limite. (Espérons que ce sera le cas; sinon nous sommes vraiment en difficulté.)

L'équation du mouvement pour $A$ implique

$$\langle 0|A_\mu(x)|0\rangle = -\mathrm{i}e_\lambda \int \mathrm{d}^4y\, D^0_{\mu\nu}(x-y)\langle 0|J^\nu(y)|0\rangle. \quad (5.50)$$

L'objet du membre de droite est l'élément de matrice d'un opérateur invariant de jauge; ainsi, en dépit de la corde de Dirac, il est invariant par rotation et par $CP$. (Notez que ce ne serait pas vrai si nous avions bêtement essayé de développer en puissance du champ monopolaire.)

En vertu de l'invariance par rotation

$$\langle 0|\bar{\psi}\gamma^0\psi|0\rangle = f(r^2), \quad (5.51)$$

$$\langle 0|\bar{\psi}\gamma^i\psi|0\rangle = r^i g(r^2), \quad (5.52)$$

où $f$ et $g$ sont deux fonctions. L'invariance par $CP$ implique que $f$ est nulle. L'équation de conservation,

$$\partial_\mu \langle 0|\bar{\psi}\gamma^\mu\psi|0\rangle = 0. \quad (5.53)$$

implique que $g$ est nulle. Ainsi, (5.49) est satisfaite trivialement et

$$eg = e_\lambda g_\lambda. \quad (5.54)$$

Ceci contraste violemment avec l'éq. (5.42), mais, comme je l'ai expliqué la contradiction n'est qu'apparente. En dépit de l'apparition de symboles similaires, les deux équations répondent à des questions différentes dans des théories différentes.

L'équation (5.54) est très satisfaisante. La théorie des monopôles dit que $eg$ doit être demi-entier, mais un théoricien des champs moderne demanderait "$eg$ défini à quelle échelle de distances?" La réponse est "à n'importe quelle échelle de distances jusqu'au coeur du monopôle. Cà n'a aucune importance".

### 5.5. *Les effets du confinement de la charge magnétique non Abélienne*

Nous avons analysé la renormalisation de la charge magnétique Abélienne, calculé le champ électromagnétique monopolaire à grandes distances en fonction du champ électromagnétique monopolaire exclusivement en dehors du coeur du monopôle. Pour complèter l'analyse, il nous faudrait maintenant étendre l'argument à la charge magnétique non Abélienne, et calculer, par exemple, le champ chromomagnétique monopolaire à grande distance des monopôles Grand-Unifiés de la section 5.3. Mais un tel calcul serait pure folie. Nous savons très bien qu'il n'y a pas de champ chromomagnétique monopolaire à grande distance; il n'y a pas de champ chromodynamique d'aucune sorte proportionnel à une puissance inverse de la distance. A cause du confinement, toutes les forces chromodynamiques décroissent exponentiellement avec la distance, avec un coefficient donné par la masse hadronique la plus légère. (Si nous ignorons les fermions, comme nous l'avons fait, ce serait la masse de la collemâne*.)

Ceci conduit à un paradoxe apparent. Pour les monopôles de Grande Unification l'existence simultanée de monopôles chromomagnétiques et électromagnétiques était nécessaire pour que la corde de Dirac reste indétectable; au cours de la diffraction d'une particule de charge fractionnaire autour de la corde, le facteur de phase chromomagnétique était nécessaire à la compensation du facteur de phase électromagnétique. (Je vous rappelle qu'à ce point le confinement des particules de charge fractionnaire est inopérant, comme nous l'avons montré, section 5.3.) Mais, à grande distance le champ chromomagnétique monopolaire disparaît, ce qui n'est pas le cas pour le champ électromagnétique. Est-ce à dire que la corde de Dirac est inobservable à grande distance?

Je pose cette question, entraîné par le courant de ma propre rhétorique. Bien entendu, la corde ne peut pas devenir observable; le confinement n'altère pas l'invariance de jauge. Néanmoins, nous aimerions comprendre comment des effets qui décroissent exponentiellement avec la distance peuvent néanmoins produire un facteur de phase indépendant de la distance. Le reste de cette section est consacré à ce problème.

*N.d.T. Traduction libre de "glueball."

Bien que le groupe approprié pour les monopôles de Grande Unification soit $SU(3)\times U(1)/Z_3$, je travaillerai ici pour simplifier avec SO(3), le plus petit groupe non Abélien. L'extension est triviale. Je donnerai d'abord des arguments généraux et les étofferai ensuite à l'aide de calculs explicites dans un modèle ultra simplifié mais exactement soluble à deux dimensions.

La construction de Lubkin associe un chemin dans la variété du groupe, $g(\tau)$ à toute sphère de l'espace ordinaire; la classe d'homotopie de ce chemin nous indique si la sphère contient un monopôle. Bien que la classe d'homotopie soit invariante de jauge, le chemin lui-même ne l'est pas. Quand on étudie des questions épineuses en théorie des champs de jauge, il est sage de travailler autant que possible avec des entités invariantes de jauge. Je travaillerai donc avec le facteur de boucle de Wilson

$$W(\tau)=\tfrac{1}{2}\operatorname{Tr} g(\tau). \tag{5.55}$$

Bien que le facteur de boucle soit indépendant de la jauge, il dépend de la représentation matricielle utilisée pour réaliser le groupe. Pour notre problème, il s'avérera avantageux de choisir la représentation de spin $1/2$ de SO(3), c'est à dire de choisir $g$ comme un élément de SU(2). Il s'agit d'une représentation bivaluée de sorte qu'il y a une ambiguité dichotomique dans le calcul de la boucle de Wilson. Nous éliminerons cette ambiguité en adoptant la convention que $g(0)=1$, de sorte que $W(0)=1$. Bien entendu, on aurait pu choisir la convention opposée, $g(0)=-1$, auquel cas $W(\tau)$ aurait été remplacé partout par $-W(\tau)$.

Le comportement de $W$ nous dit si la sphère contient un monopôle. Si $g(\tau)$ appartient à la classe d'homotopie *triviale*, $W(1)=W(0)=1$. La figure 6a montre ce comportement pour un champ typique de cette classe, le champ nul. Si $g(\tau)$ appartient à la classe d'homotopie *non triviale*, $W(1)=-W(0)=-1$. La figure 6b montre ce comportement

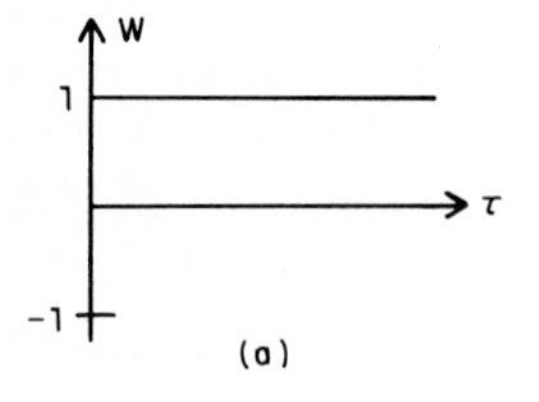


Fig. 6.

pour un champ typique de cette classe, le champ monopolaire GNO. Exprimé à l'aide de $W$, notre problème est de comprendre comment les effets qui décroissent exponentiellement avec la distance sont néanmoins capables de produire un changement indépendant de la distance dans $W$.

Mais cette formulation n'est pas tout à fait juste. Le confinement est un effet quantique, et, dans la théorie quantique, $W$ est un opérateur. Excepté dans l'approximation semiclassique dominante (qui ne montre aucun signe de confinement) l'état d'un monopôle n'est pas un état propre de $W$. L'objet que nous devons étudier est $\langle W \rangle$, la valeur moyenne de $W$.

Bien sûr, pour une petite sphère, dont le rayon est beaucoup plus petit que la longueur de confinement, l'approximation semi-classique est valable, et la fig. 6 donne des courbes précises de $\langle W \rangle$ dans le vide et dans l'état d'un monopôle. Mais pour une grande sphère, dont le rayon est beaucoup plus grand que la longueur de confinement, même la courbe pour le vide, fig. 6a, est fausse.

Quand notre courbe balaye la grande sphère, elle devient nécessairement, au milieu, une boucle de grande surface, et, par suite $\langle W \rangle$ devient très petite en vertu de la loi des aires de Wilson. Ce phénomène est ébauché sur la fig. 7a. L'ébauche n'est pas du tout à l'échelle. $\langle W(1/2) \rangle$ est $O(\exp - \pi K r^2)$ où $K$ est la tension de la corde. C'est si petit que si la courbe était tracée à l'échelle vous ne verriez pas que $\langle W(1/2) \rangle$ n'est pas nul.

Notre problème est résolu. Le monopôle n'a pas besoin de produire une variation de $\langle W \rangle$ indépendante de la distance. La loi des aires réduit $\langle W \rangle$ de 1 à une valeur positive petite. Tout ce que le champ du monopôle doit faire est de produire une petite variation additionnelle qui donne à $\langle W \rangle$ une valeur légèrement négative. La loi des aires renvoie alors $\langle W \rangle$ à $-1$. Cette situation est ébauchée sur la fig. 7b. J'insiste sur

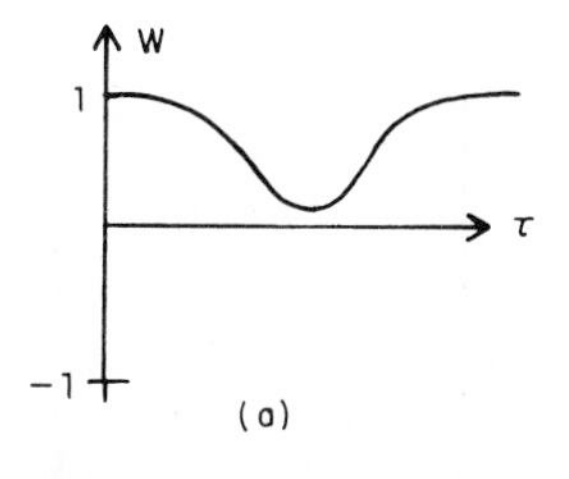


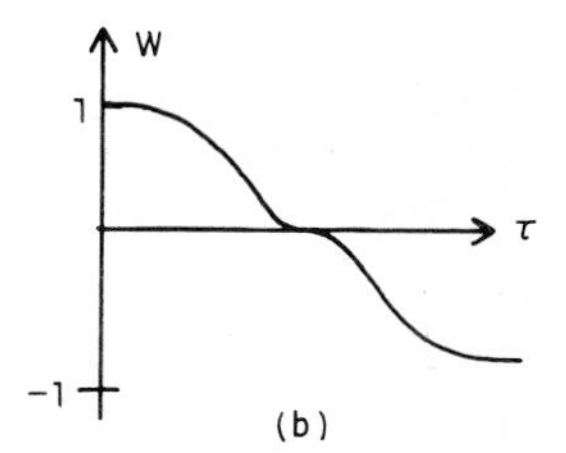


Fig. 7.

le fait que la partie droite de ce graphique représente la même physique que la partie gauche; on peut déduire l'évolution des valeurs négatives de $\langle W \rangle$ de celle des valeurs positives par un changement de convention.

Il est amusant de noter que si on avait défini $W$ par une représentation de spin entier de SO(3), on n'aurait pas eu la loi des aires mais on n'aurait pas non plus eu de test pour les monopôles.

Ainsi se concluent les arguments généraux. Maintenant, comme promis, je vais faire quelques calculs explicites dans une théorie de jauge Euclidienne à deux dimensions.

Les théories de jauge bidimensionnelles ont deux grands avantages. Elles sont d'un traitement facile. Ceci est fondamentalement dû à ce qu'elles ont très peu de degrés de liberté; dans l'espace de Minkowski, il n'y a pas de champ de rayonnement, seulement un champ coulombien. Dans une dimension d'espace, même l'électrodynamique Abélienne ordinaire donne un potentiel linéaire. Pour notre propos, elles ont un grand désavantage; elles ne possèdent pas de monopôle. Je m'en sortirai en étudiant une théorie de jauge SO(3), non dans l'espace plat à deux dimensions, mais sur une sphère bidimensionnelle. On peut considérer la théorie des champs euclidienne ordinaire à deux dimensions comme obtenue à partir de la théorie quadridimensionnelle en fixant $z$ et $t$ et en laissant varier $x$ et $y$. De même la théorie sphérique bidimensionnelle peut s'obtenir en fixant $t$ et $r$, et en laissant varier $\theta$ et $\varphi$. Le monopôle est dans la partie de l'espace que nous avons éliminée mais on en sent les effets sur la sphère que nous gardons.

Maintenant, précisons les choses. Nous traitons une théorie de jauge SO(3) sur une sphère. Exactement comme dans la section 3, nous passons à la jauge de corde où $A_\theta$ est identiquement nulle. En général cette jauge introduit une singularité de corde au pôle sud, un $A_\varphi(\pi, \varphi)$ non nul. On définit un élément de SU(2), $g(\varphi)$, par

$$\mathrm{d}g/\mathrm{d}\varphi = -A_\varphi(\pi, \varphi) g(\varphi), \tag{5.56}$$

et

$$g(0) = 1. \tag{5.57}$$

Alors, pour que la singularité de corde ne soit qu'un artifice de jauge,

$$g(2\pi) = \pm 1. \tag{5.58}$$

Le signe plus correspond à l'absence de monopôle (à l'intérieur de la sphère), le signe moins, à un monopôle. Si nous choisissons nos boucles de Lubkin comme sur la fig. 4,

$$W(\tau) = \tfrac{1}{2}\operatorname{Tr} g(2\pi\tau). \tag{5.59}$$

L'action Euclidienne $S_E$ est l'action conventionnelle restreinte à la sphère,

$$S_E = -\frac{1}{f^2}\int \operatorname{Tr} F_{\theta\varphi}F^{\theta\varphi}\sqrt{g}\, d\theta\, d\varphi = -\frac{1}{r^2 f^2}\int \operatorname{Tr}(\partial_\theta A_\varphi)^2 \frac{d\theta\, d\varphi}{\sin\theta}, \tag{5.60}$$

en vertu des éqs. (3.43) et (3.45).

Comme toujours, la valeur moyenne d'une observable invariante de jauge $\mathcal{O}$, est donnée par le rapport de deux intégrales fonctionnelles,

$$\langle \mathcal{O} \rangle = \int (dA_\varphi)\mathcal{O}e^{-S_E} \Big/ \int (dA_\varphi)e^{-S_E}. \tag{5.61}$$

Nous allons calculer $\langle \mathcal{O} \rangle$ dans l'absence (en présence) d'un monopôle en intégrant sur toutes les configurations de champs de jauge avec $g(2\pi) = +1\ (-1)$. Ceci nous garantit depuis le tout début que $\langle W \rangle$ aura le comportement approprié $\langle W(1) \rangle = \pm 1$.

Nous allons commencer par le dénominateur de l'éq. (5.61). L'extension au numérateur sera évidente. Nous écrivons $A_\varphi$ comme une somme

$$A_\varphi = \hat{A}_\varphi + \tfrac{1}{2}(1-\cos\theta)A_\varphi(\pi,\varphi). \tag{5.62}$$

Ainsi,

$$\hat{A}_\varphi(0,\theta) = \hat{A}_\varphi(\pi,\theta) = 0. \tag{5.63}$$

Cette séparation diagonalise $S_E$:

$$S_E = \hat{S} + S_1, \tag{5.64}$$

où

$$\hat{S} = \frac{\operatorname{Tr}}{r^2 f^2}\int d\theta\, d\varphi \frac{\left(\partial_\theta \hat{A}_\varphi\right)^2}{\sin\theta}, \tag{5.65}$$

et

$$\begin{aligned} S_1 &= -\frac{\operatorname{Tr}}{2r^2 f^2}\int_0^{2\pi} d\varphi \left[A_\varphi(\pi,\varphi)\right]^2 \\ &= -\frac{\operatorname{Tr}}{2r^2 f^2}\int_0^{2\pi} d\varphi \left[g^{-1} dg/d\varphi\right]^2. \end{aligned} \tag{5.66}$$

Notre théorie apparaît comme la somme de celles de deux systèmes indépendants. L'une, décrite par $\hat{S}$ est une théorie de champs bidimensionnelle, triviale; les variables dynamiques satisfont à des conditions aux limites linéaires, et l'action est une fonctionnelle quadratique de ces variables. L'autre, décrite par $S_1$, se réfère à un système non trivial, unidimensionnel. Autrement dit, ce n'est en aucune façon une théorie de champs Euclidienne à deux dimensions, mais simplement une version à temps imaginaire d'un système mécanique ordinaire, où $\varphi$ est le temps imaginaire.

En fait, il s'agit d'un système bien étudié. Les états du système sont indexés par les éléments du groupe des rotations, autrement dit, par trois angles d'Euler. Le système est un corps solide avec un point fixe, ou, plus exactement, le recouvrement double d'un corps solide, puisque $g$ est un élément de SU(2), non de SO(3); les moments angulaires demi-entiers sont permis, aussi bien que les entiers.

Si on considère les mouvements dans le voisinage de $g = 1$:

$$g = 1 - \mathrm{i}\varepsilon^a\sigma^a/2 + \mathrm{O}(\varepsilon^2), \tag{5.67}$$

alors

$$S_1 = \frac{1}{4r^2f^2}\int \frac{\mathrm{d}\varepsilon^a}{\mathrm{d}\phi}\frac{\mathrm{d}\varepsilon^a}{\mathrm{d}\phi}\mathrm{d}\phi + \mathrm{O}(\varepsilon^3), \tag{5.68}$$

d'où on déduit que le corps solide est isotrope; ses moments d'inertie principaux sont égaux,

$$I_1 = I_2 = I_3 \equiv I = 1/(2r^2f^2). \tag{5.69}$$

Ceci nous permet d'analyser le système complètement pour toutes les valeurs de $r$. Cependant, la solution est particulièrement simple quand $r$ est soit très petit soit très grand, de sorte que je me limiterai ici à ces deux cas.

Si $r$ est très petit, $r^2f^2 \ll 1$, l'intégrale fonctionnelle est dominée par les points stationnaires de l'action, les solutions des équations classiques du mouvement. Pour un corps rigide, ce sont les rotations uniformes d'un axe fixe, par exemple l'axe 3 positif,

$$g = \exp\left(-\tfrac{1}{2}\mathrm{i}\omega\sigma_3\varphi\right), \tag{5.70}$$

où $\omega$ est une constante non négative arbitraire, la vitesse angulaire. (C'est une bonne chose que les moments d'inertie se soient trouvés être tous égaux; autrement, nous aurions dû nous préoccuper de précession en l'absence de force.) Parmi ces mouvements, celui qui est dominant est celui pour lequel l'action est minimale, autrement dit pour lequel $\omega$ est minimale, compte-tenu des conditions aux limites $g(0)=g(2\pi)=1$. Le mouvement dominant est $\omega=0$ et le potentiel vecteur correspondant est $A_\varphi=0$. En présence d'un monopôle les conditions aux limites sont $g(0)=1$, $g(2\pi)=-1$. Le mouvement dominant est $\omega=1$ et le potentiel vecteur correspondant est

$$A_\varphi = -\tfrac{1}{4}\mathrm{i}\sigma_3(1-\cos\theta), \tag{5.71}$$

qui est le champ GNO. A petites distances, la théorie quantique des champs ressemble à la théorie classique comme cela doit être.

Pour $r$ grand, on a besoin des états propres et des valeurs propres du corps solide isotrope (deux fois recouvert). Les uns et les autres se trouvent dans les ouvrages standards [31], de sorte que je ne donnerai ici que les résultats. Les états propres sont de la forme $|j,m,m'\rangle$, où $j=0,1/2,1\ldots$, et $m$ et $m'$ varient indépendamment de $-j$ à $+j$ par valeurs entières. Les valeurs propres sont données par

$$H|j,m,m'\rangle = E_j|j,m,m'\rangle, \tag{5.72}$$

avec

$$E_j = j(j+1)/2I,$$

et $I$ est le moment d'inertie. Dans n'importe lequel de ces états, l'amplitude pour trouver le système dans la configuration spécifiée par l'élément $g$ du groupe est:

$$\langle g|j,m,m'\rangle = (2j+1)^{1/2}D^{(j)}_{mm'}(g), \tag{5.73}$$

où $D$ est la matrice de représentation usuelle.

Nous somme prêts. La contribution du corps solide à l'intégrale fonctionnelle est

$$\int(\mathrm{d}g)\exp(-S_1) = \langle g=\pm1|\exp(-2\pi H)|g=1\rangle, \tag{5.74}$$

d'après la formule de l'intégrale de chemin de Feynman. Si on insère un ensemble complet d'états,

$$\begin{aligned}\langle g=\pm1|\exp(-2\pi H)|g=1\rangle &\\ = \sum_{m,m',j}(2j+1)D^{(j)}_{mm'}(\pm1)D^{(j)*}_{mm'}(1)\exp(-2\pi E_j)&\\ = \sum_j(2j+1)^2(\pm1)^{2j}\exp[-\pi j(j+1)/I].&\end{aligned} \tag{5.75}$$

Ce résultat est exact. Pour $r$ grand, $f^2r^2 \gg 1$, cette expression est dominée par le premier terme de la série, $j=0$, qui est complètement insensible à la présence ou à l'absence de monopôle. Si on veut voir le monopôle, il faut au moins garder le second terme, $j=1/2$. On trouve:

$$\int (\mathrm{d}g)\exp(-S_1) = 1 \pm 4\exp(-3\pi f^2 r^2/2). \tag{5.76}$$

L'effet est négligeable.

Bien entendu, il s'agit seulement ici du dénominateur dans l'équation pour la valeur moyenne d'un opérateur éq. (5.61), mais un raisonnement analogue s'applique à l'ensemble de l'expression. Un cas particulièrement aisé à calculer concerne la valeur moyenne de l'action elle-même,

$$\begin{aligned}\langle S_{\mathrm{E}}\rangle &= \langle \hat{S}\rangle + \langle S_1\rangle = \langle \hat{S}\rangle + f^2\frac{\mathrm{d}}{\mathrm{d}f^2}\ln\int(\mathrm{d}g)\exp(-S_1)\\ &= \langle \hat{S}\rangle \mp 6\pi f^2 r^2\exp(-3\pi f^2 r^2/2),\end{aligned} \tag{5.77}$$

où, comme avant, j'ai négligé les termes exponentiellement petits par rapport à ceux que j'ai gardés. Un autre résultat facile à obtenir est

$$\langle W(\tau)\rangle = \exp(-3\pi f^2 r^2\tau/2) \pm \exp[-3\pi f^2 r^2(1-\tau)/2], \tag{5.78}$$

où je n'ai à nouveau retenu que les termes dominants. (Ici, j'ai laissé le détail du calcul en exercice). Cette expression a une interprétation physique extrêmement simple; c'est la somme des facteurs d'aires pour les deux portions dans lesquelles la boucle découpe la sphère.

Tout ceci tombe en parfait accord avec notre raisonnement général. A petite distance, tout ressemble à la physique classique; à grande distance tous les effets du monopôle sont négligeables, mais, néanmoins le monopôle fait varier $\langle W\rangle$ de $+1$ à $-1$.

Les théoriciens de la Gravitation disent, "Un trou noir n'a pas de cheveu". Ce que ceci veut dire est qu'un trou noir a des cheveux limités; les seules choses qui émergent d'un trou noir sont les champs de jauge sans masse associés aux quantités exactement conservées, les champs gravitationnels associés à l'énergie et au moment angulaire total, et les champs électromagnétiques associés à la charge électrique et magnétique totale.

Les forces chromodynamiques sont à courte portée, à cause du confinement. Néanmoins, la charge topologique est strictement conservée et peut être calculée à partir des effets chromodynamiques à des distances arbitrairement grandes. Ces deux affirmations seraient contradictoires en physique classique, mais sont parfaitement compatibles en mécanique quantique, comme nous l'avons vu. Problème pour l'étudiant: est-ce qu'un trou noir peut avoir des cheveux de couleur?

## Notes et Références

[1] P.A.M. Dirac, Proc. Roy. Soc. (London) Ser. A, 133 (1931) 60.

[2] G. 't Hooft, Nucl. Phys. B79 (1974) 276.
A.M. Polyakov, JETP Lett. 20 (1974) 194.

[3] J. Preskill, Phys. Rev. Lett. 43 (1979) 1365.

[4] S. Coleman, Classical Lumps and Their Quantum Descendants, dans: *New Phenomena in Subnuclear Physics*, éd. A. Zichichi (Plenum, New York, 1977).

[5] La réf. [4] contient une bibliographie plus développée. Voir aussi:
P. Goddard et D. Olive, Rep. Prog. Phys. 41 (1978) 1357 (revue).
E. Amaldi et N. Cabibbo, dans: *Aspects of Quantum Theory*, éds. A. Salam et E. Wigner (Cambridge Univ. Press, 1972).

[6] Notez qu'il n'y a pas de $4\pi$ dans la définition de la charge magnétique.

[7] Y. Aharonov and D. Bohm, Phys. Rev. 115 (1959) 485.

[8] Ceci est une lapalissade pour les théories de jauge non Abéliennes où tout le monde pense aux transformations de jauge comme des fonctions de l'espace–temps dans le groupe de jauge. Pour la théorie Abélienne le groupe de jauge est U(1) et la fonction dans le groupe est $\exp(-\mathrm{i}e\chi)$.

[9] T.T. Wu and C.N. Yang, Nucl. Phys. B107 (1976) 365.

[10] Ce problème a été résolu il y a très longtemps par
I. Tamm, Z. Phys. 71 (1931) 141,
et mes résultats sont les mêmes que les siens bien que ma méthode soit quelque peu différente. A ma connaissance le traitement publié le plus proche du mien est celui de H. Lipkin, W. Weisberger et M. Peshkin, Ann. Phys. 53 (1969) 203.

[11] Les commutateurs ne sont pas aussi inoffensifs qu'ils le paraissent. Si on essaye de vérifier l'identité de Jacobi pour trois $D$'s, on obtient, au lieu de zéro, un terme proportionnel à $\delta^3(r)$. Ce n'est pas un vrai problème pour deux raisons. D'abord on ne croit pas au champ monopolaire jusqu'à l'origine; ce n'est que la partie à longue portée de quelque chose de plus compliqué à courte distance. (Dans la section 4 on verra de quoi il s'agit.) Deuxièmement, même si on croyait au champ jusqu'à l'origine, il y a, comme nous le verrons, une barrière centrifuge dans *toutes* les ondes partielles, qui empêche la particule d'atteindre l'origine.

[12] L'analyse donnée ici suit celle de
A. Goldhaber, Phys. Rev. Lett. 36 (1976) 1122.
Le travail de Goldhaber a été stimulé par les recherches de
R. Jackiw et C. Rebbi, Phys. Rev. Lett. 36 (1976) 1116, et par
P. Hasenfratz et G. 't Hooft, Phys. Rev. Lett. 36 (1976) 1119.

[13] P. Goddard, J. Nuyts, and D. Olive, Nucl. Phys. B125 (1977) 1.

[14] Je suis ici très cavalier avec les singularités. Voici un argument plus rigoureux: nous définissons une configuration de champ de jauge localement non singulière dans une région, si cette région est une union d'ouverts tels que dans chaque ouvert le champ de jauge est non singulier et que dans l'intersection de n'importe quelle paire d'ouverts les champs de jauge dans les deux ouverts sont reliés par une transformation de jauge non

singulière. Il est possible de montrer qu'une configuration de champs de jauge qui est localement non singulière dans tout l'espace excepté à l'origine, est équivalente de jauge à une configuration non singulière (dans le sens ordinaire) partout excepté sur l'axe sud-polaire. (Ce théorème est démontré dans la réf. [4].) Autrement dit, si toutes les singularités, ailleurs qu'à l'origine, sont des artifices de jauge, elles peuvent toutes être repoussées sur la corde de Dirac par un choix approprié de jauge.

[15] E. Lubkin, Ann. Phys. (NY) 23 (1963) 233. Voir en particulier la section XV.

[16] Un cours quelque peu plus développé (bien que désespérément vulgaire) se trouve dans la réf. [4], en même temps que des références à la littérature mathématique.

[17] Cette analyse de la stabilité a été faite pour la première fois par
R. Brandt et F. Neri, Nucl. Phys. B161 (1979) 253.

[18] W. Nahm et D. Olive (communication privée, été 1979).

[19] Le travail décrit ici est tiré de plusieurs sources; pour les références, voir la réf. [4].

[20] H. Georgi et S.L. Glashow, Phys. Rev. Lett. 32 (1974) 438.

[21] E. Bogomol'nyi, Sov. J. Nucl. Phys. 24 (1976) 449.
S. Coleman, S. Parke, A. Neveu et C. Sommerfield, Phys. Rev. D15 (1977) 544.

[22] M. Prasad et C. Sommerfield, Phys. Rev. Lett. 35 (1975) 760.

[23] Une bonne partie de cette section est un plagiat criard de la réf. [4]. (Avis au détenteur du "copyright".)

[24] On peut en fait éviter cet horrible calcul par des astuces de théorie des groupes, comme on l'a fait pour l'opérateur de Laplace dans la section 2.

[25] Cette note est réservée aux experts. La discussion du texte laisse penser que le calcul détaillé du spectre isorotationnel est plus facile qu'il n'est en réalité. Il y a des complications techniques dûes à l'invariance de jauge. On peut toutes les maitriser, mais elles allongent le calcul davantage que si elles n'étaient pas là. Par exemple, dans la jauge temporelle, plusieurs des invariances de l'Hamiltonien ne laissent pas la solution monopolaire inchangée, à savoir, les transformations de jauge indépendantes du temps. Personne, s'il est sain d'esprit, ne s'attend à ce qu'elles conduisent à des niveaux isorotationnels. Cependant, pour s'en assurer, et pour éliminer les excitations parasites des excitations réelles, il est nécessaire de jouer avec la condition subsidiaire qui est le fléau de la quantification dans la jauge temporelle. Les conditions subsidiaires peuvent être évitées dans la jauge de Coulomb, par exemple, mais alors la forme de l'Hamiltonien est plus compliquée et cela embrouille la situation. Pour un traitement soigneux, dans la jauge de Coulomb, des dyons qui sont discutés immédiatement après, voir
E. Tomboulis et G. Woo, Nucl. Phys. B107 (1976) 221.

[26] Ces dyons ont été découverts pour la première fois par des méthodes complètement différentes des nôtres, par
B. Julia et A. Zee, Phys. Rev. D11 (1975) 2227.

[27] E. Witten, Phys. Lett. 86B (1979) 283.

[28] C. Dokos et T. Tomaras, Phys. Rev. D21 (1980) 2940.

[29] Cette astuce a été découverte par
G. 't Hooft, Nucl. Phys. B105 (1976) 538, et par
E. Corrigan, D. Olive, D. Fairlie et J. Nuyts, Nucl. Phys. B106 (1976) 475.

[30] J. Schwinger, Phys. Rev. 144 (1966) 1087; 151 (1966) 1048, 1055.

[31] Par exemple, voir
L. Landau et E. Lifshitz, *Quantum Mechanics*, 3rd ed. (Pergamon, 1977) p. 410.

5:30
S. Coleman (HARVARD)
1st LECTURE

COURSE 5

# DISCRETE QUANTUM CHROMODYNAMICS

*A pedagogical introduction to Euclidean lattice gauge theory*

Richard C. BROWER

*Physics Department – SCIPP, University of California*
*Santa Cruz, CA 95064, USA*

*M.K. Gaillard and R. Stora, eds.*
*Les Houches, Session XXXVII, 1981*
*Théories de jauge en physique des hautes énergies/Gauge theories in high energy physics*

TONIGHT

## Contents

## 1. Introduction

A decade ago in 1971, 't Hooft triggered a remarkable revolution in particle physics with his proof of renormalizability [1] for non-Abelian gauge theories (NAGT). Rapidly thereafter a consensus developed that all the interactions of elementary particles may be understood as a consequence of quantized gauge theories. However, a decade later, this consensus is *still* based more on theoretical consistency and aesthetic appeal than experimental confirmation. Crudely speaking, the problem is a mismatch between theoretical and experimental competence.

Using weak coupling perturbation theory, theorists most confidently predict phenomena beyond the energies of current experimental facilities. Undoubtedly, the next decade with its colossal accelerator projects will help to close the energy gap. For the electroweak theory of Glashow, Weinberg and Salam, the W, Z and Higgs physics in the 100 GeV range will bring theory and experiment into dramatic confrontation. However, for testing the strong interaction theory (Quantum Chromodynamics or QCD) higher energies should not be necessary, and in my opinion may not be decisive. Here the significant energies are on the few GeV scale, and we already have a plethora of experimental numbers in that range. Theorists are simply unable to deal with the obvious non-perturbative phenomena of quark confinement. These lectures present one strategy for bringing QCD predictions down to the energies where experimental results abound.

### *1.1. Scope of lectures*

In 1974, Wilson [2] suggested a specific non-perturbative formulation for QCD, regulating the Euclidean path integral via a space–time lattice cut-off in the ultraviolet (momenta $\leqslant P_{max} = \pi/a$). These lectures are intended as a pedagogical introduction to this approach to non-perturbative QCD. To keep the topic to manageable scope, the lectures will intentionally slight two closely related areas: First, spin systems on a lattice and Abelian lattice gauge theories will be avoided except as a pedagogical tool (see section 4). Much of the most beautiful and successful calculations exist for these simpler theories, so the reader is referred to an excellent review by Kogut [3] which emphasizes these results. Secondly, the Hamiltonian or spacial lattice formulation of Kogut and Susskind is omitted. Both Kogut's review [3] and Susskind's 1976 Les Houches lectures [4] give priority to the Hamiltonian or Fock space picture, and to help bridge the gap, Appendix A derives the lattice Hamiltonian from Euclidean lattice QCD.

A new feature of these lectures will be the copious reference to Monte Carlo simulations to explore qualitative features of gauge theories. This computational technique (as pioneered by Wilson [2] and extended by Creutz [5] and an increasing number of others*) has come perilously close to predicting mass ratios that can be compared with experiment. However, in my opinion, accurate tests of QCD (i.e., masses to a percent or better) require more than the faster computers of the next decade. Newer analytical techniques and a more sophisticated identification of the most relevant degrees of freedom are also needed. Computers should be viewed as laboratories for "experimental mathematics". The Monte Carlo simulations provide a cheap way to augment the experimental data, which is indispensable for stimulating and testing new theoretical ideas in physics.

Finally, any effort to understand the non-linearities of non-Abelian gauge theories also has implications for the more ambitious program to develop Grand Unified Theories (GUTs) for all the fundamental interactions. For GUTs the perturbative consequences often involve energies beyond any conceivable accelerator program; and I believe the few ingenious low energy predictions from weak coupling theory (e.g. proton decay or monopole abundance) may be insufficient to guide us unambiguously to the correct GUT. Therefore, a firmer grasp of the full non-linearities of the quantum non-Abelian gauge theory may be essential to progress in GUTs. Although the focus of these lectures will be restricted to quantum chromodynamics, it is hoped that this approach can answer more general questions of interest to GUT theorists—such as when and why confinement, chiral breaking, or Higgs-like breaking occur in a randomly chosen gauge theory.

*No attempt will be made to give a complete set of references, and the author apologizes for the inevitable injustices that result. The few citations should, however, offer the reader an entré to the relevant literature.

## 1.2. Euclidean path integral on the lattice

The problem of quantum field theory is to replace the classical Lagrangian with an associated *quantum* theory. For example, QCD has the (Euclidean) Lagrangian ($\mu = 1,2,3,4$) density:

$$\mathcal{L}_{\mathrm{E}} = \tfrac{1}{2}\mathrm{Tr}(F_{\mu\nu}F_{\mu\nu}) + \bar{\psi}_{\mathrm{f}}\gamma_{\mu}D_{\mu}\psi_{\mathrm{f}} + m_{\mathrm{f}}\bar{\psi}_{\mathrm{f}}\psi_{\mathrm{f}}, \tag{1.1}$$

where $D_\mu = \partial_\mu - \mathrm{i}g_0 A_\mu$. The gluon gauge field $A_\mu = \frac{1}{2}\lambda^\alpha A_\mu^\alpha$ is in the Lie Algebra of color SU(3), and the quark Dirac field $\psi_{\mathrm{f}}$ is a color triplet with flavors $\mathrm{f} = \mathrm{u,d,s,c,b,t,}\ldots$

$$F_{\mu\nu} = \frac{\mathrm{i}}{g_0}\left[D_\mu, D_\nu\right] = \partial_\mu A_\nu - \partial_\nu A_\mu - \mathrm{i}g_0\left[A_\mu, A_\nu\right]. \tag{1.2}$$

For $m_{\mathrm{f}} = 0$, the classical Lagrangian has a bare coupling $g_0$, but no mass scale. The renormalized perturbation series breaks scale invariance, and depends on a scale $\Lambda$. However, it is invariant under simultaneous scale ($\Lambda \to \Lambda'$) and coupling ($g_{\mathrm{R}} \to g_{\mathrm{R}}'$) redefinitions. (This non-linear invariance, called the renormalization group, will be discussed in section 2.3.)

Unlike QED, the aforementioned renormalized perturbation theory fails to give an adequate quantization prescription of quantum chromodynamics. If, as experiments indicate, there are no free quarks and gluons at energies well beyond $\Lambda(\approx 100$ MeV), no perturbative expansion around plane wave solutions can be adequate. Even the cleverly contrived "short distance" effects calculable perturbatively for quark subamplitudes, described by perturbative effects, have proven too elusive to provide accurate quantitative tests of QCD. At a more fundamental level, the series misses instanton contributions [ $\sim \exp(-8\pi^2/g_{\mathrm{R}}^2)$] altogether. The renormalized perturbation series simply does not give a complete definition of *quantum* chromodynamics.

We follow an alternate quantization procedure based on the Euclidean path integral. For simplicity, consider the path integral (or partition function) for $\phi^4$ theory:

$$Z[J] = \int_{\mathrm{paths}} \mathrm{D}\phi(x)\exp\left[-\frac{1}{\hbar g_0^2}\int_V \mathrm{d}^4x\,\mathcal{L}_{\mathrm{E}} + \int \mathrm{d}^4x\, J(x)\phi(x)\right],$$

$$\mathcal{L}_{\mathrm{E}} = \tfrac{1}{2}\partial_\mu\phi\partial_\mu\phi + (\phi^2 - v_0^2)^2, \tag{1.3}$$

where $x_4 = \mathrm{i}t$ is the Euclidean or Wick rotated "time". Complex time ($x_4 = \mathrm{i}t$) has the simultaneous effect of replacing the Schrödinger equation by the diffusion equation, and the Lorentz group O(3,1) by the rotation group O(4). We shall assume that analytic continuation to real energies ($E = \mathrm{i}p_4 \to$ real) gives the physical momentum space amplitudes, as suggested by the continuation for individual Feynman diagrams. Initially, the fields are placed in a finite space–time box of volume $V$.

Still this expression (1.3) fails to define the quantum theory because of the difficulty with defining integration over "paths", i.e. *one* integral $\int_{-\infty}^{\infty}\mathrm{d}\phi$ for *each* point $x$ in volume $V$. To remedy this, some kind of ultraviolet regularization prescription must be adopted. We choose to restrict $\phi(x) = \phi_n$ to the $L^4$ sites on a hypercubic lattice ($x_n = a(n_1, n_2, n_3, n_4)$; $n_i = 1, \ldots, L$) in our volume $V = a^4 L^4$. With the replacements

$$\int_{\text{path}} \mathrm{D}\phi(x) \to \prod_n \int_{-\infty}^{\infty} \mathrm{d}\phi_n,$$

$$\int \mathrm{d}^4x \to \sum_n a^4,$$

$$\partial_\mu \phi \to \Delta_\mu \phi_n = \frac{1}{a}\left[\phi_{n+\hat{\mu}} - \phi_n\right], \tag{1.4}$$

the path integral becomes an $L^4$ fold multiple integral.

$$Z_L[J_n] = \prod_n \int_{-\infty}^{\infty} \mathrm{d}\phi_n \exp\left(-\frac{1}{\hbar g_0^2} \sum_n a^4 \mathcal{L}[\phi_n] + \sum_n a^4 J_n \phi_n\right). \tag{1.5}$$

Now we have a discrete quantum system written as a finite statistical mechanical system. The real continuum world is recovered, we hope, by the application of the combined thermodynamic (IR) and continuum (UV) limits,

1. Thermodynamic limit: $L \to \infty$,
2. Continuum limit: $a \to 0$. (1.6)

The careful definition of these limits is a central problem of Euclidean lattice field theory.

Let us make a short notational aside: It is often convenient on the lattice to work in units $c = \hbar = a = 1$ so all quantities appear dimensionless. Hence $x = (n_1, n_2, n_3, n_4)$ is regarded as an integer valued four vector ($x_\mu$). Sites are labeled by $x$; oriented links from $x$ to $x + \mu$ are labeled by $\ell = (x, \mu)$; and oriented plaquettes (squares) are labeled by $P = (x, \mu\nu)$ where $\mu < \nu$ (see fig. 1). The unit vectors take on positive

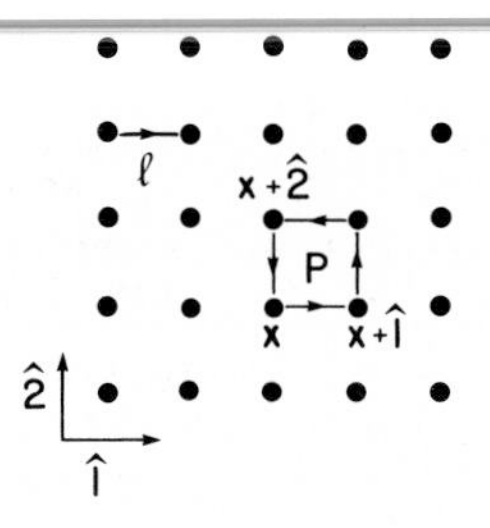


Fig. 1. Finite Euclidean lattice ($d=2, L=5$) with $L^d$ sites labeled by $x=(n_1,\ldots,n_d)$, oriented links labeled by $\ell=(x,\mu), \mu=\hat{1},\hat{2},\ldots,\hat{d}$ and oriented plaquettes labeled by $P=(x,\mu\nu), \mu<\nu$.

values only ($\mu=\hat{1},\hat{2},\hat{3},\hat{4}$). Finally, since a multiplicative constant in $Z$ has no effect on Green's functions (or thermal averages) such as

$$\langle\phi_x\phi_0\rangle_c \equiv \frac{\partial}{\partial J_x}\frac{\partial}{\partial J_0}\log Z, \tag{1.7}$$

it will be adjusted without warning to suit convenience.

### 1.3. Statistical Mechanics Connection

The Euclidean path integral emphasizes a profound isomorphism:

$$\text{Quantum field theory} \equiv \text{Equilibrium statistical mechanics.} \tag{1.8}$$

The isomorphism is seen by comparing field theory in the Euclidean path integral form with statistical mechanics in the canonical ensemble. The quantum action ($S$) corresponds to the energy functional ($E$) and the configuration probability to the Boltzmann factor.

$$\exp\left(-\frac{1}{\hbar g_0^2}S[\phi]\right) \leftrightarrow \exp\left(-\frac{1}{kT}E[s]\right). \tag{1.9}$$

The path integral is a canonical ensemble in $d$-dimensional space–time.

This connection can be either a source of linguistic confusion or a source of deeper physical insight. We would like to use commonly accepted features of statistical mechanical systems to anticipate properties of non-perturbative quantum field theory. To this end, we give a bilingual translation for this isomorphism (see table 1). With a lattice cut-off, the statistical mechanics language is more convenient. *But* it

Table 1
Bilingual dictionary

| Quantum Fieldese | Stat. Mechese |
|---|---|
| Field: $\phi(x)$ | Order Parameter (spin): $s_x$ |
| Action: $S=\int \mathrm{d}x\left[\frac{1}{2}(\partial_\mu\phi)^2+(\phi^2-v_0^2)^2\right]$ | Energy: $E=\sum_{x,\mu}\frac{1}{2}(s_{x+\mu}-s_x)^2$ |
| Coupling: $\hbar g^2$ | Temperature: $kT$ |
| Vacuum Persistence Amp: $\langle 0_{\text{out}}\vert 0_{\text{in}}\rangle=\int \mathrm{D}\phi \exp\left(-\frac{1}{\hbar g^2}S+\int J\phi\right)$ | Partition Function: $Z=\sum_{\{s_x=\pm 1\}}\exp\left(-\frac{1}{kT}E+H_x s_x\right)$ |
| Quantum Fluctuations | Thermal Fluctuations |
| Classical Limit: $\hbar g^2\to 0$ | Zero Temperature limit: $T\to 0$ |
| Lowest Action Classical Solution | Ground State Configuration |
| Connected Generating Functional: $W[J]=\frac{1}{V}\log\langle 0_{\text{out}}\vert 0_{\text{in}}\rangle$ | Free Energy Density: $F[H_x]=\frac{1}{L^d}\log Z$ |
| Vacuum Exp. Value: $\phi_{\text{cl}}=\langle 0\vert\phi\vert 0\rangle$ | Magnetization: $M=\langle s_x\rangle$ |
| Effective Potential: $V_{\text{eff}}(\phi_{\text{cl}})$ | Helmholtz Potential: $A=F+HM$ |
| Propagator: $\langle 0\vert T(\phi(x)\phi(0))\vert 0\rangle \sim \exp(-m\vert x\vert)$ | Correlation Function: $\langle s_x s_0\rangle \sim \exp(-\vert x\vert/\xi)$ |
| Mass: $m$ | Inverse Correlation length: $1/\xi$ |
| Time Evolution Op. $\hat{U}(t)=\mathrm{e}^{-t\hat{H}}$ | Transfer Matrix: $\hat{T}$ |

should be emphasized that for lattice theories the two languages are exactly, absolutely precisely equivalent!

In table 1, we choose to compare continuum $\phi^4$ field theory (1.3) with the Ising statistical system.

$$Z_{\text{Ising}}=\sum_{\{s_x=\pm 1\}}\exp\left[-\frac{1}{kT}\sum_{x,\mu}\tfrac{1}{2}(s_{x+\mu}-s_x)^2+\sum_x H_x s_x\right]. \tag{1.10}$$

Note the relationship between $\hbar g_0^2$ and $kT$, so that quantum fluctuations about the classical minima correspond to thermal fluctuations about the ground state (minimum energy).

Actually the connection between $\phi^4$ theory with deep double wells and the Ising system also occurs on the dynamical level. Magnetization at

$T \leqslant T_{\text{Curie}}$ ($\langle s_x \rangle \neq 0$) exhibits the same "mechanism" which gives spontaneous order in the Higgs potential or non-zero vev ($\langle \phi \rangle \neq 0$). Formally, lattice $\phi^4$ theory becomes the Ising model in the limit

$$\int_{-\infty}^{\infty} \mathrm{d}\phi_x \exp\left[ -\frac{1}{\hbar g_0^2} \left( \phi_x^2 - v_0^2 \right)^2 \right] \to \int \mathrm{d}\phi_x \, \delta\left( \phi_x^2 - v_0^2 \right), \tag{1.11}$$

as $g_0^2 \to 0$ for fixed $\hbar g_0^2 / v_0^2 = kT$. The integral becomes a sum over $s_x = \pm 1$ with $\phi_x = s_x v_0$. More precise connections have been made (or conjectured) under the rubric of "universality of exponents" or "hyperscaling".

For our purposes we only want to gain heuristic insight from the Statistical Mechanics Connection. As a trivial example, Ising's failure (as a graduate student!) to find spontaneous magnetization in the $d = 1$ Ising model corresponds to the well-known fact that in quantum mechanics ($d = 1$, $\phi^4$ theory) a double-well potential always has a symmetric ground state due to tunnelling. The physics and the results are the same. His advisor should have anticipated this result.

We can also anticipate some of the properties of the thermodynamic limit. First, to get finite quantities, we must, of course, divide by volume factors in extensive variables. For example, the free energy per site,

$$F[J_x] = \lim_{L \to \infty} \frac{1}{L^4} \log Z, \tag{1.12}$$

has a finite thermodynamic limit, but $Z$ does not. $F$ is also called the generating functional ($W$) for connected Green's functions (see table 1). Less trivial is the fact that the limit can be ambiguous. For example, with symmetry breaking $\langle \phi_x \rangle = \pm$const., there are two free energies ($\langle \phi_x \rangle = \partial F_{\pm} / \partial J_x$ at $J_x = 0$) which must be reached by biasing the system with applied fields $J_x \gtrless 0$ or boundary conditions prior to taking the thermodynamic limit. In field theory, one refers to this situation as multiple vacua, which cannot be connected by any transition amplitude. *Notice* that the vacuum state (or states) is the lowest eigenstate of the quantum operator Hamiltonian $\hat{H}$ and this *does not* correspond in statistical mechanics to the ground state energy ($E = E_0$). It corresponds to the lowest eigenstate of the transfer matrix ($\hat{T}$) that translates you from one lattice plane to the next along the "time" axis. (see Appendix A.)

## 2. Re-inventing gauge theories

One decade ago, at the same time that 't Hooft proved renormalizability for NAGT [1], Wegner [6] also invented the first lattice gauge theory. His motivation was to find another simple spin model, like the U(1) or XY model, that had a phase transition but no local order parameter. To this end, he considered a modification of the Ising model (1.10) so that the *global* $Z_2$ invariance ($s_x \to -s_x$ for all $x$) is replaced by a local gauge invariance at each site ($\Omega_x = \pm 1$).

*2.1. Lattice gauge theories*

Wegner's model places $Z_2$ "spins" on links $[s_\ell = s_\mu(x)]$ and four spin interactions on plaquettes $[P = (x, \mu\nu)$ with $\mu < \nu]$.

$$Z_{\text{Wegner}} = \sum_{\{s_\ell = \pm 1\}} \exp\left[\frac{1}{g_0^2}\sum_P s_\nu(x)s_\mu(x+\hat{\nu})s_\nu(x+\hat{\mu})s_\mu(x)\right]. \tag{2.1}$$

Obviously, it has a local invariance when all spins connected to site $x$ are flipped together ($\Omega_x = -1$).

$$s_\mu(x) \to \Omega_{x+\hat{\mu}} s_\mu(x)\Omega_x. \tag{2.2}$$

Thus Wegner invented a lattice gauge theory, which has no local order parameter since $\langle s_\mu(x)\rangle = 0$ for all temperatures $kT > 0$ by Elitzur's theorem [7]. His theory also has a transition to a confining phase at strong coupling (see section 4.1).

*2.1.1. Abelian lattice gauge theories*

With a little insight (actually hindsight) into the geometry, it is simple to generalize Wegner's models to Abelian groups. Link variables $(x, \mu)$ correspond to vector fields, and plaquettes express their curl. To see this clearly, reparametrize the spins, $s_\mu(x) = \exp[\mathrm{i}\theta_\mu(x)]$ for $\theta_\mu = 0, \pi$ so that

$$s_\nu(x)s_\mu(x+\hat{\nu})s_\nu(x+\hat{\mu})s_\mu(x)$$

$$= \tfrac{1}{2}\left[\exp - \{\mathrm{i}(\Delta_\mu\theta_\nu(x) - \Delta_\nu\theta_\mu(x))\} + \text{c.c.}\right]. \tag{2.3}$$

Abelian gauge theories for the $Z_N$ or U(1) groups are formulated by allowing the angular variable to take on more values $\theta_\mu = 2\pi n/N$ ($n = 1,\ldots,N$) or $-\pi < \theta_\mu < \pi$, respectively. For example, the U(1) theory (compact QED) is written as:

$$Z_{\text{QED}} = \prod_\ell \int_{-\pi}^{\pi} \mathrm{d}\theta_\ell \exp\left[\frac{1}{e_0^2}\sum_P \cos(\Delta_\mu\theta_\nu - \Delta_\nu\theta_\mu) + \mathrm{i}\sum_\ell J_\mu\theta_\mu\right], \tag{2.4}$$

with integer valued sources $J_\mu(x)$, obeying current conservation to preserve gauge invariance.

$$\mathrm{Div}\, J \equiv -\Delta_{-\mu} J_\mu(x) = \sum_{\hat{\mu}} \left[ J_\mu(x) - J_\mu(x-\hat{\mu}) \right]. \tag{2.5}$$

For small field strengths $[A_\mu(x) = e_0 a \theta_\mu \to 0]$, with appropriate gauge fixing this becomes:

$$Z \simeq \int_{\text{paths}} \mathrm{D}A \exp\left[ -\tfrac{1}{4} \int \mathrm{d}^4x \left( \tfrac{1}{4} F_{\mu\nu}^2 - \mathrm{i}eJ_\mu A_\mu \right) + \text{gauge fixing} \right], \tag{2.6}$$

which is free Maxwell theory, with zero mass photons and Coulombic forces on the sources $J_\mu$. On the lattice, the $\cos(e_0 a^2 F_{\mu\nu})$ term gives nontrivial self-interactions, which cause the Coulombic weak coupling phase to give way to a confining strong coupling phase by magnetic flux condensation (see section 4.2).

*2.1.2. Non-Abelian gauge theories*

Again we associate a group element, $U_\mu(x)$ in SU($N$), with each link $(x, \mu)$. The concept of parallel transport of the fundamental (quark) charge from $x$ to $x + \mu$ suggests the definition:

$$U_\mu(x) = P\exp\left( \mathrm{i}g_0 \int_x^{x+\mu} A_\nu \mathrm{d}\xi_\nu \right) \equiv \exp\left[ \mathrm{i}g_0 a \bar{A}_\mu(x) \right] \tag{2.7}$$

in terms of the field strength $A_\mu = \frac{1}{2}\lambda^\alpha A_\mu^\alpha$ ($\bar{A}_\mu$ is an appropriate average value) and the path ordered ($P$) exponential. By writing the standard continuum gauge transformation in the form

$$1 + \mathrm{i}gA_\mu \mathrm{d}x_\mu \to \Omega(x+dx)(1 + \mathrm{i}gA_\mu \mathrm{d}x_\mu)\Omega^\dagger(x), \tag{2.8}$$

we see that $U_\mu(x)$ depends only on gauge transformations $\Omega_x$ restricted to the sites.

$$U_\mu(x) \to \Omega_{x+\mu} U_\mu(x) \Omega_x^\dagger. \tag{2.9}$$

This is Wilson's crucial trick—path ordered products ($P$) have gauge rotations only at the end points of the path. The product around a closed path is a covariant tensor. For example, a plaquette variable

$$U_P = U_\nu^\dagger(x) U_\mu^\dagger(x+\hat{\nu}) U_\nu(x+\hat{\mu}) U_\mu(x) \tag{2.10}$$

transforms as $U_P \to \Omega_x U_P \Omega_x^\dagger$. Therefore the obvious (most local) gauge invariant choice for the action is:

$$-\frac{1}{g_0^2} S[U_\ell] = \frac{1}{g_0^2} \sum_P \operatorname{Tr}\left(U_P + U_P^\dagger - 2\right). \tag{2.11}$$

Note that the negatively oriented link from $x + \hat{\mu}$ to $x$ carries the inverse matrix

$$U_{-\mu}(x + \hat{\mu}) \equiv U_\mu^\dagger(x). \tag{2.12}$$

Finally the partition function for SU($N$) gauge theory is

$$Z_L[J_\ell]$$
$$= \prod_\ell \int \mathrm{d}U_\ell \exp\left[\frac{1}{g_0^2} \sum_P \operatorname{Tr}\left(U_P + U_P^\dagger - 2\right) + \sum_\ell \operatorname{Tr}\left(J_\ell^\dagger U_\ell + J_\ell U_\ell^\dagger\right)\right], \tag{2.13}$$

where $J$ is an arbitrary, complex matrix source and where $\mathrm{d}U_\ell$ is the invariant (or Haar) integral over the SU($N$) group. This measure can be defined abstractly by

$$\int \mathrm{d}U = 1, \qquad \int \mathrm{d}U \mathrm{f}(U_0 U) = \int \mathrm{d}U \mathrm{f}(U), \tag{2.14}$$

for any $U_0 \in \mathrm{SU}(N)$ or explicitly by the parametrization [8]

$$\int \mathrm{d}U = N \prod_{\alpha=1}^{N} \int \mathrm{d}\theta_\alpha [\det M]^{-1}, \quad U = \exp(\mathrm{i}\theta^\alpha \lambda^\alpha);$$

$$M^{\alpha\beta} = \operatorname{Tr}\left(\lambda^\alpha U^\dagger \frac{1}{\mathrm{i}} \frac{\partial}{\partial \theta^\beta} U\right), \quad [\lambda^\alpha, \lambda^\beta] = 2\mathrm{i} f^{\alpha\beta\gamma} \lambda^\gamma. \tag{2.15}$$

We again place the system in a finite box with $L^4$ sites and choose periodic boundary conditions $[U_\mu(x + \hat{\nu} L) = U_\mu(x)]$.

Thus we have introduced SU($N$) gauge theory on the lattice (without quark or Higgs matter field for now) as a $4L^4(N^2 - 1)$-fold integral. We need only compute the "thermal" average of any function $f[U_\ell]$ of the fields,

$$\langle f[U_\ell] \rangle = \frac{1}{Z_L} \prod_\ell \int \mathrm{d}U_\ell f[U_\ell] \exp\left(-\frac{1}{g_0^2} S\right), \tag{2.16}$$

in terms of one parameter $g_0^2$ and take judiciously the thermodynamic ($L \to \infty$) and continuum ($a \to 0$) limits. Voilà quantum chromodynamics! In the remainder of this section we discuss how, in principle, we take the continuum limit and what, in principle, this can teach us about QCD.

### 2.2. Continuum limit

How do we take the continuum limit ($a \to 0$) when "$a$" does not occur in our formulation of $\langle f[U_\ell]\rangle$ at *all*? There are really two types of continuum limits:

(i) The naive continuum limit or weak coupling perturbation theory which gives (cut-off) Feynman rules.

(ii) The actual continuum limit, which it is hoped gives a non-perturbative description of quark confinement.

#### 2.2.1. Weak coupling perturbation theory

The naive continuum limit is found by re-introducing the lattice spacing into $U_\mu(x)$ and expanding

$$U_\mu(x) \equiv \exp\left[\mathrm{i}ag_0\bar{A}_\mu(x)\right] \approx 1 + \mathrm{i}ag_0\bar{A}_\mu - \frac{a^2g_0^2}{2!}\bar{A}_\mu^2 + \cdots. \tag{2.17}$$

Since the action $(1/g_0^2)S[U_\ell]$ in eq. (2.11) does not really depend on $a$, this expansion is more properly regarded as a small $g_0$ expansion around the least action point $U_\mu = 1$ and its gauge copies. Using the Baker–Cambell–Hausdorff theorem, the reader is urged to show that the plaquette variable (2.10) is

$$U_P \approx \exp\left[\tfrac{1}{2}\mathrm{i}a^2g_0\lambda^\alpha\bar{F}^\alpha_{\mu\nu} + \mathrm{O}\left(a^3g_0^3\right)\right], \tag{2.18}$$

where for smoothly varying fields* $A_\mu(x)$:

$$\tfrac{1}{2}\lambda^\alpha\bar{F}^\alpha_{\mu\nu} = \Delta_\mu\bar{A}_\nu - \Delta_\nu\bar{A}_\mu - \mathrm{i}g_0\left[\bar{A}_\mu, \bar{A}_\nu\right] + \mathrm{O}(a). \tag{2.19}$$

Thus we recognize the standard continuum action emerge as $a \to 0$,

$$\frac{1}{g_0^2}S \approx \sum_x\left\{\tfrac{1}{4}a^4\bar{F}^\alpha_{\mu\nu}\bar{F}^\alpha_{\mu\nu} + \mathrm{O}(a^5)\right\}. \tag{2.20}$$

However to actually do weak coupling expansion order by order in $ag_0$ (with the "masses" $\langle\bar{A}_\mu^2\rangle$ finite) is more involved.

You must expand out *all* non-quadratic pieces in $\exp[-1/g_0^2S]$, order by order in $ag_0$ in a fixed gauge, and derive lattice cut-off Feynman rules

*The exact expression for $\bar{F}^\alpha_{\mu\nu}$ is easily written as $\frac{1}{2}\lambda^\alpha\bar{F}^\alpha_{\mu\nu} = \Sigma_l\bar{A}_l - \Sigma_{l<l'}\frac{1}{2}\mathrm{i}g_0[\bar{A}_l, \bar{A}_{l'}]$ if we label the links $U_l = \exp(\mathrm{i}ag_0\bar{A}_l)$ counter clockwise around the plaquette.

by Gaussian integration. This task has been initiated by Hasenfratz and Hasenfratz [9]. In the Feynman gauge, the lattice propagator (see appendix B) is

$$\Delta_F^{\mu\nu}(x) = \frac{1}{L^4}\sum_k \delta^{\mu\nu} e^{ik\cdot x} \Big/ \left(4a^2 \sum_\lambda \sin^2\frac{ak_\lambda}{2}\right), \tag{2.21}$$

where the sum is over integers $\ell_\mu$ in the interval $-\pi/a < k_\mu = 2\pi\ell_\mu/L \leqslant \pi/a$, or on an infinite lattice, $L \to \infty$,

$$\Delta_F^{\mu\nu}(x) = \int_{-\pi/a}^{\pi/a} \frac{d^4k}{(2\pi)^4} \delta^{\mu\nu} e^{ik\cdot x} \Big/ \left(\frac{4}{a^2}\sum_\lambda \sin^2\frac{ak_\lambda}{2}\right). \tag{2.22}$$

In addition, there are vertices with arbitrary numbers of external gluons, since the lattice action is an infinite order polynomial in $\bar{A}_\mu$. Nonetheless, it is possible in low orders in $g_0$ to add together diagrams, and introduce a fixed mass scale $\Lambda_0$ (see discussion in section 3.3) to renormalize the series. As $a \to 0$, one can see how Lorentz invariance is restored, and the standard Feynman diagrams emerge. The task is arduous, but the message is simple. The lattice can be viewed as a specific gauge invariant (but Lorentz noninvariant) momentum cut-off scheme. A difficult way to rederive renormalized perturbation theory.

#### 2.2.2. *Non-perturbative continuum limit*

Before turning to the non-perturbative method of defining the continuum limit, we need to consider several properties of the (exact) lattice Green's functions ($L \to \infty$, fixed $g_0$). Indeed we can compute exactly the two point function

$$\langle U_\mu^\dagger(x)_{ij} U_\nu(0)_{kl}\rangle = \frac{1}{N}\delta_{x,0}\,\delta_{\mu,\nu}\,\delta_{il}\,\delta_{jk}. \tag{2.23}$$

Lattice "gluons" cannot propagate! [To derive this result note that since the Haar integral averages over all gauges, we get zero for $\mu \neq \nu$ or $x \neq 0$. For $\mu = \nu$, $x = 0$ the $\delta_{il}\delta_{jk}$ tensor is the only left and right invariant tensor, normalized by $\mathrm{Tr}(U^+U) = N$.] This result is a special case of Elitzur's Theorem [7] that gauge odd ($\Omega_x = -1$) averages are zero. The only interesting quantities are averages of gauge invariant functions $f(U_\ell)$.

The simplest gauge invariant quantity is the Wilson loop operator constructed by an ordered product on a closed curve $\mathcal{C}$ (see fig. 2):

$$\psi(\mathcal{C}) = \frac{1}{N}\mathrm{Tr}\left(P \prod_{l\in\mathcal{C}} U_l\right). \tag{2.24}$$

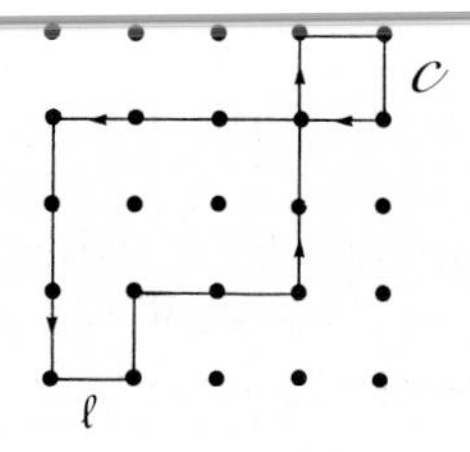


Fig. 2. Wilson loop operator, $\psi(\mathcal{C}) = (1/N)\mathrm{Tr}(P\prod_{\ell \in \mathcal{C}} U_\ell)$ on path ordered ($P$) curve $\mathcal{C}$; $W(\mathcal{C}) \equiv \langle \psi(\mathcal{C}) \rangle$.

All other gauge invariant polynomials are products over many such loops. In the continuum, the Wilson Loop,

$$\psi(\mathcal{C}) = \frac{1}{N}\mathrm{Tr}\left[P\exp\left(\mathrm{i}g_0\oint_{\mathcal{C}} A^\mu \mathrm{d}\xi_\mu\right)\right]. \tag{2.25}$$

arises in the path integral for a (spinless) quark source

$$J_\mu(x) = \int \mathrm{d}s\, Q(s)\dot{\xi}_\mu \delta^4(x_\nu - \xi_\nu(s)), \tag{2.26}$$

on a world line $\mathcal{C}$ parametrized by $\xi_\mu(s)$, where $J_\mu$ must be covariantly conserved, $[D_\mu, J_\mu] = 0$. Thus, we can consider the physically simple case of a rectangular loop, which separates quark and anti-quark charges a distance $R/a$ lattice sites for a Euclidean "time" $T/a$ (see fig. 3), as a

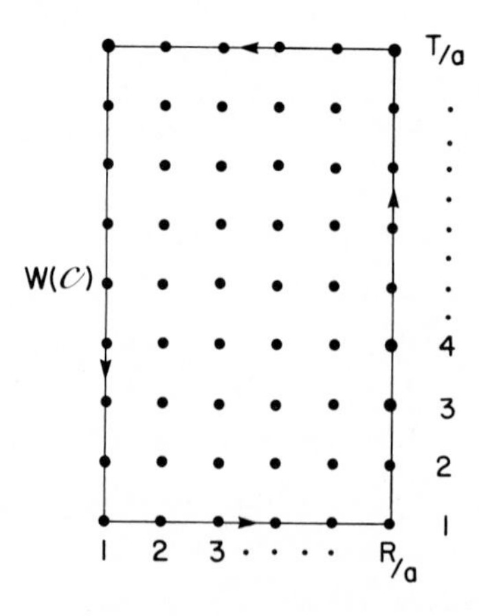


Fig. 3. The $R/a$ by $T/a$ rectangular Wilson loop used at large $T$ to define the static quark potential, $W(\mathcal{C}) \sim \exp[-TV(R)]$.

definition of the static energy of separation, $V(R)$.

$$W(\mathcal{C}) \to \exp\left[-TV\left(R, g_0^2, a\right)\right], \quad \text{as } T \to \infty. \tag{2.27}$$

Confinement on the lattice is given by a "potential" $V(R)$ that goes to infinity as $R \to \infty$. Indeed, as we shall see in the strong coupling approximations, a linearly increasing potential is expected:

$$V(R) \approx \kappa\left(g_0^2, a\right)R, \quad \text{as } R \to \infty, \tag{2.28}$$

where $\kappa$ is the surface (or string) tension. This gives Wilson's so-called *area law criterion* for quark confinement on the lattice:

$$W(\mathcal{C}) \to \exp(-\kappa RT), \tag{2.29}$$

as Area $= RT \to \infty$. Now we return to the continuum limit.

Again, since the lattice spacing does not enter into our expression for $\langle f[U_l]\rangle$ it appears difficult, if not ridiculous, to take a non-perturbative continuum limit. What we mean intuitively by the continuum limit is to interpolate more and more sites relative to a physical length, for example the Compton wave length $\lambda$ of the lowest mass bound state (called a glueball). On the Euclidean lattice, we detect such a state as an exponential fall off in the plaquette propagator with separation ($|n| =$ number of lattice sites):

$$G_{\mu\nu,\rho\sigma}(x) \equiv \frac{1}{N^2}\langle \mathrm{Tr}\left[U_{\mu\nu}(x)\right]\mathrm{Tr}\left[U_{\rho\sigma}(0)\right]\rangle_c \sim \exp(-|n|/\xi_{GB}). \tag{2.30}$$

(See fig. 4 and section 5). The correlation length $\xi_{GB}$ is a dimensionless function of the coupling. Relative to a physical distance ($T = a|n|$), there

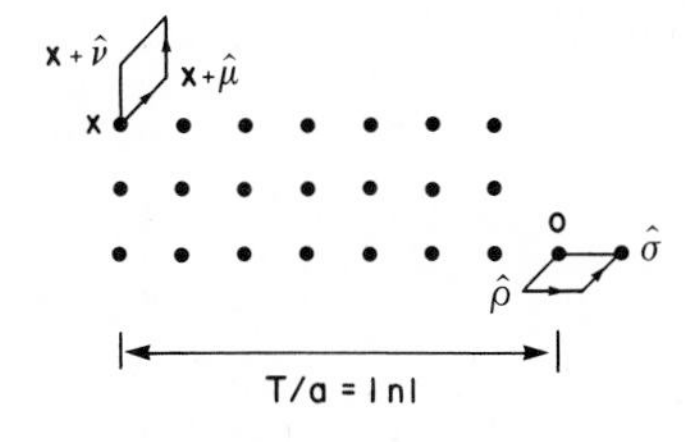


Fig. 4. The plaquette–plaquette Green's function, $G_{\mu\nu\rho\sigma}(x) = (1/N^2)\langle \mathrm{Tr}\,[U_{\mu\nu}(x)] \times \mathrm{Tr}\,[U_{\rho\sigma}(0)]\rangle$, which measures the glueball correlation length ($\xi_{GB}$) at large separation $T/a = |n| \to \infty, G \sim \exp(-|n|/\xi_{GB})$.

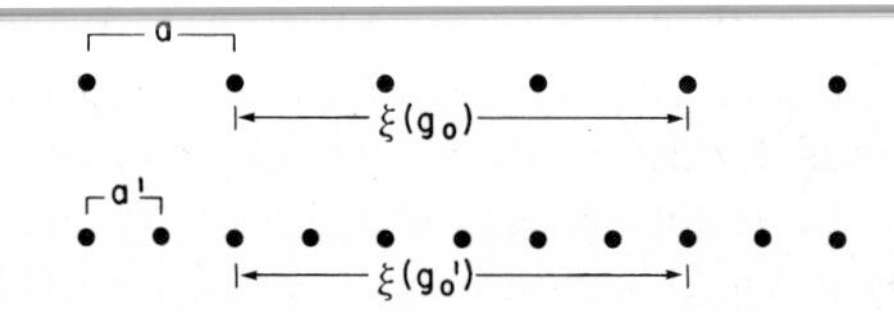


Fig. 5. The continuum limit for fixed Compton wave lengths $m^{-1} = a\xi(g_0)$ illustrated for $a \to a' = \frac{1}{2}a$ and $\xi(g_0') = 2\xi(g_0)$.

is an inverse glueball mass or Compton wavelength:

$$\lambda = (m_{\mathrm{GB}})^{-1} = a\xi_{\mathrm{GB}}(g_0^2). \tag{2.31}$$

To interpolate lattice sites, we must fix $\lambda$ as we decrease $a$ (e.g., $a \to a' = \frac{1}{2}a$ in fig. 5), and we must hope (or assume) that $g_0(a)$ can be changed continuously so that the correlation length diverges linearly in $a$, as $a \to 0$:

$$\xi(g_0) \to \text{const.}/a, \quad \text{as } g_0(a) \to g^*. \tag{2.32}$$

The fixed point ($g_0 = g^*$) at which a correlation length diverges is the trade mark of a second-order phase transition. In this limit $a \to 0$ (which is really $g_0 \to g^*$), we can hope to define a continuum theory. There may be many fixed points for a variety of correlation lengths; thus there are possibly zero, one or many continuum theories contained in a single lattice theory.

### 2.3. *Criterion for confinement*

We now wish to state sufficient conditions for confinement in the continuum limit of QCD. Clearly we need to consider the string tension. By dimensional considerations, the string tension defines a length

$$\kappa = \frac{1}{a^2}\frac{1}{\xi_{\mathrm{st}}^2(g_0^2)} \tag{2.33}$$

where the string tension correlation length $\xi_{\mathrm{st}} = 1/a\sqrt{\kappa}$ depends only on the bare coupling. Obviously we can try to define a continuum theory by just fixing $\kappa = \text{const.}$ as $a \to \infty$. Namely, look for a fixed point where $\xi_{\mathrm{st}} \to \infty$, then, by definition, the continuum theory will confine. As we will see in section 3, any gauge theory (including compact QED) has $\kappa > 0$ for strong coupling, so with a little luck this can be done. Indeed compact QED probably leads to a confining theory for $g_0 = e_0 \to e^* \approx 1.05$. This does not mean quantized Maxwell's equation confines.

We must apply another condition, which I call the *weak coupling correspondence principle*. Namely, for QCD we require that the limit $g_0 \to g^*$ corresponds to the weak coupling expansion where at short distances we have an asymptotically free theory. Presumably this "short distance" property of QCD is the correct explanation of large momentum physics. To achieve this, we require that the fixed point where $\xi_{st} \to \infty$ is the *same* ultraviolet stable fixed point at $g_0 = 0$ seen in the Gell-Mann–Low $\beta$-function:

$$-\Lambda \frac{d}{d\Lambda} g_R = \gamma_0 g_R^3 + \gamma_1 g_R^5 + \cdots, \tag{2.34}$$

for the renormalized coupling $g_R(\Lambda)$ of perturbation theory with subtraction point $p^2 = \Lambda^2$.

In other words, the principle is, that fixing the scale-breaking mass parameter $\Lambda$ (say in the $\overline{MS}$ scheme, $\Lambda_{\overline{MS}}$) must be *equivalent* to renormalizing lattice gauge theory by fixing the string tension mass $\sqrt{\kappa}$ as $a \to 0$. All masses are arbitrary but their ratios are physically measurable (e.g., $\sqrt{\kappa}/\Lambda_{\overline{MS}} \approx 400$ MeV/100 MeV $\approx 4$). We now translate these words into a scaling condition.

First we note that in the lattice cut-off approach to renormalized perturbation theory, it can be shown that $g_0(a) \to 0$, in obedience with the scaling condition

$$a \frac{d}{da} g_0(a) = \gamma_0 g_0^3 + \gamma_1 g_0^5 + \cdots, \tag{2.35}$$

where the coefficients,

$$\gamma_0 = \frac{11}{3}\left(\frac{N}{16\pi^2}\right), \qquad \gamma_1 = \frac{34}{3}\left(\frac{N}{16\pi^2}\right)^2, \tag{2.36}$$

are exactly the same as those in the one- and two-loop terms (2.34) in the Gell-Mann–Low function for the renormalized coupling. (The essence of the argument is repeated below. Let us accept it for now.) For the two-loop beta function, $\beta_2(g_0) \equiv \gamma_0 g_0^3 + \gamma_1 g_0^5$, it can be shown that:

$$a = \text{const.}\exp\left[-\int_{g_0}^{\infty} \frac{dg_0'}{\beta_2(g_0')}\right]$$

$$= \text{const.}\left(\frac{\gamma_1}{\gamma_0^2} + \frac{1}{\gamma_0 g_0^2}\right)^{\gamma_1/2\gamma_0^2} \exp(-1/2\gamma_0 g_0^2) \tag{2.37}$$

is the exact integral. As $g_0 \to 0$ it has become conventional to define the lattice integration constant $\Lambda_0$ by:

$$\Lambda_0 \equiv \frac{1}{a}\left(1/\gamma_0 g_0^2\right)^{\gamma_1/2\gamma_0^2}\exp\left(-1/2\gamma_0 g_0^2\right). \tag{2.38}$$

The lattice $\Lambda_0$ is related by a known [9] numerical coefficient to $\Lambda_{\overline{\mathrm{MS}}}$ of the modified Minimal-Subtraction scheme

$$\Lambda_{\overline{\mathrm{MS}}} = 39.0\exp\left(-3\pi^2/11N^2\right)\Lambda_0, \tag{2.39}$$

by comparing the two schemes for developing renormalized perturbation theory.

Now the *confinement criterion for the continuum limit of lattice QCD* can be stated precisely as fixing $\sqrt{\kappa}$ relative to $\Lambda_{\overline{\mathrm{MS}}}$. Hence

$$\sqrt{\kappa} \simeq \frac{c_0}{a}\left(1/\gamma_0 g_0^2\right)^{\gamma_1/2\gamma_0^2}\exp\left(-1/2\gamma_0 g_0^2\right)\left[1+\mathrm{O}\left(g_0^2\right)\right] \tag{2.40}$$

for $c_0 > 0$, and $g_0^2 \to 0$. In a quarkless world $c_0 = \sqrt{\kappa}/\Lambda_0$ would be an experimentally measurable continuum mass ratio. Note that confinement requires that the dimensionless surface tension $a^2\kappa$ which you compute on the lattice vanishes exponentially as $g_0^2 \to 0$. However, if the scaling law (2.40) is obeyed, this lattice "deconfinement" is *reinterpreted* by dimensional transmutation as fixing the physical string tension.

Now to complete our argument, we wish to show that the scaling rule (2.35) for $g_0(a)$ is correct. Consider defining $g_{\mathrm{R}}(g_0, a, \Lambda)$ by evaluating the cut-off three gluon vertex function at $p_i^2 = \Lambda^2$

$$g_{\mathrm{R}}(g_0, a, \Lambda) = \Gamma(p_1, p_2, p_3)\big|_{p_i^2=\Lambda^2}. \tag{2.41}$$

First, we require that $g_{\mathrm{R}} \to g_0$ as $g_0 \to 0$ to correspond to lowest order perturbation theory. Secondly, we know that $g_{\mathrm{R}}$ is a dimensionless function of $g_0$ and $a\Lambda$ [or equivalently $t = \log(a\Lambda)$]. Thirdly, we demand that renormalization as $a \to 0$ corresponds to adjusting $g_0(a)$ to fix $g_{\mathrm{R}}$ at each stage, $(\mathrm{d}/\mathrm{d}a)g_{\mathrm{R}}(g_0(a), t) = 0$. These three properties suffice. Rewrite the Gell–Mann–Low eq. (2.34) for $g_{\mathrm{R}}$ as

$$\frac{\mathrm{d}}{\mathrm{d}t}\frac{1}{g_{\mathrm{R}}^2} = 2\gamma_0 + 2\gamma_1 g_{\mathrm{R}}^2 + \cdots, \tag{2.42}$$

and integrate it subject to the $g_{\mathrm{R}} \to g_0$ boundary condition.

$$\frac{1}{g_{\mathrm{R}}^2} = \int_0^t \left(2\gamma_0 + 2\gamma_1 g_{\mathrm{R}}^2 + \cdots\right)\mathrm{d}t' + \frac{1}{g_0^2} + \mathrm{O}(1) + \mathrm{O}\left(g_0^2\right). \tag{2.43}$$

Then take the derivative with respect to $a$:

$$0 = a\frac{\mathrm{d}}{\mathrm{d}a}\frac{1}{g_{\mathrm{R}}^2} = 2\gamma_0 + 2\gamma_1 g_{\mathrm{R}}^2 + a\frac{\mathrm{d}}{\mathrm{d}a}\left(\frac{1}{g_0^2}\right)\left[1+\mathrm{O}(g_0^2)\right] + \mathrm{O}(g_{\mathrm{R}}^4). \tag{2.44}$$

With $g_{\mathrm{R}} \approx g_0$, the solution is

$$-a\frac{\mathrm{d}}{\mathrm{d}a}\frac{1}{g_0^2} = 2\gamma_0 + 2\gamma_1 g_0^2 + \mathrm{O}(g_0^4), \tag{2.45}$$

as promised.

To reiterate, the weak coupling correspondence principle requires the continuum limit to be taken by taking $\beta = 1/Ng_0^2(a) \to \infty$ as $a \to 0$ by the dictates of the renormalization group:

$$\beta = \frac{1}{Ng_0^2} \simeq \frac{11}{3\times 16\pi^2}\left[\ln(1/a\Lambda_0) + \tfrac{17}{121}\ln\ln 1/a\Lambda_0\right] + \mathrm{O}(1). \tag{2.46}$$

For better or worse, this is the *only limit* of lattice gauge theory consistent with asymptotically free perturbation theory. The bare coupling $g_0$ is dimensionally transmuted into a mass parameter $\Lambda_0$. Next, we stated the condition on the scaling relation for the string tension $\sqrt{\kappa}$, (2.40) if it is to be non-zero in the continuum theory.

Likewise, any mass parameter $m_i$ must also scale as

$$m_i = \frac{c_i}{a}\left(\frac{3\times 16\pi^2}{11Ng_0^2}\right)^{51/121}\exp\left(-\frac{3\times 16\pi^2}{11Ng_0^2}\right), \tag{2.47}$$

unless the physical ratio $m_i/\Lambda_0$ is either zero (e.g. chiral pions) or infinite (e.g. $Z_2$ monopoles in SO(3) lattice theory). For example, in section 5, we shall discuss further the plaquette–plaquette correlation lengths ($\xi_{\mathrm{GB}}$), [see eq. (2.30)] which measure the glueball mass ($m_{\mathrm{GB}} = 1/a\xi_{\mathrm{GB}}$). In principle, with this continuum renormalization prescription (2.46), we could extract Euclidean Green's functions for glueballs, and continue in $p_4 = \mathrm{i}E$ to find all the scattering amplitudes for the lowest mass glueballs. Lattice QCD provides a precise algorithm for the calculations of continuum physics; only time will tell if man and machines can use the algorithm to extract useful numbers.

## 3. Approximation techniques

We present three approximation schemes. The first two, strong coupling and Monte Carlo, are in direct competition, since they both are local or finite lattice approximations in essence. They remind me of the legendary American race between the coal miner John Henry and the newly invented steam-driven pile driver: brute force techniques testing the endurance of man (strong coupling) versus machine (Monte Carlo). The legend reports that, at the end of the day's race, John Henry wins but promptly dies.

The third approximation is a more ambitious and elegant theoretical program to expand in $1/N$ about the hadronic Born term at $N=\infty$. This approach gives considerable phenomenolgical insight, but its bright promise for analytical techniques is still a futuristic dream (see section 6.3). Except in two-dimensional QCD, even the leading $N=\infty$ term has not been found.

Even in their primitive state, a sort of mental interpolation between these three approximations allows us to hazard a guess at the main features of lattice QCD. Obviously, more rigorous and analytic techniques are needed. Section 4 illustrates some of the more analytical approaches in the context of Abelian theories and attempts to use them to illuminate the results of the approximations presented below.

### 3.1. *Strong coupling expansion*

In continuum quantum mechanics, the weak coupling expansion, or Feynman–Dyson series, is the standard tool for initial investigations. Correspondingly, in lattice statistical mechanics, the strong coupling (or high temperature $\beta = 1/kT = 1/\hbar g_0^2 \to 0$) expansion is the standard tool. One simply expands the exponential factor

$$\begin{aligned}\exp\left(-\frac{1}{g_0^2}S\right) &= \prod_{P,P^\dagger} \exp\left[\frac{1}{g_0^2}\operatorname{Tr}(U_P)\right] \\ &= \prod_{P,P^\dagger}\sum_{\{n_P\}}\frac{1}{n_P!}\left(\frac{1}{g_0^2}\right)^{n_P}(\operatorname{Tr}U_P)^{n_P}\end{aligned} \tag{3.1}$$

where $U_{P^\dagger} \equiv U_P^\dagger$, and performs the Haar integrals in each order of $\beta \sim 1/g_0^2$. Of course, as in the Feynman–Dyson expansion, the perfection of bookkeeping techniques to enumerate and give weights to diagrams

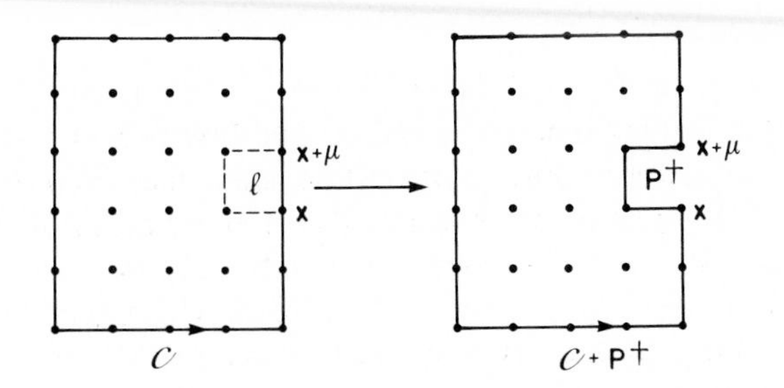


Fig. 6. Integral over link $U_\mu(x)$ on contour $\mathcal{C}$ gives rise to a "keyboard" shifted contour $\mathcal{C}' = \mathcal{C} + P^\dagger$.

can become a lifetime work. Here we will only illustrate the technique in its least sophisticated form.

An interesting quantity to consider is the Wilson loop:

$$W(\mathcal{C}) = \frac{1}{Z}\prod_l \int \mathrm{d}U_l \frac{1}{N}\operatorname{Tr}\Big(P\prod_{\ell\in\mathcal{C}} U_l\Big)\exp\Big(-\frac{1}{g_0^2}S\Big). \tag{3.2}$$

For example, let us begin with compact QED. Note that each link integral, over $U_\mu(x) = \exp[\mathrm{i}\theta_\mu(x)]$, on the boundary of the curve $\mathcal{C}$ (see fig. 6),

$$\int_{-\pi}^{\pi}\frac{\mathrm{d}\theta_\mu(x)}{2\pi}\prod_{\ell\in\mathcal{C}}\mathrm{e}^{\mathrm{i}\theta_l}\prod_{P,P^\dagger}\frac{1}{n_P!}\left(\frac{1}{e_0^2}\right)^{n_P}\exp\left[\mathrm{i}n_P(\Delta_\mu\theta_\nu - \Delta_\nu\theta_\mu)\right] \tag{3.3}$$

gives zero unless we pick up an anti-link factor, $e_0^{-2}\exp[-\mathrm{i}\theta_\mu(x)]$, from the expansion of the action. Doing this integral gives a new loop $\prod_{\ell\in\mathcal{C}+P^\dagger}\exp(\mathrm{i}\theta_l)$, where $\mathcal{C}\to\mathcal{C}+P^\dagger$ by a "keyboard" shift with a new plaquette ($P^\dagger$) inserted at that link ($x, \mu$).

Iteratively, the contour $\mathcal{C}$ can be shifted to a point. The lowest non-zero term is the minimal area $[A_{\min}(\Sigma)]$ swept out on a surface $\Sigma$ attached to the curve $\mathcal{C}(\mathcal{C} = \partial\Sigma)$.

$$W_{\mathrm{QED}} = (\mathcal{C}) \simeq \left(1/e_0^2\right)^{A_{\min}(\Sigma)/a^2}\left[1 + \mathrm{O}\left(1/e_0^6\right)\right]. \tag{3.4}$$

In particular, if we consider a rectangular $R\times T$ Wilson loop ($A_{\min} = RT$, see fig. 3), the result for compact QED as $RT\to\infty$ is:

$$W_{\mathrm{QED}}(\mathcal{C}) \sim \exp(-\kappa_{\mathrm{QED}}RT), \qquad \kappa_{\mathrm{QED}} = \frac{1}{a^2}\left[\log e_0^2 - \mathrm{O}\left(1/e_0^2\right)\right]. \tag{3.5}$$

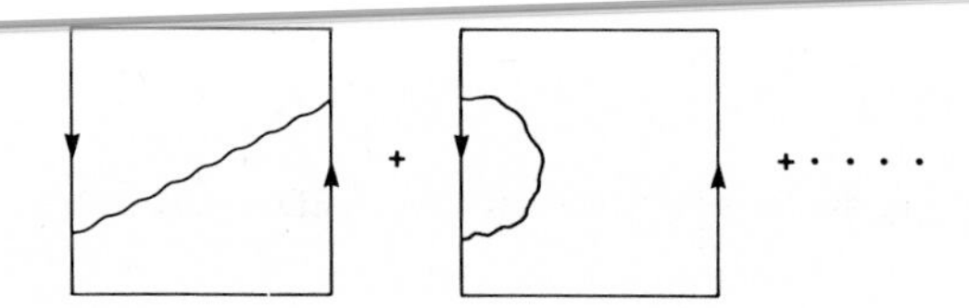


Fig. 7. First-order Feynman diagrams for a single photon exchange inside a Wilson loop.

We now contrast this linear confinement for strong coupling lattice QED, with the first-order weak coupling result from cut-off Feynman diagrams (see fig. 7):

$$W(\mathcal{C}) = \exp\left(-\frac{e_0^2}{4\pi}\left[\frac{T+R}{a} - \frac{2}{\pi}\ln\frac{RT}{a^2} - \frac{1}{\pi}\ln(2+R^2/T^2+T^2/R^2) - \frac{2}{\pi}\frac{T}{R}\tan^{-1}(T/R) - \frac{2}{\pi}\frac{R}{T}\tan^{-1}(R/T) + \frac{4}{\pi}\right]\right). \tag{3.6}$$

If we believe that these two series have non-zero radii of convergence, there must be at least two phases

$$\begin{aligned} W_{\text{QED}}(\mathcal{C}) &\sim \exp(-\text{Perimeter}), \quad \text{small } e_0^2, \\ &\sim \exp(-\text{Area}), \quad \text{large } e_0^2. \end{aligned} \tag{3.7}$$

Actually, if we look more closely at the static potential (2.27),

$$\begin{aligned} V_{\text{QED}}(R) &\simeq \frac{e_0^2}{4\pi}\left(\frac{1}{a} - \frac{1}{R}\right), \quad e_0^2 \to 0, \\ &\simeq \frac{R}{a^2}\log e_0^2, \quad e_0^2 \to \infty, \end{aligned} \tag{3.8}$$

we see that the weak coupling domain has a perimeter piece due to the self-energy of the "electron" ($1/a$) plus a Coulombic attraction ($-1/R$). Rigorous proofs have been given to prove the existence of both the linear confining phase at strong coupling and the Coulombic phase at weak coupling. Monte Carlo simulations [10], as we shall see, support this picture with an isolated second-order fixed point at $1/e_0^{*2} = 1.04$ separating the two phases (see fig. 8). Wegner's $Z_2$ gauge theory has a similar phase diagram (fig. 9) with a single fixed point, except this time

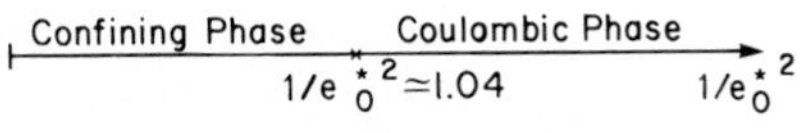

Confining Phase Coulombic Phase
$1/e_0^{*2} \simeq 1.04$ $1/e_0^{*2}$

Fig. 8. The phase diagram for U(1) (compact) lattice QED with the Monte Carlo estimate of transition [10].

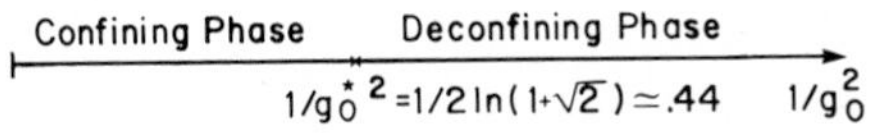

Confining Phase Deconfining Phase
$1/g_0^{*2} = 1/2 \ln(1+\sqrt{2}) \simeq .44$ $1/g_0^2$

Fig. 9. The phase diagram for the $d = 4$ Wegner $Z_2$ gauge theory.

the transition is probably first order, located at $1/g_0^2 = \frac{1}{2}\ln(\sqrt{2}+1)$, by duality arguments (section 4.1). The weak coupling $Z_2$ phase has no Coulombic piece—not too surprisingly since $Z_2$ hardly could be expected to have a photon mode with $\theta_\mu = 0, \pi$. (As $N$ is increased in $Z_N$, the single phase transition of $Z_2$ splits for $N \geqslant 5$ revealing an intermediate Coulombic phase which extends to zero coupling only for $N = \infty$ or compact QED.)

### 3.1.1. *Non-Abelian strong coupling*

Now let us return to SU($N$) gauge theory. The first order weak coupling result (3.6) is exactly the same, with $e_0^2$ replaced by $g_0^2(N^2-1)/2N$ so again there is an apparent weak coupling Coulombic phase. On the other hand, confinement requires $\kappa \sim \exp(-1/\gamma_0 g_0^2)$ as $g_0^2 \to 0$, so the string tension in linearly confining QCD (if it exists) cannot be seen in any order of weak coupling. If we are to believe the confinement hypothesis, the standard weak coupling series (which is at best an asymptotic series) must have *no range of validity*.

On the other hand, in strong coupling we obtain a non-zero series for $\kappa$ and the series is known to converge in a finite radius. To find the first non-zero contribution for the Wilson loop, the argument proceeds in analogy with QED. The zero integrals,

$$\int \mathrm{d}U\, U_{ij} = \int \mathrm{d}U\, U_{kl}^{\dagger} = 0 \tag{3.9}$$

over single links $U \equiv U_\mu(x)$ on $\mathcal{C}$, require us to "pave" the loop with plaquettes from the expansion of the exponentiated action. All non-zero diagrams are closed surfaces or surfaces bounded by $\mathcal{C}$. In lowest order, the combinatorics is easy, using the formula for a doubly occupied link,

$$\int dU\, U_{ij} U_{kl} = \int dU\, U^{\dagger}_{ij} U^{\dagger}_{kl} = \delta_{N,2} \varepsilon_{ik} \varepsilon_{jl},$$

$$\int dU\, U_{ij} U^{\dagger}_{kl} = \frac{1}{N} \delta_{il} \delta_{jk}, \tag{3.10}$$

repeatedly on links where two plaquettes intersect.

For SU(2) the quadratic invariant ($\det U = 1$) gives a special two-plaquette contribution. For $N > 2$, we have:

$$W(\mathcal{C}) = \left(1/g_0^2 N\right)^{TR/a^2} \left[1 + 2(d-2)\frac{TR}{a^2}\left(\frac{1}{g_0^2 N}\right)^4 + \cdots\right], \tag{3.11}$$

and a non-zero string tension as $RT \to \infty$:

$$\kappa = \frac{1}{a^2}\left[\log(g_0^2 N) - 2(d-2)\left(\frac{1}{g_0^2 N}\right)^4 - \cdots\right]. \tag{3.12}$$

The first term is the flat minimal surface, followed by a one-cube distortion. (see fig. 10.) For large area, the multiple distortions (acting like quasi-particles on the $d = 2$ surface) exponentiate.

These series expansions have been carried out [11, 12] to at least $(1/g_0^2)^{12}$ for $Z_2$, $Z_3$, U(1), SU(2), SU(3), and SU($\infty$) gauge theories in $d = 4$ dimensions and they are plotted in figs. 11–15. Notice that there is *no* dramatic evidence which distinguishes the Abelian theories which are known to have a decontinuing transition with $\kappa(g^*) = 0$, and the non-Abelian theories with no transition and $\kappa \sim \exp(-1/\gamma_0 g_0^2) \to 0$ as $g_0 \to 0$. The strong coupling expansions all show a rapid decrease in string tension for $g_0^2 \approx O(1)$. The polynomial corrections naturally give rise to a

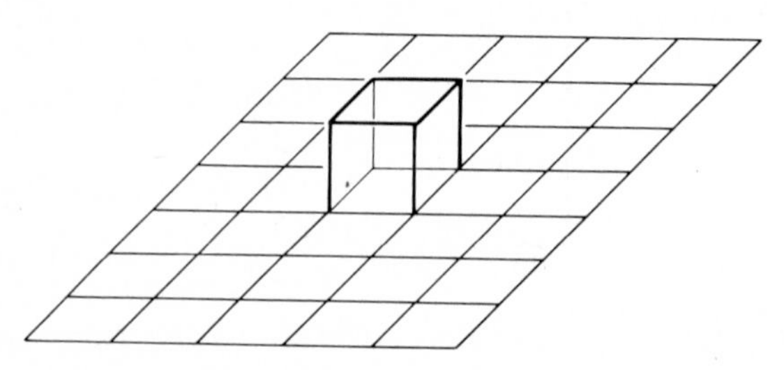

Fig. 10. The one-cube distortion as the first correction to the strong coupling series for the string tension.

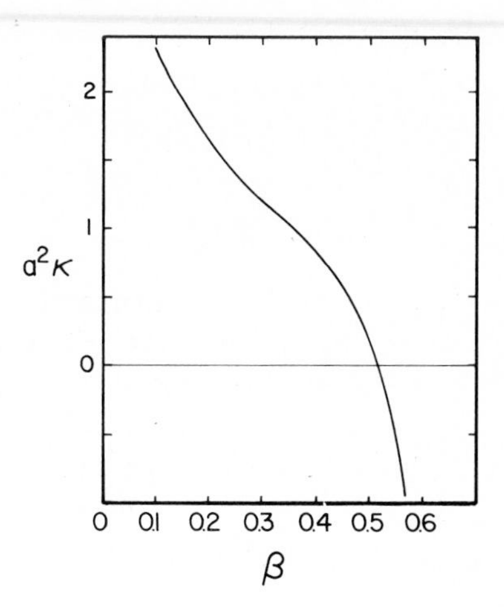


Fig. 11. Strong coupling expansion [11] for $a^2\kappa$ in $d = 4$ $Z_2$ gauge theory to twelfth order in $x = \tanh\beta$.

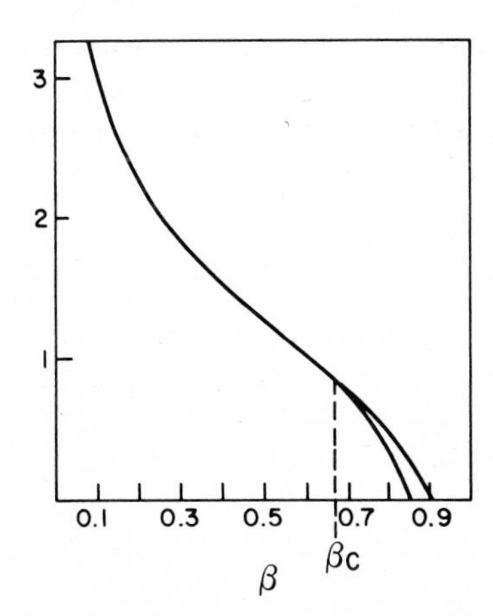


Fig. 12. Strong coupling expansion [11] for $a^2\kappa$ in $d = 4$ $Z_3$ gauge theory to ninth (upper line) and twelfth order (lower line) in $x = \tanh\beta$.

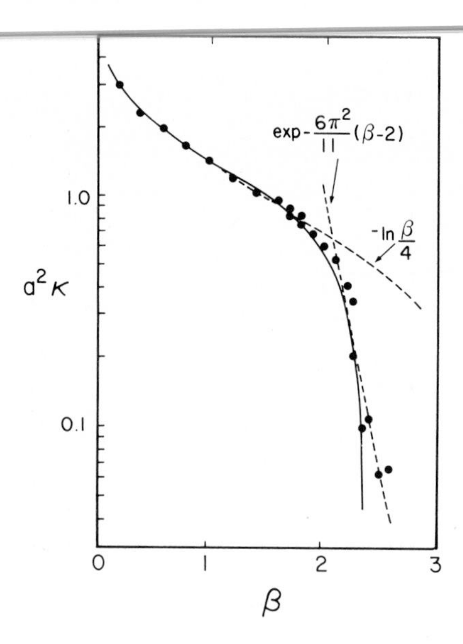


Fig. 13. Strong coupling expansion [11] for $a^2\kappa$ in $d=4$ SU(2) gauge theory to twelfth order in $\beta = 4/g_0^2$. Note comparison with the Monte Carlo points (•) of ref. [18].

nearby zero in $\kappa(g_0^2)$. Attempts to massage the series via Padés do not materially improve the situation. Probably the series diverge in this crossover region $[g_0^2 \approx O(1)]$ due to roughening phenomena [13].

Before leaving this subject, we wish to make some general comments on the technical aspects of strong coupling expansions. In general, there is the problem of calculating multi-link integrals which can be generated by differentiating the single link integral [14]:

$$Z_0(J^\dagger, J) = \int \mathrm{d}U \exp \mathrm{Tr}(J^\dagger U + J U^\dagger). \tag{3.13}$$

Extensive work on this generating function for U($N$) and SU($N$) exist in the literature. Differential equations expressing $UU^\dagger = 1$ [and det $U=1$ for SU($n$)] have been solved for all $N$ [15], including $N=\infty$ [14, 16]. For SU($N$), $Z_0(J^\dagger, J)$ becomes:

$$Z_0(J^\dagger, J) = \prod_{i=1}^{N} \int_0^{2\pi} \mathrm{d}\phi_i\, \delta\Big(\sum_i \phi_i\Big) \exp\Big[2\sum_i z_i \cos(\phi_i - \theta) + \ln G\Big], \tag{3.14}$$

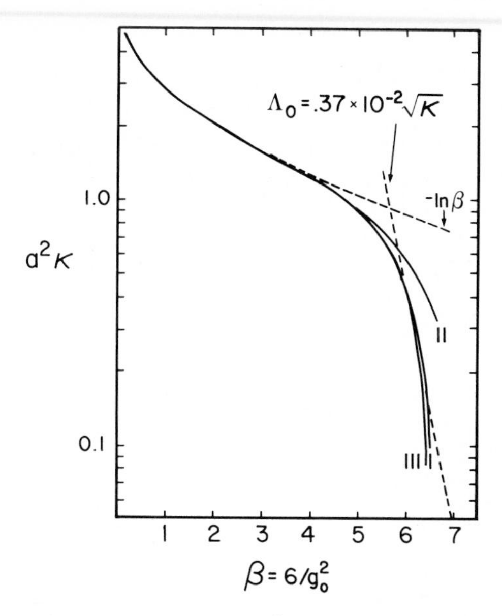


Fig. 14. Strong coupling expansion [11] for $a^2\kappa$ in $d = 4$ SU(3) gauge theory to tenth (I), eleventh (II), and twelfth (III) order in $\beta = 6/g_0^2$.

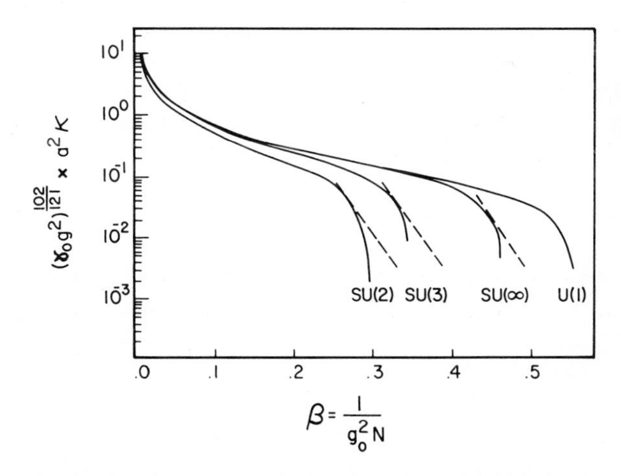


Fig. 15. Comparison of strong coupling expansions for $a^2\kappa$ in SU(2) [11], SU(3) [11], SU($\infty$) [12], and U(1) [12] gauge theories all to at least twelfth order in $\beta = 1/g_0^2 N$.

where $z_i$ are eigenvalues of $(JJ^\dagger)^{1/2}$, $\theta$ is the phase of $(\det J/\det J^\dagger)^{N/2}$ and $G$ is a Jacobian given in ref. [14]. However, the more difficult task is the combinatorics of enumerating the connected diagrams.

Another approach, which partially sums the multi-plaquette diagrams is to expand in a character expansion. This is extremely valuable for $Z_2$ theories, since $e^{\beta s_P} = \cosh\beta + s_P \sinh\beta = \cosh\beta[1+\omega(\beta)s_P]$ means that each plaquette $P$ occurs at most once in a single diagram. For the U(1) group, we have the Fourier–Bessel expansion

$$\exp(2\beta\cos\theta_P) = \sum_\nu e^{-\beta} I_\nu(\beta) e^{i\nu\theta_P}. \tag{3.15}$$

For a general non-Abelian group this is replaced by an expansion in terms of irreducible characters $\chi_\nu$.

$$\begin{aligned} &\exp\left[2\beta N \operatorname{Tr}\left(U_P + U_P^\dagger\right)\right] \\ &= \sum_\nu \omega_\nu \chi_\nu(U_P) \\ &\simeq \omega_0\left[1 + N\omega_N(\beta)\operatorname{Tr}\left(U_P + U_P^\dagger\right) + \cdots\right]. \end{aligned} \tag{3.16}$$

The first coefficient is:

$$\omega_N(\beta) = \frac{1}{2N^2}\frac{\partial}{\partial\beta}\log\left\{\int dU \exp\left[N\beta \operatorname{Tr}(U+U^\dagger)\right]\right\}. \tag{3.17}$$

So the new parameter in the expansion, $\omega_N(\beta)$, now lies in the interval $0 \leqslant \omega_N(\beta) < 1$ for $0 < \beta = 1/Ng_0^2 < \infty$ and convergence might be improved. In this form, multiple link integrals are computed using the orthogonality of characters $[d_\nu = \chi_\nu(1)]$,

$$\int dU \chi_\nu(UV)\chi_{\bar\nu}^*(UW) = \frac{1}{d_\nu}\delta_{\nu\bar\nu}\chi_\nu(W^\dagger V), \tag{3.18}$$

and other standard techniques of group representation theory. The first two terms in the character expansion for $\kappa$ are identical for the U(1) and SU($N$) groups:

$$\kappa = \frac{1}{a^2}\left[\log\frac{1}{\omega_N} - 2(d-2)\omega_N^4 - \cdots\right]. \tag{3.19}$$

To make comparisons with leading strong coupling results note that $\omega_N \simeq \frac{1}{2}e_0^2$, $1/g_0^2$, and $1/Ng_0^2$ for U(1), SU(2), and SU($N>2$) respectively.

Even with characters, the difficulty, which limits the series to around ten terms, is enumerating diagrams, not computing Haar integrals. Computer algorithms to enumerate gauge-theory diagrams are difficult to construct, and unfortunately, very high order series involving graphs with complex topologies and self-interactions, are probably necessary to really distinguish between different gauge groups. In my opinion, Monte Carlo simulations will easily win the race with strong coupling expansions. After all, for a local variable such as the free energy $F$, strong coupling to order $l=10$ only involves connected diagrams of length 2 (two adjacent cubes) which fit into a $3^4$ lattice. In effect, very small lattice simulations are competitive with strong coupling.

### 3.2. *Monte Carlo simulations*

The Monte Carlo method is an integration algorithm for direct computation of the $4(N^2-1)L^4$ integrals for SU($N$) on a finite lattice. Even with modern computers and small lattices, this is a non-trivial task. For example, if we take a $10^4$ lattice with a ten point mesh per angle for each of the three angles in SU(2) on the $4\times10^4$ links, we re-express the multiple integral as a sum with $10^{120000}$ terms! (The age of the universe is only $10^{29}$ pico seconds.)

We must avoid algorithms which exponentiate the volume factor $L^4$. Monte Carlo replaces the exponential growth with linear dependence on $L^4$ by a statistical sampling of the phase space.

Consider computing the thermal average

$$\langle f[U_l]\rangle = \frac{1}{Z}\prod_l \int \mathrm{d}U_l f[U_l]\exp(-\beta S[U]), \tag{3.20}$$

by picking a series of configurations, $C_1, C_2, C_3, \ldots$. Each configuration $C=\{U_1, U_2, \ldots\}$ gives one value to the $(N^2-1)\mathrm{d}L^d$ angles in phase space. Suppose also that the statistical probability of finding a configuration in the series reflects the equilibrium probability distribution.

$$\mathrm{d}P_{\mathrm{eq}}[C] = \frac{1}{Z}\prod_l \mathrm{d}U_l \exp(-\beta S[C]). \tag{3.21}$$

A sum over a large but finite number ($I$) of configurations directly gives an approximation to the thermal average.

$$\langle f[U_l]\rangle = \frac{1}{I}\sum_{\alpha=1}^{I} f[C_\alpha]+\mathrm{O}(1/\sqrt{I}), \tag{3.22}$$

subject only to the standard errors of the finite sample.

So now we must learn how to construct such a beautiful equilibrium sequence of configurations $\{C_1, C_2, \ldots\}$. We know that any thermal system in contact with a reservoir of temperature $T = 1/k\beta$ approaches the equilibrium distributions (3.21), if it is allowed to undergo transition $T_{C_i \to C_f}$. Introducing an equilibration (or fifth) time $t$, the Master Equation,

$$\frac{\mathrm{d}}{\mathrm{d}t} P[C_f] = \sum_{C_i} P[C_i] T_{C_i \to C_f} - P[C_f] \sum_{C_i} T_{C_f \to C_i}, \tag{3.23}$$

drives the probability distribution toward equilibrium ($P[C] \to P_{\mathrm{eq}}[C]$) as $t \to \infty$, if detailed balance is satisfied:

$$\exp(-\beta S[C_i]) T_{C_i \to C_f} = \exp(-\beta S[C_f]) T_{C_f \to C_i}. \tag{3.24}$$

By similar arguments, if we also use detailed balance to generate our sequence iteratively going from $C_\alpha \to C_{\alpha+1}$, we can demonstrate that it will *eventually* approach an equilibrium sequence. (See ref. [17], for details.)

Two common algorithms are as follows. Both build up a new configuration by changing one link at a time, $U_l \to U_l' = U_l + \Delta U_l$. The locality of the action assures you that $\Delta S = S[C] - S[C']$ depends only on the $2(d-1)$ plaquettes containing $l$ (technically the set of plaquettes in the co-boundary of $l$, $P \in \hat{\partial} l$, see fig. 16):

$$\Delta S = \sum_{P \in \hat{\partial} l} \mathrm{Tr}(\Delta U_P). \tag{3.25}$$

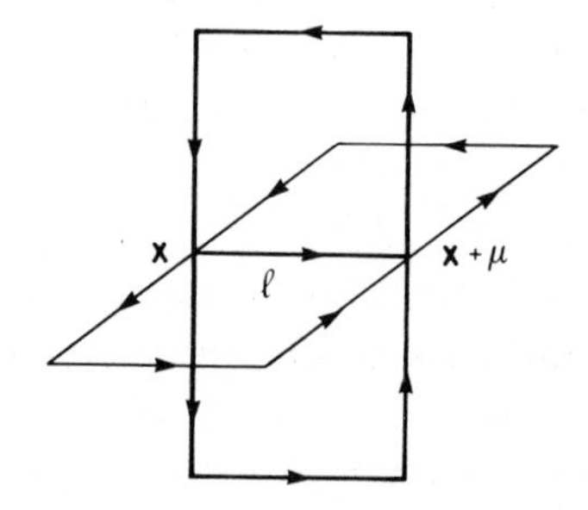


Fig. 16. The set of plaquettes $\{P\}$ containing the link $\ell$ form the co-boundary of $\ell$, ($P \in \hat{\partial}\ell$).

The first and most efficient algorithm is the heat-bath method: Knowing $S[C]$ you can choose the $U_l \to U_l'$ transition rate,

$$T_{C \to C'} \sim \exp(-\beta S[C']), \tag{3.26}$$

to satisfy detailed balance at each step. This is most useful for $Z_N$, U(1), and SU(2), where the relative probability weight $\exp(-\beta \Delta S)$ is easily parametrized.

The second algorithm is the Metropolis method. Here, if the shift $U_l \to U_l'$ decreases the action $S[C]$, the new link $U_l'$ is accepted; while if the shift increase the action, the new link $U_l'$ is conditionally accepted with probability $\exp(-\beta \Delta S)$ to satisfy detailed balance.

$$\begin{aligned} T_{C \to C'} &= 1, & S[C'] &\leqslant S[C], \\ T_{C \to C'} &= \exp(-\beta S[C] + \beta S[C']), & S[C'] &> S[C]. \end{aligned} \tag{3.27}$$

One may view the decreasing action as the classical "motion" and the random fluctuations to larger action as the quantum correction. Without this quantum correction, we would have a minimization routine for the classical action.

In practice, for both of these algorithms, while the random shift $U_l \to U_l'$ is chosen to favor small action configurations, the transition rate $T_{C \to C'}$ must still be imposed as a filter by rejecting a fraction of the trial values for $U_l'$. The art of Monte Carlo is to balance rejection rates, algebraic computation times, and convergence in the iteration number $(\alpha \to \alpha + 1)$, so as to find overall an efficient algorithm. Reasonably simple programs have been developed [18], which perform quite well for pure gauge theories, although increasingly sophisticated software and hardware, which promise substantial increases in speed, are becoming necessary for further progress. The reader is referred to Creutz's 1981 Erice lectures [18] and references therein for details.

### 3.2.1. *Monte Carlo results for the string tension*

Creutz has attempted to compute the string tension for SU(2) and SU(3) gauge theories by ratios of Wilson loops, $W(R/a, T/a)$,

$$\kappa(g_0^2) = -\frac{1}{a^2} \log \frac{W(J,J)W(J+1,J+1)}{W(J,J+1)W(J+1,J)}, \quad \text{as } J \to \infty, \tag{3.28}$$

which for large $RT$ cancel the constant and perimeter terms but not the area term. In the continuum limit, divergences along the perimeter and at the corners render the constant and perimeter terms infinite, but the area coefficient $\kappa(g_0^2)$ should be finite in the scaling limit [18]. Some results from Monte Carlo simulation are presented in figs. 17 and 18.

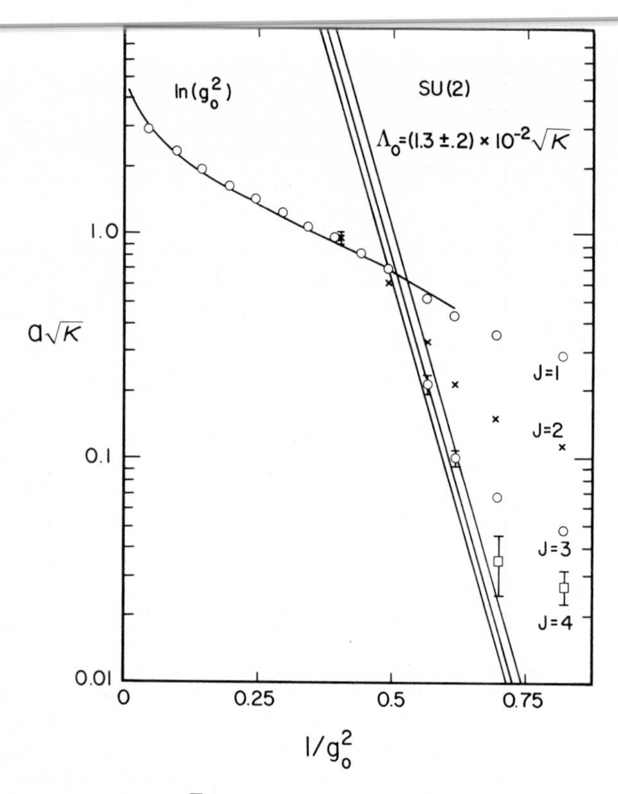


Fig. 17. The string tension $a\sqrt{\kappa}$ for SU(2) gauge theory computed by Monte Carlo simulations [18] from the ratios of loops defined in eq. (3.28).

Notice that at strong coupling the Monte Carlo agrees very well with the first term in the series. In SU(2) there is pure strong coupling behavior for $\beta = 4/g_0^2 \lesssim 2.2$*. Beyond $\beta \simeq 2.2$, it is plausible that the envelope of curves for increasing $J$ fall on the exponential required by the confinement criterion (2.40). (It is also possible that the theory is beginning to deconfine!) Monte Carlo, restricted to a finite lattice (here $L \lesssim 10$), is in effect restricted to couplings larger than those for which $\xi \simeq O(L)$. For smaller couplings, the correlation lengths can no longer diverge as $\xi \sim \exp(1/\gamma_0 g_0^2)$ as they should on the infinite lattice. Thus, we also see the onset of finite lattice effects (described by naive perturba-

*Warning: $\beta \equiv 4/g_0^2$ is the conventional definition only for SU(2).

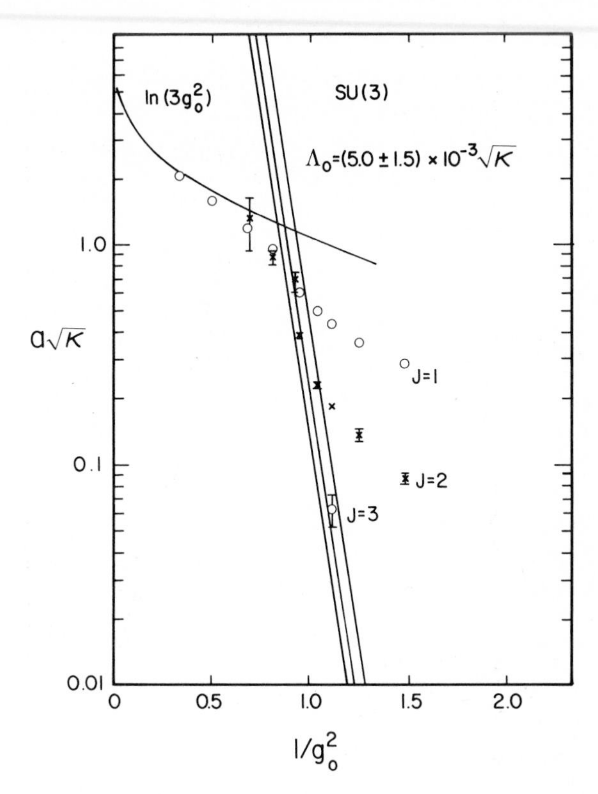


Fig. 18. The string tension $a\sqrt{\kappa}$ for SU(3) gauge theory computed by Monte Carlo simulations [18] from the ratios of loops defined in eq. (3.28).

tion theory in $g_0^2$), which are irrelevant to the non-perturbative (or confining) continuum theory.

Nonetheless, by assuming a matching condition for the Monte Carlo data and the scaling law for $\kappa$ (2.40), we apparently get reasonably reliable estimates for the ratio $\sqrt{\kappa}/\Lambda_0$,

$$\sqrt{\kappa}\,|_{\mathrm{SU(2)}} = \frac{100\Lambda_0}{1.3\pm0.2}, \qquad \sqrt{\kappa}\,|_{\mathrm{SU(3)}} = \frac{100\Lambda_0}{0.5\pm0.15}, \tag{3.29}$$

from couplings on the order of $g_0^2 \simeq 1.5$ and $g_0^2 \simeq 0.9$ for SU(2) and SU(3) respectively. Using the connection (2.39) to $\Lambda_{\overline{\mathrm{MS}}}$ for SU(3) this predicts

$$\Lambda_{\overline{\mathrm{MS}}} \simeq (60\pm20)\ \mathrm{MeV}, \tag{3.30}$$

where we have used $\sqrt{\kappa} = 1/(2\pi\alpha')^{1/2} \simeq 420$ MeV from the Regge slope parameter $\alpha' = 0.9(\text{GeV})^{-2}$. Experimental measurements from logarithmic scaling violations are very difficult, but fits have recently tended to drift to smaller values in the range $\Lambda_{\overline{\text{MS}}} \simeq 100$ to 200, in *better* agreement with Monte Carlo. Agreement is certainly adequate and may even be better if quark effects are either "removed" from the experimental measurements of $\Lambda_{\overline{\text{MS}}}$, or added to the Monte Carlo (see section 6).

In fig. 12, Münster [11] compares his SU(2) strong coupling series with Creutz's Monte Carlo, and the agreement is striking. Indeed matching strong coupling series alone to the scaling relation for $\kappa$ gives very similar values for $\Lambda_0/\sqrt{\kappa}$. We found $\Lambda_0/\sqrt{\kappa} = 1.45 \times 10^{-2}$, $3.75 \times 10^{-3}$, and $4.3 \times 10^{-4}$ for $N = 2$, 3, and $\infty$ respectively [12] by matching to the series in fig. 15.

### 3.2.2. *Precocious scaling*

These estimates of the string tension depend on the *bold* assumption that asymptotic freedom scaling sets in immediately after the first strong coupling approximation fails [i.e., $\beta > 2.2$ for SU(2)]. (This hypothesis we

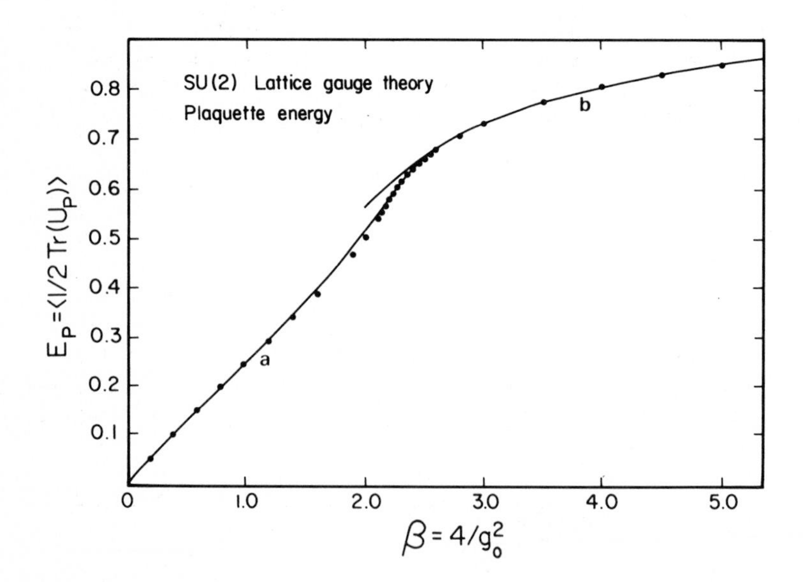


Fig. 19. The internal energy $(1 - E_p)$ in SU(2) lattice gauge theory, ($E_P = \langle \frac{1}{2}\text{Tr}(U_P)\rangle$ is the average plaquette) by Monte Carlo simulation [19], compared with strong (a) and weak (b) coupling expansions.

call precocious scaling on the lattice.) Let us look at the very accurate Monte Carlo calculations of Lautrup and Nauenberg [19] for the internal energy $E_P = \frac{1}{2}\langle \mathrm{Tr}(U_P)\rangle$ and the specific heat $C = \frac{1}{6}\beta^2\partial^2 F/\partial\beta^2$. (See fig. 19 and 20.) Notice that in the specific heat, there is a substantial peak that disagrees with either the strong coupling series (left side) or the naive weak coupling series (right side). The right side of the peak is where the string tension is computed. A further test of precocious scaling has been made by studying the scaling behavior of the finite lattice corrections [20]

$$E_P(\beta, L) \simeq E_P(\beta, L=\infty) + \frac{1}{L^4}\varepsilon(L/\xi). \tag{3.31}$$

By assuming a $\beta$-function of the form $\beta(g_0) = \varepsilon g_0 + \gamma g_0^3$ and varying the

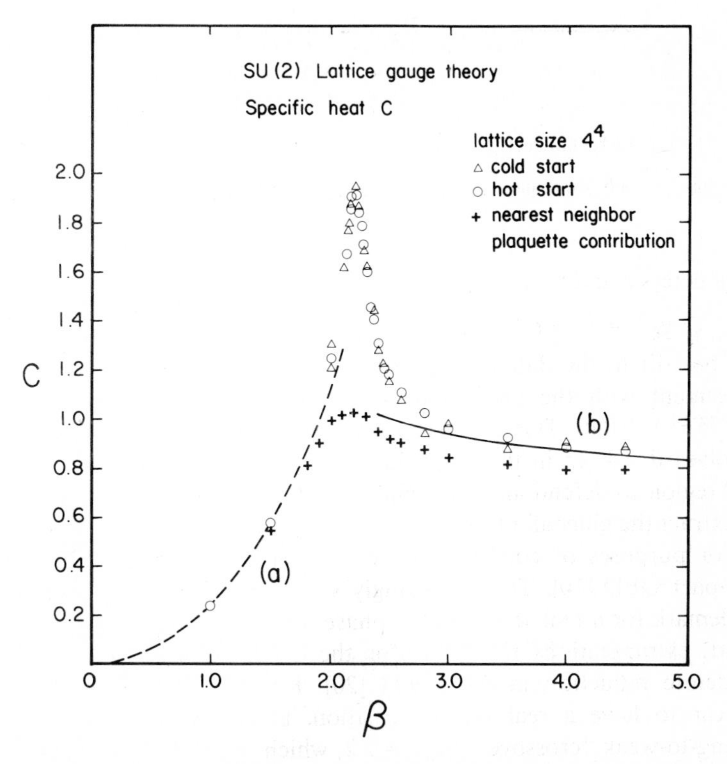


Fig. 20. The specific heat, $C = \frac{1}{6}\beta^2\partial^2 F/\partial\beta^2$, in SU(2) lattice gauge theory, by Monte Carlo simulations [19] compared with strong (a) and weak (b) coupling expansions.

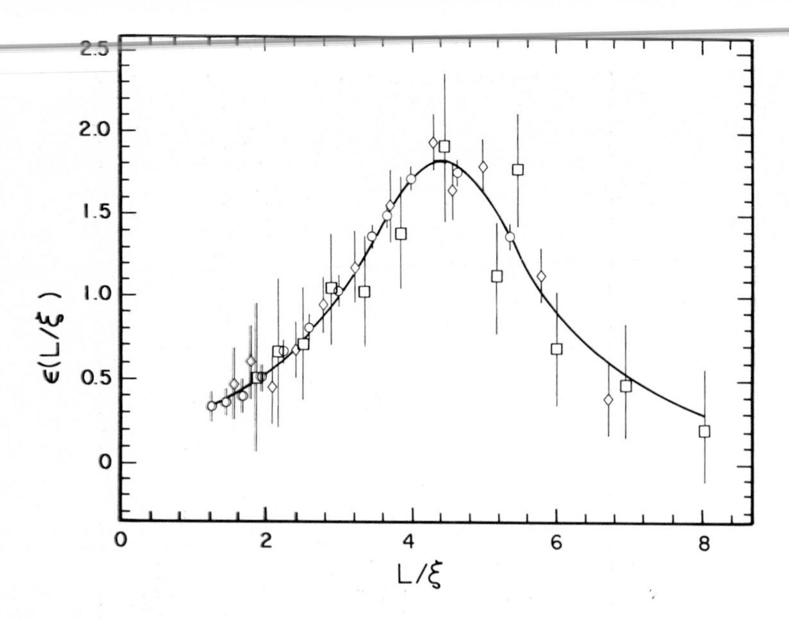


Fig. 21. The finite lattice corrections $\varepsilon(L/\xi) \simeq L^4[E_P(\beta, L) - E_P(\beta, \infty)]$ scale [20] for $\xi = \exp[-\frac{3\pi^2}{11}(\beta - L)]$ and $L = 4$ (squares), 5 (diamonds) and 6 (circles).

parameters $\varepsilon$ and $\gamma$ in the scaling law,

$$\xi(g_0) = \left(1 + \varepsilon/\gamma g_0^2\right)^{1/2\varepsilon} \tag{3.32}$$

the best fit to the data (fig. 21) gives $|\varepsilon| \lesssim 1/45$ and $\gamma \simeq 0.041$ in good agreement with the prediction of asymptotic freedom $\varepsilon = 0$ and $\gamma = 11/45\pi^2 \simeq 0.046$. This is rather astonishing, since all the data in fig. 21 involves $\beta = 4/g_0^2$ in the range 2.05–2.7. In section 4, we will return to this region to defend and "explain" precocious scaling; and in section 5, to extract the glueball mass.

For purposes of contrast I have also included the specific heat for compact QED [10]. The increasingly sharp peak, as $L$ increases, is the trademark for a real second order phase transition (fig. 22). On the other hand, extrapolations to $L = \infty$ for the SU(2) peak (fig. 23) shows a moderate reduction in the height [20]. For $\beta_c \approx 2.2$, SU(2) does not appear to have a real phase transition, but instead a rather sudden strong-to-weak "crossover" at $\beta_c = 2.2$, which, we will see, is "near" to a real transition in modified Wilson actions. Understanding this crossover to precocious scaling is the major goal of section 4.

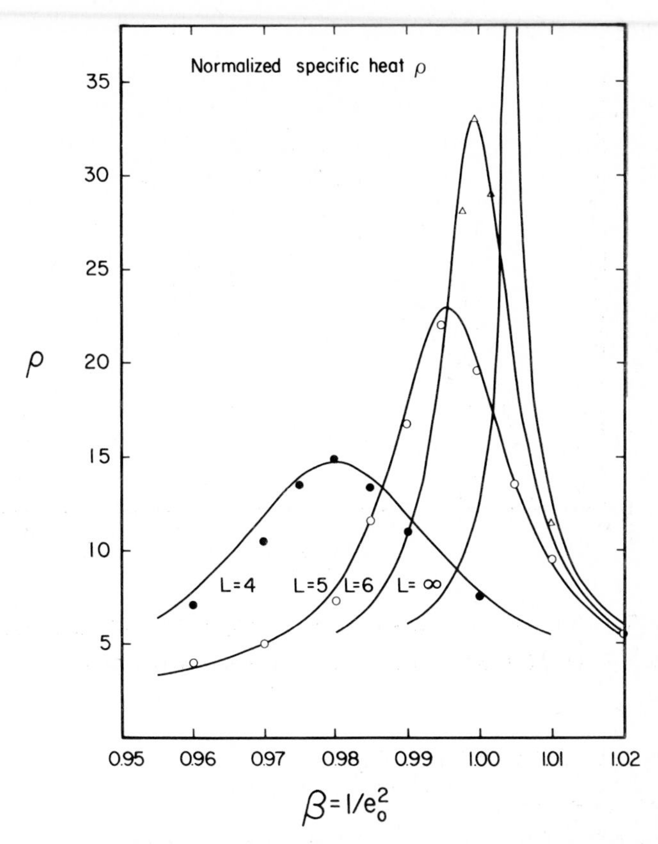


Fig. 22. The specific heat for U(1) lattice QED for increasing lattice sizes $L^d$ [10].

### 3.3. Large N expansion

In the continuum limit, QCD has no coupling constant to expand in ($g_0 = 0$). Although with massive quarks, $m_f$, one can expand around the $m_f/\Lambda = 0$ (chiral) limit or the $\Lambda/m_f = 0$ (static) limit, the gauge field dynamics still has no expansion parameter. However, as 't Hooft [21] pointed out, we can expand in $1/N$ by considering SU($N$) QCD. Of course, as in all expansions after the series is found, we set $N = 3$ colors —the physically correct value.

Here we wish to give only the central ideas of the $1/N$ expansion. No one has succeeded so far in actually performing the expansion, so its as

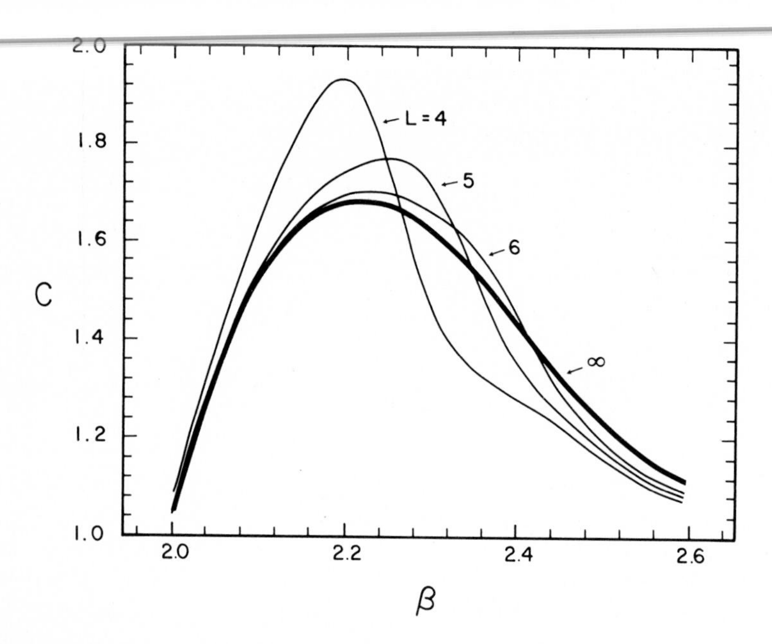


Fig. 23. The specific heat for SU(2) lattice gauge theory [20] for increasing lattice sizes $L^d$ (note expanded scale on $\beta$-axis relative to fig. 20).

yet unrealized bright promise will be discussed only as a parting remark in section 6. However, there are very useful phenomenological consequences that follow from the $1/N$ idea.

*3.3.1. Chromotopology*

First, as $N \to \infty$, with $Ng_0^2$ fixed (or equivalently $Ng_R^2$ in the renormalized perturbation series), all internal quark loops are suppressed. [This can be easily seen in terms of Feynman diagrams; for example, the $n_f$ quark loop insertions into a gluon line give $n_f g_R^2 \sim 1/N$ whereas a gluon loop gives $Ng_R^2 \sim O(1)$.] Here let us look at the strong coupling series on the lattice.

In particular, consider a connected diagram on a surface with $B$ boundaries for the quark loops, and $H$ handles and no self-intersections*

*Self-intersecting surfaces require the use of multi-link integrals [12], but I believe the Schwinger–Dyson equation (6.22) of section 6 allows the topological classification of eq. (3.34) to be generalized to these surfaces.

(see fig. 24). The factors of $N$ are easily counted. Each plaquette has a factor $1/g_0^2$ from eq (3.1); each link integral, a factor of $1/N$ from eq. (3.10); and each vertex a factor of $\mathrm{Tr}(1) = N$. Therefore the contribution to the connected correlation function is:

$$\left\langle \prod_{i=1}^{B} \mathrm{Tr}(U_{\mathcal{C}_i}) \right\rangle_c \sim (N/g_0^2 N)^F (1/N)^E (N)^V, \tag{3.33}$$

for $F$ faces (plaquettes), $E$ edges (links), and $V$ vertices (sites) on the surface. By Euler's theorem:

$$\left\langle \prod_{i=1}^{B} \mathrm{Tr}(U_{\mathcal{C}_i}) \right\rangle_c \sim \beta^F (1/N)^{B+2H-2}. \tag{3.34}$$

Indeed internal quark loops cost a factor of $1/N$, and we also see that non-planarity of gluonic handles costs a factor of $1/N^2$. This chromotopological classification is the same as that given by 't Hooft for the weak coupling series [21] or earlier for dual string models [22]. There are many phenomenological consequences.

A quark–antiquark system has a planar world sheet, so the bound state meson never dissociates into two mesons (two open strings form an internal boundary) or into a meson and a glueball (one open string and a closed string form a handle). The $N = \infty$ theory (if it confines) has absolutely stable $q\bar{q}$ mesons, and closed string glueballs (or gluehoops!). This is the hadronic Born term which dual theories desperately sought without success. In all likelihood, $N = \infty$ QCD gives a dual string theory, whether or not it is possible or even useful to construct it. Clearly the string ideas and their vast range of semi-quantitative phenomology apply to $N = \infty$ QCD. This phenomenology suggests that $1/N$ corrections are not very large, 10–20% effects in most instances. Also the $1/N$ expansion justifies the neglect of quarks in a first approximation (as we do until section 6), and the lack of internal quark loops provides a basis for Zweig's rule, and the valence quark approximations. Witten [23] has also discussed how to include the $N$ quark–baryon states as self-consistent bag-like objects with masses $O(N)$. For mesons the divergence of the bag constant like $O(N^2)$ implies zero volume objects and thus a possible mechanism for string-like flux tube formation at $N = \infty$.

This introduction to large $N$ has been very rough. Throughout the rest of the lecture, we will often lean on large $N$ ideas and therefore extend this discussion. In particular, section 5 discusses chirality at large $N$ and section 6 will pursue a little more the string-like features.

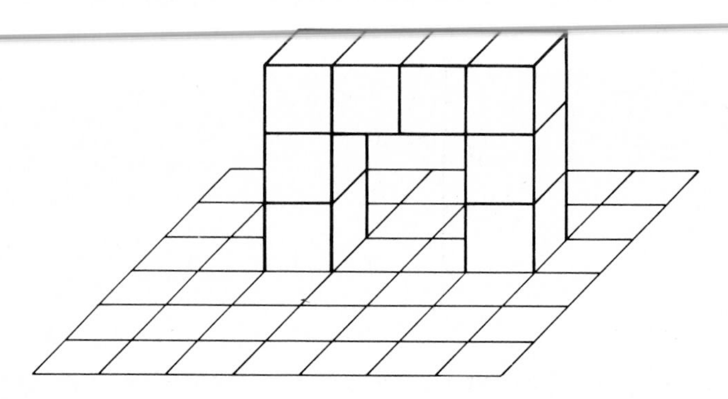

Fig. 24. A surface with one boundary ($B=1$) and one handle ($H=1$).

One feature which we will need later is the factorization property for gauge invariant operators,

$$\langle \psi_1 \psi_2 \rangle = \langle \psi_1 \rangle \langle \psi_2 \rangle + O(1/N^2), \tag{3.35}$$

for $\psi_i = \mathrm{Tr}(U_{\mathcal{C}_i})/N$. This follows from the above chromotopology—two planar surfaces with one boundary each is $O(N^2)$, whereas the connected piece with one surface and two boundaries is O(1). The $N=\infty$ theory is in this respect like a classical limit with configuration averages (apparently) replaced by a single dominant configuration, up to gauge transformations [23]. The search for this configuration, which Coleman [24] called the "Master Field", has not been fruitful to date.

## 4. Duality and disorder variables

A rather convincing picture of confinement has been achieved by looking at the strong coupling domain of a variety of lattice gauge theories [3]. The basic mechanism is a reflection of the original idea of 't Hooft and Mandelstam [25] that there is a dual version of the Meisner effect, where mobile magnetic charge (instead of a super conductor with mobile electric charge) pinches the electric (instead of magnetic) lines between quark and anti-quark sources. A narrow tube of electric flux of fixed cross-section clearly gives a static energy, which grows linearly with separation, $V(R) \sim \kappa R$. The first step is to identify dual collective variables [25].

Classical Maxwell equations have the exact symmetry ($E \to B$, $B \to -E$),

$$F_{\mu\nu} \to \tilde{F}_{\mu\nu} = \tfrac{1}{2}\varepsilon_{\mu\nu\rho\sigma} F_{\rho\sigma}, \tag{4.1}$$

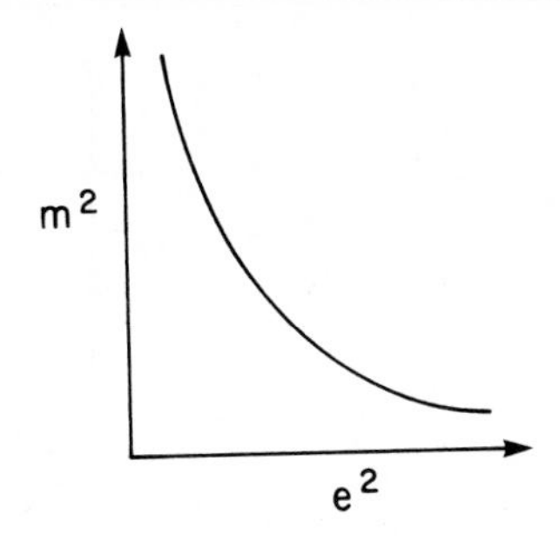


Fig. 25. The typical relationship between a charge $|e^2|$ and its dual $(\tilde{e}^2)$ or magnetic $(m^2 = \tilde{e}^2)$ charge.

if we also exchange electric $(e)$ and magnetic $(m)$ charges. As expounded in Coleman's lectures [26], quantum theory imposes the Dirac condition $(em = \text{const.})$ or

$$m^2 \equiv \tilde{e}^2 \sim \hbar / e^2 \tag{4.2}$$

so that a small electromagnetic coupling is dual to a large magnetic coupling (see fig. 25). A perturbative expansion in both electric and magnetic sources is impossible, but a very useful feature of duality is that our strong coupling series can be reinterpreted as a weak coupling series for the dual version of the theory.

In this section, we will describe some dual properties of $Z_2$ theories (section 4.1), Abelian gauge theories (section 4.2), and SU(2) gauge theory (section 4.3).

### 4.1. Duality in $Z_2$ theories

Like the classical Maxwell equations, the Ising model also enjoys an exact duality property, as we can easily see by comparing strong and weak coupling expansions. At strong coupling $(\beta \to 0)$, on a periodic lattice with $N = L^2$ sites,

$$\begin{aligned} Z &= \sum_{\{s_x\}} \prod_{\ell} \exp(\beta s_{x+\hat{\mu}} s_x) = (\cosh \beta)^{2N} \sum_{\{s_x\}} \prod_{\ell} (1 + \tanh \beta s_{x+\mu} s_x) \\ &= (2 \cosh \beta)^{2N} \big[ 1 + N(\tanh \beta)^4 + 2N(\tanh \beta)^6 \\ &\quad + \tfrac{1}{2} N(N-5)(\tanh \beta)^8 + \cdots \big], \end{aligned} \tag{4.3}$$

where the terms in the series correspond to sets of closed curves $(\Sigma_s s_x = 0$

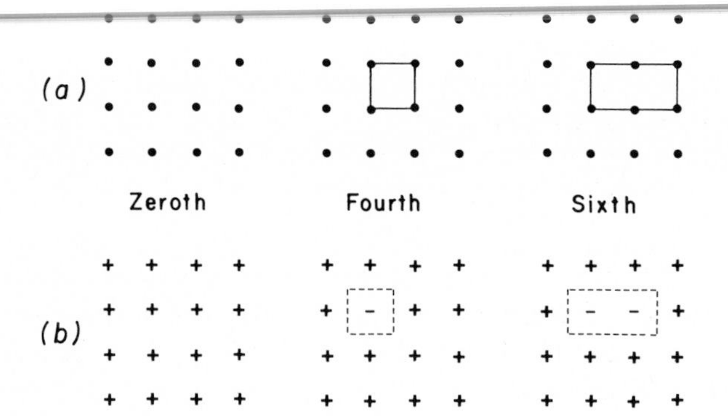


Fig. 26. (a) The strong coupling or high temperature graphs in zeroth, fourth, and sixth order respectively for the $d=2$ Ising model. (b) The corresponding Peierls contours for the weak coupling or low temperature expansion around the spin $s_x = +1$ ground state.

removes open curves) with no links doubly occupied (see fig. 26a). At weak coupling ($\tilde{\beta} \to \infty$), we expand

$$Z = \exp(2N\tilde{\beta}) \sum_{\{s_x\}} \prod_{\ell} \exp\left[-\tfrac{1}{2}\tilde{\beta}(s_{x+\hat{\mu}} - s_x)^2\right] \tag{4.4}$$

around the ordered states $s_x = 1$ and $s_x = -1$:

$$Z = 2e^{2N\hat{\beta}}\left[1 + N(e^{-2\tilde{\beta}})^4 + 2N(e^{-2\tilde{\beta}})^6 + \tfrac{1}{2}N(N-5)(e^{-2\tilde{\beta}})^8 + \cdots\right], \tag{4.5}$$

where now the terms represent domains of flipped spins enclosed by Peierls contours connecting the centers of the plaquettes (see fig. 26b). A "time" slice through a flipped domain reveals a kink–anti-kink pair at the opposite boundaries. The kinks are the dual (or disorder) variables analogous to monopoles in gauge theories. As the temperature is raised ($\tilde{\beta} = 1/kT$ decreased) the kink pairs unbind and disorder the system. The geometric descriptions of both series are in fact identical and they coincide, when $\tilde{\beta}$ obeys the duality condition,

$$\tanh \beta = e^{-2\tilde{\beta}}. \tag{4.6}$$

Therefore both the series in $\beta$ and $\tilde{\beta}$ sum to give the same partition function aside from trivial multiplicative factors.

$$Z(\beta)/(2\cosh^2\beta)^N = Z(\tilde{\beta})/2e^{2N\tilde{\beta}}. \tag{4.7}$$

This identity between the weak and strong coupling is called self-duality. Note that the duality transformation is its own inverse ($\tilde{\tilde{\beta}} = \beta$), since $\sinh 2\beta \sinh 2\tilde{\beta} = 1$. Just as for QED, the strong and weak coupling regions are interchanged under duality. By Onsager's solution, we know there is only one singular point in the free energy so this critical point must occur at the self-dual point ($\tanh \beta_c = e^{-2\beta_c}$),

$$\beta_c = \tfrac{1}{2}\ln(1+\sqrt{2}) \approx 0.44. \tag{4.8}$$

Now let us repeat this argument in a form that allows us to generalize it to the $d=3$ Ising spin model and the $d=4$ Wegner gauge theory.

The series at strong coupling can be written

$$Z = \sum_{\{s_x\}} \prod_{\ell} \sum_{k_\mu = \pm 1} \omega_{k_\mu}(\beta)(s_x s_{x+\mu})^{1/2(k_\mu + 1)}, \tag{4.9}$$

where $(\omega_{+1}, \omega_{-1}) = (\cosh\beta, \sinh\beta)$. Performing the sum over each $s_x$ imposes the constraint, $\Sigma_\mu \frac{1}{2} k_\mu(x) = 0 \ (\mathrm{mod}\, 2)$, on the four vectors ($\mu = \hat{1}, -\hat{1}, \hat{2}, -\hat{2}$) leaving each site $x$.

$$Z = (2)^N \prod_{\ell} \sum_{k_\mu = \pm 1} \omega_{k_\mu} \delta_2\left(\tfrac{1}{2}\sum_\mu k_\mu\right). \tag{4.10}$$

We may solve the constraint by viewing it as a divergence condition,

$$\Delta_\mu \phi_\mu(x) = 0 \ (\mathrm{mod}\, 2), \tag{4.11}$$

on phases $\phi_\mu = 0, 1$ for $k_\mu = \exp(i\pi\phi_\mu)$. The solution is written as a curl:

$$\phi_\mu(x) = \varepsilon_{\mu\nu} \nabla_\nu \theta_{\tilde{x}}, \tag{4.12}$$

where $\theta_{\tilde{x}}$ are angles on the dual lattice, whose sites $\tilde{x} = (n_1 + \frac{1}{2}, n_2 + \frac{1}{2})$, lie at the centers of the plaquettes (see fig. 27). We may go back to dual Ising spins $\tau_{\tilde{x}} = \exp(i\pi\theta_{\tilde{x}})$ and express

$$k_\mu(x) = -|\varepsilon_{\mu\nu}| \tau_{\tilde{x}+\nu} \tau_{\tilde{x}} \tag{4.13}$$

as Ising couplings on the dual links $\tilde{\ell} = (\tilde{x}, \nu)$; $\mu = \pm\hat{1}, \pm\hat{2}$; $\nu = \hat{1}, \hat{2}$. Substitution into eq. (4.10) gives us:

$$Z = \sum_{\{\tau_{\tilde{x}}\}} \prod_{\tilde{\ell}} \omega_{\tau_{\tilde{x}+\nu}\tau_{\tilde{x}}}(\beta) = (\sinh 2\beta)^N \sum_{\{\tau_{\tilde{x}}\}} \exp(\tilde{\beta} \tau_{\tilde{x}+\nu} \tau_{\tilde{x}}). \tag{4.14}$$

These steps can be easily generalized to $Z_2$ spin systems and $Z_2$ gauge theories in higher dimensions [27].

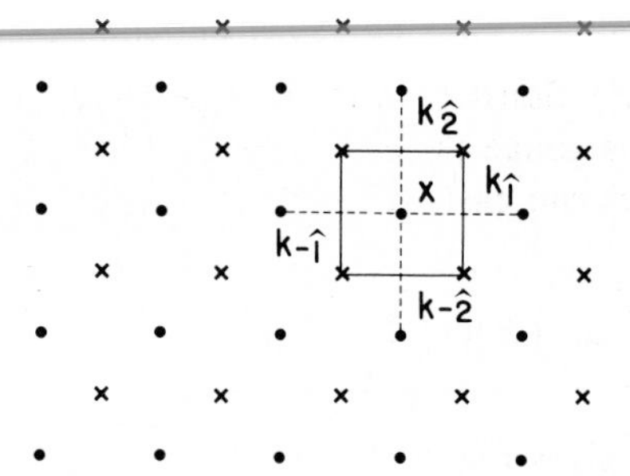


Fig. 27. The lattice (•) at $(n_1, n_2)$ and its dual lattice (×) at $\tilde{x} = (n_1 + 1/2, n_2 + 1/2)$, with a square Peierls contour surrounding the site with co-closed links $k_{\hat{1}}, k_{\hat{2}}, k_{-\hat{1}}, k_{-\hat{2}}$.

*4.1.1. A little lattice homology*

It is helpful to recognize the dual relationship between cells $c_r$ (or simplexes) on the $d$-dimensional lattice $\Lambda$ and the corresponding co-cells $\tilde{c}_{d-r}$ (or dual simplexes) on the dual lattice $\tilde{\Lambda}$. A lattice in four dimensions has $r$ rank oriented cells $c_r$, which are called sites $(x)$, links $(x, \mu)$, plaquettes $(x, \mu\nu)$, cubes $(x, \mu\nu\alpha)$, and hypercubes $(x, \mu\nu\alpha\rho)$ for $r = 0, 1, 2, 3$, and 4 respectively. To each cell $c_r$ on the lattice, there corresponds a co-cell $\tilde{c}_{d-r}$ with the same center of gravity but lying along the $d - r$ orthogonal axes. This corresponds to the standard use of the $\varepsilon_{\mu_1 \cdots \mu_d}$ symbol which takes anti-symmetric tensors (forms) into dual tensors. For example, in $d = 4$, $P$ is dual to $\tilde{P}$ as in $\tilde{F}_{\mu\nu} = \frac{1}{2}\varepsilon_{\mu\nu\rho\alpha}F_{\rho\alpha}$. The appropriate sign conventions and orientations can be determined from the $\varepsilon$-symbol. Note also that the boundary $(\partial)$ of a $c_r$ cell is made of $c_{r-1}$ cells (properly oriented of course), and the dual of the boundary of the dual cell (co-boundary[†] $\Delta = *\partial*$) is a collection of all of the $c_{r+1}$ cells which have $c_r$ on their boundary (e.g., fig. 16). The boundary of a simplex is closed $(\partial^2 = 0)$, and the co-boundary is co-closed $(\Delta^2 = 0)$.

Let us apply these ideas to the $d = 3$ Ising model. Arguments similar to those leading to eq. (4.10) now lead to the constraint $\frac{1}{2}\Sigma_\mu k_\mu(x) = 1 \pmod 2$ on the six links $\mu = \pm 1, \pm 2, \pm 3$ leaving site $x$. The solution to these constraints is given by expressing $k_\mu$ in terms of $\tau$-variables on the boundary (i.e. dual links $\tilde{\ell}$) of the plaquettes $(\tilde{P})$ dual to $k_\mu$.

$$k_\mu(x) = -\tfrac{1}{2}|\varepsilon_{\mu\nu\lambda}|\tau_\nu(\tilde{x})\tau_\lambda(\tilde{x}+\nu)\tau_\nu(x+\lambda)\tau_\lambda(\tilde{x}). \tag{4.15}$$

[†] Here we use the star (*) which is often used to denote the dual transformation (i.e. $*F \equiv \tilde{F}$).

Therefore the $d=3$ Ising model is dual to the $d=3$ Wegner gauge theory for $\tilde{\beta}=-\frac{1}{2}\ln(\tanh\beta)$. Starting with an Ising model in three dimensions, we've re-discovered a gauge theory on the dual lattice!

Finally, the reader can easily check that, for the $d=4$ Wegner lattice gauge theory,

$$Z=\sum_{\{s_\ell\}}\prod_P\sum_{k_{\mu\nu}}\omega_{k_{\mu\nu}}(\beta)(s_{\mu\nu})^{1/2(k_{\mu\nu}+1)}, \tag{4.16}$$

the sum over spins gives rise to the constraint $\frac{1}{2}\Sigma_{\mu\nu}k_{\mu\nu}(x)=0 \pmod 2$. This is solved by putting spins $\tau_\mu(\tilde{x})$ on the boundary [dual links $\tilde{\ell}=(\tilde{x},\mu)$] of the plaquettes ($\tilde{P}$) dual to $k_{\mu\nu}(x)$:

$$k_{\mu\nu}=-\tfrac{1}{2}|\varepsilon_{\mu\nu\rho\sigma}|\tau_\mu(\tilde{x})\tau_\sigma(\tilde{x}+\rho)\tau_\nu(\tilde{x}+\sigma)\tau_\sigma(\tilde{x}). \tag{4.17}$$

The $d=4$ $Z_2$ gauge theory is self-dual with $\sinh 2\beta\sinh 2\tilde{\beta}=1$, exactly like the $d=2$ Ising model. Thus if it has a single critical point, it also occurs at $\beta_c=1/g_0^2\simeq 0.44$.

### *4.1.2. Fluxon confinement mechanism*

We can now see that the Wilson loop may be disordered (area law) by magnetic flux just as Ising correlation functions are disordered by kinks. The Ising model at low temperatures ($\beta\to\infty$) is in an ordered phase so that $\langle s_x s_0\rangle\approx\langle s_x\rangle\langle s_0\rangle$ as $|x|\to\infty$. As the temperature is raised, more and more often $s_x$ and $s_0$ find themselves in domains of opposite spins (see fig. 28). To get a crude picture of the transition to the exponential regime, consider a "typical" kink orbit $\tilde{\mathfrak{C}}$ placed on a lattice with $L^2$ sites. At low temperature, the kink orbit has a small "length", $l(\beta)\ll|x|$. Consequently, out of the $L^2$ translationally equivalent kinks, only $O(l^2)$ kinks circle either $s_x$ or $s_0$, giving rise to a relative minus sign. But at higher temperature the length grows, $l(\beta)\gg|x|$, so that $O(l(\beta)|x|)$

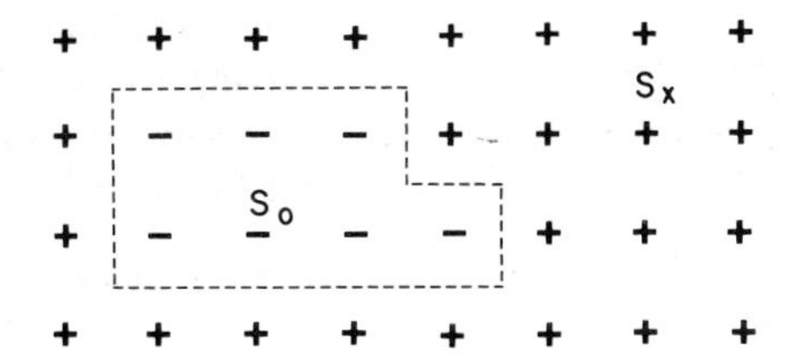


Fig. 28. Kink orbit (Peierls contour) encircles spin so as to disorder $\langle s_0 s_x\rangle$ correlation functions in the Ising model.

kinks separate $s_0$ from $s_x$. In a dilute gas approximation, we can sum over the multiple kink contributions with activity $z$,

$$\langle s_x s_0 \rangle \simeq \frac{\sum_k (L^2 - 2l|x|)^k z^k/k!}{\sum_k (L^2 z)^k/k!}$$

$$= \exp(-|x|/\xi), \quad \xi^{-1} = 2zl(\beta). \tag{4.18}$$

This result is consistent with our high temperature series, if $l(\beta) \sim \log(1/\beta)$ as $\beta \to \infty$. The above hand-waving argument, in spite of its woeful lack of rigor, expresses essentially the correct physics. The condensation of kinks ($\langle \tau_{\tilde{x}} \rangle \neq 0$) disorders Ising spin correlations.

In the $d = 3$ gauge theory, we may also look at the Wilson loop, $W(\mathcal{C}) = \langle \prod_{\ell \in \partial\mathcal{C}} s_\ell \rangle$, at weak coupling ($\beta \to \infty$) where it has perimeter behavior, i.e., the ordered phase where in an axial gauge $s_{\hat{3}}(x) = 1$, all $\langle s_\mu(x) \rangle \simeq 1$. In this phase the dual Ising model ($\tilde{\beta} \to 0$) is disordered $\langle \tau_{\tilde{x}} \rangle = 0$. Now consider the first spin flip excitation ($s_{\hat{1}}(x) = -1$), which gives a factor $(c^{-2\beta})^4 = (\tanh \tilde{\beta})^4$. This corresponds to a contour $\tilde{\mathcal{C}}$ on a dual plaquette $\tilde{P}$ of the Ising systems. As the temperature is raised ($\beta = 1/g_0^2$ decreases) larger contours $\tilde{\mathcal{C}}$ appear with a surface $\tilde{\Sigma}$ of flipped spins ($s_\mu(x) = -1$) spanning it ($\tilde{\mathcal{C}} = \partial\tilde{\Sigma}$). The particular choice of the surface $\tilde{\Sigma}$ is a gauge artifact. (Try a gauge flip $\Omega_x = -1$ at the surface $\tilde{\Sigma}$, and $\tilde{\Sigma}$ will move to the other side of $x$.)

Now we are ready to understand confinement. For small coupling ($\beta \to \infty$) only small contours $\tilde{\mathcal{C}}$ prevail, and when they link the Wilson loop $\mathcal{C}$ they give a perimeter effect—renormalizing the "quark" mass. But as $\beta$ decreases, the large contours abound (see fig. 29). Their entropy

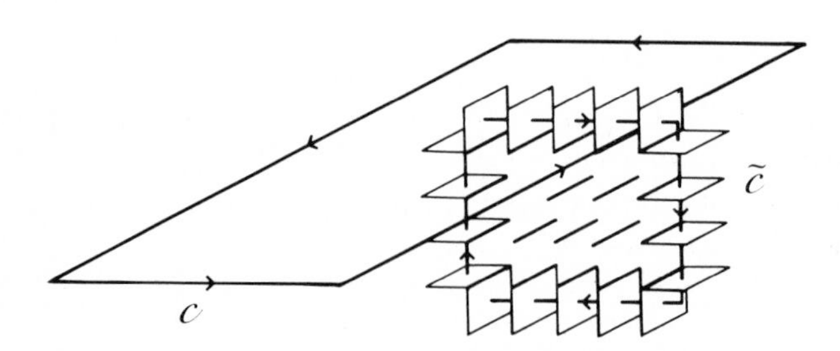


Fig. 29. The dual $Z_2$ fluxon contour $\tilde{\mathcal{C}}$ representing a sheet $\tilde{\Sigma}$ of flipped links ($s_\mu = -1$ on $\tilde{\Sigma}$, where $\partial\tilde{\Sigma} = \tilde{\mathcal{C}}$) which disorders the linked Wilson loop $\mathcal{C}$.

wins over their energetics, and we get a condensate of contours $\tilde{\mathcal{C}}(\langle\tau_{\tilde{x}}\rangle =$ const.) which flip the sign of the Wilson loop $\psi(\mathcal{C})$ each of the $\nu$ times that $\mathcal{C}$ links $\tilde{\mathcal{C}}$. In the dilute gas approximation with linking probability $\ell(\beta)RT/a^2L^3$ proportional to the area $RT$, we can estimate the Wilson loop as:

$$W(\mathcal{C}) \simeq \frac{\sum_k (L^3 - 2\ell RT/a^2)^k z^k/k!}{\sum_k (zL^3)^k/k!}$$

$$= \exp(-\kappa RT), \qquad \kappa = \frac{2}{a^2}\ell(\beta)z. \tag{4.19}$$

With $\ell \sim \log(1/\beta)$, we reproduce the earlier results of strong coupling theory.

Again the argument is shaky but the physics is real. The $Z_2$ flux condensation is the dual Meisner effect which causes confinement. In $d = 4$, we can give the same argument, except translated in time $(x_4)$. The dual loops $\tilde{\mathcal{C}}$ on a time slice correspond to the 't Hooft [25] operator $A(\tilde{\mathcal{C}})$ which in time gives rise to a world sheet of flux linking the Wilson loop. The commutation relation between 't Hooft $A(\tilde{\mathcal{C}})$ and Wilson $\psi(\mathcal{C})$ operators is:

$$\psi(\mathcal{C})A(\tilde{\mathcal{C}}) = (-1)^{\nu}A(\tilde{\mathcal{C}})\psi(\mathcal{C}), \tag{4.20}$$

where $\nu$ is the number of times $\mathcal{C}$ and $\tilde{\mathcal{C}}$ link at fixed time $x_4$. This is responsible for disordering the Wilson loop, when $A(\tilde{\mathcal{C}})$ has perimeter behavior. Magnetic flux condensation is the cause of confinement. Here, in four dimensions, it would indeed be nice to develop the operator Fock space description further, but we refrain (see appendix A).

### 4.2. *Monopoles on the lattice*

Sidney Coleman's Lectures [26] constructed and classified classical monopole configurations in NAGT. Here we attempt to define and study the dynamics of similar configurations in lattice gauge theory. Recall that the standard Dirac monopole for continuum QED in spherical coordinates and $A_4 = 0$ gauge can be written as the field from an infinitely narrow solenoid (or Dirac String),

$$A_{\phi}^{\text{Dirac}} = \frac{m}{4\pi}\{(1-\cos\theta)+\text{string}\}, \tag{4.21}$$

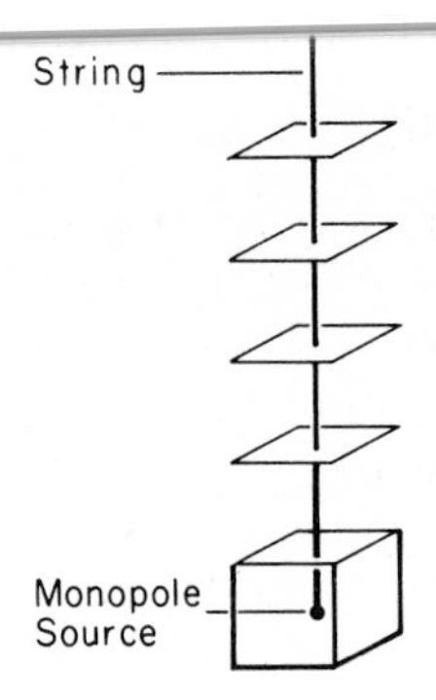


Fig. 30. The Dirac monopole in a lattice cube with its string piercing the plaquettes along the positive $z$-axis.

where the "string" $= \pi\theta(z)\,\delta(x)\,\delta(y)$ is taken along the positive $z$-axis (see fig. 30). The string is invisible if the flux through it is quantized

$$\exp\left(\mathrm{i}e\int A_\phi \,\mathrm{d}\phi\right) = \exp(\mathrm{i}e\,\mathrm{flux}) = \exp \mathrm{i}em = +1. \tag{4.22}$$

In accord with the homotopy classification, $\pi_1(\mathrm{U}(1)) = \mathrm{Z}$, we get the Dirac magnetic charge quantization condition

$$em/2\pi = 0, \pm 1, \pm 2, \ldots. \tag{4.23}$$

(In Coleman's lectures [26] the electric and magnetic potentials are $e^2/4\pi r$ and $4\pi g^2/r$ respectively so that he has $g = m/4\pi$ or $eg = 0, \pm 1/2, \pm 1, \pm 3/2 \ldots$.)

An elegant way to identify the magnetic monopoles with dual variables in lattice QED [28] is based on the periodic Gaussian or Villain form for the action, which for small coupling ($e_0^2 \to 0$) is a good approximation to compact QED (2.4).

$$Z_{\text{Villain}} = \prod_{\ell} \int_{-\pi}^{\pi} \mathrm{d}\theta_\ell \sum_{\{s_{\mu\nu}\}} \exp\left[-\frac{1}{2e_0^2}\sum_P (\theta_{\mu\nu} - 2\pi s_{\mu\nu})^2 + \mathrm{i}\sum \theta_\mu J_\mu\right]. \tag{4.24}$$

Here, $\theta_{\mu\nu} = \Delta_\mu \theta_\nu - \Delta_\nu \theta_\mu$, the integer-valued electric current obeys $\Delta_{-\mu} J_\mu = 0$ and the summation on $s_{\mu\nu} = 0, \pm 1, \pm 2$ re-instates periodicity.

The variables $s_{\mu\nu}$ may be split into an irrotational field ($s_\mu(x) = 0, \pm 1, \pm 2, \ldots$) and a solenoidal field ($G_{\mu\nu} = \frac{1}{2}\varepsilon_{\mu\nu\rho\sigma}\tilde{G}_{\rho\sigma}$, where $\tilde{\Delta}_\nu \tilde{G}_{\mu\nu}(\tilde{x}) =$

$(e_0/2\pi)m_\nu(\tilde{x})$ is the magnetic current on the dual links):

$$s_{\mu\nu} = \nabla_\mu s_\nu - \nabla_\nu s_\mu + G_{\mu\nu} \tag{4.25}$$

The $s_\mu$ variables convert the compact integrals $(-\pi < \theta_\mu \leqslant \pi)$ into non-compact integrals over $eA_\mu = \theta_\mu + 2\pi s_\mu$ $(-\infty < A_\mu < \infty)$, and the equation $\tilde{\Delta}_\mu \tilde{G}_{\mu\nu} = (e_0/2\pi)m_\mu$, can be integrated to find $\tilde{G}_{\mu\nu}$. For example a static source, $m_4(\tilde{x}_1, \tilde{x}_2, \tilde{x}_3) = (2\pi/e_0)\,\delta_{\tilde{x},0}$ up to homogeneous or irrotational contributions, gives the field:

$$G_{ij}(\tilde{x}) = \frac{e_0}{2\pi}\varepsilon_{ij} \sum_{\tilde{z} < \tilde{x}_3} m_4(\tilde{x}_1, \tilde{x}_2, \tilde{z}) \tag{4.26}$$

for a Dirac string from the cube at $\tilde{x}$ along the positive $z$ $(\hat{3})$ axis (see fig. 30). Further analysis again shows that the Dirac quantization condition (4.23) is required for the location of the string to be a gauge artifact. At long distances the electric, $J_\mu(x)$, and magnetic, $m_\mu(\tilde{x})$, currents interact as expected by Maxwell's equations [28]. The Villain approximation allows an identification of monopoles as dual variables, but it is difficult to generalize this approach to non-Abelian theories. Therefore, we return to a cruder but more direct identification of monopole configurations in compact QED.

In compact QED, de Grand and Toussaint [29] suggest the following definition for a lattice monopole. Since $\theta_\mu = ae_0 A_\mu \to 0$, in the continuum limit $(a \to 0)$, we take $\theta_\mu$ always in the interval $-\pi < \theta_\mu \leqslant \pi$. Then the test for a string through a plaquette $(P)$ is the excess curl defined by the integer $G_P \equiv G_{\mu\nu}$,

$$\Delta_\mu \theta_\nu - \Delta_\nu \theta_\mu = \bar{\theta}_{\mu\nu} + 2\pi G_{\mu\nu}, \tag{4.27}$$

where $\bar{\theta}_{\mu\nu} = a^2 e_0 \bar{F}_{\mu\nu}$ is restricted to the interval $-\pi < \theta_{\mu\nu} \leqslant \pi$. Then we define the monopole charge inside a cube, $m_4[c]$, as the sum over the total string flux leaving it:

$$m_4[c] = \frac{2\pi}{e_0} \sum_{P \in \partial c} G_P. \tag{4.28}$$

This is a sensible definition. Although individual strings $G_{\mu\nu} = G_P$ through a plaquette are not gauge invariant, $m_4(c)$ is. The monopole current $m_\mu(\tilde{x})$, found by summing over the faces of the cube $c$ dual to $\tilde{\ell} = (\tilde{x}, \mu)$, is conserved. (The cubes for $\tilde{\Delta}_\mu m_\mu(\tilde{x})$ form a co-closed set.) Finally $m_\mu(\tilde{x})$ is additive; if we sum over a volume composed of cubes the result is the same as the flux from that volume. Thus it leads to the right continuum definition of magnetic charge. As we expect in any order of perturbation theory in $e_0^2$ around the naive vacuum ($A_\mu = 0$), the magnetic charge is zero. Using Monte Carlo simulation, DeGrand and Toussaint support the thesis that the U(1) confining transition is caused by magnetic condensation [29].

### 4.3. Crossover in SU(2) gauge theory

As indicated by Monte Carlo simulation, SU(2) gauge theory does not have a phase transition, but there is a rather sudden crossover for $\beta = 4/g_0^2 \simeq 2.2$ from the strong coupling domain to a domain where precocious scaling apparently begins (see section 3.2.1).

The hope for extracting continuum mass ratios depends critically on this hypothesis of precocious scaling for intermediate couplings ($\beta \sim 2\frac{1}{2}$), so it is vital to understand the dynamics of the crossover. Here we will show how $Z_2$ monopoles and strings are responsible for the crossover.

The expectation that lattice NAGTs have no transitions separating strong coupling confining physics from weak coupling was shattered by the observation in Monte Carlo simulations of a first-order transition in SO(3) gauge theories [30]. The SO(3) action is merely the SU(2) theory, but with the trace in the adjoint representation.

$$-\frac{1}{g_0^2} S_{\mathrm{A}} = \frac{\beta_{\mathrm{A}}}{3} \sum_P \mathrm{Tr}_{\mathrm{A}}(U_P) = \frac{\beta_{\mathrm{A}}}{3} \sum_P \{[\mathrm{Tr}(U_P)]^2 - 1\}. \tag{4.29}$$

Since SU(2) is the covering group of SO(3), two points in SU(2) ($\pm U_P$) are mapped into one point in SO(3). There are (at long distances) nontrivial topological monopoles since $\pi_1(\mathrm{SO(3)}) = Z_2$. Using the fundamental representation, the monopole field can be written

$$A_\phi = \frac{m}{4\pi} \begin{pmatrix} \frac{1}{2} & 0 \\ 0 & -\frac{1}{2} \end{pmatrix} [(1 - \cos\theta) + \mathrm{string}], \tag{4.30}$$

where the Dirac quantization condition in plaquette notation is

$$\tfrac{1}{3}\mathrm{Tr}_{\mathrm{A}}(U_P) = 1, \quad \text{or} \quad \tfrac{1}{2}\mathrm{Tr}(U_P) = \pm 1. \tag{4.31}$$

Thus the monopoles have a $Z_2$ character; $m = 0$ (for no monopole) and $m = 2\pi/e_0$ (for the self-conjugate monopole or anti-monopole) are the only topologically distinct cases.

On the lattice, of course eq. (4.31) need not be exactly satisfied since the plaquette loop cannot be shrunk to zero. So we adopt as the criterion for the presence of a $Z_2$ string in SO(3) the sign of $\mathrm{Tr}(U_P)$:

$$\sigma_P = \mathrm{sign}\,\mathrm{Tr}(U_P) \tag{4.32}$$

and define the monopole charge density by $m[c] = \frac{1}{2}(1-\sigma_c)$, where

$$\sigma_c = \prod_{P\in c} \sigma_P \tag{4.33}$$

is the product of signs around the cube. As in U(1), only the total flux [$\sigma_c = \exp(\mathrm{i}\pi$ flux), flux defined mod 2] is a variable in the SO(3) group. The individual contributions through one plaquette can be flipped ($\sigma_P \to -\sigma_P$) by $U_\mu(x) \to -U_\mu(x)$. We now wish to show how these SO(3) variables naturally appear in SU(2) theories. Of course, the SO(3) monopole configurations in the SU(2) theory will be attached to physical strings (or solenoids), since they ($\sigma_P = -1$) cost $\beta/2$ in increased action relative to the naive vacuum ($U_P = 1$). Monopole pairs will have a linear, instead of a Coulombic, attracting potential; so the dynamical effects can be expected to be changed drastically.

To see the monopole and string variables, it is possible to exactly rewrite the Wilson SU(2) partition function [31],

$$Z = \prod_{\ell} \int \mathrm{d}U_\ell \exp\left[\tfrac{1}{2}\beta \sum_P \mathrm{Tr}(U_P)\right], \tag{4.34}$$

in terms of SO(3) [$\tilde{U}_\ell$] and $Z_2$ ($s_P$) variables:

$$Z = \sum_{\{s_P\}} \prod_{\ell} \int \mathrm{d}\tilde{U}_\ell \prod_c (1 + s_c\sigma_c[U_\ell]) \exp\left[\tfrac{1}{2}\beta \sum_P s_P |\mathrm{Tr}(U_P)|\right] \tag{4.35}$$

where $s_c = \prod_{P\in c} s_P$, and the variables $\sigma_c[U_\ell]$, and $|\mathrm{Tr}(U_P)|$ depend only on the SO(3) group manifold. (Namely, they are unchanged by $U_\ell \to -U_\ell$.) The equivalence of eqs. (4.34) and (4.35) is seen by a series of trivial steps: Introduce a summation ($\frac{1}{2}\Sigma_{\gamma_\ell}$) on dummy link variables $\gamma_\ell = \pm 1$, transform $U_\ell \to \gamma_\ell U_\ell$, and introduce a sum over $s_P$ with the constraint $s_P = \sigma_P \prod_{\ell\in P} \gamma_\ell$:

$$Z = \sum_{\{s_P\}} \int \mathrm{d}\tilde{U}_\ell \tfrac{1}{2} \sum_{\ell} \prod_{P} \left(1 + s_P\sigma_P \prod_{\ell\in P} \gamma_\ell\right) \exp\left[\tfrac{1}{2}\beta \sum_P s_P |\mathrm{Tr}(U_P)|\right]. \tag{4.36}$$

The sum over $\gamma_\ell = \pm 1$ replaces the constraint by the Jacobian

$$\prod_c (1 + s_c\sigma_c). \tag{4.37}$$

It is worth studying this remarkable expression (4.35) for SU(2) in detail. Without the Jacobian ($s_c = \sigma_c$), the sum can be done over $s_P$. This is a Villain version of the pure SO(3) model. The 't Hooft loop operator (4.20) in SU(2) precisely acts to flip signs in the Jacobian ($1 + s_c\sigma_c \to 1 - s_c\sigma_c$), so the $Z_2$ monopole variables $\sigma_c$ are the same as 't Hooft's disorder variables [25].

Brower, Kessler and Levine [32] have studied this expression by introducing an adjustable monopole mass ($\lambda$) and string mass ($\eta$) with the shift in the Jacobian factor:

$$\prod_c (1 + s_c\sigma_c) \to \prod_c \exp(\lambda\sigma_c)\exp(\eta s_c\sigma_c). \tag{4.38}$$

We briefly describe the conclusions of this study [32] for the pure SU(2) case ($\eta = \infty$).

The monopole density, $\rho_c = \frac{1}{2}\langle(1 - \sigma_c)\rangle$, shows a sharp rise in the Wilson model ($\lambda = 0$) for $\beta \simeq 2.2$ (fig. 31). Conversely, if we exclude monopoles (allowing only strings) by taking $\lambda \to \infty$, the average internal energy $\langle \frac{1}{2}\operatorname{Tr}(1 - U_P)\rangle$ is smooth in the crossover region until a sharp break at $\beta_c \simeq 0.965 \pm 0.005$ (fig. 32). The convergence of a mixed phase initial state, half ordered and half disordered, shows quite convincingly that the transition is first order (fig. 33). In this $\lambda = \infty$ limit, introduced

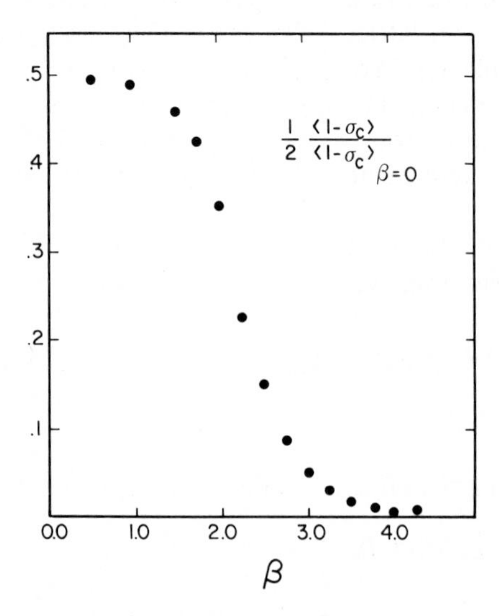


Fig. 31. The $Z_2$-monopole density in SU(2) lattice gauge theory for $\beta = 4/g_0^2$.

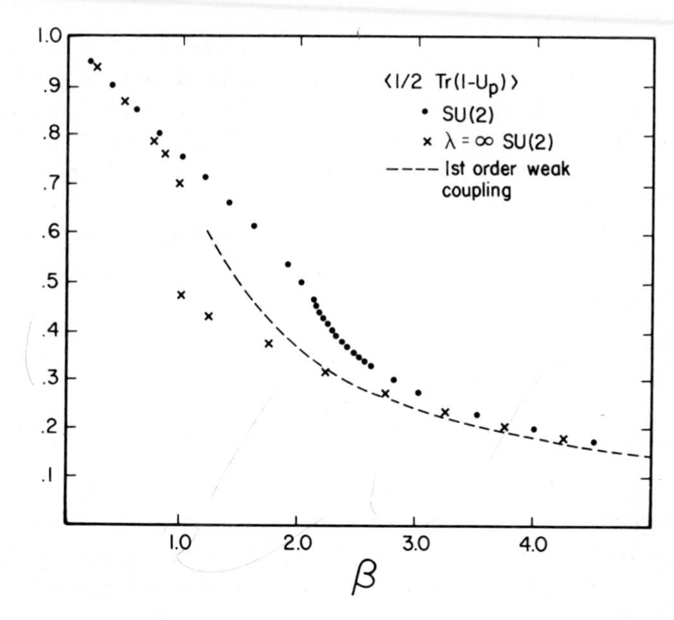


Fig. 32. The internal energy $\langle \frac{1}{2}\,\mathrm{Tr}(1-U_P)\rangle$ in SU(2) lattice gauge theory (•) compared with SU(2) with $Z_2$ monopoles excluded (×).

by Mack and Petkova [31], the dominant mechanism is $Z_2$ flux condensation as in the Wegner lattice gauge theory. Numerical support for this is marshaled by replacing $|\mathrm{Tr}(U_P)|$ by its average value at strong coupling. Thus $\beta_{\mathrm{eff}} = \beta/2\langle|\mathrm{Tr}(U_P)|\rangle = \beta(4/3\pi)$ in a $Z_2$ Wegner model, where we know the transition from duality to be at $\beta_{\mathrm{eff}} = 0.44$. This predicts $\beta_c = 1.02$ in good agreement with the SU(2) simulation value, $\beta_c = 0.965 \pm 0.005$.

Indeed, it is possible to describe the entire $\beta-\lambda$ plane as an effective $Z_2$ theory by integrating over the SU(2) variables.

$$\exp S_{\mathrm{eff}}(s_P) = \int \mathrm{d}\tilde{U}_\ell \mathrm{e}^{\lambda\sigma_c} \prod_c (1+\sigma_c s_c) \exp\left[\tfrac{1}{2}\beta \sum s_P |\mathrm{Tr}(U_P)|\right]. \tag{4.39}$$

At strong coupling the effective action is approximately

$$S_{\mathrm{eff}}(s_P) \simeq \sum_c \lambda s_c + \beta_{\mathrm{eff}} \sum_P s_P. \tag{4.40}$$

This effective theory is dual to a $Z_2$ Higgs model where $\lambda =$

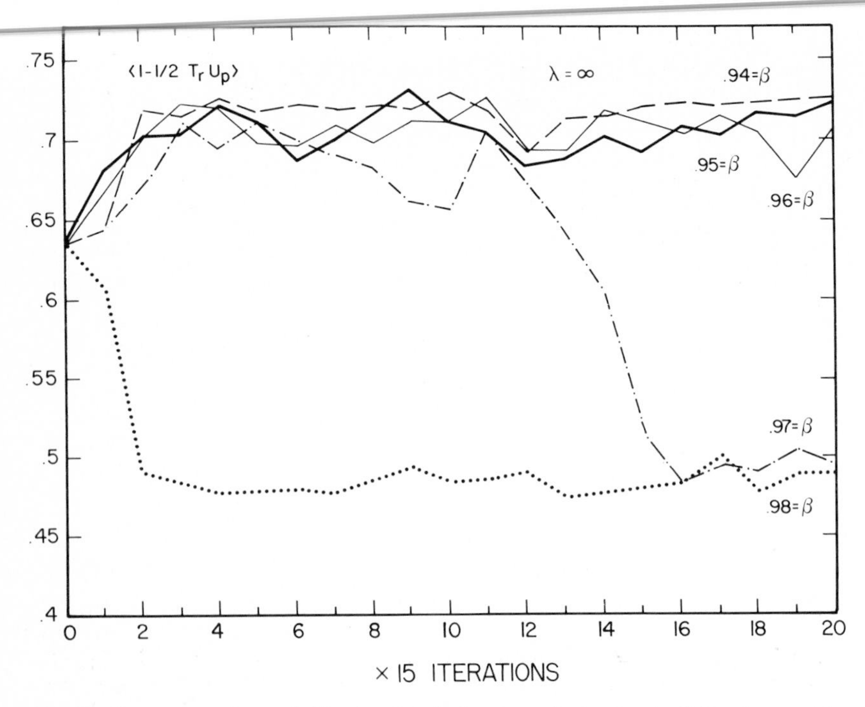


Fig. 33. Convergence from a half-ordered–half-disordered lattice for the SU(2) theory with $Z_2$ monopoles excluded.

$-\frac{1}{2}\ln(\tanh\beta_H)$ and $\beta_{\text{eff}} = -\frac{1}{2}\ln(\tanh\beta)$. On the basis of Creutz's [33] Monte Carlo simulation for this $Z_2$ Higgs theory (fig. 34), we predict the entire $\beta-\lambda$ phase plane very accurately (fig. 35). The crossover in pure SU(2) lies at the linear extension of the first-order phase line for $Z_2$ monopole and flux condensation. Moreover, by applying the Fradkin–Shenker analyticity argument [34] to our effective theory, we see why there is a path of continuation around the end point so the two "phases" of SU(2) lattice gauge theory (like the liquid–gas phases) are not precisely distinguishable. Both presumably confine.

It is interesting that in general, gauge invariant modifications of Wilson's action *do* encounter singularities separating strong and weak domains. For example, fig. 36 gives the phase structure for Bhanot and Creutz's [35] fundamental plus adjoint action $[\frac{1}{2}\beta\text{Tr}(U_P)+\frac{1}{3}\beta_A\text{Tr}_A(U_P)]$ and fig. 37 gives the phase structure on the surfaces of the fully modified

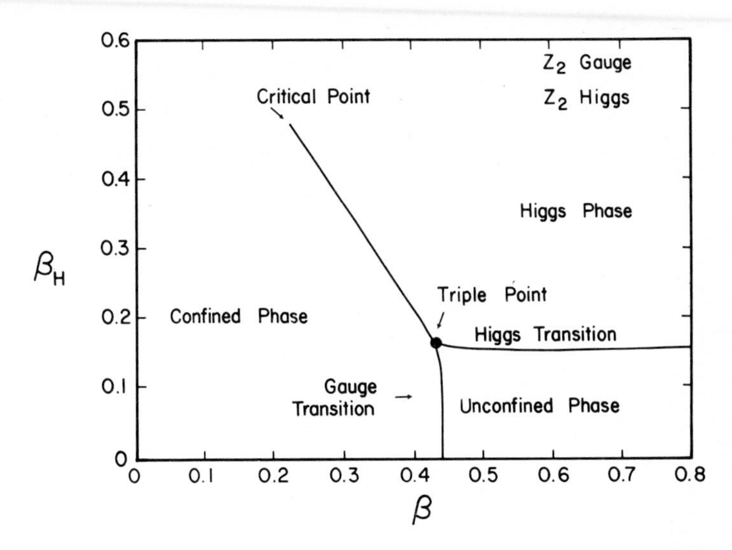


Fig. 34. The $Z_2$ Higgs phase diagram estimated by Creutz's Monte Carlo simulation [33].

$\beta, \lambda, \eta$ action found by inserting eq. (4.38) into eq. (4.35). The exploration of this rich structure has just begun (see ref. [32] for some more details).

From the success of our modified strong coupling expansion (4.40) several conclusions can be drawn. Apparently, the $Z_2$ dynamics of monopoles and strings is largely responsible for the abrupt crossover in the bulk quantities for SU(2) QCD. Brower, Nauenberg and Schalk [20] have studied scaling behavior (described in detail in section 5), and have shown that scaling works well just past $\beta_c \simeq 2.0$ for the bulk finite size effects. Consequently, it is tempting to say that $Z_2$ deconfinement initiates the onset of precocious scaling. The idea is that the flux on the single plaquette level ($\sigma_P = -1$) disappears, allowing the linear potential ($V(R) \simeq \kappa R$) to soften to a Coulombic attraction at the shortest scales. What is unanswered is how flux on larger scales survives to keep large scale confinement intact, and why the various scales come in at the proper exponential rate $\xi \sim \exp(1/2\gamma_0 g_0)$, to precociously mimic asymptotic freedom scaling for $\beta \sim 2\frac{1}{2}$!

In an Abelian theory, flux is additive (i.e. Wilson loops are multiplicative $U_{\mathcal{C}} = U_{P_1} U_{P_2} U_{P_3} U_{P_4}$ where $\mathcal{C}$ encloses the four plaquettes $P_i$), so at the transition the flux disappears simultaneously on all scales. This is our picture of Abelian deconfinement. However, the non-Abelian theory can

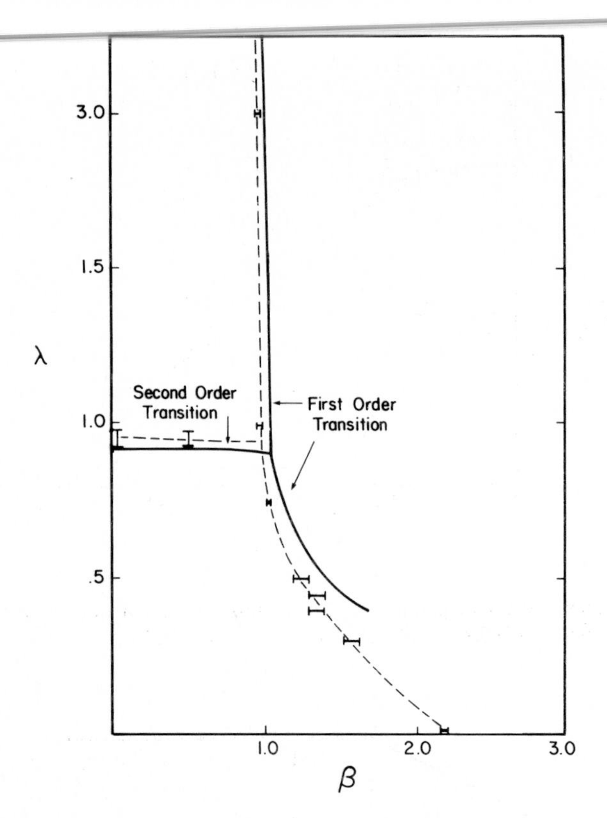


Fig. 35. A comparison of the predicted (solid line) and Monte Carlo simulations for the $\beta-\lambda$ phase plane for SU(2) gauge theory with a variable ($\lambda$) monopole activity [32].

work differently, since the large scale flux $[\mathrm{Tr}(U_{\mathcal{C}})]$ is not the sum of the smaller fluxes $[(\mathrm{Tr}(U_{\mathcal{C}}) \neq \prod_i \mathrm{Tr}(U_{P_i})]$. A more precise description of how the different scales couple is an enormously important, unanswered problem for confinement.

## 5. Masses in quarkless QCD

The proof of the recipe is in the pudding. But the lattice QCD recipe is hard to prove with only one predicted mass ratio—the SU(3) string tension [18], relative to $\Lambda_{\overline{\mathrm{MS}}}$.

$$\sqrt{\kappa}/\Lambda_{\overline{\mathrm{MS}}} \simeq 7 \pm 2. \tag{5.1}$$

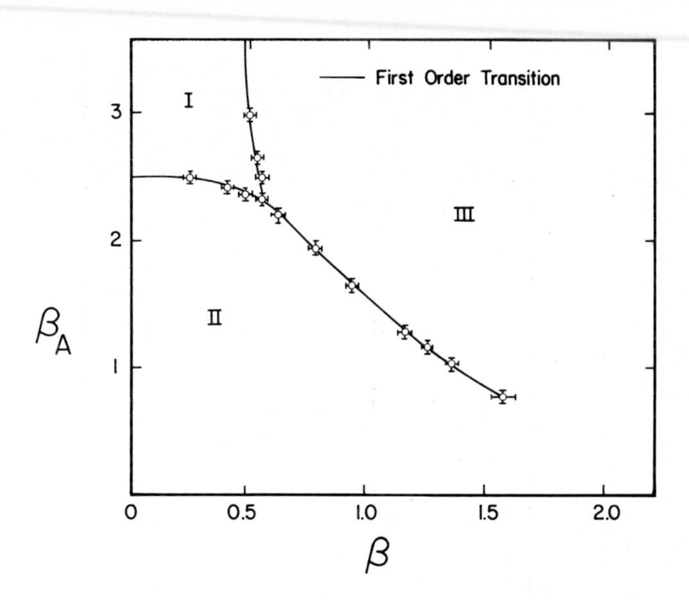


Fig. 36. Bhanot and Creutz [35] Monte Carlo results for the phase plane for SU(2) gauge theory with modified Wilson action, $\Sigma_P[\frac{1}{2}\beta \mathrm{Tr}(U_P) + \frac{1}{3}\beta_A \mathrm{Tr}_A(U_P)]$.

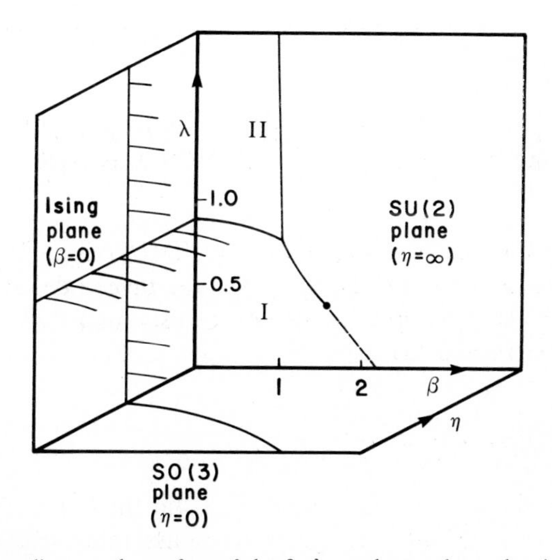


Fig. 37. The phase lines on the surface of the $\beta-\lambda-\eta$ phase volume showing some effects of a modified Wilson action with increased monopole mass ($\lambda$) or increased string mass ($\eta$) [32].

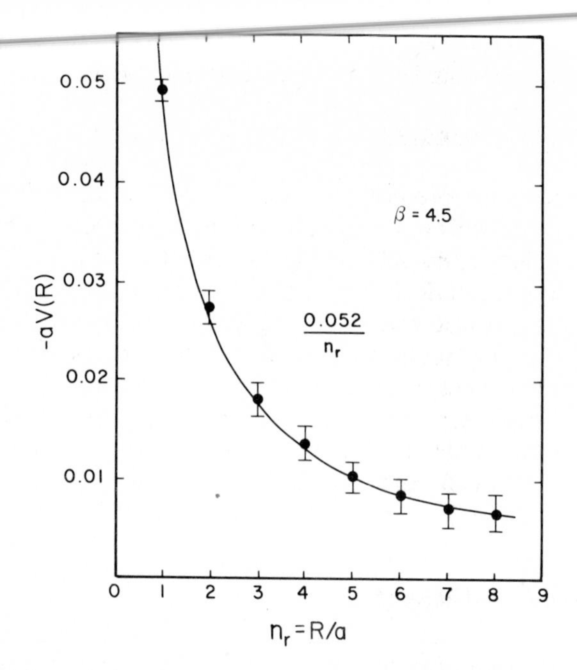


Fig. 38. The Monte Carlo estimate for the SU(2) quark–anti-quark potential by Bhanot and Rebbi [36].

Better Monte Carlo data and corrections for quark effect in $\Lambda_{\overline{\mathrm{MS}}}$ would be nice. Although in the end, $\Lambda_{\overline{\mathrm{MS}}}$ is likely never to be accurately measured experimentally (say to a percent) from logarithmic scaling violations. To suppress higher order logarithms and higher twist, one needs higher $Q^2$, which in turn means smaller corrections in more difficult experimental circumstances. To test the validity of quarkless QCD requires finding other predictions. For SU(2), Bhanot and Rebbi [36] measured the $R$ dependence in the static quark potential $V(R)$, defined by two parallel loops with $L_0$ links running around the periodic "time" axis at spacial separation $\boldsymbol{R}$,

$$\exp[-aL_0V(R)] = \langle\psi_0(\boldsymbol{x})\psi_0(\boldsymbol{x}+\boldsymbol{R})\rangle_c, \tag{5.2}$$

where

$$\psi_0 = \tfrac{1}{2}\,\mathrm{Tr}\left[\mathrm{P}\prod_{\tau=1}^{L_0} U_{\hat{4}}(\boldsymbol{x},\tau)\right].$$

For $\alpha_0 = g_0^2/4\pi \approx 0.0707$ (or $\beta = 4/g_0^2 = 4.5$) and $L_0 = 16$ in the confined domain (see section 5.2) they fit the potential of fig. 38, by

$$V(R, g_0^2) = 0.052(3.0/a - 1/R) + (91/\Lambda_0)^2 R. \tag{5.3}$$

Consistent with the precocious scaling hypothesis, they see a distinct Coulombic piece with $\alpha_{\text{eff}} = 0.0693$, very close to $\alpha_0$. In principle, the continuum limit for this static potential could be extracted by a scaling relation (2.46) at fixed $R$; however, it is not a simple matter to relate this *static* spinless quark interaction energy $V$ to the physical potential for *dynamic* quarks used in heavy quarkonium. Further work on spin- and velocity-dependent effects needs to be done.

Here we will discuss progress in calculating three mass ratios in quarkless SU(2) QCD: the glueball mass ($m_{\text{GB}}/\sqrt{\kappa}$), the deconfinement temperature ($T_c/\sqrt{\kappa}$) and the gluonic component in the $\eta$-mass ($A^{1/4}/\sqrt{\kappa}$).

### 5.1. Degeneracy and mass of glueballs

The straightforward way to measure glueball masses is to compute the plaquette $[\frac{1}{2}\operatorname{Tr} U_{\mu\nu}(x)]$–plaquette $[\frac{1}{2}\operatorname{Tr} U_{\rho\sigma}(0)]$ correlation functions at large "time" ($x = (\boldsymbol{x}, T/a), T \to \infty$ in fig. 39), and zero momentum, ($E_{\text{GB}} \simeq m_{\text{GB}} + k^2/2m_{\text{GB}} + \cdots$):

$$G_{\mu\nu\rho\sigma}(T) = \sum_{\boldsymbol{x}} \exp(\mathrm{i}\boldsymbol{x}\cdot\boldsymbol{k}) G_{\mu\nu\rho\sigma}(\boldsymbol{x}, x_4 = T) \sim \exp(-E_{\text{GB}}T). \tag{5.4}$$

Bhanot and Rebbi [36] have initiated this approach, using Monte Carlo simulations for the 120 element icosahedral subgroup of SU(2). In spite of impressive computational expertise, one sees from their graph (fig. 39) that only a very rough estimate of $m_{\text{GB}}/\sqrt{\kappa}$ is possible:

$$m_{\text{GB}}/\sqrt{\kappa} \simeq 3 \pm 1. \tag{5.5}$$

They have wisely decided not to attempt to project out spin or momentum eigenstates.

Recently Brower et al. [37] have considered another approach, which utilizes the full lattice volume by analyzing the finite lattice size scaling effects in terms of a glueball gas. The internal energy, $E_P = \langle \frac{1}{2}\operatorname{Tr}(U_P)\rangle$,

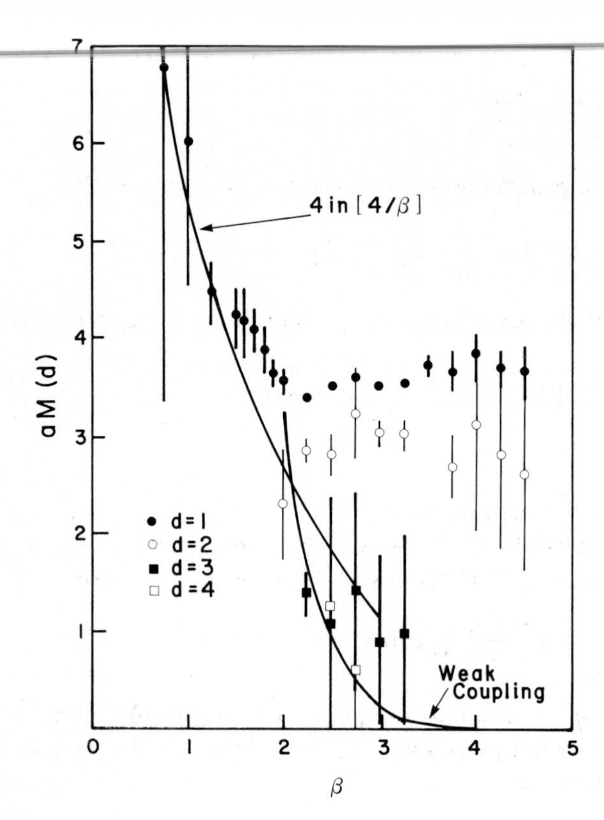


Fig. 39. Bhanot and Rebbi's [36] Monte Carlo data to estimate the SU(2) glueball mass.

in a box of size $L^4$ was shown in section 4.3 to obey the scaling law:

$$E_P(\beta, L) \simeq E_P(\beta, L=\infty) + \frac{1}{L^4}\varepsilon(L/\xi) \tag{5.6}$$

The finite size effect $\varepsilon(L/\xi)$ can be viewed as the correlation of a single plaquette $\mathrm{Tr}(U_P)$ and its periodic images at separations $L$, $L\sqrt{2}$, $L\sqrt{3}$, .... One might expect that $\varepsilon \sim \exp(-L/\xi)$ for $L \to \infty$; however, since appreciable size dependence is observed only for $L = 4$, 5, and 6, nearer images at $L\sqrt{2}$, $L\sqrt{3}$ must be included.

Consider the partition function for a single glueball field $(\phi_x)$ in a box with $L^4$ sites.

$$Z_{\mathrm{GB}} = \prod_x \int \mathrm{d}\phi_x \exp\left\{-\frac{1}{2}\sum_x \left[(\Delta_\mu \phi_x)^2 + \phi_x^2/\xi^2\right]\right\}. \tag{5.7}$$

Applying the scaling law for SU(2) to one loop ($\beta = 4/g_0^2$),

$$\frac{\partial \log \xi}{\partial \beta} = \frac{3\pi^2}{11}, \tag{5.8}$$

the glueball contribution to the internal energy is found:

$$E_{\mathrm{GB}}(\beta, L) = \frac{1}{6L^4}\frac{\partial}{\partial \beta}\log Z_{\mathrm{GB}} = \frac{\pi^2}{22\xi^2}\langle \phi_x \phi_x \rangle_L, \tag{5.9}$$

in terms of the scalar lattice propagator derived in appendix B:

$$G_{\mathrm{L}}(x, y) = \langle \phi_x \phi_y \rangle = \frac{1}{L^4}\sum_{k_\mu} \exp[\mathrm{i}(x-y)k] \Big/ \left( \frac{1}{\xi^2} + 4\sum_{\mu} \sin^2 \frac{k_\mu}{2} \right) \tag{5.10}$$

Note that the discrete sum over $k_\mu = 2\pi l_\mu / L$ ($l_\mu = 1, 2, \ldots, L$) can be exactly replaced by a sum over the periodic images at $x_n = L(n_1, n_2, n_3, n_4)$ in an infinite volume, using the identity

$$\frac{1}{L}\sum_{k_\mu} \equiv \int_{-\pi}^{\pi} \frac{\mathrm{d}k_\mu}{2\pi} \sum_{n_\mu} \exp(\mathrm{i}Ln_\mu k_\mu). \tag{5.11}$$

The scaling contribution to $\varepsilon(L/\xi)$ comes entirely from the small $k$, Lorentz invariant propagator, $4\Sigma_\mu \sin^2(\frac{1}{2}k_\mu) \simeq k^2$. For example, consider the nearest image

$$\int \frac{\mathrm{d}^4 k}{(2\pi)^4} \frac{\exp(\mathrm{i}Lk_0)}{1/\xi^2 + k_0^2 + \boldsymbol{k}^2} \simeq \int \frac{\mathrm{d}^3 k}{(2\pi)^3} \frac{\exp(-LE_{\mathrm{GB}})}{2E_{\mathrm{GB}}}, \tag{5.12}$$

where $E_{\mathrm{GB}} = (\boldsymbol{k}^2 + 1/\xi^2)^{1/2}$, which at large $L/\xi$ contributes

$$\varepsilon_0(L/\xi) = \frac{\sqrt{\pi}}{2}\frac{1}{11}(L/\xi)^{5/2}\mathrm{e}^{-L/\xi} \tag{5.13}$$

to $\varepsilon(L/\xi)$. For smaller $L/\xi$ better approximations are needed. In fig. 40, a comparison is made between: (a) the large $L/\xi$ contribution in eq. (5.13), (b) the full scaling contribution of the first image, (c) the first three images at $L$, $L\sqrt{2}$, and $L\sqrt{3}$, and (d) the full scaling contribution of the glueball gas.

Fortunately, the glueball hypothesis is used only in the region $L/\xi \gtrsim 3$, where the first three images give a good approximation. Starting with the fourth image (at $2L$), the free glueball gas also neglects the two glueball threshold (at $2M_{\mathrm{GB}}$), so this non-interacting glueball model probably cannot be trusted for $L/\xi$ less than about three, in any case.

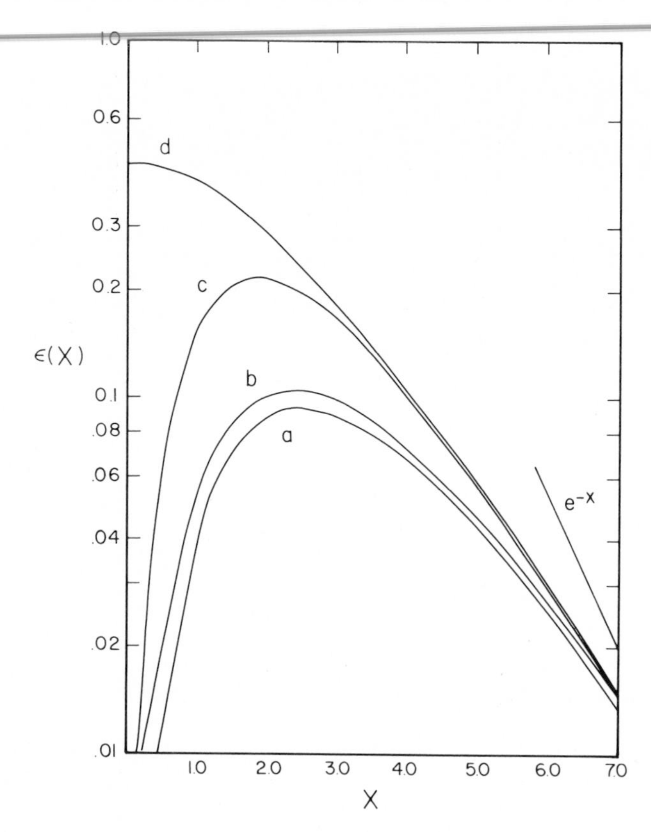


Fig. 40. Contributions to $\varepsilon(x) = L^4[E(L, \beta) - E(\infty, \beta)]$ from a glueball gas in the scaling limit at fixed $x = L/\xi$: (a) the asymptotic form of the nearest image at $L$, (b) the full scaling contribution from the nearest image, (c) the contribution from the three nearest images at $L$, $L\sqrt{2}$ and $L\sqrt{3}$, and (d) the full scaling contribution from all images.

Allowing for an unknown statistical weight or spin degeneracy $d_{GB}$ (where $d_{GB} = 2J + 1$ for spin $J$) of the lowest mass cluster of glueballs, there is a two-parameter fit ($m_{GB}$ and $d_{GB}$) to the scaling curve for $L/\xi \gtrsim 3$. Several conclusions can be drawn.

The fit has a considerable mass versus degeneracy ambiguity. However a single non-degenerate glueball state ($J = 0$) is very implausible. Degeneracies from 5 to 15 seem more reasonable (see fig. 41). The lowest mass in the cluster is quite small,

$$m_{GB}/\sqrt{\kappa} \approx 1.5 \pm 0.2, \tag{5.14}$$

and the degeneracy increases rapidly with mass (by a factor of three from

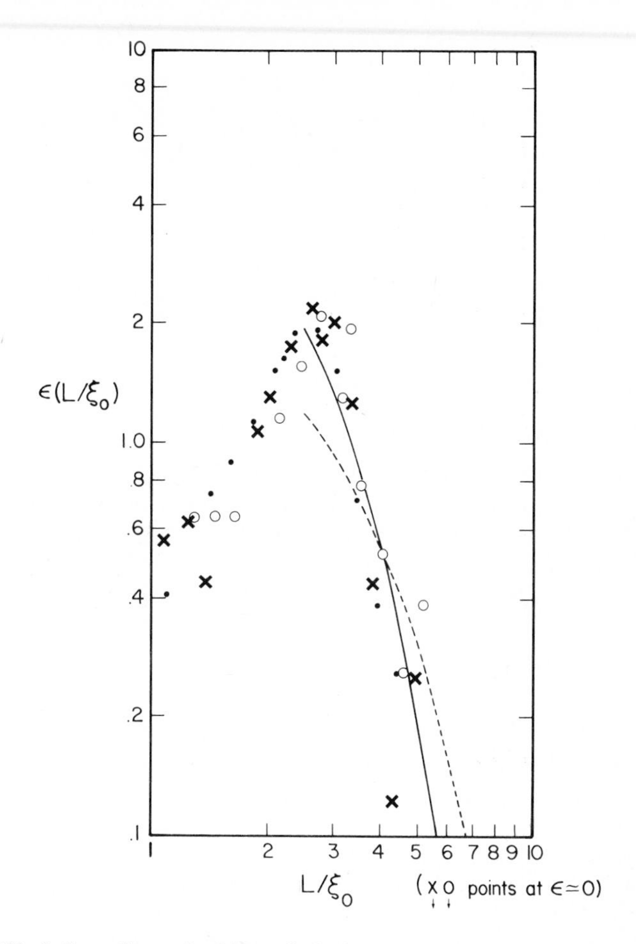


Fig. 41. Glueball gas fits to the Monte Carlo data of fig. 21 on lattices of size $L=4$ (•), $L=5(\times)$ and $L=6$ (∘): dashed line is $d_{\mathrm{GB}}=5$, $m_{\mathrm{GB}}/\sqrt{\kappa}=1.3$ and solid line is $d_{\mathrm{GB}}=15$, $m_{\mathrm{GB}}/\sqrt{\kappa}=1.9$.

$m_{\mathrm{GB}} \simeq 1.3\sqrt{\kappa}$ to $1\frac{1}{2}m_{\mathrm{GB}}$). Interestingly, the estimate of a smaller mass is consistent with Münster's [38] Padé of strong coupling:

$$m_{\mathrm{GB}} \simeq (1.8 \pm 0.8)\sqrt{\kappa} \tag{5.15}$$

for SU(2), and the high degeneracy is typical of spectra in many models of the glueball [39].

The scaling curve for $L/\xi < 2\frac{1}{2}$ is obviously not given in terms of a free gas of glueballs. It is reasonable (as we describe below) that here, at small coupling, the glueballs deconfine into a gas of gluons.

### 5.2. The deconfinement temperature

Quantum statistical mechanics is defined in terms of the partition function

$$Z_T = \mathrm{Tr}\left[\exp\left(-\frac{1}{kT}\hat{H}\right)\right] \tag{5.16}$$

where $\hat{H}$ is the quantum Hamiltonian (see appendix A) and $kT$ is the temperature (real temperature, *not* $\beta^{-1} \sim \hbar g_0^2$). In the path integral formulation this becomes

$$Z_T = \int_{\text{paths}} \mathrm{D}A \exp\left(-\int_0^{1/kT} d\tau \int \mathrm{d}^3x \mathcal{L}_E[A]\right), \tag{5.17}$$

which is just quantum field theory in a finite Euclidean time slab of length $(kT)^{-1}$ with periodic boundary conditions.

Several authors [40] have tried to estimate the critical temperature $T_c$ ($k=1$, temperature in MeV) for deconfinement in lattice SU(2) QCD. For finite temperature, you consider a slab of temporal extent $(aL_0)$ much less than the spacial extent $(aL)$, and study the quark–anti-quark potential between two static quarks on loops running around the periodic time axis introduced in eq. (5.2). In the deconfined domain the linear piece of the potential disappears and there is a "magnetization" effect $\langle \psi_0(\boldsymbol{x})\psi_0(\boldsymbol{x}+\boldsymbol{R})\rangle \to \text{const.} > 0$ as $R \to \infty$. The critical temperature relative to the string tension in SU(2) is defined as the continuum limit $g_0 \to 0$ of

$$kT_c/\sqrt{\kappa} = (1.3 \pm 0.2)/100\Lambda_0(g_0^*)aL_0^*, \tag{5.18}$$

evaluated on the critical surface $(g_0^*, L_0^*)$ for deconfinement. Of course $a\Lambda_0$ must obey the scaling law (2.38) as a function of $g_0$ as dictated by asymptotic freedom. In actual practice, McLerran and Svetitsky [40] look at the magnetization effect for $\langle|\psi_0(\boldsymbol{x})|\rangle$ at fixed $L_0 = 1,2,3$ while varying $g_0$ through the transition, and Engels et al. [40] look at the peak in the specific heat. They both obtain for SU(2)

$$kT_c/\sqrt{\kappa} \approx 0.5 \pm 0.1. \tag{5.19}$$

It is interesting to look back at the peak in the finite size scaling curve for the internal energy at

$$L_c/\xi_0 = 100aL_0\Lambda_0 \approx 2.5, \tag{5.20}$$

which translates into an effective $kT_c/\sqrt{\kappa} \simeq 0.5 \pm 0.1$. The agreement with the deconfinement transition is embarassingly good, realizing that the finite size transition from glueball to gluon gas takes place in a hypercube ($L_0 = L$) instead of a slab ($L_0 \ll L$). Alternately, we might better regard $R_c = aL_c \approx 1/0.5\sqrt{\kappa}$ as a lattice definition of the hadronic radius. It is intriguing that both $T_c$ and $R_c$ define lengths close to the pion Compton wave length.

### 5.2.1. *Higgs mechanism for deconfinement*

A lattice version of the Georgi–Glashow model can also be introduced to study the transition to deconfinement [41]. Here one adds an adjoint [or vector representation for SU(2)] Higgs field $\Phi$ and a new term in the action

$$-S_H = \sum_{x\mu} \tfrac{1}{2}\beta_H \mathrm{Tr}\left[\Phi_x U_\mu(x)\Phi_{x+\mu}U_\mu^\dagger(x)\right], \tag{5.21}$$

where $\Phi = \tau_\alpha \phi^\alpha$, $\phi^\alpha\phi^\alpha = 1$. The deconfinement transition of the U(1) lattice QED at $\beta_H = \infty$ and $\beta_c = 1/e_0^2 \simeq 1.0$ appears to pass smoothly to the Heisenberg O(3) transition [42] (see fig. 42). Further study of this transition should increase substantially our confidence in confinement in QCD. After all, the non-existence of deconfinement is a less convincing observation than the existence of that finite temperature ($T_c > 0$) or that finite Higgs coupling ($\beta_H^* > 0$) necessary to induce deconfinement. Moreover, deconfinement at $T > T_c$ was physically realized in the early universe, with definite consequences for cosmology. Likewise, although the adjoint Higgs (or scalar gluon) may be unesthetic, neither experiment nor the current trend toward gauge hierarchies argues strongly against such a hadronic constituent. The adjoint Higgs, therefore, provides a viable alternative to the standard QCD theory, which I believe should be developed as a way to parametrize our ignorance. For example, the analysis of the gluon spin in three-jet events would be more meaningfully summarized as a bound on the admixture of a spin-zero relative to the spin-one gluon component.

## 5.3. *Large N chirality and the η-mass*

At first, the problem of chirality seems to be beyond the reach of the $1/N$ expansion, where the number of quark loops is limited. After all, to build a chirally broken vacuum, we need a sea of quark–anti-quark pairs. But recent work shows that in the one-loop approximation (leading $1/N$) for $n_f$ massless quarks, $U(n_f)\times U(n_f)$ must be broken spontaneously to

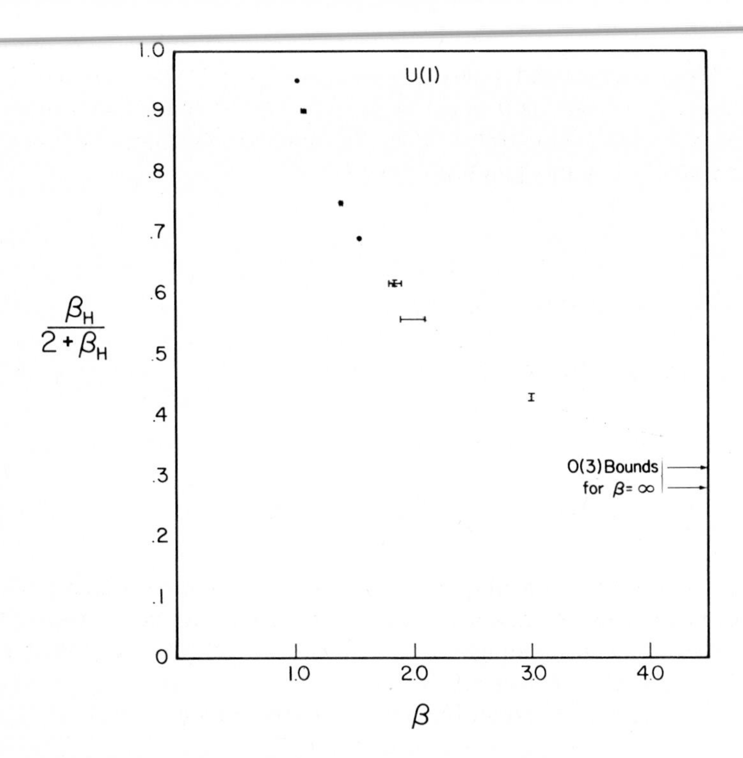


Fig. 42. The adjoint Higgs phase plane determined by Monte Carlo simulations. The upper and lower bounds at $\beta=\infty$ are rigorous results for the transition in the O(3) Heisenberg spin model in four dimensions [42].

give $U(n_f)$ with $n_f^2$ pseudoscalar Goldstein bosons [43]. The point is, that at $N=\infty$ with confinement the only singularities are meson poles, and the only way to satisfy the axial vector triangle anomaly is to have a zero mass pole. To one loop order, all $n_f^2$ states are treated symmetrically, so $n_f^2$ pseudo-scalars appear.*

Also it is hard to see how 't Hooft's resolution of the U(1) problem (no massless $\eta$) can survive the $1/N$ expansion as it apparently relies on instanton configurations, whose contribution,

$$\exp(-8\pi^2/g^2) = \exp(-8\pi^2\beta/\lambda), \tag{5.22}$$

*This picture can also be confirmed by explicit calculations with lattice fermions at large $N$, and strong coupling $\beta = 1/g_0^2 N \to 0$ (see Kluberg–Stern et al. [43]).

does not occur in any finite order in $\lambda = 1/N$. (Recall from section 3.3 that $\beta = 1/g^2N$ is fixed in the large $N$ limit.) Nonetheless, Witten [44] has argued that in two loop order the $\eta$ *does* get a mass squared of order $1/N$. Moreover, via current algebra relations (at large $N$) this mass can be related to a purely gluonic parameter,

$$A = \int \mathrm{d}^4x \exp(\mathrm{i}k\cdot x)\langle Q(x)Q(0)\rangle_\mathrm{c}\Big|_{\substack{k=0\\ \text{no quarks}}}, \tag{5.23}$$

where $Q(x)$ is the instanton density

$$Q(x) = \frac{g^2}{16\pi^2}\,\mathrm{Tr}\left(\tfrac{1}{2}\varepsilon_{\mu\nu\rho\sigma}F_{\mu\nu}F_{\rho\sigma}\right). \tag{5.24}$$

This parameter $A$ is estimated phenomenologically to be:

$$A \simeq \frac{F_\pi^2}{2n_\mathrm{f}}\left(m_{\eta'}^2 + m_\eta^2 - 2m_\mathrm{K}^2\right) \simeq (180\ \mathrm{MeV})^4, \tag{5.25}$$

taking into account small quark mass for $n_\mathrm{f} = 3$, and in principle it can be computed in our quarkless lattice QCD. Let us quickly summarize Witten's argument relating $A$ to the $\eta$-mass, before discussing the SU(2) lattice estimate of $A^{1/4}/\sqrt{\kappa}$.

Consider the $1/N$ expansion for the matrix element

$$U(k) = \int \mathrm{d}^4x \exp(\mathrm{i}kx)\langle Q(x)Q(0)\rangle \simeq U_0(k) + \frac{1}{N}U_1(k) + \cdots. \tag{5.26}$$

For massive quarks (or no quarks), $U(k)$ is non-zero as $k \to 0$ in spite of the fact that $Q(x)$ is a total divergence $[Q(x) = \partial_\mu S_\mu]$ by virtue of instanton contributions to the integer index,

$$\nu = \int_V \mathrm{d}^4x\, Q(x) = \int_{\partial V} S_\mu \mathrm{d}\Omega_\mu, \tag{5.27}$$

in the nontrivial topological sector ($\nu \neq 0$). Witten makes the plausible assumption that the leading $N = \infty$ term $U_0(k)$ is also non-zero at $k = 0$ $[A = U_0(0) \neq 0]$. This term is given by a positive sum over glueball poles, $g_n^2/(k^2 + m_n^2)$, and, having no quark loops, is independent of quark masses. This presents an apparent conflict with the decoupling of instantons and the consequent vanishing of $U(k)$ at $k \to 0$ for zero mass quarks. Witten's resolution of this conflict is to envoke an almost zero mass state $[m_\eta^2 \sim O(1/N)]$ in $U_1(k)$

$$U_1(k) \simeq -\frac{g_\eta^2}{k^2+m_\eta^2} + \cdots, \tag{5.28}$$

to cancel with $U_0(k)$ at $k=0$, as $N \to \infty$. Standard current algebra arguments, involving the anomaly in the singlet axial current,

$$\sum_f \partial_\mu(\psi_f \gamma_5 \gamma_\mu \psi_f) = \sum_f m_f \bar{\psi}_f \psi_f + 2n_f Q(x), \tag{5.29}$$

lead to eq. (5.25). Now we are prepared to discuss the calculation of $A^{1/4}/\sqrt{\kappa}$ in SU(2) lattice QCD [45].

A lattice version of $\mathrm{Tr}(F\tilde{F})$ can be introduced via the summation over all the 48 distinct and appropriately oriented "figure eight" loops,

$$Q_L(\tilde{x}) = -\frac{1}{64\pi^2} \sum \mathrm{Tr}(\varepsilon_{\mu\nu\rho\sigma} U_{\mu\nu} U_{\rho\sigma}), \tag{5.30}$$

contained in a single hypercube centered at the dual site $\tilde{x}$ (see fig. 43). Then Monte Carlo calculations [45] in SU(2) can be performed for

$$A_L = \frac{1}{a^4} \sum_{\tilde{x}} \langle Q_L(\tilde{x}) Q_L(0) \rangle. \tag{5.31}$$

Unfortunately, the lattice has broken the divergence condition $[Q(x) = \partial_\mu S_\mu]$ so $A_L$ has a weak coupling perturbative contribution, numerically

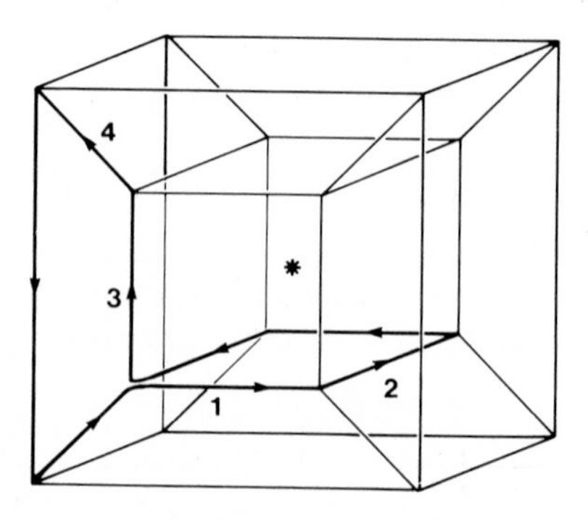


Fig. 43. A four-dimensional cube with the fourth axis represented by scale changes. Wilson loop is one contribution to a lattice approximation to $\mathrm{Tr}(F\tilde{F})$ at the dual site $\tilde{x}$.

evaluated by DiVecchia et al. [45] to be

$$10^4 a^4 A_L^{\text{pert}} \simeq \frac{5.608}{\beta^3} + \frac{2.14}{\beta^6} + \cdots . \tag{5.32}$$

Subtracting this contribution, they also see a clear non-perturbative piece (see fig. 44), which is consistent with the expected scaling for the $d = 4$ operator $Q(x)$ in the continuum limit. From this scaling piece, an estimate of the $\eta$-mass parameter in SU(2) has been given in ref. [45]:

$$A^{1/4}|_{\mathrm{SU(2)}} = (0.11 \pm 0.02)\sqrt{\kappa} . \tag{5.33}$$

Let us conclude this section with a little numerology on these SU(2) masses from quarkless discrete QCD. The ratios of $m_{\mathrm{GB}}$, $T_c$ and $A^{1/4}$ are on the order of

$$m_{\mathrm{GB}} : T_c : A^{1/4} \approx 1.5 : 0.5 : 0.1 , \tag{5.34}$$

which is qualitatively consistent with the expectation of a glueball in the 1–2 GeV range, a deconfinement temperature of order $2m_\pi \approx \frac{1}{2}$ GeV and $A^{1/4} \sim 0.2$ GeV. Beyond this, one may throw caution to the wind and multiply these numbers by $\sqrt{\kappa} \approx 420$ MeV from the experimental Regge

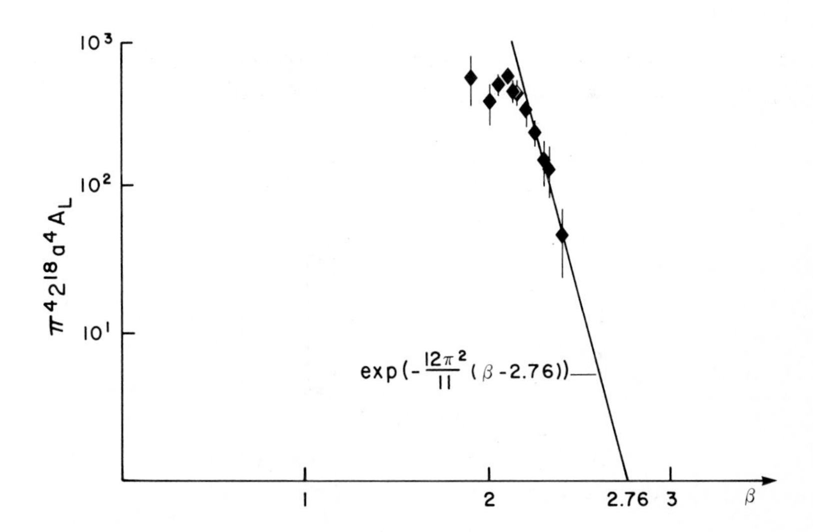


Fig. 44. The scaling contribution to $A$ by Monte Carlo simulations for ref. [45].

slope value in SU(3) with quarks. Now all these masses seem rather small

$$m_{\mathrm{GB}}: T_{\mathrm{c}}: A^{1/4} \approx 600\ \mathrm{MeV}: 200\ \mathrm{MeV}: 45\ \mathrm{MeV}. \tag{5.35}$$

I conclude, not surprisingly, that the $N$ dependence and quark effects are *not* negligible. For the glueball in strong coupling, Münster [38] sees easily a 50% increase from SU(2) to SU(3). The inclusion of quarks is the subject of section 6.

## 6. Fermions and the future

Obviously, the neglect of fermions so far has restricted us to a few questions. Almost the entire known spectrum of hadrons is classified as valence quark–anti-quark or three quark states, so confrontation with data requires them. Moreover, even confinement would be upset if virtual quarks were to modify the vacuum state dramatically.

Here I will give only the briefest of summaries of a few of the difficulties (section 6.1) and prospects (section 6.2) for including fermions on the lattice. The concluding section 6.3 gives some personal speculations on future trends involving symmetry restoration, large $N$, and strings.

### 6.1. *The doubling problem*

In nature there are nearly massless quarks (not to mention massless neutrinos), so it is reasonable to begin with a lattice action that preserves chiral symmetry:

$$\psi_x \to \exp(\mathrm{i}\gamma_5\theta)\psi_x, \qquad \bar{\psi}_x \to \bar{\psi}_x \exp(\mathrm{i}\gamma_5\theta), \tag{6.1}$$

as well as local gauge invariance:

$$\psi_x \to \Omega_x \psi_x, \qquad \bar{\psi}_x \to \bar{\psi}_x \Omega_x^{\dagger}. \tag{6.2}$$

(In the path integral $\psi_x$ and $\bar{\psi}_x$ are independent Grassman variables, $\{\psi_x, \bar{\psi}_x\} = 0$. See appendix B.)

A simple candidate for the Euclidean Fermionic action (1.1) is

$$S_{\mathrm{F}} = \tfrac{1}{2} \sum_{x,\mu} \left[ \bar{\psi}_{x+\mu}\gamma_\mu U_\mu \psi_x - \bar{\psi}\gamma_\mu U_\mu^{\dagger} \psi_{x+\mu} \right], \tag{6.3}$$

where the Euclidean $\gamma$'s obey:

$$\gamma_\mu = \gamma_\mu^{\dagger}, \qquad \{\gamma_\mu, \gamma_\nu\} = 2\,\delta_{\mu\nu}. \tag{6.4}$$

This action has the right naive continuum limit,

$$-\tfrac{1}{2}\bar{\psi}_x\{\gamma_\mu(1+\mathrm{i}ag_0A_\mu)(1-a\partial_\mu)-\mathrm{c.c.}\}\psi_x \simeq a\bar{\psi}\gamma_\mu D_\mu\psi, \tag{6.5}$$

however, there is a paradox in this observation. On the one hand, we know that this continuum theory has a triangle graph which gives an anomalous Ward identity for the divergence of the axial current (5.29), while on the other hand, the Ward identities are expected to be exactly realized on the lattice, since there are no ultraviolet or short distance singularities to invalidate their derivation from the path integral.

The lattice avoids this paradox in a curious fashion. Actually the continuum limit has sixteen massless fermions of opposite chiralities, which cancel in the triangle graph! (Doubling for each dimension, $2^4=16$.) There is no chiral anomaly in this lattice regulated theory (6.3). The details of the demonstration are involved [46], but it is easy to see the fermion "doubling" by looking at the free ($U_\mu=1$) fermionic propagator (see appendix B) for eq. (6.3):

$$\langle\psi_x\psi_0\rangle = \int_{-\pi}^{\pi}\frac{\mathrm{d}^4p}{(2\pi)^4}\frac{i\gamma_\mu \sin p_\mu \exp(\mathrm{i}p\cdot x)}{\sum\limits_\mu \sin^2 p_\mu}. \tag{6.6}$$

We see that in addition to the zero mass pole at $p_\mu=(0,0,0,0)$,

$$\sum_\mu \sin^2 p_\mu \simeq p_\mu^2 = 0, \tag{6.7}$$

there are $2^d-1=15$ other poles at the edges of the Brillouin zone, where one or more of the components $p_\mu=\pi$. Several ingenious modifications to this lattice fermion term have been suggested, but none of them escape the general conclusion: If chirality is exact on the lattice, new zero mass singularities develop in perturbation theory to cancel the anomaly. One is forced to explicitly introduce chiral symmetry breaking in the classical lattice action to obtain the desired anomaly structure in the continuum.* This is an unsatisfactory situation in the sense that spontaneous symmetry breaking of chirality by quantum effects cannot be isolated from *explicit* classical breaking.

Wilson suggests that we introduce a mass parameter $K$, which breaks chirality, removes the doubling, and can be fine tuned to give the chiral limit ($m_\pi=0$) or the physical pion mass in the continuum limit. For

*It may be possible to preserve chirality with long distance couplings in the kinetic term, however the continuum limit is still problematic [47].

QCD, this is after all an acceptable solution, since the quark masses are in fact additional phenomenological parameters. The Wilson fermion [2] term is:

$$-S_{\mathrm{F}} = K\sum_{x,\mu}\{\bar{\psi}_{x+\mu}(1+\gamma_\mu)U_\mu\psi_x + \bar{\psi}_x(1-\gamma_\mu)U_\mu^\dagger\psi_{x+\mu}\} - \sum_x \bar{\psi}_x\psi_x. \tag{6.8}$$

The projection operators $1 \pm \gamma_\mu$ remove the extra modes. In weak coupling ($U_\mu \to 1$), the poles in the quark propagator occur at

$$\sum_\mu \cos p_\mu = 1/2K. \tag{6.9}$$

For $K = 1/8$, the $\bar{\psi}_x\psi_x$ term cancels near $p \approx 0$, and we have a zero mass quark with no doubling problem at $p_\mu = \pi$. Continuing to $p_4 = -\mathrm{i}E$ at $\boldsymbol{p} = 0$ in the continuum limit, the pole can be identified as a bare quark mass $m_0^2 \simeq (1-8K)/a^2K$. Expansions in the hopping parameter $K$ can be made but they diverge at $K = 1/8$, where the interesting chiral dynamics occur.

### 6.2. *Monte Carlo with fermions*

Suppose we wish to simulate the Wilson lattice QCD with $n_{\mathrm{f}}$ flavors of fermions on a $L^d$ lattice. The partition function,

$$Z = \prod_\ell \int \mathrm{d}U_\ell \prod_i \int \mathrm{d}\psi_i \mathrm{d}\bar{\psi}_i \exp(-S_0[U] - S_{\mathrm{F}}[U,\psi,\bar{\psi}]), \tag{6.10}$$

is given in terms of the standard Wilson action,

$$S_0[U] = \frac{1}{g_0^2}\sum_P \mathrm{Tr}(U_P + U_P^\dagger), \tag{6.11}$$

plus the fermionic piece (6.8),

$$-S_{\mathrm{F}}[U] = \bar{\psi}_i D_{ij}\psi_j = \bar{\psi}_i(\delta_{ij} - KQ_{ij})\psi_j, \tag{6.12}$$

quadratic in the Grassman variables $\psi_i$, $\bar{\psi}_i$. The subscript $i$ runs through the $n = n_{\mathrm{f}}NdL^d$ values that label the flavor, color, Dirac and lattice site indices. The $2^n$ fermionic configurations easily exhaust the storage capacity of a computer, but we can perform the Grassman integral analytically (see appendix B) to obtain

$$Z = \prod_{\ell} \int dU_\ell \exp(-S_{\text{eff}}[U]),$$

$$\langle \psi_i \bar{\psi}_j \rangle = \frac{1}{Z} \prod_{\ell} \int dU_\ell D_{ij}^{-1} \exp(-S_{\text{eff}}[U]), \tag{6.13}$$

where the effective action is given by:

$$-S_{\text{eff}} = \frac{1}{g_0^2} \sum_P \text{Tr}\left(U_P + U_P^\dagger\right) + \log(\det D). \tag{6.14}$$

The determinant is positive for $0 \leqslant K < 1/8$, so the additional contribution of internal quark loops in $S_{\text{eff}}$ is well defined. Although $Q_{ij}$ is local (i.e. almost diagonal), the full expressions for $\det(1 - KQ)$ and $(1 - KQ)^{-1}$ are terribly non-local involving all Wilson loops on the lattice. An exact computation of the $n \times n$ determinant requires $n^3 \sim L^{12}$ steps on an $L^4$ lattice. Some approximation must be made to update $\det D$ and $D_{ij}^{-1}$ each time a new link $U_\ell \to U'_\ell$ is selected. Several algorithms for updating the fermionic determinant have been suggested [48]. We will discuss one approach recommended by Fucito et al. [48].

Suppose the link update is restricted to a small change, $\delta U = U'_\ell - U_\ell$, then:

$$\delta S_{\text{eff}} \simeq \delta S_0 - D_{ji}^{-1} \frac{\delta D_{ij}}{\delta U} \delta U. \tag{6.15}$$

The first term has been discussed in section 3.2. Fucito et al. [48] recommend that $D_{ij}^{-1}[U]$ be computed by an internal Monte Carlo loop on a set of fictitious bosons (called pseudofermions) $\phi_i$:

$$D_{ij}^{-1}[U] \equiv \langle \phi_i \bar{\phi}_j \rangle = \frac{\int d\phi d\bar{\phi}\, \phi_i \bar{\phi}_j \exp\left(-\bar{\phi}_i D_{ij} \phi_i\right)}{\int d\phi d\bar{\phi} \exp\left(-\bar{\phi}_i D_{ij} \phi_j\right).} \tag{6.16}$$

Now the full update algorithm is as follows: Given a configuration $C_\alpha = \{U_1, U_2, \ldots\}$ and $D_{ij}[U]$, you do $J$ Monte Carlo iterations on the $\phi$'s to compute the numbers $\langle \phi_i \bar{\phi}_j \rangle$. Next the new configuration $C_{\alpha+1} = \{U'_1, U'_2, \ldots\}$ is found by Monte Carlo, using the new *local* effective action,

$$\bar{S}_{\text{eff}} = S_0[U] - \langle \phi_i \bar{\phi}_j \rangle D_{ij}[U], \tag{6.17}$$

for fixed numbers $\langle \phi_i \phi_j \rangle$. The result is a nested loop of two Monte Carlo routines, requiring $IJ$ iterations for $I$ configurations ($C_\alpha$; $\alpha = 1, \ldots, I$).

Many other algorithms should be, and are being, investigated. The large $N$ phenomenology suggests that the internal quark loops might be neglected in first approximation. Neglecting the fermionic loops at fixed $N$ may be formulated as a quenched limit ($n_f/N \to 0$) of the $n_f$ replications. The next approximation would be to include one internal quark loop and so on. It is premature to assess the trade off between the variety of strategies. Certainly some numbers for the meson masses and vertices will be forthcoming. Most likely progress in both computer hardware and software will afford significant tests of QCD in the near future.

### 6.3. Future directions

Lurking beneath the surface of our discussion have been several subterranean themes: symmetry restoration, the $1/N$ expansion, and string variables. These will, I believe, play a bigger role in the future, but due to formidable technical difficulties, precise results are very difficult to come by.

The difficulty discussed above (section 6.1) with formulating chirality for lattice fermions, should be recognized as a more general problem of *symmetry restoration*. Many symmetries not fully realized on the lattice, are expected to be restored in the continuum limit. For example, throughout we have assumed that the cubic subgroup of the Lorentz group will be restored to the full Lorentz group. This is part of the lore of universality.

The question of symmetry restoration should also be asked about internal symmetries. Apparently, with chiral symmetry, some broken version on the lattice is necessary to avoid a chiral doublet in the continuum limit. With supersymmetry, the problem is even more acute and interesting, since the supersymmetry algebra includes the Poincaré group generators. We need to ask what portion of supersymmetry should be represented on the lattice, so we can investigate nonperturbative questions, confident that the continuum limit corresponds to a fully supersymmetric theory with a particular spectrum. Even with the gauge group itself [e.g. U(1) or SU(2)], it is not obvious that the lattice must carry the full symmetry to have it fully restored in the continuum. These questions are extremely difficult, but perhaps better guidelines for universality will emerge from the study of symmetry restoration in many different models.

The only solutions to a lattice gauge theory at $N_c = \infty$ are for QCD in $d = 2$ dimensions [49], QCD on a single tetrahedron [50], and several other toy matrix models mostly in the Hamiltonian framework [51].

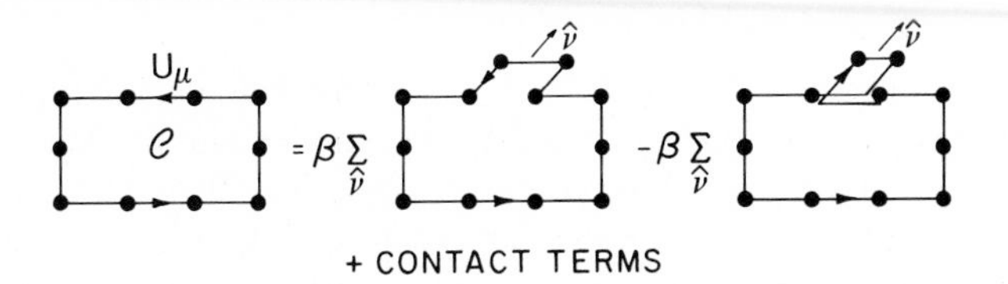


Fig. 45. The Schwinger–Dyson equations for a Wilson loop in $N = \infty$ QCD.

For $d = 2$ QCD, the Wilson loop for a simple closed curve can be expressed as

$$W(\mathcal{C}) = \exp\left\{ -\frac{\text{Area}}{a^2} \log[1/\omega_N(\beta)] \right\}, \tag{6.18}$$

and the $N \to \infty$ limit of eq. (3.17) at fixed $\beta^{-1} = g_0^2 N$ gives

$$\begin{aligned} \omega_\infty &= 1/g_0^2 N, & g_0^2 N &\geqslant 2, \\ \omega_\infty &= 1 - \tfrac{1}{4} g_0^2 N, & g_0^2 N &\leqslant 2. \end{aligned} \tag{6.19}$$

A peculiarity of the lattice is this cusp in the string tension (and free energy) that may be a $d = 2$ analogue to the sudden $d = 4$ crossover to precocious scaling. For the tetrahedron, the cusp is at $\beta = 1/g_0^2 N = \pi/8 \approx 0.393$, close to mean field theory [52] estimates of $\beta \approx 0.395$ and the strong coupling [53] estimates of $\beta \approx 0.40$ for the $N = \infty$ crossover (see fig. 15). Apparently at $N_c = \infty$, there is a weak phase transition due to the infinite group volume, which really occurs on a finite lattice. This may (or may not) be related to the monopole crossover mechanism discussed in Section 4.3, appropriately generalized to $SU(N)/Z_N$ as $N \to \infty$.

For $d = 4$ QCD, one can write down exact Schwinger–Dyson equations [14] on the lattice at $N_c = \infty$. Consider a link $U_\mu(x) = \exp(i\lambda^\alpha \overline{A}{}^\alpha_\mu)$ at the end of a curve $\mathcal{C}$ for the Wilson loop $W(\mathcal{C}) = \langle (1/N)\mathrm{Tr}(U_\mathcal{C})\rangle$. By making an infinitesimal variation in the group, $\delta_\alpha U_\mu(x) \equiv i\lambda^\alpha U_\mu(x)$, the Schwinger–Dyson equations follow from the standard arguments involving invariance of the Haar measure*:

$$\frac{1}{Z} \prod_\ell \int \mathrm{d}U_\ell\, \delta_\alpha \left[ \frac{1}{N} \mathrm{Tr}(\lambda_\alpha U_\mathcal{C}) \exp(-S_0[U]) \right] = 0. \tag{6.20}$$

[Note that in the naive continuum limit $\delta_\alpha = \delta/\delta \overline{A}{}^\alpha_\mu(x)$.] The variation of the action $S_0$ gives the "keyboard" derivative (fig. 45) and the variation

*For simplicity we use the $U(N)$ measure and sum $\alpha = 0, 1, \ldots, N^2 - 1$ where $\lambda^0 = \sqrt{2/N}$, which is equivalent to $SU(N)$ for $N \to \infty$.

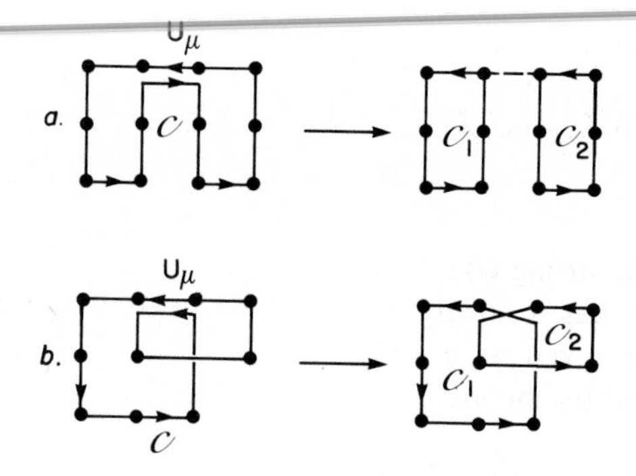


Fig. 46. The contact terms in the Schwinger–Dyson equations of fig. 45 resulting from the fissioning of a Wilson loop $W(\mathcal{C})$ at (a) an anti-link or (b) a link, into two Wilson loops $W(\mathcal{C}_1)$, $W(\mathcal{C}_2)$ in $N=\infty$ QCD.

of the $\mathrm{Tr}(\lambda_\alpha U_\mathcal{C})$ gives back the Wilson loop plus contact terms, where the Wilson loop $\mathcal{C}$ fissions into $\mathcal{C}_1+\mathcal{C}_2$ at a point of self-intersection (fig. 46a, b). By using $N=\infty$ factorization for the contact terms,

$$\frac{1}{N^2}\langle \mathrm{Tr}(U_{\mathcal{C}_1})\mathrm{Tr}(U_{\mathcal{C}_2})\rangle = \frac{1}{N}\langle \mathrm{Tr}(U_{\mathcal{C}_1})\rangle \frac{1}{N}\langle \mathrm{Tr}(U_{\mathcal{C}_2})\rangle, \tag{6.21}$$

we obtain a closed set of equations for the single Wilson loop:

$$W(\mathcal{C}) = \beta\sum_P \left[W(\mathcal{C}+P^\dagger) - W(\mathcal{C}+P)\right] + \sum_{\mathcal{C}_1+\mathcal{C}_2=\mathcal{C}} \pm W(\mathcal{C}_1)W(\mathcal{C}_2). \tag{6.22}$$

The positive terms on the right side of eq. (6.22) occur, because to preserve $UU^\dagger=1$ the anti-links $[U_\mu^\dagger(x)]$ must be counter-rotated ($\delta_\alpha U^\dagger \equiv -U^\dagger(\delta_\alpha U)U^\dagger$). Generalization to the connected part of multi-loop correlation functions is straightforward. This remarkable c-number eq. (6.22) for $W(\mathcal{C})$ with appropriate boundary conditions [14] defines uniquely the Wilson loops for lattice QCD at $N=\infty$. However, general solutions to these lattice equations are probably hopeless, with the possible exception of the continuum limit, where they resemble the Nambu string theory.

A plausibility argument for a string representation goes as follows. If we drop all the terms with minus signs in eq. (6.22), Weingarten has shown that these equations are formally equivalent to a gauge theory with a Gaussian measure,

$$\int \mathrm{d}U = \int \mathrm{d}^2U_{ij}\,\delta\big(U_{i\ell}U^{\dagger}_{\ell i} - \delta_{ij}\big) \to \int \mathrm{d}^2U_{ij}\exp\big[-N\,\mathrm{Tr}(UU^{\dagger})\big], \tag{6.23}$$

order by order in strong coupling at $N = \infty$.

Such a model has a strong coupling expansion for the Wilson loop $W(\mathcal{C})$ at $N = \infty$ given as a sum over *all* planar surfaces ($\Sigma$ with no handles) weighted by the area $A(\Sigma)$ (or number of plaquettes $A/a^2$)

$$W(\mathcal{C}) = \sum_{\partial\Sigma = \mathcal{C}} \exp\left[-\frac{\kappa}{a^2}A(\Sigma)\right]. \tag{6.24}$$

This is a naive lattice version of the Nambu string. In the continuum, the quantum string Hamiltonian has a tachyon state ($m_0^2 < 0$), which may reflect the fact that Weingarten's lattice partition function is actually divergent. [The Gaussian factor is insufficient to control the quartic terms in the action, $\beta\,\mathrm{Tr}(U_P + U_P^{\dagger})$.]

For $N = \infty$ QCD, strong coupling is again a sum over planar surfaces, but the relative minus signs (due to the $UU^{\dagger} = 1$ constraint) reflect themselves in complicated weights for self-intersecting surfaces. Migdal and Makeenko [54] have tried to define continuum versions of the Schwinger–Dyson equations by introducing a functional version of the keyboard derivative at $x$ (see fig. 45):

$$\sum_{P^{\dagger}} W(\mathcal{C} + P^{\dagger}) \simeq a^2 \sum_{\nu} \frac{\delta W(\mathcal{C})}{\delta\sigma_{\mu\nu}(x)} + W(\mathcal{C}). \tag{6.25}$$

Combined with the negative (twisted) keyboard terms, they "derive" the functional equation,

$$\partial_{\nu}\frac{\delta W}{\delta\sigma_{\mu\nu}} = \frac{1}{\beta}\int_{\mathcal{C}} \mathrm{d}s\,\dot{\xi}_{\mu}\delta^4(\xi - x)W(\mathcal{C}_{x\xi})W(\mathcal{C}_{\xi x}), \tag{6.26}$$

where $\xi_{\mu}(s)$ parametrizes the curve $\mathcal{C}$ from $x$ to $\xi(\mathcal{C}_{\xi x})$ and from $\xi$ back to $x(\mathcal{C}_{x\xi})$ so that $\mathcal{C} = \mathcal{C}_{x\xi} + \mathcal{C}_{\xi x}$. Of course, giving any concrete meaning to a finite renormalized theory based on this divergent functional expression is extremely subtle. They then conjecture an equivalence to a fermionic variation of the Nambu string; natural enough in view of the relative minus signs (6.22), but difficult to really prove. Perhaps Polyakov's recent reformulation of the $d = 4$ quantum string in path integral form [55], with the conformal anomaly seen earlier in the quantum

Hamiltonian form [56], will bring strings closer to Euclidean QCD so a connection to Migdal–Makeenko can be found. In my opinion, the application of string variables to reformulate gauge theories in the confined phase is a goal of great importance.

A decade ago, the renormalizability of non-Abelian gauge theories [1], the first lattice gauge models [6], the quantized string models [57], and super symmetric models [58] were just beginning to be formulated. We have subsequently witnessed an explosion of applications and conjectured non-perturbative consequences of these ideas. Perhaps, in the next decade, the lattice will provide a secure (although temporary) scaffolding on which these non-perturbative effects can be reliably computed.

**Acknowledgements**

To all the students and organizers of the 1980 Zakopane and 1981 Les Houches summer schools, I express my gratitude for providing an atmosphere of free and open enquiry. From their enthusiastic questions I learned more than I taught. I am also indebted to Roscoe Giles and Candace Arnott at Santa Cruz for editorial help in preparing the final draft.

## Appendix A. Transfer matrix and continuous time

To introduce the operator formalism for lattice QCD, go to the $A_0 = 0$ gauge $[U_4(x) = 1]$, so that each "time" slice $x_4 = \varepsilon n$ has eigenstates dependent on the spatial links $U_\ell \equiv U_s(\boldsymbol{x})$, $s = \hat{1}, \hat{2}, \hat{3}$.

$$|\{U_\ell\}\rangle = \otimes \prod_{s,\boldsymbol{x}} |U_s(\boldsymbol{x})\rangle. \tag{A.1}$$

The temporal lattice spacing is set to $\varepsilon$ on the $x_4$ axis. The state at each link $(\boldsymbol{x}, s)$ obeys completeness and orthogonality conditions relative to the Haar measure $\mathrm{d}U$ in SU($N$).

$$\int dU|U\rangle\langle U| = 1, \qquad \int \mathrm{d}U f(U)\langle U_0|U\rangle = f(U_0^\dagger). \tag{A.2}$$

On an asymmetric lattice with temporal and spacial extent $\varepsilon L_0$ and $aL$ respectively, the partition function can be expressed as a trace of the transfer matrix $\hat{T}$

$$Z = \mathrm{Tr}(\hat{T}^{L_0}), \tag{A.3}$$

where $\hat{T}$ takes you from one "time" slice, $|\{U_\ell\}\rangle$ at $x_4 = \varepsilon n$, to another, $|\{U'_\ell\}\rangle$ at $x_4 = \varepsilon(n+1)$.

$$\begin{aligned}\langle\{U'_\ell\}|\hat{T}|\{U_\ell\}\rangle &= \prod_{P_s} \exp\left[\frac{\varepsilon}{2g_0^2 a}\operatorname{Tr}\left(U'_{P_s} + U'^{\dagger}_{P_s}\right)\right] \\ &\times \prod_{r,x} \exp\left[\frac{a}{\varepsilon g_0^2}\operatorname{Tr}\left(U'^{\dagger}_s U_s + U^{\dagger}_s U'_s - 2\right)\right] \\ &\times \prod_{P_s} \exp\left[\frac{\varepsilon}{2g_0^2 a}\operatorname{Tr}\left(U_{P_s} + U^{\dagger}_{P_s}\right)\right]. \end{aligned} \tag{A.4}$$

The matrix $\hat{T}$ has been factored into a kinetic term $\hat{T}_1$ (for the space–time plaquettes) and two potential factors $\hat{T}_2$ (for the spatial plaquettes $U_{P_s}$ and $U'_{P_s}$), so that

$$\hat{T} = \hat{T}_2 \hat{T}_1 \hat{T}_2 \tag{A.5}$$

is a real symmetric matrix. With operators $\hat{U}_s(\boldsymbol{x})$ on each link,

$$\hat{U}_s(\boldsymbol{x})|\{U_\ell\}\rangle = U_s(\boldsymbol{x})|\{U_\ell\}\rangle, \tag{A.6}$$

it is trivial to find the potential factors:

$$\hat{T}_2 = \prod_{P_s} \exp\left[\frac{\varepsilon}{2g_0^2 a}\operatorname{Tr}\left(\hat{U}_{P_s} + \hat{U}^{\dagger}_{P_s}\right)\right]. \tag{A.7}$$

The kinetic factor is a product over single link operators, $\hat{T}_1 = \prod_\ell \hat{T}_{1\ell}$ satisfying:

$$\langle U'_\ell|\hat{T}_{1\ell}|U_\ell\rangle = \exp\left[\frac{a}{\varepsilon g_0^2}\operatorname{Tr}\left(U'^{\dagger}_\ell U_\ell + U^{\dagger}_\ell U'_\ell - 2\right)\right]. \tag{A.8}$$

Formally introducing the *right* rotation operator at spatial links, $\ell = (\boldsymbol{x}, r)$,

$$\hat{R}_\ell(G)|U_\ell\rangle = |U_\ell G\rangle, \tag{A.9}$$

by the orthogonality relation, we have

$$\hat{T}_{1\ell} = \int \mathrm{d}G \exp\left[\frac{a}{\varepsilon g_0^2}\operatorname{Tr}(G + G^\dagger - 2)\right]\hat{R}_\ell(G). \tag{A.10}$$

Thus the exact transfer matrix is constructed for the Wilson lattice gauge theory.

In the time continuous limit (holding $\varepsilon L_0$ fixed as $\varepsilon \to 0$), we can do Gaussian integration near $G \simeq 1$. Parametrizing the operator,

$$\hat{R}(G) \simeq \exp(-\mathrm{i}\theta_\alpha \hat{L}^\alpha), \quad \text{for } G = \mathrm{e}^{-\mathrm{i}\theta_\alpha\lambda^\alpha/2}, \tag{A.11}$$

the integral is approximately evaluated.

$$\hat{T}_\ell \sim \prod_\alpha \int \frac{\mathrm{d}\theta_\alpha}{2\pi} \exp\left(-\frac{a}{2\varepsilon g_0^2}\theta_\alpha^2\right) \exp(-\mathrm{i}\theta_\alpha \hat{L}^\alpha)$$

$$\sim \prod_\alpha \exp\left[-\frac{\varepsilon g_0}{2a}(\hat{L}^\alpha)^2\right]. \tag{A.12}$$

Thus we have the Hamiltonian $\hat{H}$ operator by the identification

$$(\hat{T})^{L_0} \sim \exp(-L_0 \varepsilon \hat{H}) \tag{A.13}$$

to $O(\varepsilon^2)$, where

$$a\hat{H} = \sum_x \left\{ \frac{g_0^2}{2} \hat{L}_s^\alpha \hat{L}_s^\alpha - \frac{1}{2g_0^2} \sum_{P_s} \mathrm{Tr}\left[\hat{U}_{P_s}(x) + U_{P_s}^\dagger(x)\right] \right\}. \tag{A.14}$$

Each spacial link $\ell = (x, s)$ carries the algebra

$$[\hat{L}^\alpha, \hat{L}^\beta] = \mathrm{i} f^{\alpha\beta\gamma} \hat{L}^\gamma, \qquad \left[\hat{L}^\alpha, \hat{U}_{ij}\right] = (\hat{U}^\alpha \lambda^\alpha/2)_{ij} \tag{A.15}$$

for right generators.

The residual gauge invariance $[\hat{U}_s(x) \to \Omega_x \hat{U}_s(x) \Omega_x^\dagger]$ is generated by:

$$\hat{Q}^\alpha(x) = \sum_{s = \pm\hat{1}, \pm\hat{2}, \pm\hat{3}} \hat{L}_s^\alpha(x), \tag{A.16}$$

where $\hat{L}_{-s}^\alpha = \hat{U}_s^\dagger \hat{L}_s^\alpha \hat{U}_s$ are the left generators. Gauss' law $[\hat{Q}(x) = 0]$ can be imposed as a weak constraint on the physical states.

The naive continuum Hamiltonian is easily found by the identification of the $E_s(x)$ field on the link $\ell = (x, s)$ and the $B_s(x)$ field on the links dual to $P_s$.

$$\hat{L}_s^\alpha = \frac{a^2}{g_0} E_s^\alpha, \qquad \hat{U}_{P_s} = \exp\left(\mathrm{i}a^2 g^2 \frac{\lambda^\alpha}{2} B_s^\alpha\right). \tag{A.17}$$

Expanding the plaquette variable $U_{P_s}$ to quadratic order, we obtain:

$$H \simeq a^3 \sum_{x,s} \left(\tfrac{1}{2} E_s^\alpha E_s^\alpha + \tfrac{1}{2} B_s^\alpha B_s^\alpha\right), \tag{A.18}$$

where the algebra (A.15) for $\hat{U}_s = \exp[\mathrm{i}ag(\lambda^\alpha/2)A_s^\alpha]$ becomes the canonical commutators

$$\left[A_r^\alpha(x), E_s^\beta(y)\right] = \mathrm{i}\,\delta_{rs}\,\delta_{\alpha\beta} \frac{\delta_{x,y}}{a^3}. \tag{A.19}$$

Continuum QCD requires the replacement

$$a^3\sum_x \to \int \mathrm{d}^3x, \qquad \frac{\delta_{x,y}}{a^3} \to \delta^3(\boldsymbol{x}-\boldsymbol{y}), \tag{A.20}$$

in eq. (A.18) and eq. (A.19) with the standard renormalization difficulties to remove the short distance infinities.

## Appendix B. Finite lattice propagators

### *B.1. Boson propagator*

The multiple Gaussian integral for a real symmetric matrix $K_{yx}$ $(x, y = 1, 2, \ldots, n)$ is easily found by completing the square $(\phi_x = K^{-1}_{yx}J_x + \phi'_x)$ and diagonalizing $K$.

$$\begin{aligned}\prod\int_{-\infty}^{\infty} &\mathrm{d}\phi_x \exp\left(-\tfrac{1}{2}\phi_y K_{yx}\phi_x + J_x\phi_x\right)\\ &= (2\pi)^{n/2}\exp\left(+\tfrac{1}{2}J_y K^{-1}_{yx}J_x\right)/(\det K)^{1/2}.\end{aligned} \tag{B.1}$$

In the case of a free particle, we have

$$\begin{aligned}\phi_y K_{yx}\phi_x &= (\Delta_\mu\phi_x)^2 + (am)^2\phi_x^2\\ &= \phi_y\left(\Delta_{-\mu}\Delta_\mu + a^2m^2\right)\delta_{yx}\phi_x.\end{aligned} \tag{B.2}$$

The last step (summation by parts) has no surface terms, if we work on a periodic lattice. The inverse $G(x-y) = K^{-1}_{xy}$ is found by going to Fourier space,

$$\tilde{\phi}_k = \frac{1}{L^2}\sum_x \exp(\mathrm{i}x\cdot k)\phi_x, \quad Lk_\mu/2\pi = 0, 1, \ldots, L-1, \tag{B.3}$$

so that $K_{xy}G_{yz} = \delta_{xz}$ becomes

$$L^2\left[\sum_\mu (1-\mathrm{e}^{+\mathrm{i}k_\mu})(1-\mathrm{e}^{-\mathrm{i}k_\mu}) + (ma)^2\right]\tilde{G}_k = 1, \tag{B.4}$$

and the solution for $G(x)$ is

$$G(x) = \frac{1}{L^4}\sum_k \exp(-\mathrm{i}k\cdot x)\Big/\left[(ma)^2 + 4\sum_\mu \sin^2\frac{k_\mu}{2}\right]. \tag{B.5}$$

In the $L\to\infty$ limit, reintroducing the $a$ dependence, we obtain the infinite lattice propagator,

$$G(x) = \int_{-\pi/a}^{\pi/a} \frac{\mathrm{d}^4 k}{(2\pi)^4} \mathrm{e}^{ikx} \bigg/ \left( m^2 + \frac{4}{a^2} \sum_\mu \sin^2 \frac{ak_\mu}{2} \right). \tag{B.6}$$

The particle is given by the pole for real $E = \mathrm{i}k_4$,

$$\frac{4}{a^2} \sinh^2 \frac{aE}{2} - m^2 - \frac{4}{a^2} \sum_i \sin^2 \frac{ak_i}{2} = 0, \tag{B.7}$$

which in the continuum limit $(a \to 0)$ occurs at $E^2 \approx m^2 + \boldsymbol{k}^2 + \mathrm{O}(a^2 k^4)$.

### B.2. *Fermionic propagator*

The fermionic propagator requires doing the integral over Grassman variables $\bar{\psi}_i, \psi_i$, for $i = 1, \ldots, n$.

$$Z_\psi(\bar{J}_i, J_i) = \prod_i \int \mathrm{d}\psi_i \mathrm{d}\bar{\psi}_i \exp\left( \bar{\psi}_i D_{ij} \psi_j + \bar{J}_i \psi_i + \bar{\psi}_i J_i \right). \tag{B.8}$$

All Grassman variables $\psi_i$, $\bar{\psi}_i$, $J_i$ and $\bar{J}_i$ anti-commute with each other. The rule of integration for each variable is:

$$\int \mathrm{d}\psi f(\psi) = f_1, \tag{B.9}$$

where $f(\psi) = f_0 + f_1 \psi + f_2 \psi^2 + \cdots = f_0 + f_1 \psi$.

Again we may complete the square by $\psi_i = D_{ij}^{-1} J_i + \psi_i'$ and $\bar{\psi}_i = \bar{J}_j D_{ji}^{-1} + \bar{\psi}_i'$, so that:

$$Z_\psi = \exp\left( \bar{J}_i D_{ij}^{-1} J_j \right) \prod_i \int \mathrm{d}\psi_i' \mathrm{d}\bar{\psi}_i' \exp\left( \bar{\psi}_i' D_{ij} \psi_j' \right), \tag{B.10}$$

and the Gaussian integral is easily evaluated explicitly by expanding, since the only non-zero contribution occurs in order $D^n$.

$$Z_\psi = \det D \exp\left( \bar{J}_i D_{ij}^{-1} J_j \right). \tag{B.11}$$

In the text, we need the expression

$$\langle \psi_i \bar{\psi}_j \rangle = \frac{\partial}{\partial \bar{J}_i} \frac{\partial}{\partial J_j} \log Z_\psi = D_{ij}^{-1} \tag{B.12}$$

as well.

Applying (B.12) to the Wilson fermions (6.8),

$$\bar{\psi}_i D_{ij} \psi_j = \bar{\psi}_y \left[ K \delta_{y,x+\mu}(1+\gamma_\mu) + K \delta_{y,x-\mu}(1-\gamma_\mu) - \delta_{y,x} \right] \psi_x, \tag{B.13}$$

we get the propagator

$$G_F(x) = \frac{1}{L^4} \sum_{p_\mu} \frac{\exp(-\mathrm{i} p \cdot x)}{2K \sum_\mu (\cos p_\mu - \mathrm{i}\gamma_\mu \sin p_\mu) - 1} \tag{B.13}$$

for fermions on a finite lattice, $Lp_\mu/2\pi = 0, 1, \ldots, L-1$.

## References


[1] G. 't Hooft, Nucl. Phys. B35 (1971) 167.

[2] K.G. Wilson, Phys. Rev. D14 (1974) 2455; Phys. Rep. 23C (1976) 331; in: New Developments in Quantum Field Theory and Statistical Mechanics, Cargese (1976), eds. M. Lévy and P. Mitter (Plenum, New York, 1977) p. 143.

[3] J.B. Kogut, Rev. Mod. Phys. 51 (1979) 659.

[4] L. Susskind, Coarse Grained Quantum Chromodynamics, in: Weak and Electromagnetic Interactions at High Energy, Les Houches (1976), session 29, eds. R. Balian and C.H. Llewellyn–Smith (North-Holland, Amsterdam, 1977) p. 207.

[5] M. Creutz, Phys. Rev. D21 (1980) 2308.

[6] F. Wegner, J. Math. Phys. 12 (1971) 2259.

[7] S. Elitzur, Phys. Rev. D12 (1975) 3978.

[8] V. Baluni and J.F. Willemsen, Phys. Rev. D13 (1976) 3342.

[9] A. Hasenfratz and P. Hasenfratz, Phys. Lett. 93B (1980) 165, R. Dashen and D. Gross, Phys. Rev. 23D (1981) 2340.

[10] B. Lautrup and M. Nauenberg, Phys. Lett. B95 (1980) 63.

[11] G. Münster, Phys. Lett. 95B (1980) 59,
G. Münster and P. Weiz, Phys. Lett. 96B (1980) 119; See also
R. Balian, J.M. Drouffe, and C. Itzykson, Phys. Rev. D11 (1975) 2104 (Errata, Phys. Rev. D19 (1979) 2514).

[12] R. Brower, G. Cristofano and J. Rudnick, Strong Coupling Expansion for QCD for $N_c = \infty$, UCSC preprint 135 (1980), unpublished.

[13] A. Hasenfratz, E. Hasenfratz and P. Hasenfratz, Nucl. Phys. B180 (1981) 353.

[14] R. Brower and M. Nauenberg, Nucl. Phys. B180 (1980) 221.

[15] R. Brower, P. Rossi and C.-I. Tan, The External Field Problem for QCD, Brown preprint HET-445 (1981).

[16] E. Brezin and D. Gross, Phys. Lett. B97 (1981) 120.

[17] K. Binder, Monte Carlo Investigations of Phase Transitions and Critical Phenomena, in: Phase Transitions and Critical Phenomena, vol. 5B, eds. C. Domb and M.S. Green (Academic Press, New York, 1976) p. 2,
V. Metropolis, A.W. Rosenbluth, A.H. Teller and E. Teller, J. Chem. Phys. 21 (1953) 1087.

[18] M. Creutz, Phys. Rev. Lett. 45 (1980) 313; Numerical Studies of Gauge Field Theories, in: 19th International School of Subnuclear Physics, Erice, Italy (1981), Brookhaven preprint BNL-29840, 1981.
[19] B. Lautrup and M. Nauenberg, Phys. Rev. Lett. 45 (1980) 410.
[20] R. Brower, M. Nauenberg and T. Schalk, Phys. Rev. D24 (1981) 548.
[21] G. 't Hooft, Nucl. Phys. B72 (1974) 461.
[22] M. Jacob, ed., Dual Theory, Physics Reports Reprint Book Series, vol. 1 (North-Holland, Amsterdam, 1974).
[23] E. Witten, Nucl. Phys. B160 (1979) 57.
[24] S. Coleman, $1/N$, in: Point like Structures Inside and Outside Hadrons, 17th International School of Subnuclear Physics, Erice, Italy (1979), SLAC-PUB 2484 (1980).
[25] G. 't Hooft, Nucl. Phys. B138 (1978) 1,
S. Mandelstam, Phys. Rep. 23C (1976) 245.
[26] S. Coleman, Monopoles Revisited, in: 19th International School of Subnuclear Physics, Erice, Italy (1981).
[27] R. Balien, J. Drouffe and C. Itzykson, Phys. Rev. D11 (1975) 2098.
[28] T. Banks, R. Myerson and J. Kogut, Nucl. Phys. B129 (1977) 493,
A. Ukawa, A. Guth and P. Windey, Phys. Rev. D21 (1980) 1013.
[29] T. DeGrand and D. Toussaint, Phys. Rev. D22 (1980) 2478.
[30] J. Greensite and B. Lautrup, Phys. Rev. Lett. 47 (1981) 9,
I. Halliday and A. Schwimmer, Phys. Lett. 101B (1981); Phys. Lett. 102B (1980) 337.
[31] G. Mack and V. Petkova, Ann. Phys. 123 (1979) 442; Ann. Phys. 125 (1980) 117.
[32] R. Brower, D. Kessler and H. Levine, Phys. Rev. Lett. 47 (1981) 621; Dynamics of SU(2) Lattice Gauge Theories, Harvard preprint HUTP-81/A040 (1981).
[33] M. Creutz, Phys. Rev. D21 (1980) 1006.
[34] E. Fradkin and S. Shenker, Phys. Rev. D19 (1979) 3682.
[35] G. Bhanot and M. Creutz, Variant Actions and Phase Structure in Lattice Gauge Theory, Brookhaven preprint BNL-29640 (1981).
[36] G. Bhanot and C. Rebbi, Nucl. Phys. B180 (1981) 469.
[37] R. Brower, M. Creutz and M. Nauenberg, Finite Size Scaling and Low Mass Glueballs, Santa Cruz preprint UCSC-TH-144 (1981).
[38] G. Münster, Nucl. Phys. B190 (1981) 439.
[39] J. Donoghue, K. Johnson, and B. Li, Phys. Lett. 99B (1981) 416.
[40] L.D. McLerran and B. Svetitsky, Phys. Lett. 98B (1981) 195,
J. Engels, F. Karsch, I. Montvay and H. Satz, Phys. Lett. 101B (1981) 89.
[41] H. Georgi and S. Glashow, Phys. Rev. Lett. 28 (1972) 1494,
A. De Rújula, R. Giles and R.L. Jaffe, Phys. Rev. D17 (1978) 285,
H. Georgi, Phys. Rev. D22 (1980) 225.
[42] J. Fröhlich, B. Simon and T. Spencer, Commun. Math. Phys. 50 (1976) 79,
B. Simon, J. Stat. Phys. 20 (1980) 491.
[43] S. Coleman and E. Witten, Phys. Rev. Lett. 45 (1980) 100,
H. Kluberg-Stern, A. Morel, O. Napoly and B. Peterson, Nucl. Phys. B190 (1981) 504.
[44] E. Witten, Nucl. Phys. B156 (1979) 269.
[45] P. DiVecchia, K. Fabricius, G. Rossi and G. Veneziano, Preliminary Evidence for $U_A(1)$ Breaking in QCD from Lattice Calculations, CERN-TH-3091 (1981).
[46] J. Kogut and L. Susskind, Phys. Rev. D11 (1975) 395,
L. Susskind, Phys. Rev. D16 (1977) 3031, See also
P. Ginsparg, Aspects of Symmetry Behavior in Quantum Field Theory, Ph.D. thesis, Cornell Univ. (1981), unpublished.

[47] S. Drell, M. Weinstein and S. Yankielowicz, Phys. Rev. D14 (1976) 1927,
L. Karten and J. Smit, Nucl. Phys. B183 (1981) 103.
[48] F. Fucito, E. Marinari, G. Parisi and C. Rebbi, Nucl. Phys. B180 (1981) 369,
D.H. Weingarten and D.N. Petcher, Phys. Lett. 99B (1981) 333,
D.J. Scalapino and R.L. Sugar, Phys. Rev. Lett. 46 (1981) 519,
R. Blankenbecler, D.J. Scalapino and R.L. Sugar, Phys. Rev. D24 (1981) 2278,
A. Duncan and M. Furman, Monte Carlo Calculations with Fermions: The Schwinger Model, Columbia preprint CU-TP-194 (1981).
[49] D. Gross and E. Witten, Phys. Rev. D21 (1980) 446,
V.A. Kazakov and I.K. Kostov, Nonlinear Strings in Two-dimensional U($\infty$) Gauge Theory, Landau Inst. preprint 9 (1980).
[50] R. Brower, P. Rossi and C.-I. Tan, Phys. Rev. D23 (1981) 953.
[51] A. Jevicki, Classical and Quantum Dynamics of Two-dimensional Non-linear Field Theories: A Review, Brown preprint HET-401 (1979).
[52] J. Greensite and B. Lautrup, Phys. Lett. 104B (1981) 41.
[53] F. Green and S. Samuel, Phys. Lett. 103B (1981) 48, Nucl. Phys. B190 (1981) 113, See also ref. [12].
[54] A.A. Migdal and Yu. Makeenko, Phys. Lett. 88B (1979) 135 (Erratum, Phys. Lett. 89B (1980) 437).
[55] A.M. Polyakov, Phys. Lett. 103B (1981) 207; Phys. Lett. 103B (1981) 211.
[56] R. Brower, Phys. Rev. D6 (1972) 1655.
R. Brower and K. Friedman, Phys. Rev. D7 (1973) 535.
[57] P. Goddard, J. Goldstone, C. REbbi and C.B. Thorn, Nucl. Phys. B56 (1973) 109,
S. Mandelstam, Dual-Resonance Models, Phys. Rep. 13C (1974) 259.
[58] Yu. A. Gol'fand and E.P. Lichtman, JETP Lett. 13 (1971) 323.

I
RIEN NE FONCTIONNE
DE CE QU'ON NOUS DONNE
ET SES COUCHES
II
PROFESSEUR DE L'AMERIQUE
UNE MALADIE TRES TYPIQUE
IL EXPLIQUE
THEORIQUE
BIEN FORT MELANCOLIQUE
ON ESSAYAIT TOUS LES MOYENS
RIEN NE VALAIT
NI DU LAIT
SAUF UNE FORTE DOSE DU SEX
DE BELGIQUE
LA MECHANIQUE STATISTIQUE
QU'IL DISAIT
FAUX NI VRAI
ETAITEN TOUS SYMBOLIQUES

**Seminars related to Courses 4 and 5**

1. A random lattice gauge theory, by T.D. Lee (Columbia Univ.).

2. The mass gap in QCD, by I.M. Singer.

3. Population explosion in the vacuum, by V.F. Weisskopf (MIT).

4. The phase structure of vector-like gauge theories with massless fermions, by A. Zaks (Tel-Aviv).